Advanced Organic Chemistry

Reinhard Bruckner

Advanced Organic Chemistry

Reaction Mechanisms

A Harcourt Science and Technology Company
San Diego • San Francisco • New York • Boston • London • Sydney • Tokyo

Sponsoring Editor	Jeremy Hayhurst
Production Managers	Joanna Dinsmore and Andre Cuello
Editorial Coordinator	Nora Donaghy
Promotions Manager	Stephanie Stevens
Copyeditor	Brenda Griffing
Proofreader	Cheryl Uppling
Composition	TechBooks
Printer	Maple Press

Front cover image: Computer animation of the catalytic hydrogenation of an olefin (J. Brickmann and W. Sachs, Physical Chemistry, TH Darmstadt).

This book is printed on acid-free paper. ∞

Academic Press
A Harcourt Science and Technology Company
525 B Street, Suite 1900, San Diego, California 92101-4495, U.S.A.
http://www.academicpress.com

Academic Press
Harcourt Place, 32 Jamestown Road, London NW1 7BY, UK
http://www.academicpress.com

Harcourt/Academic Press
A Harcourt Science and Technology Company
200 Wheeler Road, Burlington, Massachusetts 01803
http://www.harcourt-ap.com

Library of Congress Catalog Card Number: 2001088748

International Standard Book Number: 0-12-138110-2

PRINTED IN THE UNITED STATES OF AMERICA
01 02 03 04 05 06 MM 9 8 7 6 5 4 3 2 1

Contents

Foreword

We are at the start of a revolution in molecular science that will more profoundly change our lives, our culture, indeed, our world than did the Industrial Revolution a century ago. From the human genome project, the largest natural product characterization effort ever, to the search for the molecular signatures of life on other planets, this molecular revolution is creating an ever-expanding view of ourselves and our universe. At the core of this revolution is chemistry, the quintessential molecular science within which is organic chemistry, a discipline that will surely be the source of many of the major advances in chemistry, biology, medicine, materials science, and environmental science in the 21st century.

In his text on organic chemistry, the translation of which has been impressively led by Professors Harmata and Glaser, Professor Bruckner has masterfully addressed the core concepts of the discipline, providing a rich tapestry of information and insight. The student of contemporary organic chemistry will be well-served by the depth and quality of this treatment. The underlying philosophy of this text is that much of chemistry can be understood in terms of structure, which in turn influences reactivity, ultimately defining the higher order activities of synthesis. Whether one seeks to understand nature or to create the new materials and medicines of the future, a key starting point is thus understanding structure and mechanism.

Professor Bruckner addresses the interrelationship of structure and mechanism with the rich insight of one schooled at the interface of physical organic chemistry and synthesis. His treatment is impressively rigorous, integrated, and broad. He achieves breadth through the careful selection of representative and fundamental reactive intermediates and reactions. Rigor and integration derive from his disciplined adherence to structure, orbital theory, and mechanism. The result is a powerfully coherent treatment that enables the student to address the rich subject matter at hand and importantly by analogy the far-ranging aspects of the field that lie beyond the scope of the book. Extending from his treatment of radicals, nucleophiles, carbenium ions, and organometallic agents to concerted reactions and redox chemistry, Bruckner provides an analysis that effectively merges theory and mechanism with examples and applications. His selection of examples is superb and is further enhanced by the contemporary references to the literature. The text provides clarity that is essential for facilitating the educational process.

This is a wonderfully rich treatment of organic chemistry that will be a great value to students at any level. Education should enable and empower. This text does both, providing the student with the insights and tools needed to address the tremendous challenges and opportunities in the field. Congratulations to Professors Bruckner, Harmata, and Glaser for providing such a rich and clear path for those embarking on an understanding of the richly rewarding field of organic chemistry.

Paul A. Wender
Stanford University

Preface to the English Edition

Writing a textbook at any level is always a challenge. In organic chemistry, exciting new discoveries are being made at an ever-increasing pace. However, students of the subject still arrive in the classroom knowing only what they have been taught, often less. The challenge is to present appropriate review material, present venerable, classic chemistry while dealing with the latest results, and, most importantly, provoke thought and discussion. At the time this book was written, there was a need for an advanced text that incorporated these aspects of our science.

The German version of the text was designed for second- and third-year chemistry majors: 60–70% of the contents of this book address students before the "Diplom-chemiker-Vorexamen," while the remaining 30–40% address them thereafter. The German book is typically used one year after a standard introductory textbook such as that by Vollhardt and Schore, Streitweiser and Heathcock, or McMurry. Accordingly, in the United States this text can be used in a class for advanced undergraduates or beginning graduate students. Curricula of other English-speaking countries should allow the use of this text with optimum benefit at a similar point of progress. A good understanding of the fundamentals of organic and physical chemistry will suffice as a foundation for using this textbook to advantage.

The approach taken in this book conveys the message that the underlying theory of organic chemistry pervades the entire science. It is not necessary at this level to restrict the learning of reactions and mechanisms to any particular order. MO theory and formalisms such as electron pushing with arrows are powerful tools that can be applied not only to the classic chemistry that led to their development but also to the most recently developed reactions and methods, even those that use transition metals.

Theory, mechanism, synthesis, structure, and stereochemistry are discussed throughout the book in a qualitative to semiquantitative fashion. Fundamental principles such as the Hammond postulate that can be applied in the most varied contexts are reinforced throughout the book. Equations such as the Erying equation or the rate laws of all kinds of reactions are introduced with the view that they have context and meaning and are not merely formulas into which numbers are plugged.

The present text, to the best of our knowledge, does not duplicate the approach of any other treatment at a comparable level. We are convinced that this book, which has already filled a niche in the educational systems of German- and the French-speaking countries (a French translation appeared in 1999), will do the same in the textbook market of English-speaking countries now that an English edition has become available.

We hope that you enjoy many fruitful hours of insight in the course of studying this book, and we welcome your constructive comments on its content and approach.

Michael Harmata
Norman Rabjohn Distinguished Professor of Organic Chemistry
Department of Chemistry
University of Missouri
Columbia, Missouri 65211
(*for feedback:* HarmataM@missouri.edu)

Reinhard Bruckner
Professor of Organic Chemistry
Institut für Organische Chemie und Biochemie
der Albert-Ludwigs-Universität
Albertstrasse 21
79104 Freiburg, Germany
(*for feedback:* reinhard.brueckner@organik.chemie.uni-freiburg.de)

April 16, 2001

Preface to the German Edition

To really understand organic chemistry requires three passes. First, one must familiarize oneself with the physical and chemical properties of organic chemical compounds. Then one needs to understand their reactivities and their options for reactions. Finally, one must develop the ability to design syntheses. A typical schedule of courses for chemistry students clearly incorporates these three components. Introductory courses focus on compounds, a course on reaction mechanisms follows, and a course on advanced organic chemistry provides more specialized knowledge and an introduction to retrosynthesis.

Experience shows that the *second* pass, the presentation of the material organized according to reaction mechanisms, is of central significance to students of organic chemistry. This systematic presentation reassures students not only that they can master the subject but also that they might enjoy studying organic chemistry.

I taught the reaction mechanisms course at the University of Göttingen in the winter semester of 1994, and by the end of the semester the students had acquired a competence in organic chemistry that was gratifying to all concerned. Later, I taught the same course again—I still liked its outline—and I began to wonder whether I should write a textbook based on this course. A text *of this kind* was not then available, so I presented the idea to Björn Gondesen, the editor of *Spektrum*. Björn Gondesen enthusiastically welcomed the book proposal and asked me to write the "little booklet" as soon as possible. I gave up my private life and wrote for just about two years. I am grateful to my wife that we are still married; thank you, Jutta!

To this day, it remains unclear whether Björn Gondesen used the term "little booklet" in earnest or merely to indicate that he expected *one* book rather than a series of volumes. In any case, I am grateful to him for having endured patiently the mutations of the "little booklet" first to a "book" and then to a "mature textbook." In fact, the editor demonstrated an indestructible enthusiasm, and he remained supportive when I presented him repeatedly with increases in the manuscript of yet another 50 pages. The reader has Björn Gondesen to thank for the two-color production of this book. All "curved arrows" that indicate electron shifts are shown in red so that the student can easily grasp the reaction. Definitions and important statements also are graphically highlighted.

In comparison to the preceding generation, students of today study chemistry with a big handicap: an explosive growth of knowledge in all the sciences has been accompanied in particular by the need for students of organic chemistry to learn a greater number of reactions than was required previously. The omission of older knowledge is possible only if that knowledge has become less relevant and, for this reason, the following reactions were omitted: Darzens glycidic ester synthesis, Cope elimination, S_Ni reaction, iodoform reaction, Reimer–Tiemann reaction, Stobble condensation, Perkin synthesis, benzoin condensation, Favorskii rearrangement, benzil–benzilic acid rearrangement, Hofmann and Lossen degradation, Meerwein–Ponndorf reduction, and Cannizzaro reaction. A few other reactions were omitted because they did not fit into the current presentation (nitrile and alkyne chemistry, cyanohydrin formation, reductive amination, Mannich reaction, enol and enamine reactions).

This book is a highly modern text. All the mechanisms described concern reactions that are used today. The mechanisms are not just *l'art pour l'art*. Rather, they present a conceptual tool to facilitate the learning of reactions that one needs to know in any case. Among the modern reactions included in the present text are the following: Barton–McCombie reaction, Mitsunobu reaction, Mukaiyama redox condensations, asymmetric hydroboration, halolactonizations, Sharpless epoxidation, Julia–Lythgoe and Peterson olefination, *ortho*-lithiation, *in situ* activation of carboxylic acids, preparations and reactions of Gilman, Normant, and Knochel cuprates, alkylation of chiral enolates (with the methods by Evans, Helmchen, and Enders), diastereoselective aldol additions (Heathcock method, Zimmerman–Traxler model), Claisen–Ireland rearrangements, transition metal–mediated C,C-coupling reactions, Swern and Dess-Martin oxidations, reductive lithiations, enantioselective carbonyl reductions (Noyori, Brown, and Corey–Itsuno methods), and asymmetrical olefin hydrogenations.

The presentations of many reactions integrate discussions of stereochemical aspects. Syntheses of mixtures of stereoisomers of the target molecule no longer are viewed as valuable—indeed such mixtures are considered to be worthless—and the control of the stereoselectivity of organic chemical reactions is of paramount significance. Hence, suitable examples were chosen to present aspects of modern stereochemistry, and these include the following: control of stereoselectivity by the substrate, the reagent, or an ancilliary reagent; double stereodifferentiation; induced and simple diastereoselectivity; Cram, Cram chelate, and Felkin–Anh selectivity; asymmetric synthesis; kinetic resolution; and mutual kinetic resolution.

B

Indicates relevance for undergraduate students

Indicates relevance for graduate students

You might ask how then, for heaven's sake, is one to remember all of this extensive material? Well, the present text contains only about 70% of the knowledge that I would expect from a *really well-trained* undergraduate student; the remaining 30% presents material for graduate students. To ensure the best orientation of the reader, the sections that are most relevant for optimal undergraduate studies are marked in the margin with a B on a gray background, and sections relevant primarily to graduate students are marked with an A on a red background. I have worked most diligently to show the reactions in reaction diagrams that include every intermediate—and in which the flow of the valence electrons is highlighted in color—and, whenever necessary, to further discuss the reactions in the text. It has been my aim to describe all reactions so well, that in hindsight—because the course of every reaction will seem so plausible—the readers feel that they might even have *predicted* their outcome. I tried especially hard to realize this aim in the presentation of the chemistry of carbonyl compounds. These mechanisms are presented in four chapters (Chapters 7–11), while other authors usually cover all these reactions in one chapter. I hope this pedagogical approach will render organic chemistry more comprehensible to the reader.

Finally, it is my pleasure to thank—in addition to my untiring editor—everybody who contributed to the preparation of this book. I thank my wife, Jutta, for typing "version 1.0" of most of the chapters, a task that was difficult because she is not a chemist and that at times became downright "hair raising" because of the inadequacy of my dictation. I thank my co-workers Matthias Eckhardt (University of Göttingen, Dr. Eckhardt by now) and Kathrin Brüschke (chemistry student at the University of Leipzig) for their careful reviews of the much later "version .10" of the chapters. Their comments and corrections resulted in "version .11" of the manuscript, which was then edited professionally by Dr. Barbara Elvers (Oslo). In particular, Dr. Elvers polished the language of sections that had remained unclear, and I am very grateful for her editing. Dr. Wolfgang Zettelmeier (Laaber-Waldetzenberg) prepared the drawings for the resulting "version .12," demonstrating great sensitivity to my aesthetic wishes. The typsesetting was accomplished essentially error-free by Konrad Triltsch (Würzburg), and my final review of the galley pages led to the publication of "version .13" in book form. The production department was turned upside-down by all the "last minute" changes—thank you very much, Mrs. Notacker! Readers who note any errors, awkward formulations, or inconsistencies are heartily encouraged to contact me. One of these days, there will be a "version .14."

It is my hope that your reading of this text will be enjoyable and useful, and that it might convince many of you to specialize in organic chemistry.

Reinhard Brückner
Göttingen, August 8, 1996

Acknowledgments

My part in this endeavor is over. Now, it is entirely up to the staff at Harcourt/Academic Press to take charge of the final countdown that will launch *Advanced Organic Chemistry: Reaction Mechanisms* onto the English-speaking market. After three years of intense trans-Atlantic cooperation, it is my sincere desire to thank those individuals in the United States who made this enterprise possible. I am extremely obliged to Professor Michael Harmata from the University of Missouri at Columbia for the great determination he exhibited at *all* phases of the project. It was he who doggedly did the legwork at the 1997 ACS meeting in San Francisco, that is, cruised from one science publisher's stand to the next, dropped complimentary copies of the German edition on various desks, and talked fervently to the responsibles. David Phanco from Academic Press was immediately intrigued and quickly set up an agreement with the German publisher. David Phanco was farsighted enough to include Mike Harmata on board as a "language polisher" (of the translation) before he passed on the torch to Jeremy Hayhurst in what then was to become Harcourt/Academic Press. The latter's sympathetic understanding and constant support in the year to follow were absolutely essential to the final success of the project: Mike Harmata, at that time a Humboldt Fellow at the University of Göttingen, and I needed to develop a very Prussian sense of discipline when doing our best to match the first part of the translation to the quality of the original. I am very much indebted to Professor Rainer Glaser, who reinforced the Missouri team and, being bilingual, finished the second half of the translation skillfully and with amazing speed. He also contributed very valuably to improving the galley proofs, as did Joanna Dinsmore, Production Manager at Harcourt/Academic Press. It is she who deserves a great deal of gratitude for her diligence in countless hours of proofreading, and for her patience with an author who even at the page proof stage felt that it was never too late to make all sorts of small amendments for the future reader's sake. It is my sincere hope, Ms. Dinsmore, that in the end you, too, feel that this immense effort was worth the trials and tribulations that accompanied it.

Reinhard Bruckner
Freiburg, April 25, 2001

Radical Substitution Reactions at the Saturated C Atom

1

B

In a substitution reaction a part X of a molecule R—X is replaced by a group Y (Figure 1.1). The subject of this chapter is substitution reactions in which a part X that is bound to an sp^3-hybridized C atom is replaced by a group Y via radical intermediates. Radicals are valence-unsaturated and therefore usually short-lived atoms or molecules. They contain one or more unpaired ("lone") electrons. From inorganic chemistry you are familiar with at least two radicals, which by the way are quite stable: NO and O_2. NO contains one lone electron; it is therefore a monoradical or simply a "radical." O_2 contains two lone electrons and is therefore a biradical.

R_{sp^3}—X

+ Reagent,
– By-products

R_{sp^3}—Y

X = H, Hal, O—C(=S)—SMe, O—C(=S)—N (imidazole)

Y = H, Hal, OOH, CH_2—CH_2(CO_2R), C—CH

Fig. 1.1. Some substrates and products of radical substitution reactions.

1.1 Bonding and Preferred Geometries in C Radicals, Carbenium Ions and Carbanions

B

At the so-called radical center an organic radical R· has an electron septet, which is an electron deficiency in comparison to the electron octet of valence-saturated compounds. Carbon atoms are the most frequently found radical centers and most often have three neighbors (see below). Carbon-centered radicals with their electron septet occupy an intermediate position between the carbenium ions, which have one electron less (electron sextet at the valence-unsaturated C atom), and the carbanions, which have one electron more (electron octet at the valence-unsaturated C atom). Since there is an electron deficiency present both in C radicals and in carbenium ions, the latter are more closely related to each other than C radicals are related to carbanions. Because of this, C radicals and carbenium ions are also stabilized or destabilized by the same substituents.

Nitrogen-centered radicals $(R_{sp^3})_2N\cdot$ or oxygen-centered radicals $(R_{sp^3})O\cdot$ are less stable than C-centered radicals $(R_{sp^3})_3C\cdot$. They are higher in energy because of the higher electronegativity of these elements relative to carbon. Nitrogen- or oxygen-centered radicals of the cited substitution pattern consequently have only a limited chance to exist.

Which geometries are preferred at the valence-unsaturated C atom of C radicals, and how do they differ from those of carbenium ions or carbanions? And what types of bonding are found at the valence-unsaturated C atoms of these three species? It is simplest to clarify the preferred geometries first (Section 1.1.1). As soon as these geometries are known, molecular orbital (MO) theory will be used to provide a description of the bonding (Section 1.1.2).

We will discuss the preferred geometries and the MO descriptions of C radicals and the corresponding carbenium ions or carbanions in two parts. In the first part we will examine C radicals, carbenium ions, and carbanions with a trivalent central C atom. The second part treats the analogous species with a divalent central C atom. A third part (species with a monovalent central C atom) can be dispensed with because the only species of this type that is important in organic chemistry is the alkynyl anion, which, however, is of no interest here.

1.1.1 Preferred Geometries

B

The preferred geometries of carbenium ions and carbanions are correctly predicted by the valence shell electron pair repulsion (VSEPR) theory. The VSEPR theory, which comes from inorganic chemistry, explains the stereostructure of covalent compounds of the nonmetals and the main group metals. It makes no difference whether these compounds are charged or not.

The VSEPR theory analyzes the stereostructure of these compounds in the environment of the central atom. This stereostructure depends mainly on (a) the number n of atoms or atom groups (in inorganic chemical terminology, referred to as ligands) linked to the central atom and (b) the number m of nonbonding valence electron pairs localized at the central atom. If the central atom under consideration is a C atom, $n + m \leq 4$. In this case, the VSEPR theory holds in the following shorthand version, which makes it possible to determine the preferred geometries: the compound considered has the stereostructure in which the repulsion between the n bonding partners and the m nonbonding valence electron pairs on the C atom is as small as possible. This is the case when the orbitals that accommodate the bonding and the nonbonding electron pairs are as far apart from each other as possible.

For carbenium ions this means that the n substituents of the valence-unsaturated central atom should be at the greatest possible distance from each other:

- In alkyl cations R_3C^+, $n = 3$ and $m = 0$. The substituents of the trivalent central atom lie in the same plane as the central atom and form bond angles of 120° with each other (trigonal planar arrangement). This arrangement was confirmed experimentally by means of a crystal structural analysis of the *tert*-butyl cation.
- In alkenyl cations $=C^+—R$, $n = 2$ and $m = 0$. The substituents of the divalent central atom lie on a common axis with the central atom and form a bond angle of 180°. Alkenyl cations have not been isolated yet because of their low stability (Section 1.2). However, calculations support the preference for the linear structure.

According to the VSEPR theory, in carbanions the *n* substituents at the carbanionic C atom and the nonbonding electron pair must move as far away from each other as possible:

- In alkyl anions R_3C^-, $n = 3$ and $m = 1$. The substituents lie in one plane, and the central atom lies outside it. The carbanion center has a trigonal pyramidal geometry. The bond angles are similar to the tetrahedral angle (109° 28′). This stereostructure may be called pseudotetrahedral when the carbanionic electron pair is counted as a pseudosubstituent.
- In alkenyl anions $=C^-—R$, $n = 2$ and $m = 1$. The substituents and the divalent central atom prefer a bent structure. The bond angle in alkenyl anions is approximately 120°. When the nonbonding valence electron pair is considered as a pseudosubstituent of the carbanion center, this preferred geometry may also be called pseudotrigonal planar.

The most stable structures of alkyl and alkenyl anions predicted with the VSEPR theory are supported by reliable calculations. There are no known experimental structural data. In fact, up to recently, one would have cited the many known geometries of the lithium derivatives of these carbanions as evidence for the structure. One would simply have "dropped" the C—Li bond(s) from these geometries. However, it is now known that the considerable covalent character of most C—Li bonds makes organolithium compounds unsuitable models for carbanions.

Since the VSEPR theory describes the mutual repulsion of valence electron *pairs*, it can hardly be used to make statements about the preferred geometries of C radicals. It is intuitively expected that C radicals assume a middle position between their carbenium ion and carbanion analogs. In agreement with this, alkyl radicals are either planar (methyl radical) or slightly pyramidal but able to swing rapidly through the planar form (inversion) to another near-planar structure (*tert*-butyl radical). In addition, some carbon-centered radicals are considerably pyramidalized (e.g., those whose carbon center is substituted with several heteroatoms). Alkenyl radicals are bent, but they can undergo *cis/trans* isomerization through the linear form very rapidly. Because they are constrained in a ring, aryl radicals are necessarily bent.

1.1.2 Bonding

The type of bonding at the valence-unsaturated C atom of carbenium ions, carbanions, and C-centered radicals follows from the geometries described in Section 1.1.1. From the bond angles at the central C atom, it is possible to derive its hybridization. Bond angles of 109° 28′ correspond to sp^3, bond angles of 120° correspond to sp^2, and bond angles of 180° correspond to *sp* hybridization. From this hybridization it follows which atomic orbitals (AOs) of the valence-unsaturated C atom are used to form the molecular orbitals (MOs). The latter can, on the one hand, be used as bonding MOs. Each one of them then contains a valence electron pair and represents the bond to a substituent of the central atom. On the other hand, one AO of the central atom represents a nonbonding MO, which is empty in the carbenium ion, contains an electron in the radical, and contains the nonbonding electron pair in the carbanion. How the valence electrons are distributed over the respective MO set follows from the *Aufbau* principle: they are placed, one after the other, in the MOs, in the order of increasing energy. The Pauli principle is also observed: any MO can take up only two electrons and only on the condition that they have opposite spins.

The bonding at the valence-unsaturated C atom of carbenium ions R_3C^+ is therefore described by the MO diagram in Figure 1.2 (left), and the bonding of the valence-unsaturated C atom of carbenium ions of type $=C^+—R$ is described by the MO diagram in Figure 1.3 (left). The MO description of R_3C^- carbanions is shown in Figure 1.2 (right), and the MO description of carbanions of type $=C^-—R$ is shown in Figure 1.3 (right). The MO description of the radicals R· or =CR· employs the MO picture for the analogous carbenium ions or carbanions, depending on which of these species the geometry of the radical is similar to. In each case only seven instead of six or eight valence electrons must be accommodated (Figures 1.2 and 1.3, left).

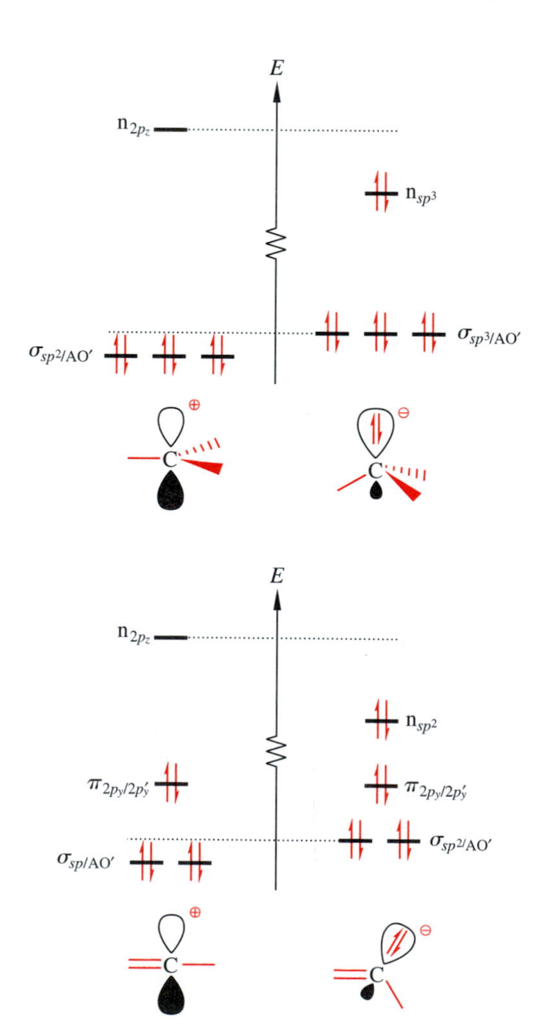

Fig. 1.2. Energy levels and occupancies (red) of the MOs at the trivalent C atom of planar carbenium ions R_3C^+ (left) and pyramidal carbanions R_3C^- (right).

Fig. 1.3. Energy levels and occupancies (red) of the MOs at the divalent C atom of linear carbenium ions $=C^+—R$ (left) and bent carbanions $=C^-—R$ (right).

1.2 Stability of Radicals

B

Stability in chemistry is not an absolute but a relative concept. It always refers to a stability difference with respect to a reference compound. Let us consider the standard heats of reaction ΔH^0 of the dissociation reaction R—H → R· + H·, that is, the dissociation enthalpy (DE) of the broken C—H bond. It reflects, on the one hand, the strength of this C—H bond and, on the other hand, the stability of the radical R· produced. As you see immediately, the dissociation enthalpy of the R—H bond depends in many ways on the structure of R. But it is not possible to tell clearly whether this is due to an effect on the bond energy of the broken R—H bond and/or an effect on the stability of the radical R· that is formed.

	$HC{\equiv}C_{sp}{-}H$	$C_6H_5{-}C_{sp^2}{-}H$ (phenyl)	$H_2C{=}C_{sp^2}H{-}H$	$H_3C{-}C_{sp^3}H_2{-}H$
DE/kcal/mol	131	111	110	98

To what must one ascribe, for example, the fact that the dissociation enthalpy of a C_{sp^n}—H bond depends essentially on n alone and increases in the order n = 3,2, and 1?

To help answer this question it is worthwhile considering the following: the dissociation enthalpies of bonds such as C_{sp^n}—C, C_{sp^n}—O, C_{sp^n}—Cl, and C_{sp^n}—Br also depend heavily on n and increase in the same order, n = 3, 2, and 1. The extent of the n–dependence of the dissocation energies depends on the element which is cleaved off. This is only possible if the n–dependence reflects, at least in part, an n–dependence of the respective C_{sp^n}–element bond. (Bond enthalpy tables in all textbooks ignore this and assign a bond enthalpy to each C_{sp^n}–element bond that is dependent on the element but not on the value of n!) Hence, the bond enthalpy of every C_{sp^n}–element bond increases in the order n = *3, 2, and 1*. This is so because all C_{sp^n}–element bonds become shorter in the same order. This in turn is due to the s character of the C_{sp^n}–element bond, which increases in the same direction.

An immediate consequence of the different ease with which C_{sp^n}–element bonds dissociate is that in radical substitution reactions, alkyl radicals are preferentially formed. Only in exceptional cases are vinyl or aryl radicals formed. Alkynyl radicals do not appear at all in radical substitution reactions. In the following we therefore limit ourselves to a discussion of substitution reactions that take place via radicals of the general structure $R^1R^2R^3C\cdot$.

1.2.1 Reactive Radicals

If radicals $R^1R^2R^3C\cdot$ are produced by breaking the C—H bond in molecules of the type $R^1R^2R^3C$—H, one finds that the dissociation enthalpies of such C—H bonds differ with molecular structure. Experience shows that these differences can be explained completely by the effects of the substituents R^1, R^2, and R^3 on the stability of the radicals $R^1R^2R^3C\cdot$ formed.

Table 1.1 shows one substituent effect, which influences the stability of radicals. The dissociation enthalpies of reactions that lead to $R{-}CH_2\cdot$ radicals are listed. The substituent R varies from C_2H_5 through $H_2C{=}CH{-}$(vinyl substituent, vin) to $C_6H_5{-}$ (phenyl substituent, Ph). The dissociation enthalpy is greatest for R = H. This shows that a radical center is stabilized by 9 kcal/mol by the neighboring C=C double bond of an alkenyl or aryl substituent.

Table 1.1. Stabilization of Radicals by Unsaturated Substituents

	$\frac{DE}{\text{kcal/mol}}$	VB formulation of the radical R •
H	98	
H	89	
H	89	

In the valence-bond (VB) model this effect results from the fact that radicals of this type can be described by superpositioning several resonance forms (Table 1.1, right). In the MO model, the stabilization of radical centers of this type is due to the overlap of the π system of the unsaturated substituent with the $2p_z$ AO at the radical center (Figure 1.4). This overlap is called conjugation.

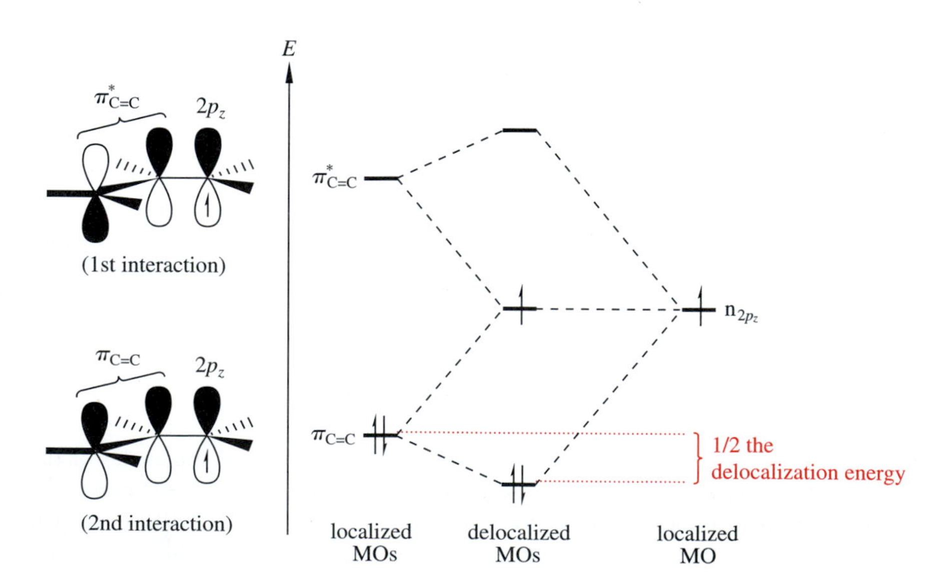

Fig. 1.4. Stabilization by overlap of a singly occupied $2p_z$ AO with adjacent parallel $\pi_{C=C}$- or $\pi^*_{C=C}$- MOs.

Table 1.2 illustrates an additional substituent effect on radical stability. Here the dissociation enthalpies of reactions that lead to polyalkylated radicals $(Alk)_{3-n}H_nC\cdot$ are listed ("Alk" stands for alkyl group). From these dissociation enthalpies it can be seen that alkyl substituents stabilize radicals. A primary radical is 6 kcal/mol more stable, a secondary radical is 9 kcal/mol more stable, and a tertiary radical is 12 kcal/mol more stable than the methyl radical.

Table 1.2. Stabilization of Radicals by Alkyl Substituents

	$\frac{DE}{\text{kcal/mol}}$	VB formulation of the radical R •
$H_3C{-}H$	104	
$H_3C{-}H_2C{-}H$	98	i.e., 1 no-bond resonance form per H_β
$(H_3C)_2HC{-}H$	95	6 no-bond resonance forms
$(H_3C)_3C{-}H$	92	9 no-bond resonance forms

In the VB model, this effect is explained by the fact that radicals of this type, too, can be described by the superpositioning of several resonance forms. These are the somewhat exotic no-bond resonance forms (Table 1.2, right). From the point of view of the MO model, radical centers with alkyl substituents have the opportunity to interact with these substituents. This interaction involves the C—H bonds that are in the position α to the radical center and lie in the same plane as the $2p_z$ AO at the radical center. Specifically, σ_{C-H} MOs of these C—H bonds are able to overlap with the radical $2p_z$ orbital (Figure 1.5). This overlap represents the rare case of lateral overlap between a σ bond and a p orbital. It is referred to as hyperconjugation to distinguish it from lateral overlap between π bonds and p orbitals, which is referred to as conjugation. When the σ_{C-H} bond and the $2p_z$ AO enclose a dihedral angle χ that is different from that required for optimum overlap (0°), the stabilization of the radical center by hyperconjugation decreases. In fact, it decreases by the square of the cosine of the dihedral angle χ.

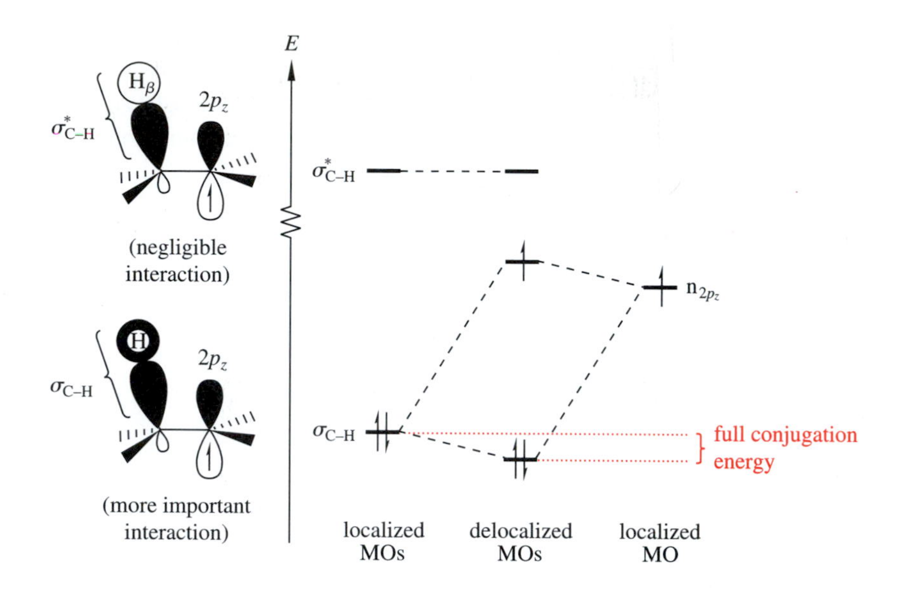

Fig. 1.5. Stabilization by overlap of a singly occupied $2p_z$ AO with vicinal nonorthogonal σ_{C-H} MOs.

1.2.2 Unreactive Radicals

B

Just as several alkyl substituents increasingly stabilize a radical center (Table 1.2), two phenyl substituents stabilize a radical center more than one does. The diphenylmethyl radical ("benzhydryl radical") is therefore more stable than the benzyl radical. The triphenylmethyl radical ("trityl radical") is even more stable because of the three phenyl substituents. They actually stabilize the trityl radical to such an extent that it forms by homolysis from the so-called Gomberg hydrocarbon even at room temperature (Figure 1.6). Although this reaction is reversible, the trityl radical is present in equilibrium quantities of about 2 mol%.

Gomberg hydrocarbon → H ⇌ 2 $Ph_3C\cdot$

Fig. 1.6. Reversible formation reaction of the triphenylmethyl radical. The equilibrium lies on the side of the Gomberg hydrocarbon.

A

Starting from the structure of the trityl radical, radicals were designed that can be obtained even *in pure form* as "stable radicals" (Figure 1.7). There are two reasons why these radicals are so stable. For one thing, they are exceptionally well resonance-stabilized. In addition, their dimerization to valence-saturated species has a considerably reduced driving force. In the case of the trityl radical, for example, dimerization leads to the Gomberg hydrocarbon in which an aromatic sextet is lost. The trityl radical can not dimerize giving hexaphenylethane, because too severe van der Waals repulsions between the substituents would occur. There are also stable N- or O-centered radicals. The driving force for their dimerization is weak because relatively weak N—N or O—O bonds would be formed.

By the way, the destabilization of the dimerization product of a radical is often more important for the existence of stable radicals than optimum resonance stabilization. This is shown by comparison of the trityl radical derivatives **A** and **B** (Figure 1.7). In radical **A** the inclusion of the radical center in the polycycle makes optimum resonance stabilization possible because the dihedral angle χ between the $2p_z$ AO at the central atom and the neighboring π orbitals of the three surrounding benzene rings is exactly 0°. And yet radical **A** dimerizes! In contrast, the trityl radical derivative **B** is distorted like a propeller, to minimize the interaction between the methoxy substituents on the adjacent rings. The $2p_z$ AO at the central atom of radical **B** and the π orbitals of the surrounding benzene rings therefore form a dihedral angle χ of a little more than 45°. The resonance stabilization of radical **B** is therefore only one half as great—$\cos^2 45° = 0.50$—as that of radical **A**. In spite of this, radical **B** does not dimerize at all.

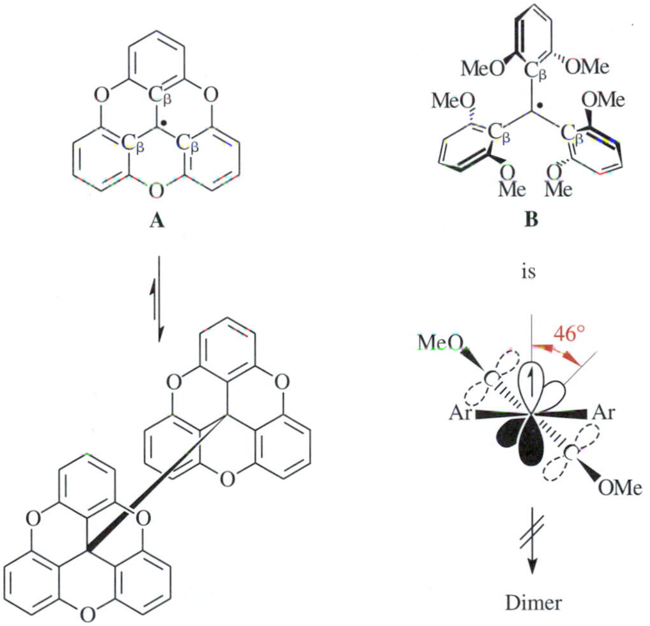

Fig. 1.7. Comparison of the trityl radical derivatives **A** and **B**; **A** dimerizes, **B** does not.

1.3 Relative Rates of Analogous Radical Reactions

In Section 1.2.1 we discussed the stabilities of reactive radicals. It is interesting that they make an evaluation of the relative rates of formation of these radicals possible.

This follows from the **Bell–Evans–Polanyi principle** (Section 1.3.1) or the **Hammond postulate** (Section 1.3.2).

1.3.1 The Bell–Evans–Polanyi Principle

In thermolyses of aliphatic azo compounds, two alkyl radicals and one equivalent of N_2 are produced according to the reaction at the bottom of Figure 1.8. A whole series of such reactions was carried out, and their heats of reaction, that is, their reaction enthalpies ΔH_r, were determined. Heat was taken up in all thermolyses. They were thus endothermic reactions (ΔH_r has a positive sign). Each substrate was thermolyzed at several different temperatures T and the associated rate constants k_r were determined. The temperature dependence of the k_r values for each individual reaction was analyzed by using the **Eyring equation** (Equation 1.1).

$$k_r = \frac{k_B \cdot T}{h} \exp\left(-\frac{\Delta G^{\ddagger}}{RT}\right) = \frac{k_B \cdot T}{h} \exp\left(-\frac{\Delta H^{\ddagger}}{RT}\right) \exp\left(+\frac{\Delta S^{\ddagger}}{R}\right) \tag{1.1}$$

k_B: Boltzmann constant (3.298×10^{-24} cal/K)
T: absolute temperature (K)
h: Planck's constant (1.583×10^{-34} cal·s)
$\Delta G^{\ddagger}$: Gibbs free energy (kcal/mol)
$\Delta H^{\ddagger}$: enthalpy of activation (kcal/mol)
$\Delta S^{\ddagger}$: entropy of activation (cal mol^{-1} K^{-1})
R: gas constant (1.986 cal mol^{-1} K^{-1})

Equation 1.1 becomes Equation 1.2 after (a) dividing by T, (b) taking the logarithm, and (c) differentiating with respect to T.

$$\Delta H^{\ddagger} = RT \frac{\partial \ln\left(\frac{k_r}{T}\right)}{\partial T} \tag{1.2}$$

With Equation 1.2 it was possible to calculate the activation enthalpy $\Delta H^{\ddagger}$ for each individual reaction.

The pairs of values $\Delta H_r/\Delta H^{\ddagger}$, which were now available for each thermolysis, were plotted on the diagram in Figure 1.8, with the enthalpy change ΔH on the vertical axis and the reaction progress on the horizontal axis. The horizontal axis is referred to as the **reaction coordinate** (RC). Among "practicing organic chemists" it is not accurately calibrated. It is implied that on the reaction coordinate one has moved by x% toward the reaction product(s) when all the structural changes that are necessary *en route* from the starting material(s) to the product(s) have been x% completed.

For five out of the six reactions investigated, Figure 1.8 shows an increase in the activation enthalpy $\Delta H^{\ddagger}$ with increasing positive reaction enthalpy ΔH_r. Only for the sixth reaction—drawn in red in Figure 1.8—is this not true. Accordingly, except for this one reaction $\Delta H^{\ddagger}$ and ΔH_r are proportional for this series of radical-producing thermolyses. This proportionality is known as the Bell–Evans–Polanyi principle and is described by Equation 1.3.

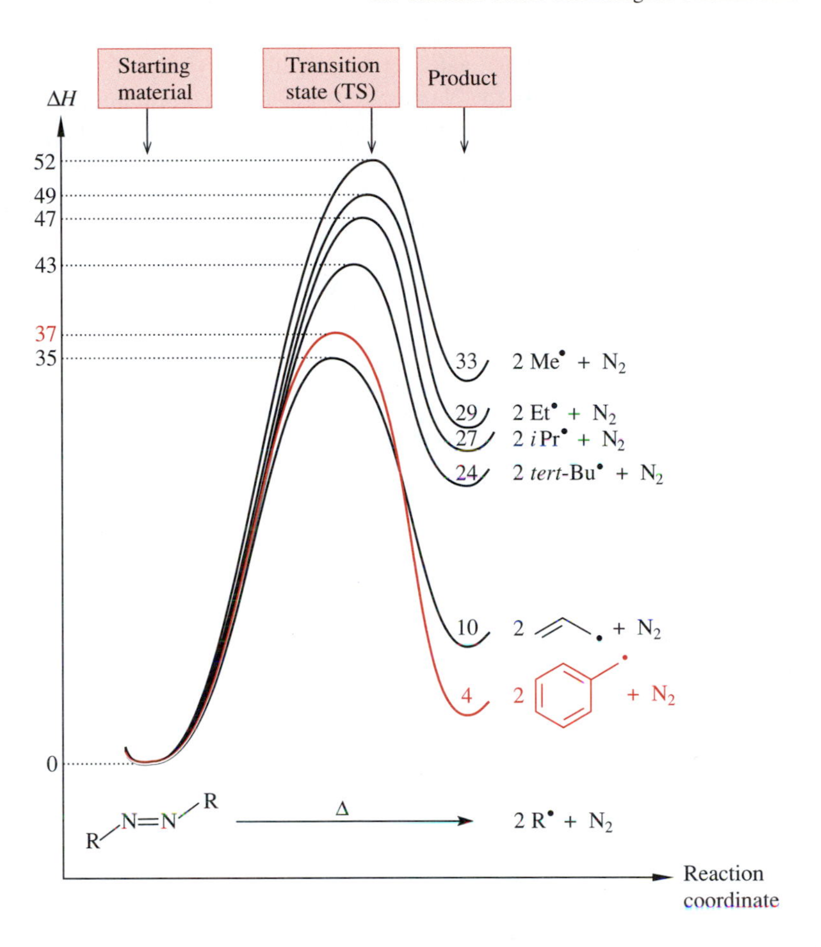

Fig. 1.8. Enthalpy change along the reaction coordinate in a series of thermolyses of aliphatic azo compounds. All thermolyses in this series except the one highlighted in color follow the Bell–Evans–Polanyi principle.

$$\Delta H^{\ddagger} = \text{const.} + \text{const.}' \cdot \Delta H_r \tag{1.3}$$

The thermolyses presented in this chapter are *one* example of a series of analogous reactions. The Bell–Evans–Polanyi relationship of Equation 1.3 also holds for many other series of analogous reactions.

1.3.2 The Hammond Postulate

B

In many series of analogous reactions a second proportionality is found experimentally, namely, between the free energy change (ΔG_r; a thermodynamic quantity) and the free energy of activation ($\Delta G^{\ddagger}$, a kinetic quantity). In series of analogous reactions, a third parameter besides $\Delta H^{\ddagger}$ and $\Delta G^{\ddagger}$, no doubt also depends on the ΔH_r and ΔG_r values, respectively, namely, the structure of the transition state. This relationship is generally assumed or postulated, and only in a few cases has it been verified by calculations (albeit usually only in the form of the so-called "transition structures"; they are likely to resemble the structures of the transition state, however). This relationship is therefore not stated as a law or a principle but as a postulate, the so-called Hammond postulate.

The Hammond Postulate

The Hammond postulate can be stated in several different ways. For *individual reactions* the following form of the Hammond postulate applies. In an endergonic reaction the transition state (TS) is similar to the *product(s)* with respect to energy and structure. Endergonic reactions thus take place through so-called late transition states. (A reaction is endergonic when the free energy change ΔG_r, is greater than zero.) Conversely, in an exergonic reaction the transition state is similar to the *starting material(s)* with respect to energy and structure. Exergonic reactions thus take place via so-called early transition states. (A reaction is called exergonic when the change in the free energy ΔG_r is less than zero.)

For series of analogous reactions this results in the following form of the Hammond postulate: in a series of *increasingly endergonic* analogous reactions the transition state is *increasingly* similar to the product(s), i.e., increasingly late. On the other hand, in a series of *increasingly exergonic* analogous reactions, the transition state is *increasingly* similar to the starting material(s), i.e., increasingly early.

What does the statement that increasingly endergonic reactions take place via increasingly product-like transition states mean for the special case of two irreversible endergonic analogous reactions, which occur as competitive reactions? With help from the foregoing statement, the outcome of this competition can often be predicted. The energy of the competing transition states should be ordered in the same way as the energy of the potential reaction products. This means that the more stable reaction product is formed via the lower-energy transition state. It is therefore produced more rapidly or, in other words, in a higher yield than the less stable reaction product.

The form of the Hammond postulate just presented is very important in the analysis of the selectivity of many of the reactions we will discuss in this book in connection with chemoselectivity (definition in Section 1.7.2; also see Section 3.2.2), stereoselectivity (definition in Section 3.2.2), diastereoselectivity (definition in Section 3.2.2), enantioselectivity (definition in Section 3.2.2), and regioselectivity (definition in Section 1.7.2).

Selectivity

Selectivity means that one of several reaction products is formed preferentially or exclusively. In the simplest case, for example, reaction product 1 is formed at the expense of reaction product 2. Selectivities of this type are usually the result of a *kinetically controlled reaction process,* or "kinetic control." This means that they are usually not the consequence of an equilibrium being established under the reaction conditions between the alternative reaction products 1 and 2. In this latter case one would have a *thermodynamically controlled reaction process,* or "thermodynamic control."

Hammond Postulate and Kinetically Determined Selectivities

With reference to the occurrence of this type of kinetically determined selectivities of organic chemical reactions, the Hammond postulate now states that:

- If the reactions leading to the alternative reaction products are one step, the most stable product is produced most rapidly, that is, more or less selectively. This type of selectivity is called product-development control.

- If these reactions are two step, the product that is derived from the more stable intermediate is produced more rapidly, that is, more or less selectively.
- If these reactions are more than two step, one must identify the least stable intermediate in each of the alternative pathways. Of these high-energy intermediates, the least energy-rich is formed most rapidly and leads to a product that, therefore, is then formed more or less selectively. The selectivity in cases 2 and 3 is therefore also due to "product development control."

1.4 Radical Substitution Reactions: Chain Reactions

B

Radical substitution reactions can be written in the following form:

$$R_{sp^3}-X \xrightarrow[\text{radical initiator (cat.)}]{\text{reagent}} R_{sp^3}-Y$$

All radical substitution reactions are chain reactions. Every chain reaction starts with an initiation reaction. In one or more steps, this reaction converts a valence-saturated compound into a radical, which is sometimes called an initiating radical (the reaction arrow with the circle means that the reaction takes place through several intermediates, which are not shown here):

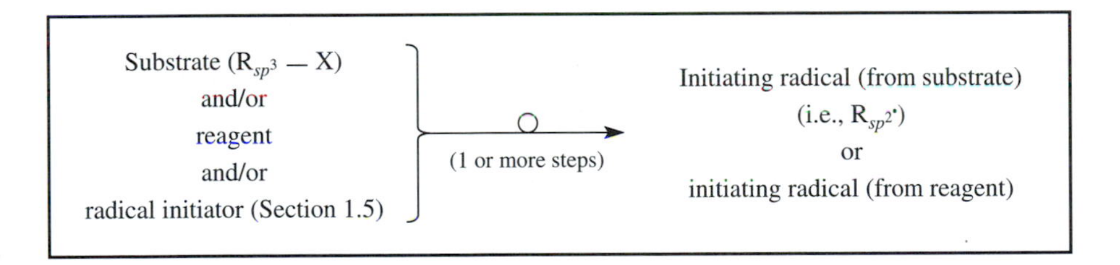

The initiating radical is the radical that initiates a sequence of two, three, or more so-called propagation steps:

$$\text{Initiating radical (from substrate) + reagent} \xrightarrow{k_{\text{prop},1}} \ldots$$

$$\ldots \xrightarrow{k_{\text{prop},n}} \ldots$$

$$\ldots \xrightarrow{k_{\text{prop},\omega}} \text{Initiating radical (from substrate)} + \ldots$$

$$\Sigma_{\text{Propagation steps}}: \quad R_{sp^3}-X + \text{Reagent} \longrightarrow R_{sp^3}-Y + \text{By-product(s)}$$

Depending on whether the initiating radical comes from the substrate or the reagent, the propagation steps must be formulated as above or as follows:

$$\text{Initiating radical (from reagent) } R_{sp^3}-X \xrightarrow{k_{\text{prop},1}} \ldots$$

$$\ldots \xrightarrow{k_{\text{prop},n}} \ldots$$

$$\ldots \xrightarrow{k_{\text{prop},\omega}} \text{Initiating radical (from reagent)} + \ldots$$

$$\Sigma_{\text{Propagation steps}}: \quad R_{sp^3}-X + \text{Reagent} \longrightarrow R_{sp^3}-Y + \text{By-product(s)}$$

As the reaction equations show, the last propagation step supplies the initiating radical consumed in the first propagation step. From this you also see that the mass conversion of a chain reaction is described by an equation that results from the propagation steps alone: they are added up, and species that occur on both sides of the reaction arrow are dropped.

If the radical intermediates of the propagation steps did nothing other than always enter into the next propagation step of the chain again, even a single initiating radical could initiate a complete starting material(s) → product(s) conversion. However, radical intermediates may also react with each other or with different radicals. This makes them disappear, and the chain reaction comes to a stop.

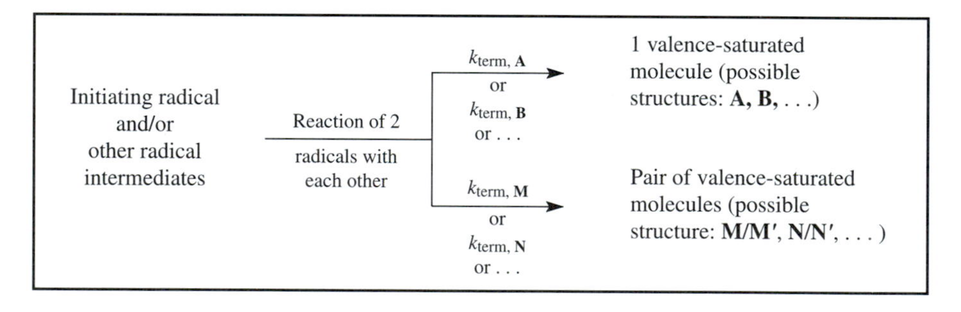

B

Reactions of the latter type therefore represent chain-terminating reactions or termination steps of the radical chain. A continuation of the starting material(s) → product(s) conversion becomes possible again only when a new initiating radical is made available via the starting reaction(s). Thus, for radical substitutions via chain reactions to take place *completely*, new initiating radicals must be produced continuously.

The ratio of the rate constants of the propagation and the termination steps determines how many times the sequence of the propagation steps is run through before a termination step ends the conversion of starting material(s) to product(s). The rate constants of the propagation steps (k_{prop} in the second- and third-to-last boxes of the present section) are greater than those of the termination steps (k_{term} in the fourth box), frequently by several orders of magnitude. An initiating radical can therefore initiate from 1000 to 100,000 passes through the propagation steps of the chain.

How does this order of the rate constants $k_{prop} \gg k_{term}$ come about? As high-energy species, radical intermediates react exergonically with most reaction partners. According to the Hammond postulate, they do this very rapidly. Radicals actually often react with the first reaction partner they encounter. Their average lifetime is therefore very short. The probability of a termination step in which *two* such short-lived radicals must meet is consequently low.

There is a great diversity of starting reaction(s) and propagation steps for radical substitution reactions. Bond homolyses, fragmentations, atom abstraction reactions, and addition reactions to C=C double bonds are among the possibilities. All of these reactions can be observed with substituted alkylmercury(II) hydrides as starting materials. For this reason, we will examine these reactions as the first radical reactions in Section 1.6.

1.5 Radical Initiators

Only for some of the radical reactions discussed in Sections 1.6–1.9 is the initiating radical produced immediately from the starting material or the reagent. In all other radical substitution reactions an auxiliary substance, the radical initiator, added in a substoichiometric amount, is responsible for producing the initiating radical.

Radical initiators are thermally labile compounds, which decompose into radicals upon moderate heating. These radicals initiate the actual radical chain through the formation of the initiating radical. The most frequently used radical initiators are azobisisobutyronitrile (AIBN) and dibenzoyl peroxide (Figure 1.9). After AIBN has been heated for only 1 h at 80°C, it is half-decomposed, and after dibenzoyl peroxide has been heated for only 1 h at 95°C, it is half-decomposed as well.

Azobisisobutyronitrile (AIBN) as radical initiator:

NC–C(CH₃)₂–N=N–C(CH₃)₂–CN ⟶ NC–C(CH₃)₂• + N≡N + •C(CH₃)₂–CN

Dibenzoyl peroxide as radical initiator:

PhC(=O)O–OC(=O)Ph ⟶ 2 PhC(=O)O•

2 PhC(=O)O• ⟶ 2 Ph• + 2 O=C=O

Fig. 1.9. Radical initiators and their mode of action (in the "arrow formalism" for showing reaction mechanisms used in organic chemistry, arrows with half-heads show where *single* electrons are shifted, whereas arrows with full heads show where electron *pairs* are shifted).

Side Note 1.1 Decomposition of Ozone in the Upper Stratosphere

Reactions that take place via radical intermediates are occasionally also begun by radical initiators, which are present unintentionally. Examples are the autooxidation of ethers (see later: Figure 1.28) or one of the ways in which ozone is decomposed in the upper stratosphere. This decomposition is initiated by, among other things, the **f**luoro**c**hloro**h**ydro**c**arbons (FCHCs), which have risen up there and form chlorine radicals under the influence of the short-wave UV light from the sun (Figure 1.10). They function as initiating radicals for the decomposition of ozone, which takes place via a radical chain. However, this does not involve a radical substitution reaction.

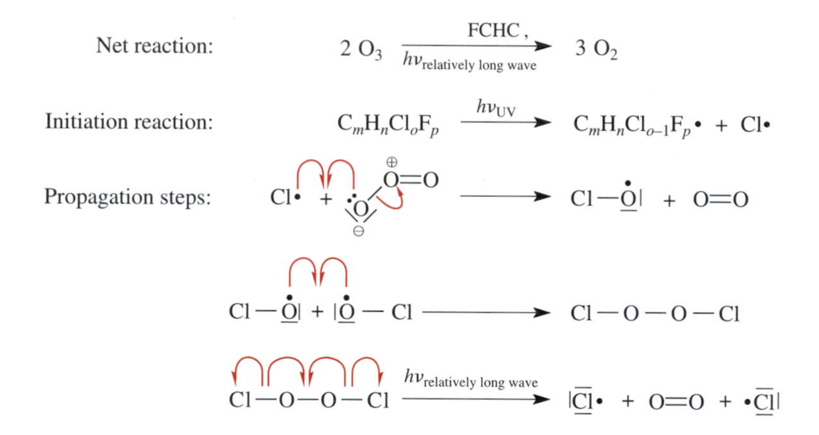

Fig. 1.10. FCHC-initiated decomposition of stratospheric ozone.

1.6 Radical Chemistry of Alkylmercury(II) Hydrides

B

Alkyl derivatives of mercury in the oxidation state +2 are produced during the solvomercuration of olefins (the first partial reaction of the reaction sequence in Figure 1.11).

$$\text{R-CH=CH}_2 \xrightarrow[\text{R''OH}]{\text{Hg(O-C(=O)-R')}_2,} \text{R-CH(OR'')-CH}_2\text{-Hg-O-C(=O)-R'} \xrightarrow{\text{NaBH}_4} \text{R-CH(OR'')-CH}_2\text{-H}$$

Fig. 1.11. Net reaction (a) for the hydration of olefins ($R' = CH_3$, $R'' = H$) or (b) for the addition of alcohol to olefins ($R' = CF_3$, $R'' =$ alkyl) via the reaction sequence (1) solvomercuration of the olefin (for mechanism, see Figure 3.37; regioselectivity: Figure 3.38); (2) reduction of the alkylmercury compound obtained (for mechanism, see Figure 1.12).

Oxymercuration provides (β-hydroxyalkyl)mercury(II) carboxylates while alkoxymercuration gives (β-alkoxyalkyl)mercury(II) carboxylates. These compounds can be reduced with $NaBH_4$ to (β-hydroxyalkyl)- or (β-alkoxyalkyl)mercury(II) hydrides. A ligand exchange takes place at the mercury: a carboxylate substituent is replaced by hydrogen. The β-oxygenated alkylmercury(II) hydrides obtained in this way are so unstable that they react further immediately. These reactions take place via radical intermediates. The latter can be transformed into various kinds of products by adjusting .the reaction conditions appropriately. The most important radical reactions of alkylmercury(II) hydrides are fragmentation to an alcohol (Figure 1.12), addition to a C=C double bond (Figure 1.13), and oxidation to a glycol derivative (Figure 1.14). The mechanisms for these reactions will be discussed below.

When (β-hydroxyalkyl)mercury(II) acetates are treated with $NaBH_4$ and no additional reagent, they first form (β-hydroxyalkyl)mercury(II) hydrides. These decompose via the chain reaction shown in Figure 1.12 to give a mercury-free alcohol. Overall, a substitution reaction R—Hg(OAc) → R—H takes place. The initiation step for the chain reaction participating in this transformation is a homolysis of the C—Hg bond. This takes place rapidly at room temperature and produces the radical ·Hg—H and a β-hydroxylated alkyl radical. As the initiating radical, it starts the first of the two

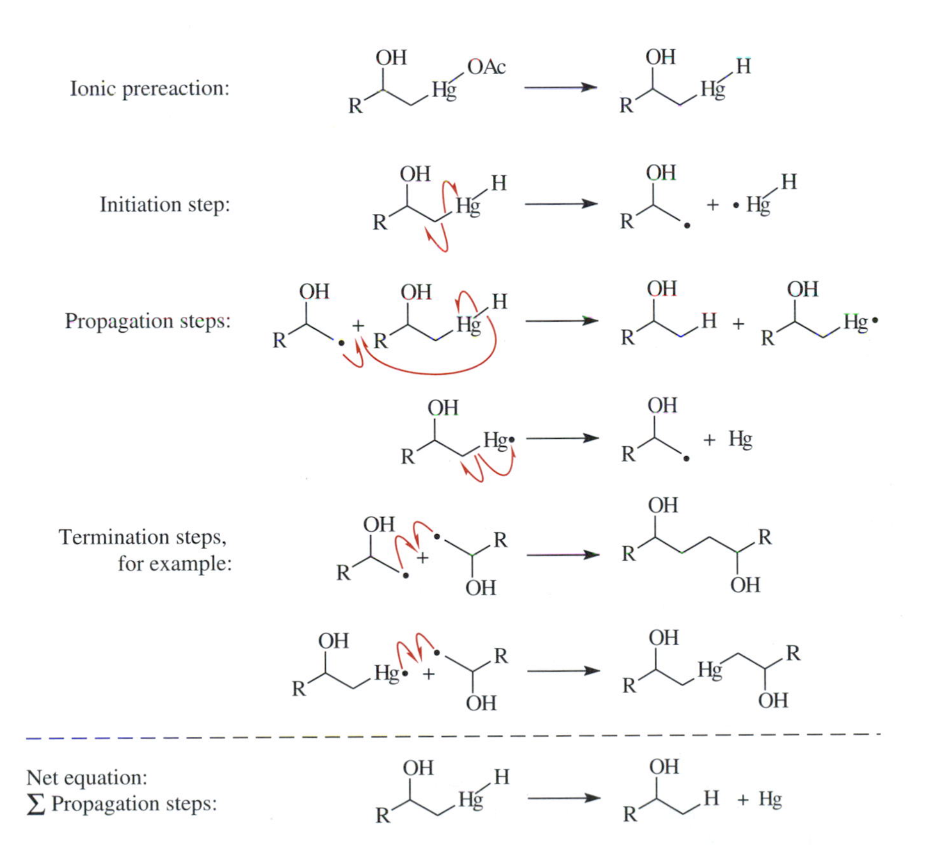

Fig. 1.12. $NaBH_4$ reduction of (β-hydroxyalkyl)mercury(II) acetates to alcohols and radical fragmentation of (β-hydroxyalkyl)mercury (II) hydrides.

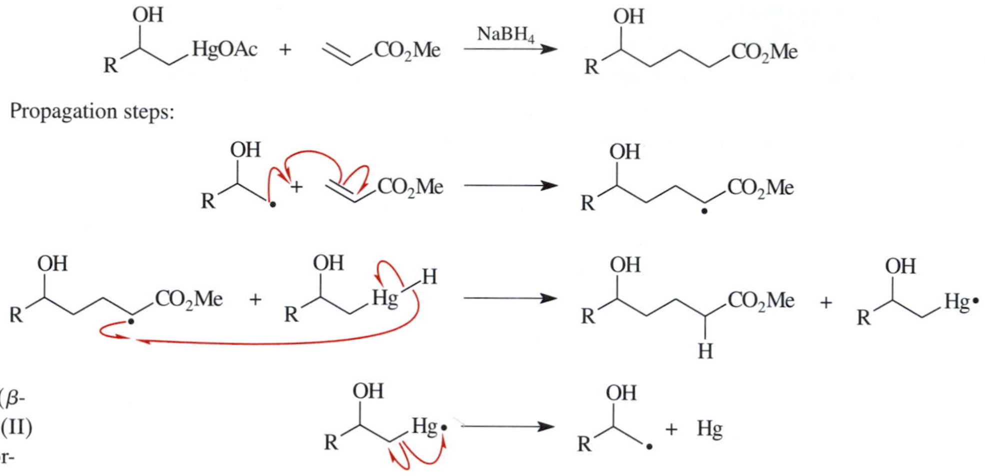

Propagation steps:

Fig. 1.13. $NaBH_4$-mediated addition of (β-hydroxyalkyl)mercury(II) acetates to an acceptor-substituted olefin.

propagation steps. This first step is an atom transfer reaction or, more specifically, a hydrogen atom transfer reaction. The second propagation step involves a radical fragmentation. This is the decomposition of a radical into a radical with a lower molecular weight and at least one valence-saturated compound (in this case, elemental mercury). The net reaction equation is obtained according to Section 1.4 by adding up the two propagation steps.

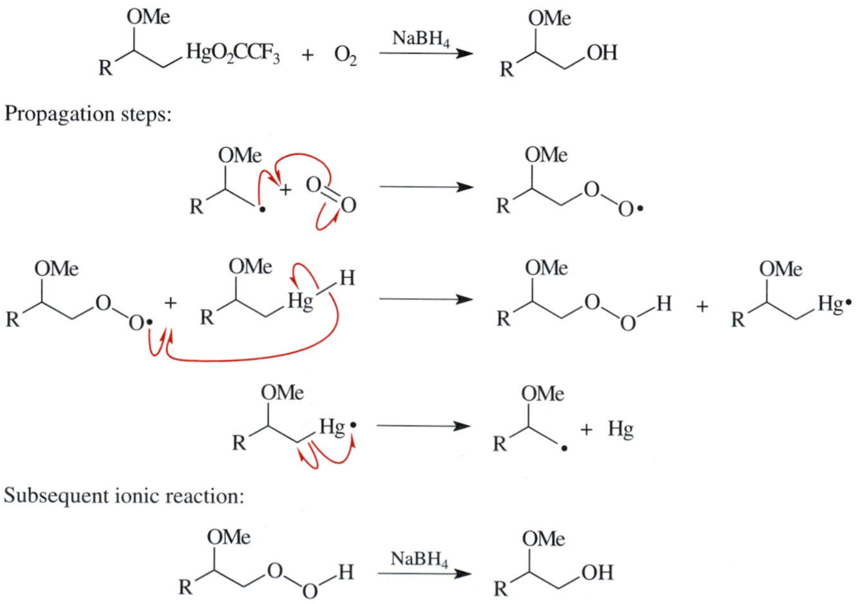

Propagation steps:

Subsequent ionic reaction:

Fig. 1.14. $NaBH_4$-induced air oxidation of a (β-alkoxyalkyl)mercury (II) trifluoroacetate (see Figure 3.39) to a glycol derivative.

These propagation steps are repeated many times while the organic mercury compound is consumed and alcohol and elemental mercury are released. This process is interrupted only by termination steps (Figure 1.12). Thus, for example, two mercury-free radicals can combine to form one dimer, or a mercury-free and a mercury-containing radical can combine to form a dialkylmercury compound.

(β-Hydroxyalkyl) mercury(II) acetates and $NaBH_4$ react to form C-centered radicals through the reaction steps shown in Figure 1.12 also when methyl acrylate is present in the reaction mixture. Under these conditions, these radicals can add to the C=C double bond of the ester (Figure 1.13). The addition takes place via a reaction chain, which comprises three propagation steps. The reaction product is a methyl ester, which has a carbon chain that is three C atoms longer than the carbon chain of the organomercury compound.

A

The radicals produced during the decomposition of alkylmercury(II) hydrides can also be added to molecular oxygen (Figure 1.14). A hydroperoxide is first produced in a chain reaction, which again comprises three propagation steps. However, it is unstable in the presence of $NaBH_4$ and is reduced to an alcohol.

1.7 Radical Halogenation of Hydrocarbons

Many hydrocarbons can be halogenated with elemental chlorine or bromine while being heated and/or irradiated together.

B

$$\geqslant C_{sp^3}-H + Cl_2\,(Br_2) \xrightarrow{h\nu \text{ or } \Delta} \geqslant C_{sp^3}-Cl\,(Br) + HCl\,(HBr)$$

The result is simple or multiple halogenations.

1.7.1 Simple and Multiple Chlorinations

Presumably you are already familiar with the mechanism for the thermal chlorination of methane. We will use Figure 1.15 to review briefly the net equation, the initiation step, and the propagation steps of the monochlorination of methane. Figure 1.16 shows the energy profile of the propagation steps of this reaction.

B

$$CH_4 \text{ (large excess)} + Cl_2 \xrightarrow{400\,°C} CH_3Cl + HCl$$

Initiation step: $$Cl-Cl \xrightarrow{\Delta} 2\ Cl\bullet$$

Propagation steps: $$Cl\bullet + H-CH_3 \longrightarrow Cl-H + \bullet CH_3$$

$$\bullet CH_3 + Cl-Cl \longrightarrow CH_3-Cl + Cl\bullet$$

Fig. 1.15. Mechanism for monochlorination of methane with Cl_2.

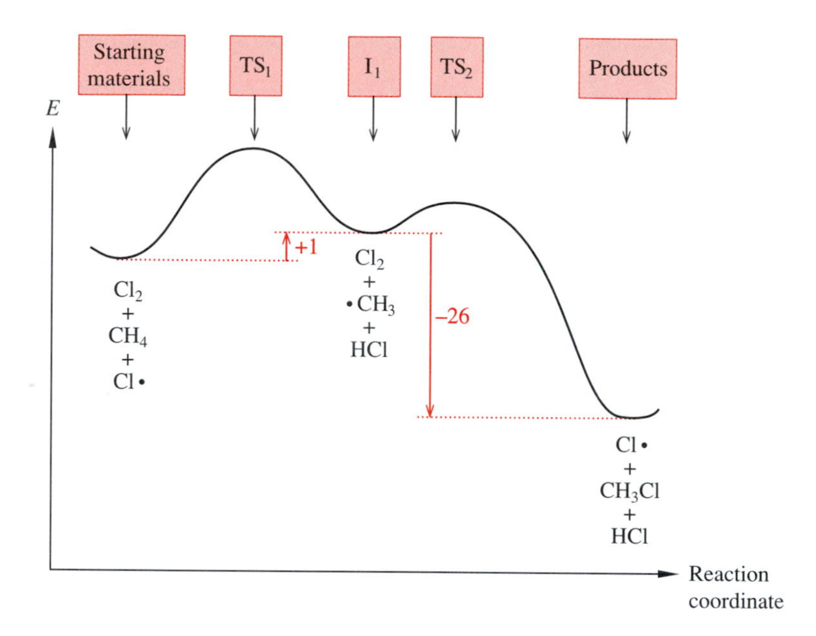

Fig. 1.16. Energy profile of the propagation steps of the monochlorination of methane with Cl_2 (enthalpies in kcal/mol).

In the energy profile, each of the two propagation steps is represented as a transition from one energy minimum over an energy maximum into a new energy minimum. Energy minima in an energy profile characterize either long-lived species [starting material(s), product(s)] or short-lived intermediates. On the other hand, energy maxima in an energy profile (transition states) are snapshots of the geometry of a molecular system, whose lifetime corresponds to the duration of a molecular vibration (ca. 10^{-13} s).

A chemical transformation that takes place via exactly one transition state is called an **elementary reaction**. This holds regardless of whether it leads to a short-lived intermediate or to a product that can be isolated. According to the definition, an n-step reaction consists of a sequence of n elementary reactions. It takes place via n transition states and $(n-1)$ intermediates.

In the reaction of a 1:1 mixture of methane and chlorine one does not obtain the monochlorination product selectively, but a 46:23:21:9:1 mixture of unreacted methane, mono-, di-, tri-, and tetrachloromethane. Thus all conceivable multiple chlorination products are also produced. Multiple chlorinations, like monochlorinations, occur as radical chain substitutions. They are based on completely analogous propagation steps (Figure 1.17).

According to Figure 1.18, *analogous* propagation steps possess the same heat of reaction, independent of the degree of chlorination. With the help of Hammond's postulate, one concludes from this that the associated free activation energies should also be independent of the degree of chlorination. This means that the monochlorination of methane and each of the subsequent multiple chlorinations should take place with one and the same rate constant. This is essentially the case. The relative chlorination rates for $CH_{4-n}Cl_n$ in the temperature range in question are 1 ($n = 0$), 3 ($n = 1$), 2

$$CH_4 + Cl_2 \text{ (comparable molar amount)} \xrightarrow{400\,°C} \left\{ \begin{array}{l} CH_3Cl \\ CH_2Cl_2 \\ CHCl_3 \\ CCl_4 \end{array} \right. + HCl$$

Initiation step: $Cl{-}Cl \xrightarrow{\Delta} 2\ Cl\bullet$

Propagation steps: $Cl\bullet + H{-}CH_{3-n}Cl_n \longrightarrow Cl{-}H + \bullet CH_{3-n}Cl_n$

$\bullet CH_{3-n}Cl_n + Cl{-}Cl \longrightarrow CH_{3-n}Cl_{n+1} + Cl\bullet$

Fig. 1.17. Mechanism for the polychlorination of methane.

($n = 2$), and 0.7 ($n = 3$). The resulting lack of selectivity, fortunately, is of no concern in the industrial reactions of CH_4 with Cl_2. The four chlorinated derivatives of methane are readily separated from each other by distillation; each one is needed in large amounts.

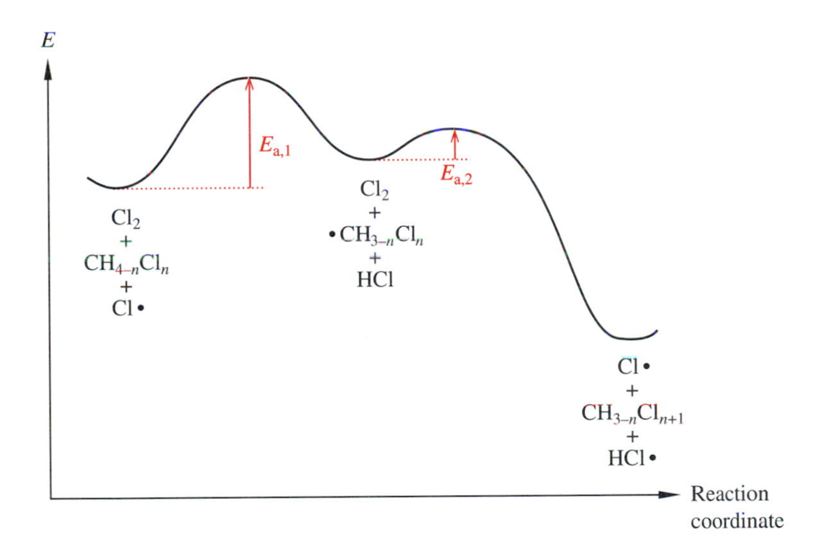

Fig. 1.18. Energy profile of the propagation steps of the polychlorinations $CH_3Cl \rightarrow CH_2Cl_2$, $CH_2Cl_2 \rightarrow CHCl_3$, and $CHCl_3 \rightarrow CCl_4$ of methane (n = 1–3 in the diagram), and of the monochlorination $CH_4 \rightarrow CH_3Cl$ ($n = 0$ in the diagram).

If only methyl chloride is needed, it can be produced essentially free of multiple chlorination products only if a large excess of methane is reacted with chlorine. In this case, there is always more unreacted methane available for more monochlorination to occur than there is methyl chloride available for a second chlorination.

A

Another preparatively valuable multiple chlorination is the photochemical perchlorination of methyl chloroformate, which leads to diphosgene:

$$Cl{-}C(=O){-}O{-}CH_3 \xrightarrow{Cl_2,\,h\nu} Cl{-}C(=O){-}O{-}CCl_3$$

Diphosgene

1.7.2 Regioselectivity of Radical Chlorinations

B

Clean monochlorinations can be achieved only with hydrocarbons that react via resonance-stabilized radicals. They also exhibit high regioselectivity. This follows from the structure of these resonance-stabilized radicals.

Regioselectivity

A given molecular transformation, for example, the reaction C—H → C—Cl, is called **regioselective** when it takes place preferentially or exclusively at one place on a substrate. Resonance-stabilized radicals are produced regioselectively as a consequence of product-development control in the radical-forming step.

In the industrial synthesis of benzyl chloride (Figure 1.19), only the H atoms in the benzyl position are replaced by Cl because the reaction takes place via resonance-stabilized benzyl radicals (cf. Table 1.1, bottom line) as intermediates. At a reaction temperature of 100°C, the first H atom in the benzyl position is substituted a little less than 10 times faster (→ benzyl chloride) than the second (→ benzal chloride) and this is again 10 times faster than the third (→ benzotrichloride).

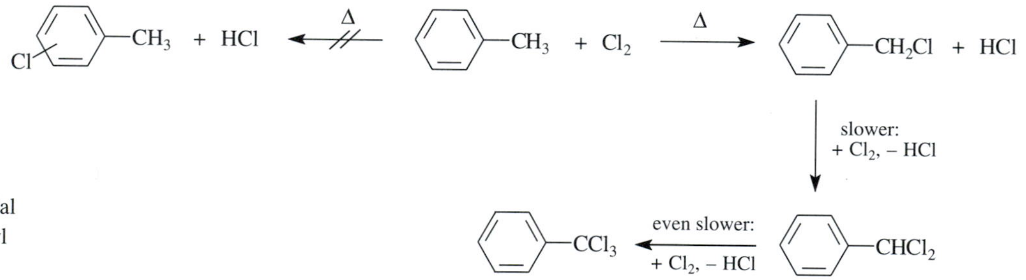

Cl–C₆H₄–CH_3 + HCl ←(Δ, not formed)— C_6H_5–CH_3 + Cl_2 —Δ→ C_6H_5–CH_2Cl + HCl

slower: + Cl_2, – HCl

even slower: + Cl_2, – HCl

C_6H_5–CCl_3 ← C_6H_5–$CHCl_2$

Fig. 1.19. Industrial synthesis of benzyl chloride.

In the reaction example of Figure 1.20, the industrial synthesis of allyl chloride, only an H atom in the allylic position is substituted. Its precursor is a resonance-stabilized (Table 1.1, center line) allylic radical.

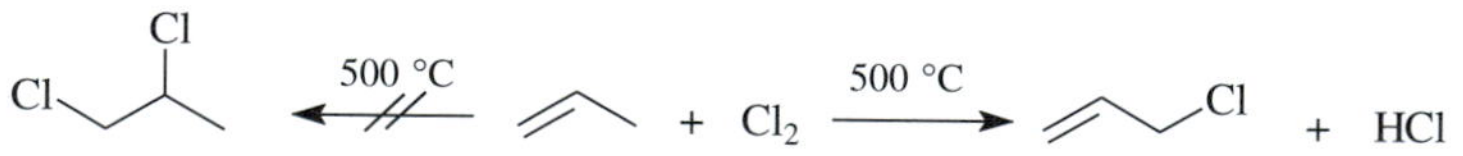

Cl / Cl ←(500 °C, not formed)— propene + Cl_2 —500 °C→ allyl–Cl + HCl

Fig. 1.20. Industrial synthesis of allyl chloride.

Incidentally, this reaction of chlorine with propene is also chemoselective:

Chemoselectivity

Reactions in which the reagent effects preferentially or exclusively one out of several types of possible transformations are **chemoselective.**

In the present case the only transformation that results is C—H → C—Cl, that is, a substitution, and not a transformation C=C + Cl_2 → Cl—C—C—Cl, which would be an addition.

Let us summarize. *Large* differences in stability between potential radical intermediates guarantee high regioselectivity of radical substitution reactions. *Small* differences in stability between potential radical intermediates no longer guarantee regioselective chlorinations. Interestingly, however, they do not yet rule out a considerable measure of regiocontrol in analogous brominations. This is illustrated in the following by a comparison of the chlorination (below) and bromination (Section 1.7.3) of isopentane.

The chlorination of isopentane gives multiply chlorinated compounds as well as all four conceivable monochlorination products (Table 1.3). These monochlorination products are obtained with relative yields of 22% (substitution of C_{tert}—H), 33% (substitution of C_{sec}—H), 30 and 15% (in each case substitution of C_{prim}—H). Consequently, one cannot talk about the occurrence of regioselectivity.

Two factors are responsible for this. The first one is a statistical factor. Isopentane contains a single H atom, which is part of a C_{tert}—H bond. There are two H atoms that are part of C_{sec}—H bonds, and 6 + 3 = 9 H atoms as part of C_{prim}—H bonds. If each H atom were substituted at the same rate, the cited monochlorides would be produced in a ratio of 1:2:6:3. This would correspond to relative yields of 8, 17, 50, and 25%.

The discrepancy from the experimental values is due to the fact that H atoms bound to different types of C atoms are replaced by chlorine at different rates. The substitu-

Table 1.3. Regioselectivity of Radical Chlorination of Isopentane

Isopentane —(Cl_2, Δ)→ Cl-substituted (tert) + Cl (sec) + Cl (prim) + Cl (prim) + multiple chlorinated compounds

The relative yields of the above monochlorination products are...	... 22%	33%	30%	15%
In order to produce the above compounds in the individual case...	... 1	2	6	3 ...
...H atoms were available for the substitution. Yields on a per-H-atom basis were...	... 22%	16.5%	5%	5% ...
... for the monochlorination product. In other words: $k_{C-H \rightarrow C-Cl, rel}$ in the position concerned is ...	... 4.4	3.3	≡ 1	≡ 1
..., that is, generally for ...	... C_{tert}—H	C_{sec}—H	C_{prim}—H	

tion of C_{tert}—H takes place via a tertiary radical. The substitution of C_{sec}—H takes place via the somewhat less stable secondary radical, and the substitution of C_{prim}—H takes place via even less stable primary radicals (for the stability of radicals, see Table 1.2). According to Hammond's postulate, the rate of formation of these radicals should decrease in the direction indicated. Hydrogen atoms bound to C_{tert} should thus be substituted more rapidly than H atoms bound to C_{sec}, and these should in turn be substituted by Cl more rapidly than H atoms bound to C_{prim}. As the analysis of the regioselectivity of the monochlorination of isopentane carried out by means of Table 1.3 shows, the relative chlorination rates of C_{tert}—H, C_{sec}—H, and C_{prim}—H are 4.4:3.3:1, in agreement with this expectation.

1.7.3 Regioselectivity of Radical Brominations Compared to Chlorinations

B

In sharp contrast to chlorine, bromine and isopentane form monosubstitution products with pronounced regioselectivity (Table 1.4). The predominant monobromination product is produced in 92.2% relative yield through the substitution of C_{tert}—H. The second most abundant monobromination product (7.4% relative yield) comes from the substitution of C_{sec}—H. The two monobromination products in which a primary H atom is replaced by Br occur only in trace quantities. The analysis of these regioselectivities illustrated in Table 1.4 gives relative rates of 2000, 79, and 1 for the bromination of C_{tert}—H, C_{sec}—H, and C_{prim}—H, respectively.

The low regioselectivity of the radical chain chlorination in Table 1.3 and the high regioselectivity of the analogous radical chain bromination in Table 1.4 are typical: bromine is generally considerably more suitable than chlorine for the regioselective halogenation of saturated hydrocarbons. Still, even the 93:7 regioselectivity in the bromination of isopentane is only somewhat attractive from a synthetic perspective. In the following we will explain mechanistically why the regioselectivity for chlorination is so much lower than for bromination.

How the enthalpy ΔH of the substrate/reagent pair R—H/Cl· changes when R· and H—Cl are produced from it is plotted for four radical chlorinations in Figure 1.21 (left). These give differently alkylated C radicals, which are the methyl, a primary, a secondary, and a tertiary radical. The reaction enthalpies ΔH_r for all four reactions are known and are plotted in the figure. Only the methyl radical is formed slightly endothermically ($\Delta H_r = +1$ kcal/mol). The primary radical, which is 6 kcal/mol more stable (cf. Table 1.2), is already formed exothermically with $\Delta H_r = -5$ kcal/mol. Secondary and tertiary radicals, which are 3 and 6 kcal/mol more stable than primary radicals are formed even more exothermically.

Hammond's postulate can be applied to this series of the selectivity-determining steps of the radical chlorination shown in Figure 1.21. They all take place via early transition states, that is, via transition states that are similar to the starting materials. The more stable the resulting radical, the more similar to the starting materials is the transition state. The stability differences between these radicals are therefore manifested only to a very small extent as stability differences between the transition states that lead to them. All transition states are therefore very similar in energy, and thus

Table 1.4. Regioselectivity of Radical Bromination of Isopentane

$\xrightarrow[\Delta]{Br_2,}$	Br– (tertiary bromide) +	Br (secondary bromide) +	Br (primary bromide) +	Br (primary bromide) + multiple brominated compounds
The relative yields of the above monobromination products are...	... 92.2%	7.38%	0.28%	0.14%
In order to produce the above compounds in the individual case...	... 1	2	6	3 ...
...H atoms were available for the substitution. Yields on a per-H-atom basis were...	... 92.2%	3.69%	0.047%	0.047% ...
... for the monobromination product above. In other words: $k_{C–H \rightarrow C–Br, rel}$ in the position concerned is ...	... 2000	79	≡ 1	≡ 1
..., that is, generally for ...	... C_{tert}—H	C_{sec}—H	C_{prim}—H	

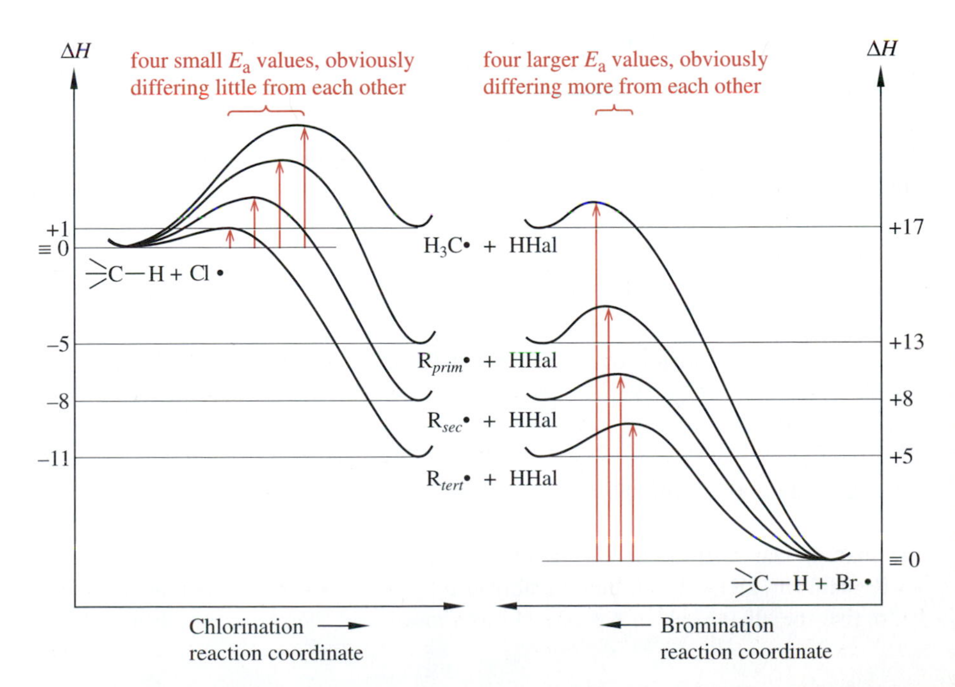

Fig. 1.21. Thermochemical analysis of that propagation step of radical chlorination (left) and bromination (right) of alkanes that determines the regioselectivity of the overall reaction. The ΔH_r values were determined experimentally; the $\Delta H^{\ddagger}$ values are estimated.

are passed through with very similar reaction rates. This means that the regioselectivity of the radical chlorination under consideration is low.

In radical brominations the energy profiles of the selectivity-determining step are completely different from what they are in analogous chlorinations. This is shown on the right side of Figure 1.21 and is rationalized as follows. The abstraction of a C-bound H atom by Cl atoms leads to the formation of an H-Cl bond with a bond enthalpy of 103 kcal/mol. In contrast, the abstraction of a C-bound H atom by a Br atom leads to the formation of a C-Br bond. Its bond enthalpy is 88 kcal/mol, which is 15 kcal/mol below the bond enthalpy of an H-Cl bond. Accordingly, even the most stable radical considered in Figure 1.21, the tertiary radical, is formed endothermically ($\Delta H_r =$ +5 kcal/mol). From the secondary through the primary to the methyl radical increasingly less stable radicals are produced in the analogous brominations of this figure. They are therefore formed increasingly endothermically and consequently probably also increasingly endergonically. According to Hammond's postulate, the selectivity-determining step of radical brominations thus proceeds via late, that is, product-like, transition states. Consequently, the substituent effects on the free energy changes of the selectivity-determing step appear almost undiminished as substituent effects on the free energies of the respective transition states. These transition states are therefore passed through with very different rate constants. The regioselectivity of radical brominations is consequently considerably higher than the regioselectivity of analogous chlorinations.

At the end of Section 1.7.4 we will talk about an additional aspect of Figure 1.21. To understand this aspect, however, we must first determine the rate law according to which radical halogenations take place.

1.7.4 Rate Law for Radical Halogenations; Reactivity/Selectivity Principle

A

A simplified reaction scheme for the kinetic analysis of radical chain halogenations can be formulated as follows:

$$\mathrm{Hal_2} \underset{k_2}{\overset{k_1}{\rightleftharpoons}} 2\,\mathrm{Hal}\cdot \quad \text{with } K_{dis} = \frac{k_1}{k_2}$$

$$\mathrm{Hal}\cdot + \mathrm{RH} \xrightarrow{k_3} \mathrm{Hal\,H} + \mathrm{R}\cdot$$

$$\mathrm{R}\cdot + \mathrm{Hal_2} \xrightarrow{k_4} \mathrm{R\,Hal} + \mathrm{Hal}\cdot$$

Let us assume that only the reaction steps listed in this scheme participate in the radical chain halogenations of hydrocarbons. Let us thus disregard the fact that chain termination can also occur owing to the radical-consuming reactions R· + Hal· → R—Hal and 2R· → R—R and possibly also by disproportionation of alkyl radicals R· to give the alkane, which has one H atom more, and the olefin, which has one H atom less. According to this scheme, the thermolysis of halogen molecules gives halogen atoms with the rate constant k_1. On the one hand, these recombine with the rate constant k_2 to form the halogen molecule again. On the other hand, the halogen atoms participate as the initiating radical in the first propagation step, which takes place with the rate constant k_3. The second and last propagation step follows with the rate constant k_4.

Explicit termination steps do not have to be considered in this approximate kinetic analysis. A termination step has already been *implicitly* considered as the reversal of the starting reaction (rate constant k_2). As soon as all halogen atoms have been converted back into halogen molecules, the chain reaction comes to a stop.

The rate law for the halogenation reaction shown above is derived step by step in Equations 1.4–1.8. We will learn to set up derivations of this type in Section 2.4.1. There we will use a much simpler example. We will not discuss Bodenstein's steady-state approximation used in Equations 1.6 and 1.7 in more detail until later (Section 2.5.1). What will be explained there and in the derivation of additional rate laws in this book is sufficient to enable you to follow the derivation of Equations 1.4–1.8 in detail in a second pass through this book.

$$\frac{d[\text{RHal}]}{dt} = k_4[\text{R}\bullet][\text{Hal}_2] \tag{1.4}$$

$$\frac{d[\text{R}\bullet]}{dt} = k_3[\text{Hal}\bullet][\text{RH}] - k_4[\text{R}\bullet][\text{Hal}_2] = 0 \text{ in the framework of Bodenstein's steady-state approximation} \tag{1.5}$$

$$\frac{d[\text{Hal}\bullet]}{dt} = k_1[\text{Hal}_2] - k_2[\text{Hal}\bullet]^2 = 0 \text{ in the framework of Bodenstein's steady-state approximation} \tag{1.6}$$

$$\text{Eq.}(1.6) \Rightarrow [\text{Hal}\bullet] = \sqrt{\frac{k_1}{k_2}[\text{Hal}_2]} = \sqrt{K_{\text{dis}}[\text{Hal}_2]} \tag{1.7}$$

$$\text{Eq.}(1.7) \text{ in Eq.}(1.5) \Rightarrow k_4[\text{R}\bullet][\text{Hal}_2] = k_3[\text{RH}]\sqrt{K_{\text{dis}}[\text{Hal}_2]} \tag{1.8}$$

$$\text{Eq.}(1.8) \text{ in Eq.}(1.4) \Rightarrow \frac{d[\text{RHal}]}{dt} = k_3[\text{RH}]\sqrt{K_{\text{dis}}[\text{Hal}_2]}$$

At this stage, it is sufficient to know the result of this derivation, which is given as Equation 1.9:

$$\text{Gross reaction rate} = k_3[\text{RH}]\sqrt{K_{\text{dis}}[\text{Hal}]_2} \tag{1.9}$$

It says: the substitution product R—X is produced at a rate that is determined by two constants and two concentration terms. For given initial concentrations of the substrate R—H and the halogen and for a given reaction temperature, the rate of formation of the substitution product is *directly proportional* to the rate constant k_3, k_3 being the rate constant of the propagation step in which the radical R· is produced from the hydrocarbon R—H.

Let us recall the energy profiles from Figure 1.21. They represent precisely the step of chlorination (left side) and bromination (right side), which determines the regioselectivity and takes place with the rate constant k_3. According to Section 1.7.3, this step is faster for chlorination than for bromination.

If we look at the reaction scheme we set up at the beginning of this section, then this means that

$$k_3 \text{ (chlorination)} > k_3 \text{ (bromination)}$$

B

Also, according to Equation 1.9, the *overall reaction* "radical chlorination" takes place on a given substrate considerably faster than the overall reaction "radical bromination." If we consider this and the observation from Section 1.7.3, which states that radical chlorinations on a given substrate proceed with considerably lower regioselectivity than radical brominations, we have a good example of the so-called reactivity/selectivity principle:

Reactivity/Selectivity Principle

A highly reactive reagent generally reacts with lower selectivity than a less reactive reagent.

1.7.5 Chemoselectivity of Radical Brominations

B

Let us go back to radical brominations (cf. Section. 1.7.3). The bromination of alkyl aromatics takes place completely regioselectively: only the benzylic position is brominated. The intermediates are the most stable radicals that are available from alkyl aromatics, namely, benzylic radicals. Refluxing *ortho*-xylene reacts with 2 equiv. of bromine to give one monosubstitution per benzylic position. The same transformation occurs when the reactants are irradiated at room temperature in a 1:2 ratio (Figure 1.22, right). The rule of thumb "SSS" applies to the reaction conditions that afford these benzylic substitutions chemoselectively. SSS stands for "searing heat + sunlight → side chain substitution."

Fig. 1.22. Competing chemoselectivities during the reaction of bromine with *ortho*-xylene by a polar mechanism (left) and a radical mechanism (right).

Starting from the same reagents, one can also effect a double substitution *on the aromatic ring* (Figure 1.22, left). However, the mechanism is completely different (Figure 5.11 and following figures). This substitution takes place under reaction conditions in which no radical intermediates are formed. (Further discussion of this process will be presented in Section 5.2.1.) Under these reaction conditions, the rule of thumb "CCC" applies. CCC stands for "catalyst + cold → core substitution."

Hydrogen atoms in the benzylic position can be replaced by elemental bromine as shown. This is not true for hydrogen atoms in the allylic position. With elemental bromine they react less rapidly than the adjacent olefinic C=C double bond does. As a consequence, bromine adds to olefins chemoselectively and does not affect allylic hydrogen (Figure 1.23, left). A chemoselective allylic bromination of olefins succeeds only according to the **Wohl–Ziegler process** (Figure 1.23, right), that is, with *N*-bromosuccinimide (NBS).

Fig. 1.23. Bromine addition and bromine substitution on cyclohexene.

Figure 1.24 gives a mechanistic analysis of this reaction. NBS is used in a stoichiometric amount, and the radical initiator AIBN (cf. Figure 1.9) is used in a catalytic amount. The starting of the chain comprises several reactions, which in the end deliver Br· as the initiating radical. Figure 1.24 shows one of several possible starting reaction sequences. Next follow three propagation steps. The second propagation step—something new in comparison to the reactions discussed before—is an ionic reaction between NBS and HBr. This produces succinimide along with the elemental bromine, which is required for the third propagation step.

In the first propagation step of the Wohl–Ziegler bromination, the bromine atom abstracts a hydrogen atom from the allylic position of the olefin and thereby initiates a **substitution.** This is not the only reaction mode conceivable under these conditions. As an alternative, the bromine atom could attack the C=C double bond and thereby start a radical **addition** to it (Figure 1.25). Such an addition is indeed observed when cyclohexene is reacted with a Br_2/AIBN mixture.

The difference is that in the Wohl–Ziegler process there is always a much lower Br_2 concentration than in the reaction of cyclohexene with bromine itself. Figure 1.25 shows *qualitatively* how the Br_2 concentration controls whether the combined effect of Br·/Br_2 on cyclohexene is an addition or a substitution. The decisive factor is that the addition takes place via a reversible step and the substitution does not. During the addition, a bromocyclohexyl radical forms from cyclohexene and Br· in an equilibrium reaction. This radical is intercepted by forming dibromocyclohexane only when a *high* concentration of Br_2 is present. However, if the concentration of Br_2 is *low,* there is no such

Fig. 1.24. Mechanism for the allylic bromination of cyclohexene according to the Wohl–Ziegler process.

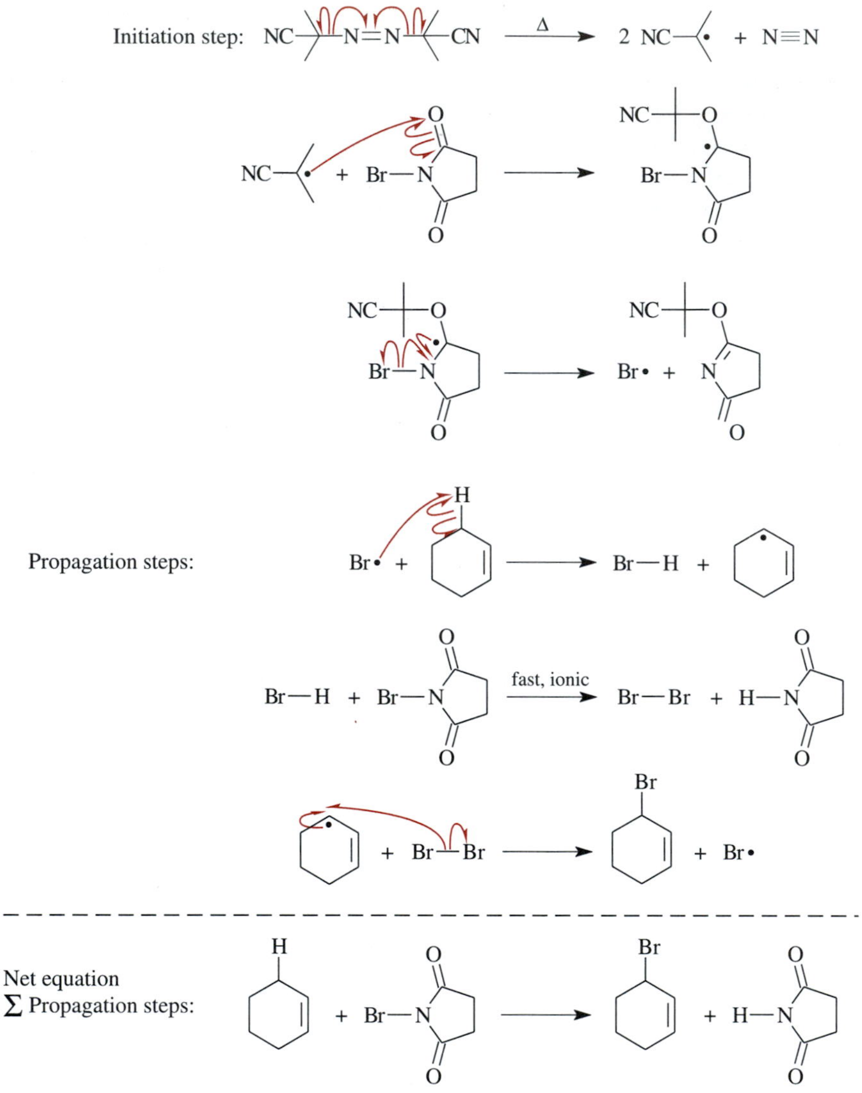

Initiation step:

Propagation steps:

Net equation
Σ Propagation steps:

Fig. 1.25. Reaction scheme for the action of $Br\cdot/Br_2$ on cyclohexene and the kinetic analysis of the resulting competition between allylic substitution (right) and addition (left) (in $k_{\sim X}$, ~X means homolytic cleavage of a bond to atom X).

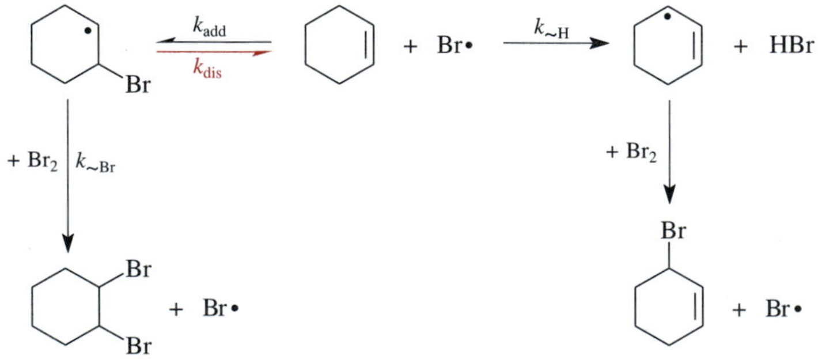

reaction. The bromocyclohexyl radical is then produced only in an unproductive equilibrium reaction. In this case the irreversible substitution therefore determines the course of the reaction.

Figure 1.26 gives a *quantitative* analysis of the outcome of this competition. Equation 1.14 provides the following decisive statement: The ratio of the rate of formation of the substitution product to the rate of formation of the addition product—which equals the ratio of the yield of the substitution product to the yield of the addition product—is inversely proportional to the concentration of Br_2.

$$d\left[\text{3-bromocyclohexene}\right]/dt = k_{\sim H}\left[\text{cyclohexene}\right]\left[Br\bullet\right] \qquad (1.10)$$

because $k_{\sim H}$ describes the rate-determining step of the allylic substitution

$$d\left[\text{1,2-dibromocyclohexane}\right]/dt = k_{\sim Br}\left[Br_2\right]\left[\text{2-bromocyclohexyl radical}\right] \qquad (1.11)$$

because $k_{\sim Br}$ describes the rate-determining step of the addition reaction

$$\left[\text{2-bromocyclohexyl radical}\right] = \frac{k_{add}}{k_{dis}}\left[\text{cyclohexene}\right]\left[Br\bullet\right] \qquad (1.12)$$

because the equilibrium condition is met

Equation (1.12) in Equation (1.11) $\Rightarrow$

$$d\left[\text{1,2-dibromocyclohexane}\right]/dt = \frac{k_{add}\cdot k_{\sim Br}}{k_{dis}}\left[\text{cyclohexene}\right]\left[Br\bullet\right]\left[Br_2\right] \qquad (1.13)$$

Divide Equation (1.10) by Equation (1.13) $\Rightarrow$

$$\frac{d\left[\text{3-bromocyclohexene}\right]/dt}{d\left[\text{1,2-dibromocyclohexane}\right]/dt} = \frac{k_{\sim H}\cdot k_{dis}}{k_{add}\cdot k_{\sim Br}}\cdot\frac{1}{\left[Br_2\right]} \qquad (1.14)$$

Fig. 1.26. Derivation of the kinetic expression for the chemoselectivity of allylic substitution versus bromine addition in the system $Br\cdot/Br_2/$cyclohexene. The rate constants are defined in Figure 1.25.

1.8 Autoxidations

B

Reactions of compounds with oxygen with the development of flames are called combustions. In addition, flameless reactions of organic compounds with oxygen are known. They are referred to as autoxidations. Of the autoxidations, only those that take place via sufficiently stable radical intermediates can deliver pure compounds and at the same time appealing yields. Preparatively valuable autoxidations are therefore limited to substitution reactions of hydrogen atoms that are bound to tertiary, allylic, or benzylic carbon atoms. An example can be found in Figure 1.27. *Unintentional autoxidations* can unfortunately occur at the O—C_{prim}—H of ethers such as diethyl ether or tetrahydrofuran (THF) (Figure 1.28).

Side Note 1.2 Autoxidations in Practice

An Industrially Important Autoxidation

The most important autoxidation used industrially is the synthesis of cumene hydroperoxide from cumene and air (i.e., "diluted" oxygen) (Figure 1.27). It is initiated by catalytic amounts of dibenzoyl peroxide as the radical initiator (cf. Figure 1.9). The cumyl radical is produced as the initiating radical from a sequence of three starting reactions. It is a tertiary radical, which is additionally stabilized by resonance. The cumyl radical is consumed in the first propagation step of this autoxidation and is regenerated in the second propagation step. These steps alternate until there is a chain termination. The autoxidation product, cumene hydroperoxide, is not isolated in pure form because all hydroperoxides are explosive. Instead the crude product is usually subjected to the cumene hydroperoxide rearrangement (see Section 11.4.1), which produces phenol and acetone. Worldwide, 90% of each of these industrially important chemicals is synthesized by this process.

Unintentional Autoxidation of Ethers

Two ethers that are frequently used as solvents are relatively easy to autoxidize—unfortunately, because this reaction is not carried out intentionally. Diethyl ether and THF form hydroperoxides through a substitution reaction in the α position to the oxygen atom (Figure 1.28). These hydroperoxides, incorrectly but popularly also referred to as **"ether peroxides,"** are fairly stable in dilute solution. However, they are highly explosive in more concentrated solutions or in pure form. Diethyl ether or THF must therefore be used only when free of peroxide.

What makes the α position of these ethers autoxidizable? α-Oxygenated radicals are stabilized by the free electron pairs on the heteroatom. When the sp^3 orbitals that accommodate the latter form a sufficiently small dihedral angle—0° would be best—with the plane in which the half-occupied $2p_z$ orbital at the radical center is located, the occupied orbitals can overlap with the half-occupied orbital (Figure 1.29). This results in a small but definite energy gain. It is similar to the energy gain the half-occupied $2p_z$ orbital of a radical center experiences as a result of overlap with a suitably oriented $\sigma_{C—H}$ MO (see Figure 1.5).

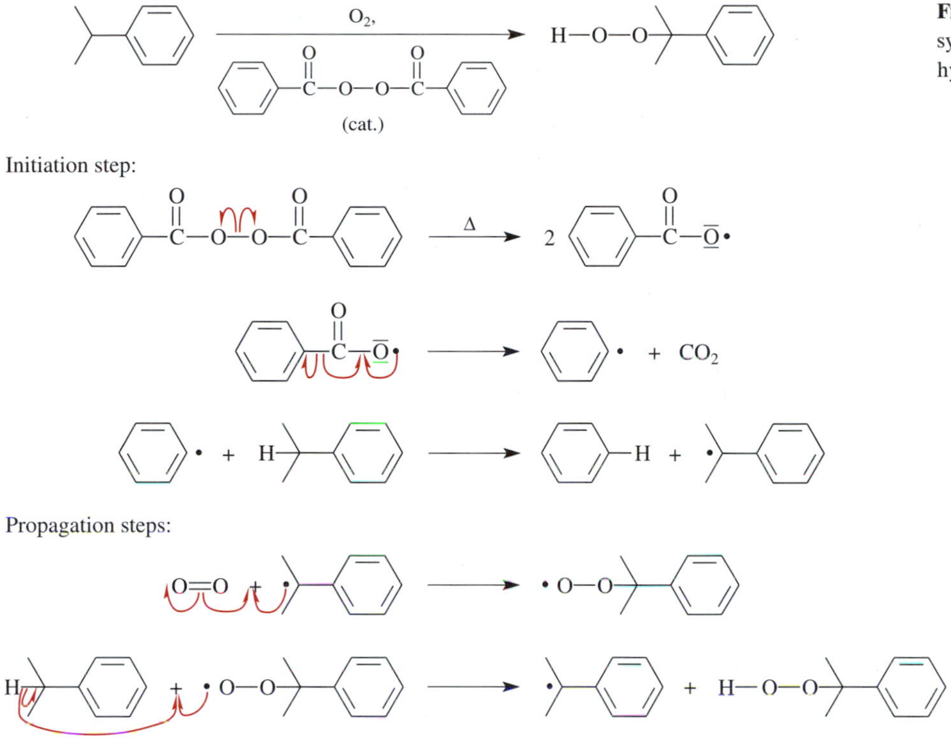

O_2, (cat.)

H—O—O

Fig. 1.27. Industrial synthesis of cumene hydroperoxide.

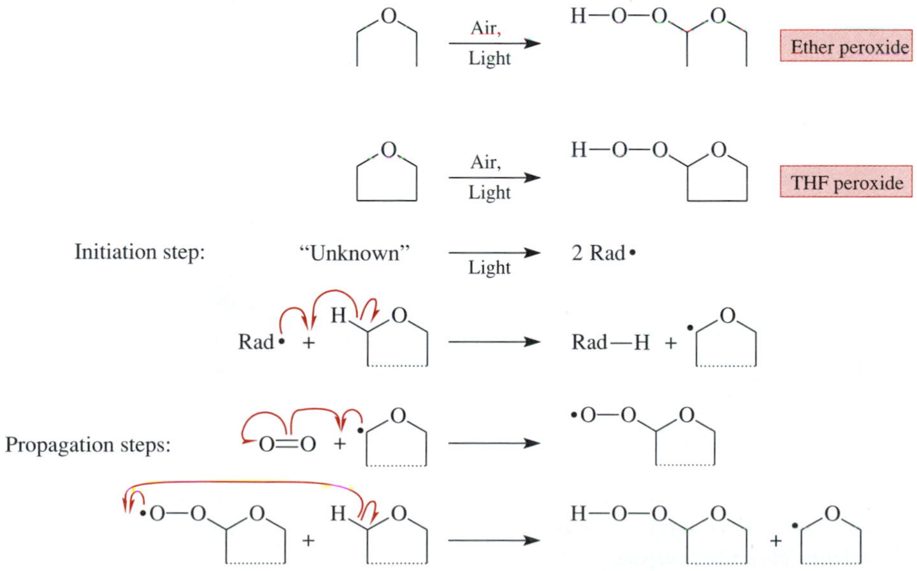

Initiation step:

Δ

2

$+ CO_2$

+ H

H +

Propagation steps:

O=O +

• O—O

H + • O—O

+ H—O—O

Air, Light

H—O—O

Ether peroxide

Air, Light

H—O—O

THF peroxide

Initiation step: "Unknown" Light 2 Rad•

Rad• + H

Rad—H +

Propagation steps: O=O +

•O—O

•O—O + H

H—O—O +

Fig. 1.28. Autoxidation of diethyl ether and THF: net equations (top) and mechanism (bottom).

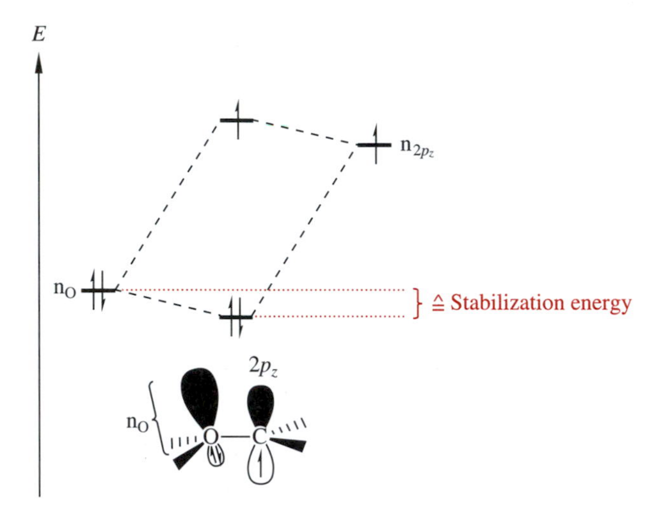

Fig. 1.29. MO diagram of α-oxygenated alkyl radicals.

tert-Butyl methyl ether is used routinely, especially in industry, as a substitute for diethyl ether. There are two reasons why it is not easy to autoxidize. In the *tert*-butyl group there is no H atom α to the heteroatom. In the methyl group such H atoms in the α position would be available but would have to be substituted via a radical, which, being unalkylated, is not very stable.

1.9 Defunctionalizations via Radical Substitution Reactions

1.9.1 Simple Defunctionalizations

A series of functionalized hydrocarbons can be defunctionalized with the help of radical substitution reactions. The functional group is then replaced by a hydrogen atom. The groups that can be removed in this way include iodide, bromide, and some sulfur-containing alcohol derivatives. Compounds with a homolysis-sensitive element–hydrogen bond serve as hydrogen atom donors. The standard reagent is Bu_3Sn-H. A substitute that does not contain tin and is considerably less toxic is $(Me_3Si)_3Si$—H. Defunctionalizations of this type are usually carried out for the synthesis of a hydrocarbon or to produce a hydrocarbon-like substructure. Figure 1.30 illustrates the possibilities of this method using a deiodination and a debromination as examples. These reactions represent general synthetic methods for obtaining cyclic esters or ethers. We will see later how easy it is to prepare halides like those shown in Figure 3.36 from olefin precursors.

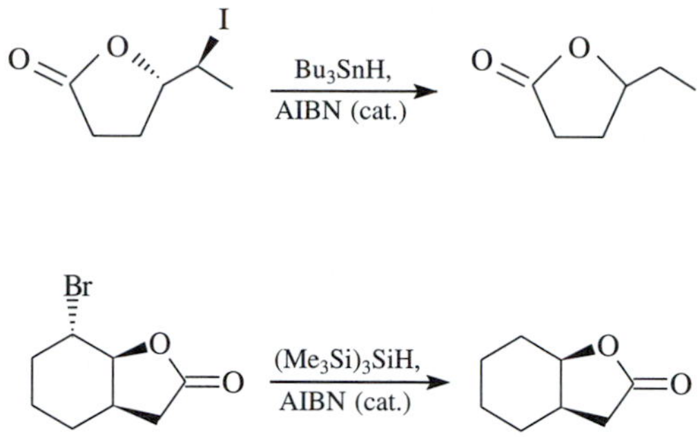

Fig. 1.30. Dehalogenations through radical substitution reactions.

Both in radical defunctionalizations effected with Bu_3SnH and in those carried out with $(Me_3Si)_3SiH$, the radical formation is initiated by the radical initiator AIBN (Figure 1.31). The initiation sequence begins with the decomposition of AIBN, which is triggered by heating or by irradiation with light, into the cyanated isopropyl radical. In the second step of the initiation sequence, the cyanated isopropyl radical produces the respective initiating radical; that is, it converts Bu_3SnH into $Bu_3Sn\cdot$ and $(Me_3Si)_3SiH$ into $(Me_3Si)_3Si\cdot$. The initiating radical gets the actual reaction chain going, which in each case comprises two propagation steps.

Both Bu_3SnH and $(Me_3Si)_3SiH$ are able to defunctionalize alkyl iodides or bromides but not alcohols. On the other hand, in the so-called Barton–McCombie reaction they can defunctionalize certain alcohol *derivatives,* namely, ones that contain a C═S double bond (e.g., thiocarboxylic esters or thiocarbonic esters). Figure 1.32 shows how the OH group of cholesterol can be removed by means of a Barton–McCombie reaction. The C═S-containing alcohol derivative used there is a xanthate (for the mechanism of the formation reaction of xanthates, see Figure 7.4).

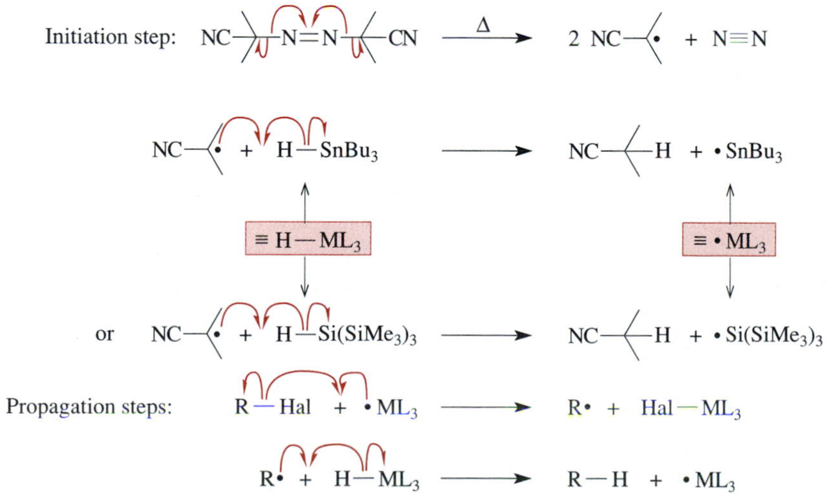

Fig. 1.31. Mechanism of the radical dehalogenations of Figure 1.30.

Fig. 1.32. Defunctionalization of an alcohol by means of the radical substitution reaction of Barton and McCombie.

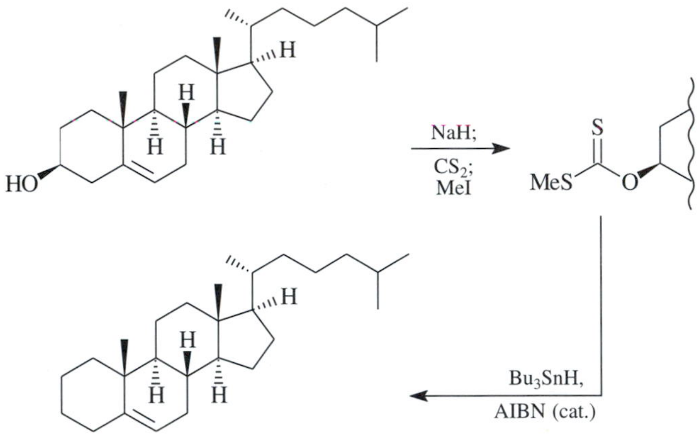

The starting sequence of this defunctionalization is identical to the one that was used in Figure 1.31 in connection with the dehalogenations. It thus leads via the initially formed $Me_2(NC)C\cdot$ radicals to $Bu_3Sn\cdot$ radicals. These radicals enter into the actual reaction chain, which consists of three propagation steps (Figure 1.33). The $Bu_3Sn\cdot$ radical is a thiophile; that is, it likes to combine with sulfur. Thus the point of attack of the first propagation step is the double-bonded sulfur atom of the xanthate. The second propagation step is a radical fragmentation.

Fig. 1.33. Propagation steps of the Barton–McCombie reaction in Figure 1.32.

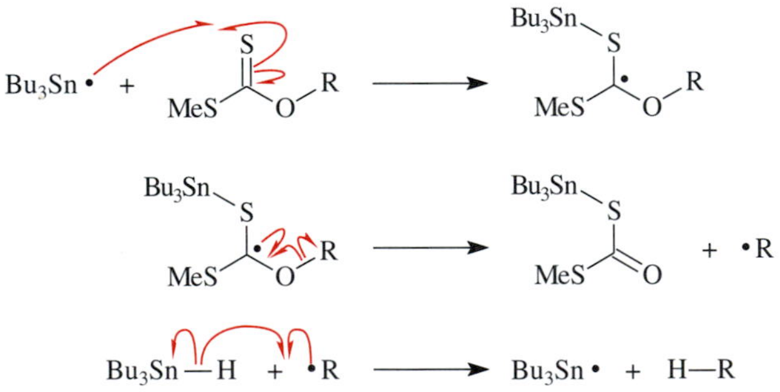

Other C=S-containing esters derived from alcohols can also be defunctionalized according to Barton and McCombie. Imidazolylthiocarbonic esters, which in contrast to xanthates can be synthesized under neutral conditions, are one example. This is important for the defunctionalization of base-sensitive alcohols. Figure 1.34 shows a reaction of this type. If the corresponding xanthate was prepared through consecutive reactions with NaH, CS_2, and MeI (for mechanism, see Figure 7.4), this compound would lose its *cis* configuration. This would occur because of the presence of the keto group. It would undergo a reversible deprotonation giving the enolate (cf. Figure 1.34). The latter would be reprotonated in part—actually even preferentially—such as to form the *trans* isomer.

The starting sequence of the Barton–McCombie reaction in Figure 1.34 is again identical to the one in Figure 1.31. However, because now Bu_3SnD is the reducing agent instead of Bu_3SnH, D is incorporated into the product instead of H.

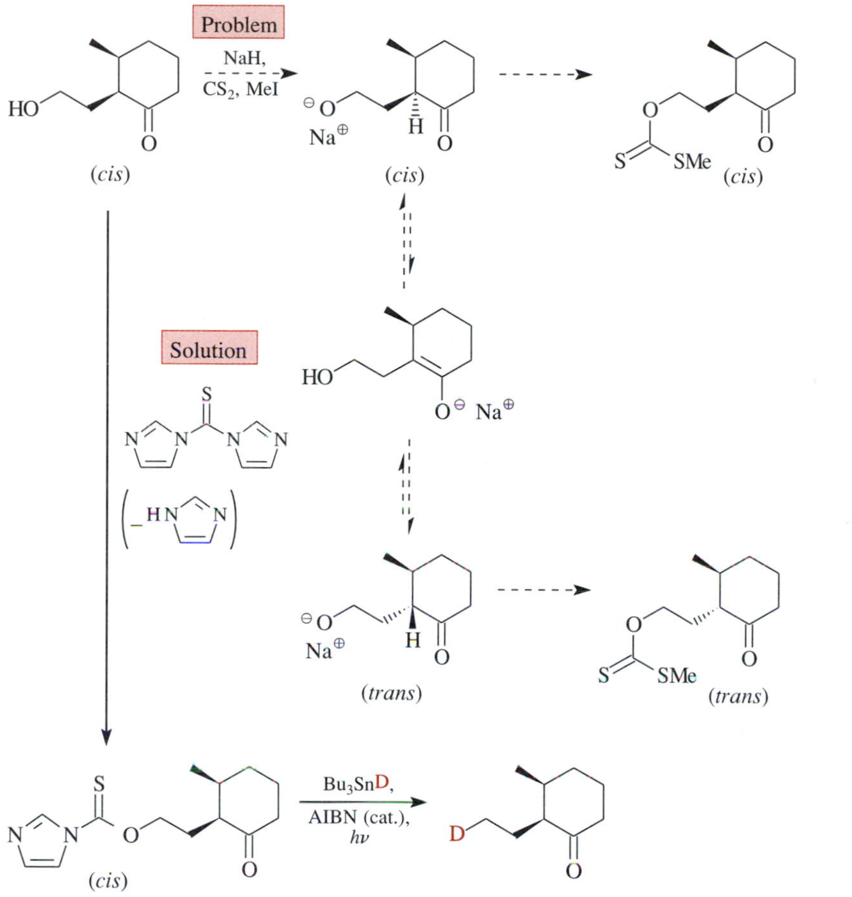

Fig. 1.34. Deoxygenation/deuteration of alcohols via a radical substitution reaction.

1.9.2 Defunctionalization via 5-Hexenyl Radicals: Competing Cyclopentane Formation

A

The defunctionalizations discussed in Section 1.9.1 considered in a different way, are also reactions for producing radicals. Certain radicals ordinarily cannot be reduced to the corresponding hydrocarbon: the 5-hexenyl radical and its derivatives. These radicals often cyclize (irreversibly) before they abstract a hydrogen atom from the reducing agent. By this cyclization, the cyclopentylmethyl radical or a derivative of it is produced selectively. An isomeric cyclohexyl radical or the corresponding derivative is practically never obtained. Often only the cyclized radical abstracts a hydrogen atom from the reducing agent. Figure 1.35 gives an example of a reaction of this type in the form of a cyclopentane annulation.

The precursor to the cyclizable radical is again (cf. Figure 1.34) a thiocarbonic acid imidazolide, as shown in Figure 1.35. The reducing agent is $(Me_3Si)_3SiH$, but one also could have used Bu_3SnH. AIBN functions as the radical initiator. After the usual ini-

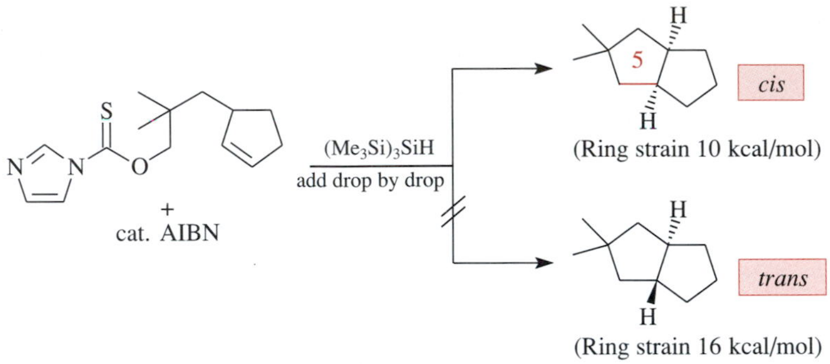

Fig. 1.35. Cyclization of a 5-hexenyl radical intermediate from a Barton–McCombie defunctionalization as a method for cyclopentane annulation.

tiation sequence (Figure 1.31, formula lines 1 and 3) the initiating radical $(Me_3Si)_3Si\cdot$ is available, and a sequence of four propagation steps is run through (Figure 1.36). In the second propagation step the 5-hexenyl radical is produced, and in the third step it is cyclized. The cyclization leads stereoselectively to a *cis*- instead of a *trans*-annulated bicyclic system. The reason for this is that the preferentially formed *cis* radical has less ring strain than its *trans* isomer. Yet, the *cis*-selectivity is due to kinetic rather than thermodynamic control. Therefore, it is a consequence of product-development control.

It is interesting that one can also use the same reagents to defunctionalize the same thiocarbonic acid derivative *without cyclization* (Figure 1.37). To do this, one simply changes the sequence in which the reagents are added according to Figure 1.37. There is no cyclization when the substrate/AIBN mixture is added dropwise to an excess of the reducing agent. A relatively concentrated solution of the reducing agent is then al-

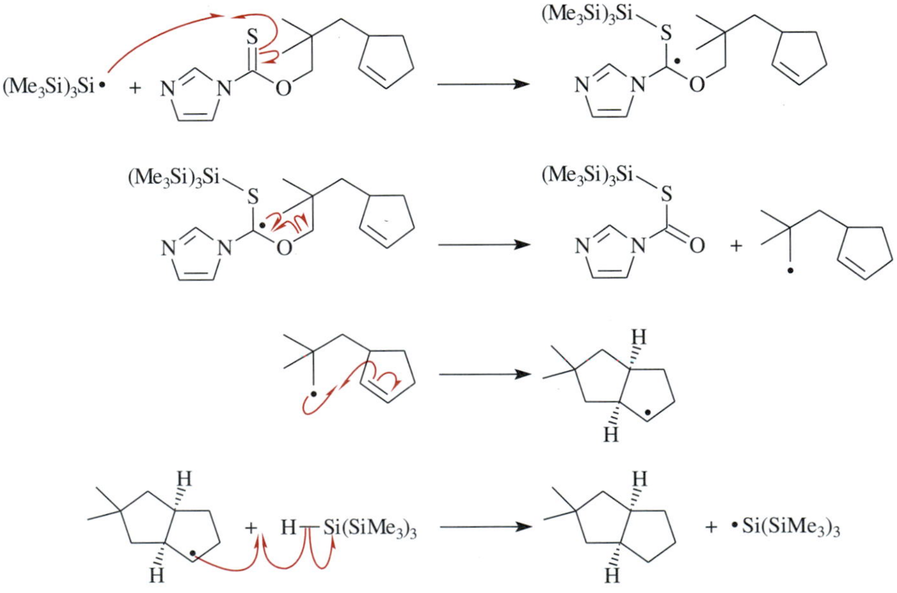

Fig. 1.36. Propagation steps in the radical substitution/cyclization of Figure 1.35.

ways available as a reaction partner for the substrate and the radicals derived from it. According to Figure 1.35, on the other hand, cyclization predominates when the reducing agent/AIBN mixture is added to the substrate dropwise over several hours. In this way the substrate and the derived radicals are exposed to only an extremely dilute solution of the reducing agent during the entire reaction.

add drop by drop to $(Me_3Si)_3SiH$; + cat. AIBN

Fig. 1.37. Cyclization-free defunctionalization of the thiocarbonic acid derivative from Figure 1.35.

According to Figure 1.38 the question whether cyclization takes place or not is decided as soon as the C—O bond of the substrate is cleaved and radical **A** is present. Radical **A** either cyclizes (at a rate that is equal to the product of the rate constant k_{cycl} and the concentration of **A**) or it reacts with the silane. In the latter case, the rate is the product of the rate constant $k_{\sim H \rightarrow prim}$, the concentration of the radical **A,** and the concentration of the silane. Let us assume that the concentration of the silane does not change during the reduction. This assumption is not correct, but sufficiently accurate for the present purpose. Then the ratio of the rates of the two alternative reactions of the hexenyl radical **A** is equal to the yield ratio of the cyclized and the noncyclized reaction products.

$$\frac{\%\ \text{Cyclization to the diquinane}}{\%\ \text{Reduction to cyclopentene}} = \frac{k_{cycl}}{k_{\sim H \rightarrow prim}} \cdot \frac{1}{[(Me_3Si)_3SiH]_{\text{quasistationary}}} \qquad (1.15)$$

According to Equation 1.15, the yield ratio of the two reduction products depends on a single variable, namely, the concentration of the reducing agent. This is in the denominator of Equation 1.15, which means two things: (1) when the radical intermediate **A** encounters a small amount of reducing agent, the diquinane is produced preferentially (see Figure 1.35) and (2) when the same radical encounters a large amount of reducing agent, the cyclopentane is preferentially formed (see Figure 1.37).

k_{cycl}; $(Me_3Si)_3SiH$; **A**; $(Me_3Si)_3SiH$, $k_{\sim H \rightarrow prim}$

Fig. 1.38. Reaction scheme for the kinetic analysis of the complementary chemoselectivity of the cyclizing defunctionalization (top reaction) of Figure 1.35 and the noncyclizing defunctionalization (bottom reaction) of Figure 1.37.

References

B. Giese, "C-Radicals: General Introduction," in *Methoden Org. Chem. (Houben-Weyl) 4th ed. 1952-, C-Radicals* (M. Regitz, B. Giese, Eds.), Vol. E19a, 1, Georg Thieme Verlag, Stuttgart, **1989.**

W. B. Motherwell and D. Crich, "Free-Radical Chain Reactions in Organic Chemistry," Academic Press, San Diego, CA, **1991.**

J. E. Leffler, "An Introduction to Free Radicals," Wiley, New York, **1993.**

M. J. Perkins, "Radical Chemistry," Ellis Horwood, London, **1994.**

J. Fossey, D. Lefort, J. Sorba, "Free Radicals in Organic Chemistry," Wiley, Chichester, U.K., **1995.**

Z. B. Alfassi, (Ed.), "General Aspects of the Chemistry of Radicals," Wiley, Chichester, U.K., **1999.**

Z. B. Alfassi, "The Chemistry of N-Centered Radicals," Wiley, New York, **1998.**

Z. B. Alfassi, (Ed.), "S-Centered Radicals," Wiley, Chichester, U.K., **1999.**

1.2

D. D. M. Wayner and D. Griller, "Free Radical Thermochemistry," in *Adv. Free Radical Chem.* (D. D. Tanner, Ed.), Vol. 1, Jai Press, Inc., Greenwich, CT, **1990.**

J. A. Martinho Simoes, A. Greenberg, J. F. Liebman (Eds.), "Energetics of Organic Free Radicals," In: *Struct. Energ. React. Chem. Ser.* **1996**, *4,* Blackie, Glascow, U.K., **1996.**

D. Gutman, "The controversial heat of formation of the *tert*-C_4H_9 radical and the tertiary carbon-hydrogen bond energy," *Acc. Chem. Res.* **1990,** *23*, 375–380.

J. C. Walton, "Bridgehead radicals," *Chem. Soc. Rev.* **1992**, *21* ,105–112.

1.3

E. Grunwald, "Reaction coordinates and structure/energy relationships," *Progr. Phys. Org. Chem.* **1990**, *17,* 55–105.

A. L. J. Beckwith, "The pursuit of selectivity in radical reactions," *Chem. Soc. Rev.* **1993,** *22,* 143–161.

1.5

J. O. Metzger, "Generation of Radicals," in *Methoden Org. Chem. (Houben-Weyl) 4th ed. 1952-, C-Radicals* (M. Regitz, B. Giese, Eds.), Vol. E19a, 60, Georg Thieme Verlag, Stuttgart, **1989.**

H. Sidebottom and J. Franklin, "The atmospheric fate and impact of hydrochlorofluorocarbons and chlorinated solvents," *Pure Appl. Chem.* **1996**, *68*, 1757–1769.

1.6

G. A. Russell, "Free radical chain reactions involving alkyl- and alkenylmercurials," *Acc. Chem. Res.* **1989**, *22*, 1–8.

G. A. Russell, "Free Radical Reactions Involving Saturated and Unsaturated Alkylmercurials," in *Advances in Free Radical Chemistry* (D. D. Tanner, Ed.), **1990**, *1*, Jai Press, Greenwich, CT.

1.7

J. O. Metzger, "Reactions of Radicals with Formation of C,Halogen-Bond," in *Methoden Org. Chem. (Houben-Weyl) 4th ed. 1952-, C-Radicals* (M. Regitz, B. Giese, Eds.), Vol. E19a, 268, Georg Thieme Verlag, Stuttgart, **1989.**

K. U. Ingold, J. Lusztyk, K. D. Raner, "The unusual and the unexpected in an old reaction. The photochlorination of alkanes with molecular chlorine in solution," *Acc. Chem. Res.* **1990**, *23*, 219–225.

1.8

J. O. Metzger, "Reactions of Radicals with Formation of C,O-Bond," in *Methoden Org. Chem. (Houben-Weyl) 4th ed. 1952-, C-Radicals* (M. Regitz, B. Giese, Eds.), Vol. E19a, 383, Georg Thieme Verlag, Stuttgart, **1989.**

W. W. Pritzkow and V. Y. Suprun, "Reactivity of hydrocarbons and their individual C-H bonds in respect to oxidation processes including peroxy radicals," *Russ. Chem. Rev.* **1996**, *65*, 497–503.

Z. Alfassi, "Peroxy Radicals," Wiley, New York, **1997.**

1.9

J. O. Metzger, "Reactions of Radicals with Formation of C,H-Bond," in *Methoden Org. Chem. (Houben-Weyl) 4th ed. 1952-, C-Radicals* (M. Regitz, B. Giese, Eds.), Vol. E19a, 147, Georg Thieme Verlag, Stuttgart, **1989.**

S. W. McCombie, "Reduction of Saturated Alcohols and Amines to Alkanes," in *Comprehensive Organic Synthesis* (B. M. Trost, I. Fleming, Eds.), Vol. 8, 811, Pergamon Press, Oxford, **1991.**

B. Giese, A. Ghosez, T. Göbel, H. Zipse, "Formation of C-H Bonds by Radical Reactions," in *Methoden Org. Chem. (Houben-Weyl) 4th ed. 1952-, Stereoselective Synthesis* (G. Helmchen, R. W. Hoffmann, J. Mulzer, and E. Schaumann, Eds.), Vol. E21d, 3913, Georg Thieme Verlag, Stuttgart, **1995.**

C. Chatgilialoglu, "Organosilanes as radical-based reducing agents in synthesis," *Acc. Chem. Res.* **1992**, *25*, 180–194.

V. Ponec, "Selective de-oxygenation of organic compounds," *Rec. Trav. Chim. Pays-Bas* **1996**, *115*, 451–455.

S. Z. Zard, "On the trail of xanthates: Some new chemistry from an old functional group," *Angew. Chem.* **1997**, *109*, 724–737; *Angew. Chem. Int. Ed. Engl.* **1997**, *36*, 672–685.

C. Chatgilialoglu and M. Newcomb, "Hydrogen donor abilities of the group 14 hydrides," *Adv. Organomet. Chem.* **1999,** *44,* 67–112.

P. A. Baguley and J. C. Walton, "Flight from the tyranny of tin: The quest for practical radical sources free from metal encumbrances," *Angew. Chem.* **1998,** *110,* 3272–3283; *Angew. Chem. Int. Ed. Engl.* **1998,** *37,* 3072–3082.

Further Reading

A. Ghosez, B. Giese, H. Zipse, W. Mehl, "Reactions of Radicals with Formation of a C,C-Bond," in *Methoden Org. Chem. (Houben-Weyl) 4th ed. 1952-, C-Radicals* (M. Regitz, B. Giese, Eds.), Vol. E19a, 533, Georg Thieme Verlag, Stuttgart, **1989.**

M. Braun, "Radical Reactions for Carbon-Carbon Bond Formation," in *Organic Synthesis Highlights* (J. Mulzer, H.-J. Altenbach, M. Braun, K. Krohn, H.-U. Reißig, Eds.), VCH, Weinheim, New York, etc., **1991**, 126–130.

B. Giese, B. Kopping, T. Gobel, J. Dickhaut, G. Thoma, K. J. Kulicke, F. Trach, "Radical cyclization reactions," *Org. React.* **1996**, *48*, 301–866.

D. P. Curran, "The design and application of free radical chain reactions in organic synthesis," *Synthesis* **1988**, 489.

T. V. RajanBabu, "Stereochemistry of intramolecular free-radical cyclization reactions," *Acc. Chem. Res.* **1991**, *24*, 139–45.

D. P. Curran, N. A. Porter, B. Giese (Eds.), "Stereochemistry of Radical Reactions: Concepts, Guidelines, and Synthetic Applications," VCH, Weinheim, Germany, **1995.**

C. P. Jasperse, D. P. Curran, T. L. Fevig, "Radical reactions in natural product synthesis," *Chem. Rev.* **1991**, *91*, 1237–1286.

G. Mehta and A. Srikrishna, "Synthesis of polyquinane natural products: An update," *Chem. Rev.* **1997**, *97*, 671–720.

V. K. Singh and B. Thomas, "Recent developments in general methodologies for the synthesis of linear triquinanes," *Tetrahedron* **1998**, *54*, 3647–3692.

S. Handa and G. Pattenden, "Free radical-mediated macrocyclizations and transannular cyclizations in synthesis," *Contemp. Org. Synth.* **1997**, *4*, 196–215.

G. Descotes, "Radical Functionalization of the Anomeric Center of Carbohydrates and Synthetic Applications," in *Carbohydrates* (H. Ogura, A. Hasegawa, T. Suami, Eds.), 89, Kodansha Ltd, Tokyo, Japan, **1992.**

C. Walling and E. S. Huyser, "Free radical addition to olefins to form carbon-carbon bonds," *Org. React.* **1963,** *13,* 92–149.

B. Giese, T. Göbel, B. Kopping, H. Zipse, "Formation of C—C bonds by addition of free radicals to olefinic double bonds," in *Stereoselective Synthesis* (Houben-Weyl) 4th ed., (G. Helmchen, R. W. Hoffmann, J. Mulzer, E. Schaumann, Eds.), 1996, Vol. E21 (Workbench Edition), *4,* 2003–2287, Georg Thieme Verlag, Stuttgart.

L. Yet, "Free radicals in the synthesis of medium-sized rings," *Tetrahedron* **1999,** *55,* 9349–9403.

B. K. Banik, "Tributyltin hydride induced intramolecular aryl radical cyclizations: Synthesis of biologically interesting organic compounds," *Curr. Org. Chem.* **1999,** *3,* 469–496.

F. W. Stacey and J. F. Harris, Jr., "Formation of carbon-heteroatom bonds by free radical chain additions to carbon-carbon multiple bonds," *Org. React.* **1963,** *13,* 150–376.

O. Touster, "The nitrosation of aliphatic carbon atoms," *Org. React.* **1953,** *7,* 327–377.

C. V. Wilson, "The reaction of halogens with silver salts of carboxylic acids," *Org. React.* **1957,** *9,* 332–387.

R. A. Sheldon and J. K Kochi, "Oxidative decarboxylation of acids by lead tetraacetate," *Org. React.* **1972,** *19,* 279–421.

Nucleophilic Substitution Reactions at the Saturated C Atom

2

2.1 Nucleophiles and Electrophiles; Leaving Groups

B

Stated with some exaggeration, organic chemistry is comparatively simple to learn because most organic chemical reactions follow a single pattern. This pattern is

$$\text{nucleophile} + \text{electrophile} \xrightarrow[\text{pair shift(s)}]{\text{valence electron}} \text{product(s)}$$

A nucleophile is a species that attacks the reaction partner by making a pair of electrons available to it; it is thus an electron pair donor. An electrophile is a species that reacts by accepting a pair of electrons from the reaction partner so that it can be shared between them. An electrophile is thus an electron pair acceptor.

Electrophiles and Nucleophiles

As **electron pair donors,** nucleophiles must either contain an electron pair that is easily available because it is nonbonding or they must contain a bonding electron pair that can be donated from the bond involved and thus be made available to the reaction partner. From this it follows that nucleophiles are usually anions or neutral species but not cations. In this book nucleophile is abbreviated as "Nu^-," regardless of charge.

According to the definition, electrophiles are **electron pair acceptors.** They therefore contain either a deficiency in the valence electron shell of one of the atoms they consist of or they are indeed valence-saturated but contain an atom from which a bonding electron pair can be removed as part of a leaving group. Concomitantly this atom accepts the electron pair of the nucleophile. Electrophiles are therefore, as a rule, cations or neutral compounds but not anions. In this book electrophile is abbreviated as "E^+," regardless of charge.

For most organic chemical reactions, the pattern just specified can thus be written more briefly as follows:

$$Nu{:}^{\ominus} + E^{\oplus} \longrightarrow Nu{-}E \quad (+\text{ by-products})$$

In this chapter we deal with nucleophilic substitution reactions at the saturated, that is, the sp^3-hybridized C atom (abbreviated "S_N reactions"). In these reactions, alkyl groups are transferred to the nucleophiles. Organic electrophiles of this type are referred to as **alkylating agents.** They have the structure $(R_{3-n}H_n)\,C_{sp^3}{-}X$. The group X is displaced by the nucleophile according to the equation

$$\text{Nu:}^{\ominus} + -\overset{|}{\underset{|}{C}}_{sp^3}-X \xrightarrow{k} \text{Nu}-\overset{|}{\underset{|}{C}}_{sp^3}- + :X^{\ominus}$$

as X^-. Consequently, both the bound group X and the departing entity X^- are called **leaving groups.** Some uncharged and a few positively charged three-membered heterocycles also react as alkylating agents. Instead of simple alkyl groups, they transfer alkyl groups with a heteroatom in the β position. The most important heterocyclic alkylating agents of this type are the epoxides. When there are no Brønsted or Lewis acids present, epoxides act as β-hydroxy alkylating agents with respect to nucleophiles:

$$\text{Nu:}^{\ominus} + \text{epoxide}(R_x) \xrightarrow{k} \text{Nu}\text{–}C(R_x)\text{–}C\text{–}O^{\ominus} \left(\xrightarrow[\text{workup}]{\text{aqueous}} \text{Nu}\text{–}C(R_x)\text{–}C\text{–}OH \right)$$

According to this equation, the product is indeed produced by an S_N reaction because the nucleophile displaces an oxyanion as a leaving group from the attacked C atom. Nonetheless this oxyanion is still a part of the reaction product. In this respect this reaction can also be considered to be an addition reaction. An intermolecular addition reaction is one that involves the combination of two molecules to form one new molecule. An intramolecular addition reaction is one that involves the combination of two moieties within a molecule to form one new molecule.

2.2 Good and Poor Nucleophiles

B

Which nucleophiles can be alkylated rapidly and are thus called "good nucleophiles"? Or, in other words, which nucleophiles have "high nucleophilicity"? And which nucleophiles can be alkylated only slowly and are thus called "poor nucleophiles"? Or, in other words, which nucleophiles have "low nucleophilicity"? Or let us ask from the point of view of the alkylating agent: Which alkylating agents react rapidly in S_N reactions and thus are "good alkylating agents" (good electrophiles)? Which alkylating agents react slowly in S_N reactions and thus are "poor alkylating agents" (poor electrophiles)? As emerges from these definitions, good and poor nucleophiles, high and low nucleophilicity, good and poor alkylating agents, good and poor electrophiles, and high and low electrophilicity are kinetically determined concepts.

Answers to all these questions are obtained via pairs of S_N reactions, which are carried out as competition experiments. In a competition experiment two reagents react simultaneously with one substrate (or two substrates react simultaneously with one reagent). Two reaction products can then be produced. The main product is the compound that results from the *more reactive* (synonymous with "faster reacting") reaction partner.

Accordingly, it is possible to distinguish between good and poor *nucleophiles* when S_N reactions are carried out as competition experiments. There the nucleophiles are made available as mixtures to a standard alkylating agent. The nucleophile that reacts to form the main product is then the "better" nucleophile. As has been observed in the investigation of a large number of competition experiments of this type, gradations of the nucleophilicity exist that are essentially independent of the substrate.

What are the causes behind the recurring gradation of this nucleophilicity series? Nucleophilicity obviously measures the ability of the nucleophile to make an electron pair available to the electrophile (i.e., the alkylating agent or the epoxide). With this as the basic idea, the experimentally observable nucleophilicity gradations can be interpreted as follows.

Nucleophilicity Gradations

- Within a group of nucleophiles that attack at the electrophile with the same atom, the nucleophilicity decreases with *decreasing basicity of the nucleophile* (Figure 2.1). Decreasing basicity is equivalent to decreasing affinity of an electron pair for a *proton*, which to a certain extent, is a model electrophile for the electrophiles of S_N reactions.

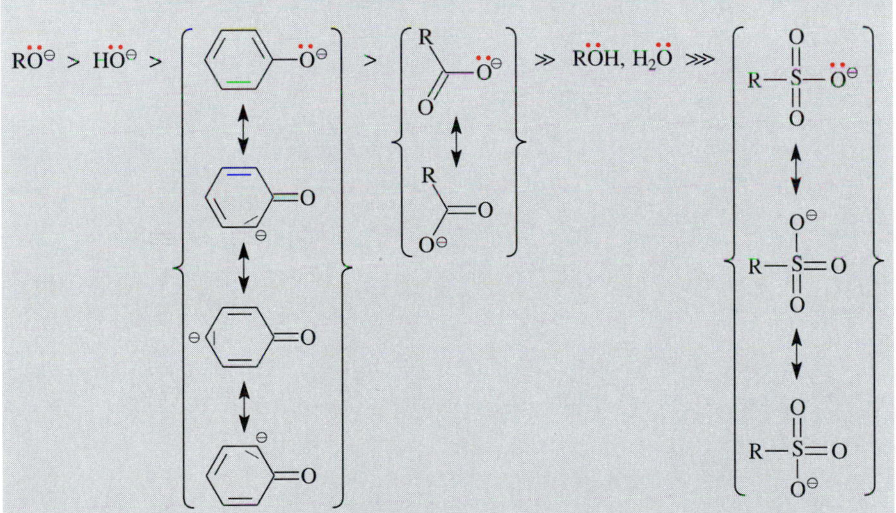

$RO^{\ominus} > HO^{\ominus} >$ [phenolate resonance structures] $>$ [carboxylate resonance structures] $\gg ROH, H_2O \ggg$ [sulfonate resonance structures]

Fig. 2.1. Nucleophilicity of O nucleophiles with different basicities.

- This parallel between nucleophilicity and basicity can be reversed by steric effects. Less basic but sterically unhindered nucleophiles therefore have a higher nucleophilicity than strongly basic but sterically hindered nucleophiles (Figure 2.2). This is most noticeable in reactions with sterically demanding alkylating agents or sterically demanding epoxides.

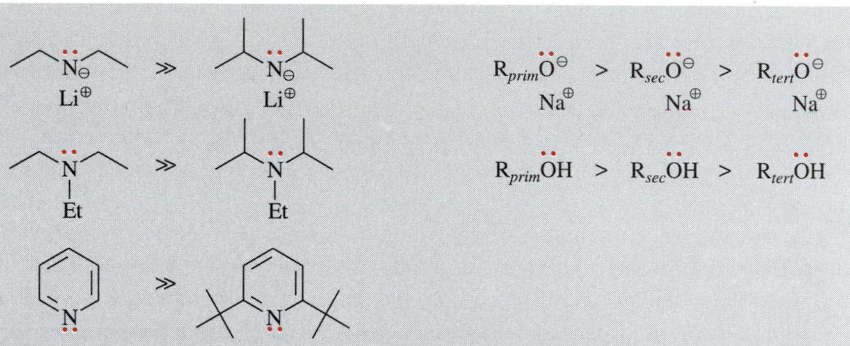

Fig. 2.2. Nucleophilicity of N and O nucleophiles that are sterically hindered to different degrees.

$R_{prim}O^{\ominus}Na^{\oplus} > R_{sec}O^{\ominus}Na^{\oplus} > R_{tert}O^{\ominus}Na^{\oplus}$

$R_{prim}OH > R_{sec}OH > R_{tert}OH$

- *Nucleophilicity decreases with increasing electronegativity of the attacking atom.* This is true both in comparisons of atomic centers that belong to the same *period* of the periodic table of the elements

$$R_2N^{\ominus} \gg RO^{\ominus} \gg F^{\ominus} \qquad RS^{\ominus} \gg Cl^{\ominus}$$

$$Et_3N \gg Et_2O$$

- and in comparisons of atomic centers from the same *group* of the periodic table:

$$RS^{\ominus} > RO^{\ominus} \qquad I^{\ominus} > Br^{\ominus} > Cl^{\ominus} \gg F^{\ominus}$$

$$RSH > ROH$$

- The nucleophilicity of a given nucleophilic center is increased by attached heteroatoms that possess free electron pairs (α-effect):

$$HO{-}O^{\ominus} > H{-}O^{\ominus}$$

$$H_2N{-}NH_2 > H{-}NH_2$$

The reason for this is the unavoidable overlap of the orbitals that accommodate the free electron pairs at the nucleophilic center and its neighboring atom.

2.3 Leaving Groups and the Quality of Leaving Groups

B

In Figure 2.3 substructures have been listed in the order of their suitability as leaving groups in S_N reactions. Substrates with good leaving groups are listed on top and substrates with increasingly poor leaving groups follow. At the bottom of Figure 2.3 are

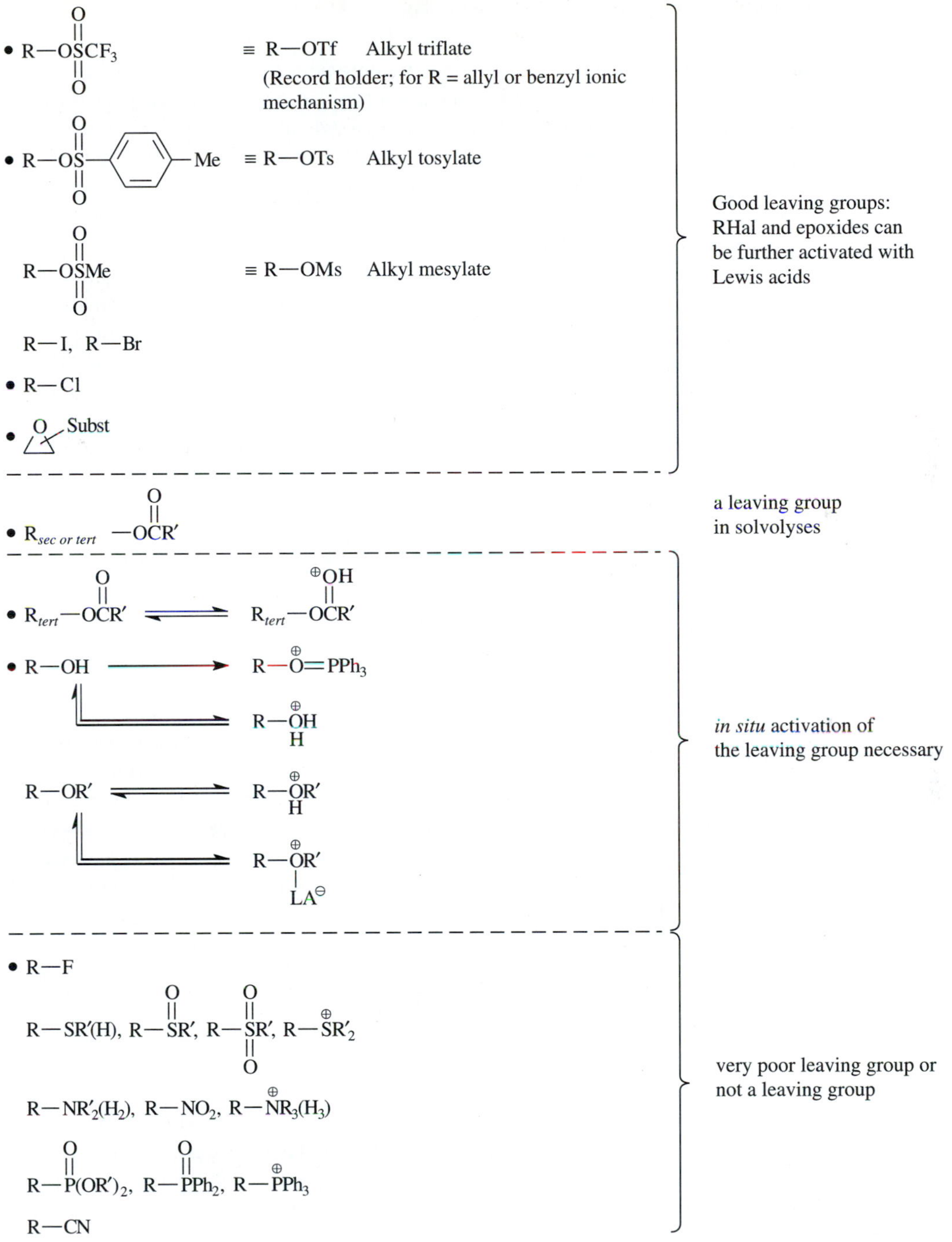

• $R—OSO_2CF_3$ ≡ R—OTf Alkyl triflate
(Record holder; for R = allyl or benzyl ionic mechanism)

• $R—OSO_2—C_6H_4—Me$ ≡ R—OTs Alkyl tosylate

$R—OSO_2Me$ ≡ R—OMs Alkyl mesylate

R—I, R—Br

• R—Cl

• epoxide (O, Subst)

Good leaving groups: RHal and epoxides can be further activated with Lewis acids

---

• $R_{sec\ or\ tert}$ —OC(=O)R′

a leaving group in solvolyses

• R_{tert}—OC(=O)R′ ⇌ R_{tert}—OC(=$^{\oplus}$OH)R′

• R—OH → $R—\overset{\oplus}{O}{=}PPh_3$

R—OH → $R—\overset{\oplus}{O}H_2$

R—OR′ ⇌ $R—\overset{\oplus}{O}R'(H)$

R—OR′ → $R—\overset{\oplus}{O}R'—LA^{\ominus}$

in situ activation of the leaving group necessary

• R—F

R—SR′(H), R—S(=O)R′, $R—SO_2R'$, $R—\overset{\oplus}{S}R'_2$

$R—NR'_2(H_2)$, $R—NO_2$, $R—\overset{\oplus}{N}R_3(H_3)$

$R—P(=O)(OR')_2$, $R—P(=O)Ph_2$, $R—\overset{\oplus}{P}Ph_3$

R—CN

very poor leaving group or not a leaving group

Fig. 2.3. Leaving group ability of various functional groups; LA = Lewis acid.

substrates whose functional group is an extremely poor leaving group. In part the effect of nucleophiles on substrates of the latter type does result in a reaction, but it is not an S_N reaction. This is, for example, the case when the nucleophile abstracts an acidic proton in the α position to the functional group instead of replacing the functional group. For example, S_N reactions with ammonium salts, nitro compounds, sulfoxides, sulfones, sulfonium salts, phosphonic acid esters, phosphine oxides, and phosphonium salts usually fail as a result of such a deprotonation. Another reaction competing with the substitution of a functional group by a nucleophile is an attack on the functional group by the nucleophile. For example, S_N reactions of nitriles, phosphonic acid esters, and phosphonium salts often fail because of this problem.

Alcohols, ethers, and carboxylic acid esters occupy an intermediate position. These compounds as such—except for the special cases shown in Figure 2.3—do not enter into any S_N reactions with nucleophiles. The reason for this is that poor leaving groups would have to be released (^-OH, ^-OR, ^-O_2CR; see below for details). However, these compounds can enter into S_N reactions with nucleophiles when they are activated as oxonium ions, for example via a reversible protonation, via bonding of a Lewis acid (LA in Figure 2.3), or via a phosphorylation. Thus, upon attack by the nucleophile, better leaving groups (e.g., HOH, HOR, HO_2CR, $O{=}PPh_3$) can be released.

Only special carboxylic acid esters and special ethers, namely epoxides, enter into S_N reactions as such, that is, without derivatization to an oxonium ion (Figure 2.3, center). In carboxylic acid esters of secondary and tertiary alcohols, the carboxylate group ^-O_2CR can become a leaving group, namely in solvolyses (see below). With epoxides as the substrate, an alkoxide ion is also an acceptable leaving group. Its release, which is actually disadvantageous (see below) because of the high basicity, is in this case coupled with the compensating release of part of the 26 kcal/mol epoxide ring strain. Product-development control therefore makes this reaction path feasible. (Part of the epoxide strain is of course also released in the transition state of an S_N reaction of an epoxide, which has been protonated or activated by a Lewis acid. Consequently, even in the presence of acids, epoxides react more rapidly than other ethers with nucleophiles.)

What makes a leaving group good or bad in substrates that react with nucleophiles as alkylating agents? The Hammond postulate implies that a good leaving group is a stabilized species, not a high-energy species. Therefore good leaving groups are usually weak bases, not strong bases. This can be rationalized as follows: A 1:1 mixture of a strong base with protons would be high in energy relative to the corresponding conjugate acid. From this we can conclude that a mixture of a strongly basic leaving group with the product of an S_N reaction is also relatively high in energy. Very basic leaving groups are produced relatively slowly according to Hammond. In other words, strong Brønsted bases are poor leaving groups; weak Brønsted bases are good leaving groups.

The suitability of halide ions as leaving groups is predicted correctly based on this reasoning alone, where $I^- > Br^- > Cl^- > F^-$. The trifluoromethanesulfonate anion (triflate anion) $F_3C{-}SO_3^-$ is for the same reason a far better leaving group than the *p*-toluenesulfonate anion (tosylate anion) $Me{-}C_6H_4{-}SO_3^-$ or the methanesulfonate anion (mesylate anion) $H_3C{-}SO_3^-$. For this reason, HOH and ROH can leave *protonated* alcohols or ethers as leaving groups, but neither the OH^- group (from alcohols) nor the OR^- group (from ethers, except for epoxides, see above) can leave.

The lower the bond enthalpy of the bond between the C atom and the leaving group, the better the leaving group. This again follows from the Hammond postulate. For this reason as well, the suitability of the halide ions as leaving groups is predicted as $I^- > Br^- > Cl^- > F^-$.

Phosphoric acid derivatives (different from biochemistry!) and sulfuric acid derivatives are not useful alkylating agents in organic chemistry. Exceptions are dimethyl sulfate and diethyl sulfate, which can be obtained commercially, as well as five-membered cyclic sulfates. These compounds contain two transferable alkyl groups. But because of the mechanism, the second alkyl group is transferred more slowly than the first one, if it is transferred at all. For these reasons, sulfates are not popular alkylating agents in organic synthesis. Furthermore, they do not even transfer the first alkyl moiety more rapidly than the alkyl sulfonates and alkyl halides ranked at the top in Figure 2.3.

2.4 S_N2 Reactions: Kinetic and Stereochemical Analysis—Substituent Effects on Reactivity

2.4.1 Energy Profile and Rate Law for S_N2 Reactions: Reaction Order

An S_N2 reaction refers to an S_N reaction

$$\text{Nu:}^{\ominus} + \text{R—X} \xrightarrow{k} \text{Nu—R} + \text{:X}^{\ominus}$$

in which the nucleophile and the alkylating agent are converted into the substitution product in one step, that is, via one transition state (Figure 2.4).

Do you remember the definition of an elementary reaction (Section 1.7.1)? The S_N2 reaction *is* an elementary reaction. Recognizing this is a prerequisite for deriving the

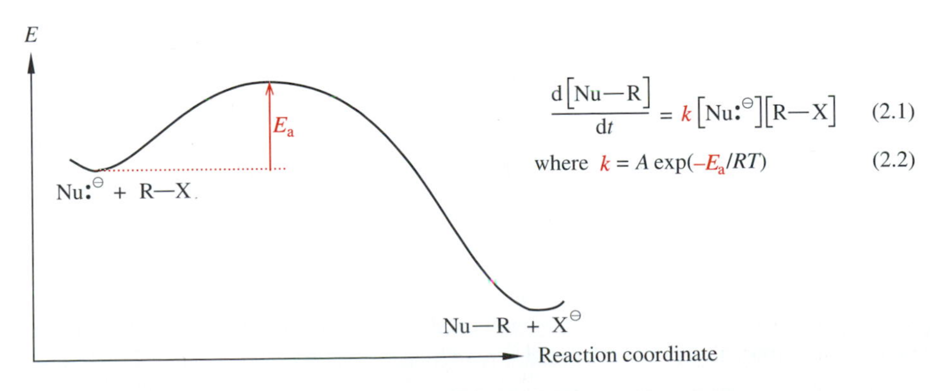

Fig. 2.4. Energy profile and rate law for S_N2 reactions.

rate law for the S_N2 reaction because the rate law for any elementary reaction can be written down *immediately.*

Side Note 2.1
Rate Laws

Rate laws establish a relationship between

- the change in the concentration of a product, an intermediate, or a starting material as a function of time, on the one hand
- and the concentrations of the starting material(s) and possibly the catalyst
- as well as the rate constants of the elementary reactions that are involved in the overall reaction, on the other hand.

By "gross reaction rate" one understands either a product formation rate *d*[product]/*dt* or a starting material consumption rate −*d*[starting material]/*dt*. The following applies unless the stoichiometry requires an additional multiplier:

$$\frac{d[\text{final product}]}{dt} = \frac{-d[\text{starting material(s)}]}{dt} \tag{2.3}$$

With the help of the rate laws that describe the elementary reactions involved, it is possible to derive Equation 2.4:

$$\frac{d[\text{final product}]}{dt} = f\{[\text{starting material(s) and optionally catalyst}], \text{rate constants}_{\text{ER}}\} \tag{2.4}$$

where the subscript ER refers to the elementary reactions participating in the overall reaction. If the right-hand side of Equation 2.4 does not contain any sums or differences, the sum of the powers of the concentration(s) of the starting material(s) in this expression is called the order m of the reaction. It is also said that the reaction is of the mth order. A reaction of order $m = 1$ is a first-order reaction, or unimolecular. A reaction of order $m = 2$ (or 3) is a second- (or third-) order reaction or a bimolecular (or trimolecular) reaction. A reaction of order m, where m is not an integer, is a reaction of a mixed order. The rate laws for elementary reactions are especially easy to set up. The recipe for this is

$$\frac{d[\text{product of an elementary reaction}]}{dt} = \frac{d[\text{starting material(s) of the elementary reaction}]}{dt}$$

The rate of product formation or starting material consumption is equal to the product of the rate constant k for this elementary reaction and the concentration of *all* starting materials involved, including any catalyst. It turns out that all elementary reactions are either first- or second-order reactions.

Because the reactions we consider in this section are single-step and therefore elementary reactions, the rate law specified in Section 2.3 as Equation 2.1 is obtained for the rate of formation of the substitution product Nu—R. It says that these reactions are bimolecular substitutions. They are consequently referred to as **S_N2 reactions.** The bimolecularity makes it possible to distinguish between this type of substitution and S_N1 reactions, which we will examine in Section 2.5: for a given concentration of the substrate, an increased concentration of the nucleophile increases the rate of formation of the S_N2 product according to Equation 2.1 but not the rate of formation of the S_N1 product(cf. Equation 2.9 in Section 2.5.1).

The rate constant of each elementary reaction is related to its activation energy E_a by the Arrhenius equation. This of course also holds for the rate constant of S_N2 reactions (see Equation 2.2 in Figure 2.4).

2.4.2 Stereochemistry of S_N2 Substitutions

B

In sophomore organic chemistry you most likely heard that S_N2 reactions take place stereoselectively. Let us consider Figure 2.5 as an example: the attack by potassium acetate on the *trans*-tosylate **A** gives exclusively the cyclohexyl acetate *cis*-**B**. No *trans*-isomer is formed. In the starting material, the leaving group is equatorial and the C—H bond at the attacked C atom is axial. In the substitution product *cis*-**B** the acetate is axial and the adjacent H atom is equatorial. Thus a 100% inversion of the configuration has taken place in this S_N2 reaction. This is also true for all other S_N2 reactions investigated stereochemically.

tert-Bu trans-A OTs H; $K^{\oplus}$ acetate, $- K^{\oplus}{}^{\ominus}OTs$ → tert-Bu cis-B (OAc axial, H); but no tert-Bu trans-B

Fig. 2.5. Proof of the inversion of configuration at the attacked C atom in an S_N2 reaction.

The reason for the inversion of configuration is that S_N2 reactions take place with a backside attack by the nucleophile on the bond between the C atom and the leaving group. In the transition state of the S_N2 reaction, the attacked C atom has five bonds. The three substituents at the attacked C atom not participating in the S_N reaction and this C atom itself are for a short time located in one plane:

$$Nu:^{\ominus} + C(a)(b)(c)X \longrightarrow \left[\overset{\delta\ominus}{Nu}\cdots C(a)(b)(c) \cdots \overset{\delta\ominus}{X} \right]^{\ddagger} \longrightarrow Nu{-}C(a)(b)(c) + :X$$

The S_N2 mechanism is casually also referred to as an "umbrella mechanism." The nucleophile enters in the direction of the umbrella handle and displaces the leaving group, which was originally lying above the tip of the umbrella. The geometry of the transition state corresponds to the geometry of the umbrella, which is just flipping over. The geometry of the substitution product corresponds to the geometry of the flipped-over umbrella. The former nucleophile is located at the handle of the flipped-over umbrella.

2.4.3 A Refined Transition State Model for the S_N2 Reaction; Crossover Experiment and Endocyclic Restriction Test

B

Figure 2.6 shows several methylations, which in each case take place as one-step S_N reactions. The nucleophile is in each case a sulfonyl anion; a methyl(arenesulfonate) reacts as electrophile. The experiments from Figure 2.6 were carried out to clarify whether these methylations take place inter- or intramolecularly.

In experiment 1 the perprotio-sulfonyl anion $[H_6]$-**A** reacts to form the methylated perprotio-sulfone $[H_6]$-**B.** It is not known whether this is the result of an intra- or an intermolecular S_N reaction. In experiment 2 of Figure 2.6, the sulfonyl anion $[D_6]$-**A,** which is perdeuterated in both methyl groups, reacts to form the hexadeuterated methylsulfone $[D_6]$-**B.** Even this result does not clarify whether the methylation is intra- or intermolecular. An explanation is not provided until the third experiment in Figure 2.6, a so-called **crossover experiment.**

The purpose of every crossover experiment is to determine whether reactions take place intra- or intermolecularly. In a crossover experiment two substrates differing from each other by a *double* substituent variation are reacted as a *mixture.* This substrate mixture is subjected to precisely the same reaction conditions in the crossover experiment that the two individual substrates had been exposed to in separate experiments. This double substituent variation allows one to determine from the structures of the reaction products their origin, i.e., from which parts of which starting materials they were formed (see below for details).

The product mixture is then analyzed. There are two possible outcomes. It can contain nothing other than the two products that were already obtained in the individual experiments. In this case, each substrate would have reacted only with itself. With the substrate mixture of the crossover experiment, this is possible only for an intramolecular reaction. The product mixture of a crossover experiment could alternatively consist of four compounds. Two of them could not have arisen from the individual experiments. They could have been produced only by "crossover reactions" *between* the two components of the mixture. A crossover reaction of this type can only be intermolecular.

In the third crossover experiment of Figure 2.6, a 1:1 mixture of the sulfonyl anions $[H_6]$-**A** and $[D_6]$-**A** was methylated. The result did *not* correspond to the sum of the individual reactions. Besides a 1:1 mixture of the methylsulfone $[H_6]$-**B** obtained in experiment 1 and the methylsulfone $[D_6]$-**B** obtained in experiment 2, a 1:1 mixture of the two crossover products $[H_3D_3]$-**B** and $[D_3H_3]$-**B** was isolated, in the same yield. The fact that both $[H_3D_3]$-**B** and $[D_3H_3]$-**B** occurred proves that the methylation was intermolecular. The crossover product $[H_3D_3]$-**B** can only have been produced because a CD_3 group was transferred from the sulfonyl anion $[D_6]$-**A** to the deuterium-free sulfonyl anion $[H_6]$-**A.** The crossover product $[D_3H_3]$-**B** can only have been produced

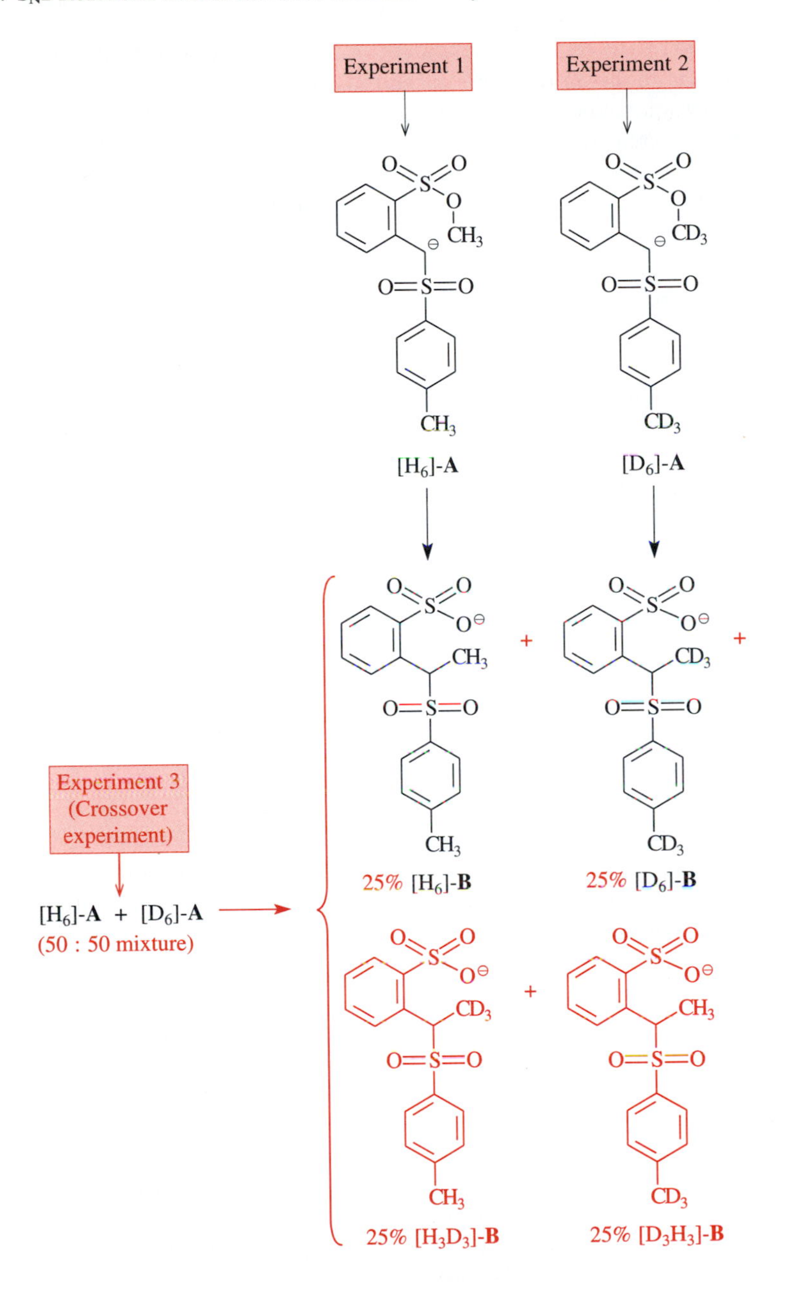

Fig. 2.6. Determination of the mechanism of one-step S_N reactions on methyl (arenesulfonates): intra- or intermolecularity.

because a CH_3 group was transferred from the sulfonyl anion $[H_6]$-**A** to the deuterium-containing sulfonyl anion $[D_6]$-**A.**

The *intramolecular* methylation of the substrate of Figure 2.6, which was not observed, would have had to take place through a six-membered cyclic transition state. In other cases cyclic six-membered transition states of intramolecular reactions are so favored that intermolecular reactions usually do not occur. It simply takes too long for the two reaction partners to find each other. Why then is a cyclic transition state not able to compete in the S_N reactions in Figure 2.6?

The conformational degrees of freedom of cyclic transition states are considerably limited or "restricted" relative to the conformational degrees of freedom of noncyclic transition states (cf. the fewer conformational degrees of freedom of cyclohexane vs *n*-hexane). Mechanistic investigations of this type are therefore also referred to as **endocyclic restriction tests.** They prove or refute in a very simple way certain transition state geometries because of this conformational restriction. The endocyclic restriction imposes limitations of the conceivable geometries on cyclic transition states. These geometries do not comprise all the possibilities that could be realized for intermolecular reactions proceeding through acyclic transition states.

In the S_N reactions of Figure 2.6, the endocyclic restriction would therefore impose a geometry in an intramolecular substitution that is energetically disfavored relative to the geometry that can be obtained in an intermolecular substitution. As shown on the right in Figure 2.7, in an intramolecular substitution the reason for this is that the nucleophile, the attacked C atom, and the leaving group cannot lie on a common axis. However, such a geometry can be realized in an intermolecular reaction (see Figure 2.7, left). From this, one concludes that in an S_N2 reaction the approach path of the nucleophile must be *collinear* with the bond between the attached C atom and the leaving group.

This approach path is preferred in order to achieve a transition state with optimum bonding interactions. Let us assume that in the transition state, the distance between the nucleophile and the attacked C atom and between the leaving group and this C atom are exactly the same. The geometry of the transition state would then correspond precisely to the geometry of an umbrella that is just flipping over. The attacked C atom would be—as shown in Figure 2.7 (left)—sp^2-hybridized and at the center of a trigonal bipyramid. The nucleophile and the leaving group would be bound to this C atom via σ bonds. Both would come about by overlap with one lobe of the $2p_z$ AO. *For this reason,* a linear arrangement of the nucleophile, the attacked C atom, and the leaving group is preferred in the transition state of S_N reactions.

Figure 2.7 (right) shows that in a bent transition state of the S_N reaction neither the nucleophile nor the leaving group can form similarly stable σ bonds by overlap with the $2p_z$ AO of the attacked C atom. Because the orbital lobes under consideration are not parallel, both the $Nu\cdots C_{sp^2}$ and the $C_{sp^2}\cdots$ leaving group bonds would be bent. **Bent bonds** are weaker than linear bonds because of the smaller orbital overlap. This is known from the special case of bent C—C bonds (Figure 2.8) as encountered, for example, in the very strained C—C bonds of cyclopropane. Bent bonds are also used in the "banana bond" model to describe the C=C double bond in olefins. (In this model, the double bond is represented by two bent single bonds between sp^3-hybridized C atoms; cf. introduction to Chapter 3.) Both types of bent C—C bonds are less stable than linear C—C bonds, such as in ethane.

2.4.4 Substituent Effects on S_N2 Reactivity

B

How substituents in the alkylating agent influence the rate constants of S_N2 reactions can be explained by means of the transition state model developed in Section 2.4.3. This model makes it possible to understand both the steric and the electronic substituent effects.

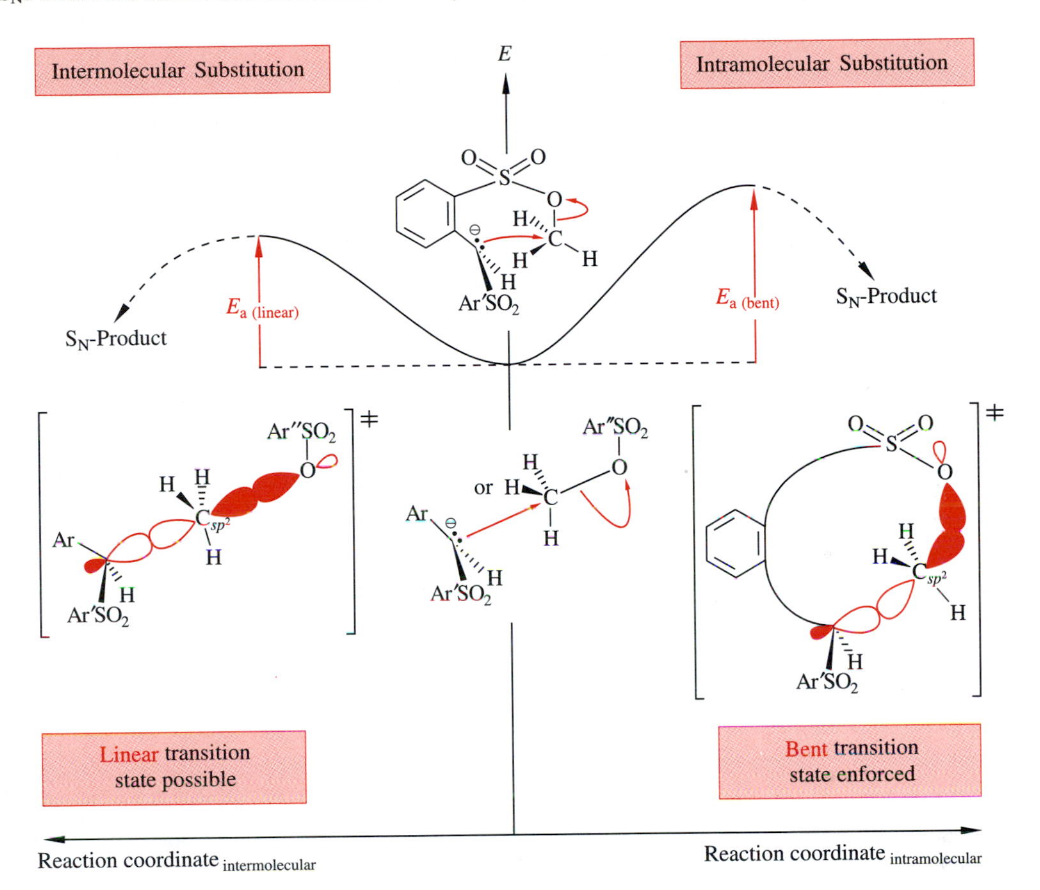

Fig. 2.7. Illustration of the intermolecular course of the S_N reactions of Figure 2.6; energy profiles and associated transition state geometries.

When an S_N2 alkylating agent is attacked by a nucleophile, the steric interactions become larger in the vicinity of the attacked C atom (Figure 2.9). On the one hand, the inert substituents come closer to the leaving group X. The bond angle between these substituents and the leaving group decreases from approximately the tetrahedral angle 109° 28′ to approximately 90°. On the other hand, the attacking nucleophile approaches the inert substituents until the bond angle that separates it from them is also approximately 90°. The resulting increased steric interactions destabilize the transition

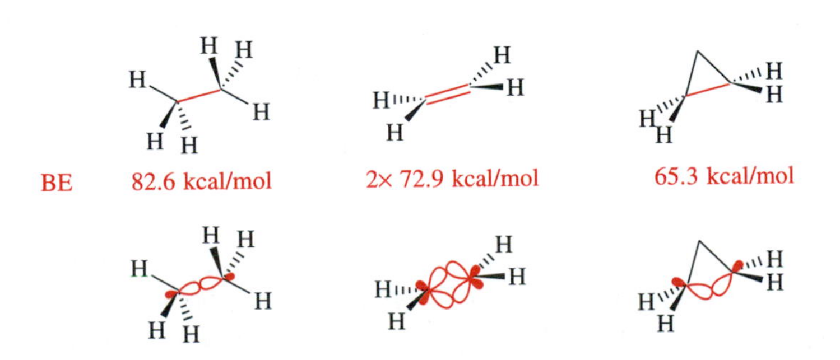

Fig. 2.8. Bond enthalpy (BE) of linear (left) and bent (middle and right; cf. explanatory text) C—C bonds.

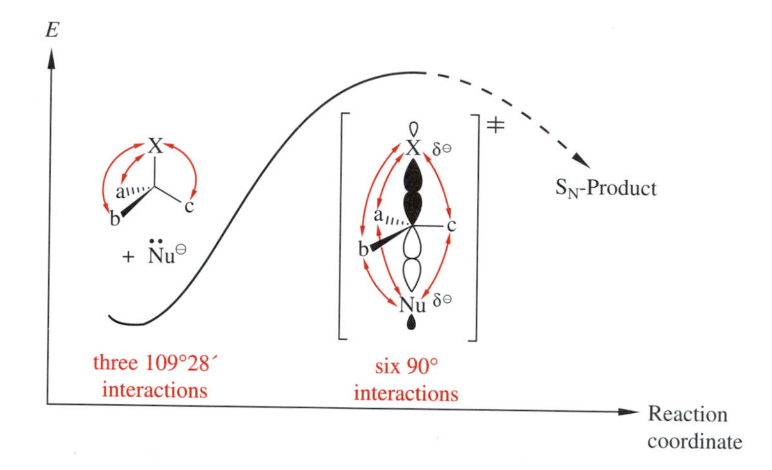

Fig. 2.9. Steric effects on S_N2 reactivity: substituent compression in the transition state.

state. Consequently, the activation energy E_a increases and the rate constant k_{S_N2} decreases. It should be noted that this destabilization is not compensated for by the simultaneous increase in the bond angle *between* the inert substituents from approximately the tetrahedral angle to approximately 120°.

This has two consequences:

Tendencies and Rules

- The S_N2 reactivity of an alkylating agent decreases with an increasing **number** of the alkyl substituents at the attacked C atom. In other words, α branching at the C atom of the alkylating agent reduces its S_N2 reactivity. This reduces the reactivity so much that tertiary C atoms can no longer be attacked according to an S_N2 mechanism at all:

$Nu^{\ominus}$ +	Me—X	Et—X	*i*Pr—X	*tert*-Bu—X
$k_{S_N2,\,rel}$ =	30	1	0.025	tiny

Generally stated, for S_N2 reactivity we have $k(\text{Me—X}) > k(\text{R}_{prim}\text{—X}) >> k(\text{R}_{sec}\text{—X})$; $k(\text{R}_{tert}\text{—X}) \approx 0$ (unit: $1\ \text{mol}^{-1}\ \text{s}^{-1}$).

- The S_N2 reactivity of an alkylating agent decreases with an increase in size of the alkyl substituents at the attacked C atom. In other words, β branching in the alkylating agent reduces its S_N2 reactivity. This reduces the reactivity so much that a C atom with a tertiary C atom in the β position can no longer be attacked at all according to an S_N2 mechanism:

$Nu^{\ominus}$ +	$MeCH_2$—X	$EtCH_2$—X	*i*PrCH_2—X	*tert*-BuCH_2—X
$k_{S_N2,\,rel}$ =	1	0.4	0.03	tiny

Generally stated, for S_N2 reactivity we have $k(\text{MeCH}_2\text{—X}) > k(\text{R}_{prim}\text{CH}_2\text{—X}) >> k(\text{R}_{sec}\text{CH}_2\text{—X})$; $k(\text{R}_{tert}\text{CH}_2\text{—X}) \approx 0$ (unit: $1\ \text{mol}^{-1}\ \text{s}^{-1}$).

Besides these rate-reducing steric substituent effects, in S_N2 reactions there is a rate-increasing electronic substituent effect. It is due to facilitation of the rehybridization

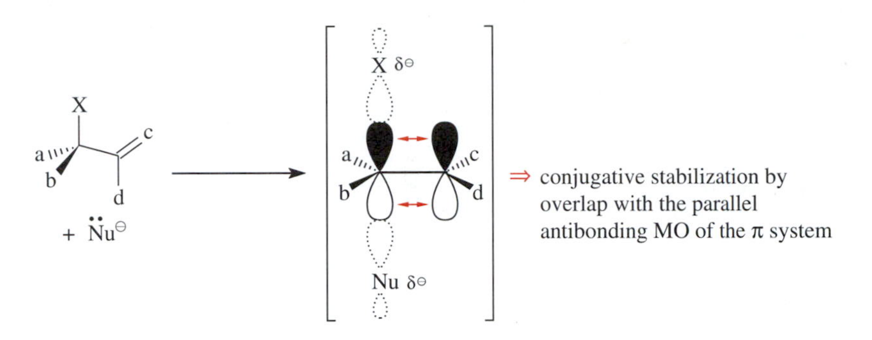

Fig. 2.10. Electronic effects on S_N2 reactivity: conjugative stabilization of the transition state by suitably aligned unsaturated substituents.

of the attacked C atom from sp^3 to sp^2 (Figure 2.10). This effect is exerted by unsaturated substituents bound to the attacked C atom. These include substituents, such as alkenyl, aryl, or the C=O double bond of ketones or esters. When it is not prevented by the occurrence of strain, the π-electron system of these substituents can line up in the transition state parallel to the $2p_z$ AO at the attacked C atom. This orbital thereby becomes part of a *delocalized* π-electron system. Consequently, there is a reduction in energy and a corresponding increase in the S_N2 reaction rate. Allyl and benzyl halides are therefore just as good alkylating agents as methyl iodide:

$Nu^{\ominus}$ +	$MeCH_2$ — X	vinylCH_2 — X	PhCH_2 — X
$k_{S_N2,\,rel}$ =	1	40	120

Because of the substituent effect just described, allyl and benzyl halides generally react with nucleophiles according to an S_N2 mechanism. This occurs even though the S_N1 reactivity of allyl and benzyl halides is *higher* than that of nonconjugated alkylating agents (see Section 2.5.4).

α-Halogenated ketones and α-halogenated acetic acid esters also react with nucleophiles according to the S_N2 mechanism. However, for them the alternative of an S_N1 mechanism is completely out of the question. This is because it would have to take place via a carbenium ion, which would be extremely destabilized by the strongly electron-withdrawing acyl or alkoxyacyl substituent.

2.5 S_N1 Reactions: Kinetic and Stereochemical Analysis; Substituent Effects on Reactivity

2.5.1 Energy Profile and Rate Law of S_N1 Reactions; Steady State Approximation

Substitution reactions according to the S_N1 mechanism take place in two steps (Figure 2.11). In the first and slower step, heterolysis of the bond between the C atom and the leaving group takes place. A carbenium ion is produced as a high-energy, and con-

B

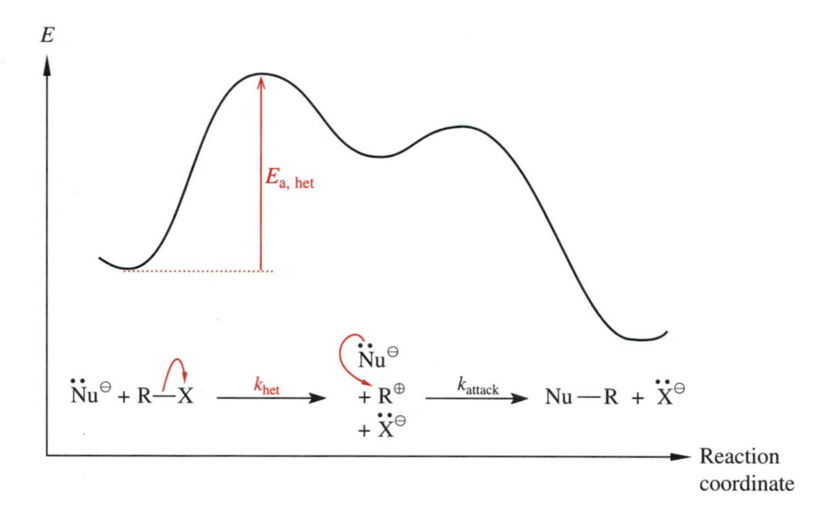

Fig. 2.11. Mechanism and energy profile of S_N1 reactions: $E_{a,het}$ designates the activation energy of the heterolysis, k_{het} and k_{attack} designate the rate constants for the heterolysis and the nucleophilic attack on the carbenium ion, respectively.

sequently short-lived, intermediate. In a considerably faster second step, it combines with the nucleophile to form the substitution product Nu—R.

In a substitution according to the S_N1 mechanism, the nucleophile does not actively attack the alkylating agent. The reaction mechanism consists of the alkylating agent dissociating *by itself* into a carbenium ion and the leaving group. Only then does the nucleophile change from a "spectator" into an active participant. Specifically, it intercepts the carbenium ion to form the substitution product.

What does the rate law for the substitution mechanism of Figure 2.11 look like? The rate of formation of the substitution product Nu—R in the second step can immediately be written as Equation 2.5 because this step represents an elementary reaction. Here, as in Figure 2.11, k_{het} and k_{attack} designate the rate constants for the heterolysis and the nucleophilic attack, respectively.

However, Equation 2.5 cannot be correlated with experimentally determined data. The reason for this is that the concentration of the carbenium ion intermediate appears in it. This concentration is extremely small during the entire reaction and consequently cannot be measured. However, one cannot set it equal to zero, either. In that case, Equation 2.5 would mean that the rate of product formation is also equal to zero and thus that the reaction does not take place at all. Accordingly, we must have a better approximation, one that is based on the following consideration:

$$\frac{d[\mathrm{Nu-R}]}{dt} = k_{attack}[\mathrm{R}^{\oplus}][\mathrm{Nu}^{\ominus}] \qquad (2.5)$$

$$\frac{d[\mathrm{R}^{\oplus}]}{dt} = 0 \text{ within the limits of the Bodenstein approximation} \qquad (2.6)$$

$$= k_{het}[\mathrm{R-X}] - k_{attack}[\mathrm{R}^{\oplus}][\mathrm{Nu}^{\ominus}] \qquad (2.7)$$

$$\Rightarrow [R^{\oplus}] = \frac{k_{het}}{k_{attack}} \cdot \frac{[R-X]}{[Nu^{\ominus}]} \tag{2.8}$$

Equation 2.8 in Equation 2.5 $\Rightarrow$

$$\frac{d\,[Nu-R]}{dt} = k_{het} \cdot [R-X] \tag{2.9}$$

Bodenstein's Steady State Approximation

The concentration of an intermediate in a multistep reaction is always very low when it reacts faster than it is produced. If this concentration is set equal to zero in the derivation of the rate law, unreasonable results may be obtained. In such a case, one resorts to a different approximation. One sets the change of the concentration of this intermediate as a function of time equal to zero. This is equivalent to saying that the concentration of the intermediate during the reaction takes a value slightly different from zero. This value can be considered to be invariant with time, i.e., steady. Consequently this approximation is called Bodenstein's steady state approximation.

Equipped with the Bodenstein principle, let us now continue the derivation of the rate law for S_N reactions that take place according to Figure 2.11. The completely inadequate approximation [carbenium ion] = 0 must be replaced by Equation 2.6. Let us now set the left-hand side of Equation 2.6, the change of the carbenium ion concentration with time, equal to the difference between the rate of formation of the carbenium ion and its consumption. Because the formation and consumption of the carbenium ion are elementary reactions, Equation 2.7 can immediately be set up. If we now set the right-hand sides of Equations 2.6 and 2.7 equal and solve for the concentration of the carbenium ion, we get Equation 2.8. With this equation, it is possible to rewrite the previously unusable Equation 2.5 as Equation 2.9. The only concentration term that appears in this equation is the concentration of the alkylating agent. In contrast to the carbenium ion concentration, it can be readily measured.

The rate law of Equation 2.9 identifies the S_N reactions of Figure 2.11 as unimolecular reactions. They are therefore referred to as S_N1 reactions. The rate of product formation thus depends only on the concentration of the alkylating agent and not on the concentration of the nucleophile. This is the key experimental criterion for distinguishing the S_N1 from the S_N2 mechanism.

From Equation 2.9 we can also derive the following: the S_N1 product is produced with the rate constant k_{het} of the first reaction step. Thus the rate of product formation does not depend on the rate constant k_{attack} of the second reaction step. In a multistep reaction, a particular step may be solely responsible for the rate of product formation. This is referred to as the rate-determining step. In the S_N1 reaction, this step is the heterolysis of the alkylating agent. The energy profile of Figure 2.11 shows that here—as everywhere else—the rate-determining step of a multistep sequence is the step in which the highest activation barrier must be overcome.

Equation 2.9 can also be interpreted as follows. Regardless of which nucleophile enters into an S_N1 reaction with a given alkylating agent, the substitution product is produced with the same rate constant. Figure 2.12 illustrates this using as an example S_N1 reactions of two different nucleophiles with $Ph_2CH—Cl$. Both take place via the benzhydryl cation Ph_2CH^+. In the first experiment, this cation is intercepted by pyridine as the nucleophile to form a pyridinium salt. In the second experiment, it is intercepted by triethylamine to form a tetraalkylammonium salt. Both reactions take place with

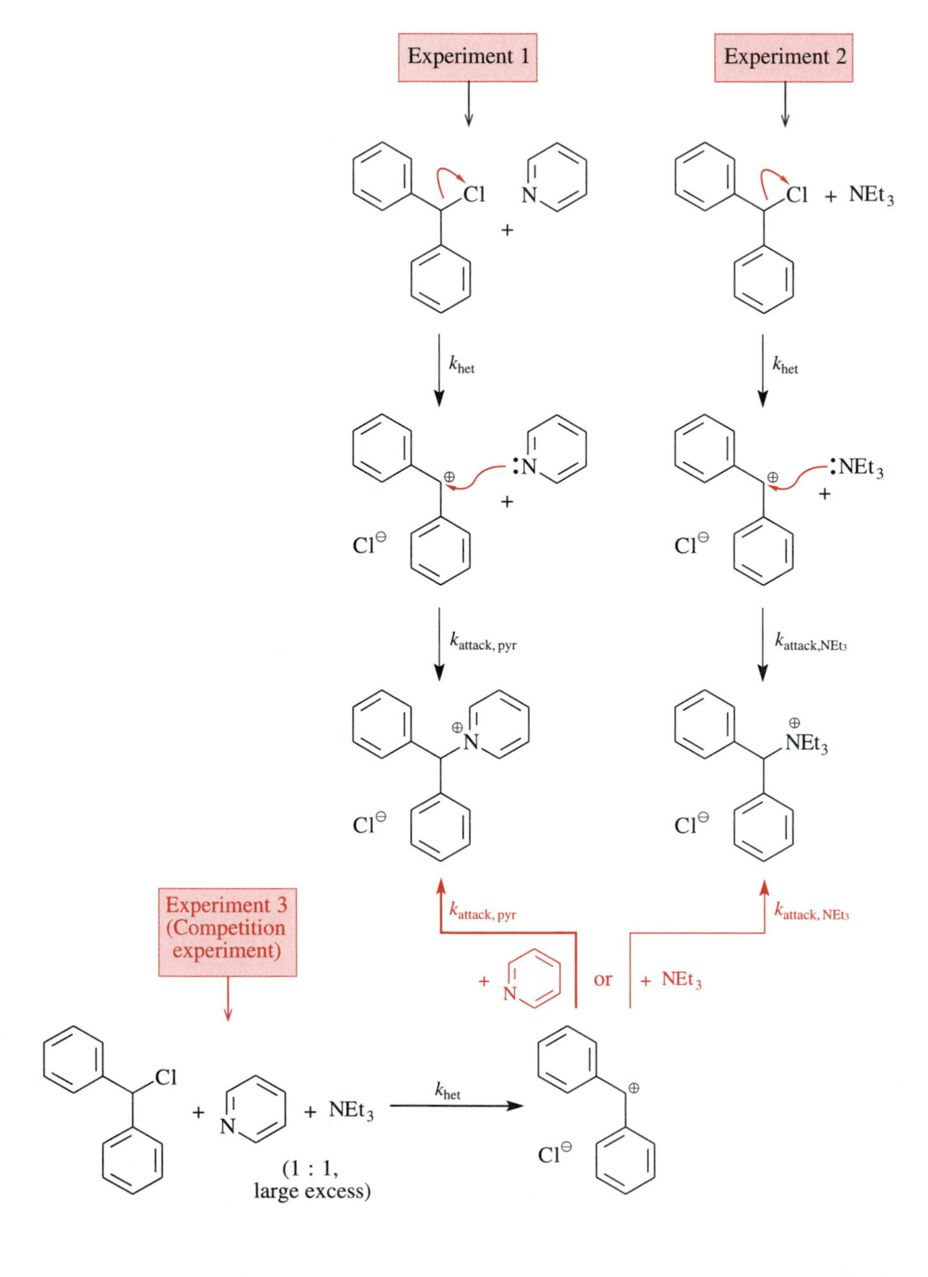

Fig. 2.12. Rate and selectivity of S_N1 reactions. Both reactions 1 and 2 take place via the benzhydryl cation. Both reactions (with pyridine and triethylamine) take place with the same rate constant. The higher nucleophilicity of pyridine does not become noticeable until experiment 3, the "competition experiment": The pyridinium salt is by far the major product.

the same rate constant, namely the heterolysis constant k_{het} of benzhydryl chloride. The fact that pyridine is a better nucleophile than triethylamine therefore does not become noticeable at all under these conditions.

The situation is different in experiment 3 of Figure 2.12. There, both S_N1 reactions are carried out as competitive reactions: benzhydryl chloride heterolyzes (just as fast as before) in the presence of equal amounts of both amines. The benzhydryl cation is now intercepted faster by the more nucleophilic pyridine than by the less nucleophilic triethylamine. By far the major product is the pyridinium salt.

2.5.2 Stereochemistry of S_N1 Reactions; Ion Pairs

B

What can be said about the stereochemistry of S_N1 reactions? In the carbenium ion intermediates $R^1R^2R^3C^+$, the valence-unsaturated C atom has a trigonal planar geometry (cf. Figure 1.2). These intermediates are therefore achiral (if the substituents R^i themselves do not contain chiral centers).

When in carbenium ions of this type $R^1 \neq R^2 \neq R^3$, these carbenium ions react with achiral nucleophiles to form chiral substitution products $R^1R^2R^3C—Nu$. These must be produced as a 1:1 mixture of both enantiomers (i.e., as a racemic mixture). Achiral reaction partners alone can never deliver an optically active product. But in apparent contradiction to what has just been explained, not all S_N1 reactions that start from enantiomerically pure alkylating agents and take place via achiral carbenium ions produce a racemic substitution product. Let us consider, for example, Figure 2.13.

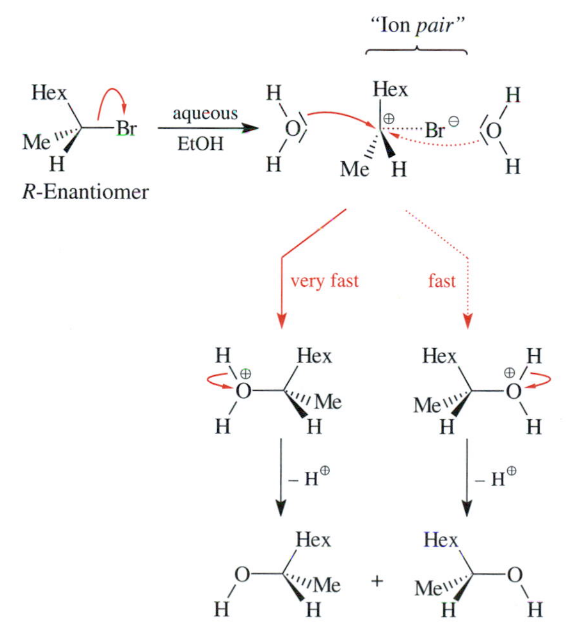

Fig. 2.13. Stereochemistry of an S_N1 reaction that takes place via a contact ion pair. The reaction proceeds with 66% inversion of configuration and 34% racemization.

Figure 2.13 shows an S_N1 reaction with optically pure *R*-2-bromooctane carried out as a solvolysis. By solvolysis one means an S_N1 reaction performed in a polar solvent that also functions as the nucleophile. The solvolysis reaction in Figure 2.13 takes place in a water/ethanol mixture. In the rate-determining step a secondary carbenium ion is produced. This ion must be planar and therefore achiral. However, it is so unstable (Section 2.5.4) that it is highly reactive. Consequently, it reacts so quickly with the solvent that at this point in time it has still not completely "separated" from the bromide ion, which was released when it was formed. In other words, the reacting carbenium ion is still almost in contact with this bromide ion. Thus it exists as part of a so-called contact ion pair (the emphasis is on ion *pair*) $R^1R^2HC^+\cdots Br^-$.

The contact ion pair $R^1R^2HC^+\cdots Br^-$, in contrast to a free carbenium ion $R^1R^2HC^+$, is *chiral.* Starting from enantiomerically pure *R*-2-bromooctane, the contact ion pair first produced is also a pure enantiomer. In this ion pair, the bromide ion adjacent to the carbenium ion center partially protects one side of the carbenium ion from the attack of the nucleophile. Consequently, the nucleophile preferentially attacks from the side that lies opposite the bromide ion. Thus the solvolysis product in which the configuration at the attacked C atom has been inverted is the major product. To a minor extent the solvolysis product with retention of configuration at the attacked C atom is formed.

It was actually found that 83% of the solvolysis product was formed with inversion of configuration and 17% with retention. This result is equivalent to the occurrence of 66% inversion of configuration and 34% racemization. We can therefore generalize and state: When in an S_N1 reaction the nucleophile attacks the contact ion pair, the reaction proceeds with partial inversion of configuration (or alternatively, with partial but not complete racemization).

On the other hand, when in an S_N1 reaction the nucleophile attacks the carbenium ion after it has separated from the leaving group, the reaction takes place with complete racemization. This is the case with more stable and consequently longer-lived carbenium ions. For example, the α-methylbenzyl cation, which is produced in the rate-determining step of the solvolysis of *R*-phenethyl bromide in a water/ethanol mixture is such a cation (Figure 2.14).

The following should be noted for the sake of completeness. The bromide ion in Figure 2.14 moves so far away from the α-methylbenzyl cation intermediate that it allows the solvent to attack both sides of the carbenium ion with equal probability. However, the bromide ion does not move away from the carbenium ion to an arbitrary distance. The electrostatic attraction between oppositely charged particles holds the carbenium ion and the bromide ion together at a distance large enough for solvent molecules to fit in between. This is the so-called solvent-separated ion pair. The emphasis lies on *solvent-separated.*

2.5.3 Solvent Effects on S_N1 Reactivity

B

In contrast to the S_N2 mechanism (Section 2.4.3), the structure of the transition state of the rate-determining step in the S_N1 mechanism cannot be depicted in a simple way. As an aid, one therefore uses a transition state model. According to the Hammond postu-

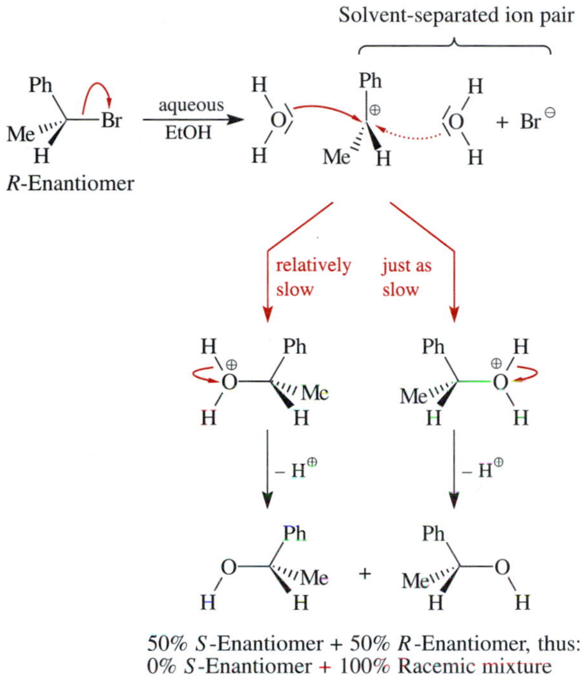

Fig. 2.14. Stereochemistry of an S_N1 reaction that takes place via a solvent-separated ion pair. The reaction proceeds with 0% inversion and 100% racemization.

late (here in the form: "The transition state of an endothermic reaction is similar to the product"), a suitable model for the transition state of the rate-determining step of an S_N1 reaction is the corresponding carbenium ion. The rate constant k_{het} of S_N1 reactions should therefore be greater the more stable the carbenium ion produced in the heterolysis is:

$$R{-}X \xrightarrow{k_{Het}} R^{\oplus} + :X^{\ominus}$$

As a measure of the rate of formation of this carbenium ion, one can take the free energy of heterolysis of the alkylating agent from which it arises. Low free energies of heterolysis are related to the formation of a stable carbenium ion and thus a high S_N1 reactivity. Because S_N1 reactions are carried out in solution, the free energies of heterolysis in solution are the suitable stability measure. These values are not immediately available, but they can be calculated from other experimentally available data (Table 2.1) by means of the thermodynamic cycle in Figure 2.15.

Heterolyses of alkyl bromides in the gas phase always require more energy than the corresponding homolyses (cf. Table 2.1). In fact, the ΔG values of these heterolyses are considerably more positive than the ΔH values of the homolyses (even when it is taken into consideration that ΔG and ΔH values cannot be compared directly with each other). The reason for this is that extra energy is required for separating the charges. As Table

Table 2.1. Free Energy Values from Gas Phase Studies (Lines 1–3). Free Energies of Heterolysis in Water (Line 4) Calculated Therefrom According to Figure 2.15*

$R{-}Br \longrightarrow R^{\oplus} + Br^{\ominus}$	Me	Et	*i*Pr	*tert*-Bu	$PhCH_2$
$\Delta G_{het,g}$ / kcal/mol	+214	+179	+157	+140	+141
$\Delta G_{hyd\ (R^{\oplus})}$ / kcal/mol	–96	–78	–59	–54	–59
$\Delta G_{hyd\ (Br^{\ominus})}$ / kcal/mol	–72	–72	–72	–72	–72
$\Delta G_{het,\ H_2O}$ / kcal/mol	+47	+30	+27	+14	+11
(cf. $\Delta H_{hom,\ any\ medium}$ / kcal/mol)	(+71)	(+68)	(+69)	(+63)	(+51)
$\tau_{1/2,\ het,\ 298\ K}$	≥ 10^{16} yr	≥ 10^5 yr	≥ 220 yr	≥ 0.7 s	≥ 0.007 s

*From these energies and using the Eyring equation (Equation 1.1), one can calculate minimum half-lives for the pertinent heterolyses in water (line 6). These are minimum values because the $\Delta G_{het,H_2O}$ values were used for $\Delta G^{\ddagger}$ in the Eyring equation, whereas actually $\Delta G^{\ddagger} > * \Delta G_{het,\ H_2O}$. ΔG_{hyd} = free energy of hydration, ΔG_{het} = free energy of heterolysis, and ΔH_{hom} = enthalpy of homolysis.

2.1 shows, this situation is reversed in polar solvents such as water. There the heterolyses require less energy than the homolyses.

You see the reason for this solvent effect when you look at lines 2 and 3 in Table 2.1. Ions are stabilized quite considerably by solvation. Heterolyses of alkylating agents, and consequently S_N1 reactions, therefore, succeed only in highly solvating media. These include the polar protic solvents such as methanol, ethanol, acetic acid, and aqueous acetone as well as the polar aprotic solvents acetone, acetonitrile, DMF, NMP, DMSO, and DMPU (Figure 2.16). Unfortunately, DMPU does not solvate as well as HMPA, which is a carcinogen.

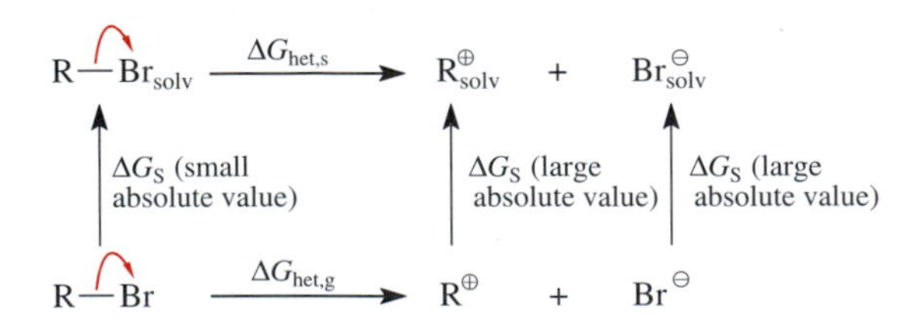

Fig. 2.15. Thermodynamic cycle for calculating the free energy of heterolysis of alkyl bromides in solution ($\Delta G_{het,s}$) from free energies in the gas phase: ΔG_s refers to the free energy of solvation, $\Delta G_{het,g}$ refers to the free energy of heterolysis in the gas phase.

DMF = **Dim**ethyl**f**ormamide: H—C(=O)—NMe_2

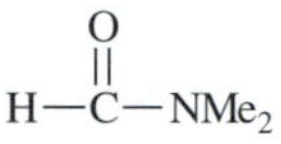

NMP = *N*-**M**ethyl**p**yrrolidone: (cyclic structure with O and NMe)

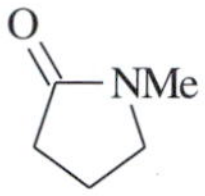

DMSO = **Dim**ethyl**s**ulf**o**xide: Me—S(=O)—Me

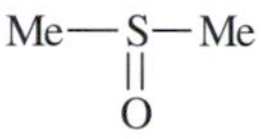

DMPU = *N,N*′-**Dim**ethyl-*N,N*′-**p**ropylene **u**rea: (cyclic urea with Me—N and N—Me)

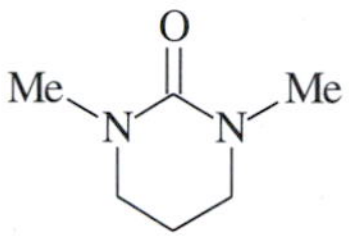

HMPA = **Hexam**ethyl **p**hosphoric acid tri**amide**: $(Me_2N)_3P{=}O$

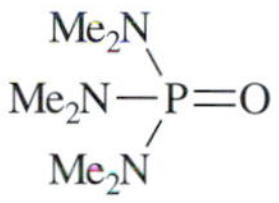

Fig. 2.16. Polar aprotic solvents and the abbreviations that can be used for them.

It is important to note that the use of polar solvents is only a necessary but not a sufficient condition for the feasibility of S_N1 substitutions or for the preference of the S_N1 over the S_N2 mechanism.

2.5.4 Substituent Effects on S_N1 Reactivity

B

The stabilities of five carbenium ions formed in the heterolyses of alkyl bromides in water (as an example of a polar solvent) can be compared by means of the associated free energy values $\Delta G_{het,\,H_2O}$ (Table 2.1, line 4). From these free energy values we get the following order of stability in polar solvents: $CH_3^+ < MeCH_2^+ < Me_2CH^+ < Me_3C^+ < Ph{-}CH_2^+$. One can thus draw two conclusions: (1) a phenyl substituent stabilizes a carbenium ion center, and (2) alkyl substituents also stabilize a carbenium ion center, and they do so better the more alkyl groups are bound to the center. The stability order just cited is similar to the stability order for the same carbenium ions in the gas phase (Table 2.1, row 1). However, in the gas phase considerably larger energy differences occur. They correspond to the inherent stability differences (i.e., without solvation effects).

We observed quite similar substituent effects in Section 1.2.1 in connection with the stability of radicals of type $R^1R^2R^3C\cdot$. These effects were interpreted both in VB and in MO terms. For the carbenium ions, completely analogous explanations apply, which are shown in Tables 2.2 (effect of conjugating groups) and 2.3 (effect of alkyl groups). These tables use resonance theory for the explanation. Figures 2.17 and 2.18 explain the same effects by means of MO theory. Both approaches explain how substituents can stabilize adjacent carbenium ion centers. Conjugating substituents, which are electron rich, stabilize a neighboring carbenium ion center via a resonance effect, and alkyl substituents do the same via an inductive effect.

Table 2.2. Stabilization of a Trivalent Carbenium Ion Center by Conjugating Substituents: Experimental Findings and Their Explanation by Means of Resonance Theory

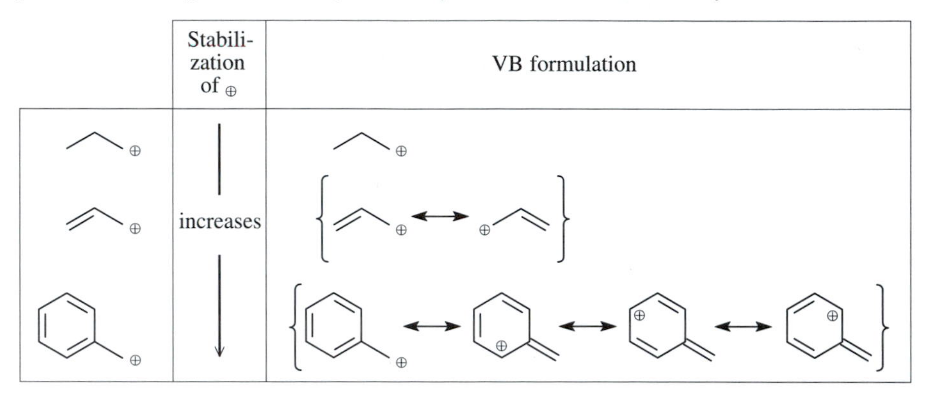

The very large inherent differences in the stability of carbenium ions are reduced in solution—because of a solvent effect—but they are not eliminated. This solvent effect arises because of the dependence of the free energy of hydration $\Delta G_{\text{hyd}(\text{R}^+)}$ on the

Table 2.3. Stabilization of Trivalent Carbenium Ion Centers by Methyl Substituents: Experimental Findings and Their Explanation by Means of Resonance Theory

	Stabilization of ⊕	VB formulation
$H_3C^{\oplus}$	increases ↓	$H_3C^{\oplus}$
$H_3C{-}H_2C^{\oplus}$		{ H_β-substituted ethyl cation ⟷ no-bond resonance forms with $H_\beta^{\oplus}$ } i.e., 1 no-bond resonance form per H_β
$(H_3C)_2HC^{\oplus}$		{ $(H_\beta)_3C$–$C^{\oplus}H$–$C(H_\beta)_3$ ⟷ 6 no-bond resonance forms }
$(H_3C)_3C^{\oplus}$		{ $((H_\beta)_3C)_3C^{\oplus}$ ⟷ 9 no-bond resonance forms }

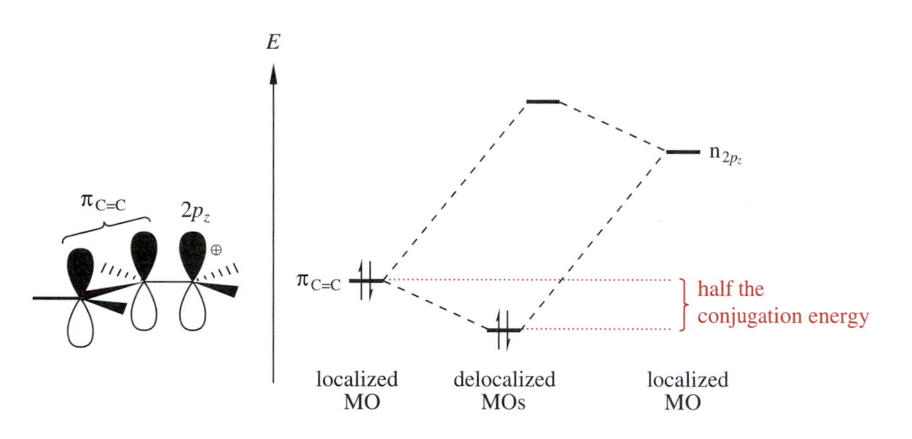

Fig. 2.17. MO interactions responsible for the stabilization of trivalent carbenium ion centers by suitably oriented unsaturated substituents ("conjugation").

structure of the carbenium ions (Table 2.1, row 2). This energy becomes less negative going from Me^+ to $Ph—CH_2^+$, as well as in the series $Me^+ \rightarrow Et^+ \rightarrow i Pr^+ \rightarrow tert\text{-}Bu^+$. The reason for this is hindrance of solvation. It increases with increasing size or number of the substituents at the carbenium ion center.

Allyl halides heterolyze just as easily as benzyl halides because they produce a resonance-stabilized carbenium ion just as the benzyl halides do. Even faster heterolyses are possible when the charge of the resulting carbenium ion can be delocalized by more than one unsaturated substituent and can thereby be stabilized especially well. This explains the remarkably high S_N1 reactivities of the benzhydryl halides (via the benzhydryl cation) and especially of the triphenylmethyl halides (via the trityl cation):

	$PhCH_2—X$	$Ph_2CH—X$	$Ph_3C—X$
$k_{S_N1,\,rel}$	1	1000	10^8

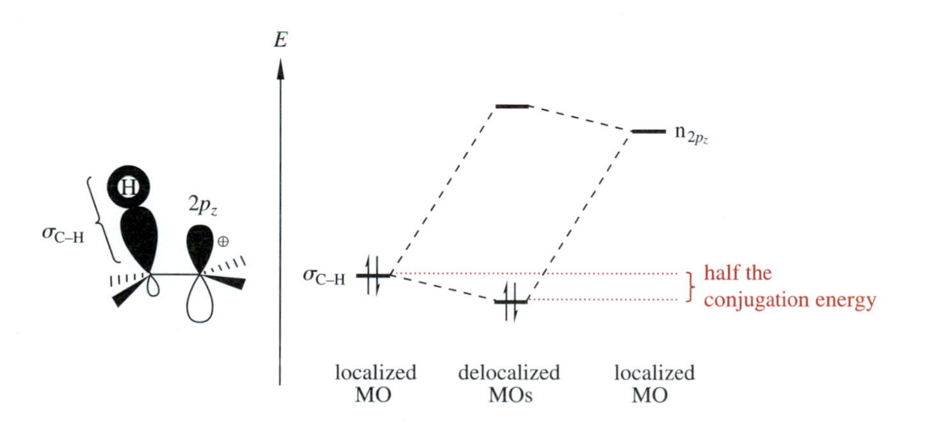

Fig. 2.18. MO interactions responsible for the stabilization of trivalent carbenium ion centers by suitably oriented C—H bonds in the β position ("hyperconjugation").

Side Note 2.2
***para*-Methoxylated Trityl Cations in Nucleotide Synthesis**

A

One methoxy group in the *para* position of the trityl cation contributes to further stabilization due to its electron-donating resonance effect. The same is true for each additional methoxy group in the *para* position of the remaining phenyl rings. The resulting increase in the S_N1 reactivity of multiply *para*-methoxylated trityl ethers is used in nucleotide synthesis (Figure 2.19), where these ethers serve as acid-labile protecting groups. The more stable the resulting trityl cation (i.e., the more methoxy groups it contains in *para* positions), the faster it is formed under acidic conditions.

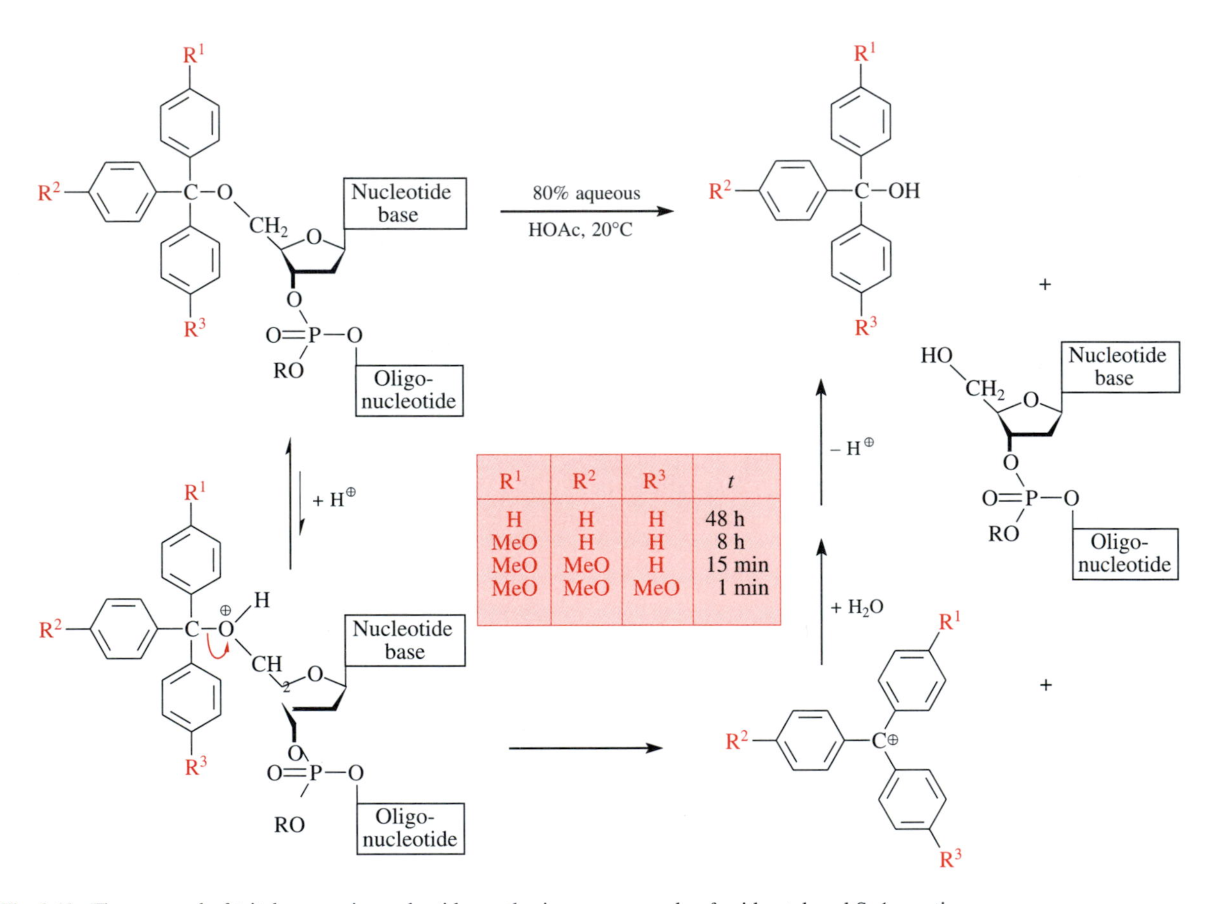

R¹	R²	R³	t
H	H	H	48 h
MeO	H	H	8 h
MeO	MeO	H	15 min
MeO	MeO	MeO	1 min

Fig. 2.19. The removal of trityl groups in nucleotide synthesis as an example of acid-catalyzed S_N1 reactions.

B

Finally, two or even three amino or dimethylamino groups in the *para* position stabilize trityl cations so efficiently because of their pronounced electron-donating resonance effect that the associated non-ionized neutral compounds are no longer able to exist at all but heterolyze quantitatively to salts. These salts are the well-known triphenylmethane dyes.

2.6 When Do S_N Reactions at Saturated C Atoms Take Place According to the S_N1 Mechanism and When Do They Take Place According to the S_N2 Mechanism?

From Sections 2.4.4 and 2.5.4, the following rules of thumb can be derived:

Rules of Thumb

S_N1 reactions are observed

- always in substitutions on R_{tert}—X, Ar_2HC—X, and Ar_3C—X;
- always in substitutions on substituted and unsubstituted benzyl and allyl triflates;
- in substitutions on R_{sec}—X when poor nucleophiles are used (e.g., in solvolyses);
- in substitutions on R_{sec}—X that are carried out in the presence of strong Lewis acids such as in the substitution by aromatics ("Friedel–Crafts alkylation;" Figure 5.21);
- almost never in substitutions on R_{prim}—X (exception: R_{prim}—$N^+{\equiv}N$).

S_N2 reactions take place

- almost always in substitutions in sterically unhindered benzyl and allyl positions (exception: benzyl and allyl triflates react according to S_N1);
- always in substitutions in MeX and R_{prim}—X;
- in substitutions in R_{sec}—X, provided a reasonably good nucleophile is used;
- never in substitutions in substrates of the type R_{tert}—X or R_{tert}—C—X.

2.7 Unimolecular S_N Reactions That Do Not Take Place via Simple Carbenium Ion Intermediates: Neighboring Group Participation

2.7.1 Conditions for and Features of S_N Reactions with Neighboring Group Participation

B

A leaving group in an alkylating agent can be displaced not only by a nucleophile added to the reaction mixture but also by one in the alkylating agent itself. This holds for alkylating agents that contain a nucleophilic electron pair at a suitable distance from the leaving group. The structural element on which this electron pair is localized is called a neighboring group. It displaces the leaving group stereoselectively through a backside attack. This attack thus corresponds to that of an S_N2 reaction. But because the substitution through the neighboring group takes place intramolecularly, it represents a unimolecular process. In spite of this, organic chemists, who want to emphasize the mechanistic relationship and not the rate law, classify substitution reactions with neighboring group participation as S_N2 reactions.

Because of this neighboring group participation, a cyclic and possibly strained (depending on the ring size) intermediate **A** is formed from the alkylating agent:

$$\text{C–X} \xrightarrow[\text{rate-determining}]{-:X^{\ominus}} \mathbf{A} \xrightarrow{:Nu^{\ominus}} \mathbf{B}\ (\text{C–Nu})$$

A (three- or five-membered) **B**

This intermediate generally contains a positively charged center, which represents a new leaving group. This group is displaced in a second step by the external nucleophile through another backside attack. This step is now clearly an S_N2 reaction. In the reaction product **B** the external nucleophile occupies the same position the leaving group X originally had. Reactions of this type thus take place with complete retention of configuration at the attacked C atom. This distinguishes them both from substitutions according to the S_N2 mechanism and from substitutions according to the S_N1 mechanism.

For an S_N reaction to take place with neighboring group participation, a neighboring group not only must be present, but it must be sufficiently reactive. The attack of the "neighboring group" must take place faster than the attack by the external nucleophile. Otherwise, the latter would initiate a normal S_N2 reaction. In addition, the neighboring group must actively displace the leaving group before this group leaves the alkylating agent on its own. Otherwise, the external nucleophile would enter via an S_N1 mechanism. Therefore, reactions with neighboring group participation take place more rapidly than comparable reactions without such participation. This fact can be used to distinguish between S_N reactions with neighboring group participation and normal S_N1 and S_N2 reactions.

The nucleophilic electron pairs of the neighboring group can be nonbonding, or they can be in a π bond or, in special cases, in a σ bond (Figure 2.20). Generally they can displace the leaving group only when this produces a three- or five-membered cyclic intermediate. The formation of rings of a different size is almost always too slow for neighboring group participations to compete with reactions that proceed by simple S_N1 or S_N2 mechanisms.

Fig. 2.20. Alkylating agents with structural elements that can act as neighboring groups in S_N reactions. (In the example listed in parentheses, the neighboring group participation initiates an alkylation of the aromatic compound, in which an external nucleophile is not participating at all.)

The "range" of possible neighboring group participations mentioned above thus reflects the almost universal ring closure rates of five-membered ≥ three-membered > six-membered >> other ring sizes.

In general, one speaks of a neighboring group effect only when the cyclic and usually positively charged intermediate cannot be isolated but is subject to an S_N2-like ring opening by the external nucleophile. This ring opening always happens for three-membered ring intermediates. There it profits kinetically (Hammond postulate!) from the considerable reduction in ring strain. In the case of five-membered intermediates, the situation is different. They not only are subject to attack by the external nucleophile, but they can also react to form stable five-membered rings by elimination of a cation—usually a proton.

2.7.2 Increased Rate through Neighboring Group Participation

B

The sulfur-containing dichloride "mustard gas," a high-boiling liquid used in World War I as a combat gas in the form of an aerosol, hydrolyzes much more rapidly to give HCl and a diol (Figure 2.21) than its sulfur-free analog 1,5-dichloropentane. Therefore, mustard gas released HCl especially efficiently into the lungs of soldiers who had inhaled it (and thus killed them very painfully). The reason for the higher rate of hydrolysis of mustard gas is a neighboring group effect. It is due to the availability of a free electron pair in a nonbonding orbital on the sulfur atom.

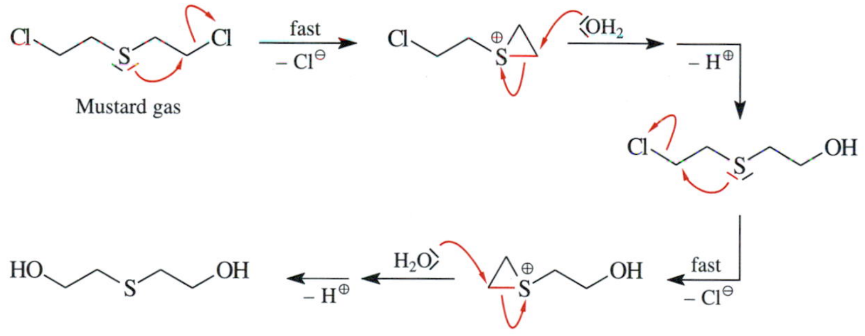

Fig. 2.21. Acceleration of the hydrolysis of mustard gas by neighboring group participation.

Phenethyl tosylate solvolyzes in CF_3CO_2H orders of magnitude faster than ethyl tosylate (Figure 2.22). Because the neighboring phenyl ring can make a π electron pair available, a phenonium ion intermediate is formed. Phenonium ions are derivatives of the spirooctadienyl cation shown. They are therefore related to the Wheland complexes of electrophilic aromatic substitution (see Chapter 5). The intermediacy of a phenonium ion in the solvolysis of phenethyl tosylate was proven by isotope labeling. A deuterium label located exclusively α to the leaving group in the tosylate

Fig. 2.22. Acceleration of the trifluoroacetolysis of phenethyl tosylate by neighboring group participation.

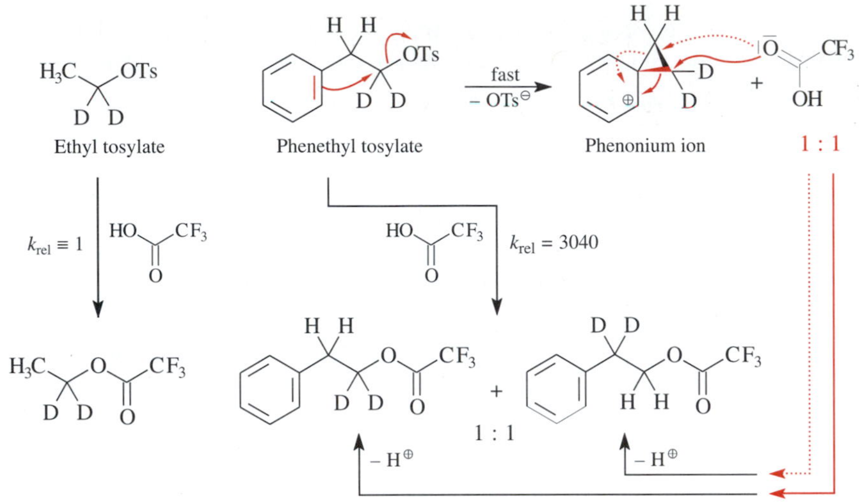

was scrambled in the reaction product—it emerged half in the α position and half in the β position.

2.7.3 Stereoselectivity through Neighboring Group Participation

A

A carboxylic ester can participate in an S_N reaction as a neighboring group. In this case, a nonbonding electron pair on the double-bonded oxygen displaces the leaving group in the first reaction step. Let us consider, for example, the glycosyl bromide **B** in Figure 2.23. There a bromide ion is displaced by an acetate group in the β position. The departure of the bromide ion is facilitated because it is immediately intercepted by Ag(I) ions and converted into insoluble AgBr. The five-membered ring of the resonance-stabilized carboxonium ion **E** is formed. At this stage, an axially oriented C—O bond has emerged from the equatorial C—Br bond of substrate **B** via an inversion of the configuration. In the second reaction step this C—O bond is broken by the solvent methanol via another inversion of the configuration at the attacked C atom. In the resulting substitution product **C,** the methoxy group is equatorial just like the bromine in the starting material **B.** The saponification of the acetate groups of this compound leads to isomerically pure β-D-methylglucopyranoside (Figure 2.23, top right). This is of preparative interest because the direct acetalization of glucose leads to the diastereomeric α-methylglucoside (Figure 2.23).

Bromide **A,** which is likewise derived from glucose, is a stereoisomer of bromide **B.** It is interesting that in silver(I)-containing methanol, **A** stereoselectively produces the same substitution product **C** as produced by **B.** In glycosyl bromide **A,** however, the C—Br bond cannot be broken with the help of the acetate group in the β position. In that case, a *trans*-annulated and consequently overly strained carboxonium intermediate would have been produced. Glycosyl bromide **A** therefore reacts with-

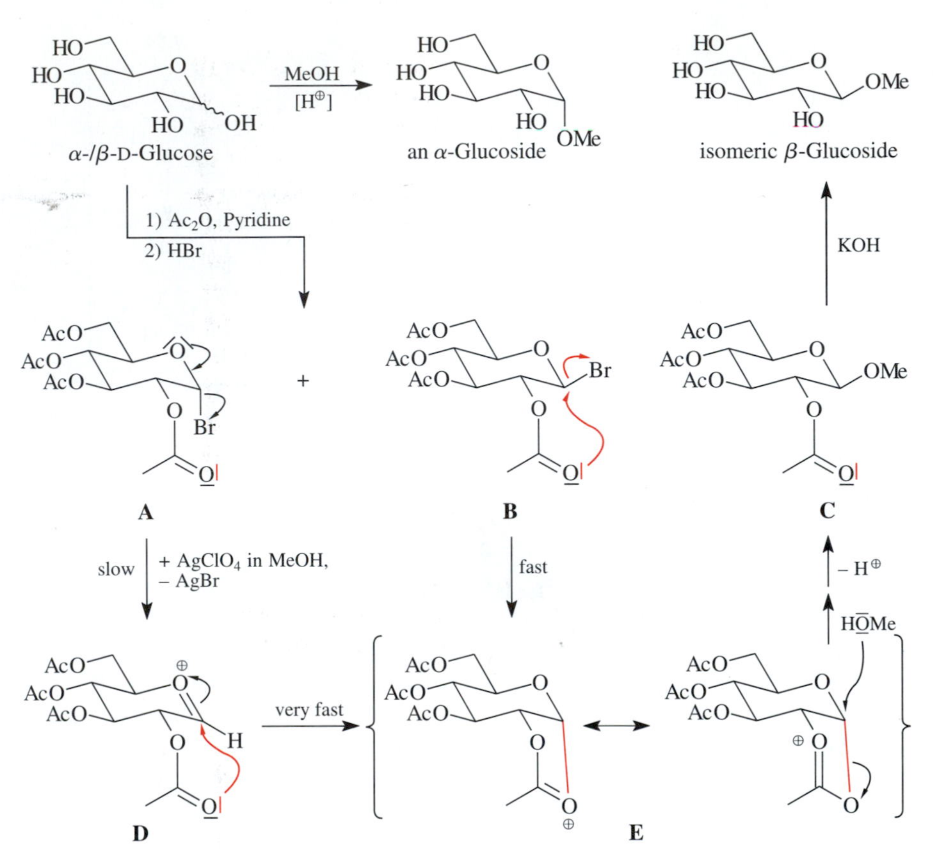

Fig. 2.23. Stereoselectivity due to neighboring group participation in glycoside syntheses.

out neighboring group participation and consequently more slowly than its diastereomer **B.** The first step is a normal S_N1 reaction with heterolysis of the C—Br bond to form the carboxonium ion **D.** However, as soon as this is present, the acetate group in the β position still exerts a neighboring group effect. With its nucleophilic electron pair, it closes the ring to the same cyclic carboxonium ion **E** that was produced in a single step from the diastereomeric glycosyl bromide **B.** The subsequent ring opening of **E** by methanol gives the methyl glucoside **C,** regardless of whether it was derived from **A** or **B.**

S_N reactions of enantiomerically pure diazotized α amino acids are also preparatively important (Figure 2.24). There the carboxyl group acts as a neighboring group. It has a double effect. For one thing, it slows down the S_N1-like decomposition of the diazonium salt, which for normal aliphatic diazonium salts takes place extremely fast (leading to carbenium ions with their abundant nonselective secondary chemistry: see Section 11.3.1). The analogous heterolysis of a diazotized α amino acid is slowed down in agreement with the Hammond postulate. This is because a carbenium ion would have to be produced, which would be strongly destabilized by the carboxyl group acting as an electron acceptor. In addition, the carboxylic acid group enters actively into the reaction as a neighboring group and displaces the leaving group (N_2) with the nu-

B

Fig. 2.24. Stereoselectivity due to neighboring group participation in the synthesis of α-functionalized carboxylic acids.

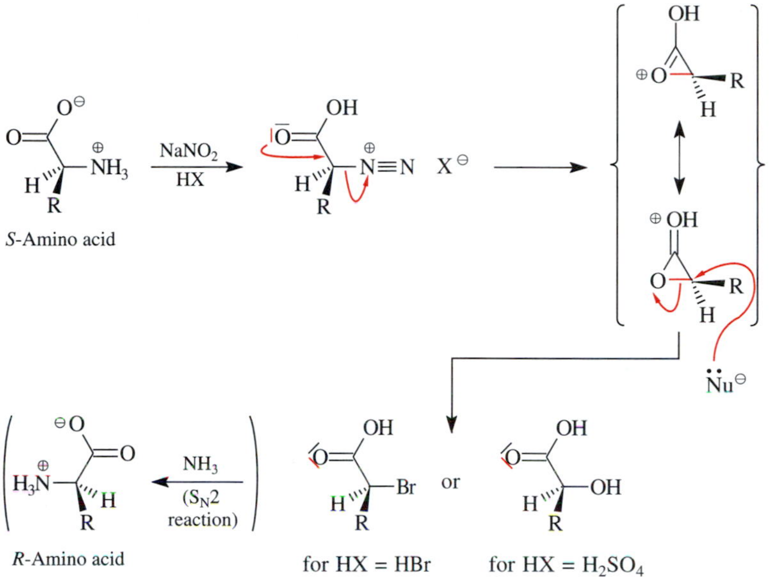

cleophilic, free electron pair of its double-bonded O atom. A highly strained protonated three-membered ring lactone is produced, in which the C—O_2C bond is attacked from the back side by the external nucleophile. The nucleophile therefore assumes the position of the original amino group, i.e., appears with retention of the configuration. According to Figure 2.24, H_2O is the entering nucleophile when the diazotization is carried out with aqueous H_2SO_4 and $NaNO_2$. But if the diazotization is carried out with aqueous HBr and $NaNO_2$, then a bromide ion acts as the external nucleophile. These chemoselectivities are explained by the nucleophilicity order $HSO_4^- < H_2O < Br^-$ (cf. Section 2.2).

A

A bonding electron pair that is contained in a C—C single bond is normally not a neighboring group. However, one famous exception is a particular bonding electron pair in the norbornane ring system (Figure 2.25). It is fixed in precisely such a way that it can interact with a reaction center on C2.

This electron pair speeds up the departure of a suitably oriented leaving group. It also determines the orientation of the nucleophile in the reaction product. The formulas in Figure 2.25 describe both of these aspects better than many words.

Nontheless, Figure 2.25 requires two comments. (1) The six-membered ring substructure of all other molecules including the cations possesses a boat conformation and not a twist–boat conformation, as the two-dimensional structures of the figure imply. If they had been drawn with the correct boat conformation, the crucial orbital interaction could not have been drawn with so few C—C bonds crossing each other. The ball-and-stick stereoformulas at the bottom of Figure 2.25 show the actual geometry of both cations. (2) The carbenium ion shown in Figure 2.25 is an isomer of, and less stable than, the carbonium ion shown there. It has slightly longer C1–C6 and C1–C2 bonds, and no C2–C6 bond (the C2/C6 separation is 2.5 Å in the carbenium ion and 1.8 Å in the carbonium ion). Whether the acetolysis of the *endo*-bornyl brosylate takes place via the

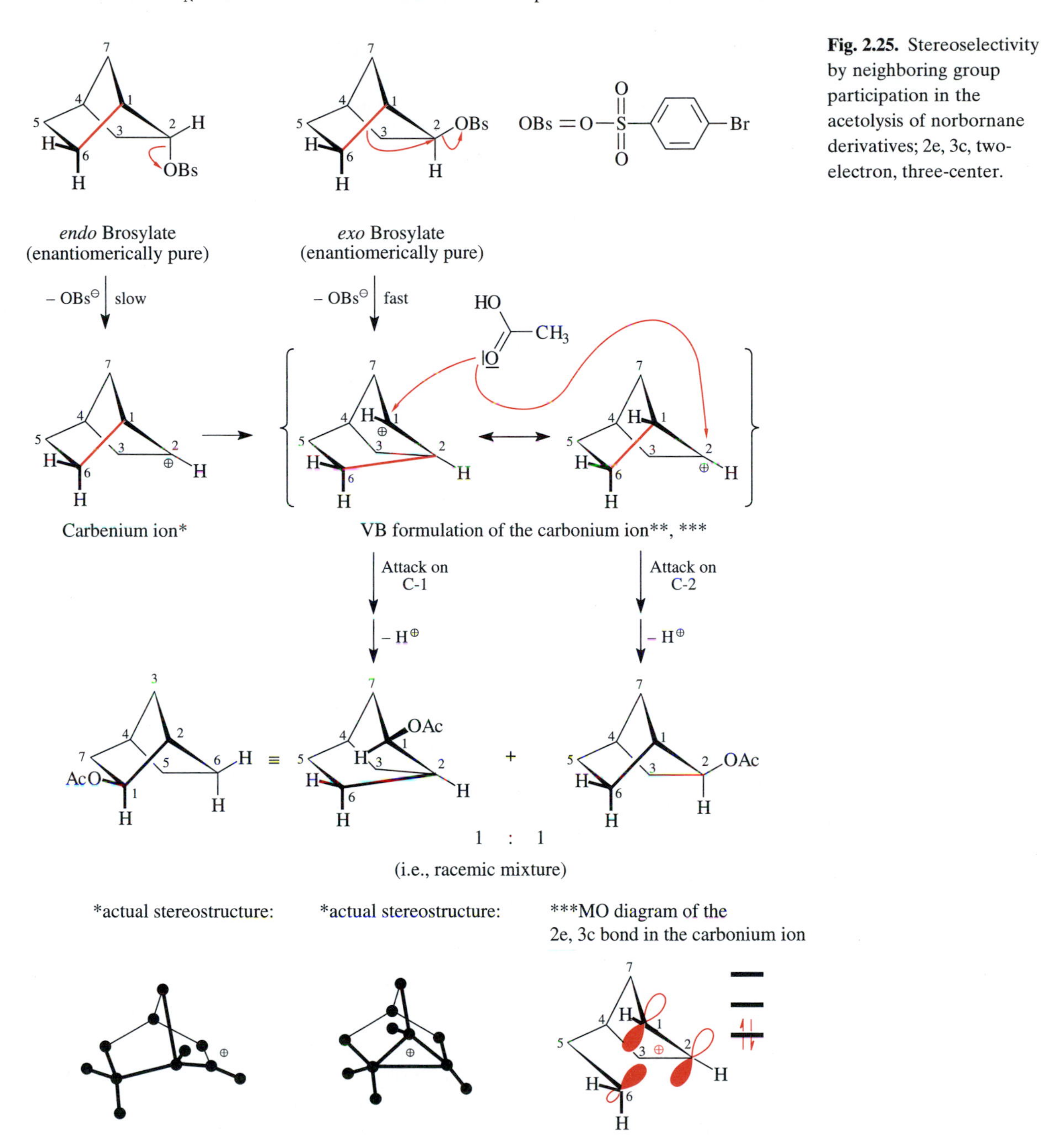

Fig. 2.25. Stereoselectivity by neighboring group participation in the acetolysis of norbornane derivatives; 2e, 3c, two-electron, three-center.

fully formed carbenium ion or follows a reaction path that leads to the carbonium ion without actually reaching the carbenium ion has not been clarified experimentally.

For many decades chemists had been interested in whether the positively charged intermediate of the S_N reaction in Figure 2.25 was a carbenium or a carbonium ion. Also the existence of a rapidly equilibrating mixture of two carbenium ions was considered. It is now known with certainty that this intermediate is a carbonium ion; it is known as a nonclassical carbocation. In the carbonium ion, between the centers C1, C2, and C6, there is a bond that is built up from two sp^2 AOs and one sp^3 AO (see MO diagram, lower right, Figure 2.25). It accommodates two electrons.

2.8 Preparatively Useful S_N2 Reactions: Alkylations

B

The three-part Figure 2.26, with a list of synthetically important reactions, shows the various nucleophiles that can be alkylated according to the S_N2 mechanism.

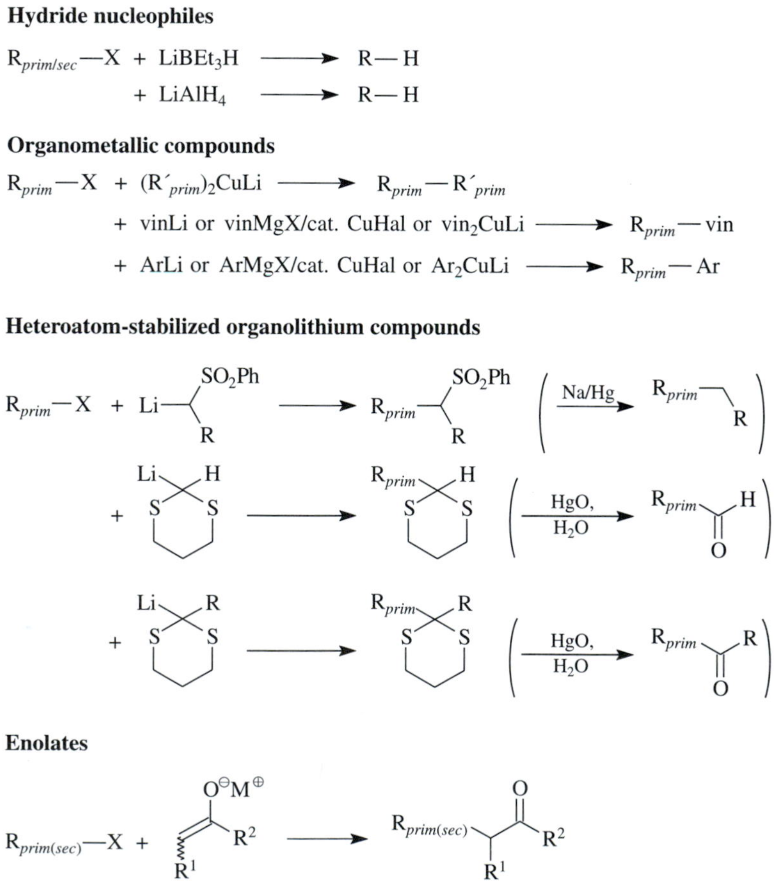

Hydride nucleophiles

$R_{prim/sec}$—X + $LiBEt_3H$ ⟶ R—H

+ $LiAlH_4$ ⟶ R—H

Organometallic compounds

R_{prim}—X + $(R'_{prim})_2CuLi$ ⟶ R_{prim}—R'_{prim}

+ vinLi or vinMgX/cat. CuHal or vin_2CuLi ⟶ R_{prim}—vin

+ ArLi or ArMgX/cat. CuHal or Ar_2CuLi ⟶ R_{prim}—Ar

Heteroatom-stabilized organolithium compounds

R_{prim}—X + Li—CH(SO_2Ph)(R) ⟶ R_{prim}—CH(SO_2Ph)(R) (Na/Hg ⟶ R_{prim}—CH_2—R)

+ 2-Li-2-H-1,3-dithiane ⟶ 2-R_{prim}-2-H-1,3-dithiane (HgO, H_2O ⟶ R_{prim}—CHO)

+ 2-Li-2-R-1,3-dithiane ⟶ 2-R_{prim}-2-R-1,3-dithiane (HgO, H_2O ⟶ R_{prim}—C(=O)—R)

Enolates

$R_{prim(sec)}$—X + R^1CH=C($O^{\ominus}M^{\oplus}$)R^2 ⟶ $R_{prim(sec)}$—CH(R^1)—C(=O)—R^2

Fig. 2.26. Preparatively important S_N2 reactions.

Fig. 2.26. (Continued)

$$+ \; R^1CH{=}C(O^{\ominus}M^{\oplus})OR^2 \longrightarrow R_{prim(sec)}{-}CH(R^1){-}C(=O)OR^2$$

$$+ \; RO_2C{-}CH{=}C(O^{\ominus}M^{\oplus})CH_3 \longrightarrow R_{prim(sec)}{-}CH(CO_2R){-}C(=O)CH_3 \quad \left(\xrightarrow[\Delta]{H_3O^{\oplus},} R{-}CH_2{-}C(=O)CH_3\right)$$

$$+ \; RO_2C{-}CH{=}C(O^{\ominus}M^{\oplus})OR \longrightarrow R_{prim(sec)}{-}CH(CO_2R){-}C(=O)OR \quad \left(\xrightarrow[\Delta]{H_3O^{\oplus},} R{-}CH_2{-}C(=O)OH\right)$$

Further C-nucleophiles

$$R_{prim}{-}X \; + \; M^{\oplus\ominus}C{\equiv}C{-}H \quad \text{or} \quad M^{\oplus\ominus}C{\equiv}C{-}R \longrightarrow R_{prim}{-}C{\equiv}C{-}H\;(R)$$

$$R_{prim(sec)}{-}X \; + \; K^{\oplus}CN^{\ominus} \longrightarrow R_{prim(sec)}{-}CN$$

Kolbe nitrile synthesis

MeI or $CH_2{=}CH{-}CH_2{-}$Br or BnBr + 1-(pyrrolidin-1-yl)cyclohexene $\longrightarrow$ 2-R-cyclohexylidene-pyrrolidinium ($\oplus$) $\xrightarrow{H_2O}$ 2-R-cyclohexanone

N-nucleophiles

$$R{-}X \; + \; K^{\oplus\ominus}N(\text{phthalimide}) \longrightarrow R{-}N(\text{phthalimide}) \xrightarrow{N_2H_4} \left(R{-}NH_2 \; + \; \text{phthalhydrazide (HN–HN)}\right)$$

Gabriel synthesis of primary amines

$$+ \; Na^{\oplus}N_3^{\ominus} \longrightarrow R{-}N_3 \quad \left(\xrightarrow[H_2/Pd-C]{LiAlH_4 \text{ or } Bu_3SnH \text{ or}} R{-}NH_2\right)$$

$$+ \; K^{\oplus}NCO^{\ominus} \longrightarrow R{-}N{=}C{=}O$$

S-nucleophiles

$$R{-}X \; + \; Na_2S_2O_3 \longrightarrow R{-}S{-}SO_3^{\ominus}\,Na^{\oplus} \quad \left(\xrightarrow{HCl} R{-}SH\right)$$

$$+ \; Na^{\oplus\ominus}S{-}C(=O)CH_3 \longrightarrow R{-}S{-}C(=O)CH_3 \quad \left(\xrightarrow{OH^{\ominus}} R{-}SH\right)$$

Fig. 2.26. (Continued)

$$+ \; S{=}C(NH_2)_2 \longrightarrow R{-}S{-}C(NH_2)_2^{\oplus} \; X^{\ominus} \left(\xrightarrow{OH^{\ominus}} R{-}SH \right)$$

$$+ \; Na^{\oplus}\,{}^{\ominus}O_2SR' \longrightarrow R{-}S(=O)_2{-}R'$$

Hal-nucleophiles

$$R_{prim(sec)}{-}X + Na^{\oplus}I^{\ominus} \xrightarrow[\text{methyl ethyl ketone}]{\text{acetone or}} R_{prim(sec)}{-}I$$

Finkelstein reaction

P-nucleophiles (⟶ **precursors for the Wittig or Korner–Wadsworth–Emmons reaction)**

$$R_{prim}{-}X \text{ or } iPr{-}I + PPh_3 \longrightarrow R{-}PPh_3^{\oplus} X^{\ominus}$$

$$R_{prim}{-}X + P(OEt)_3 \longrightarrow \left[R_{prim}{-}P(OEt)_2{=}{}^{\oplus}O{-}Et \; X^{\ominus} \right] \xrightarrow{-\,EtX} R_{prim}{-}P(OEt)_2{=}O$$

Arbuzov Reaction

O-nucleophiles

$$R_{prim}{-}X + Na^{\oplus}\,{}^{\ominus}OR \longrightarrow R_{prim}{-}OR \text{ or}$$

Williamson ether synthesis

$$R_{prim(sec)}{-}X + Na^{\oplus}\,{}^{\ominus}OAr \longrightarrow R_{prim(sec)}{-}OAr$$

$$+ Cs^{\oplus}\,{}^{\ominus}OAc \longrightarrow R_{prim(sec)}{-}OAc$$

Whereas all the alkylations in Figure 2.26 take place in basic or neutral solutions, carboxylic acids can be methylated as such with diazomethane (Figure 2.27). The actual nucleophile (the carboxylate ion) and the actual methylating agent ($H_3C{-}N^{+}{\equiv}N$) are then produced from the reaction partners by proton transfer.

Fig. 2.27. Preparation of methyl esters from carboxylic acids and diazomethane.

$$R{-}CH(OH){-}C(=O){-}O{-}H + {}^{\ominus}|CH_2{-}\overset{\oplus}{N}{\equiv}N \longrightarrow R{-}CH(OH){-}C(=O){-}O^{\ominus} + CH_3{-}\overset{\oplus}{N}{\equiv}N \xrightarrow{-\,N{\equiv}N} R{-}CH(OH){-}C(=O){-}O{-}CH_3$$

S-Enantiomer *S*-Enantiomer

Several S_N reactions of alcohols are initiated by positively charged phosphorus(V) compounds (Figures 2.28–2.31). In these cases, alcohols are treated with phosphonium salts Ph_3P^+—X that contain a leaving group X. This leaving group is displaced by the O atom of the OH group of the alcohol. In this way this O atom is bound to the P atom and becomes part of a good leaving group, Ph_3P═O. If

- the species Ph_3P^+—X, which is attacked by the OH group, is produced from Ph_3P and EtO_2C—N═N—CO_2Et, and
- the nucleophile is a carboxylic acid or the carboxylate ion formed from it *in situ,* and
- in addition, the OH group is located at a stereocenter,

then the reaction is called a **Mitsunobu inversion.** With its help it is possible to invert the configuration of a stereocenter equipped with an OH group (Figure 2.28). The enantiomers of optically pure alcohols, which contain a single stereocenter that bears an OH group, are easily accessible through a Mitsunobu inversion process.

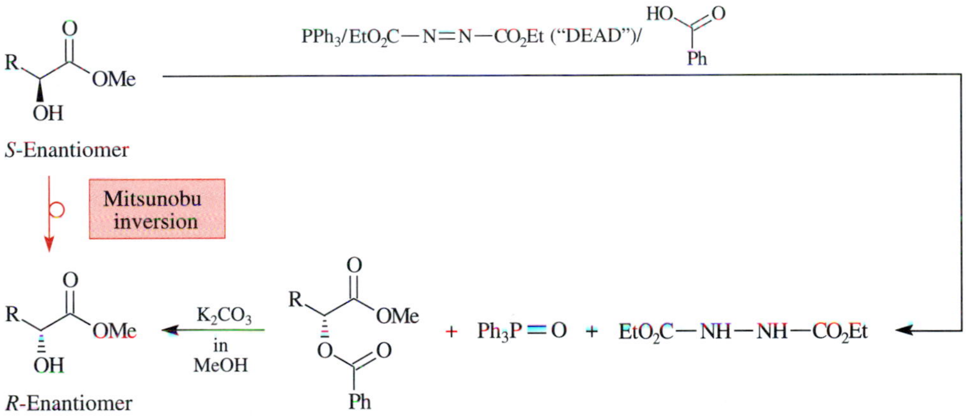

Fig. 2.28. Mitsunobu inversion: a typical substrate, the reagents, and products. (Possible preparation of the substrate: Figure 2.27.)

Figure 2.29 shows the mechanism of this reaction. A key intermediate is the alkylated phosphine oxide **A,** into which the carboxylate ion enters in a backside attack and displaces the leaving group O═PPh_3.

If the Mitsunobu inversion is carried out intramolecularly (i.e., in a hydroxycarboxylic acid), a lactone is produced with inversion of the configuration at the OH-bearing stereocenter (Figure 2.30). This lactonization is stereochemically complementary to the paths via activated hydroxycarboxylic acids, which lead to lactones with retention of the configuration at the OH-bearing C atom (Section 6.4.2).

Let us finally consider the reactions shown in Figure 2.31. There the oxophilic species Ph_3P^+—X are produced from Ph_3P and reagents of the form X—Y. While one half of such a reagent is transferred as X^+ to Ph_3P, an anion Y^- is produced from the other half of the molecule. The phosphonium ion now converts the OH group of the alcohol—through a nucleophilic attack of the oxygen at the phosphorus—to the same leaving group Ph_3P═O that was also present in the Mitsunobu

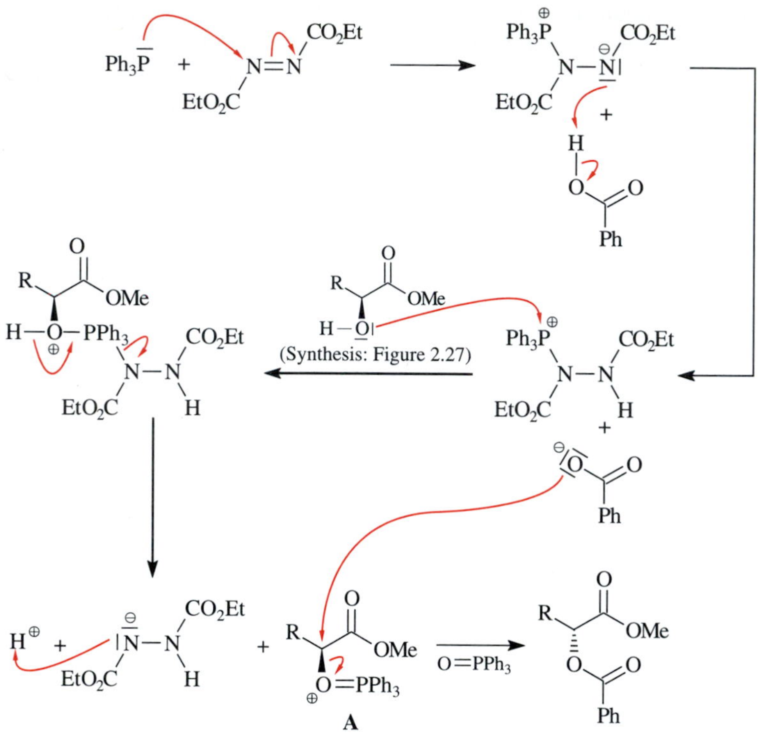

Fig. 2.29. Mechanism of the Mitsunobu inversion in Figure 2.28.

inversion. The group X, which had been bound to the phosphorus, is released in the form of a nucleophile X^-, which ultimately displaces the leaving group $Ph_3P{=}O$. Consequently, from three molecules of starting materials (alcohol, Ph_3P, X—Y) one substitution product, one equivalent of $Ph_3P{=}O$, and one equivalent of H—Y are produced. At the same time the phosphorus, which was originally present in the oxidation state +3, is oxidized to P(V). This type of reaction is called a **redox condensation according to Mukaiyama.**

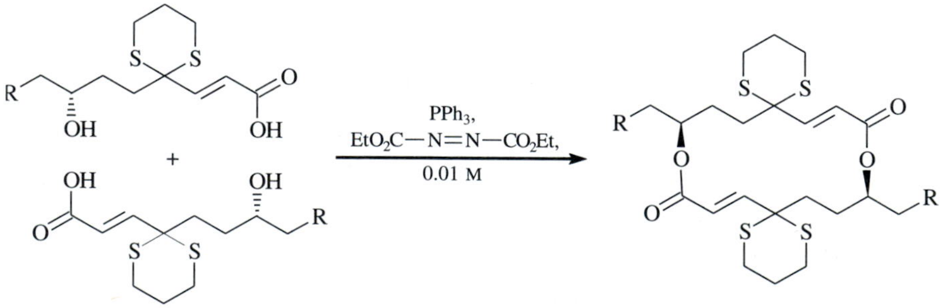

Fig. 2.30. Stereoselective lactonization by means of a Mitsunobu inversion.

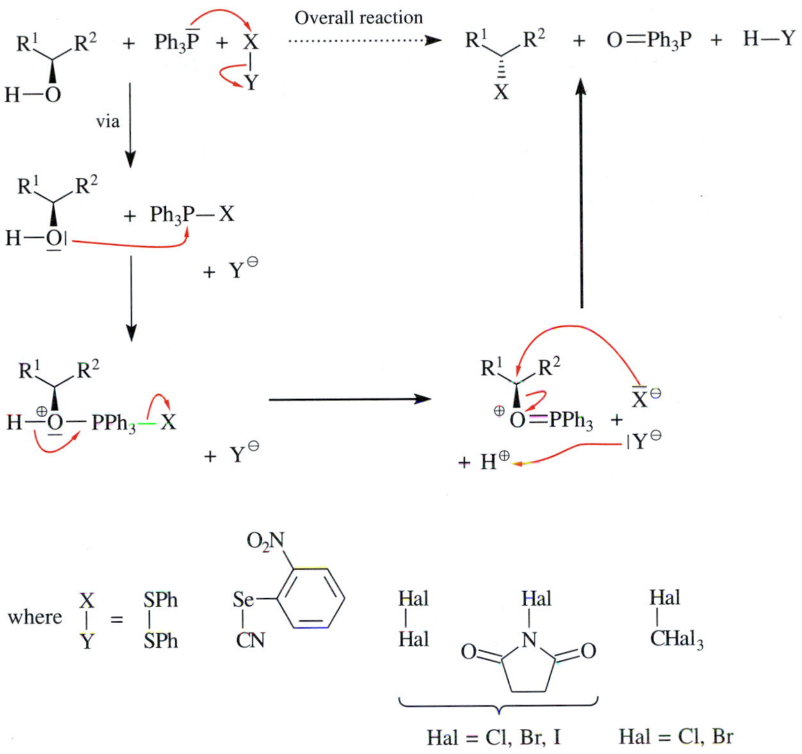

Fig. 2.31. Preparatively important redox condensations according to Mukaiyama, and related reactions.

References

G. L. Edwards, "One or More CC Bond(s) Formed by Substitution: Substitution of Halogen," in *Comprehensive Organic Functional Group Transformations* (A. R. Katritzky, O. Meth-Cohn, and C. W. Rees, Eds.), Vol. 1, 105, Elsevier Science, Oxford, U.K., **1995.**

2.1

H. Mayr and M. Patz, "Scales of nucleophilicity and electrophilicity: A system for ordering polar organic and organometallic reactions," *Angew. Chem.* **1994,** *106,* 990–1010; *Angew. Chem. Int. Ed. Engl.* **1994,** *33*, 938–958.

2.4

P. Beak, "Mechanisms of reactions at nonstereogenic heteroatoms: evaluation of reaction geometries by the endocyclic restriction test," *Pure Appl. Chem.* **1993,** *65,* 611–615.

2.5

T. Baer, C.-Y. Ng, I. Powis (Eds.), "The Structure, Energetics and Dynamics of Organic Ions," Wiley, Chichester, U.K., **1996.**

M. Saunders and H. A. Jiménez-Vázquez, "Recent studies of carbocations," *Chem. Rev.* **1991,** *91,* 375–397.

V. D. Nefedov, E. N. Sinotova, V. P. Lebedev, "Vinyl cations," *Russ. Chem. Rev.* **1992,** *61,* 283–296.
T. T. Tidwell, "Destabilized carbocations," *Angew. Chem. Int. Ed. Engl.* **1984,** *23,* 20.
G. A. Olah, "My search for carbocations and their role in chemistry (Nobel Lecture)," *Angew. Chem.* **1995**, *107,* 1519–1532; *Angew. Chem. Int. Ed. Engl.* **1995,** *34,* 1393–1405.
K. Okamoto, "Generation and Ion-Pair Structures of Unstable Carbocation Intermediates in Solvolytic Reactions," in *Advances in Carbocation Chemistry* (X. Creary, Ed.) **1989,** *1,* Jai Press, Greenwich, CT.
P. E. Dietze, "Nucleophilic Substitution and Solvolysis of Simple Secondary Carbon Substrates," in *Advances in Carbocation Chemistry* (J. M. Coxon, Ed.) **1995,** *2,* JAI, Greenwich, CT.
R. F. Langler, "Ionic reactions of sulfonic acid esters," *Sulfur Rep.* **1996,** 19, 1–59.

2.6

A. R. Katritzky and B. E. Brycki, "Nucleophilic substitution at saturated carbon atoms. Mechanisms and mechanistic borderlines: evidence from studies with neutral leaving groups," *J. Phys. Org. Chem.* **1988,** *1,* 1–20.
J. P. Richard, "Simple Relationships between Carbocation Lifetime and the Mechanism for Nucleophilic Substitution at Saturated Carbon," in *Advances in Carbocation Chemistry* (X. Creary, Ed.) **1989,** *1,* Jai Press, Greenwich, CT.
A. R. Katritzky and B. E. Brycki, "The mechanisms of nucleophilic substitution in aliphatic compounds," *Chem. Soc. Rev.* **1990,** *19,* 83–105.
H. Lund, K. Daasbjerg, T. Lund, S. U. Pedersen, "On electron transfer in aliphatic nucleophilic substitution," *Acc. Chem. Res.* **1995,** *28,* 313–319.

2.7

G. M. Kramer and C. G. Scouten, "The 2-Norbornyl Carbonium Ion Stabilizing Conditions: An Assessment of Structural Probes," in *Advances in Carbocation Chemistry* (X. Creary, Ed.) **1989,** *1,* Jai Press, Greenwich, CT.

2.8

J. M. Klunder and G. H. Posner, "Alkylations of Nonstabilized Carbanions," in *Comprehensive Organic Synthesis* (B. M. Trost, I. Fleming, Eds.), Vol. 3, 207, Pergamon Press, Oxford, **1991.**
H. Ahlbrecht, "Formation of C—C Bonds by Alkylation of σ-Type Organometallic Compounds," in *Stereoselective Synthesis* (Houben-Weyl) 4th ed., 1996, (G. Helmchen, R. W. Hoffmann, J. Mulzer, E. Schaumann, Eds.), **1996,** Vol. E21 (Workbench Edition), *2,* 645–663, Georg Thieme Verlag, Stuttgart.
D. W. Knight, "Alkylations of Vinyl Carbanions," in *Comprehensive Organic Synthesis* (B. M. Trost, I. Fleming, Eds.), Vol. 3, 241, Pergamon Press, Oxford, **1991.**
H. Ahlbrecht, "Formation of C—C Bonds by Alkylation of π-Type Organometallic Compounds," in *Stereoselective Synthesis* (Houben-Weyl) 4th ed., 1996, (G. Helmchen, R. W. Hoffmann, J. Mulzer, E. Schaumann, Eds.), **1996,** Vol. E21 (Workbench Edition), *2,* 664–696, Georg Thieme Verlag, Stuttgart.
P. J. Garratt, "Alkylations of Alkynyl Carbanions," in *Comprehensive Organic Synthesis* (B. M. Trost, I. Fleming, Eds.), Vol. 3, 271, Pergamon Press, Oxford, **1991.**
G. H. Posner, "Substitution reactions using organocopper reagents," *Org. React.* **1975,** *22,* 253–400.
B. H. Lipshutz and S. Sengupta, "Organocopper reagents: Substitution, conjugate addition, carbo/metallocupration, and other reactions," *Org. React.* **1992,** *41,* 135–631.
M. V. Bhatt and S. U. Kulkarni, "Cleavage of Ethers," *Synthesis* **1983,** 249.
C. Bonini and G. Righi, "Regio- and chemoselective synthesis of halohydrins by cleavage of oxiranes with metal halides," *Synthesis* **1994,** 225–38.
U. Raguarsson and L. Grehn, "Novel Gabriel reagents," *Acc. Chem. Res.* **1991,** *24,* 285–289.
T. H. Black, "The preparation and reactions of diazomethane," *Aldrichimica Acta* **1983,** *16,* 3.

B. R. Castro, "Replacement of alcoholic hydroxyl groups by halogens and other nucleophiles via oxyphosphonium intermediates," *Org. React. (N.Y.)* **1983,** *29,* 1.

D. L. Hughes, "Progress in the Mitsunobu reaction. A review," *Org. Prep. Proced. Int.* **1996,** *28,* 127–164.

D. L. Hughes, "The Mitsunobu reaction," *Org. React.* **1992,** *42,* 335–656.

A. Krief and A.-M. Laval, "σ-Nitrophenyl selenocyanate, a valuable reagent in organic synthesis: Application to one of the most powerful routes to terminal olefins from primary-alcohols (the Grieco-Sharpless olefination reaction) and to the regioselective isomerization of allyl alcohols," *Bull. Soc. Chim. Fr.* **1997,** *13,* 869–874.

C. Simon, S. Hosztafi, S. Makleit, "Application of the Mitsunobu reaction in the field of alkaloids," *J. Heterocycl. Chem.* **1997,** *34,* 349–365.

Further Reading

R. M. Magid, "Nucleophilic and organometallic displacement reactions of allylic compounds: Stereo- and regiochemistry," *Tetrahedron* **1980,** *36,* 1901.

Y. Yamamoto, "Formation of C—C Bonds by Reactions Involving Olefinic Double Bonds, Vinylogous Substitution Reactions," in *Methoden Org. Chem. (Houben-Weyl) 4th ed. 1952-, Stereoselective Synthesis* (G. Helmchen, R. W. Hoffmann, J. Mulzer, E. Schaumann, Eds.), Vol. E21b, 2011, Georg Thieme Verlag, Stuttgart, **1995.**

L. A. Paquette and C. J. M. Stirling, "The intramolecular SN′ reaction," *Tetrahedron* **1992,** *48,* 7383.

R. M. Hanson, "The synthetic methodology of nonracemic glycidol and related 2,3-epoxy alcohols," *Chem. Rev.* **1991,** *91,* 437–476.

B. B. Lohray, "Cyclic sulfites and cyclic sulfates: Epoxide like synthons," *Synthesis* **1992,** 1035–1052.

S. C. Eyley, "The Aliphatic Friedel-Crafts Reaction," in *Comprehensive Organic Synthesis* (B. M. Trost, I. Fleming, Eds.), Vol. 2, 707, Pergamon Press, Oxford, **1991.**

Y. Ono, "Dimethyl carbonate for environmentally benign reactions," *Pure Appl. Chem.* **1996,** *68,* 367–376.

C. M. Sharts and W. A. Sheppard, "Modern methods to prepare monofluoroaliphatic compounds," *Org. React.* **1974,** *21,* 125–406.

J. E. McMurry, "Ester cleavages via S_N2-type dealkylation," *Org. React.* **1976,** *24,* 187–224.

S. K. Taylor, "Reactions of epoxides with ester, ketone and amide enolates," *Tetrahedron* **2000,** *56,* 1149–1163.

M. Schelhaas and H. Waldmann, "Protecting group strategies in organic synthesis," *Angew. Chem.* **1996,** *108,* 2192–2219; *Angew. Chem. Int. Ed. Engl.* **1996,** *35,* 2056–2083.

K. Jarowicki and P. Kocienski, "Protecting groups," *Contemp. Org. Synth.* **1996,** *3,* 397–431.

3 Additions to the Olefinic C═C Double Bond

B

Olefins contain a C═C double bond. The C═C double bond can be described with two different models. According to the less frequently used model briefly mentioned in Section 2.4.3 (Figure 2.8), C═C double bonds consist of two bent C—C single bonds. With a bond energy of 73 kcal/mol, each of these bonds is 10 kcal/mol less stable than the linear C—C bond of an aliphatic compound. Normally, the second model is used to describe the bonding in olefins. According to this model, a C═C double bond consists of a σ and a π bond. The σ bond with 83 kcal/mol has 20 kcal/mol more bond energy than the π bond (63 kcal/mol). The higher stability of σ- in comparison to π-C—C bonds is due to the difference in the overlap between the AOs that form these bonds. σ-C—C bonds are produced by the overlap of two sp^n atomic orbitals (n = 1, 2, 3), which is quite effective because it is *frontal*. π-C—C bonds are based on the overlap of $2p_z$ atomic orbitals, which is not as good because it is *lateral*.

Regardless of which bond model is utilized, it is clear that upon exposure to a suitable reagent the relatively weak C═C double bond of an olefin will be given up and a relatively stable C—C single bond conserved instead. From the point of view of the usual bonding model, this means that the respective reactions of olefins are addition reactions, in which the C—C π bond is converted into σ bonds to two new substituents, a and b. One C atom of the C═C double bond picks up fragment a from the reagent and the other C atom picks up fragment b.

$$(R^1)(R^2)C_{sp^2} \overset{\pi}{=} C_{sp^2}(R^3)(R^4) + \text{Reagent(s)} \xrightarrow{\text{Addition}} a \overset{\sigma}{-} C_{sp^3}(R^1)(R^2) - C_{sp^3}(R^3)(R^4) \overset{\sigma}{-} b$$

Suitable reagents have the structure a—b in the most simple case. However, they may also have a different structure. The fragments a and b, which are bound to the olefinic C═C double bond, do not necessarily constitute the *entire* reagent (e.g., epoxidation with percarboxylic acids: see Figure 3.14) nor must they originate from a single reagent (e.g., formation of bromohydrins: see Figure 3.33). In these cases also one talks about an addition reaction.

In addition reactions to olefinic double bonds, two new sp^3-hybridized C atoms are produced. Each of them can be a stereogenic center (stereocenter) and obviously *is* one if it possesses four different substituents. If stereocenters are produced, their configuration must be specified. Stated more accurately, the first question is which *absolute* configuration is produced at any new stereocenter. Then the question arises about the configuration of the new stereocenters *relative* (a) to each other or (b) to additional stereocenters the molecule contained before the addition reaction occurred.

Stereochemical aspects are therefore an important part of the chemistry of addition reactions to the olefinic double bond. We will therefore investigate them in detail in this chapter. In fact, the content of Chapter 3 is arranged according to the stereochemical characteristics of the (addition) reactions.

3.1 The Concept of *cis* and *trans* Addition

B

In a number of additions of two fragments a and b to an olefinic C=C double bond, a new stereocenter is produced at *both* attacked C atoms (Figure 3.1). For describing the configurations of these new stereocenters relative to each other the following nomenclature has come into use: If the stereostructure (not the preferred conformation!) of the addition product arises because fragments a and b have added to the C=C double bond of the substrate from the same side, a *cis* addition has taken place. Conversely, if the stereostructure of the addition product is obtained because fragments a and b were bonded to the C=C double bond of the substrate from opposite sides, we have a *trans* addition. The terms *cis* and *trans* addition are also used when the addition products are acyclic and thus their configuration cannot be described with the *cis/trans* nomenclature.

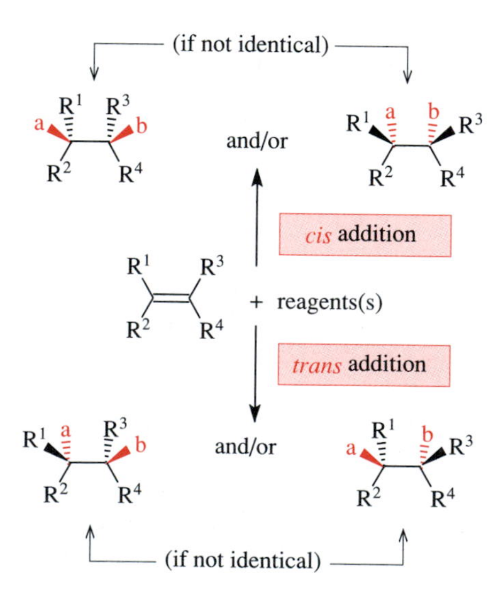

Fig. 3.1. The *cis* and *trans* additions of two fragments a and b to C=C double bonds.

3.2 Vocabulary of Stereochemistry and Stereoselective Synthesis I

3.2.1 Isomerism, Diastereomers/Enantiomers, Chirality

B

Two molecules that have the same empirical formula can exhibit four different degrees of relationship, which are shown in Figure 3.2.

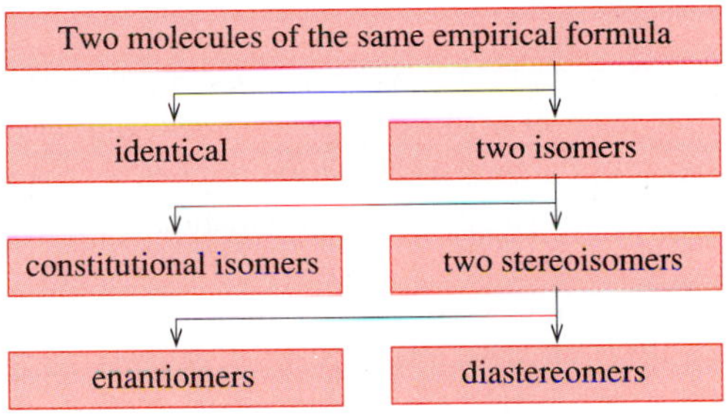

Two molecules of the same empirical formula

identical | two isomers

constitutional isomers | two stereoisomers

enantiomers | diastereomers

Fig. 3.2. Possible relationships between molecules of the same empirical formula.

According to Figure 3.2, molecules of the same empirical formula are either identical or isomers. Isomers either differ in the connectivity of their constituent atoms—this then involves **constitutional isomers** (structural isomers)—or they do not differ in this way; then they are **stereoisomers.** Stereoisomers that are related to each other as an image and a mirror image are **enantiomers;** otherwise they are **diastereomers.** Enantiomers are **enantiomorphic** to each other; diastereomers are **diastereomorphic.** Diastereomers that contain several stereocenters but differ in the configuration of only one stereocenter are called **epimers.** Finally, epimeric glycosides that are configured differently at the anomeric (i.e., the *gem*-dioxygenated) C atom are designated **anomers.**

According to the foregoing analysis, conformers such as *gauche*- and *anti*-butane or chair and twist–boat cyclohexane would be considered to be diastereomers of each other. However, under most conditions these conformers interconvert so rapidly that butane and cyclohexane are considered to be single species and not mixtures of stereoisomers.

Let us also visualize the important concept of **chirality.** Only molecules that differ from their mirror image have enantiomers. Molecules of this type are called **chiral.** For the occurrence of enantiomers or for the presence of the "property of differing from its mirror image"—that is, chirality—there is a necessary and sufficient condition: *it is the absence of an intramolecular rotation/reflection axis.*

The most frequently occurring rotation/reflection axis in organic chemistry is S_1, the intramolecular mirror plane. Its presence makes *cis*-1,2-dibromocyclohexane (structure **A** in Figure 3.3) as well as *meso*-2,3-dibromosuccinic acid (structure **B**) achiral. This is true even though there are two stereocenters in each of these compounds. But

Fig. 3.3. Molecules that contain several stereocenters as well as one (**A–C**) or no (**D, E**) rotation/reflection axis and consequently are achiral (**A–C**) or chiral (**D, E**).

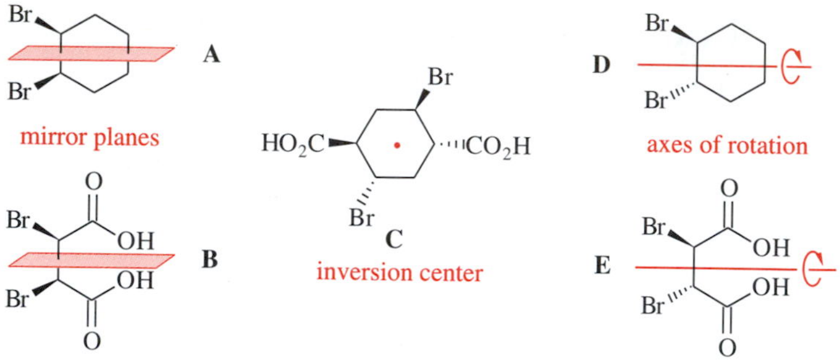

why is the dibromocyclohexane dicarboxylic acid **C** (Figure 3.3) achiral? The answer is that it contains an S_2 axis, an inversion center, which occurs in organic chemistry rarely. This compound is thus achiral although it contains four stereocenters.

Note that according to the foregoing definition, chirality occurs only in molecules that do not have a *rotation/reflection axis.* However, if the molecule has only (!) an *axis of rotation,* it is chiral. For example, both *trans*-1,2-dibromocyclohexane (**D** in Figure 3.3) and the dibromosuccinic acid **E** have a two-fold axis of rotation (C_2) as the only symmetry element. In spite of that, these compounds *are* chiral because the presence of an axis of rotation, in contrast to the presence of a rotation/reflection axis, is not a criterion for achirality.

3.2.2 Chemoselectivity, Diastereoselectivity/Enantioselectivity, Stereospecificity/Stereoconvergence

B

Preparatively useful reactions should deliver a *single* product or at least a *preferred* product and not a product mixture. If they give one product exclusively or preferentially, they are called selective reactions. One usually specifies the type of selectivity involved, depending on which of the conceivable side products do not occur at all (highly selective reaction) or occur only to a minor degree (moderately selective reaction) (Figure 3.4).

A reaction that preferentially or exclusively gives one of several conceivable reaction products with different empirical formulas takes place with **chemoselectivity.** A reaction that preferentially or exclusively gives one of several conceivable constitutional isomers is also chemoselective. A reaction that preferentially or exclusively gives one of several conceivable stereoisomers is referred to as moderately or highly **stereoselective.** When these conceivable stereoisomers are diastereomers, one also talks about a diastereoselective reaction or about the occurrence of **diastereoselectivity.** When the conceivable stereoisomers are enantiomers, we have an enantioselective reaction, or **enantioselectivity.**

The objective of **stereoselective synthesis** is to produce compounds as pure diastereomers (diastereomerically pure) and/or pure enantiomers (enantiomerically pure). To be able to evaluate the success of efforts of this type, one needs quantitative measures for diastereoselectivity and enantioselectivity.

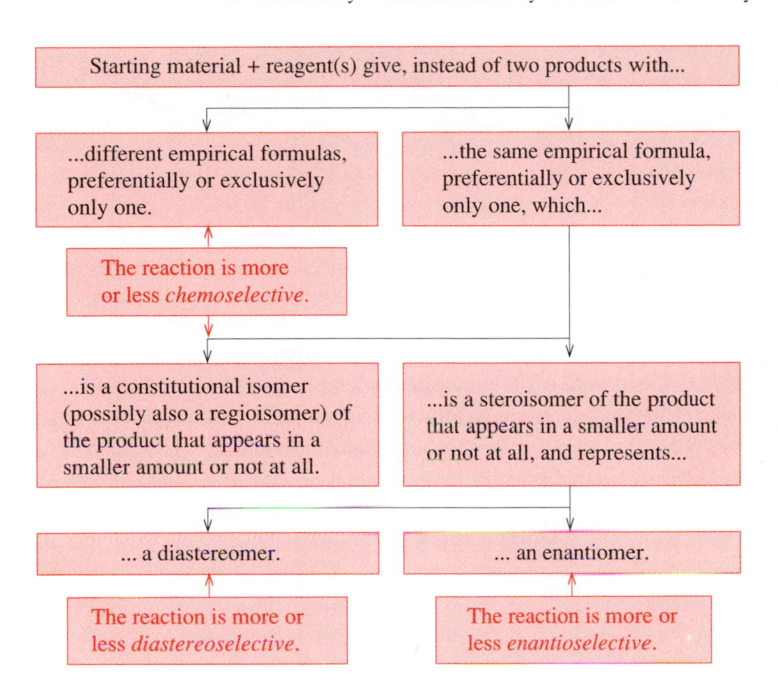

Fig. 3.4. Chemo-, diastereo-, and enantioselectivity.

The stereochemical quality of diastereoselective reactions is expressed as a numeric ratio $ds = x{:}y\ ({:}z{:}\ldots)$ (from **dia**stereoselectivity). It is the ratio of the yields of the two (or more) diastereomers formed normalized to 100. "Normalized to 100" means that we must have $x + y\ (+z + \cdots) = 100$. Figures 3.5 and 3.6 illustrate the use of the measure *ds* as applied to a moderately diastereoselective bromination. The complete description of the stereochemical result is shown in Figure 3.5, and the common short form is shown in Figure 3.6.

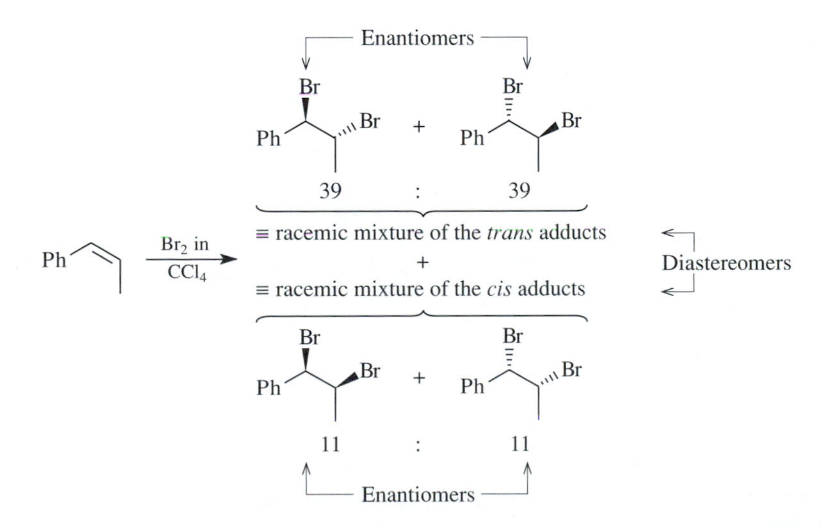

Fig. 3.5. An unusually detailed representation of a moderately diastereoselective reaction. (Figure 3.6 gives a short form of the same reaction.)

Ph–CH=CH–CH$_3$ $\xrightarrow{Br_2 \text{ in } CCl_4}$ (trans dibromide) + (cis dibromide)

78 : 22

i.e., ds = 78 : 22

Fig. 3.6. Short way of writing the stereochemical result of the bromination of Figure 3.5.

Of the dibromides obtained, 78% is the *trans* adduct, which is produced as a racemic mixture (i.e., a 1:1 mixture) of the two enantiomers. This is because the individual yield of each *trans* enantiomer was $0.5 \times 78 = 39\%$ of the total amount of dibromides. The remaining 22% dibromide is the *cis* adduct, which is also produced as a racemic (i.e., a 1:1) mixture of the two enantiomers. This is because the individual yield of each *cis* enantiomer was $0.5 \times 22 = 11\%$ of the total amount of dibromides. The diastereoselectivity with which the *trans* (vs the *cis*) addition product is produced is consequently $ds = 78{:}22$.

Inevitably, in *all* reactions between achiral reaction partners—hence also in many additions in this chapter—chiral products occur as racemic mixtures. That part of Figure 3.5, which states just this, is therefore trivial. Moreover, it tends to distract from the essential information. In order to focus on the latter, in general, equations for such reactions are written so that only different diastereomers are depicted, though it is *understood* that these diastereomers occur as racemic mixtures. Thus Figure 3.6 is the shorthand documentation of what Figure 3.5 presented in much more detail. The same shorthand notations will be used in this chapter and in the remainder of the book.

The quality of enantioselective reactions is numerically expressed as the so-called **enantiomeric excess** (*ee*). It is equal to the yield of the major enantiomer minus the yield of the minor enantiomer in the product whose total yield is normalized to 100%. For example, in the Sharpless epoxidation of allyl alcohol (see Figure 3.7) *S*- and *R*-glycidol are formed in a ratio of 19:1. For a total glycidol yield standardized to 100%, the *S*-glycidol fraction (95% yield) thus exceeds the *R*-glycidol fraction (5% yield) by 90%. Consequently, *S*-glycidol is produced with an *ee* of 90%.

allyl alcohol (OH) $\xrightarrow[\text{L-(+)-DET}]{tert\text{-BuOOH, } Ti(OiPr)_4,}$ (*S*-glycidol, OH) + (*R*-glycidol, OH)

95 : 5

i.e., *ee* = 90%

Fig. 3.7. Definition of the enantiomeric excess *ee* using the Sharpless epoxidation of allyl alcohol as an example. The chiral auxiliary is tartaric acid diethyl ester (**die**thyl **t**artrate, DET).

The concept of **stereospecificity** has been introduced to characterize the stereochemical course of *pairs* of highly diastereoselective reactions. Another term, **stereoconvergence,** is also very useful in this connection.

Definition: Stereospecific Reaction

Here we note that if in a *pair* of analogous reactions *one* starting material is converted highly diastereoselectively into a *first* product and a *stereoisomer* of the starting material is converted highly diastereoselectively into a *diastereomer* of the first product, these reactions are called **stereospecific.** An example of this can be found in Figure 3.8.

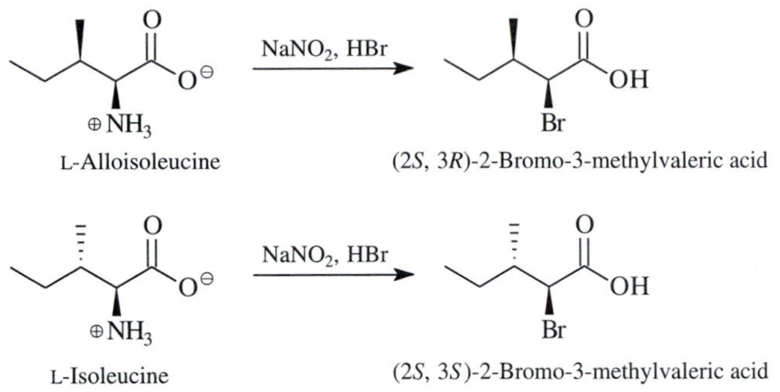

L-Alloisoleucine → ($NaNO_2$, HBr) → (2*S*, 3*R*)-2-Bromo-3-methylvaleric acid

L-Isoleucine → ($NaNO_2$, HBr) → (2*S*, 3*S*)-2-Bromo-3-methylvaleric acid

Fig. 3.8. A pair of stereospecific reactions (using the type of reaction in Figure 2.24 as an example).

Definition: Stereoconvergent Reaction

It is possible that in a pair of analogous reactions, *one* starting material is converted highly diastereoselectively into one product, and a *stereoisomer* of the starting material is converted highly diastereoselectively into *the same* product. Reactions of this type can be called **stereoconvergent** (for an example, see Figure 3.9), but this term has not been generally accepted.

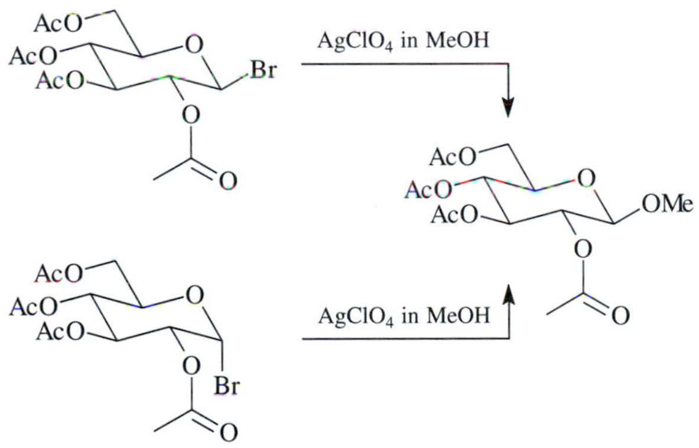

$AgClO_4$ in MeOH

$AgClO_4$ in MeOH

Fig. 3.9. A pair of stereoconvergent reactions (using the glucoside syntheses from Figure 2.23 as an example).

3.3 Additions That Take Place Diastereoselectively as *cis* Additions

B

All one-step additions to C=C double bonds are mechanistically required to take place *cis* selectively (*ds* > 99:1) (Sections 3.3.1–3.3.3). Moreover, the heterogeneously catalyzed hydrogenation of olefins also usually takes place with very high *cis* selectivity, in spite of its being a multistep reaction (Section 3.3.4).

3.3.1 A Cycloaddition Forming Three-Membered Rings

B

Cycloadditions are ring-forming addition reactions in which the product, the so-called cycloadduct, possesses an empirical formula that corresponds to the sum of the empirical formulas of the starting material and the reagent. All one-step cycloadditions take place with *cis* selectivity. Three-, four-, five-, or six-membered but not larger rings can be produced by cycloadditions to monoolefins (Figure 3.10).

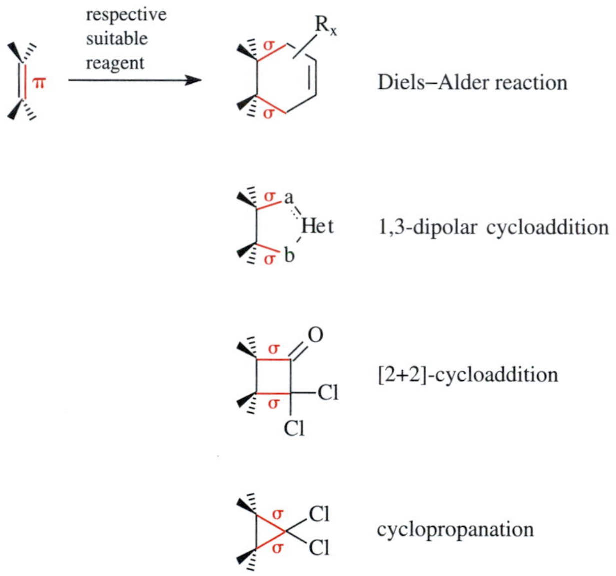

Fig. 3.10. *cis*-Selective cycloadditions.

You will become familiar with selected cycloadditions that lead to four-, five-, or six-membered rings in Chapter 12. Two more cycloadditions, which are also oxidations, will be examined in Chapter 14, which deals with oxidations and reductions: the ozonolysis reaction can be found in Section 14.3.2 (as well as in Section 12.5.5) and the *cis-vic* dihydroxylation with OsO_4 can be found in Section 14.3.2. Here we discuss only the addition of dichlorocarbene to olefins as an example of a *cis* addition of the cycloaddition type (Figure 3.11).

Dichlorocarbene cannot be isolated, but it can be produced in the presence of an olefin and then reacted with it immediately. The best dichlorocarbene precursor is the anion Cl_3C^-, which easily eliminates a chloride ion. This anion is obtained from Cl_3CH and fairly strong bases. The OH^- ion is sufficiently basic for effecting this deprotonation provided that it comes into contact with the Cl_3CH molecules. Consequently, when chloroform is stirred with potassium hydroxide solution, there is only moderate conversion into the anion Cl_3C^-. This is because Cl_3CH and aqueous KOH are not miscible so that KOH cannot efficiently migrate from the aqueous phase into the Cl_3CH. The solvation of the K^+ and OH^- ions in the organic phase would be far too low.

The transfer of only the OH^- ions into the chloroform, however, succeeds very well in the presence of a **phase transfer catalyst.** The most frequently used phase transfer

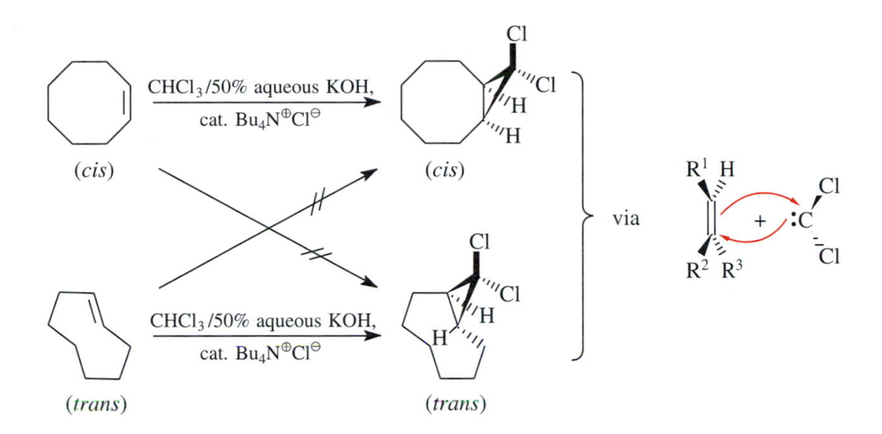

Fig. 3.11. Two reactions that demonstrate the stereospecificity of the *cis* addition of dichlorocarbene to olefins.

catalysts are tetraalkylammonium chlorides or tetraalkylammonium hydrogensulfates. With a large excess of OH^- ions, they give the corresponding tetraalkylammonium hydroxides in an equilibrium reaction. These dissolve in chloroform because the cation is so large that it requires practically no solvation. Via the Coulomb attraction, these cations pull the nonsolvatable OH^- ions with them into the chloroform phase.

The dichlorocarbene produced in this way *in* chloroform adds not only stereoselectively but also stereospecifically to olefins (Figure 3.11). Doubly chlorinated cyclopropanes are produced. These are not only of preparative interest as such, but they can also be used as starting materials for the synthesis of chlorine-free cyclopropanes. Chlorine can successfully be removed, for example, with Bu_3SnH according to the mechanism discussed in Section 1.9.1 for bromides and iodides. Being able to prepare dichlorocarbene as conveniently as described makes the two-step process attractive for producing chlorine-free cyclopropanes.

3.3.2 Additions to C═C Double Bonds That Are Related to Cycloadditions and Form Three-Membered Rings, Too

A

In contrast to the dichlorocyclopropanations from Section 3.3.1, the reactions discussed in this section are not cycloadditions in a strict sense. The reason is that the empirical formula of the addition products presented here is not equal to the sum of the empirical formulas of the reaction partners. Accordingly, the reaction products here are not cycloadducts in the strict sense.

The addition of the **Simmons–Smith reagent** to olefins leads in a single step to chlorine-free cyclopropanes (Figure 3.12). This addition also runs stereoselectively and stereospecifically. A Simmons–Smith reagent can be produced from diiodomethane and Zn/Cu couple, which in turn is produced from zinc dust and catalytic amounts of $CuSO_4$, $CuCl_2$, or $Cu(OAc)_2$. This Simmons–Smith reagent is usually assigned the structural formula $I—CH_2—ZnI$. This species, however, is subject to an equilibrium similar

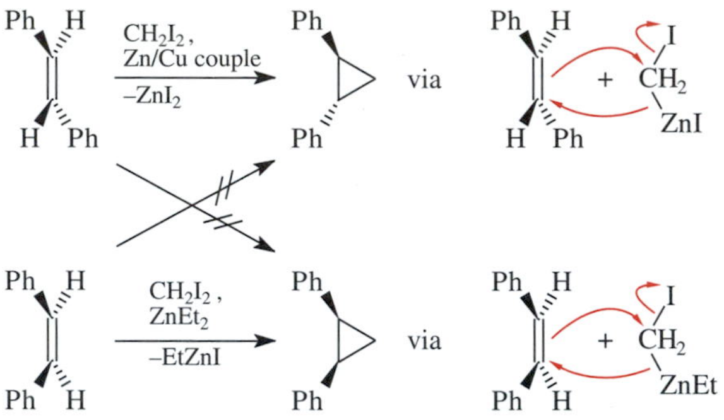

Fig. 3.12. Two reactions that demonstrate the stereospecificity of *cis* cyclopropanations with the Simmons–Smith reagent. In the first reaction the zinc carbenoid is produced according to the original method, and in the second it is produced by the Furukawa variant.

to the one that is known as the Schlenk equilibrium for Grignard compounds (cf. Figure 8.1):

$$2\,I-CH_2-ZnI \rightleftharpoons I-CH_2-Zn-CH_2-I + ZnI_2$$

In the Simmons–Smith reaction, as formulated in Figure 3.12, it has seemingly been clarified that the attacking species is I—CH_2—ZnI rather than I—CH_2—Zn—CH_2—I, which is in equilibrium with it in small amounts. A reagent that has properties quite similar to those of the actual Simmons–Smith reagent (but is advantageously produced in a homogeneous reaction) results from a halogen/metal exchange between diiodomethane and $ZnEt_2$. It can be written as I—CH_2—ZnEt.

Simmons–Smith reagents would be ordinary organic zinc compounds with a covalent C—Zn bond if the carbon that binds the zinc were not also bound to iodine as a leaving group. Because of this arrangement, both the C—I bond and the adjacent C—Zn bond are highly stretched. As a result of this distortion, the geometry of the Simmons–Smith reagents is similar to the geometry of the carbene complex of ZnI_2 or IZnEt. The fact that these carbene complexes are not the same as a mixture of free carbene (CH_2) and ZnI_2 or EtZnI is indicated by designating the Simmons–Smith reagents as **carbenoids.** These carbenoids have the ability to transfer a CH_2 group to olefins.

A $C(CO_2Me)_2$ unit can be added to C=C double bonds via another carbenoid, namely a rhodium–carbene complex (Figure 3.13). Again, these additions are not

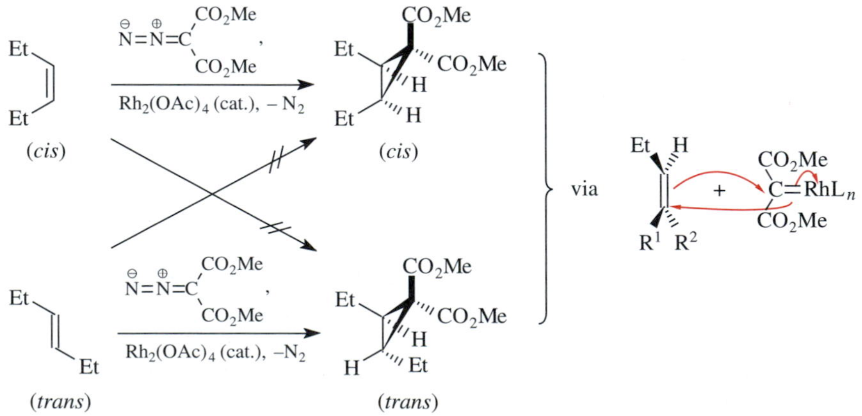

Fig. 3.13. Two reactions that demonstrate the stereospecificity of Rh-catalyzed *cis* cyclopropanations of electron-rich olefins.

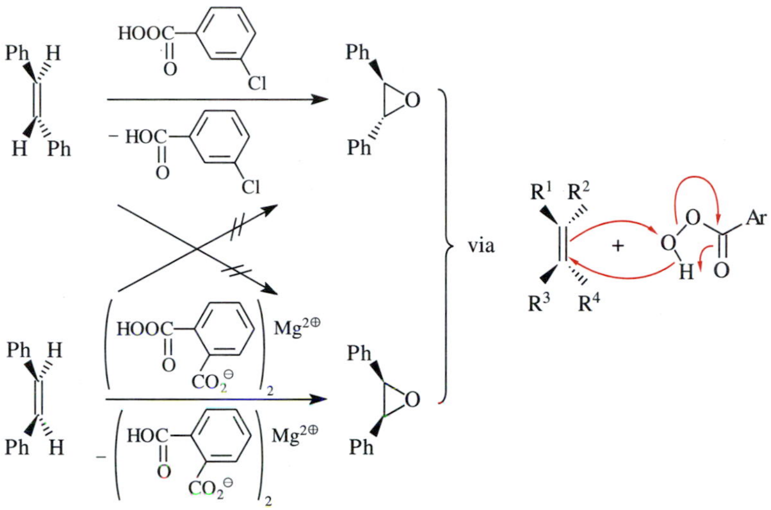

Fig. 3.14. Stereospecific *cis* epoxidations of olefins with percarboxylic acids.

only stereoselective but also stereospecific. The rhodium–carbene complex mentioned forms from diazomalonic ester (for a preparation, see Figure 12.39) and dimeric rhodium(II)acetate while eliminating molecular nitrogen. Diazoacetic ester and a number of transition metal salts—including dimeric rhodium(II)acetate—form similar transition metal–carbene complexes while eliminating N_2. They react with C=C double bonds to transfer a $CH(CO_2Me)$ group.

A third one-step addition reaction to C=C double bonds that forms three-membered rings is the epoxidation of olefins with **percarboxylic acids** (Figure 3.14). Suitable percarboxylic acids must, however, not be (too) explosive. Thus, aromatic percarboxylic acids are preferable. Until recently one epoxidized almost exclusively with *meta*-chloroperbenzoic acid (MCPBA). An alternative has become magnesium monoperoxyphthalate (MMPP). In the transition state of this type of epoxidation, four electron pairs are shifted simultaneously (which is a "record" in this book except for the Corey–Winter elimination in Figure 4.42).

3.3.3 *cis*-Hydration of Olefins via the Hydroboration/Oxidation/Hydrolysis Reaction Sequence

Boranes

B

In monoborane (BH_3), monoalkylboranes RBH_2, or dialkylboranes R_2BH there is only an electron sextet at the boron atom. In comparison to the more stable electron octet, the boron atom thus lacks two valence electrons. It "obtains" them by bonding with a suitable electron pair donor. When no better donor is available, the bonding electron pair of the B—H bond of a second borane molecule acts as the donor so that a "two-electron, three-center bond" is produced. Under these conditions, boranes are consequently present as dimers: "BH_3," for example, as B_2H_6. Still, small fractions of the monomers appear as minor components in the dissociation equilibrium of the dimer: B_2H_6, for example, thus contains some BH_3.

The Lewis bases Me_2S or THF are better electron pair donors than the boranes themselves. Therefore, they convert dimeric boranes into Me_2S or THF complexes of the monomers such as $Me_2S \cdots BH_3$ or THF $\cdots BH_3$. This type of complex also dissociates to a small extent and thus contains a small equilibrium concentration of free monomeric boranes.

Only *monomeric* boranes can undergo a *cis* addition to olefins. Specifically, *every* B—H bond of a monomeric borane can in principle be involved in this kind of reaction. It effects a one-step addition of formerly joined fragments BR_2 and H to the C=C double bond. This addition is therefore called the **hydroboration** of a C=C double bond.

Because BH_3 contains three B—H bonds, it can add to as many as three C=C double bonds. Monoalkylboranes RBH_2, with their two B—H bonds, add to as many as two C=C double bonds. Dialkylboranes R_2BH naturally can add to only one. The steric hindrance of a B—H bond increases in the series $BH_3 < RBH_2 < R_2BH$. Therefore its reactivity decreases. Consequently, BH_3 does not necessarily react to produce trialkylboranes (but may do so as shown shortly: see Figure 3.16). This failure to produce trialkylboranes is especially common with sterically demanding olefins, which may therefore only lead to dialkyl- or even to monoalkylboranes (examples: Figure 3.15). The 9-BBN (for 9-borabicyclo[3.3.1]nonane), which is produced from 1,5-cyclooctadiene and B_2H_6, illustrates the selective formation of a dialkylborane; 9-BBN itself is a valuable hydroborating agent (synthesis applications are given later: Figures 3.18–3.21 and Table 3.1).

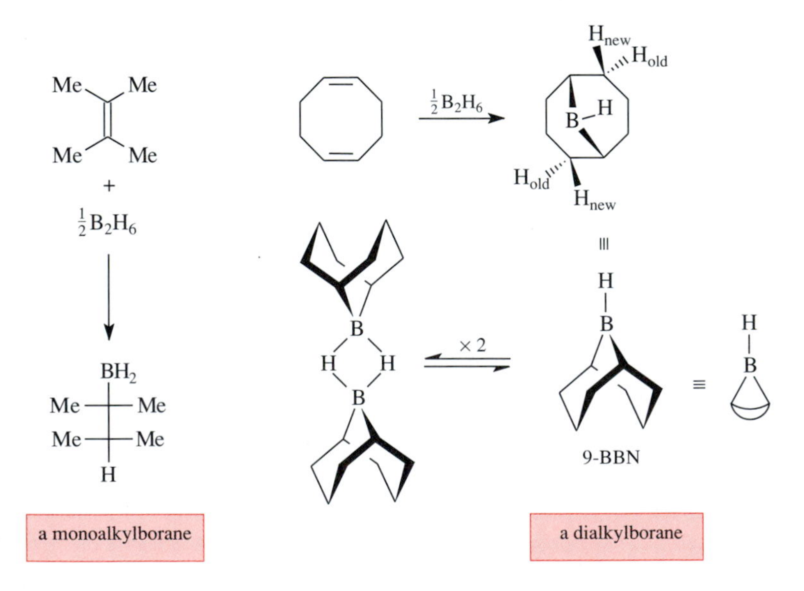

Fig. 3.15. Synthesis of monoalkyl- and dialkylboranes through incomplete hydroboration of multiply substituted C=C double bonds.

Hydration of Cyclohexene

B

Hydroboration of olefins has become very important in organic chemistry, mainly because of a secondary reaction, which is almost always carried out after it. In this reaction, the initially obtained trialkylboranes are first oxidized with sodium hydroxide solution/H_2O_2 to boric acid trialkyl esters (trialkyl borates), which are then hydrolyzed to alcohol and sodium borate. Thus, through the reaction sequence of (1) hydroboration and (2) $NaOH/H_2O_2$ treatment, one achieves an H_2O addition to ("hydration" of) C=C double bonds. The reaction example in Figure 3.16 illustrates this starting from $Me_2S \cdots BH_3$ or THF $\cdots BH_3$ and cyclohexene.

Because the C=C double bond of the cyclohexene used in Figure 3.16 is labeled with deuterium, it is possible to follow the stereochemistry of the whole reaction sequence. First there is a *cis*-selective hydroboration. Two diastereomeric racemic trialkylboranes are produced. Without isolation, these are oxidized/hydrolyzed with sodium hydroxide solution/H_2O_2. The reaction product is the sterically homogeneous

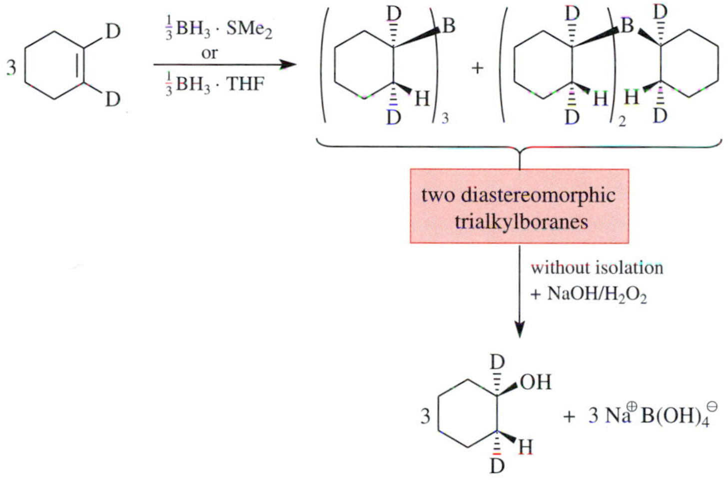

Fig. 3.16. *cis*-Hydration of an olefin via the reaction sequence hydroboration/oxidation/hydrolysis.

but, of course, racemic dideuteriocyclohexanol. The stereochemistry of the product proves the *cis* selectivity of this hydration.

Three types of reactions participate in the reaction sequence of Figure 3.16 (cf. Figure 3.17):

(1) The hydroboration of the C=C double bond to form the trialkylboranes **A** and **B.**
(2) Their oxidation to the trialkylborates **D** and **E,** respectively.
(3) Their hydrolyses to the dideuteriocyclohexanol **C.**

The mechanism of partial reaction 1 is detailed in Figure 3.17 (top), and the mechanism of partial reaction 3 is detailed in the same figure at the bottom. We will discuss partial step 2 using the example **B** → **E** in Figure 11.36. The upper part of Figure 3.17 illustrates the details of the consecutive hydroboration of C=C double bonds by BH_3, mono-, and dialkylboranes, respectively. The second and the third hydroboration steps take place without any diastereoselectivity with respect to the stereocenters, which are contained already in the attacking mono- or dialkylborane. The isotopomeric trialkylboranes **A** and **B** are therefore produced in the statistical (i.e., 3:1) ratio. During the oxidation of these trialkylboranes with $NaOH/H_2O_2$, a 3:1 mixture of the isotopomeric

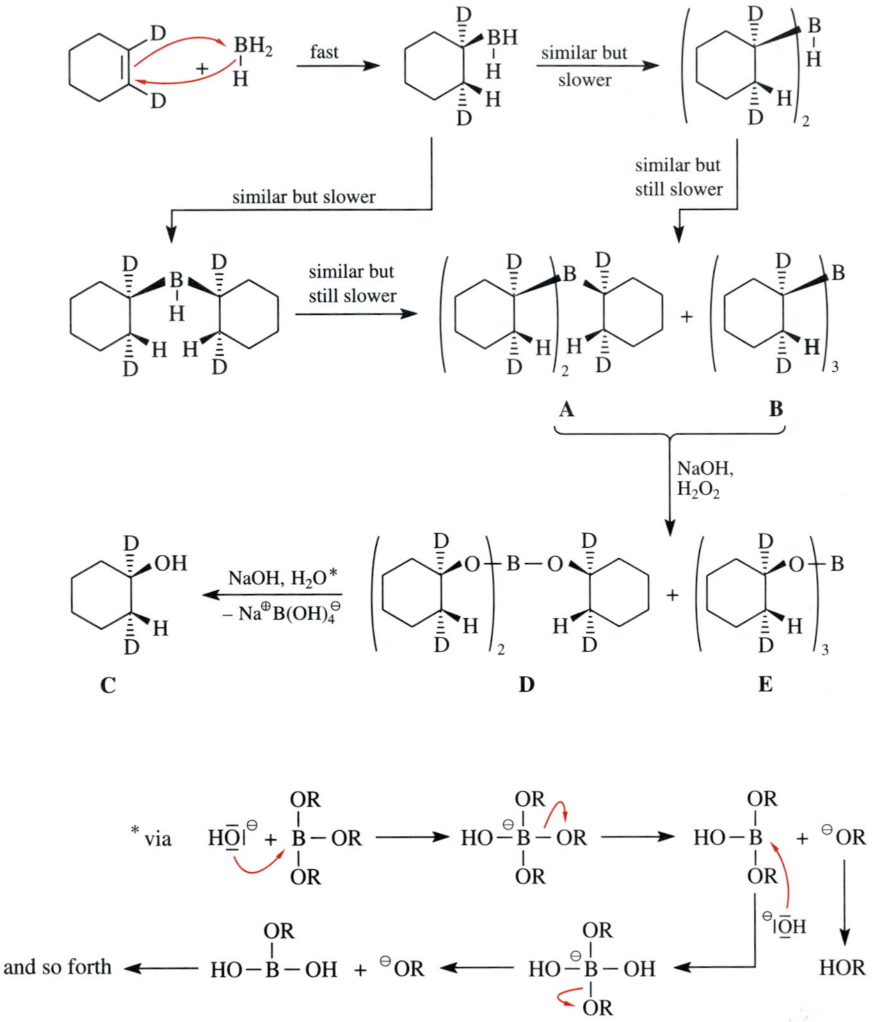

Fig. 3.17. Mechanistic details of the hydroboration/oxidation/hydrolysis sequence of Figure 3.16.

trialkyl-borates **D** and **E** is produced. In this step, the C—B bonds are converted to C—O bonds with retention of configuration. During the subsequent alkaline hydrolysis of the borates **D** and **E** (Figure 3.17, bottom), the three O—B bonds are broken one after the other while the O—C bonds remain intact. In each case the intermediate is an anion with a tetracoordinate B atom (i.e., a borate complex). Because the sodium borate is produced in an alkaline rather than acidic solution, the hydrolysis is irreversibe.

Regioselective Hydroboration of Asymmetric Olefins

B

The hydroboration of olefins, in which in contrast to tetramethylethylene, 1,5-cyclooctadiene (both in Figure 3.15), or cyclohexene (cf. Figure 3.17) the $C_\alpha{=}C_\beta$ is not symmetrically substituted, can lead to constitutionally isomeric trialkylboranes. This is because the new C—B bond can form either at the C_α or at the C_β of the

$C_\alpha{=}C_\beta$ double bond. In the oxidation/hydrolysis sequence that follows, constitutionally isomeric alcohols are produced. In one of them, the OH group binds to C_α and in the other it binds to C_β. If only one constitutional isomer of the trialkylborane and consequently only one constitutional isomer of the alcohol shall be produced, the hydroboration step must take place regioselectively. Whether regioselectivity occurs is determined by steric and electronic effects.

The two reactions in Figure 3.18 prove that the regioselectivity of the hydroboration of asymmetrical olefins is influenced by steric effects. The substrate is an olefin whose attacked centers C_α and C_β differ only in the size of the alkyl fragment bound to them, Me or *i*Pr (while they do not differ with respect to the small positive partial charges located there; see below). The upper hydroboration of this olefin shown in Figure 3.18 takes place with BH_3 itself. It exhibits no regioselectivity at all. This highly reactive and sterically undemanding reagent forms C_α—B bonds as rapidly as C_β—B bonds. The monoalkylboranes thus formed become the reagents for the second hydroboration step. In their attack on the next $C_\alpha{=}C_\beta$ double bond, they are subject to a very small steric hindrance. They therefore attack the less hindered C_α center with a very slight preference. The dialkylboranes thus formed are the most hindered hydroboronating agents in this mixture and react with a slight preference for attacking C_α. Nonetheless, these steric effects are small. After oxidation/hydrolysis, one therefore finds only a 57:43 regioselectivity in favor of the product hydroxylated at C_α.

The bottom half of Figure 3.18 shows an almost perfect solution of this regioselectivity problem, namely, by hydroborating with 9-BBN (see Figure 3.15 for the formula).

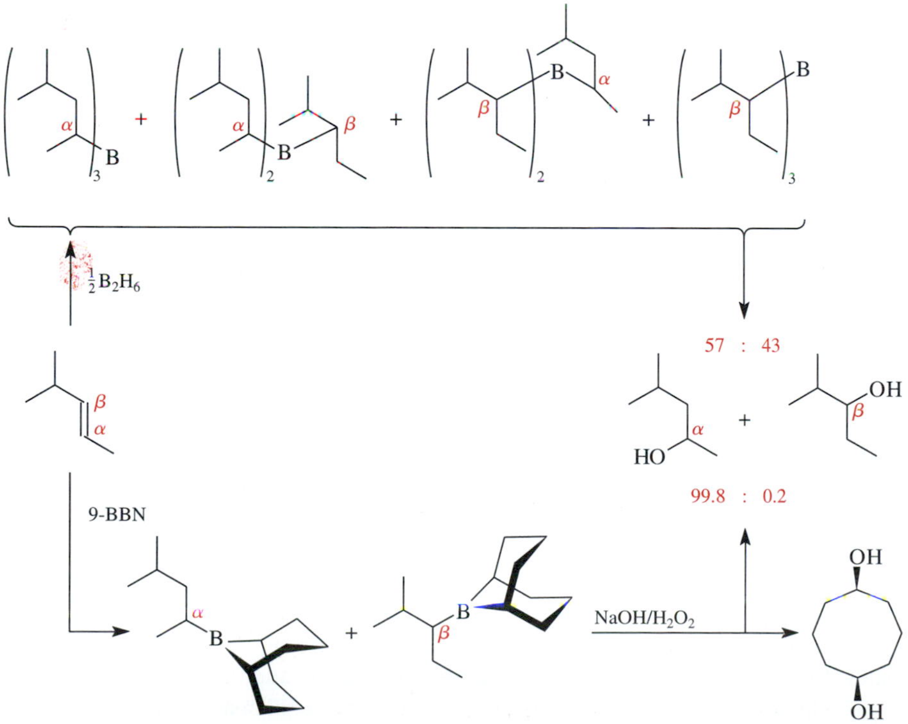

Fig. 3.18. Regioselectivity of the hydroboration of an α,β-dialkylated ethylene.

The B—H bond of 9-BBN is much less readily accessible than the B—H bond in BH_3 and the resulting primary products. Thus, the hydroboration with 9-BBN takes place much more slowly than the one with BH_3 but at the same time with much higher regioselectivity. The less hindered hydroboration product and the alcohol derived from it are produced with a considerable regioselectivity of 99.8:0.2.

Asymmetric olefins, which carry more alkyl substituents at the C_β center than at the C_α center, are also hydroborated by the unhindered BH_3 with considerable regioselectivity (Table 3.1). After oxidative workup, one isolates the alcohol in the α position almost exclusively. According to what has already been stated, as the more bulky reagent, 9-BBN reacts with more sensitivity to steric effects than BH_3 and its secondary products. It therefore makes possible olefin hydrations with almost perfect regiocontrol.

Table 3.1. Regioselectivity for the Hydration (via Hydroboration/Oxidation/Hydrolysis) of Ethylene Derivatives That Contain More Alkyl Substituents at C_β Than at C_α

Fraction of α-alcohol in the hydroboration/oxidation of ...	R^1, β, α	R^1, R^2, β, α, R^3	R^1, R^2, β, α
... with B_2H_6:	94%	100%	99%
... with 9-BBN:	99.9%		99.8%

The regioselectivities in Table 3.1 are probably also caused by an electronic effect, which acts in the same direction as the steric effect. The electronic effect can be explained in various ways: for example, by considering a resonance formula of the type

for each H atom in the allylic position. When more H atoms are allylic next to C_β than to C_α—precisely this condition is satisfied for the olefins in Table 3.1—a negative partial charge appears at C_α relative to C_β. Because the B atom in all boranes is an *electrophilic* center, it attacks preferentially at the more electron-rich C_α.

Addition reactions to unsymmetric olefins in which a reagent with the structure H—X transfers the H atom to the less-substituted C atom and the "X group" to the more-substituted C atom give, according to an old nomenclature, a so-called **Markovnikov addition product.** On the other hand, addition reactions of reagents H—X in which the H atom is transferred to the more substituted C atom of an asymmetrically substituted C=C double bond and the "X group" is transferred to the less substituted C atom lead to a so-called **anti-Markovnikov addition product.**

The hydroboration of asymmetrical olefins thus gives monoalkylboranes (addition of H—BH_2), dialkylboranes (addition of H—BHR), or trialkylboranes (addition of H—BR_2), which are typical anti-Markovnikov products. Therefore, expressed in the terminology discussed above, the reaction sequence hydroboration/oxidation/hydrolysis brings about the anti-Markovnikov addition of H_2O to asymmetrically substituted olefins.

The hydroboration/oxidation/hydrolysis of trisubstituted olefins also takes place as a *cis* addition. The reaction equation from Figure 3.19 shows this using 1-methylcyclohexene as an example. 9-BBN adds to both sides of its C=C double bond at the same reaction rate. In the energy profile this fact means that the activation barriers for both modes of addition are equally high. The reason for this equality is that the associated transition states are enantiomorphic. They thus have the same energy—just as enantiomorphic molecules.

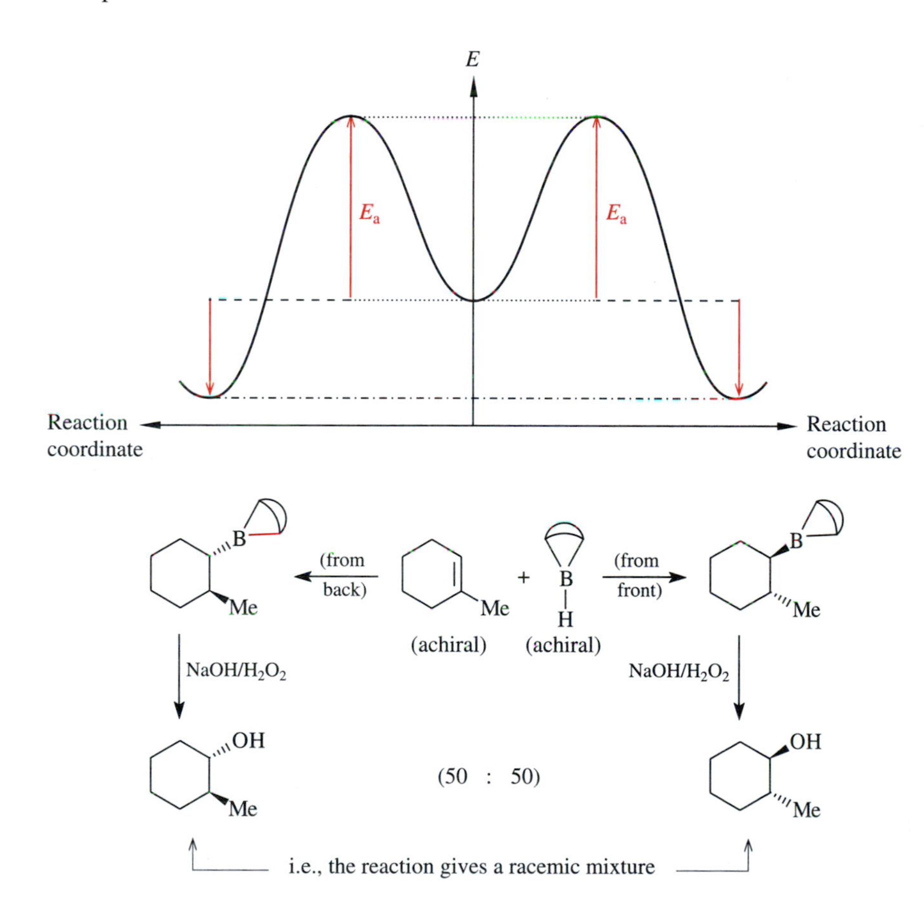

Fig. 3.19. *cis*-Selective hydration of an achiral trisubstituted olefin.

Stereoselective Hydration of Asymmetric Olefins and Substrate Control of the Stereoselectivity

Let us now consider the analogous hydroborations/oxidations of chiral derivatives of 1-methylcyclohexene, namely of racemic 3-ethyl-1-methylcyclohexene (Figure 3.20) and of enantiomerically pure α-pinene (Figure 3.21).

Fig. 3.20. *cis*-Selective hydration of a chiral, racemic, trisubstituted olefin with induced diastereoselectivity; ΔE_a corresponds to the extent of the substrate control of diastereoselectivity.

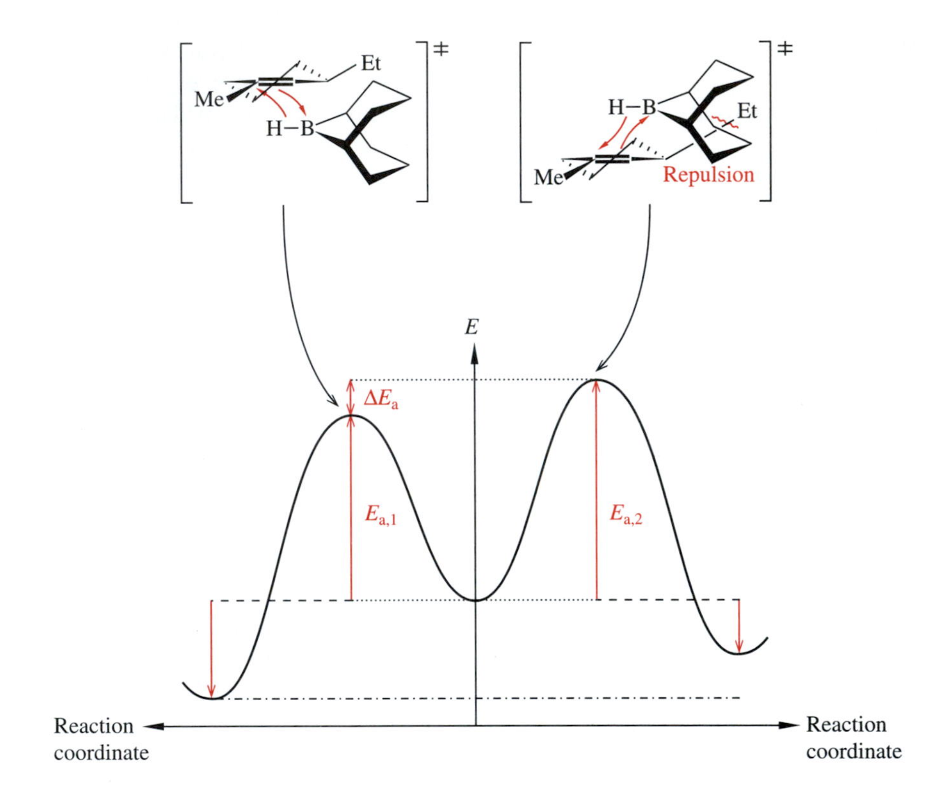

Et, B, Me — **A** (+ 50% *ent*-**A**) ← k_1 (from back) — Et, Me (+ 50% of the enantiomer) + B–H → k_2 (from front) → Et, B, Me — **B** (+ 50% *ent*-**B**)

NaOH/H_2O_2 ↓ NaOH/H_2O_2 ↓

Et, OH, Me — *trans,trans* (+ 50% of the enantiomer)

Et, OH, Me — *cis,trans* (+ 50% of the enantiomer)

B

The stereochemical result is no longer characterized solely by the fact that the newly formed stereocenters have a uniform configuration *relative to each other.* This was the only type of stereocontrol possible in the reference reaction 9-BBN + 1-

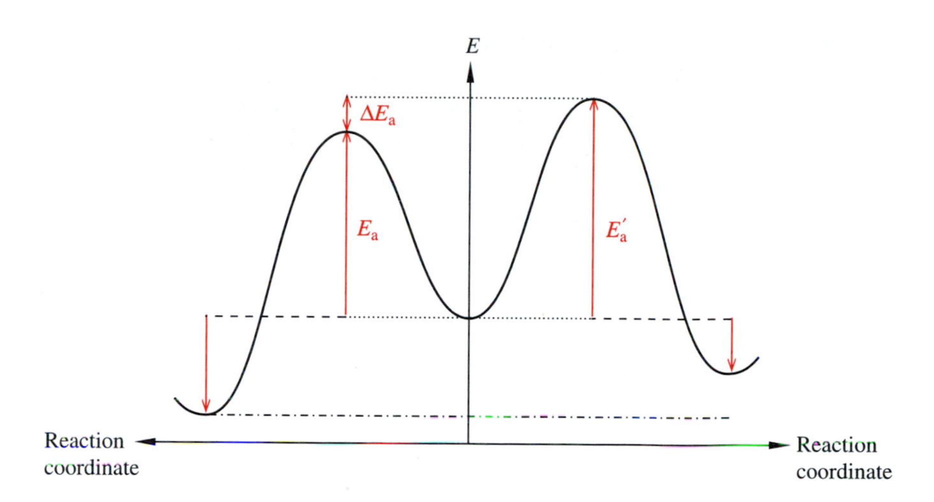

Fig. 3.21. *cis*-Selective hydration of a chiral, enantiomerically pure, trisubstituted olefin with induced diastereoselectivity; ΔE_a corresponds to the degree of substrate control of diastereoselectivity.

(from back) · (pure enantiomer) · + · B–H · (from front) · (pure enantiomer)

B · Me · (pure enantiomer) — $NaOH/H_2O_2$ → OH · Me · (the only product) · *cis,trans* (pure enantiomer)

B · Me · (pure enantiomer) — $NaOH/H_2O_2$ → OH · Me · *trans,trans* (pure enantiomer)

methylcyclohexene (Figure 3.19), that is, in the hydroboration/oxidation/hydrolysis of an *achiral* olefin with an achiral borane. In the hydroborations of the cited *chiral* olefins with 9-BBN, an additional question arises. What is the relationship between the new stereocenters and the stereocenter(s) present in the olefin at the start? When a uniform relationship between the old and the new stereocenters arises, a type of diastereoselectivity not mentioned previously is present. It is called **induced diastereoselectivity.** It is based on the fact that the substituents on the stereocenter(s) of the chiral olefin hinder one face of the chiral olefin while not affecting the other. This is an example of what is called **substrate control of stereoselectivity.** Accordingly, in the hydroborations/oxidations of Figures 3.20 and 3.21, 9-BBN does *not* add to the top and the bottom sides of the olefins with the same reaction rate. The transition states of the two modes of addition are not equivalent with respect to energy. The reason for this inequality is that the associated transition states are diastereomorphic. They thus have different energies—just as diastereomorphic molecules.

When 9-BBN attacks 3-ethyl-1-methylcyclohexene from the side on which the ethyl group is located (Figure 3.20), the transition state involved is more energy rich than the transition state of the 9-BBN addition to methylcyclohexene. This is due to a repulsion between the reagent and the ethyl group (Figure 3.19). On the other hand, the attack by 9-BBN from the other side of 3-ethyl-1-methylcyclohexene is not sterically hindered. Therefore, the corresponding transition state has approximately the same energy as that of the 9-BBN addition to methylcyclohexene. This circumstance makes the "left" transition state in Figure 3.20 the more energetically favorable, and it can therefore be passed through faster. As a consequence, one obtains a single, diastereomerically pure trialkylborane, or after oxidation and hydrolysis, a single alcohol. Because the olefin is used as a racemic mixture, this alcohol is also racemic. Otherwise it would have been a pure enantiomer.

Figure 3.21 shows reaction equations and the energy relationships of the hydroboration of enantiomerically pure α-pinene with 9-BBN. The reagent approaches only the side of the C=C double bond that lies opposite the isopropylidene bridge. The addition is thus completely diastereoselective. Moreover, the trialkylborane obtained is a pure enantiomer. It is used as Alpine-Borane® for the enantioselective reduction of carbonyl compounds (Section 8.4).

3.3.4 Heterogeneously Catalyzed Hydrogenation

B

There is no known one-step addition of molecular hydrogen to C=C double bonds (hydrogenation).

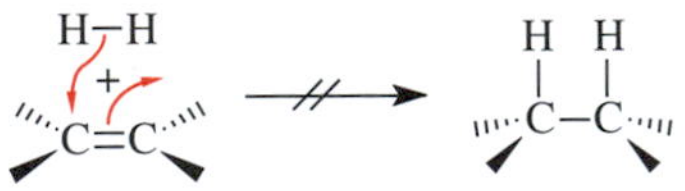

The addition of hydrogen to olefins is made possible only with noble metal catalysts. They allow for a multistep, low-energy reaction pathway. The noble metal catalyst may be soluble in the reaction mixture; in that case we have a homogeneously catalyzed hydrogenation (cf. Section 14.4.7). But the catalyst may also be insoluble; then we deal with a heterogeneously catalyzed hydrogenation.

Heterogeneously catalyzed hydrogenations are by far the most common. They take place as surface reactions (see Section 14.4.7 for a discussion of the mechanism). Their rate is higher, the greater the surface of the catalyst. Therefore the catalyst is precipitated on a carrier with a high specific surface such that it is very finely distributed. Suitable carriers are activated carbon or aluminum oxide. To bring the olefin, the catalyst, and the hydrogen, which is dissolved continuously from the gas phase, into intimate contact, the reaction mixture is either shaken or stirred. This procedure makes it possible to carry out heterogeneously catalyzed hydrogenations at room tempera-

Ph, Ph, Me, Me (*Z*) → H_2 (only the dissolved part reacts), cat. Pd/C, AcOEt or EtOH, shake → (*meso*)

Ph, Me, Me, Ph (*E*) → the same → (racemic mixture)

Fig. 3.22. Stereoselectivity and stereospecificity of a pair of heterogeneously catalyzed hydrogenations.

ture and under normal or often only slightly increased pressure. Most heterogeneously catalyzed hydrogenations take place with high *cis*-selectivity. This implies that stereoisomeric olefins are frequently hydrogenated with stereospecificity (Figure 3.22). For steric reasons, monosubstituted ethylenes react more rapidly with hydrogen than 1,1-disubstituted ethylenes, which react faster than *cis*-disubstituted ethylenes, which react faster than *trans*-disubstituted ethylenes, which react faster than trisubstituted ethylenes, which in turn react faster than tetrasubstituted ethylene:

Therefore, in polyenes that contain differently substituted C=C double bonds, it is often possible to hydrogenate chemoselectively the least hindered C=C double bond:

H_2, PtO_2

3.4 Enantioselective *cis* Additions to C=C Double Bonds

B

In the discussion of the hydroboration in Figure 3.20, you saw that one principle of stereoselective synthesis is the use of *substrate control* of stereoselectivity. But there are quite a few problems in stereoselective synthesis that cannot be solved in this way. Let us illustrate these problems by means of two hydroborations from Section 3.3.3:

- 9-BBN effects the hydration of the C=C double bond of 1-methylcyclohexene according to Figure 3.19 in such a way that after the oxidative workup, *racemic*

2-methyl-1-cyclohexanol is obtained. This brings up the question: Is an *enantioselective* H_2O addition to the same olefin possible? The answer is yes, but only with the help of the so-called *reagent control* of stereoselectivity (cf. Section 3.4.2).

A

- 9-BBN attacks the C=C double bond of 3-ethyl-1-methylcyclohexene according to Figure 3.20 exclusively from the side that lies opposite the ethyl group at the stereocenter. Consequently, after oxidation and hydrolysis, a *trans,trans*-configured alcohol is produced. The question that arises is: Can this diastereoselectivity be reversed in favor of the *cis,trans* isomer? The answer is possibly, but, if so, only by using *reagent control* of stereoselectivity (cf. Section 3.4.4).

Before these questions are answered, we want to consider a basic concept for stereoselective synthesis—topicity—in the following subsection.

3.4.1 Vocabulary of Stereochemistry and Stereoselective Synthesis II: Topicity, Asymmetric Synthesis

The "faces" on both sides of a C=C, C=O, or C=N double bond can have different geometrical relationships to each other (Figure 3.23). If they can be converted into each other by means of a rotation of 180° about a two-fold intramolecular axis of rotation, then they are **homotopic.** If they cannot be superimposed on each other in this way, they are **stereoheterotopic.** Stereoheterotopic faces of a double bond are **enantiotopic** if they can be related to each other by a reflection; otherwise they are **diastereotopic.** Let us illustrate this with examples. The faces of the C=C double bond of cyclohexene are homotopic, those in 1-methylcyclohexene are enantiotopic, and those in 3-ethyl-1-methylcyclohexene are diastereotopic.

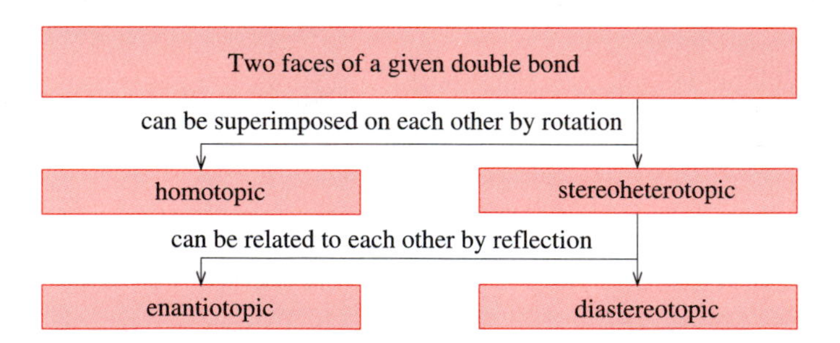

Fig. 3.23. Topicities of the faces of a double bond.

Let us consider an addition reaction to one of the cited C=X double bonds during which at least one stereocenter is produced. The topicity of the faces of the C=X double bond in question allows one to predict whether such an addition can, in principle, take place stereoselectively. This is because depending on the topicity of the faces of the reacting C=X double bond, the transition states that result from the attack by reagents of one kind or another (see below) from one or the other face are enantiomorphic or diastereomorphic. If the transition states are enantiomorphic, they have the same energy. The addition proceeds through each of them to the same extent and we thus get a 50:50 ratio of enantiomers (i.e., no enantioselectivity). If the transition states are diastereomorphic, they may have different energies. The addition preferentially takes place via the lower energy transition state and preferentially results in one stereoisomer. It is thus diastereoselective (i.e., $ds \neq 50{:}50$) or enantioselective (i.e., $ee \neq 0\%$).

When all these results are analyzed accurately, it is seen that addition reactions to the C=X double bond

- of achiral compounds (i.e., to C=X double bonds with homotopic or enantiotopic faces) cannot take place enantioselectively; they always give racemic mixtures (e.g., hydroboration in Figure 3.19);
- of chiral compounds (i.e., to C=X double bonds with diastereotopic faces) can take place diastereoselectively with achiral reagents. In this case we have substrate control of stereoselectivity (e.g., hydroborations in Figures 3.20 and 3.21); or
- of achiral compounds (i.e., to C=X double bonds with homotopic or enantiotopic faces) with chiral reagents can lead to enantiomerically enriched or enantiomerically pure compounds.

An enantioselective addition of the latter type (here) or, in general, the successful conversion of achiral starting materials into enantiomerically enriched or enantiomerically pure products is referred to as **asymmetric synthesis** (for examples, see Sections 3.4.2, 3.4.6, 8.4, 8.5.2, 14.4.7).

3.4.2 Asymmetric Hydroboration of Achiral Olefins

B

The conclusion drawn from Section 3.4.1 for the hydroborations to be discussed here is this: an addition reaction of an enantiomerically pure chiral reagent to a C=X double bond with enantiotopic faces can take place via two transition states that are diastereomorphic and thus generally differing from one another in energy. In agreement with this statement, there *are* diastereoselective additions of enantiomerically pure mono- or dialkylboranes to C=C double bonds that possess enantiotopic faces. Consequently, when one subsequently oxidizes all C—B bonds to C—OH bonds, one has realized an enantioselective hydration of the respective olefin.

An especially efficient reagent of this type is the boron-containing five-membered ring compound shown in Figure 3.24. Since this reagent is quite difficult to synthesize, it has not been used much in asymmetric synthesis. Nonetheless, this reagent will be presented here simply because it is particularly easy to see which face of a C=C double bond it will attack.

In the structure shown in Figure 3.24 the top side attack of this borane on the C=C double bond of 1-methylcyclohexene prevails kinetically over the bottom side attack. This is because only the top side attack of the boranes avoids steric interactions between the methyl substituents on the borane and the six-membered ring. In other words, the *reagent* determines the face to which it adds. We thus have **reagent control of stereoselectivity.** As a result, the mixture of the diastereomeric trialkylboranes **C** and **D**, both of which are pure enantiomers, is produced with $ds = 97.8{:}2.2$. After the normal $NaOH/H_2O_2$ treatment, they give a 97.8:2.2 mixture of the enantiomeric *trans*-2-

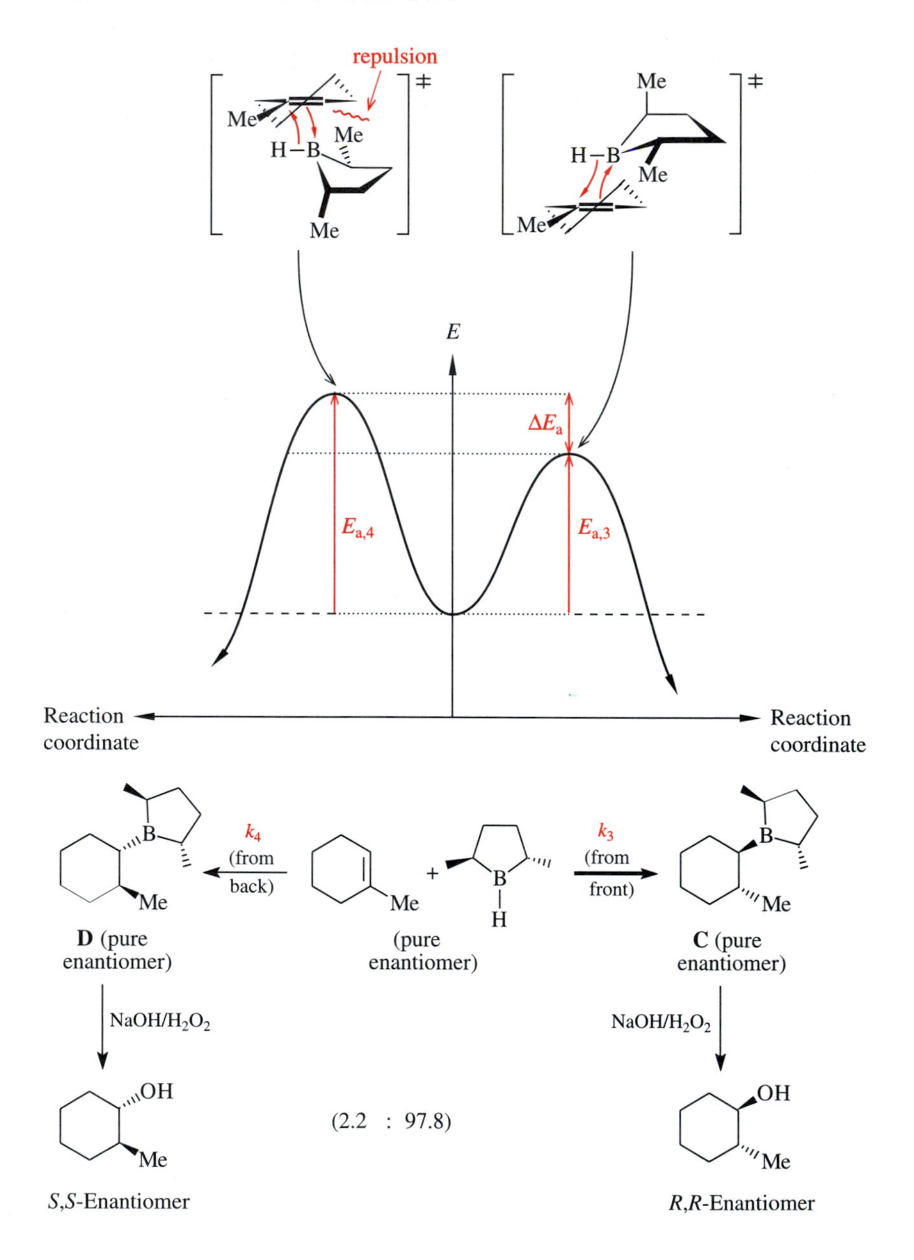

Fig. 3.24. Asymmetric hydration of an achiral olefin via hydroboration/oxidation/hydrolysis; ΔE_a corresponds to the extent of reagent control of diastereoselectivity.

methylcyclohexanols. The 1*R*,2*R* alcohol will therefore have an *ee* value of 97.8% − 2.2% = 95.6%.

The *S*,*S* enantiomer of this alcohol is obtained with the same *ee* value of 95.6% when the enantiomer of the borane shown in Figure 3.24 is used for the hydroboration of 1-methylcyclohexene. The first problem we ran into in the introduction to Section 3.4 is thereby solved!

3.4.3 Thought Experiment I on the Hydroboration of Chiral Olefins with Chiral Boranes: Mutual Kinetic Resolution

During the addition of a racemic chiral dialkylborane to a racemic chiral olefin a maximum of four diastereomeric racemic trialkylboranes can be produced. Figure 3.25 il-

A

Et, Me, (racemic), B, H, k_1, k_2, Substrate control of stereoselectivity

50% **A** (+ 50% *ent*-**A**) 50% **B** (+ 50% *ent*-**B**)

50% **C** (+ 50% *ent*-**C**) 50% **E** (+ 50% *ent*-**E**) 50% **F** (+ 50% *ent*-**F**)

Reagent control of stereo-selectivity k_3 k_4 k_5 k_6 k_7 k_8 (racemic)

50% **D** (+ 50% *ent*-**D**) 50% **G** (+ 50% *ent*-**G**) 50% **H** (+ 50% *ent*-**H**)

Fig. 3.25. Thought experiment I: Products from the addition of a racemic chiral dialkylborane to a racemic chiral olefin. Rectangular boxes: previously discussed reference reactions for the effect of substrate control (top box: reaction from Figure 3.20) or reagent control of stereoselectivity [leftmost box: reaction from Figure 3.24 (rewritten for racemic instead of enantiomerically pure reagent)]. Solid reaction arrows, reagent control of stereoselectivity; dashed reaction arrows, substrate control of stereoselectivity; red reaction arrows (kinetically favored reactions), reactions proceeding with substrate control (solid lines) or reagent control (dashed lines) of stereoselectivity; black reaction arrows (kinetically disfavored reactions), reactions proceeding opposite to substrate control (solid lines) or reagent control (dashed lines) of stereoselectivity.

lustrates this using the example of the hydroboration of 3-ethyl-1-methylcyclohexene with the cyclic borane from Figure 3.24. This hydroboration, however, was never carried out experimentally. This should not prevent us from considering what would happen if it were carried out.

Our earlier statements on substrate and reagent control of stereoselectivity during hydroborations are incorporated in Figure 3.25. Because of the obvious analogies between the old and the new reactions, the following can be predicted about the product distribution shown:

- As the main product we expect the racemic trialkylborane **E;** the substrate and the reagent controls work together to promote its formation.
- The racemic trialkylborane **H** should be produced in trace quantities only. Its formation is disfavored by both the reagent and the substrate control.
- As minor products we expect the racemic trialkylboranes **F** and/or **G; F** is favored by reagent and disfavored by substrate control of stereoselectivity, whereas for **G** it is exactly the opposite.

We thus summarize: the yield ratios of the conceivable hydroboration products **E, F, G,** and **H** should be much:little:little:none. One enantiomer of **E** comes from the reaction of the *S*-olefin with the *S,S*-borane; the other enantiomer of **E** comes from the reaction of the *R*-olefin with the *R,R*-borane. Thus each enantiomer of the reagent has preferentially reacted with *one* enantiomer of the substrate. The diastereoselectivity of this reaction thus corresponds to a *mutual kinetic resolution.*

The condition for the occurrence of a mutual kinetic resolution is thus that considerable substrate control of stereoselectivity and considerable reagent control of stereoselectivity occur simultaneously.

From the diastereoselectivities in Figure 3.25 one concludes the following for the rate constants: $k_5 > k_6$, $k_7 > k_8$ (which implies $k_5 \gg k_8$). It is not known whether k_6 or k_7 is greater.

For the discussion in Sections 3.4.4 and 3.4.5, we will assume(!) that $k_6 > k_7$; that is, the reagent control of stereoselectivity is more effective than the substrate control of stereoselectivity. The justification for this assumption is simply that it makes additional thought experiments possible. These are useful for explaining interesting phenomena associated with stereoselective synthesis, which are known from other reactions. Because the thought experiments are much easier to understand than many of the actual experiments, their presentation is given preference.

3.4.4 Thought Experiments II and III on the Hydroboration of Chiral Olefins with Chiral Boranes: Reagent Control of Diastereoselectivity, Matched/Mismatched Pairs, Double Stereodifferentiation

A

At the beginning of Section 3.4, we wondered whether 3-ethyl-1-methylcyclohexene could also be hydroborated/oxidized/hydrolyzed to furnish the *cis,trans*-configured alcohol. There is a solution (Figure 3.26) if two requirements are fulfilled. First, we must rely on the assumption made in Section 3.4.3 that this olefin is attacked by the cyclic *ent*-**G** (pure enantiomer) ← k_7 (from back) — (pure *R*-enantiomer) + (pure enantiomer) — k_6 (from front) → **F** (pure enantiomer)

NaOH/H_2O_2

trans,trans (pure enantiomer) — *cis,trans* (pure enantiomer)

Fig. 3.26. Thought experiment II: Reagent control of stereoselectivity as a method for imposing on the substrate a diastereoselectivity that is alien to it (mismatched pair situation).

borane in such a way that the reagent control of stereoselectivity exceeds the substrate control of the stereoselectivity. Second, both the olefin and the borane must be used in enantiomerically pure form.

Figure 3.25 already contains all the information that is necessary to deduce the product distribution of the reaction in Figure 3.26. The reaction partners in Figure 3.26 can provide nothing but the *cis,trans* adduct as the pure enantiomer **F** and the *trans,trans* adduct as the pure enantiomer *ent*-**G.** The adduct **F** is produced with the rate constant k_6 and the adduct *ent*-**G** is produced with the rate constant k_7. After oxidative cleavage and hydrolysis with NaOH/H_2O_2, the adduct **F** results in the *cis,trans* alcohol. The diastereoselectivity of this reaction is given by $ds = (k_6/k_7){:}1$. When $k_6 > k_7$, the *cis,trans* alcohol is produced in excess. If on the other hand, in contrast to our assumption, k_7 were to exceed k_6, the *cis,trans* alcohol would still be produced although only in less than 50% yield. It would be present in the product mixture—as before—in a (k_6/k_7):1 ratio, but now only as a minor component.

If, as in the reaction example in Figure 3.26, during the addition to enantiomerically pure chiral olefins, substrate and reagent control of diastereoselectivity act in opposite directions, we have a so-called **mismatched pair.** For obvious reasons it reacts with relatively little diastereoselectivity and also relatively slowly. Side reactions and, as a consequence, reduced yields are not unusual in this type of reaction.

Conversely, the addition of enantiomerically pure chiral dialkylboranes to enantiomerically pure chiral olefins can also take place in such a way that substrate control and reagent control of diastereoselectivity act in the same direction. Then we have a **matched pair.** It reacts faster than the corresponding mismatched pair and with especially high diastereoselectivity. This approach to stereoselective synthesis is also referred to as **double stereodifferentiation.**

Thought experiment III in Figure 3.27 provides an example of how such a double stereodifferentiation can be used to increase stereoselectivity. The two competing hy-

Fig. 3.27. Thought experiment III: Reagent control of stereoselectivity as a method for enhancing the substrate control of stereoselectivity (matched pair situation).

droborations take place with $ds = (k_5/k_8){:}1$, as a comparison with what the rate constants in Figure 3.25 reveal. After oxidation/hydrolysis, the *trans,trans* alcohol is obtained besides the *cis,trans* alcohol in the same ratio $(k_5/k_8){:}1$. According to the results from Section 3.4.3, it was certain that $k_5 \gg k_8$. The diastereoselectivity of the thought experiment in Figure 3.27 is therefore considerably higher than the one in Figure 3.26.

3.4.5 Thought Experiment IV on the Hydroboration of Chiral Olefins with Chiral Dialkylboranes: Kinetic Resolution

Figure 3.26 showed the reaction of our enantiomerically pure chiral cyclic dialkylborane with *R*-3-ethyl-1-methylcyclohexene. It took place relatively slowly with the rate constant k_6. The reaction of the same dialkylborane with the isomeric *S*-olefin was shown in Figure 3.27. It took place considerably faster with the rate constant k_5. The combination of the two reactions is nothing other than the reaction of Figure 3.28. There the same enantiomerically pure borane is reacted simultaneously with both olefin enantiomers (i.e., with the racemic olefin). What is happening? In the *first* moment of the reaction the *R*- and the *S*-olefin react in the ratio k_6 (small)/k_5 (big). The matched pair thus reacts faster than the mismatched pair. This means that at low conversions ($\ll 50\%$) the trialkylborane produced is essentially derived from the *S*-olefin only. It has the stereostructure **E.** Therefore, relative to the main by-product **F,** compound **E** is produced with a diastereoselectivity of almost k_5/k_6. However, this selectivity decreases as the reaction progresses because the reaction mixture is depleted of the more reactive *S*-olefin. When the reaction is almost complete, the less reactive *R*-olefin has been enriched to a ratio k_5/k_6. Then it consequently has an *ee* value of almost:

$$\frac{k_5 - k_6}{k_5 + k_6} \cdot 100\%$$

Et B Me k_6 (slow) Et Me + B H k_5 (fast) Et B Me

F (pure enantiomer) (racemic) (pure enantiomer) **E** (pure enantiomer)

k_7 (slowest)

Et B Me

ent-**G** (pure enantiomer)

Fig. 3.28. Thought experiment IV: Reaction of ≤ 0.5 equiv of an enantiomerically pure chiral dialkylborane with a racemic chiral olefin to effect a kinetic resolution of the latter.

Of the greatest preparative interest is a point in time at which there is approximately 50% conversion in the reaction. Only then can the yield of the trialkylborane of one enantiomeric series or the yield of the olefin of the other enantiomeric series reach its theoretical maximum value of 50%. In other words, in the ideal case, at a conversion of exactly 50%, the olefin enantiomer of the matched pair has reacted completely (owing to the high rate constant k_5), and the olefin enantiomer from the mismatched pair has not yet been attacked (owing to the much smaller rate constant k_6). If the hydroboration were stopped at this time by adding H_2O_2/NaOH, the following desirable result would be obtained: 50% of the alcohol derived almost exclusively from the *S*-olefin and 50% of almost exclusively *R*-configured unreacted olefin would be isolated. The reaction in Figure 3.28 would thus have been a kinetic resolution. The kinetic requirements for the success of a kinetic resolution are obviously more severe than the requirements for the success of a mutual kinetic resolution (Section 3.4.3). For the latter, both substrate and reagent control of stereoselectivity must be high. For a good kinetic resolution to occur, it is additionally required that one type of stereocontrol clearly dominate the other type (i.e., $k_5 \gg k_6$).

3.4.6 Catalytic Asymmetric Synthesis: Sharpless Oxidations of Allyl Alcohols

B

Asymmetric syntheses can even be carried out more easily and more elegantly than by reacting achiral substrates with enantiomerically pure chiral reagents via diastereomorphic transition states. Instead, one can allow the substrate to react exclusively and via diastereomorphic transition states with an enantiomerically pure species formed *in situ* from an achiral reagent and an enantiomerically pure chiral additive. The exclusive attack of this species on the substrate implies that the reagent itself reacts substantially slower with the substrate than its adduct with the chiral additive. If high stereoselectivity is observed, it is due exclusively to the presence of the additive.

This type of **additive** (or ligand) **control of stereoselectivity** has three advantages: First of all, after the reaction has been completed, the chiral additive can be separated from the product with physical methods, for example, chromatographically. In the second place, the chiral additive is therefore also easier to recover than if it had to be first liberated from the product by means of a chemical reaction. The third advantage of additive control of enantioselectivity is that the enantiomerically pure chiral additive does not necessarily have to be used in stoichiometric amounts, but that catalytic amounts may be sufficient. This type of **catalytic asymmetric synthesis,** especially on an industrial scale, will be increasingly important in the future.

The most important catalytic asymmetric syntheses include addition reactions to C=C double bonds. The best known of them are the Sharpless epoxidations. Sharpless epoxidations cannot be carried out on all olefins but only on primary or secondary allyl alcohols. Nonetheless, nowadays they represent the most frequently used synthetic access to enantiomerically pure target molecules.

There are two reasons for this. First, the Sharpless epoxidation can be applied to almost *all* primary and secondary allyl alcohols. Second, it makes trifunctional compounds accessible in the form of enantiomerically pure α,β-epoxy alcohols. These can react with a wide variety of nucleophiles to produce enantiomerically pure "second-generation" products. Further transformations can lead to other enantiomerically pure species that ultimately may bear little structural resemblance to the starting α,β-epoxy alcohols.

Sharpless epoxidations are discussed in the plural because primary (Figures 3.29, and 3.30) and secondary allylic alcohols (Figure 3.31) are reacted in different manners. Primary allyl alcohols are reacted to completion. Secondary allyl alcohols—if they are racemic—are usually reacted only to 50% conversion (the reason for this will become clear shortly). The oxidation agent is always a hydroperoxide, usually *tert*-BuOOH. The chiral additive used is 6–12 mol% of an enantiomerically pure dialkyl ester of tartaric acid, usually the diethyl ester (**die**thyl**t**artrate, DET). Ti(O*i*Pr)$_4$ is added so that the oxidizing agent, the chiral ligand, and the substrate can assemble to form an enantiomerically pure chiral complex. No epoxidation at all takes place in the absence of Ti(O*i*Pr)$_4$.

The transition states of the Sharpless epoxidations have not been unambiguously established. In any case, the structural proposals for such transition states are too com-

Fig. 3.29. Enantioselective Sharpless epoxidation of achiral primary allyl alcohols. If the substrates are drawn as shown, the direction of the attack of the complexes derived for L−(+) and D−(−)−DET can be remembered with the following mnemonic: **L,** from **l**ower face; **D, d**oesn't attack from **d**ownface.

HO, CO_2Et / HO, CO_2Et ≡ L-(+)-DET (6–12 mol%), *tert*-BuOOH (standard) or $PhCMe_2OOH$, Ti(O*i*Pr)$_4$ (5–10 mol%), 3 Å molecular sieves, CH_2Cl_2

D-(−)-DET, ROOH, Ti(O*i*Pr)$_4$, 3 Å molecular sieves, CH_2Cl_2

R_x OH → O R_x OH

with or without additive; *tert*-BuOOH, $Ti(OiPr)_4$

Additive				
D-(−)-DET	→	90	:	1
L-(+)-DET	→	1	:	22
none	→	2.3	:	1

Fig. 3.30. Diastereoselective Sharpless epoxidation of chiral primary allyl alcohols. (For preparation of starting materials, see Figure 14.54.)

plex to be discussed in this chapter. However, the nature of the stereocontrol is not based on any effect other than the one discussed in connection with asymmetric hydroboration (Section 3.4.2). Consequently, in the following we present only experimental findings on the Sharpless epoxidation; no interpretations are offered.

Achiral primary allyl alcohols are epoxidized *enantioselectively,* as Figure 3.29 shows. On the other hand, *chiral primary* allyl alcohols are epoxidized *diastereoselectively* according to Figure 3.30. The additive control of stereoselectivity is here much higher than the substrate control (third line in Figure 3.30). Thus the diastereoselectivity in the mismatched pair (middle line in Figure 3.30) is only slightly smaller than in the matched pair (top line in Figure 3.30).

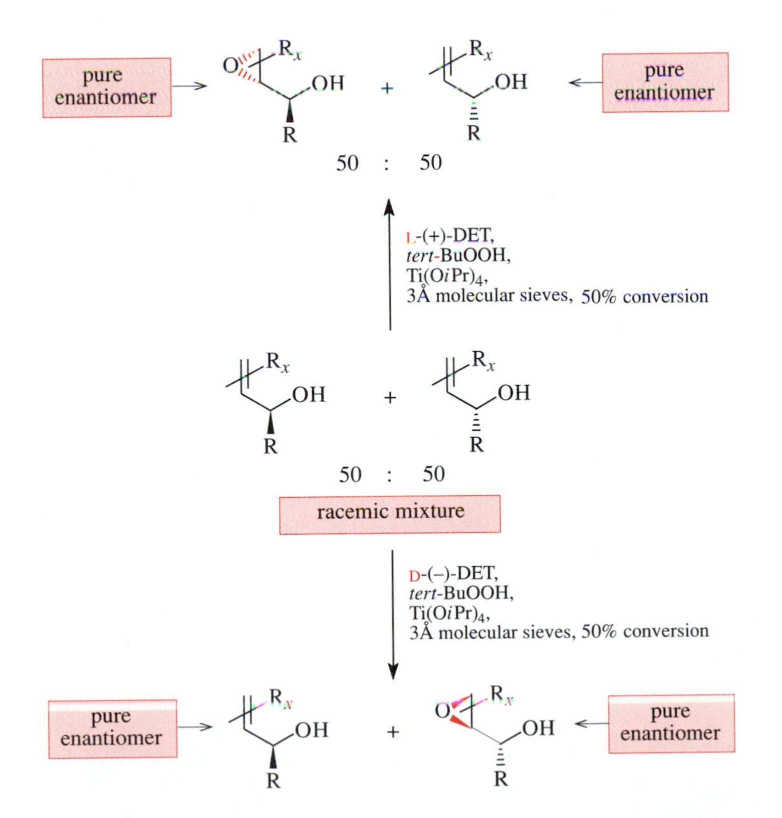

Fig. 3.31. Sharpless kinetic resolution of chiral racemic secondary allyl alcohols.

Racemic chiral secondary allyl alcohols can be subjected to a kinetic resolution by means of the Sharpless epoxidation (Figure 3.31). The reagent mixture reacts with both enantiomers of the allyl alcohol with very different rates. The unreactive enantiomer is therefore isolated with enantiomer excesses close to *ee* = 100% in almost 50% yield at approximately 50% conversion. The other enantiomer is epoxidized much more rapidly (i.e., almost completely). The epoxide is obtained from it with high diastereoselectivity and can also be isolated with *ee* values of up to 100% in almost 50% yield.

Note: Other examples of very efficient catalytic asymmetric syntheses are certain homogeneously catalyzed hydrogenations. They are discussed in connection with the redox reactions in Section 14.4.7.

3.5 Additions That Take Place Diastereoselectively as *trans* Additions (Additions via Onium Intermediates)

B

trans Additions to olefins are two-step reactions. They occur only for reagents that for certain reasons are unable to form the *final* σ bonds in a one-step addition. Instead they form two *preliminary* σ bonds (Figure 3.32), which are incorporated into a three-membered heterocyclic intermediate. This intermediate carries a positive formal charge on the heteroatom. Thus it contains an "onium" center there. The fact that this intermediate contains two new σ bonds provides a better energy balance for breaking the π bond of the olefin than if an acyclic intermediate were produced, which, of course, would contain only one new σ bond. During the formation of the cyclic in-

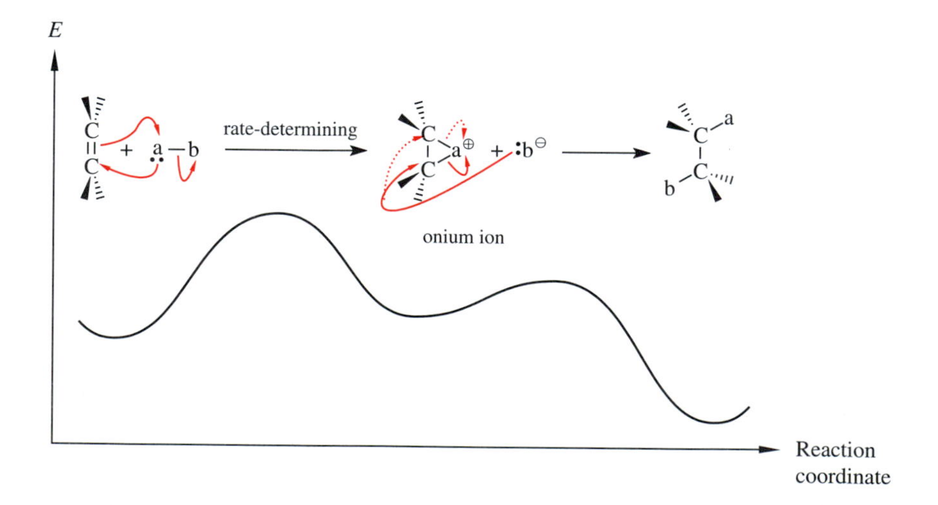

Fig. 3.32. Mechanism and energy profile of *trans*-selective additions to C=C double bonds. The first step, the formation of the onium ion, is rate-determining. The second step corresponds to the S_N2 opening of an epoxide.

termediate, however, it is necessary to compensate for the energy that (a) is required to separate the charges that result when the onium center is formed and (b) must be invested as ring strain. Therefore the first step of *trans* additions to the C═C double bond, i.e., the one that leads to the onium ion is endothermic. The Hammond postulate thereby identifies it as the rate-determining step of the overall reaction (Figure 3.32).

The onium centers of the three-membered intermediates considered in this section are Cl^+, Br^+, I^+, and $Hg^+(OAc)$. In each of these onium ions the ring strain is lower than in an epoxide. This is because the C—Cl^+, C—Br^+, C—I^+, and C—$Hg^+(OAc)$ bonds are considerably longer than the C—O bonds in an epoxide. The reason for this is that the heteroatoms of these onium ions have larger atomic radii than oxygen. At the same time, the C—C bond within the ring of each onium ion is approximately just as long as both the C—C and the C—O bonds of an epoxide ring. The ratio of C-heteroatom bond *length* to the endocyclic C—C bond *length* now defines the bond *angle* in these heterocycles. In an epoxide, the C—C—O and C—O—C angles measure almost exactly 60° and thereby cause the well-known 20 kcal/mol ring strain. On the other hand, in symmetrically substituted onium ions derived from Cl^+, Br^+, I^+, or $Hg^+(OAc)$, the C—C—Het^+ bond angles must be greater than 60° and the C—Het^+—C bond angles must be smaller:

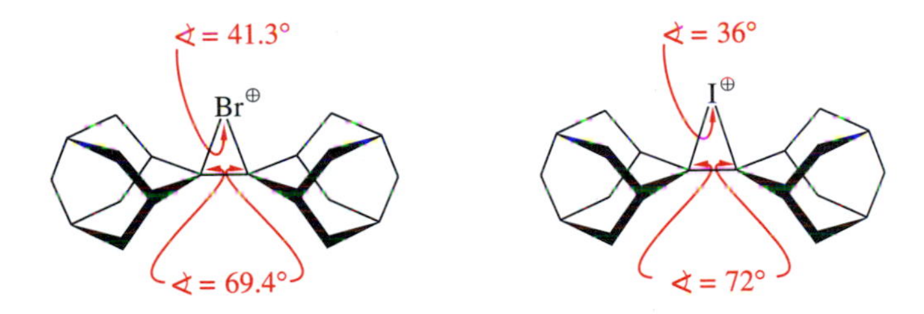

The increase in the C—C—Het^+ angle reduces the angular strain and thus increases the stability. On the other hand, the simultaneous decrease in the C—Het^+—C bond angle hardly costs any energy. According to the VSEPR theory, the larger an atom is, the smaller the bond angle can be between bonds emanating from that atom. Consequently, for atoms such as Cl, Br, I, or Hg, relative to C and O, a smaller bond angle does not result in a significant increase in strain.

In the second step of *trans* additions the onium intermediates react with a nucleophile. Such a nucleophile can be either an anion released from the reagent in the first step or a better nucleophile present in the reaction mixture. It opens the onium intermediate in an S_N2 reaction with backside attack. In this way, the addition product is produced *trans*-selectively (Figure 3.32).

3.5.1 Addition of Bromine

The *trans* addition of bromine to C═C double bonds follows the onium mechanism generally formulated in Figure 3.32. Cyclohexene reacts stereoselectively to give (racemic!) *trans*-1,2-dibromocyclohexane:

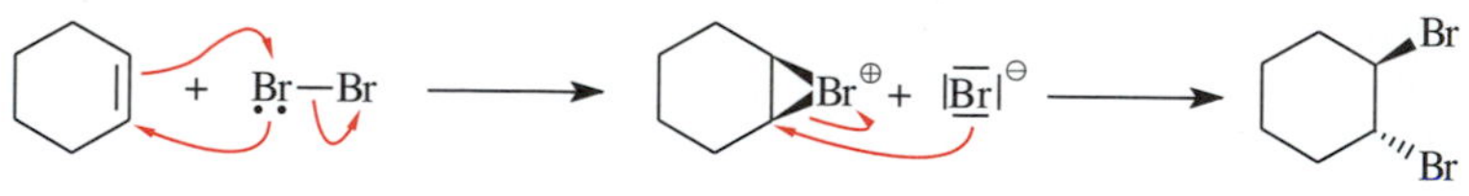

trans-Selectivity is also exhibited by the additions of bromine to fumaric or maleic acid, which follow the same mechanism:

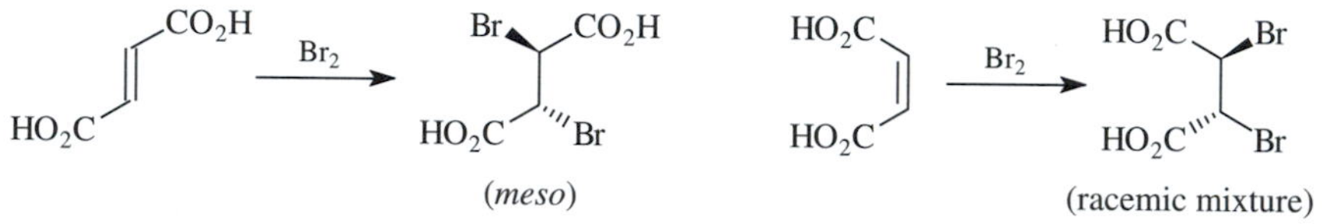

(*meso*)

(racemic mixture)

This is proven by the fact that the reaction with fumaric acid gives *meso*-dibromosuccinic acid, whereas maleic acid gives the *chiral* dibromosuccinic acid, of course as a racemic mixture. Note that these reactions are stereospecific.

Instead of a bromonium ion, in certain cases an isomeric acyclic and sufficiently stable cation can occur as an intermediate of the bromine addition to olefins. This holds true for bromine-containing benzyl cations. Therefore, the bromine addition to β-methyl styrene shown in Figures 3.5 and 3.6 takes place *without* stereocontrol.

The addition of Cl_2 to C=C double bonds is *trans*-selective only when it takes place via three-membered chloronium ions. But it can also take place without stereocontrol, namely, when carbenium ion intermediates appear instead of chloronium ions. This is observed in Cl_2 additions, which can take place via benzyl or *tert*-alkyl cations.

In general, an addition of I_2 to C=C double bonds is thermodynamically impossible, although an iodonium ion can still form.

3.5.2 The Formation of Halohydrins; Halolactonization and Haloetherification

Halohydrins are β-halogenated alcohols. They can be obtained in H_2O-containing solvents from olefins and reagents, which transfer Hal^+ ions. *N*-Bromosuccinimide (transfers Br^+; Figures 3.33 and 3.34 as well as 3.36), chloramine-T (transfers Cl^+; Figure 3.35), and elemental iodine (transfers I^+; Figure 3.36) have this ability. Bromonium and chloronium ions are then attacked by H_2O according to an S_N2 mechanism. This furnishes the protonated bromo- or chlorohydrins, which are subsequently deprotonated.

A

Instead of H_2O, COOH or OH groups of the substrate located at a suitable distance can also open the halonium ion intermediate through a nucleophilic backside attack. In this way, cyclic halohydrin derivatives are produced (Figure 3.36). They are referred to as halolactones or haloethers.

B

With NBS in aqueous DMSO, cyclohexene gives racemic *trans*-2-bromo-1-cyclohexanol. This stereochemical result means that we have a *trans* addition. In the analogous bromohydrin formation from 3,3-dimethylcyclohexene, the analogous dimethylated 2-bromo-1-cyclohexanol is also produced *trans*-selectively as well as regioselectively

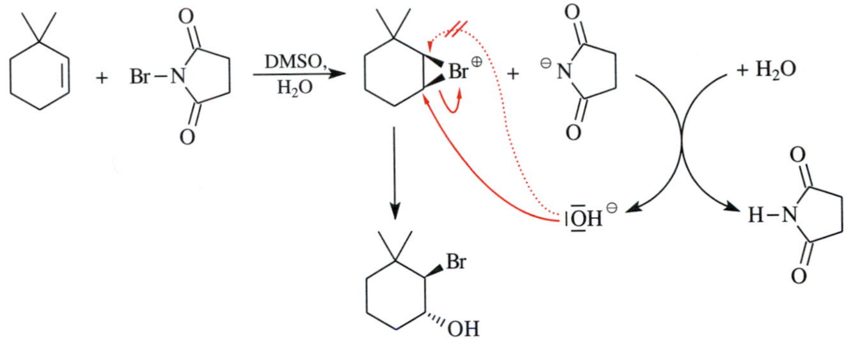

Fig. 3.33. Stereo- and regioselective formation of a bromohydrin from a 1,2-disubstituted ethylene.

(Figure 3.33). In the bromonium ion intermediate the backside attack by the H_2O molecule does not take place at the hindered neopentyl center according to the rules for S_N2 reactivity (Section 2.4.4).

As shown in Figure 3.34, trisubstituted olefins are also converted to bromohydrins by NBS in an aqueous organic solvent. This reaction takes place via a *trans* addition and therefore must take place via a bromonium ion. Once again, the reaction is also regioselective. At first glance, the regioselectivity of this reaction might seem surprising: The H_2O molecule attacks the bromonium ion intermediate at the tertiary instead of at the secondary C atom. One might have expected the backside (S_N2) attack of the nucleophile to be directed at the secondary C atom of the bromonium ion (cf. Section 2.4.4). However, this is not the case. The bromonium ion is very distorted: The C_{sec}—Br^+ bond at 1.9 Å is considerably shorter than and consequently *considerably stronger* than the C_{tert}—Br^+ bond. The latter is stretched to 2.6 Å and thereby weakened. This distortion reduces the ring strain of the bromonium ion. The distortion becomes possible because the stretching of the C_{tert}—Br^+ bond produces a partial positive charge on the tertiary C atom, which is stabilized by the alkyl substituents located there. In this bromonium ion, the bromine atom has *almost* separated from the tertiary ring C atom (but only almost, because otherwise the result would not be 100% *trans* addition).

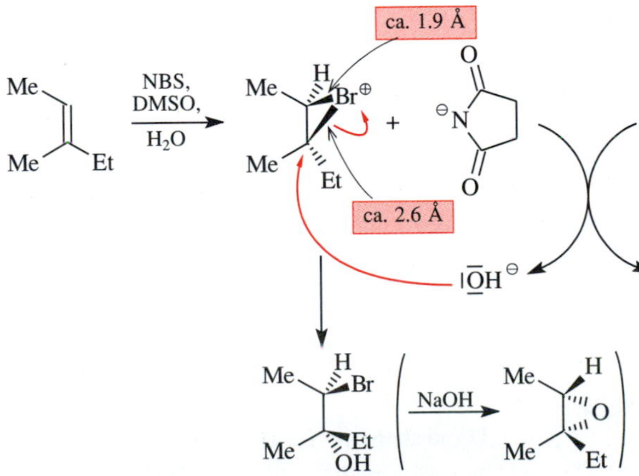

Fig. 3.34. Stereo- and regioselective formation of a bromohydrin from a trisubstituted ethylene; conversion to an epoxide.

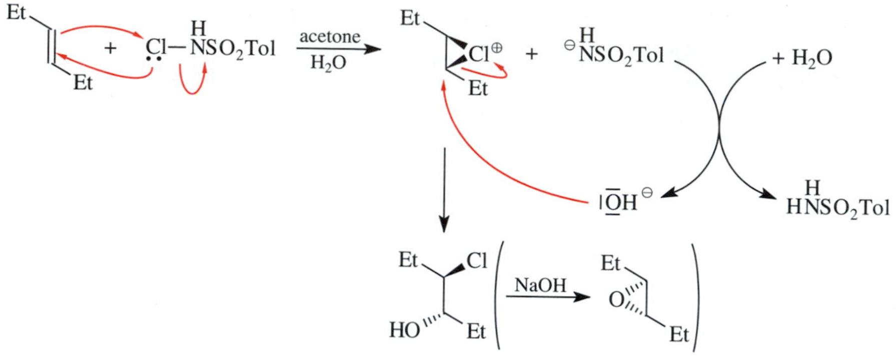

Fig. 3.35. Stereoselective formation of a chlorohydrin; conversion to an epoxide.

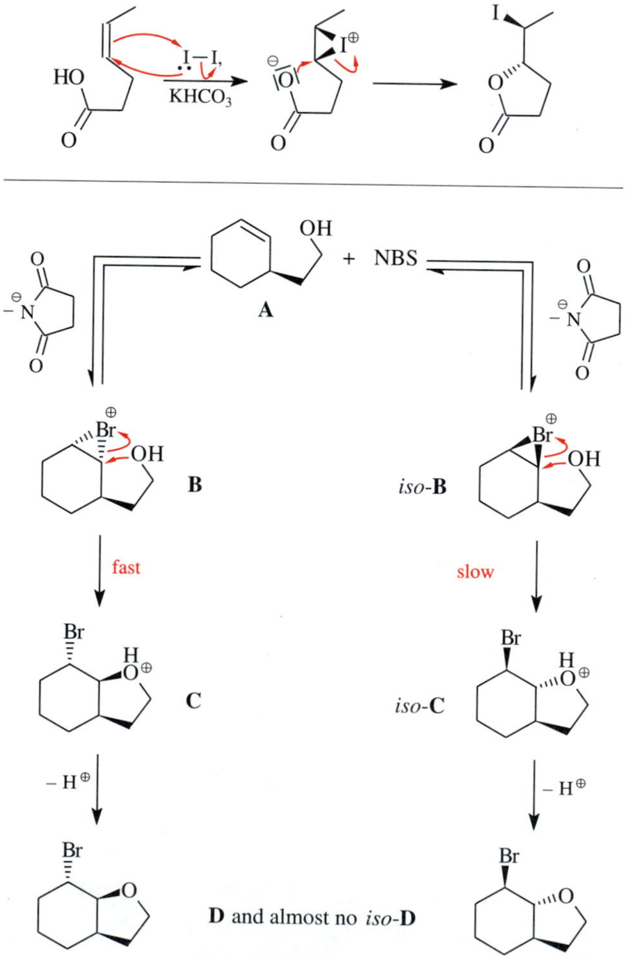

Fig. 3.36. Stereoselective iodolactonization (top) and stereoselective bromoetherification (bottom). See Figure 1.30 for the dehalogenations of the iodolactone and of the bromoether shown.

Olefins and chloramine-T in aqueous acetone form chlorohydrins through another *trans* addition (Figure 3.35).

Halohydrin formation is not only important in and of itself. Bromohydrins (see Figure 3.34) as well as chlorohydrins (Figure 3.35) can be cyclized stereoselectively to epoxides by deprotonation with NaH or NaOH via an intramolecular S_N2 reaction. In this way, epoxides can be obtained from olefins in two steps. Though this route is longer than single-step epoxidation of olefins with percarboxylic acids (Figure 3.14), it has merits: it avoids the use of reagents that in the extreme case might explode.

Similar addition mechanisms explain the so-called halolactonization and the related haloetherification (Figure 3.36). With the help of these reactions one can produce halogenated five- and six-membered ring lactones or ethers stereoselectively. Dehalogenation afterward is possible (Figure 1.30).

The formation of the halonium ion intermediate in halolactonizations and haloetherifications is a reversible step. Therefore, initially, comparable amounts of the diastereomeric bromonium ions **B** and *iso*-**B** are produced from the unsaturated alcohol **A** of Figure 3.36. However, essentially only one of them—namely, **B**—undergoes a nucleophilic backside attack by the alcoholic OH group. The brominated tetrahydrofuran **D** is produced via the oxonium ion **C.** An analogous intramolecular backside attack by the alcoholic OH group in the bromonium ion *iso*-**B** is energetically disfavored and hardly observed. The result is that the bromonium ion *iso*-**B** can revert to the starting material **A,** whereby the overall reaction takes place almost exclusively via the more reactive bromonium ion **B.**

3.5.3 Solvomercuration of Olefins: Hydration of C=C Double Bonds through Subsequent Reduction

Mercury(II) salts add to C=C double bonds (Figure 3.37) in nucleophilic solvents according to the onium mechanism of Figure 3.32. However, the heterocyclic primary product is not called an onium, but rather a mercurinium ion. Its ring opening in an H_2O-containing solvent gives a *trans*-configured alcohol, which can then be demercurated with $NaBH_4$. The hydration product of the original olefin is obtained and is, in our case, cyclohexanol.

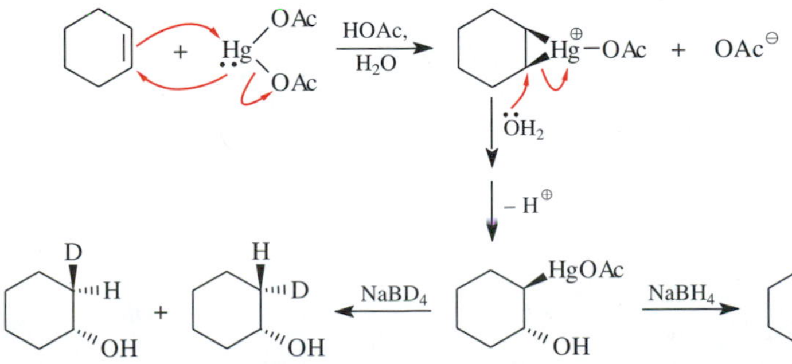

Fig. 3.37. Hydration of a symmetric olefin through solvomercuration/reduction.

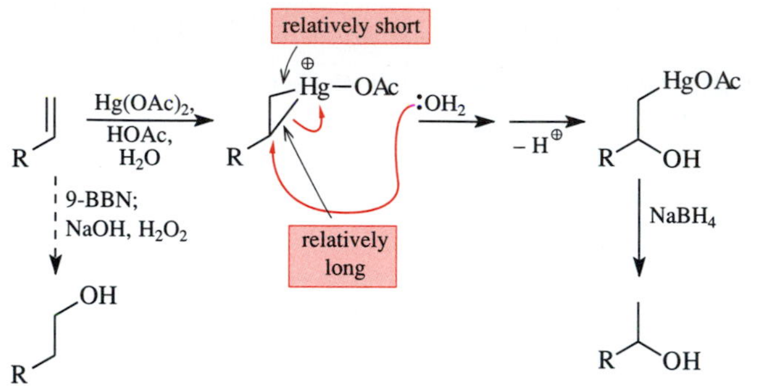

Fig. 3.38. Regioselective hydration of an asymmetric olefin via solvomercuration/reduction. The regioselectivity of the solvomercuration/reduction sequence is complementary to that of hydroboration/oxidation/hydrolysis.

The mechanism of this defunctionalization was discussed in connection with Figure 1.12. It took place via approximately planar radical intermediates. This is why in the reduction of alkyl mercury(II) acetates, the C—Hg bond changes to the C—H bond without stereocontrol. The stereochemical integrity of the mercury-bearing stereocenter is thus lost. When the mercurated alcohol in Figure 3.37 is reduced with $NaBD_4$ rather than $NaBH_4$, the deuterated cyclohexanol is therefore produced as a mixture of diastereomers.

Asymmetrically substituted C=C double bonds are hydrated according to the same mechanism (Figure 3.38). The regioselectivity is high, and the explanation for this is that the mercurinium ion intermediate is distorted in the same way as the bromonium ion in Figure 3.34. The H_2O preferentially breaks the stretched and therefore weakened C_{sec}—Hg^+ bond by a backside attack and does not affect the shorter and therefore more stable C_{prim}—Hg^+ bond.

From the Hg-containing alcohol in Figure 3.38 and $NaBH_4$ one can obtain the Hg-free alcohol. The overall result is a hydration of the C=C double bond. According to the nomenclature of Section 3.3.3, its regioselectivity corresponds to a Markovnikov addition. It is complementary to the regioselectivity of the reaction sequence hydroboration/oxidation/hydrolysis (Table 3.1). The latter sequence would have converted the same olefin regioselectively into the primary instead of the secondary alcohol.

Besides H_2O, simple alcohols or acetic acid can also be added to olefins by solvomercuration/reduction. Figure 3.39 shows MeOH addition as an example. The regioselectivities of this reaction and of the H_2O addition in Figure 3.38 are identical.

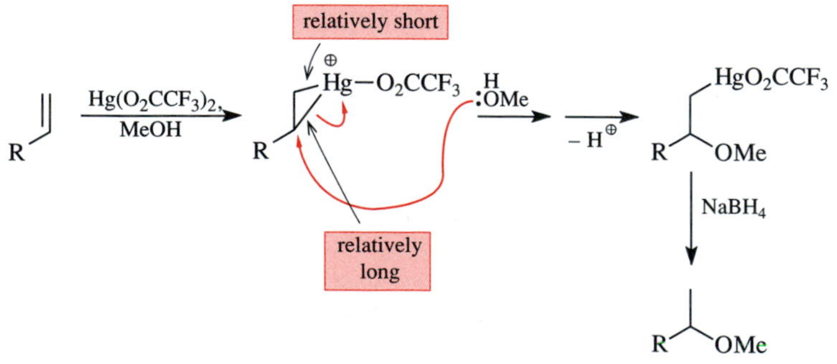

Fig. 3.39. Regioselective methanol addition to an asymmetric olefin via solvomercuration/reduction.

3.6 Additions That Take Place or Can Take Place without Stereocontrol Depending on the Mechanism

3.6.1 Additions via Carbenium Ion Intermediates

Carbenium ions become intermediates in two-step additions to olefins when they are more stable than the corresponding cyclic onium intermediates. With many electrophiles (e.g., Br_2 and Cl_2), this occasionally is the case. With others, such as protons and carbenium ions, it is always the case. (Figure 3.40).

In Section 3.5.1 it was mentioned that Br_2 and Cl_2 form resonance-stabilized benzyl cation intermediates with styrene derivatives and that *gem*-dialkylated olefins react with Br_2 but not with Cl_2 via halonium ions. Because C—Cl bonds are shorter than C—Br bonds, chloronium ions presumably have a higher ring strain than bromonium ions. Accordingly, a β-chlorinated tertiary carbenium ion is more stable than the isomeric chloronium ion, but a β-brominated tertiary carbenium ion is less stable than the isomeric bromonium ion.

Protons and carbenium ions *always* add to C═C double bonds via carbenium ion intermediates simply because no energetically favorable onium ions are available from an attack by these electrophiles. An onium intermediate formed by the attack of a proton would contain a divalent, positively charged H atom. An onium intermediate produced by the attack of a carbenium ion would be a carbonium ion and would thus contain a pentavalent, positively charged C atom. Species of this type are at best detectable under exotic reaction conditions or in the norbornyl cation (Figure 2.25).

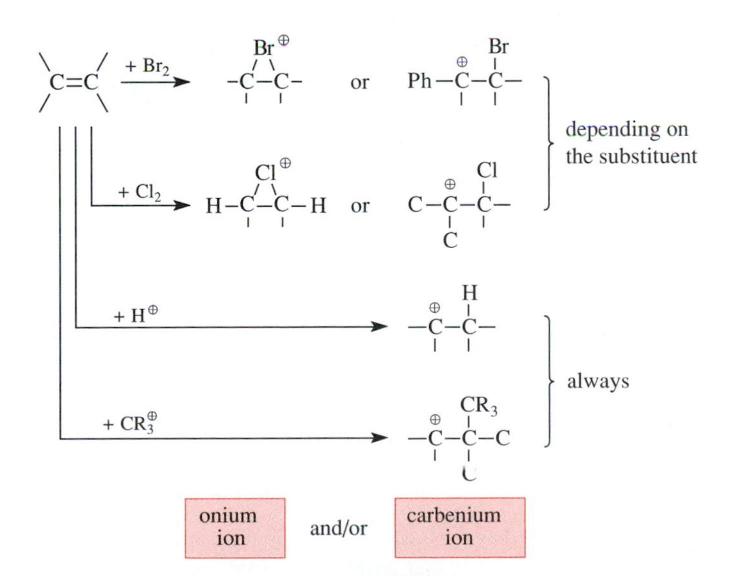

Fig. 3.40. Competition between onium and carbenium ion intermediates in electrophilic additions to C═C double bonds.

Preparatively it is important that mineral acids, carboxylic acids, and *tert*-carbenium ions can be added to olefins via carbenium ion intermediates. Because of their relatively low stability, primary carbenium ions form more slowly in the course of such reactions than the more stable secondary carbenium ions, and these form more slowly than the even more stable tertiary carbenium ions (Hammond postulate!). Therefore, mineral and carboxylic acids add to asymmetric olefins regioselectively to give Markovnikov products (see Section 3.3.3 for an explanation of this term). In addition, these electrophiles add most rapidly to those olefins from which tertiary carbenium ion intermediates can be derived.

An example of an addition of a mineral acid to an olefin that takes place via a tertiary carbenium ion is the formation of a tertiary alkyl bromide from a 1,1-dialkylethylene and HBr (Figure 3.41).

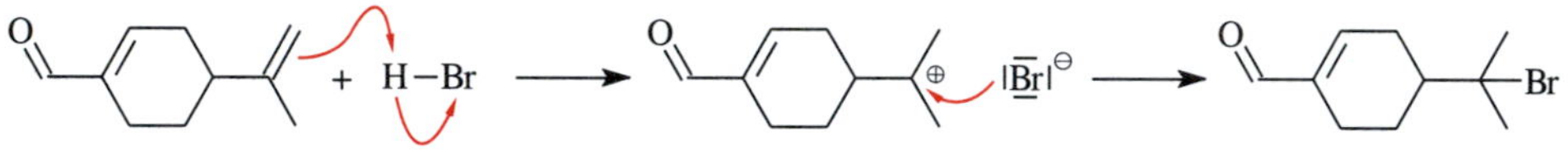

Fig. 3.41. Addition of a mineral acid to an olefin.

Figure 3.42 shows an addition of a carboxylic acid to isobutene, which takes place via the *tert*-butyl cation. This reaction is a method for forming *tert*-butyl esters. Because the acid shown in Figure 3.42 is a β-hydroxycarboxylic acid whose alcohol group adds to an additional isobutene molecule, this also shows an addition of a primary alcohol to isobutene, which takes place via the *tert*-butyl cation. Because neither an ordinary carboxylic acid nor, of course, an alcohol is sufficiently acidic to protonate the olefin to give a carbenium ion, catalytic amounts of a mineral or sulfonic acid are also required here.

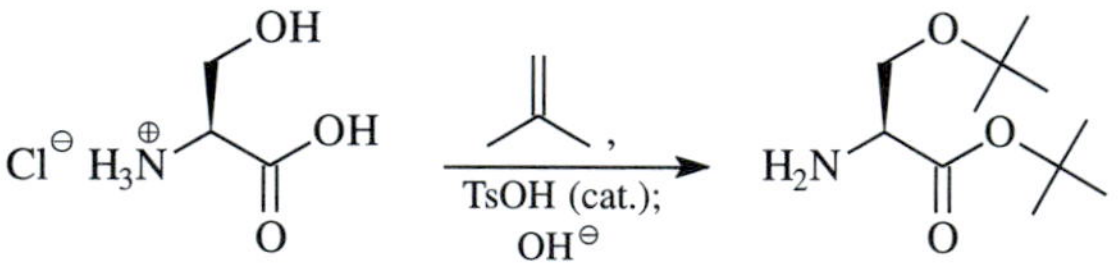

Fig. 3.42. Addition of a carboxylic acid and an alcohol to an olefin.

Side Note 3.1 Catalytic Polymerization of Isobutene

Intermolecular additions of carbenium ions to olefins give polymers. Such a reaction is used in industry, for example, in the cationic polymerization of isobutene (Figure 3.43). One of the rare cases of an intermolecular carbenium ion addition to an olefin without polymer formation occurs in the industrial synthesis of isooctane (Figure 3.44).

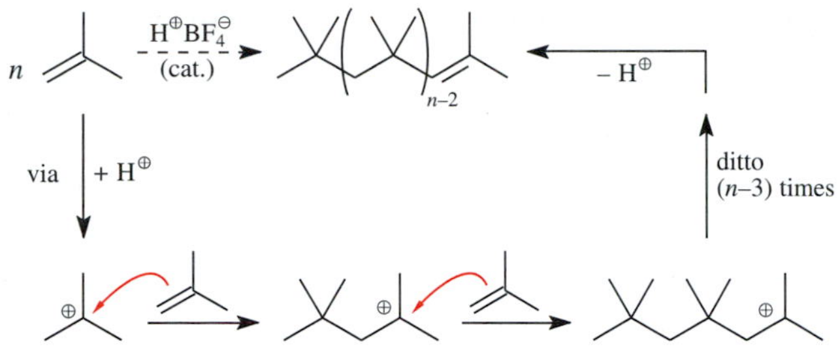

Fig. 3.43. Carbenium ion additions to isobutene as key steps in the cationic polymerization of isobutene. The dashed arrow corresponds to the overall reaction.

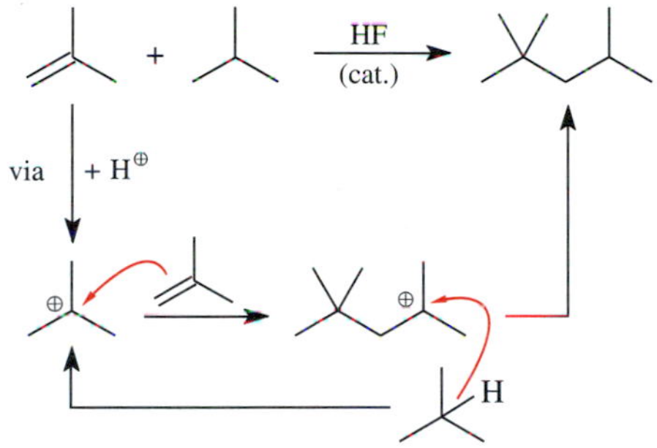

Fig. 3.44. Carbenium ion addition to isobutene as a partial step in the Ipatiev synthesis of isooctane. The upper line shows the overall reaction.

A

Carbenium ion additions to C=C double bonds are of greater preparative importance when they take place intramolecularly as ring closure reactions (see Figure 3.45).

In polyenes even tandem additions are possible. The best known and the most impressive example is the biomimetic synthesis of the steroid structure.

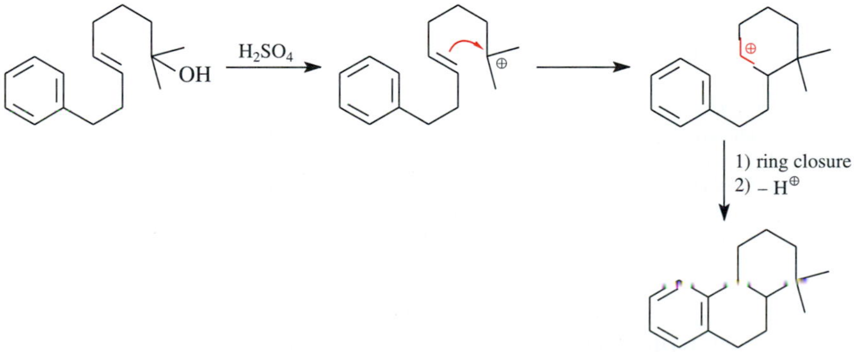

Fig. 3.45. Ring closure reaction by addition of a carbenium ion to a C=C double bond (for the mechanism of the second ring closure, see Section 5.2.5).

3.6.2 Additions via "Carbanion" Intermediates

B

Nucleophiles can be added to acceptor-substituted olefins. In that case, enolates and other "stabilized carbanions" occur as intermediates. Reactions of this type are discussed in this book only in connection with 1,4-additions of ylids (Section 9.2.2), organometallics (Section 8.6), or enolates (Section 10.6) to α,β-unsaturated carbonyl and carboxyl compounds.

References

G. Melloni, G. Modena, U. Tonellato, "Relative Reactivities of C—C Double and Triple Bonds Towards Electrophiles," *Acc. Chem. Res.* **1981,** *14,* 227.

3.3

G. Poli and C. Scolastico, "Formation of C—O Bonds by 1,2-Dihydroxylation of Olefinic Double Bonds," in *Methoden Org. Chem. (Houben-Weyl) 4th ed. 1952-, Stereoselective Synthesis* (G. Helmchen, R. W. Hoffmann, J. Mulzer, E. Schaumann, Eds.), Vol. E21e, 4547–4599, Georg Thieme Verlag, Stuttgart, 1995.

W. Sander, G. Bucher, S. Wierlacher, "Carbenes in matrixes: Spectroscopy, structure, and reactivity," *Chem. Rev.* **1993,** *93,* 1583–1621.

R. R. Kostikov, A. P. Molchanov, A. F. Khlebnikov, "Halogen-containing carbenes," *Russ. Chem. Rev.* **1989,** *58,* 654–666.

M. G. Banwell and M. E. Reum, "gem-Dihalocyclopropanes in Chemical Synthesis," in *Advances in Strain in Organic Chemistry* (B. Halton, Ed.) **1991,** *1,* Jai Press, Greenwich, CT.

E. V. Dehmlow and S. S. Dehmlow, "Phase Transfer Catalysis," 3rd ed., VCH, New York, **1993.**

Y. Goldberg, "Phase Transfer Catalysis: Selected Problems and Applications," Gordon and Bresch, Philadelphia, PA, **1992.**

W. B. Motherwell and C. J. Nutley, "The role of zinc carbenoids in organic synthesis," *Contemporary Organic Synthesis* **1994,** *1,* 219.

J. Adams and D. M. Spero, "Rhodium(II) catalyzed reactions of diazo-carbonyl compounds," *Tetrahedron* **1991,** *47,* 1765–1808.

T. Ye and M. A. McKervey, "Organic Synthesis with α-Diazo Carbonyl Compounds," *Chem. Rev.* **1994,** *94,* 1091–1160.

M. P. Doyle, M. A. McKervey, T. Ye, "Reactions and Syntheses with α-Diazocarbonyl Compounds," Wiley, New York, **1997.**

J. Aube, "Epoxidation and Related Processes," in *Comprehensive Organic Synthesis* (B. M. Trost, I. Fleming, Eds.), Vol. 1, 843, Pergamon Press, Oxford, **1991.**

A. S. Rao, "Addition Reactions with Formation of Carbon-Oxygen Bonds: (i) General Methods of Epoxidation," in *Comprehensive Organic Synthesis* (B. M. Trost, I. Fleming, Eds.), Vol. 7, 357, Pergamon Press, Oxford, **1991.**

R. Schwesinger, J. Willaredt, T. Bauer, A. C. Oehlschlager, "Formation of C—O Bonds by Epoxidation of Olefinic Double Bonds," in *Methoden Org. Chem. (Houben-Weyl) 4th ed. 1952-, Stereoselective Synthesis* (G. Helmchen, R. W. Hoffmann, J. Mulzer, E. Schaumann, Eds.), Vol. E21e, 4599, Georg Thieme Verlag, Stuttgart, **1995.**

H. Heaney, "Oxidation reactions using magnesium monoperphthalate and urea hydrogen peroxide," *Aldrichimica Acta* **1993,** *26,* 35–45.

K. Smith and A. Pelter, "Hydroboration of C=C and Alkynes," in *Comprehensive Organic Synthesis* (B. M. Trost, I. Fleming, Eds.), Vol. 8, 703, Pergamon Press, Oxford, **1991.**

M. Zaidlewicz, "Formation of C—H Bonds by Reduction of Olefinic Double Bonds—Hydroboration and Hydroalumination," in *Methoden Org. Chem. (Houben-Weyl) 4th ed. 1952-, Stere-*

oselective Synthesis (G. Helmchen, R. W. Hoffmann, J. Mulzer, E. Schaumann, Eds.), Vol. E21d, 4396, Georg Thieme Verlag, Stuttgart, **1995.**

M. Zaidlewicz, "Formation of C—O Bonds by Hydroboration of Olefinic Double Bonds Followed by Oxidation," in *Methoden Org. Chem. (Houben-Weyl) 4th ed. 1952-, Stereoselective Synthesis* (G. Helmchen, R. W. Hoffmann, J. Mulzer, E. Schaumann, Eds.), Vol. E21e, 4519, Georg Thieme Verlag, Stuttgart, **1995.**

B. Carboni and M. Vaultier, "Useful synthetic transformations via organoboranes. 1. Amination Reactions," *Bull. Soc. Chim. Fr.* **1995,** *132,* 1003–1008.

J. S. Siegel, "Heterogeneous Catalytic Hydrogenation of C═C and Alkynes," in *Comprehensive Organic Synthesis* (B. M. Trost, I. Fleming, Eds.), Vol. 8, 417, Pergamon Press, Oxford, **1991.**

U. Kazmaier, J. M. Brown, A. Pfaltz, P. K. Matzinger, H. G. W. Leuenberger, "Formation of C—H Bonds by Reduction of Olefinic Double Bonds: Hydrogenation," in *Methoden Org. Chem. (Houben-Weyl) 4th ed. 1952-, Stereoselective Synthesis* (G. Helmchen, R. W. Hoffmann, J. Mulzer, E. Schaumann, Eds.), Vol. E21d, 4239, Georg Thieme Verlag, Stuttgart, **1995.**

3.4

A. Pelter, K. Smith, H. C. Brown, "Borane Reagents," Academic Press, London, **1988.**

H. C. Brown and B. Singaram, "Development of a simple procedure for synthesis of pure enantiomers via chiral organoboranes," *Acc. Chem. Res.* **1988,** *21,* 287.

H. C. Brown and P. V. Ramachandra, "Asymmetric Syntheses via Chiral Organoboranes Based on α-Pinene," in *Advances in Asymmetric Synthesis* (A. Hassner, Ed.), **1995,** *1,* JAI, Greenwich, CT.

H. B. Kagan and J. C. Fiaud, "Kinetic resolution," *Top. Stereochem.* **1988,** *18,* 249.

A. Pfenniger, "Asymmetric epoxidation of allylic alcohols: The Sharpless epoxidation," *Synthesis* **1986,** 89.

R. A. Johnson and K. B. Sharpless, "Addition Reactions with Formation of Carbon-Oxygen Bonds: (ii) Asymmetric Methods of Epoxidation," in *Comprehensive Organic Synthesis* (B. M. Trost, I. Fleming, Eds.), Vol. 7, 389, Pergamon Press, Oxford, **1991.**

R. A. Johnson and K. B. Sharpless, "Asymmetric Oxidation: Catalytic Asymmetric Epoxidation of Allylic Alcohols," in *Catalytic Asymmetric Synthesis* (I. Ojima, Ed.), 103, VCH, New York, **1993.**

T. Katsuki and V. S. Martin, "Asymmetric epoxidation of allylic alcohols: The Katsuki-Sharpless epoxidation reaction," *Org. React.* **1996,** *48,* 1–299.

T. Linker, "The Jacobsen-Katsuki epoxidation and its controversial mechanism," *Angew. Chem.* **1997,** 109, 2150–2152; *Angew. Chem. Int. Ed. Engl.* **1997,** *36,* 2060–2062.

R. M. Hanson, "The synthetic methodology of nonracemic glycidol and related 2,3-epoxy alcohols," *Chem. Rev.* **1991,** *91,* 437–476.

3.5

M. F. Ruasse, "Bromonium ions or β-bromocarbocations in olefin bromination: A kinetic approach to product selectivities," *Acc. Chem. Res.* **1990,** *23,* 87–93.

M.-F. Ruasse, "Electrophilic bromination of carbon–carbon double bonds: structure, solvent and mechanism," *Adv. Phys. Org. Chem.* **1993,** *28,* 207.

R. S. Brown, "Investigation of the early steps in electrophilic bromination through the study of the reaction with sterically encumbered olefins," *Acc. Chem. Res.* **1997,** *30,* 131–138.

S. Torii and T. Inokuchi, "Addition Reactions with Formation of Carbon-Halogen Bonds," in *Comprehensive Organic Synthesis* (B. M. Trost, I. Fleming, Eds.), Vol. 7, 527, Pergamon Press, Oxford, **1991.**

J. Mulzer, "Halolactonization: The Career of a Reaction," in *Organic Synthesis Highlights* (J. Mulzer, H.-J. Altenbach, M. Braun, K. Krohn, H.-U. Reißig, Eds.), VCH, Weinheim, New York, **1991,** 158–164.

A. N. Mirskova, T. I. Drozdova, G. G. Levkovskaya, M. G. Voronkov, "Reactions of N-chloramines and N-haloamides with unsaturated compounds," *Russ. Chem. Rev.* **1989,** *58,* 250–271.

E. Block and A. L. Schwan, "Electrophilic Addition of X-Y Reagents to Alkenes and Alkynes," in *Comprehensive Organic Synthesis* (B. M. Trost, I. Fleming, Eds.), Vol. 4, 329, Pergamon Press, Oxford, **1991.**

T. Hosokawa and S. Murahashi, "New aspects of oxypalladation of alkenes," *Acc. Chem. Res.* **1990,** *23,* 49–54.

3.6

R. C. Larock and W. W. Leong, "Addition of H-X Reagents to Alkenes and Alkynes," in *Comprehensive Organic Synthesis* (B. M. Trost, I. Fleming, Eds.), Vol. 4, 269, Pergamon Press, Oxford, **1991.**

U. Nubbemeyer, "Formation of C—C Bonds by Reactions Involving Olefinic Double Bonds, Addition of Carbenium Ions to Olefinic Double Bonds and Allylic Systems," in *Methoden Org. Chem. (Houben-Weyl) 4th ed. 1952-, Stereoselective Synthesis* (G. Helmchen, R. W. Hoffmann, J. Mulzer, E. Schaumann, Eds.), Vol. E21c, 2288, Georg Thieme Verlag, Stuttgart, **1995.**

H. Mayr, "Carbon-carbon bond formation by addition of carbenium ions to alkenes: Kinetics and mechanism," *Angew. Chem.* **1990,** *102,* 1415–1428; *Angew. Chem. Int. Ed. Engl.* **1990,** *29,* 1371–1384.

R. Bohlmann, "The folding of squalene: An old problem has new results," *Angew. Chem.* **1992,** *104,* 596–598; *Angew. Chem. Int. Ed. Engl.* **1992,** *31,* 582–584.

I. Abe, M. Rohmer, G. D. Prestwich, "Enzymatic cyclization of squalene and oxidosqualene to sterols and triterpenes," *Chem. Rev.* **1993,** *93,* 2189–2206.

S. R. Angle and H. L. Mattson-Arnaiz, "The Formation of Carbon-Carbon Bonds via Benzylic-Cation-Initiated Cyclization Reactions," in *Advances in Carbocation Chemistry* (J. M. Coxon, Ed.) **1995,** *2,* JAI, Greenwich, CT.

D. Schinzer, "Electrophilic Cyclizations to Heterocycles: Iminium Systems," in *Organic Synthesis Highlights II* (H. Waldmann, Ed.), VCH, Weinheim, New York, **1995,** 167–172.

D. Schinzer, "Electrophilic Cyclizations to Heterocycles: Oxonium Systems," in *Organic Synthesis Highlights II* (H. Waldmann, Ed.), VCH, Weinheim, New York, **1995,** 173–179.

D. Schinzer, "Electrophilic Cyclizations to Heterocycles: Sulfonium Systems," in *Organic Synthesis Highlights II* (H. Waldmann, Ed.), VCH, Weinheim, New York, **1995,** 181–185.

C. F. Bernasconi, "Nucleophilic addition to olefins: Kinetics and mechanism," *Tetrahedron* **1989,** *45,* 4017–4090.

Further Reading

K. B. Wilberg, "Bent bonds in organic compounds," *Acc. Chem. Res.* **1996,** *29,* 229–234.

T. Arai and K. Tokumaru, "Present status of the photoisomerization about ethylenic bonds," *Adv. Photochem.* **1996,** *20,* 1–57.

B. B. Lohray, "Recent advances in the asymmetric dihydroxylation of alkenes," *Tetrahedron: Asymmetry* **1992,** *3,* 1317–1349.

V. K. Singh, A. DattaGupta, G. Sekar, "Catalytic enantioselective cyclopropanation of olefins using carbenoid chemistry," *Synthesis* **1997,** 137–149.

I. P. Beletskaya and A. Pelter, "Hydroborations catalyzed by transition metal complexes," *Tetrahedron* **1997,** *53,* 4957—5026.

V. K. Khristov, K. M. Angelov, A. A. Petrov, "1,3-Alkadienes and their derivatives in reactions with electrophilic reagents," *Russ. Chem. Rev.* **1991,** *60,* 39–56.

A. Hirsch, "Addition reactions of Buckminsterfullerene," *Synthesis* **1995,** 895–913.

4

β-Eliminations

4.1 Concepts of Elimination Reactions

4.1.1 The Concepts of α,β- and 1,n-Elimination

Reactions in which two atoms or atom groups X and Y are removed from a compound are referred to as **eliminations** (Figure 4.1). In many eliminations X and Y are removed in such a way that they do not become constituents of one and the same molecule. In other eliminations they become attached to one another such that they leave as a molecule of type X—Y or X═Y or as N≡N. The atoms or groups X and Y can be bound to C atoms and/or to heteroatoms in the substrate. These atoms can be sp^3 or sp^2 hybridized.

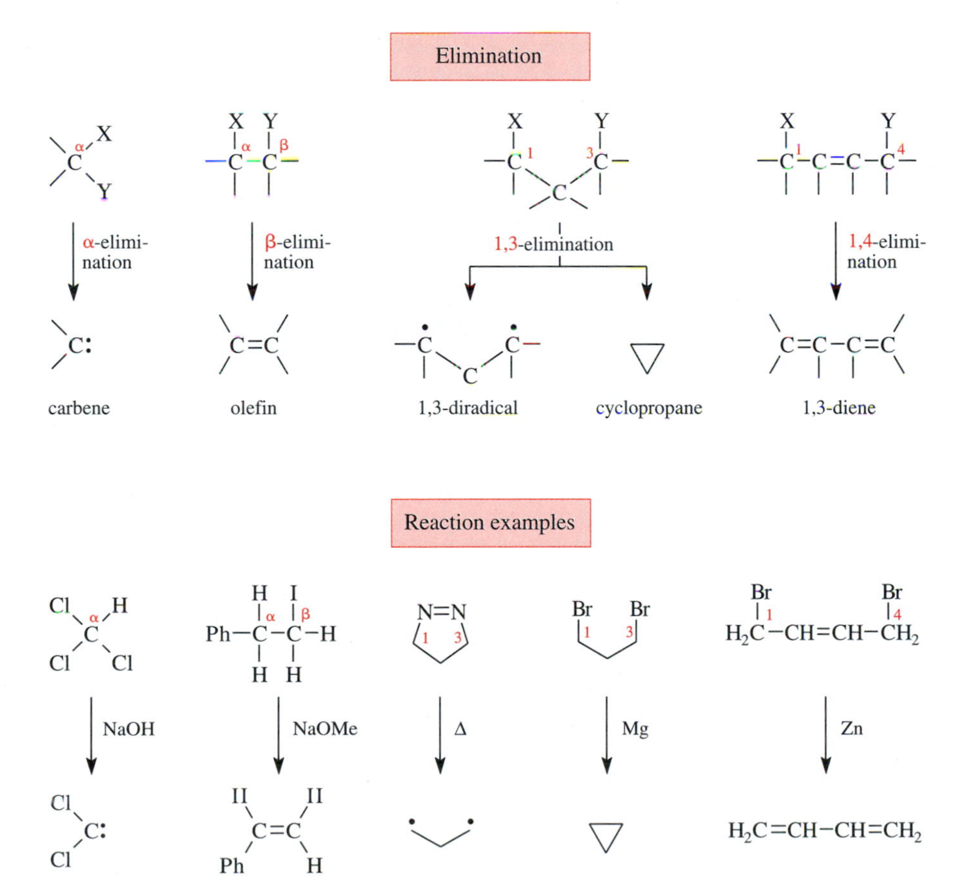

Fig. 4.1. 1,n-Eliminations (n = 1–4) of two atoms or groups X and Y, which are bound to sp^3-hybridized C atoms.

Depending on the distance between the atoms or groups X and Y removed from the substrate, their elimination has a distinct designation. If X and Y are geminal, their removal is an α-elimination. If they are vicinal, it is a β-elimination. If X and Y are separated from each other by *n* atoms, their removal is called **1,*n*-elimination,** that is, 1,3-, 1,4-elimination, and so on (Figure 4.1).

Chapter 4 is limited to a discussion of the most important eliminations, which are the **olefin-forming β-eliminations.**

In this book, **α-eliminations** are described only in connection with their applications: the α-elimination of HCl from $CHCl_3$ + base (Figure 3.11) and the α-elimination of XZnI from carbenoids X—CH_2—ZnI (Figure 3.12) as well as the α-elimination of N_2 from diazomalonic ester (Figure 3.13) in cyclopropanations, the α-elimination of LiBr from carbenoids Br—CR^1R^2—Li in thermal rearrangements (Figures 11.23, 11.29), the α-elimination of N_2 from α-diazoketones in the photochemical Wolff rearrangement (Section 11.3.2), and the related α-elimination of N_2 from azides in the Curtius degradation (Section 11.4.5).

β-Eliminations in which at least one of the leaving groups is removed from a heteroatom are considered to be oxidations. Eliminations of this type are therefore not treated here but in the redox chapter (mainly in Section 14.3.1).

1,3-Eliminations are mentioned in the preparation of 1,3-dipoles such as diazoalkanes or α-diazoketones (Section 12.5.3) and nitrile oxides (Section 12.5.4), in connection with the decomposition of primary ozonides to carbonyl oxides (Section 12.5.5) and the decomposition of phenylpentazole to phenyl azide (Section 12.5.6).

4.1.2 The Terms *syn*- and *anti*-Elimination

B

In various eliminations the mechanism implies a well-defined stereorelationship between the eliminated atoms or groups X and Y and the plane of the resulting C=C double bond (Figure 4.2). For example, X and Y may leave into one and the same half-space flanking this double bond. Their removal is then called a ***syn*- or *cis*-elimination.** (Be careful when using the second term: *cis*-eliminations can also give *trans*- or *E*-olefins!) There are other eliminations where group X leaves the substrate in the direction of one half-space and group Y leaves in the direction of the other half-space, both flanking the C=C double bond produced. These are so-called ***anti*-** or

syn-Elimination

is identical with

anti-Elimination

Fig. 4.2. *Syn*- (= *cis*-) and *anti*- (= *trans*-) eliminations.

***trans*-eliminations.** (Again, be careful about using the second term: *trans*-eliminations also can led to *cis*- or *Z*-olefins!) The third possibility concerns eliminations in which there is no need of an unambiguous spatial relationship between the groups X and Y to be removed and the plane of the resulting double bond.

4.1.3 When Are Stereogenic *syn*- and *anti*-Selective Eliminations Stereoselective?

An elimination in which a *cis,* a *trans,* an *E,* or a *Z* double bond is produced can be called a **stereogenic elimination.** Strictly *syn*- and *anti*-selective stereogenic β-eliminations of X and Y may be, but need not be, stereoselective. They are necessarily stereoselective when in the substrate X and Y are bound to stereocenters. Examples of stereoselectivities of this type are provided by the eliminations in Figure 4.2 for $X \neq R^1 \neq R^2$ and $Y \neq R^3 \neq R^4$. On the other hand, stereoselectivity is not guaranteed when X and Y are removed from substrates possessing the structure $X—CR^1R^2—CY_2—R^3$, that is, from substrates in which Y is not bound to a stereocenter but only X is (because $R^1 \neq R^2 \neq X$).

This last situation characterizes, among others, eliminations from the following kinds of substrates: (1) the entity X to be removed is either a halogen atom or a group linked via a heteroatom to the developing olefinic carbon atom (for the latter case, the abbreviation "Het" will be used for X from now on in Chapter 4), and (2) the entity Y to be removed is an H atom. Accordingly, the substrates now considered possess the structure $Het—CR^1R^2—CH_2—R^3$. If R^1 is different from R^2, H/Het-eliminations from such substrates are stereogenic and *may* be stereoselective (quite independent of the elimination mechanism). As an illustration, we want to discuss three representative examples: stereogenic eliminations from $Het—C(Ph)H—CH_2—R$ (Figure 4.3), from $Het—C(Ph)Me—CH_2—R$ (Figure 4.5), and from $Het—C(Et)Me—CH_2—R$ (Figure 4.6).

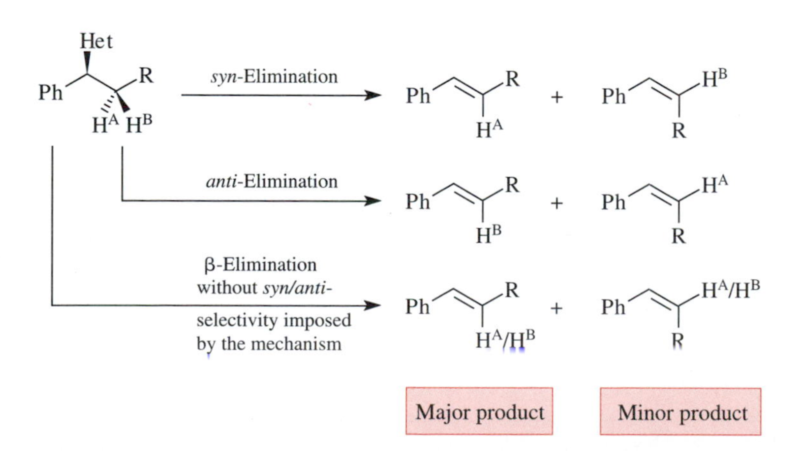

Fig. 4.3. Mechanism-independent occurrence of considerable stereocontrol in β-eliminations of Het and H from a substrate $Het—C(Ph)H—CH_2—R$.

β-Eliminations from the benzyl derivatives of Figure 4.3 take place preferentially to form the *trans*- and only to a lesser extent the *cis*-olefin. This is independent of whether the elimination mechanism is *syn*-selective or *anti*-selective or neither.

This *trans*-selectivity ultimately results from the fact that *trans*-olefins are more stable than their *cis* isomers. This energy difference is especially pronounced for the olefins in Figure 4.3 because they are styrene derivatives. Styrenes with one alkyl group in the *trans* position on the olefinic C═C double bond enjoy the approximately 3 kcal/mol styrene resonance stabilization. This is essentially lacking in *cis*-styrenes because there the phenyl ring is rotated out of the plane of the olefinic C═C double bond to avoid the *cis*-alkyl substituent. However, the *trans*-selectivity documented in Figure 4.3 is not the consequence of thermodynamic control. This could occur only for a reversible elimination or if the olefins could interconvert under the reaction conditions in some other way. Yet, under the conditions of Figure 4.3, olefins are almost always produced irreversibly and without the possibility of a subsequent *cis/trans* isomerization. Therefore, the observed *trans*-selectivity is the result of kinetic control.

Figure 4.4 explains how this takes place. Whether a *cis*- or a *trans*-olefin is formed is determined in the C═C-forming step of each elimination mechanism. For a given mechanism the (immediate) precursor of each of these olefins is one and the same species. This is to be converted either to the *cis*-olefin via one transition state or to the *trans*-olefin via another transition state. Because acyclic *cis*-olefins without a heteroatom substituent always have a higher energy than *trans*-olefins, according to the Hammond postulate the *cis*-transition state must always have a higher energy than the *trans*-transition state. Of course, olefin formation takes place preferentially via the lower energy pathway, and thus it leads to the *trans* isomer. In other words, the *trans*-selectivities of the eliminations of Figure 4.3 are caused by product-development control.

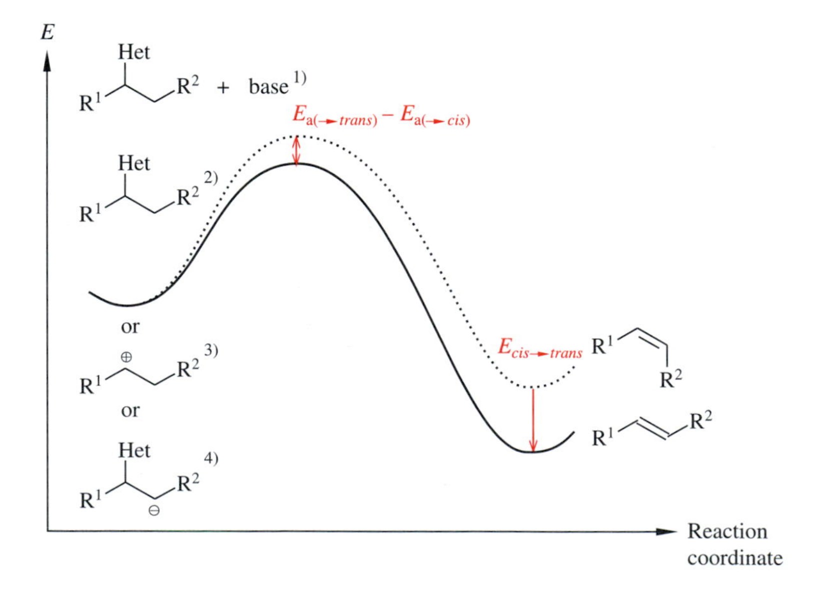

Fig. 4.4. Energy profile of the C═C-forming step of the four mechanisms according to which the β-eliminations of Figure 4.3 can take place in principle as a function of the chemical nature of the substituent Het and the reaction conditions. The conceivable starting materials for this step are, depending on the mechanism, the four species depicted on the left, where [1] is for E2-elimination, [2] is for β-elimination via a cyclic transition state, [3] is for E1-elimination, and [4] is for $E1_{cb}$-elimination.

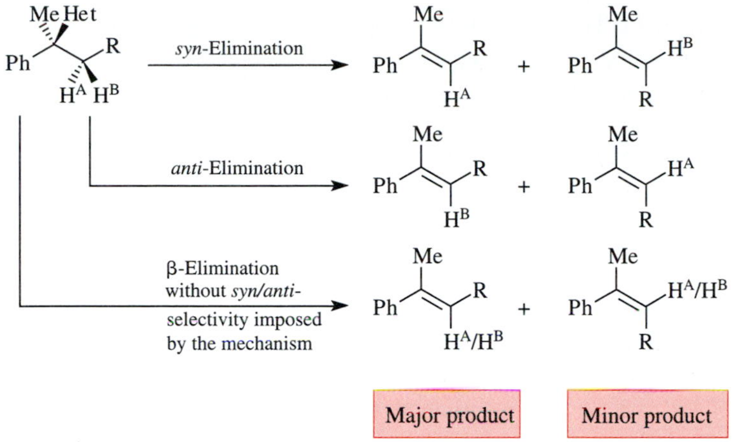

Fig. 4.5. Mechanism-independent occurrence of some stereocontrol in β-eliminations of Het and H from a substrate Het—C(Ph)Me—CH_2—R. Only eliminations in which the removed H atom does not come from the methyl group are considered.

All β-eliminations from the benzyl derivative in Figure 4.5 exhibit a certain stereoselectivity, in this case *E*-stereoselectivity. This is true regardless of whether the elimination is *syn*- or *anti*-selective or neither. The reason for the preferred formation of the *E*-product is again product-development control. This comes about because there is a significant energy difference between the isomeric elimination products due to the presence (*E* isomers) or absence (*Z* isomers) of styrene resonance stabilization.

The basis for the occurrence of product-development control vanishes if there is only a marginal energy difference between *E*,*Z*-isomeric trisubstituted olefins, as for those shown in Figure 4.6. Therefore the corresponding β-eliminations proceed without any stereocontrol.

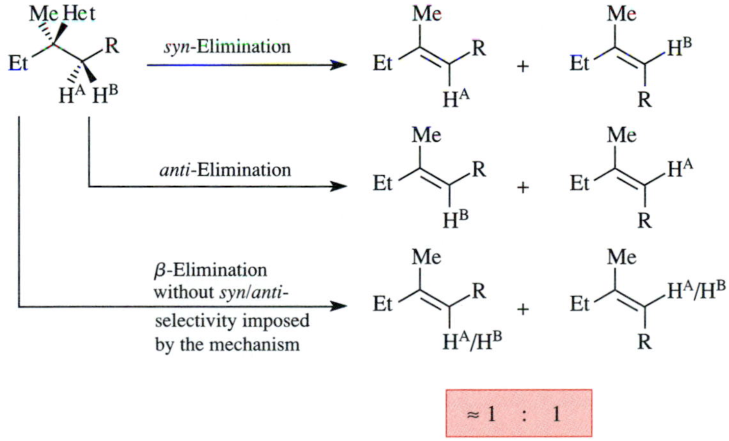

Fig. 4.6. Mechanism-independent absence of stereocontrol in β-eliminations of Het and H from a substrate Het—C(Et)Me—CH_2—R. Only eliminations in which the eliminated H atom comes neither from the ethyl group nor from the methyl group are discussed.

If we consider what has just been said from a different standpoint, we arrive at the following conclusion:

Rule of Thumb for Stereocontrol in Eliminations

Eliminations of H/Het from substrates in which the leaving group (Het) is bound to a stereocenter and the H atom is not, may be suited for the stereoselective synthesis of *trans*- or of *E*-olefins. However, even this kind of stereocontrol is limited to olefins that are *considerably* more stable than their *cis*- or *Z* isomers, respectively. On the other hand, eliminations of this type are never suitable for the synthesis of *cis*- or *Z*-olefins.

4.1.4 Formation of Regioisomeric Olefins by β-Elimination: Saytzeff and Hofmann Product(s)

Substrates in which the leaving group "Het" is bound to a primary C atom can form just a single olefin by the β-elimination of H/Het:

$$\mathrm{H{-}C^{\alpha}_{prim}(Het)(H){-}C^{\beta}{-}H} \xrightarrow{\beta\text{-Elimination}} \mathrm{(H)C^{\alpha}{=}C^{\beta}(H)}$$

On the other hand, in β-eliminations where the leaving group Het is bound to a secondary or a tertiary C atom, a neighboring H atom can be removed from up to two (Figure 4.7) or even up to three (Figure 4.8) constitutionally different positions. In this way up to as many as two or three constitutionally isomeric olefins can be produced. In spite of this, it may be possible to form a single olefin selectively. In such a case the elimination takes place only in one particular direction. One therefore says that it takes place **regioselectively** or that it gives only one of the **regioisomeric olefins.**

If the β elimination of H/Het from R_{sec}—Het can, in principle, deliver (two) regioisomeric olefins whose C=C double bonds (Figure 4.7) contain a different number of alkyl substituents, they are differentiated as Hofmann and Saytzeff products: the Hofmann product is the olefin with the less alkylated double bond, and the Saytzeff product is the olefin with the more alkylated double bond. Because C=C double bonds are stabilized by alkyl substituents, a Hofmann product is in general less stable than its Saytzeff isomer. Accordingly, eliminations of H/Het from R_{sec}—Het, which exhibit product-development control, furnish a Saytzeff product with more or less regioselectivity.

Alternatively, the (two) regioisomeric elimination products of H/Het from R_{sec}—Het could also have the same number of alkyl substituents on their C=C double bonds. In such a case the regioselective production of one olefin at the expense of the other

$$\underset{\mathbf{A}}{\mathrm{C^{\beta'}{=}C^{\alpha}_{sec}(H){-}C^{\beta}{-}H}} \xleftarrow{\beta'\text{-elimination}} \mathrm{H{-}C^{\beta'}{-}C^{\alpha}_{sec}(Het)(H){-}C^{\beta}{-}H} \xrightarrow{\beta\text{-elimination}} \underset{\mathbf{B}}{\mathrm{H{-}C^{\beta'}{-}C^{\alpha}_{sec}(H){=}C^{\beta}}}$$

Fig. 4.7. Regioselectivity of the elimination of H/Het from R_{sec}—Het. When $C^{\beta'}$ has fewer alkyl substituents than C^{β}, one distinguishes between the constitutionally isomeric products as Hofmann product (**A**) and Saytzeff product (**B**).

Me Et Pr + Et Me Pr ← β′-Elimination — H Het H Me C Et H H H H Pr (β′, α, β, β″) — β-Elimination → Me Et Pr + Me Pr Et

↓ β″-Elimination

Me Et Me Et Pr + Pr

Fig. 4.8. An unsolvable regio- and stereoselectivity problem: β-elimination of H/Het from EtPrBuC—Het.

is generally not possible because the basis for product-development control is missing, namely, a clear stability difference between the regioisomers.

Tertiary substrates R_{tert}—Het can give a maximum of three regioisomeric olefins by β-elimination of H/Het and achieving regiocontrol can become an unsolvable problem. For example, it is impossible to eliminate H/Het regioselectively by any method from the substrate shown in Figure 4.8. Moreover, the discussion in Section 4.1.3 showed that none of the three regioisomers would be obtained stereoselectively. Consequently none of the six olefins can be produced selectively from the substrate specified in Figure 4.8.

4.1.5 The Synthetic Value of Het^1/Het^2 in Comparison to H/Het Eliminations

B

If one wants to obtain olefins by an H/Het elimination from R—Het, there are evidently unpleasant limitations with respect to regiocontrol (Section 4.1.4) and stereocontrol (Section 4.1.3). Most of these limitations disappear when the same olefin is synthesized by a β-elimination of Het^1/Het^2:

$$-C^{\beta'}-C^{\alpha}(\text{Het}^1)(-C^{\beta''}-)-C^{\beta}(\text{Het}^2)- \xrightarrow{\beta\text{-Elimination}} -C^{\beta'}-C^{\alpha}(-C^{\beta''}-)=C^{\beta}$$

Rigorous regiocontrol in β-elimination of Het^1/Het^2 is available because these functional groups occur in the substrate only *once* in the β position to each other.

In addition, many Het^1/Het^2 eliminations have the advantage that their mechanism allows for *syn*-selectivity (Sections 4.7.3–4.7.5) or *anti*-selectivity (Section 4.7.3). When such a β-elimination is carried out starting from a substrate in which both Het^1 and Het^2 are bound to a stereocenter, 100% stereoselectivity is observed.

Figure 4.9 shows Het[1]/Het[2] eliminations in the same carbon framework in which the H/Het eliminations of Figure 4.8 led to a mixture of six isomeric olefins. In Figure 4.9 *every* Het[1]/Het[2] elimination leads to a single olefin isomer. In this way, each component of the olefin mixture in Figure 4.8 can be obtained as a pure isomer! Still, one difficulty must not be ignored: it is often more laborious to prepare the substrate for a Het[1]/Het[2] elimination than for an H/Het elimination. If one compares the olefin precursors from Figure 4.9 with those from Figure 4.8, one realizes that clearly.

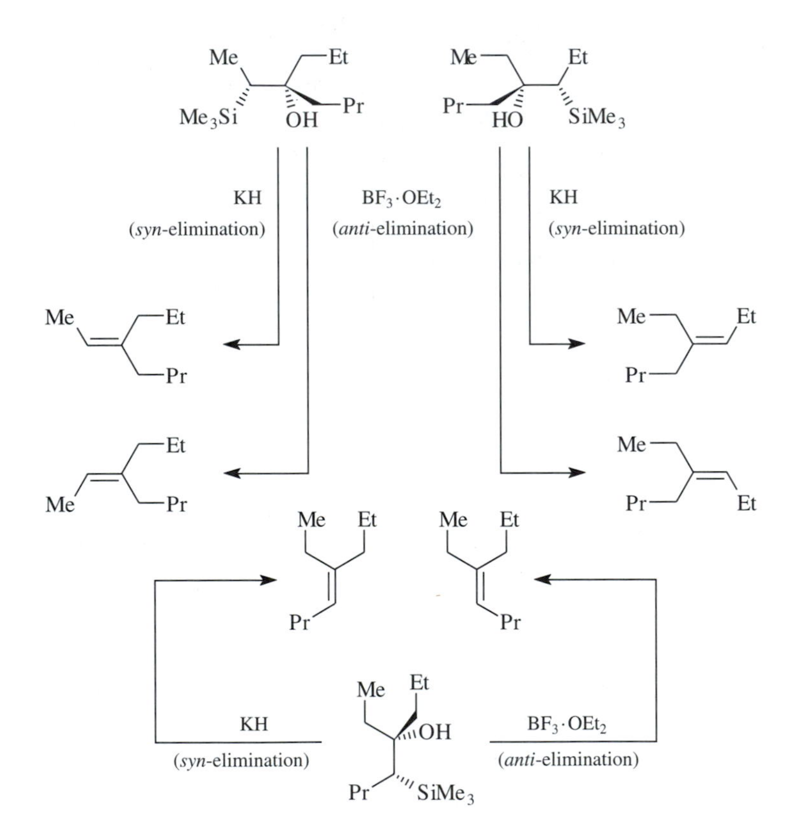

Fig. 4.9. A solution to the regio- and stereoselectivity problems depicted in Figure 4.8 with the help of the Peterson elimination (for the mechanism, see Figure 4.38) as an example of either a *syn*- or an *anti*-selective Het[1]/Het[2] elimination, depending on the reaction conditions.

4.2 β-Eliminations of H/Het via Cyclic Transition States

B

The thermal decomposition (pyrolysis) of alkylaryl selenoxides **(selenoxide pyrolysis)** to an olefin and an aryl selenic acid Ar—Se—OH often takes place even at room temperature (Figure 4.10). This reaction is one of the mildest methods for introducing a C=C double bond by means of a β-elimination. The mechanism is described by the simultaneous shift of three electron pairs in a five-membered cyclic

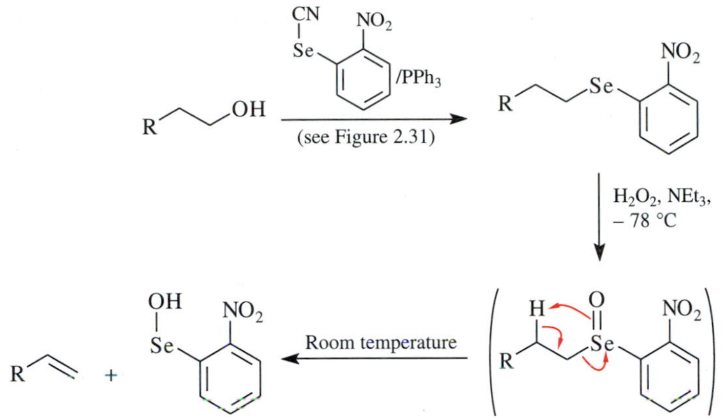

Fig. 4.10. Selenoxide pyrolysis for the dehydration of primary alcohols.

transition state. One of these electron pairs becomes a nonbonding electron pair on the Se atom. The selenium is consequently reduced in the course of the pyrolysis.

Conveniently, the required alkylaryl selenoxide does not have to be isolated. Instead it is produced in solution by a low-temperature oxidation of the corresponding alkylaryl selenide and eliminated in the same synthetic operation (i.e., through a "one-pot reaction") by thawing or gently heating the reaction mixture.

The reaction from Figure 4.11 proves that a selenoxide pyrolysis is a *syn*-elimination. The cyclohexylphenyl selenoxide shown reacts regioselectively to produce the less stable Hofmann product **(D)**. The Saytzeff product **(E)** is not produced at all, although it is more stable than **D.** This observation shows that the transition state **C** of an *anti-*

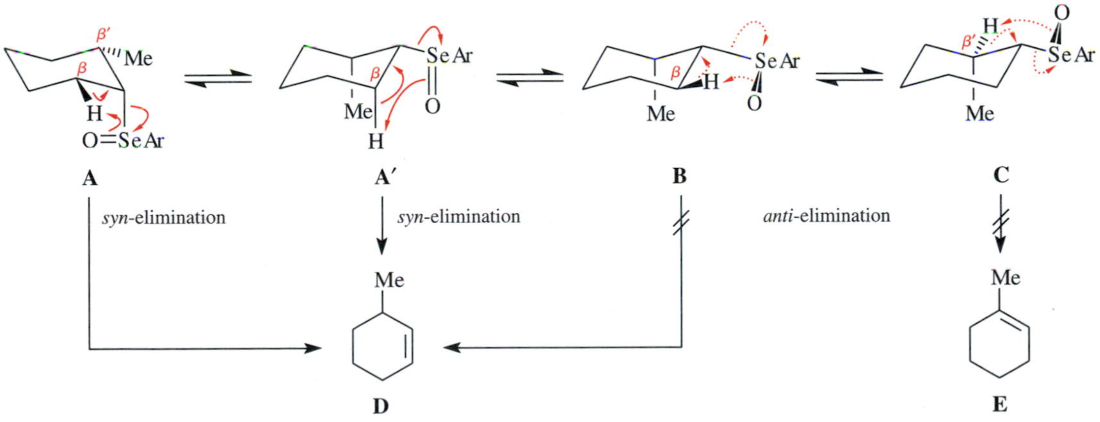

Fig. 4.11. Regiocontrol in a selenoxide pyrolysis based on its mechanism-imposed *syn* selectivity.

elimination (→**E**) has a higher energy than the transition states **A** or **A′** of the conceivable *syn*-eliminations leading to **D.** With the *anti*-transition state **C** being so disfavored, the Hofmann product **D** should, of course, not form via transition state **B** either, because in that case it would also originate from an *anti*-elimination.

Certain alkylaryl selenides can be prepared by the electrophilic selenylation of enolates (Figure 4.12; also see Table 10.4). With a subsequent H_2O_2 oxidation to produce the selenoxide followed by the elimination of Ph—Se—OH, one proceeds in a total of two synthetic steps from a carbonyl or carboxyl compound to its α,β-unsaturated analogue.

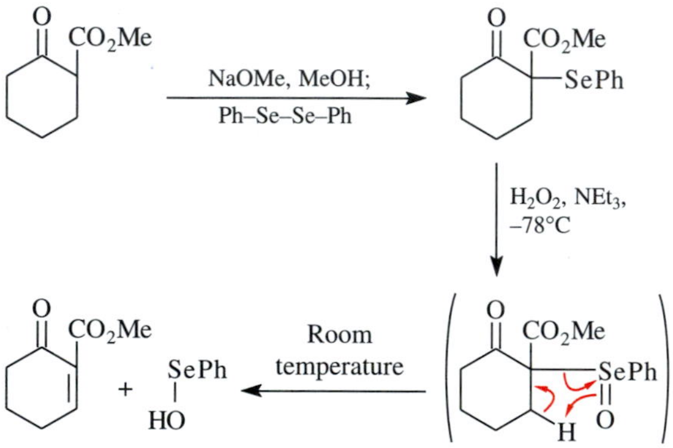

NaOMe, MeOH; Ph–Se–Se–Ph

H_2O_2, NEt_3, −78°C

Room temperature

Fig. 4.12. "Dehydrogenation" of carbonyl or carboxyl compounds via their α-phenylselenyl derivatives.

In the **Chugaev reaction** (Figure 4.13), *O*-alkyl-*S*-methyl xanthates are pyrolyzed to an olefin, carbon oxysulfide, and methanethiol at 200°C.

In contrast to selenoxide pyrolysis, xanthate pyrolysis takes place via a *six-membered* transition state. As in selenoxide pyrolysis, in xanthate pyrolysis three electron pairs are shifted at the same time. However, the additional atom in the ring provides a stereochemical flexibility which the selenoxide pyrolysis does not have: The Chugaev reaction is not necessarily *syn*-selective (Figure 4.14). Still, the favored transition state is the one in which the leaving group and the *syn* H atom in the β-position leave the substrate, that is, a *syn*-transition state. However, in the transition state of the Chugaev reaction all bonds of the partial structure $C\alpha \cdots O \cdots C(SMe) \cdots S \cdots H \cdots C_\beta$ are long enough that the leaving group and an *anti* H atom in the β position can leave the substrate, jointly as a thiocarboxylic acid half-ester.

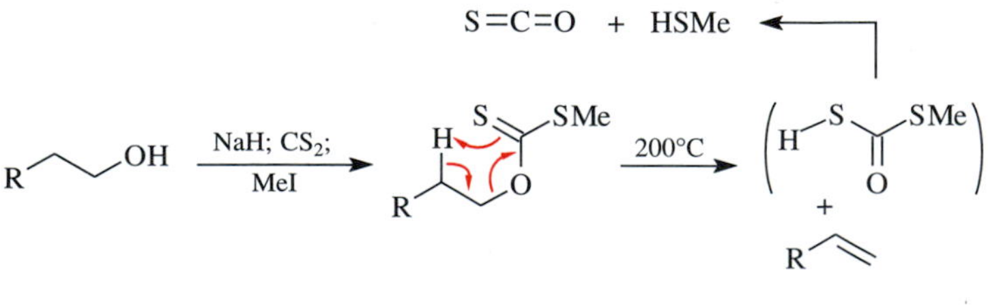

S=C=O + HSMe

NaH; CS_2; MeI

200°C

Fig. 4.13. The Chugaev reaction for the dehydration of alcohols.

Fig. 4.14. Competition of *syn*- and *anti*-elimination in the Chugaev reaction and corresponding regioselectivities.

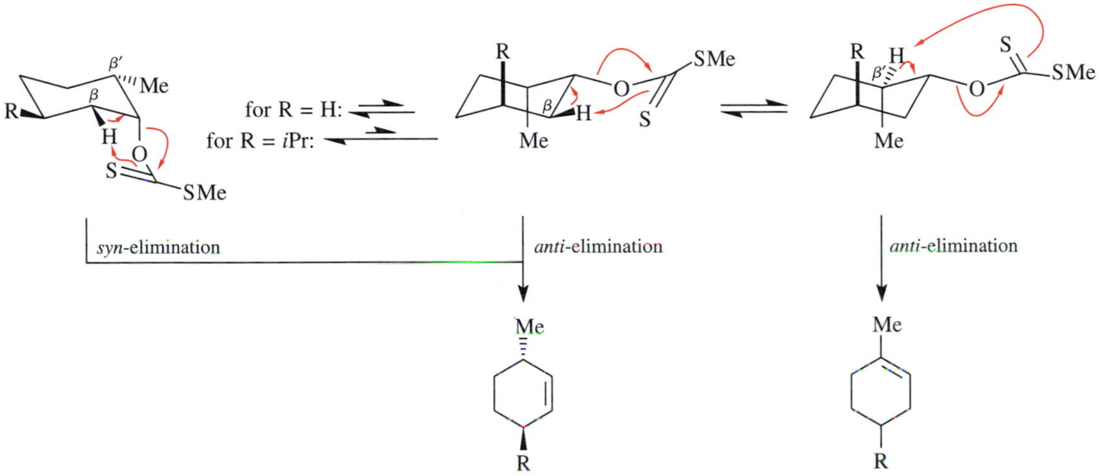

Figure 4.14 illustrates the competition of such *syn*- vs *anti*-pathways starting from two differently substituted cyclohexyl xanthates. The one with the substituent R = isopropyl contains a so-called **conformational lock,** at least to a certain extent. This term refers to a substituent that fixes a molecule predominantly in a single conformation while this molecule could assume several preferred conformations in the absence of this substituent. For example, an α-branched alkyl group such as an isopropyl group (cf. Figure 4.14) or, even better, a *tert*-butyl group acts as a conformational lock on the cyclohexane ring. This group fixes cyclohexane preferentially in the chair conformation in which the group is oriented equatorially. Due to the presence of this conformational lock R = isopropyl, the xanthate group of the second substrate of Figure 4.14 is oriented axially. As a consequence, the H atoms in the β- or β'-positions that would have to participate in an *anti*-elimination are so far removed that they cannot be reached in a cyclic transition state. Therefore, only a highly *syn*-selective elimination is possible. This steric course explains why the thermodynamically less stable Hofmann product is produced regioselectively in the second elimination in Figure 4.14 (R = isopropyl). The first elimination in Figure 4.14 (R = hydrogen) starts from a cyclohexyl xanthate devoid of a conformational lock. Fifty percent of the resulting olefin mixture represents the thermodynamically more stable Saytzeff product. It stems from an *anti*-elimination in a substrate conformer with an equatorially disposed xanthate group because the equally equatorially disposed *anti*-H at $C_{\beta'}$ *is* within reach in the corresponding cyclic transition state.

4.3 β-Eliminations of H/Het via Acyclic Transition States: The Mechanistic Alternatives

B

Let us now turn to β-eliminations that take place via acyclic transition states. One distinguishes between three elimination mechanisms (Figure 4.15) depending on the order in which the C—H and the C—Het bonds of the substrate are broken. If both bonds are broken at the same time, one deals with one-step E2 elimination. When first one and then the other bond is broken, we have two-step eliminations. These can take place according to the E1 or the $E1_{cb}$ mechanism. In the E1 mechanism the C—Het bond is broken first and the C—H bond is broken second. Conversely, in the $E1_{cb}$ mechanism the C—H bond is broken first, by deprotonation with a base. In this way the conjugate base (cb) of the substrate is produced. Subsequently, the C—Het bond breaks.

According to the definition given above, E2 eliminations are one-step eliminations. Still, in an E2 transition state the C—H bond can be broken to a different extent than the C—Het bond. If the C—H bond is broken to a greater extent than the C—Het bond, we have an E2 elimination with an $E1_{cb}$-like distortion of the transition state geometry. Such transition states exhibit characteristic partial charges. In the $E1_{cb}$-like distorted E2 transition state a small negative charge develops on the C atom connected to the attacked H atom. In the E1-like distorted E2 transition state, the C atom linked to the leaving group carries a small positive charge. E2 eliminations of good leaving groups with weak bases are candidates for E1-like transition states. Conversely, E2 eliminations of poor leaving groups with a strong base tend to proceed via an $E1_{cb}$-like transition state.

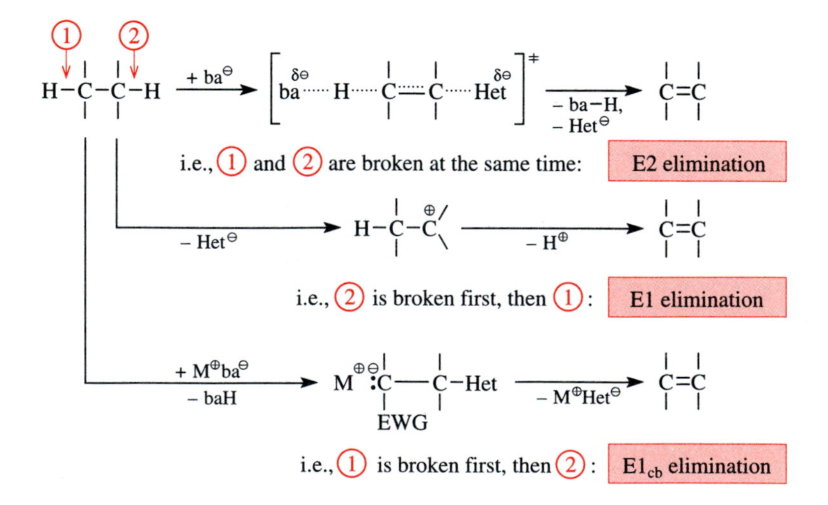

Fig. 4.15. The three mechanisms of H/Het elimination via acyclic transition states: ba^-, the attacking base; baH, the protonated base; EWG, electron-withdrawing group.

4.4 E2 Eliminations of H/Het and the E2/S_N2 Competition

B

The rate of formation of the olefin in E2 eliminations is described by Equation 4.1. It shows the *bi*molecularity of this reaction, which is responsible for the short-hand notation of its mechanism. Typical substrates for E2 eliminations are alkyl halides and sulfonates:

$$\text{H–C–C–Y} + X^{\ominus} \xleftarrow[\text{(one-step)}]{k_{S_N2}} \text{H–C–C–X} + \text{Y:}^{\ominus} \xrightarrow[\text{(one-step)}]{k_{E2}} \text{C=C} + \text{H–Y} + X^{\ominus}$$

The removal of H—Hal or H—O_3S—R from these substrates is effected by bases Y^-. They make their electron pair available to the H atom located in the β position to the leaving group. This allows the H atom to be transferred to the base as a proton.

In principle, the bases Y^- are also nucleophiles, and, hence, they can attack the same alkyl halides and sulfonates via the S_N2 mechanism. The point of attack is the C atom that bears the leaving group. In order to carry out E2 eliminations chemoselectively, competing S_N2 reactions must be excluded. To understand the outcome of the competition (E2 elimination vs S_N2 reaction), it is analyzed kinetically with Equations 4.1–4.3.

$$\frac{\text{d [elimination product]}}{\text{d}t} = k_{E2}\,[\text{RX}][\text{Y}^-] \qquad (4.1)$$

$$\frac{\text{d [substitution product]}}{\text{d}t} = k_{S_N2}\,[\text{RX}][\text{Y}^-] \qquad (4.2)$$

Division of Eq. 4.1 by Eq. 4.2 yields

$$\frac{\text{d [elimination product]}}{\text{d [substitution product]}} = \frac{k_{E2}}{k_{S_N2}} = \frac{\text{\% elimination product}}{\text{\% substitution product}} \qquad (4.3)$$

According to Equation 4.3, the yield ratio of E2 to S_N2 product equals the ratio of the rate constants k_{E2}:k_{S_N2}. This ratio depends on the substrate structure, the nature of the added base, and stereoelectronic factors. The individual influences will be investigated in more detail in Sections 4.4.1–4.4.3.

4.4.1 Substrate Effects on the E2/S_N2 Competition

B

Tables 4.1 and 4.2 summarize the typical substrate effects on the chemoselectivity of E2 vs S_N2 reactions. These substrate effects are so pronounced because NaOEt was used as base. As a reasonably strong base and a quite good nucleophile, NaOEt is able to convert a fair portion of many elimination substrates into S_N2 products (see Section 4.4.2).

Table 4.1. Effect of Alkyl Groups in the α Position of the Substrate on the Result of the E2/S_N2 Competition

$$\text{R—Br} \xrightarrow[\text{in EtOH, 55°C}]{\geq 1\ \text{M NaOEt}} \text{R—OEt} + \text{Olefin}$$

Substrate	k_{S_N2} [10^{-5} l mol^{-1} s^{-1}]	k_{E2} [10^{-5} l mol^{-1} s^{-1}]	k_{E2} (per β-H) [10^{-5} l mol^{-1} s^{-1}]	Olefin formed	Olefin fraction
EtBr	118	1.2	0.4	ethene	1%
*i*PrBr	2.1	7.6	1.3	propene	79%
tert-BuBr	≪ 2.1	79	8.8	isobutene	100%

Table 4.1 gives the chemoselectivity of E2 eliminations from representative bromides of the type R_{prim}—Br, R_{sec}—Br, and R_{tert}—B r. The fraction of E2 product increases in this sequence from 1 to 79 and to 100% and allows for the following generalization: E2 eliminations with sterically unhindered bases can be carried out chemoselectively (i.e., without a competing S_N2 reaction) only starting from tertiary alkyl halides and sulfonates. To obtain an E2 product from primary alkyl halides and sulfonates at all or to obtain an E2 product from secondary alkyl halides and sulfonates exclusively, one must change the base (see Section 4.4.2).

Table 4.1 also allows one to identify reasons for the chemoselectivities listed therein. According to Equation 4.3, they equal the ratio k_{E2}:k_{S_N2} of the two competing reactions. According to Table 4.1, this ratio increases rapidly in the sequence R_{prim}—Br, R_{sec}—Br, and R_{tert}—Br because the k_{S_N2} values become smaller and smaller in this order (for the reason, see Section 2.4.4) while the k_{E2} values become larger and larger.

There are two reasons for this increase in the k_{E2} values. The first reason is a statistical factor: the number of H atoms located in β-position(s) relative to the leaving group and that can be eliminated together with it is 3, 6, and 9, respectively, in the three bromides discussed. One can adjust the gross k_{E2} values of Table 4.1 for this factor by converting them to k_{E2}-values per *single* H atom in a β-position. However, these numbers—"k_{E2} (per β H atom)" in Table 4.1—still increase in the series R_{prim}—Br → R_{sec}—Br → R_{tert}—Br. This second effect is due to the fact that the E2 eliminations considered lead to C=C double bonds with an increasing number of alkyl substituents: EtBr results in an unsubstituted, *i*PrBr in a monosubstituted, and *tert*-BuBr in a disubstituted alkene. The stability of olefins, as is well known, increases with the degree of alkylation. A certain fraction of this stability increase becomes noticeable in the transition state of E2 eliminations in the form of product-development control: The rate constant k_{E2} (per β H atom) is therefore smallest for the most unstable E2 product (ethene) and largest for the most stable E2 product (isobutene).

Table 4.2. Effect of Alkyl Groups in the β Position in the Substrate on the Result of the E2/S_N2 Competition

$$\text{R—Br} \xrightarrow[\text{in EtOH, 55°C}]{\text{0.1 M NaOEt}} \text{R—OEt} + \text{Olefin}$$

Substrate	k_{S_N2} [10^{-5} l mol^{-1} s^{-1}]	k_{E2} [10^{-5} l mol^{-1} s^{-1}]	k_{E2} (per β-H) [10^{-5} l mol^{-1} s^{-1}]	Olefin formed	Olefin fraction
Br	172	1.6	0.53		1%
Br	54.7	5.3	2.7		9%
Br	5.8	8.5	8.5		60%

Table 4.2 compiles the chemoselectivities of E2 eliminations starting from Me—CH_2Br and the simplest bromides of types R_{prim}—CH_2Br and R_{sec}—CH_2Br. It shows how an increasing number of β substituents influences the E2/S_N2 competition: the more numerous they are, the greater is the fraction of E2 product. The reasons for this finding are the same as the reasons given in the discussion of Table 4.1 for the influence of the increasing number of α substituents on the result of the E2/S_N2 competition. All of the corresponding quantities—k_{S_N2}-, gross k_{E2}- and k_{E2} (per β H atom) values—are contained in Table 4.2.

4.4.2 Base Effects on the E2/S_N2 Competition

Chemoselective E2 eliminations can be carried out with sterically hindered, sufficiently strong bases. Their bulkiness causes them to attack at an H atom at the periphery of the molecule rather than at a C atom deep within the molecule. These bases are therefore called **nonnucleophilic bases.** The weaker nonnucleophilic bases include the bicyclic amidines DBN (diazabicyclononene) and DBU (diazabicycloundecene). These can be used to carry out chemoselective E2 eliminations even starting from primary and secondary alkyl halides and sulfonates (Figure 4.16).

β-Eliminations of epoxides lead to allyl alcohols after release of a high-energy (cf. Section 2.3) alkoxide ion as a leaving group. For this reaction to take place, the strongly

DBN DBU Cl DBN

Fig. 4.16. Relatively weak nonnucleophilic bases, used in a chemoselective E2 elimination from a primary alkyl halide.

basic bulky lithium dialkyl amides LDA (lithium diisopropyl amide), LTMP (lithium tetramethyl piperidide) or LiHMDS (lithium hexamethyl disilazide) shown in Figure 4.17 are used. As for the amidine bases shown in Figure 4.16, the bulkiness of these amides guarantees that they are nonnucleophilic. They react, for example, with epoxides in chemoselective E2 reactions even when the epoxide contains a primary C atom that is easily attacked by nucleophiles (see, e.g., Figure 4.17).

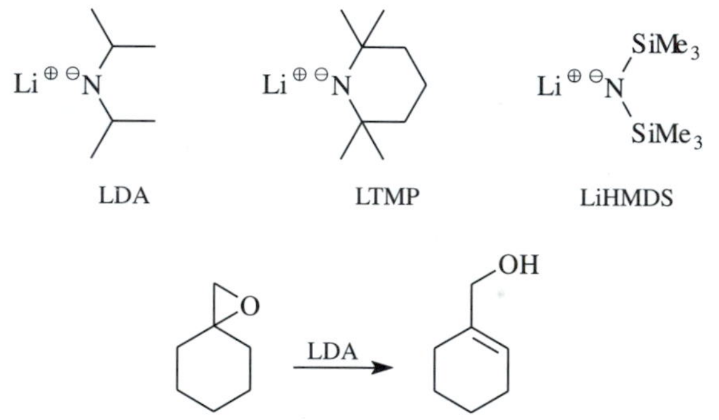

Fig. 4.17. Strong nonnucleophilic bases; use in a chemoselective E2 elimination from an epoxide.

4.4.3 A Stereoelectronic Effect on the E2/S_N2 Competition

(4-*tert*-Butylcyclohexyl)trimethylammonium iodide and potassium-*tert*-butoxide (KO*t*Bu) can undergo both an E2 and an S_N2 reaction (Figure 4.18). However, the reactivity and chemoselectivity differ drastically depending on whether the *cis*- or the *trans*-configured cyclohexane derivative is the substrate. The *cis* isomer gives a 90:10 mixture of E2 and S_N2 products in a fast reaction. The *trans* isomer reacts much more slowly and gives only the substitution product. According to Equation 4.3, these findings mean that $k_{E2,cis}:k_{S_N2,\,cis} = 90:10$ and $k_{E2,trans}:k_{S_N2,\,trans} < 1:99$. A plausible assumption is that for the respective substitution reactions $k_{S_N2,\,cis} \approx k_{S_N2,\,trans}$. This, in turn, means that the opposite chemoselectivities of the two reactions in Figure 4.18 arise almost exclusively from the fact that $k_{E2,cis} \gg k_{E2,trans}$.

How this gradation of the k_{E2} values comes about is easily understood by considering Figure 4.19 and realizing the following: in the most stable transition state of an E2 elimination the half-broken $^{\alpha}$C $\cdots$ Het bond and the likewise half-broken $^{\beta}$C $\cdots$ H bond are oriented *parallel.* This is because the hybridization change at $^{\alpha}$C and $^{\beta}$C from sp^3 to sp^2 hybridization has started in the transition state. As a result, $2p_z$-like AOs are being formed as $^{\alpha}$C and $^{\beta}$C and their increasingly effective overlap establishes the π bond. In other words, coplanar $^{\alpha}$C $\cdots$ Het and $^{\beta}$C $\cdots$ H bonds stabilize the transition state of an E2 elimination through a π-like interaction. Interactions of this type are possible in the most stable transition state of *syn*-eliminations as well as in the most stable transition state of *anti*-eliminations. When a substrate can assume both of these transition state geometries, the *anti*-elimination is always preferred. This is a consequence of the lower steric hindrance of the *anti*-transition state, in which the substituents at $^{\alpha}$C vs $^{\beta}$C are nearly staggered. The sterically more hindered *syn* transition

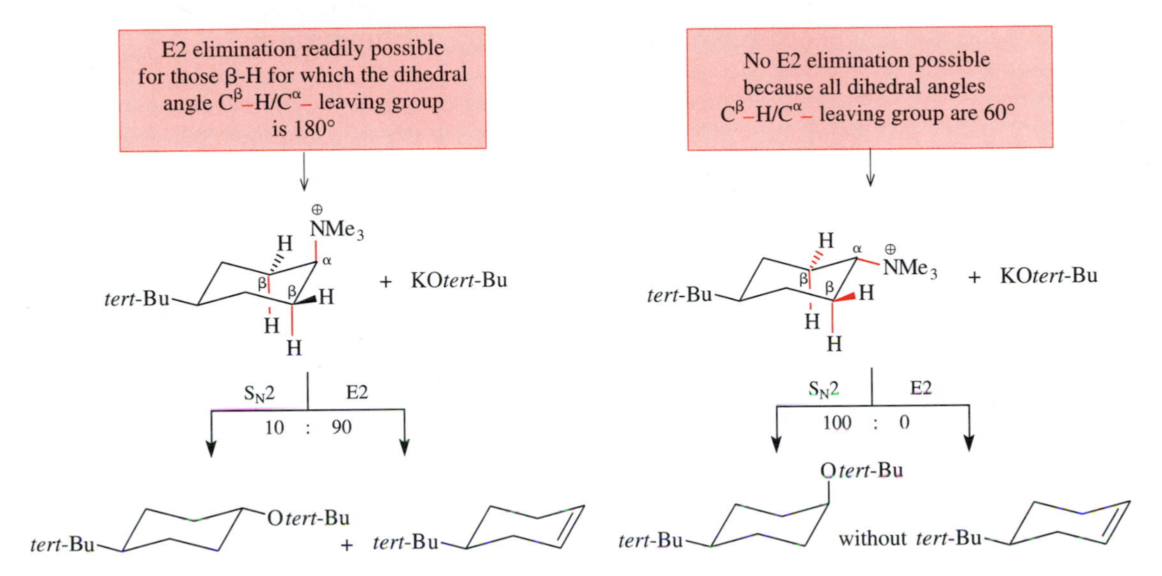

Fig. 4.18. E2/S_N2 competition in cyclohexanes with poor leaving groups locked in axial (left reaction) or equatorial (right reaction) position.

state has a nearly eclipsed structure. In addition, there are reasons to believe that orbital overlap is better in the transition state of an *anti*-elimination.

Let us return to the finding that $k_{E2,cis} > k_{E2,trans}$ in the E2 eliminations from the isomeric tetraalkylammonium salts in Figure 4.18. The transition state geometries of these reactions are derived from the chair conformers shown in Figure 4.18. The stereostructure is in each case fixed by the equatorially disposed *tert*-butyl group, which acts as a conformational lock. The much higher elimination rate of the *cis* isomer is due to its ability to undergo an *anti*-elimination, because the $^{\beta}$C—H and C—NMe_3^+ bonds exhibit a dihedral angle of 180°. In contrast, in the *trans*-configured ammonium salt, the corresponding dihedral angle is 60°, which is suitable neither for a *syn*- nor for an *anti*-elimination. The *trans* substrate and KO*t*Bu can therefore react only via the S_N2 mode.

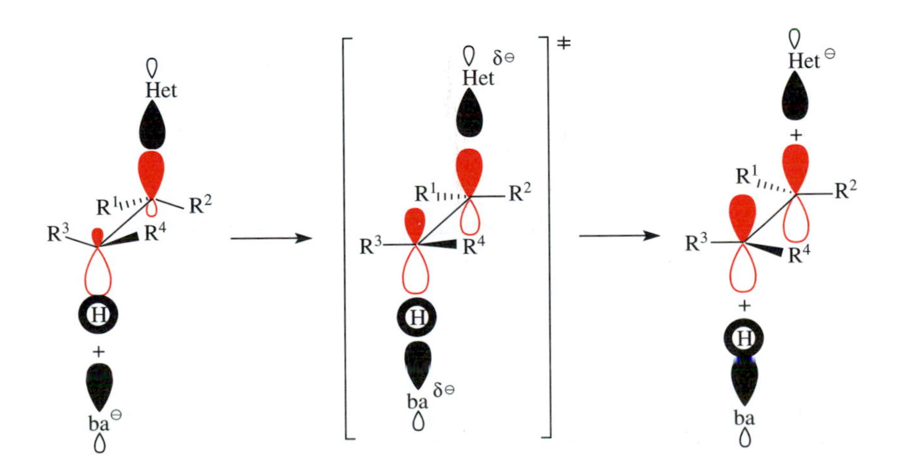

Fig. 4.19. π-Like bonding interactions in the transition state of *anti*-selective E2 eliminations; ba^- refers to the attacking base.

4.4.4 The Regioselectivity of E2 Eliminations

B

When E2 eliminations from secondary or tertiary substrates can give regioisomeric olefins, which differ from each other as Saytzeff and Hofmann products (i.e., in the degree of alkylation of their C=C bond), the Saytzeff product predominates usually because of product-development control (see Section 4.1.4). The fraction of Hofmann product can be increased, however, by using bulky bases (Figure 4.20) or equipping the starting materials with poorer leaving groups than halides or sulfonates (Figure 4.21). On the other hand, *completely* Hofmann-selective E2 eliminations can be achieved almost only in cyclic systems if the requirement of *anti*-selectivity gives rise to such a regiocontrol (Figures 4.22 and 4.23). We want to explore this in detail in the following.

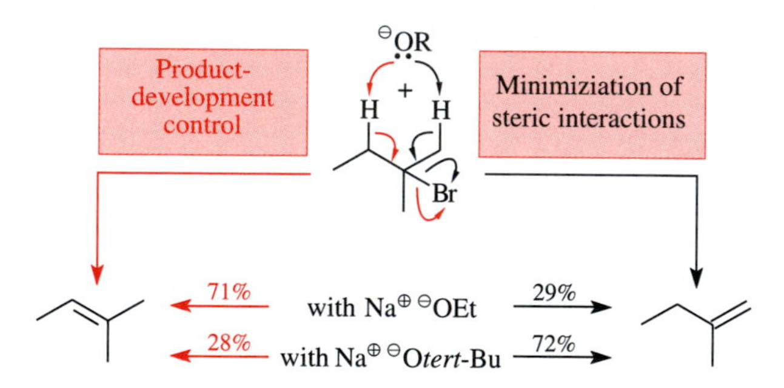

Fig. 4.20. Steric base effects on the Saytzeff/Hofmann selectivity of an E2 elimination. The small base EtO^- can attack the H atoms in both positions β to the leaving group, i.e., it does not matter whether the H atom is bound to a primary or secondary C atom. The regioselectivity therefore results *only* from product-development control: the thermodynamically more stable Saytzeff product is produced preferentially although not exclusively. The attack of the sterically hindered base $tBuO^-$ is directed preferentially toward the more readily accessible H atoms at the primary C atom. Therefore, it provides mainly the Hofmann product.

The sterically hindered base *tert*-BuO^- reacts with *tert*-amyl bromide in a different manner than EtO^- and in such a way that steric interactions are minimized (Figure 4.20). This makes the substructure C_{prim}—H a more suitable point of attack than the substructure C_{sec}—H. If a C_{tert}—H were present, it would not be attacked at all. As a consequence, the Hofmann product is produced predominantly but not exclusively.

Using one and the same base, one can reverse the normal Saytzeff preference of the E2 elimination into a preference for the Hofmann elimination through a variation of the leaving group. This is especially true for eliminations from tetraalkylammonium salts $R—NMe_3^+I^-$ and trialkylsulfonium salts $R—SMe_2^+I^-$. Their positively

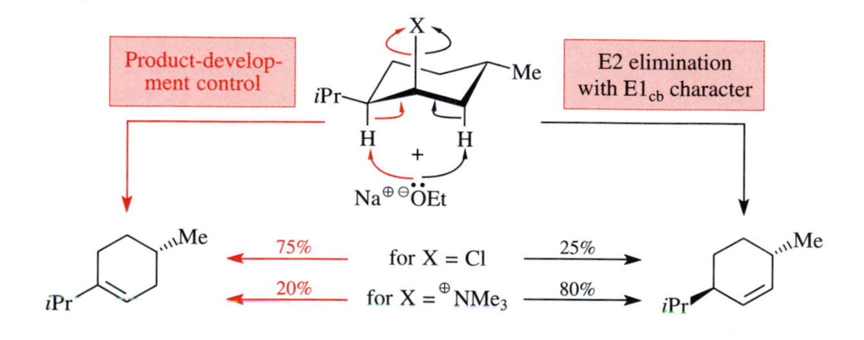

Fig. 4.21. Leaving-group effect on the Saytzeff/Hofmann selectivity of an E2 elimination. Poor, positively charged leaving groups react preferentially to give the Hofmann product via an E2 elimination with $E1_{cb}$ character.

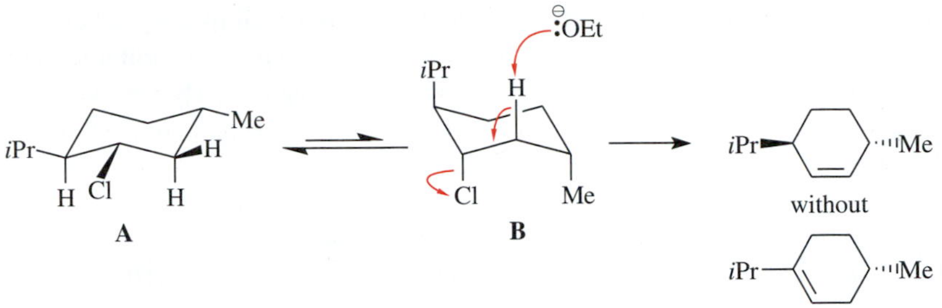

Fig. 4.22. *anti*-Selectivity of an E2 elimination as a reason for Hofmann selectivity.

charged leaving groups primarily acidify the H atoms in the α position, but do the same to a lesser extent with the H atoms in the β position. The latter effect is decisive. In addition, NMe_3 and SMe_2 are just as poor leaving groups in eliminations as in S_N reactions (cf. Section 2.3). According to Section 4.3, both propensities favor an E2 elimination via an $E1_{cb}$-like transition state. Obviously, a β H atom participating in such an elimination must be sufficiently acidified. As is well known, the C,H acidity decreases in the series C_{prim}—H > C_{sec}—H > C_{tert}—H. Therefore, in $E1_{cb}$-like E2 eliminations, the H atom is preferentially removed from the least alkylated C atom β to the leaving group. This regioselectivity corresponds to a preference for the Hofmann product, as shown in the bottom example in Figure 4.21. In addition, a steric effect probably acts in the same direction: the NMe_3^+ or SMe_2^+ groups in the substrate are relatively large because they are branched. Therefore, they allow an attacking base more ready access to the structural element C^{β}_{prim}—H relative to C^{β}_{sec}—H and to the structural element C^{β}_{sec}—H relative to C^{β}_{tert}—H.

Many cyclohexane derivatives undergo—in apparent contradiction to what has been said before—*perfectly* Hofmann selective E2 eliminations even when they contain good leaving groups and even when sterically modest bases are used. Figure 4.22 demonstrates this using menthyl chloride as an example. In substrates of this type, a stereoelectronic effect, which has already been discussed in another connection (Section 4.4.3), takes over the regiocontrol: Many E2 eliminations exhibit pronounced *anti*-selectivity. The only H/Cl elimination from menthyl chloride that is *anti*-selective leads to the Hofmann product. It takes place through the all-axial conformer **B** of substrate **A**, even though this conformer is present in only small equilibrium quantities.

Fig. 4.23. *anti*-Selectivity of an E2 elimination as the reason for the chemoselective formation of a 1,3-diene instead of a bromo-olefin.

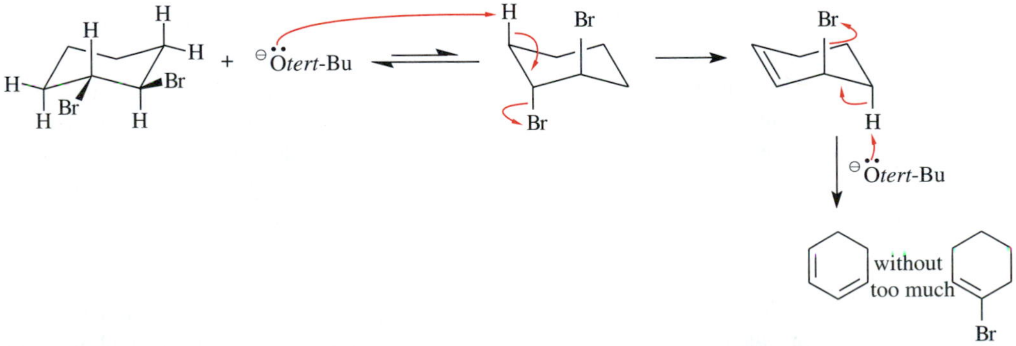

Because of the same necessity for *anti*-selectivity, *trans*-1,2-dibromocyclohexane does not react to form 1-bromocyclohexene in the KO*t*Bu-promoted HBr elimination (Figure 4.23). Instead 3-bromocyclohexene is produced through a fairly regioselective E2 elimination. In a subsequent fast 1,2-elimination, it loses a second equivalent of HBr. In this way 1,3-cyclohexadiene is produced.

4.4.5 One-Pot Conversion of an Alcohol to an Olefin

A

The dehydration of alcohols to olefins is possible when their OH group is converted into a good leaving group by protonation or by binding to a Lewis acid. However, eliminations in esters (e.g., sulfonates, halides) derived from the same alcohols would usually take place under milder (i.e., nonacidic) conditions than the dehydrations of the alcohols themselves. Sometimes it can therefore be advantageous to perform such dehydrations via the corresponding esters. Gratifyingly, it may be possible to prepare suitable esters from the alcohols for that purpose and subject them, without working them up, directly to the elimination. Through a mesylate produced in this way *in situ* it was possible, for example, to generate the highly sensitive dienediyne in Figure 4.24.

Fig. 4.24. One-pot procedure for the dehydration of alcohols: activation/elimination with $MsCl/NEt_3$.

In accord with the goal of "faster, milder, more selective" in organic synthesis, special reagents were also developed for the one-pot dehydration of particularly sensitive alcohols. One of these is the Burgess reagent (Figure 4.25). It activates the alcohol in the form of an aminosulfuric acid ester. The latter decomposes either via a cyclic transition state (i.e., by a *syn*-elimination, shown as an example in Figure 4.25), or through a NEt_3-induced and *anti*-selective process via an E2 mechanism, depending on the substrate.

Another special reagent for the one-pot dehydration of alcohols after an *in situ* activation is Martin's persulfurane (Figure 4.26). It attacks the OH group so that its O atom becomes part of a leaving group $Ph_2S{=}O$. For Martin's persulfurane the elimination mechanism depends on the exact substrate structure, too. Starting from secondary alcohols, *anti*-selective E2 eliminations take place. Tertiary alcohols can eliminate according to the E2 (see example in Figure 4.26) or according to the E1 mechanism.

MeO_2C O ⊖N—S=O ⊕NEt_3 H O

MeO_2C O HN—S=O ⊕NEt_3 + O⊖

$\overset{\oplus}{H}NEt_3$ + MeO_2C O HN—S=O + O⊖

$\overset{\oplus}{H}NEt_3$ + MeO_2C O ⊖N—S=O H O

NEt_3 + MeO_2C O HN—S=O O

Fig. 4.25. Dehydration of an alcohol with the Burgess reagent. The aminosulfuric acid ester intermediate decomposes via a cyclic transition state (*syn*-elimination).

R_F—O Ph—S—Ph R_F—O

R_F—O Ph—S⊕—Ph

OH + $R_FO^\ominus$ O⊖

R_F—O Ph—S—Ph O

R_F—OH + Ph—S—Ph O +

R_F—O⊖ + Ph—S⊕—Ph H O

$R_F = PhC(CF_3)_2$

Fig. 4.26. Dehydration of an alcohol with Martin's persulfurane. The tertiary alcohol shown reacts via the E2 mechanism

4.5 E1 Elimination of H/Het from R_{tert}—X and the E1/S_N1 Competition

B

E1 eliminations take place via carbenium ion intermediates. Consequently, they are encountered in substrates that are prone to undergo heterolyses to a carbenium ion. E1 eliminations are also possible from substrates in which Brønsted or Lewis acids facilitate the formation of carbenium ions. In other words: E1 eliminations are possible with substrates and under conditions under which S_N1 reactions can also take place (Section 2.5).

The outcome of this competition depends only on how the carbenium ion intermediate reacts: if it encounters a poor nucleophile, it neglects it and the E1 product is produced; if it encounters a good nucleophile, it reacts with it to form the S_N1 product.

In general, a high reaction temperature favors the E1 at the expense of the S_N1 reaction. Only the elimination has a strong positive reaction entropy because a greater number of species are produced than were available at the start. Thus the translational entropy ΔS increases during the reaction and the $T\Delta S$ term in the driving force $\Delta G = \Delta H - T\Delta S$ becomes more and more negative at high temperatures and thereby more and more noticeable. According to the Hammond postulate, this thermodynamic effect is felt to some extent in the transition state of the E1 elimination. Through this lowering of $\Delta G^{\ddagger}$ the elimination rate is increased.

B

4.5.1 Energy Profiles and Rate Laws for E1 Eliminations

tert-Butyl bromide undergoes an E1 elimination (Figure 4.27) when it is heated in a polar medium in the absence of a good nucleophile; it leads to isobutene and HBr in two steps. The first step, the heterolysis to carbenium and bromide ions, is rate-determining and essentially irreversible. The second step is the deprotonation of the carbenium ion to the olefin. In Figure 4.27 and in the associated kinetic equations (Equations 4.4–4.6) it is shown as a unimolecular reaction. The possibility of a bimolecular deprotonation step, in which either the bromide ion or the solvent participates as a base, should not be ignored a priori. Taking into account Figure 4.29, a related E1 elimination in which the carbenium ion is definitely not subject to a bimolecular deprotonation by Br^-, we can conclude that in the present case the deprotonation is unimolecular. By the way, from a preparative point of view, the E1 elimination in Figure 4.27 can also be used to obtain anhydrous hydrogen bromide.

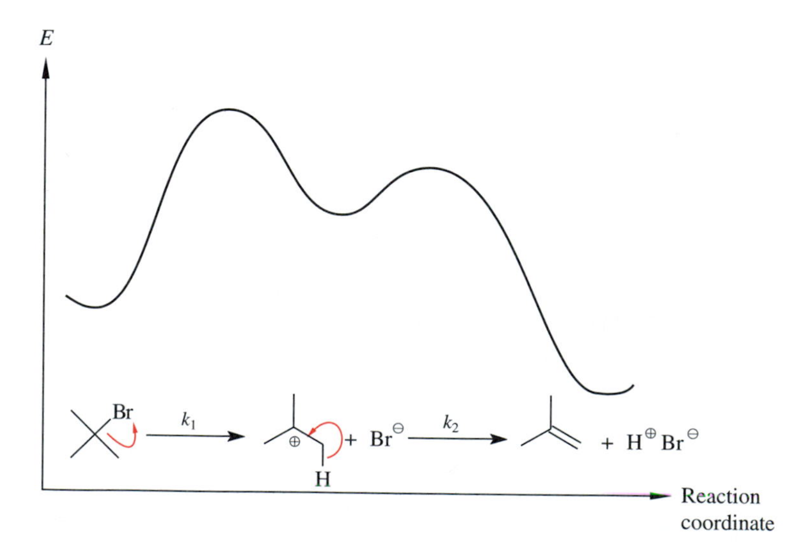

Fig. 4.27. The E1 elimination *tert*-BuBr → isobutene + HBr and its energy profile.

Take a look at the derivation of the rate law for this E1 elimination:

$$\frac{\mathrm{d}\,[\text{olefin}]}{\mathrm{d}t} = k_2\,[\mathrm{R}^+] \tag{4.4}$$

$$\frac{\mathrm{d}\,[\mathrm{R}^+]}{\mathrm{d}t} = k_1\,[\mathrm{RBr}] - k_1\,[\mathrm{Y}^-]$$

$= 0$ because of the steady state approximation

$$\Rightarrow [\mathrm{R}^+] = \frac{k_1}{k_2}\,[\mathrm{RBr}] \tag{4.5}$$

Inserting Eq. 4.5 into Eq. 4.4 yields

$$\frac{\mathrm{d}\,[\text{olefin}]}{\mathrm{d}t} = k_1\,[\mathrm{RBr}] \tag{4.6}$$

Equation 4.6 proves the unimolecularity of this reaction (from which the designation "E1 elimination" is derived).

Tertiary alcohols, tertiary ethers, or carboxylic acid esters of tertiary alcohols can undergo E1 eliminations, but only in the presence of Brønsted or Lewis acids. Anyone who has prepared a tertiary alkoxide by a Grignard reaction and treated the crude reaction mixture with HCl and obtained the olefin knows that tertiary alcohols can be converted into olefins even with dilute hydrochloric acid.

A similar E1 elimination is the cleavage of *tert*-butyl hexyl ether in CF_3CO_2H/CH_2Cl_2 to isobutene and hexanol (Figure 4.28). This reaction proceeds in three steps. In an equilibrium reaction, the ether is protonated to give the oxonium ion. The second step

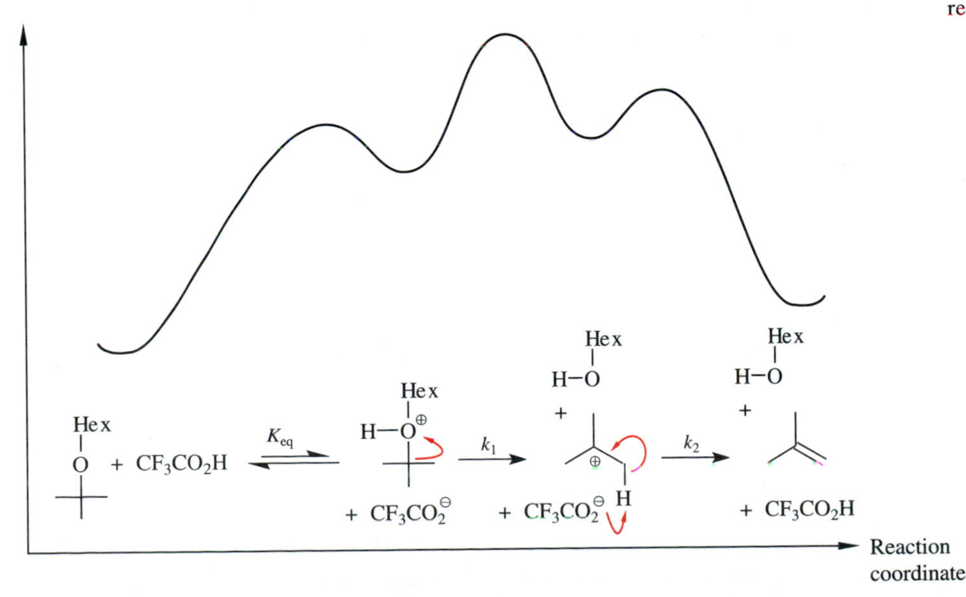

Fig. 4.28. The acid-catalyzed E1 elimination of *tert*-butyl hexyl ether to isobutene and hexanol and its energy profile. K_{eq} refers to the equilibrium constant of the acid/base reaction.

is rate-determining and essentially irreversible: it is a regioselective heterolysis to the *tert*-butyl cation, the more stable of the two carbenium ions that might, in principle, form. In addition, 1-hexanol is released. In the third step, a fast and probably bimolecular deprotonation of the carbenium ion to the olefin takes place. The ether cleavage in Figure 4.28 exemplifies how one can release primary or secondary alcohols from their *tert*-butyl ethers under mild conditions.

The kinetic analysis of the E1 elimination of Figure 4.28 can be carried out as follows:

$$\frac{\mathrm{d\,[olefin]}}{\mathrm{d}t} = k_2\,[\mathrm{R}^+][\mathrm{F_3CCO_2^-}] \tag{4.7}$$

$$\frac{\mathrm{d\,[R^+]}}{\mathrm{d}t} = k_1\,[\mathrm{R_2^+OH}] - k_2[\mathrm{R}^+][\mathrm{F_3CCO_2^-}]$$

$$= 0 \text{ because of the steady state approximation}$$

$$\Rightarrow [\mathrm{R}^+] = \frac{k_1}{k_2}\,\frac{[\mathrm{R_2^+OH}]}{[\mathrm{F_3CCO_2^-}]} \tag{4.8}$$

Because the oxonium ion forms in a fast equilibrium reaction, its concentration is equal to the equilibrium concentration:

$$[\mathrm{R_2^+OH}] = K_{\mathrm{eq}}[\mathrm{R_2O}]\,[\mathrm{CF_3CO_2H}] \tag{4.9}$$

Inserting Eq. 4.8 and 4.9 into Eq. 4.7 yields

$$\frac{\mathrm{d\,[olefin]}}{\mathrm{d}t} = k_1 \bullet K_{\mathrm{eq}}[\mathrm{R_2O}]\,\underbrace{[\mathrm{CF_3CO_2H}]}_{\text{is not consumed}} \tag{4.10}$$

$$= \text{const.} \bullet [\mathrm{R_2O}] \tag{4.11}$$

The rate law of this E1 reaction in the form of Equation 4.11 reveals unimolecularity once more. However, it hides what the more detailed form of Equation 4.10 discloses: namely, that the elimination rate increases with increasing CF_3CO_2H concentration, which means that this is a *bimolecular* E1 elimination. From Equation 4.10 we can also see that the E1 rate increases at a given concentration of the acid when a more acidic acid is used.

4.5.2 The Regioselectivity of E1 Eliminations

B

When the carbenium ion intermediate of an E1 elimination can be deprotonated to give two regioisomeric olefins, generally both of them are produced. If these olefin isomers are Saytzeff and Hofmann products, the first one is produced preferentially because of product-development control. This is illustrated in Figure 4.29 by the E1 elimination from *tert*-amyl iodide, bromide, and chloride. The different halides form the *tert*-amyl cation and consequently the olefin, too, at *different* rates. Still, in each reaction the *same* 82:18 ratio of Saytzeff to Hofmann product is obtained. This can be explained most simply if the deprotonation of the *tert*-amyl cation occurs unimolecularly in all cases, that is, without participation of the halide ion as a base.

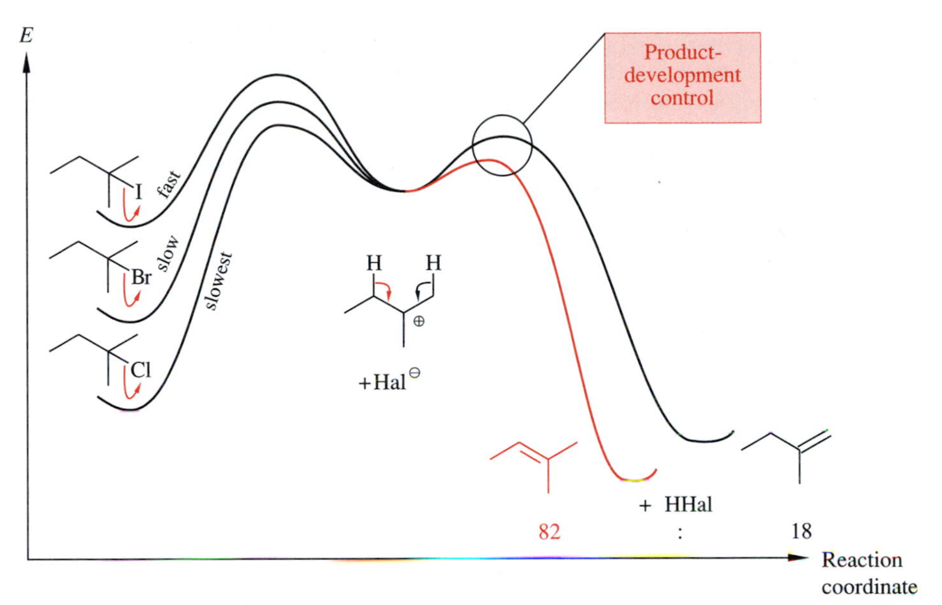

Fig. 4.29. Saytzeff preference of E1 eliminations from *tert*-amyl halides and energy profile. The Saytzeff: Hofmann preference amounts to 82:18 regardless of which halide ion accompanies the carbenium ion. This observation is explained most simply by the assumption that this halide ion is not involved in the deprotonation step forming the C=C double bond.

A

It is interesting that E1 eliminations from tertiary alcohols carried out in acidic media give more Saytzeff product than E1 eliminations from the corresponding tertiary alkyl halides. As an example, compare the E1 eliminations from Figure 4.30 (alkylated *tert*-amyl alcohols) with those in Figure 4.29 (*tert*-amyl halides). The E1 eliminations via the alkylated *tert*-amyl cations from Figure 4.30 give a 95:5 ratio of Saytzeff to Hofmann product. From the (unsubstituted) *tert*-amyl cations in Figure 4.29, the same compounds were produced (when we set R′ = H) in the less biased ratio of 82:18.

The reason for the increased regioselectivity is that with tertiary alcohols as substrates there is no longer exclusively kinetic control. This is because the regioisomeric olefins are no longer formed irreversibly. Instead they can be reprotonated, deprotonated again, and thereby finally equilibrated. In this way, the greatest part of the initially formed Hofmann product is converted to the more stable Saytzeff isomer. Product formation is thus

R OH

+ $BF_3 \cdot OEt_2$

R ⊕

R + R + $H^{\oplus}\,{}^{\ominus}BF_3(OH)$

95 : 5

isomerizes the primary products!

Fig. 4.30. Explanation for the particularly high Saytzeff selectivity of the E1 elimination from *tert*-amyl alcohol and its derivatives.

the result of thermodynamic control. The fact that a greater regioselec-tivity results with thermodynamic control than with kinetic control through product-development control is precisely what one would expect, considering the Hammond postulate.

The elimination in Figure 4.31 provides proof that the olefins initially formed from tertiary alcohols under E1 conditions can be reprotonated. The Saytzeff *and* the Hofmann products shown there can be protonated to provide the tertiary carbenium ion through which they were formed and also to a different tertiary carbenium ion. The consequence of this is that in the major product obtained after the final deprotonation, the C=C double bond is no longer located at the C atom that carried the OH group in the starting alcohol, but is moved one center away.

OH

+ $BF_3 \cdot OEt_2$

+ + + $H^{\oplus}{}^{\ominus}BF_3(OH)$

isomerizes the primary products!

10 : 90 : 0

Fig. 4.31. E1 elimination with subsequent C=C migration.

As discussed, the major product of the elimination in Figure 4.31 is evidently produced under thermodynamic control. In this respect it is surprising that it contains a trisubstituted C=C double bond, while the minor product has a tetrasubstituted one. The stability of a C=C double bond normally increases with each additional alkyl substituent. However, here an opposite effect dominates. When we compare the two isomers methylenecyclohexane (which contains a disubstituted C=C double bond) and 1-methylcyclohexene (which contains a trisubstituted C=C double bond) as model olefins, then methylenecyclohexane has 2 kcal/mol more energy than 1-methylcyclohexene. This difference in stability is too great to be accounted for only by the difference between di- and trisubstitution of the C=C double bond. A second factor is that methylene-substituted cyclohexane has a greater ring strain than methyl-substituted cyclohexene. The same ring strain effect, possibly reinforced by greater steric hindrance, is probably responsible for the surprising difference in stability between the olefin isomers in Figure 4.31.

4.5.3 E1 Eliminations in Protecting Group Chemistry

B

Acid treatment of *tert*-butyl ethers of primary and secondary alcohols produces alcohols by E1 eliminations. We saw this in the discussion of Figure 4.28. In a quite analogous way and under the same acidic conditions, *tert*-butyl esters and O-*tert*-butyl carbamates are cleaved. In accordance with that, *tert*-butyl ethers, esters, and carbamates are frequently used as protecting groups. An example of such an application in peptide synthesis is given in Figure 4.32. Of course, the elimination product (isobutene) is of no interest in this case, but the leaving group is, because it is the unprotected peptide.

H_2N H N O O N H O O O H N O O

CF_3CO_2H

OH ⊕ H_2N H N O O N H OH ⊕ O ⊕ H O H N O O

OH H_2N H N O O N H O OH H O H N O

a carbamic acid + 3 ⊕ H

$-CO_2$, − 3

Fig. 4.32. Three E1 eliminations in the deprotection of a protected tripeptide. For the sake of brevity, a *single* formula in the second row of the scheme shows how the three *tert*-Bu—O bonds heterolyze; of course, they are activated and broken *one after the other*. In the deprotection of the *tert*-butylated lysine side chain, the leaving group is a carbamic acid. Carbamic acids decarboxylate spontaneously (Figure 7.12), which explains the final transformation. The preparation of the protected tripeptide is shown in Figure 4.35.

Each of the E1 eliminations from the three *tert*-butyl groups shown takes place in three steps. In the first step the most basic O atom is protonated in an equilibrium reaction. Oxonium ions or carboxonium ions are produced. The second step is the heterolysis of the O—*tert*-Bu bond. Each leaving group is uncharged and therefore energetically acceptable: the *tert*-butyl ether of the serine side chain gives an alcohol, the *tert*-butylcarbamate of the lysine side chain gives a carbamic acid (which decarboxylates subsequently), and the *tert*-butyl ester of the glycine moiety gives a carboxylic acid. The other product of the heterolyses is the well-stabilized *tert*-butyl cation, which is deprotonated in the last step to give isobutene.

4.6 $E1_{cb}$ Eliminations

4.6.1 Unimolecular $E1_{cb}$ Eliminations: Energy Profile and Rate Law

B

Knoevenagel reactions (for their mechanism, see Section 10.4.2) end with H_2O being eliminated from the initially formed alcohol in the basic medium. The elimination takes place via an $E1_{cb}$ mechanism, which is shown in Figure 4.33 via the example of the elimination of H_2O from a β-nitro alcohol. In a fast exergonic reaction,

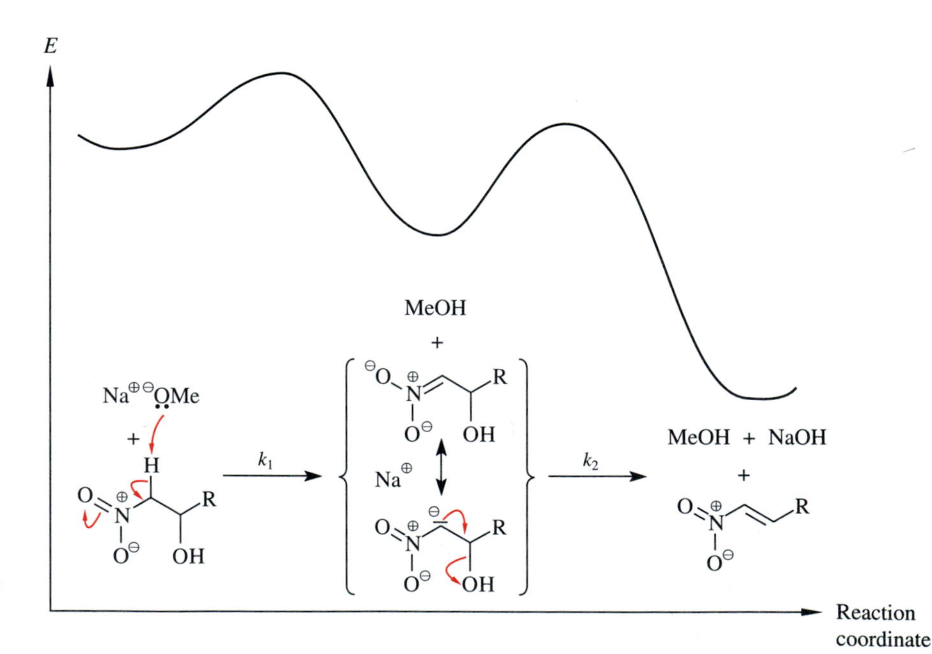

Fig. 4.33. A unimolecular $E1_{cb}$ elimination and its energy profile.

this substrate is deprotonated quantitatively to give a nitronate. In the following slower second reaction step, an OH^- group leaves, and a nitroolefin is produced. It is preferentially formed as the *trans* isomer because of product-development control.

The following derivation delivers the rate law for this elimination as Equation 4.14. The term $[\text{nitroalcohol}]_0$ refers to the initial concentration of the nitro-alcohol.

$$\frac{d\,[\text{olefin}]}{dt} = k_2[\text{nitronate}] \tag{4.12}$$

Because $k_1 \gg k_2$ and because of the stoichiometry, we have:

$$[\text{nitronate}] = [\text{nitroalcohol}]_0 - [\text{olefin}] \tag{4.13}$$

Inserting Eq. 4.13 into Eq. 4.12 yields

$$\frac{d\,[\text{olefin}]}{dt} = k_2([\text{nitroalcohol}]_0 - [\text{olefin}]) \tag{4.14}$$

This rate law contains only one concentration variable—the term $[\text{nitroalcohol}]_0$ is, of course, a constant and not a variable—and thus refers to a **uni**molecular process. This elimination is therefore designated E**1**$_{cb}$.

4.6.2 Nonunimolecular $E1_{cb}$ Eliminations: Energy Profile and Rate Law

B

At the end of an aldol condensation (for the mechanism, see Section 10.4.1), water is eliminated from a β-hydroxy carbonyl compound. This takes place according to the mechanism given in Figure 4.34. In a fast but endergonic and consequently reversible reaction, the β-hydroxy carbonyl compound is deprotonated to give an enolate. In the subsequent rate-determining reaction step an OH^- group is expelled, and an α,β-unsaturated carbonyl compound, the elimination product, is produced. If it were formed irreversibly, it would be produced preferentially as the *trans* isomer because of product-development control. However, in aldol condensations, this stereoselectivity is reinforced by thermodynamic control. Under the reaction conditions, any initially formed *cis* isomer would be isomerized subsequently to the *trans* isomer. Therefore, the *trans* selectivity of aldol condensations is in general due to thermodynamic control.

The rate law for the $E1_{cb}$ mechanism in Figure 4.34 results as Equation 4.17 from the following derivation.

$$\frac{d\,[\text{olefin}]}{dt} = k[\text{enolate}] \quad (4.15)$$

The enolate concentration follows from the equilibrium condition

$$[\text{enolate}] = K_{eq}\,[\text{aldol}]\,[\text{NaOEt}] \quad (4.16)$$

Inserting Eq. 4.16 and Eq. 4.15 yields

$$\frac{d\,[\text{olefin}]}{dt} = k \bullet K_{eq}\,[\text{aldol}]\,[\text{NaOEt}] \quad (4.17)$$

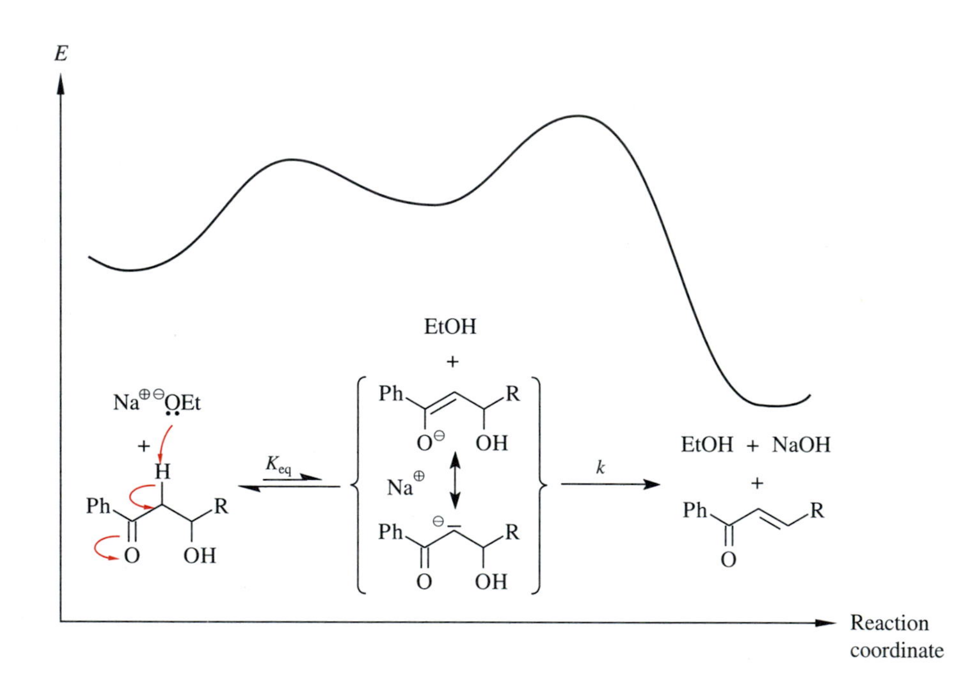

Fig. 4.34. A bimolecular $E1_{cb}$ elimination and its energy profile; K_{eq} refers to the constant of the acid/base equilibrium.

Thus, this elimination takes place *bimolecularly*. Therefore, the designation $E1_{cb}$ for this mechanism is justified only if one thinks of the unimolecularity of the rate-determining step.

4.6.3 $E1_{cb}$ Eliminations in Protecting Group Chemistry

A

In peptide synthesis functional groups in the amino acid side chains are often protected with acid-labile protecting groups (Section 4.5.3). The tripeptide in Figure 4.35 contains, for example, a serine *tert*-butyl ether and an L-lysine ε-protected as an O-*tert*-butyl carbamate. In the standard strategy of synthesizing oligopeptides from the C- to the N-terminus (cf. Section 6.4.3) the C-terminus is either connected to the acid-labile

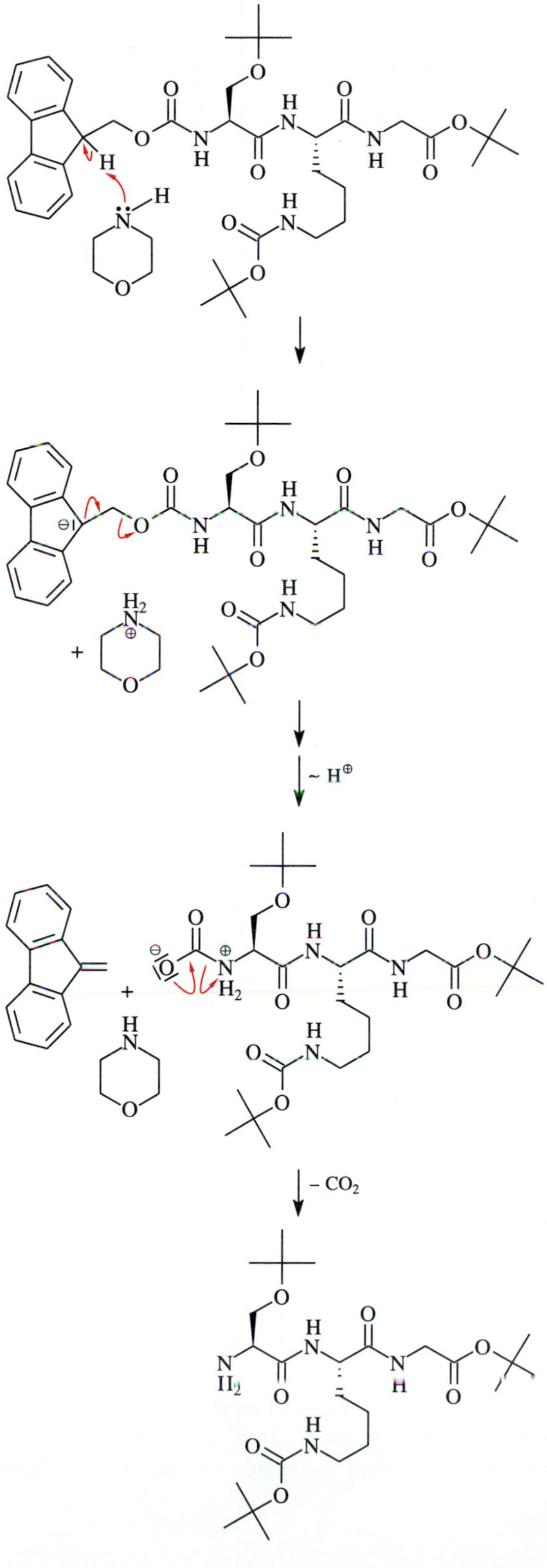

Fig. 4.35. Elimination of a fluorenylmethoxycarbonyl group (Fmoc group) from the terminal N atom of a protected tripeptide according to the $E1_{cb}$ mechanism.

Merrifield resin or to an acid-labile protecting group of lower molecular mass. For illustration, Figure 4.35 shows a glycine *tert*-butyl ester as the C-terminus of the mentioned tripeptide.

At the N-terminus of the growing peptide chain there is initially also a protecting group. This group must be removed as soon as the free NH_2 group is needed for condensation with the next activated amino acid. However, when the protecting group of the terminal nitrogen is removed, all side chain protecting groups as well as the C-terminal protecting group must remain intact. Because these are (only) acid-labile, the nitrogen protecting group should be baselabile. An N-bound fluorenylmethoxycarbonyl (Fmoc) group is ideal for this purpose (Figure 4.35): it can be removed from the N atom by the weakly basic reagent morpholine. This elimination takes place according to an $E1_{cb}$ mechanism. This is because the carbanion intermediate is easily produced because as a fluorenyl anion (i.e., a dibenzoannulated cyclopentadienyl anion) it has the benefit of considerable aromatic stabilization.

In the synthesis of oligonucleotides and oligodeoxynucleotides, the 5′ ends of the nucleotide building blocks are protected with the acid-labile dimethoxytrityl group mentioned in Section 2.5.4. The phosphorous acid portion (when the synthesis takes place through phosphorous acid derivatives) or the phosphoric acid portion (which are present in all syntheses at least at the end) and the nucleotide bases are protected with base-sensitive protecting groups because these survive the acid-catalyzed removal of the dimethoxytrityl group undamaged. A well-established base-labile protecting group for the phosphorus moiety (Figure 4.36) is the β-cyanoethyl ester. Its cleavage is due to an $E1_{cb}$ elimination via a cyano-stabilized "carbanion."

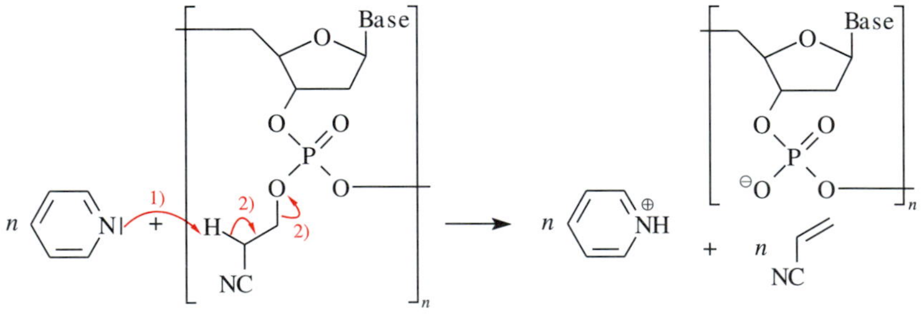

Fig. 4.36. Cleavage of a poly[(β-cyanoethyl)phosphate] according to the $E1_{cb}$ mechanism.

4.7 β-Eliminations of Het^1/Het^2

B

It was indicated already in Section 4.1.5 that Het^1/Het^2 eliminations offer possibilities for regio- and stereocontrol in the synthesis of olefins that are quite different from those of H/Het eliminations. It was also indicated that the value of these reactions for olefin synthesis depends on, among other things, how laborious it is to obtain the Het^1- and Het^2-containing substrates. Accordingly, in the next sections we will discuss such eliminations and the preparation of the respective elimination precursors as well.

4.7.1 Fragmentation of β-Heterosubstituted Organometallic Compounds

B

Haloalkanes and Li, Mg, or Zn can form organometallic compounds. How these conversions take place mechanistically will be discussed in Section 14.4.1. What is of interest here, however, is that if in organometallics of this type a leaving group is located in the β position of the metal, the leaving group will almost unavoidably be eliminated quickly together with the metal. In general, only O^-- or $–N^-C(=O)R$-groups resist this elimination; they are simply too poor as leaving groups.

A β-elimination of this type—often also referred to as fragmentation—takes place in the intiation of a Grignard reaction using ethylene bromide:

Mg + Br⁄⁄Br ⟶ (Br–Mg⁄⁄Br) ⟶ Br_2Mg + $H_2C=CH_2$

The purpose of this operation is not to release ethylene but to etch the Mg shavings.

A second example of this type of elimination is the generation of dichloroketene from trichloroacetyl chloride and zinc:

$$Zn + Cl_3C-C(=O)-Cl \longrightarrow (Cl-Zn-CCl_2-C(=O)-Cl) \longrightarrow ZnCl_2 + Cl_2C=C=O$$

One should therefore always keep in mind that *β-heterosubstituted saturated organometallic compounds hardly ever exist.* On the contrary, they almost inevitably fragment through M^+/Het^- elimination to an olefin:

Mg + Br–CH₂–C(R_n)–OR ⟶ (Br–Mg–CH₂–C(R_n)–OR) ⟶ BrMgOR + olefin(R_n)

Li + Br–CH₂–C(R_n)–OR ⟶ (Li–CH₂–C(R_n)–OR) ⟶ LiOR + olefin(R_n)

4.7.2 Julia–Lythgoe Synthesis of *trans*-Olefins

A

α-Lithiated primary alkylphenylsulfones can be added to aldehydes (Figure 4.37). A lithium alkoxide is produced as a primary product just as upon addition of other C nucleophiles (Section 8.5). If acetic anhydride is added to it, one obtains a (β-acetoxyalkyl) phenylsulfone as a mixture of two diastereomers. If one treats this mixture with sodium amalgam in methanol, the result is a Het[1]/Het[2] elimination. Each diastereomer of the

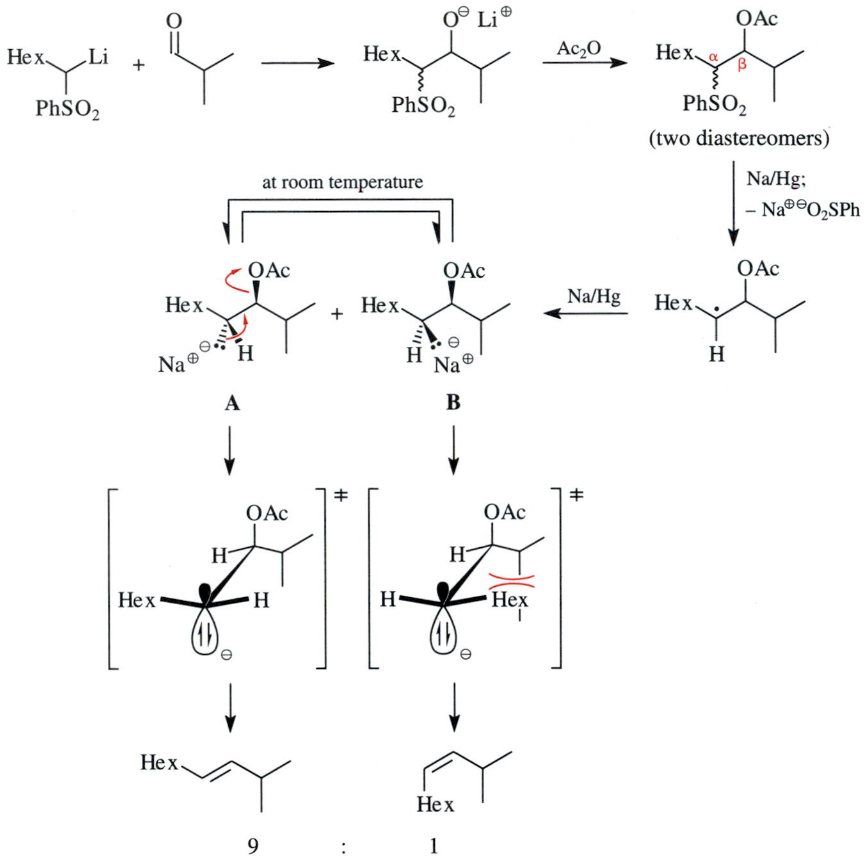

Fig. 4.37. Julia–Lythgoe synthesis of *trans*-olefins involving a Het[1]/Het[2] elimination.

acetoxysulfone gives the same mixture of a *trans*- and a *cis*-olefin. The *trans*-isomer predominates distinctly so that this so-called Julia–Lythgoe olefination has become an important synthesis for *trans*-olefins.

The mechanism of this reductive elimination is not known with certainty. However, a plausible first step is that $NaHg_x$ dissolves by liberating an electron which is transferred to the sulfonylated phenyl ring so that a radical anion is produced (Figure 4.37). One of its C—S bonds is broken regioselectively so that a benzenesulfinate anion $PhSO_2^-$ and a secondary alkyl radical form. This radical has one stereocenter less than its $PhSO_2^-$ containing precursor because the septet center is essentially planar (Section 1.2.1). In the second electron transfer step, this radical is reduced to a sodium organometallic and a new stereocenter is produced. Since no stereocontrol can be expected, most probably both diastereomeric reduction products **A** and **B** are formed. If these species lose Na^+OAc^-, they must be able to be converted into each other before the elimination. Because of product-development control, the *anti*-elimination of diastereomer **A** leading to the *trans*-olefin would take place more rapidly than the *anti*-elimination of diastereomer B, which leads to the less stable *cis*-olefin. Due to the epimerization reaction **A** $\rightleftharpoons$ **B,** the mixture of the sodium organometallics **A** and **B** could then react to the olefin exclusively via the more reactive diastereomer **A.**

4.7.3 Peterson Olefination

A

A β-hydroxysilane, like the one shown in Figure 4.38 (top, left), can be prepared stereoselectively (e.g., via the Cram-selective reduction of an α-silylated ketone according to the reactions in Figure 8.9 or via the Cram-selective addition of organometallic compounds to α-silylated aldehydes similar to what is shown in Table 8.3). These compounds undergo a stereoselective *anti*-elimination in the presence of acid and a stereoselective *syn*-elimination in the presence of a base (Figure 4.38). Both reactions are referred to as Peterson olefination. The stereochemical flexibility of the Peterson elimination is unmatched by any other Het[1]/Het[2] elimination discussed in this section.

Fig. 4.38. Stereoselective Het[1]/Het[2] eliminations in the Peterson olefination. The base-induced Peterson olefination (top reaction) takes place as a *syn*-elimination, and the acid-catalyzed Peterson olefination takes place as an *anti*-elimination.

The *acid-catalyzed* Peterson olefination is presumably an E2 elimination, that is, a one-step reaction. On the other hand, the base-induced Peterson olefination probably takes place via an intermediate. This intermediate probably is a four-membered heterocycle with a pentavalent, negatively charged Si atom. This heterocycle probably decomposes in the same way as the oxaphosphetane intermediate of the Wittig reaction (Section 4.7.4) does: by a [2+2]-cycloreversion.

The same acidic or basic medium is also suitable for the stereoselective conversion of α-silylated tertiary alcohols into trisubstituted olefins. Figure 4.9 showed an impressive series of examples of this.

4.7.4 Oxaphosphetane Fragmentation, Last Step of Wittig and Horner–Wadsworth–Emmons Reactions

B

According to Section 9.3.2, P-ylides and aldehydes first react in a [2+2]-cycloaddition to form a heterocycle, which is referred to as oxaphosphetane (Figure 4.39).

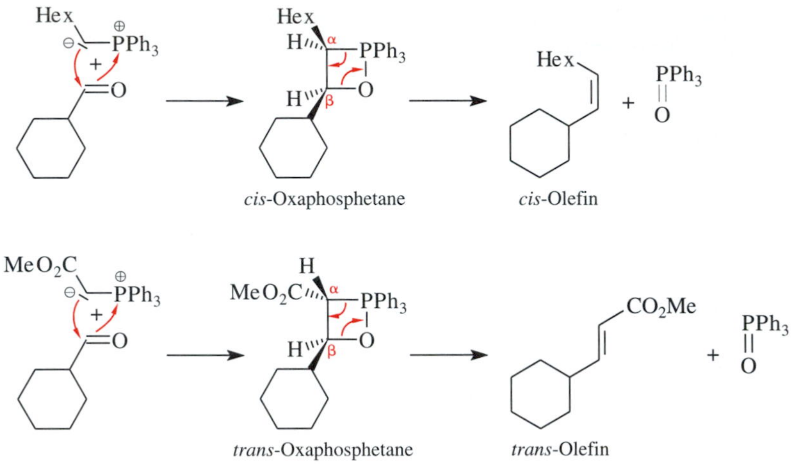

Fig. 4.39. *syn*-Selective eliminations from oxaphosphetanes in Wittig olefinations with unstabilized (upper row; gives *cis*-olefin) and stabilized P-ylides (bottom row; gives *trans*-olefin).

When the negative formal charge on the ylide C atom is resonance-stabilized by conjugating substituents, *trans*-configured oxaphosphetanes are produced (cf. Section 9.3.2). On the other hand, if the negative formal charge on the ylide C atom is not resonance-stabilized, *cis*-configured oxaphosphetanes are produced (best under "salt-free conditions," as described in Section 9.3.2). Regardless of their stereochemistry, such oxaphosphetanes decompose rapidly and stereoselectively to give $Ph_3P{=}O$ and an olefin by a *syn*-elimination. In this way *trans*-oxaphosphetanes give conjugated *trans*-olefins including, for example, *trans*-configured α,β-unsaturated esters. *cis*-Oxaphosphetanes lead to *cis*-olefins.

A *syn*-elimination of $Ph_2MeP{=}O$ and simultaneous stereoselective olefin formation from an oxaphosphetane are shown in Figure 4.40 (note that this oxaphosphetane is

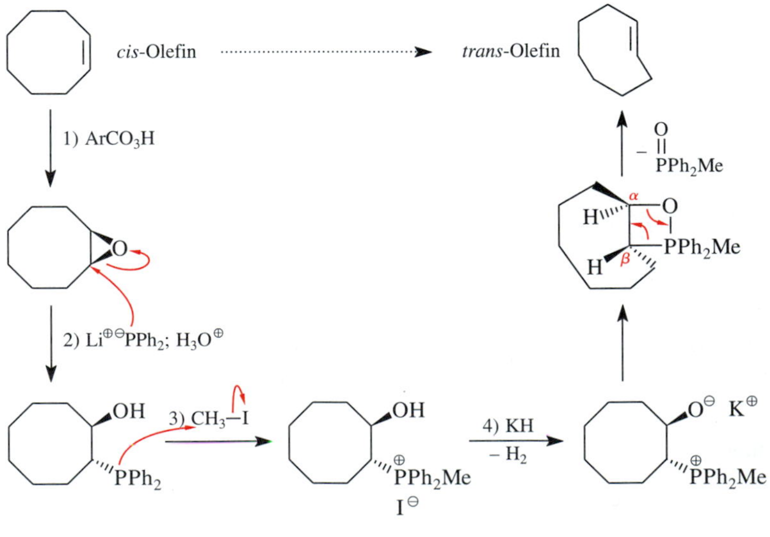

Fig. 4.40. *syn*-Selective $Ph_2MeP{=}O$ elimination as a key step in the preparation of *trans*-cyclooctene.

not produced via a P-ylide). In Figure 4.40, this elimination is part of a so-called olefin inversion in which, via a four-step reaction sequence, an olefin such as *cis*-cyclooctene, a molecule with little strain, can be converted into its *trans*- and, in this case, more highly strained, isomer.

The first steps in this sequence are (1) *cis*-selective epoxidation (for the mechanism, see Section 3.3.2), (2) ring opening of the epoxide with lithium diphenylphosphide in an S_N2 reaction with backside attack, (3) S_N2 reaction of the resulting alkyldiarylphosphine with MeI to furnish a hydroxyphosphonium salt. In the fourth step, this salt is deprotonated to give a zwitterion (internal salt, betaine) with the substructure $^{-}O—C—C—P^{+}Ph_2Me$. This zwitterion is less stable than the charge-free isomeric oxaphosphetane and consequently cyclizes to provide the latter. By way of a decomposition by [2+2]-cycloreversion as known from the Wittig olefination, $Ph_2MeP=O$ and the stereoisomer of the starting olefin are now produced. In the example in Figure 4.40, *trans*-cyclooctene is produced, which is unlikely to be accessible in another way. It is even more impressive that *cis,cis*-1,5-cyclooctadiene can be converted into *trans,trans*-1,5-cyclooctadiene in a completely analogous manner.

You will learn about the reaction of α-metallated phosphonic acid esters with aldehydes in Section 9.4. This reaction also seems to give a *trans*-configured oxaphosphetane (Figure 4.41). A *syn*-selective β-elimination by a [2+2]-cycloreversion of a compound containing a P=O double bond follows; this compound is $(EtO)_2P(=O)O^{-}$. The second elimination product is an olefin. It is predominantly or exclusively *trans*-configured.

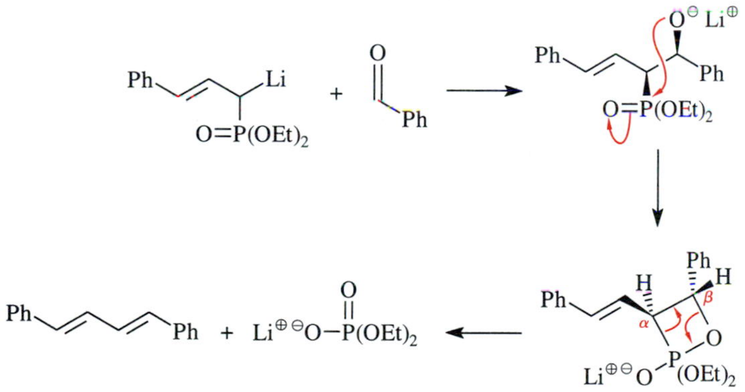

Fig. 4.41. A Het[1]/Het[2] elimination from the presumed oxaphosphetane intermediate of a Horner–Wadsworth–Emmons reaction

4.7.5 Corey–Winter Reaction

Vicinal *syn*- and *anti*-diols, as shown in Figure 4.42, can be prepared diastereoselectively (cf. Figures 8.10, 8.13–8.15, 8.32). In the Corey–Winter process they are first converted into cyclic thiocarbonates (cf. Section 6.4.4 for a similar reaction mechanism). Upon heating in trimethyl phosphite, these thiocarbonates furnish olefins. In what is evidently a one-step reaction, phosphorus and sulfur combine with one another and the five-membered heterocycle fragments. CO_2 is released and the olefin results from a *syn*-elimination. Because of the latter, a *syn*-diol gives the *trans*-olefin and an *anti*- diol gives the *cis*-olefin in the Corey–Winter sequence.

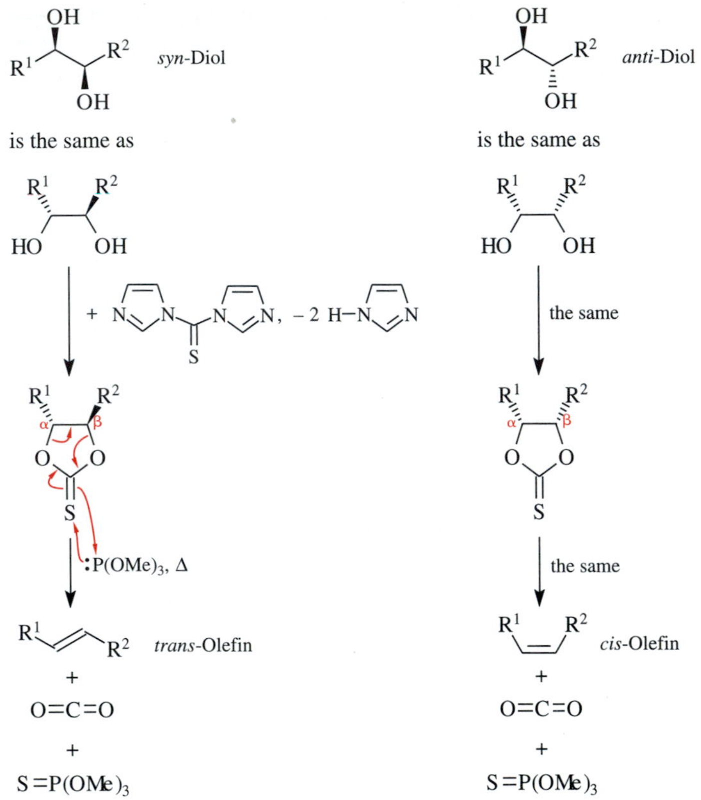

Fig. 4.42. Stereoselectivity and stereospecificity in Corey–Winter eliminations from epimeric 1,2-diols.

References

S. E. Kelly, "Alkene Synthesis," in *Comprehensive Organic Synthesis* (B. M. Trost, I. Fleming, Eds.), Vol. 1, 729, Pergamon Press, Oxford, **1991.**

A. Krebs and J. Swienty-Busch, "Eliminations to Form Alkenes, Allenes and Alkynes and Related Reactions," in *Comprehensive Organic Synthesis* (B. M. Trost, I. Fleming, Eds.), Vol. 6, 949, Pergamon Press, Oxford, **1991.**

J. M. Percy, "One or More C=C Bond(s) by Elimination of Hydrogen, Carbon, Halogen or Oxygen Functions," in *Comprehensive Organic Functional Group Transformations* (A. R. Katritzky, O. Meth-Cohn, C. W. Rees, Eds.), Vol. 1, 553, Elsevier Science, Oxford, U.K., **1995.**

J. M. J. Williams (Ed.), "Preparation of Alkenes: A Practical Approach," Oxford University Press, Oxford, U.K., **1996.**

4.2

H. J. Reich and S. Wollowitz, "Preparation of α,β-unsaturated carbonyl compounds and nitriles by selenoxide elimination," *Org. React. (N.Y.)* **1993,** *44,* 1.

A. Krief and A.-M. Laval, "*o*-Nitrophenyl selenocyanate, a valuable reagent in organic synthesis: Application to one of the most powerful routes to terminal olefins from primary-alcohols (the Grieco-Sharpless olefination reaction) and to the regioselective isomerization of allyl alcohols)," *Bull. Soc. Chim. Fr.* **1997,** *134,* 869–874.

Y. Nishibayashi and S. Uemura, "Selenoxide elimination and [2,3]-sigmatropic rearrangement," *Top. Curr. Chem.* **2000,** *208,* 201–234.

H. R. Nace, "The preparation of olefins by the pyrolysis of xanthates. The Chugaev reaction," *Org. React.* **1962,** *12,* 57–100.

S. Z. Zard, "On the trail of xanthates: Some new chemistry from an old functional group," *Angew. Chem.* **1997,** *109,* 724–737; *Angew. Chem. Int. Ed. Engl.* **1997,** *36,* 672–685.

4.4

E. Baciocchi, "Base dependence of transition-state structure in alkene-forming E2 reactions," *Acc. Chem. Res.* **1979,** *12,* 430.

J. K. Crandall and M. Apparu, "Base-promoted isomerizations of epoxides," *Org. React.* **1983,** *29,* 345–443.

A. C. Cope and E. R. Trumbull, "Olefins from amines: The Hofmann elimination reaction and amine oxide pyrolysis," *Org. React.* **1960,** *11,* 317–493.

4.5

M. F. Vinnik and P. A. Obraztsov, "The mechanism of the dehydration of alcohols and hydration of alkenes in acid solution," *Russ. Chem. Rev.* **1990,** *59,* 106–131.

A. Thibblin, "Mechanisms of solvolytic alkene-forming elimination-reactions," *Chem. Soc. Rev.* **1993,** *22,* 427.

R. F. Langler, "Ionic reactions of sulfonic acid esters," *Sulfur Rep.* **1996,** *19,* 1–59.

4.7

E. Block, "Olefin synthesis via deoxygenation of vicinal diols," *Org. React.* **1984,** *30,* 457–566.

P. Kocienski, "Reductive Elimination, Vicinal Deoxygenation and Vicinal Desilylation," in *Comprehensive Organic Synthesis* (B. M. Trost, I. Fleming, Eds.), Vol. 6, 975, Pergamon Press, Oxford, **1991.**

D. J. Ager, "The Peterson olefination reaction," *Org. React.* **1990,** *38,* 1–223.

A. G. M. Barrett, J. M. Hill, E. M. Wallace, J. A. Flygare, "Recent studies on the Peterson olefination reaction," *Synlett* **1991,** 764–770.

T. Kawashima and R. Okazaki, "Synthesis and reactions of the intermediates of the Wittig, Peterson, and their related reactions," *Synlett* **1996,** 600.

M. Julia, "Recent advances in double bond formation," *Pure Appl. Chem.* **1985,** *57,* 763.

Further Reading

H. N. C. Wong, C. C. M. Fok, T. Wong, "Stereospecific deoxygenation of epoxides to olefins," *Heterocycles* **1987,** *26,* 1345.

M. Schelhaas and H. Waldmann, "Protecting group srategies in organic synthesis," *Angew. Chem.* **1996,** *108,* 2192–2219; *Angew. Chem. Int. Ed. Engl.* **1996,** *35,* 2056–2083.

K. Jarowicki and P. Kocienski, "Protecting groups," *Contemp. Org. Synth.* **1996,** *3,* 397–431.

Substitution Reactions on Aromatic Compounds

5

Substitution reactions on aromatic compounds are the most important methods for the preparation of aromatic compounds. Synthesizing them from nonaromatic precursors is considerably less important. Via substitution reactions, electrophiles and nucleophiles can be introduced into aromatics. A series of mechanisms is available for this. Those that are discussed in this chapter are listed in Table 5.1.

B

5.1 Electrophilic Aromatic Substitutions via Wheland Complexes (“Ar-S_E Reactions”)

The electrophilic aromatic substitution via Wheland complexes, or the Ar-S_E reaction, is the classical method for functionalizing aromatic compounds. In this section we will focus on the mechanistic foundations as well as the preparative possibilities of this process.

B

5.1.1 Mechanism: Substitution of H^+ vs *ipso*-Substitution

For an Ar-S_E reaction to be able to occur, first the actual electrophile must be produced from the reagent (mixture) used. Then this electrophile initiates the aromatic substitution. It takes place, independently of the chemical nature of the electrophile, essentially according to a two-step mechanism (Figure 5.1). A third step, namely, the initial formation of a π complex from the electrophile and the substrate, is generally of minor importance for understanding the reaction event.

B

In the first step of the actual Ar-S_E reaction, a substituted cyclohexadienyl cation is formed from the electrophile and the aromatic compound. This cation and its derivatives are generally referred to as a σ or Wheland complex. Wheland complexes are described in the language of the VB method by superpositioning mentally at least three carbenium ion resonance forms (Figure 5.1). In the following, these resonance forms are referred to briefly as **“sextet formulas.”** There is an additional resonance form for each substituent, which can stabilize the positive charge of the Wheland complex by a +M effect (see Section 5.1.3). This resonance form is an all-octet formula.

Wheland complexes are high-energy intermediates because they do not contain the conjugated aromatic electron sextet present in the product and in the starting material. Consequently, the formation of these complexes is the rate-determining step of

Table 5.1. Substitution Reactions of Aromatic Compounds: Mechanistic Alternatives*

Section	Type of substitution using a benzene derivative as an example	Substitution type also known for: naphthalene	Substitution type also known for: five-membered ring aromatic compounds	Mechanistic designation
5.1–5.2	R_x–C₆H₄–H $\xrightarrow{E^\oplus}$ R_x–C₆H₄–E	yes	yes	"Classic Ar-S_E"
5.3.2	R_x–C₆H₄–Br $\xrightarrow[\text{or BuLi; } E^\oplus]{\text{Mg or Li}}$ R_x–C₆H₄–E	yes	yes	Ar-S_E via organometallic compounds
5.3.1	R_x–C₆H₃(MDG)–H $\xrightarrow[E^\oplus]{sec\text{-BuLi;}}$ R_x–C₆H₃(MDG)–E	yes	yes	
5.3.3/13.3.2	R_x–C₆H₄–$B(OR)_2$ $\xrightarrow[Pd(PPh_3)_4]{E^\oplus\text{, cat.}}$ R_x–C₆H₄–E	yes	yes	The same; also transition metal-mediated C,C coupling
5.4	R_x–C₆H₄–$\overset{\oplus}{N}\equiv N$ $\xrightarrow{Nu^\ominus}$ R_x–C₆H₄–Nu	yes	no	Ar-S_N1
	R_x–C₆H₄–$\overset{\oplus}{N}\equiv N$ $\xrightarrow{Nu^\ominus/Cu(I)}$ R_x–C₆H₄–Nu	yes	no	
5.5	EWG–C₆H₄–Hal $\xrightarrow{Nu^\ominus}$ EWG–C₆H₄–Nu	yes	no	Ar-S_N via Meisenheimer complexes
–	R_x–C₆H₄–Hal $\xrightarrow[Nu^\ominus/Pd(0)]{Nu^\ominus/Cu(I)\text{ or}}$ R_x–C₆H₄–Nu	yes	no	Ar-S_N of the Ullmann type
13.1–13.3		yes	yes	Transition metal-mediated C,C coupling
5.6	R_x–C₆H₄–Hal $\xrightarrow{Nu^\ominus}$ R_x–C₆H₄–Nu (ortho) + R_x–C₆H₄–Nu (meta)	yes	no	Ar-S_N via arynes

*MDG, metallation-directing group; EWG, electron-withdrawing group.

Wheland complex (or σ complex)

Fig. 5.1. The common reaction mechanism for all Ar-S_E reactions. Almost always we have X = H and rarely X = *tert*-Bu or X = SO_3H.

Ar-S_E reactions (cf. Figure 5.1). This, in turn, means that Wheland complexes are also a good—even the best—model for the transition state of Ar-S_E reactions.

In the second step of the Ar-S_E reaction, an aromatic compound is regenerated by cleaving off a cation from the C atom which was attacked by the electrophile. Most often the eliminated cation is a proton (Figure 5.1, X = H or $X^+ = H^+$).

In a few cases, cations other than the proton are eliminated from the Wheland complex to reconstitute the aromatic system. The *tert*-butyl cation (Figure 5.1, X=*tert*-Bu) and protonated SO_3 (Figure 5.1, X=SO_3H) are suitable for such an elimination. When the latter groups are replaced in an Ar-S_E reaction, we have the special case of an *ipso* substitution. Among other things, *ipso* substitutions play a role in the few Ar-S_E reactions that are reversible (Section 5.1.2).

5.1.2 Thermodynamic Aspects of Ar-S_E Reactions

Substitution and Addition Compared: Heats of Reaction

As you know, Br_2 *adds* to olefinic C=C bonds (Section 3.5.1). On the other hand, Br_2 *replaces* an sp^2-bonded H atom on the formal C=C double bond of aromatic compounds. Why is it not the other way around? That is, why do cyclohexene and Br_2 not give 1-bromocyclohexene via a substitution reaction, and why do benzene and Br_2 not give a dibromocyclohexadiene via an addition reaction?

If one compares the heats of reaction for these potentially competing reactions (Figure 5.2), one arrives at the following:

1) The substitution reaction C_{sp^2}—H + Br-Br → C_{sp^2}—Br + H-Br is exothermic by approximately −11 kcal/mol. It is irrelevant whether the attacked sp^2-hybridized carbon atom is part of an olefin or an aromatic compound.
2) In an addition reaction C=C + Br—Br → Br—C—C—Br there is a decrease in enthalpy of 27 kcal/mol *in the substructure shown,* that is, a reaction enthalpy of −27 kcal/mol. This enthalpy decrease equals the heat of reaction liberated when Br_2 is added to cyclohexene. However, it does not equal the heat of reaction for the addition of Br_2 to benzene.

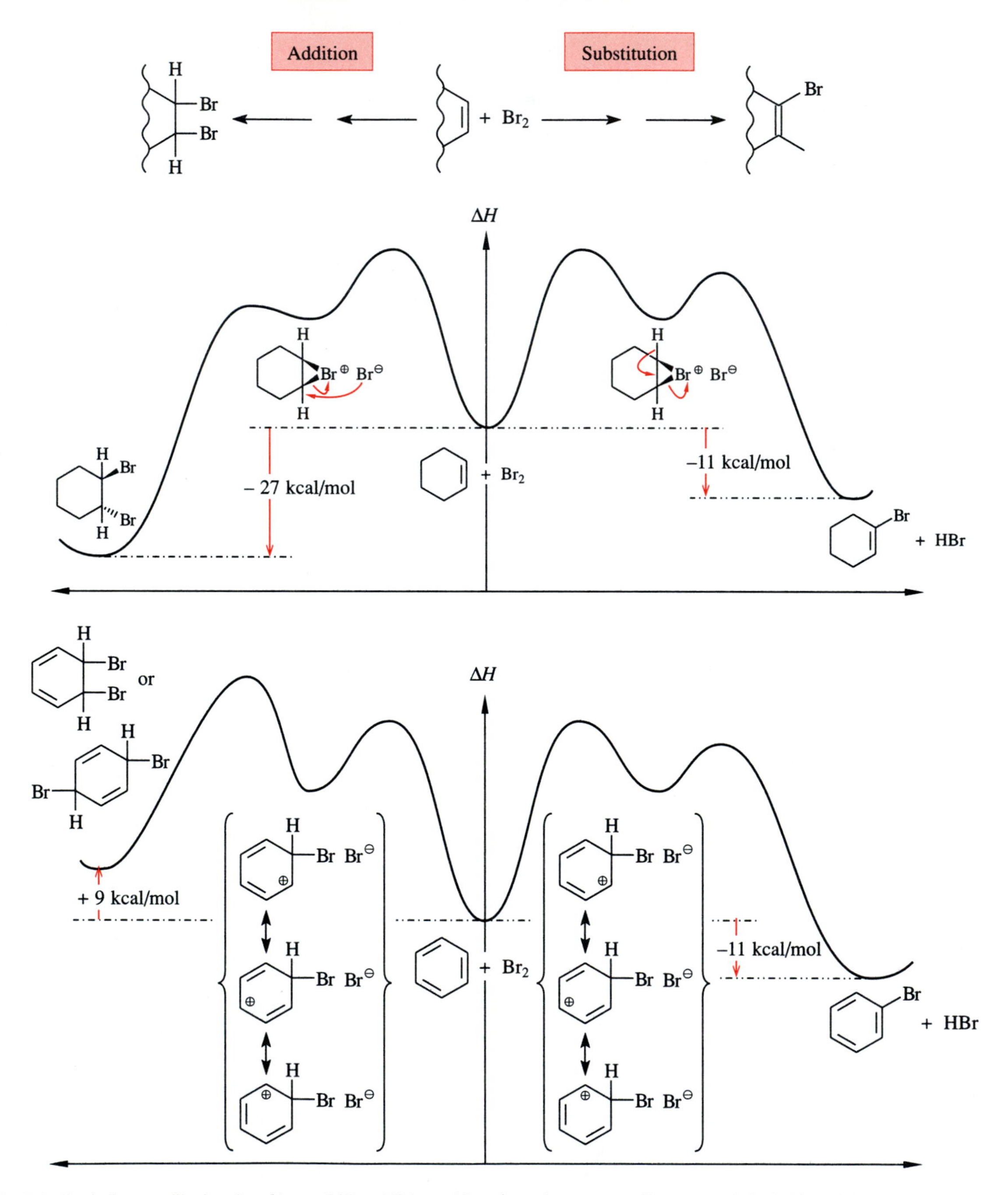

Fig. 5.2. Enthalpy profile for the electrophilic addition of Br_2 (reactions proceeding toward the left) and for the electrophilic substitution by Br_2 (reactions proceeding toward the right) of cyclohexene (top) and of benzene (bottom).

3) When Br_2 is added to benzene, the above-mentioned −27 kcal/mol must be balanced with the simultaneous loss of the benzene conjugation, which is +36 kcal/mol. All in all, this makes the addition of Br_2 to benzene by approximately +9 kcal/mol endothermic. Moreover, when Br_2 is added to benzene, the entropy decreases. Consequently, the addition of bromine to benzene would not only be endothermic but also endergonic. The latter means that such an addition is thermodynamically impossible.
4) The exothermic substitution reaction (see above) on benzene is also exergonic because no significant entropy change occurs. This substitution is therefore thermodynamically possible and actually takes place under suitable reaction conditions (Section 5.2.1).
5) Finally, because the addition of Br_2 to cyclohexene is 27 kcal/mol − 11 kcal/mol = 16 kcal/mol more exothermic than the substitution of Br_2 on cyclohexene can we conclude that the first reaction also takes place more rapidly? Not necessarily! The (fictitious) substitution reaction of Br_2 on cyclohexene should be a multi-step reaction and proceed via a bromonium ion formed in the first and also rate-determining reaction step. This bromonium ion has been demonstrated to be the intermediate in the known addition reaction of Br_2 to cyclohexene (Section 3.5.1). Thus, one would expect that the outcome of the competition of substitution vs addition depends on whether the bromonium ion is converted—*in each case in an elementary reaction*—to the substitution or to the addition product. The Hammond postulate suggests that the bromonium ion undergoes the more exothermic (exergonic) reaction more rapidly. In other words, the addition reaction is expected to win not only thermodynamically but also kinetically.

A

ipso Substitutions and the Reversibility of Ar-S_E Reactions

According to Section 5.1.1 electrophilic *ipso* substitutions via Wheland complexes occur, for example, when a proton attacks at the substructure C_{sp^2}—*tert*-Bu or C_{sp^2}—SO_3H of appropriately substituted aromatic compounds. After expulsion of a *tert*-butyl cation or an HSO_3^+ ion, an aromatic compound is obtained, which has been defunctionalized in the respective position.

Figure 5.3 shows an example of a de-*tert*-butylation by a reaction of this type. In fact, *both tert*-butyl groups of the aromatic compounds shown in the figure are removed by *ipso* substitutions. The fate of the *tert*-butyl cations released depends on whether a reactive solvent such as benzene (Figure 5.3 shows this case) or an inert solvent (not shown) is used. The benzene is *tert*-butylated by *tert*-butyl cations in a Friedel–Crafts alkylation (cf. Section 5.2.5). From the perspective of the *tert*-butyl groups, this *ipso* substitution represents an S_N1 reaction with benzene acting as the nucleophile. The driving force for this reaction is based on thermodynamic control: the *tert*-butyl groups leave a sterically hindered substrate and enter into a sterically unhindered product. If the de-*tert*-butylation of the same substrate is carried out in an inert solvent instead of in benzene, it becomes an E1 elimination of Ar—H from the perspective of the *tert*-butyl groups: the released *tert*-butyl cations are deproto-

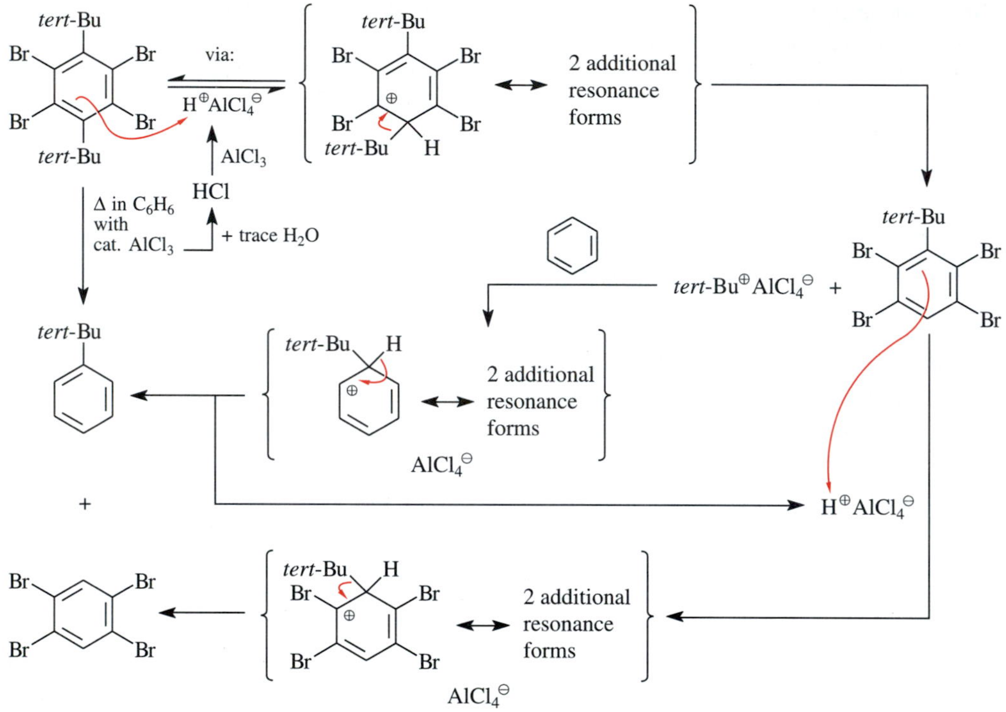

Fig. 5.3. De-*tert*-butylation via Ar–S_E reaction.

nated to give isobutene. If the latter is continuously distilled off the reaction mixture, the de-*tert*-butylation equilibrium is also shifted toward the defunctionalized aromatic compounds.

The *tert*-butyl cations liberated from compounds Ar—*tert*-Bu upon *ipso*-attack of a proton may also reattack the aromatic compound from which they stemmed. If this course is taken, the *tert*-butyl groups are ultimately bound to the aromatic nucleus with a regioselectivity that is dictated by thermodynamic control. Figure 5.4 shows how in this way 1,2,4-tri-*tert*-butylbenzene is smoothly isomerized to give 1,3,5-tri-*tert*-butylbenzene.

Fig. 5.4. De-*tert*-butylation/re-*tert*-butylation as a possibility for isomerizing *tert*-butylated aromatic compounds via Ar-S_E reactions.

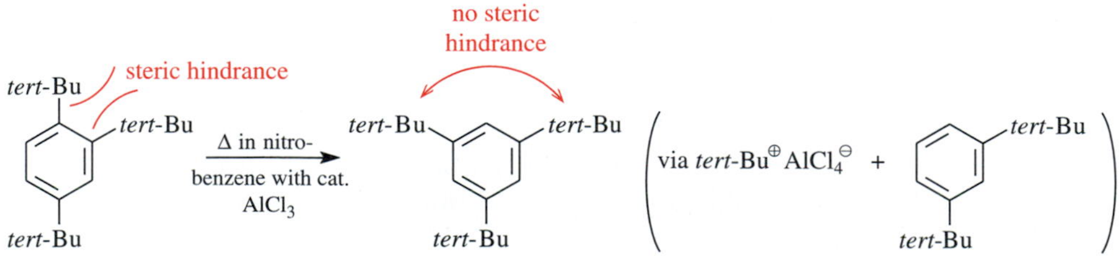

Side Note 5.1 Large-Scale Preparation of a Schäffer Acid

The other important *ipso* substitution by the Ar-S_E mechanism is the protodesulfonylation of aromatic sulfonic acids. It is used, for example, in the industrial synthesis of 2-hydroxynaphthalene-6 sulfonic acid (Schäffer acid), an important coupling component for the production of azo dyes. Schäffer acid is produced by sulfonylating 2-naphthol twice (for the mechanism, see Section 5.2.2) and then desulfonylating once. The first SO_3H group is introduced into the activated 1-position of 2-naphthol. The second SO_3H group is introduced into the 6-position of the initially formed 2-hydroxynaphthalene-1-sulfonic acid. The second step of the industrial synthesis of Schäffer acid is therefore the regioselective monodesulfonylation of the resulting 2-hydroxynaphthalene-1,6-disulfonic acid at the 1-position (Figure 5.5). Of course, the expelled electrophile HSO_3^+ must not re-attack the monosulfonylated product. Therefore, it must be scavenged by another reagent. The simplest way to do this is by desulfonylation with dilute sulfuric acid. The water contained therein scavenges the HSO_3^+ cations to form HSO_4^- and H_3O^+ or SO_4^{2-} and H_3O^+.

Δ in dilute H_2SO_4:

5 sextet resonance forms

in dilute H_2SO_4 irreversible

Fig. 5.5. Desulfonylation via Ar-S_E reaction (see Figure 5.10 for an explanation of this regioselectivity).

If the released electrophile HSO_3^+ is *not* intercepted during the protodesulfonylation as in Figure 5.5, it attacks the defunctionalized aromatic compound again. In this way an isomer of the original sulfonic acid may be obtained. The best-known example of such an isomerization is the conversion of naphthalene-1-sulfonic acid into naphthalene-2-sulfonic acid (Figure 5.6). Naphthalene-1-sulfonic acid is destabilized by the so-called *peri*-interaction, that is, the steric interaction between the C^8—H bond of the naphthalene and the substituent on C1. The *peri*-interaction is thus a *cis*-olefin strain. Because naphthalene-2-sulfonic acid does not suffer from this interaction, it becomes the only reaction product under conditions of thermodynamic control.

In conclusion we can make the following statement: most Ar-S_E reactions are irreversible because they have a sufficiently strong driving force and, at the same time, because they can be carried out under sufficiently mild conditions. The most important reversible Ar-S_E reactions are *tert*-alkylation and sulfonylation.

B

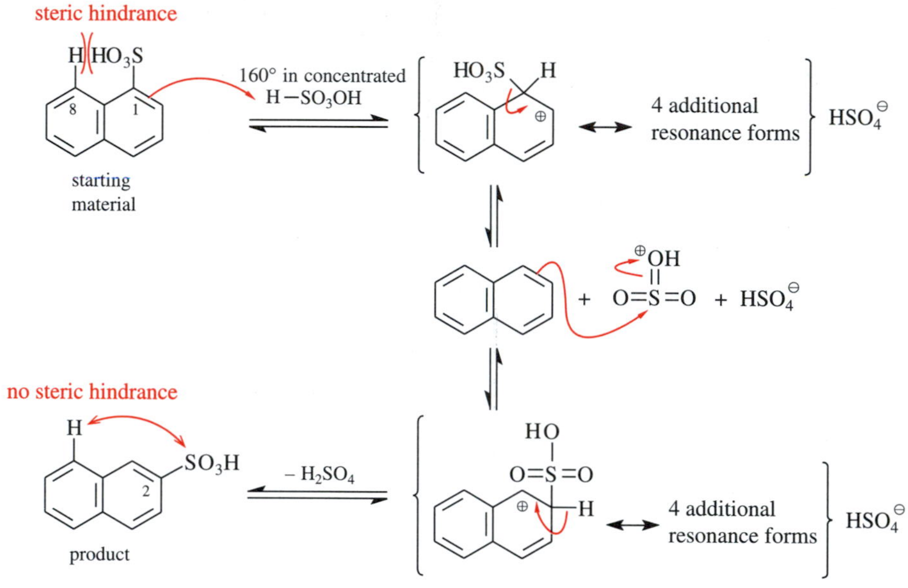

Fig. 5.6. Desulfonylation/resulfonylation as a possibility for isomerizing aromatic sulfonic acids via Ar-S_E reactions.

5.1.3 Kinetic Aspects of Ar-S_E Reactions: Reactivity and Regioselectivity in Reactions of Electrophiles with Substituted Benzenes

B

In the rate-determining step of an Ar-S_E reaction, a monosubstituted benzene and an electrophile can form three isomeric Wheland complexes. By the subsequent elimination of a proton, one Wheland complex gives the *ortho*-disubstituted, the second Wheland complex gives the *meta*-disubstituted, and the third Wheland complex gives the *para*-disubstituted benzene. According to Section 5.1.1, Wheland complexes are also excellent transition state models for the rate-determining step of Ar-S_E reactions. According to the Hammond postulate the more stable Wheland complexes are, the faster they form. To the extent that they form irreversibly, it is inferred that the major substitution product is the one that forms from the most stable Wheland complex.

Stabilization and Destabilization of Wheland Complexes through Substituent Effects

B

Which Wheland complexes are the most stable? This is determined to a small extent by steric effects and to a considerably greater extent by electronic effects: As a carbocation, a substituted Wheland complex is considerably more stable than an unsubstituted one only when it carries one or more donor substituents, and unsubstituted Wheland complexes $E—C_6H_6^+$ are still considerably more stable than Wheland complexes that contain one or more acceptor substituents. Therefore, donor-substituted benzenes are attacked by electrophiles more rapidly than benzene, and acceptor-substituted benzenes are attacked more slowly.

A more detailed analysis of the stabilizing effect of donor substituents and the destabilizing effect of acceptor substituents (both are referred to as "Subst" in the following) on Wheland complexes E-$C_6H_5^+$-Subst explains, moreover, the regioselectivity of an Ar-S_E attack on a monosubstituted benzene. *Isomeric donor-containing Wheland complexes and acceptor-containing Wheland complexes have different stabilities.* This follows from the uneven charge distributions in the Wheland complexes.

Figure 5.7 shows on the left side the "primitive model" of the charge distribution in the Wheland complex. Therein the positive charges only appear *ortho* and *para* to the attacked C atom, and in each case they equal +0.33. This charge distribution is obtained by superimposing the three resonance forms of Figure 5.1.

most simple model: + 0.33, + 0.33, + 0.33 (H, E)

refined model: + 0.25, + 0.10, + 0.30, + 0.10, + 0.25 (H, E)

Fig. 5.7. Charge distribution in Wheland complexes E-$C_6H_6^+$ (refined model, on the right: calculation for E = H).

The formula on the right in Figure 5.7 gives a more subtle model of the charge distribution in the Wheland complex. There positive partial charges exist on all five sp^2-hybridized ring atoms. The greatest charge (+0.30) can be found in the position *para* to the attacked C atom, a somewhat smaller charge (+0.25) in the *ortho* position, and a much smaller charge (+0.10) in the *meta* position.

Equipped with this refined charge distribution model, it is now possible to compare (1) the stabilities of isomeric donor-substituted Wheland complexes with each other, (2) the stabilities of isomeric acceptor-substituted Wheland complexes with each other, and (3) the stabilities of each Wheland complex already mentioned with the stability of the unsubstituted Wheland complex E-$C_6H_6^+$. The results can be found in Figure 5.8 for donor-substituted and in Figure 5.9 for acceptor-substituted Wheland complexes.

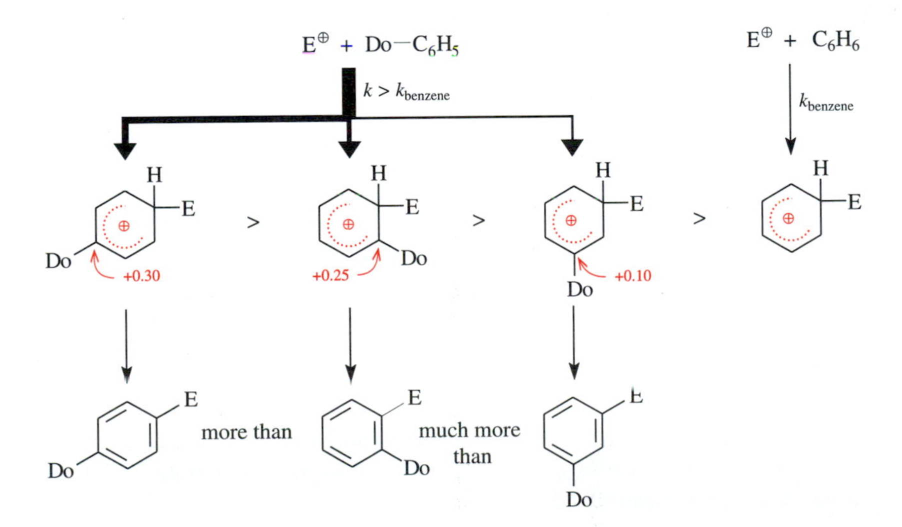

Fig. 5.8. Ar-S_E reactions with donor-substituted benzenes (Do, donor substituent); comparing the regioselectivity and the reactivity with benzene. The thicknesses of the initial arrows show qualitatively to what extent the reaction takes place via the corresponding transition state.

Every donor-substituted Wheland complex E-C_6H_5-Do^+ is more stable than the reference compound E-$C_6H_6^+$ (Figure 5.8). Regardless of its position relative to the attacked C atom, the donor turns out to be always located on a partially positively charged C atom, which it stabilizes by donating electrons. Of course, the donor provides the greatest possible stabilization when it is bound to the C atom with the greatest positive charge (+0.30), and it effects the smallest stabilization when it is bound to the C atom with the smallest positive charge (+0.10). If the donor is bound to the C atom with the +0.25 charge, the result is stabilization of intermediate magnitude. This gives the following stability order for the Wheland complexes of interest: *para*-E—$C_6H_5^+$—Do > *ortho*-E—$C_6H_5^+$— Do > *meta*-E—$C_6H_5^+$—Do > E—$C_6H_6^+$.
Taking into account the Hammond postulate, this means two things for Ar-S_E reactions:

Reactivity and Regioselectivity in Ar-S_E Reactions of Donor-Substituted Aromatic Compounds

1) Each H atom of a donor-substituted aromatic compound should be substituted faster by an electrophile than an H atom in benzene.
2) Donor-substituted benzenes and electrophiles should produce mixtures of *para*- and *ortho*-disubstituted aromatic compounds, in which the *para*-disubstituted product is formed in a greater amount. Only traces of the *meta*-disubstitution product are expected, although even at the *meta*-C atom a donor-substituted benzene is substituted faster than benzene itself at any of its C atoms.

Fig. 5.9. Ar-S_E reactions with acceptor-substituted benzenes (EWG, electron-withdrawing group); comparing the regioselectivity and the reactivity with benzene. The thicknesses of the initial arrows again indicate qualitatively to what extent the reaction takes place via the respective transition states (but there is no relationship to the thicknesses of the arrows in Figure 5.8).

Because of completely analogous considerations, every acceptor-substituted Wheland complex E—C_6H_5—EWG^+ is less stable than the reference compound E—$C_6H_6^+$ (Figure 5.9). From this analysis, one derives the following expectations for Ar-S_E reactions of acceptor-substituted benzenes:

Reactivity and Regioselectivity in Ar-S_E Reactions of Acceptor-Substituted Aromatic Compounds

1) Each H atom of an acceptor-substituted aromatic compound should be substituted more slowly by an electrophile than an H atom in benzene.
2) The substitution product should primarily be the *meta*-disubstituted aromatic compound. Among the by-products, the *ortho*-disubstituted benzene should predominate, while practically no *para* product is expected.

Substituent Effects on Reactivity and Regioselectivity of Ar-S_E Reactions of Monosubstituted Benzenes

B

The vast majority of reactivities and regioselectivities observed in an attack by electrophiles on monosubstituted benzenes (Table 5.2) is in agreement with the preceding generalizations (columns 2 and 4). The very few substituents that are not in agreement (column 3) *deactivate* the aromatic compound as do electron acceptors, but they are *para*- > *ortho*-directing as are electron donors.

Table 5.2. Experimental Results Concerning the Regioselectivity and the Reactivity of Ar-S_E Reactions of Monosubstituted Benzenes

	Do	"Chameleon"-R	EWG
Substitution rate for each H atom of this aromatic compound compared with C_6H_5-H	greater	somewhat smaller	smaller
Regioselectivity	*para* > *ortho*	*para* > *ortho*	*meta**
The following can be introduced ...	... practically every $E^{\oplus}$	... majority of the $E^{\oplus}$	... only a good $E^{\oplus}$

*For the byproducts *ortho* > *para*

Therefore, the reactivity and the regioselectivity of Ar-S_E reactions with substituted benzenes can be predicted reliably. According to what has been stated above, one only has to identify the electron donating and withdrawing substituents in the substrate. The electronic effects of the most important functional groups are listed in Table 5.3, where they are ordered semiquantitatively: the best electron donors are on top and the best

Table 5.3. Inductive and Resonance Substituent Effects of Various Functional Groups

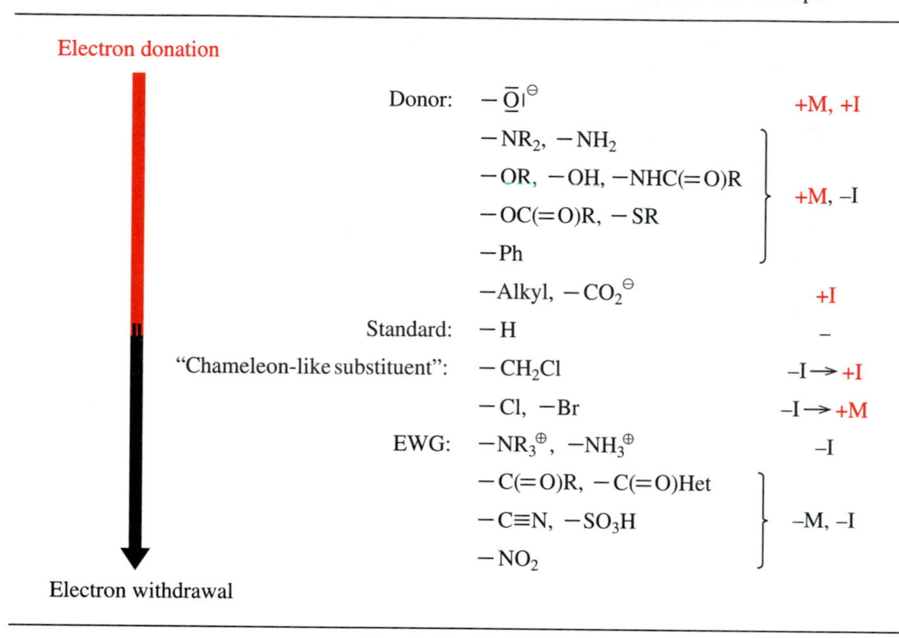

Electron donation ↓ Electron withdrawal		
Donor:	$-\overline{O}|^{\ominus}$	+M, +I
	$-NR_2$, $-NH_2$	+M, –I
	$-OR$, $-OH$, $-NHC(=O)R$	+M, –I
	$-OC(=O)R$, $-SR$	+M, –I
	$-Ph$	+M, –I
	$-$Alkyl, $-CO_2^{\ominus}$	+I
Standard:	$-H$	–
"Chameleon-like substituent":	$-CH_2Cl$	–I → +I
	$-Cl$, $-Br$	–I → +M
EWG:	$-NR_3^{\oplus}$, $-NH_3^{\oplus}$	–I
	$-C(=O)R$, $-C(=O)Het$	–M, –I
	$-C{\equiv}N$, $-SO_3H$	–M, –I
	$-NO_2$	–M, –I

electron acceptors are at the bottom. It should be emphasized, though, that the significance of these substituent effects for the understanding of organic reactions goes far beyond the scope of this chapter.

To complete this discussion, let us go back to those substituents (Table 5.2, column 3) that deactivate an aromatic compound like an electron acceptor but direct the electrophile like an electron donor in the *para* and *ortho* positions. The electronic effects of these substituents are obviously as variable as the color of a chameleon. These substituents destabilize each of the isomeric Wheland complexes through inductive electron withdrawal. However, the extent of electron withdrawal depends on the magnitude of the positive charge in the position of this substituent. If the positive charge is small—in the *meta* position with respect to the entering electrophile—electron withdrawal is important. Conversely, if the positive charge in the position of the substituent in question is large—in the *para* > *ortho* position with respect to the entering electrophile—then electron withdrawal is reduced. There are two reasons for this dependence on the substituents. (1) The "chameleon-like" $-$I substituents Cl and—to a lesser extent—Br lend some electron density to the Wheland complex through their +M effect during a *para* or an *ortho* attack of an electrophile. (2) The "chameleon-like" $-$I substituents CH_2Cl, Br, and I exert a diminished $-$I-effect during a *para* or an *ortho* attack of an electrophile, which results in a large positive partial charge in the α-position of the substituents, whereas they exert their full $-$I-effect during an analogous *meta* attack, which gives rise to a small positive partial change in the vicinity of these substituents. It is often observed that the ability of a substituent to withdraw electron density is smaller with respect to positively charged centers than with respect to centers without a charge.

B

Regioselectivity for Ar-S_E Reactions of Naphthalene

Although what was stated in Section 5.1.3 is indeed completely correct, one might think that the analysis of the partial charge distribution in the (not yet attacked) aromatic substrate could allow, too, an at least qualitatively correct prediction of the reactivity and the regioselectivity in Ar-S_E reactions. This opinion cannot be upheld, however, as an analysis of the rate and regioselectivity of Ar-S_E reactions with naphthalene clearly shows (Figure 5.10).

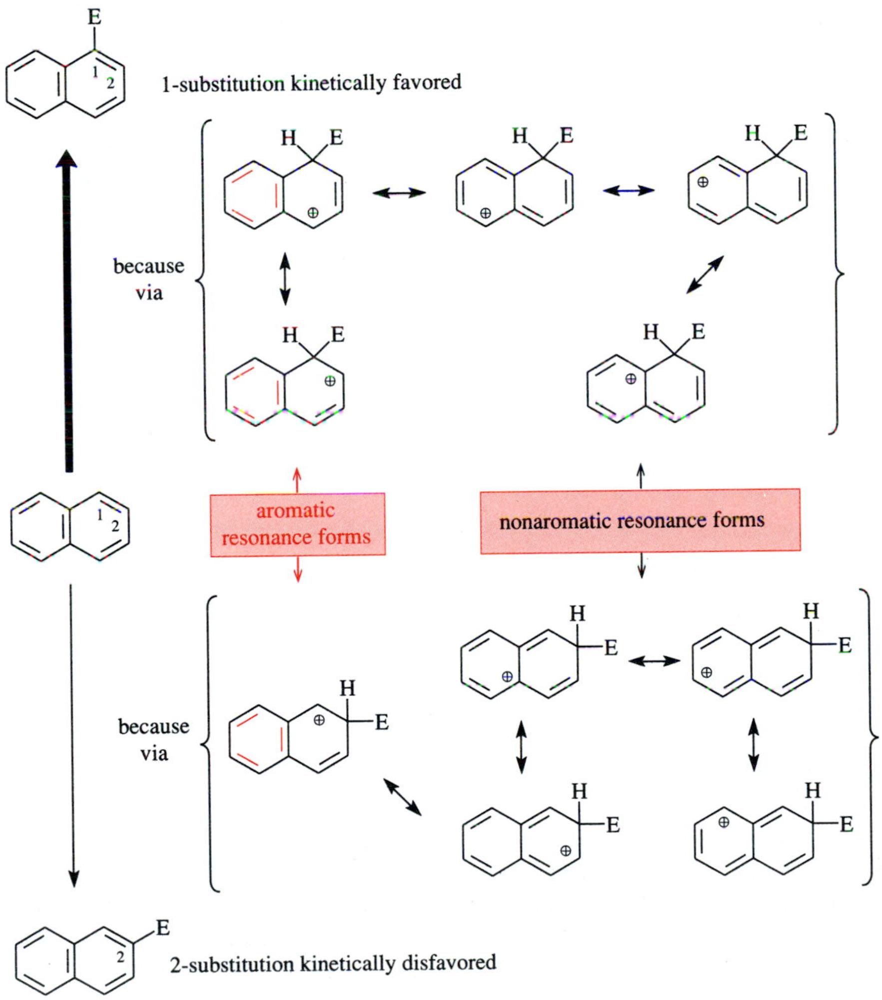

1-substitution kinetically favored

because via

aromatic resonance forms

nonaromatic resonance forms

because via

2-substitution kinetically disfavored

Fig. 5.10. Kinetically controlled Ar-S_E reactions of naphthalene.

There is neither a partial positive nor a partial negative charge on the two nonequivalent positions 1 and 2 of naphthalene, which are poised for electrophilic substitution. Based on the above-mentioned incorrect mechanistic model, one would consequently predict that electrophiles attack naphthalene without regiocontrol. Furthermore, this should occur with the same reaction rate with which benzene is attacked. Both predictions contradict the experimental results! For example, naphthalene is brominated with a 99:1 selectivity in the 1-position in comparison to the 2-position. The bromination at C1 takes place 12,000 times faster and the bromination at C2 120 times faster than the bromination of benzene.

The regioselectivity and reactivity of Ar-S_E reactions of naphthalene are explained *correctly* by comparing the free activation enthalpies for the formation of the Wheland complexes 1-E—$C_{10}H_{10}^+$ and 2-E—$C_{10}H_{10}^+$ from the electrophile and naphthalene and for the formation of the Wheland complex E—$C_6H_6^+$ from the electrophile and benzene, respectively.

With respect to the *reactivities,* the decisive effect is the following: in the formation of either of the two isomeric Wheland complexes from naphthalene, the difference between the naphthalene resonance energy (66 kcal/mol)—which is uplifted—and the benzene resonance energy (36 kcal/mol)—which is maintained—is lost, that is, an amount of only 30 kcal/mol. By contrast, the formation of a Wheland complex from benzene costs the full 36 kcal/mol of the benzene resonance energy. This explains why $k_{\text{naphthalene}} > k_{\text{benzene}}$.

The *regioselectivity* of Ar-S_E reactions with naphthalene follows from the different stabilities of the Wheland complex intermediate of the 1-attack (Figure 5.10, top) compared with that of the 2-attack (Figure 5.10, bottom). For the Wheland complex with the electrophile at C1 these are five sextet resonance forms. In two of them the aromaticity of one ring is retained. The latter forms are thus considerably more stable than the other three. The Wheland complex with the electrophile at C2 can also be described with five sextet resonance forms. However, only one of them represents an aromatic species. The first Wheland complex is thus more stable than the second. The 1-attack is consequently preferred over the 2-attack.

5.2 Ar-S_E Reactions via Wheland Complexes: Individual Reactions

5.2.1 Ar—Hal Bond Formation by Ar-S_E Reaction

B

Cl_2 and Br_2 react with donor-substituted aromatic compounds (aniline, acetanilide, phenol) without a catalyst. With nonactivated or deactivated aromatic compounds, these halogen molecules react only in the presence of a Lewis acid catalyst ($AlCl_3$ or $FeBr_3$). In the reaction example in Figure 5.11, as in other electrophilic substitution reactions, the following is true: if all substituents in multiply substituted aromatic compounds activate the same position (*cooperative substitution effect*), only that one position is attacked. On the other hand, if these substituents activate different positions (*competitive substitution effect*), regioselectivity can be achieved only when the directing effect of one substituent predominates. Consequently, in the reaction example in Figure 5.12 the stronger donor methoxy, not the weaker donor methyl, determines the course of the reaction and directs the electrophile to the position *ortho to itself.*

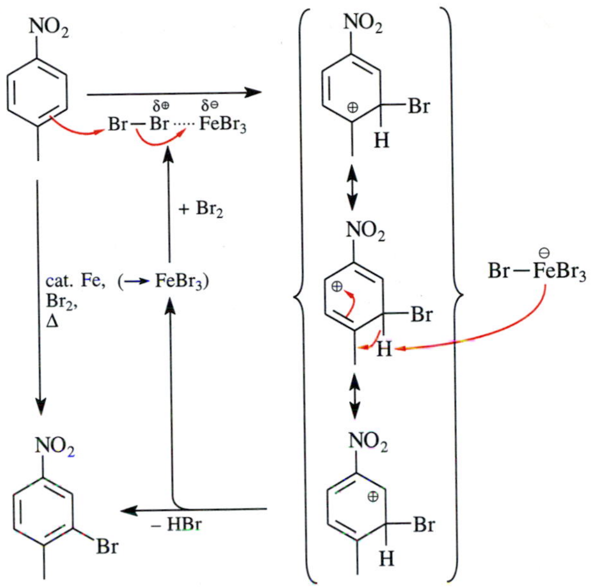

Fig. 5.11. Regioselective bromination of a benzene derivative that contains a strong acceptor and a weak donor substituent.

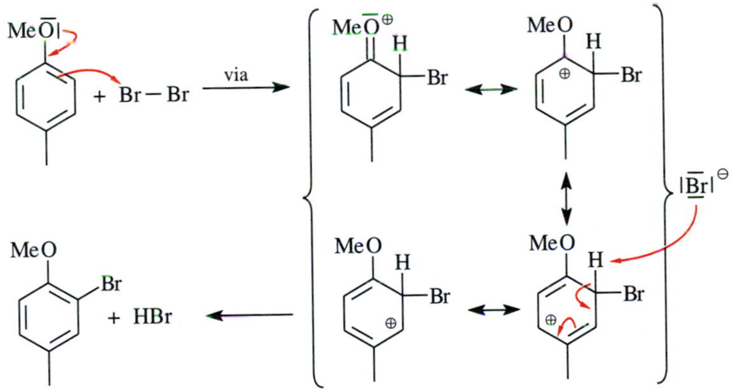

Fig. 5.12. Regioselective bromination of a benzene derivative that contains two donor substituents of different strength. The stronger donor methoxy directs to the position *ortho* to itself.

To obtain single or multiple electrophilic brominations chemoselectively, one can simply vary the stoichiometry:

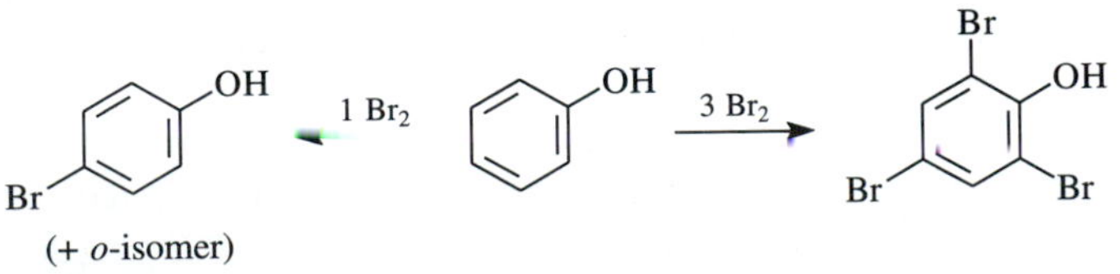

Or, to the same end, different brominating reagents can be used:

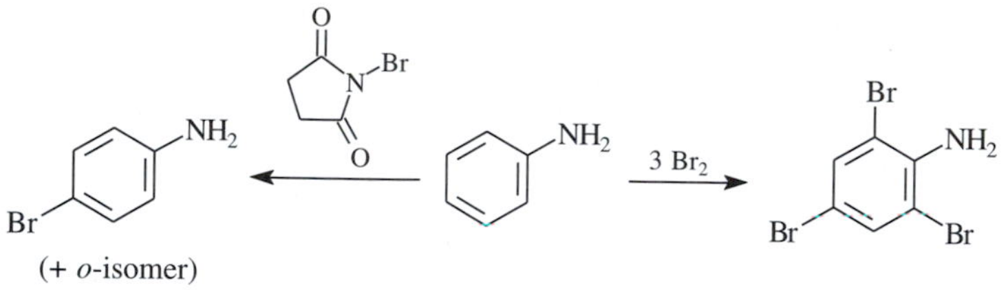

(+ *o*-isomer)

In alkylbenzenes the *ortho* and *para* H atoms or the benzyl CH atoms can be replaced by Br_2. Chemoselective substitutions in the *ortho* and *para* positions succeed at low temperature in the presence of a catalyst. Figure 5.13 shows an example; Figure 1.22 gave another example. Chemoselective substitutions with Br_2 in the benzyl

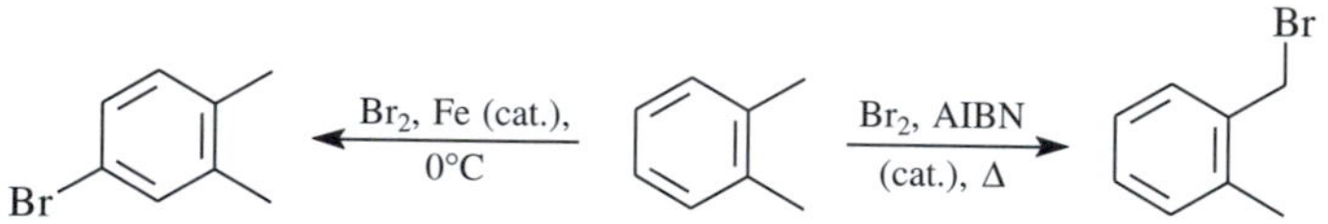

Fig. 5.13. Side chain and aromatic substitution in the reaction of *ortho*-xylene with Br_2.

position are possible using radical chemistry (for the mechanism, see the discussion of Figure 1.22), that is, by heating the reagents in the presence of a radical initiator or by irradiating them. Examples can be found in Figures 5.13 and 1.22.

When acetophenone is complexed with stoichiometric or greater amounts of $AlCl_3$, it is brominated by Br_2 in the position *meta* to the acceptor, that is, on the benzene ring.

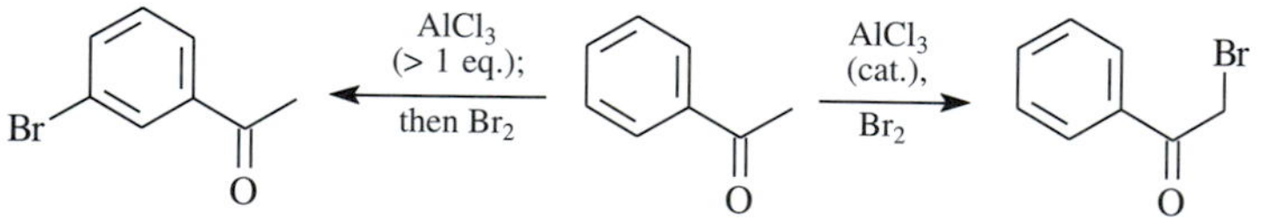

On the other hand, if only catalytic amounts of $AlCl_3$ are added, the acetyl group of the acetophenone is brominated. Under these conditions a catalytic amount of HCl is first formed. It allows the acetophenone to equilibrate with the tautomeric enol. This is a better nucleophile than the aromatic compound because it is brominated electrophilically without intermediate loss of the aromaticity. HBr is the stoichiometric byproduct of this substitution. Just like the HCl that is formed initially, it catalyzes the enolization of unreacted acetophenone and thus keeps the reaction going.

I_2 is a very weak electrophile. It is just reactive enough to attack the *para* position of aniline. Phenol ethers are attacked by iodine only in the presence of silver(I) salts ($\rightarrow$ AgI + I_3^+). Benzene and alkyl benzenes react with I_2 only when the iodine is activated still more strongly by oxidation with iodic or nitric acid.

5.2.2 Ar—SO_3H Bond Formation by Ar-S_E Reaction

B

An SO_3H group can be introduced into aromatic compounds through an electrophilic substitution reaction, which is referred to as **sulfonation.** Suitable reagents are dilute, concentrated, or fuming sulfuric acid. The H_2SO_4 molecule as such probably functions as the actual electrophile only in reactions with dilute sulfuric acid. If the sulfuric acid is used concentrated or even fuming, it contains much better electrophiles (Figure 5.14). One such electrophile, the protonated sulfuric acid ($H_3SO_4^+$), is produced by autoprotolysis of sulfuric acid, and another, disulfuric acid ($H_2S_2O_7$), by condensation of two molecules of sulfuric acid. The dehydration of protonated sulfuric acid ($H_3SO_4^+$) gives a third good electrophile (HSO_3^+), which can alternatively be generated by the protonation of the SO_3 fraction of fuming sulfuric acid.

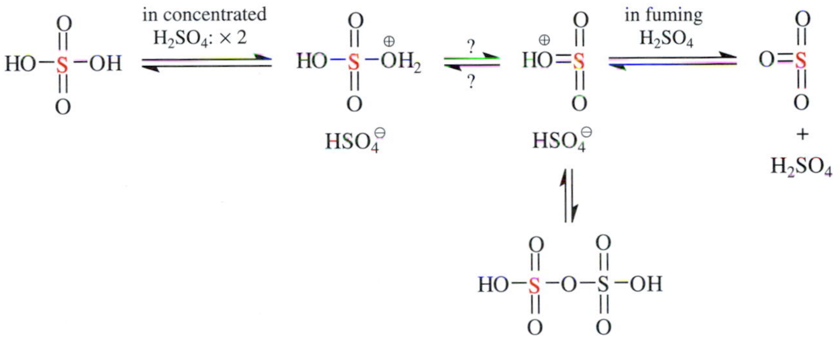

Fig. 5.14. Effective electrophiles in sulfonations of aromatic compounds with sulfuric acids of different concentrations.

Initially, an arylamine does not react as an activated aromatic compound with sulfuric acid but as a base instead: anilinium hydrogensulfates are produced (Figure 5.15). However, when the latter are heated, they decompose to give the starting materials reversibly. Only via the very small equilibrium amounts of free amine and free sulfuric acid does one observe the slow formation of the substitution product, an aromatic aminosulfonic acid. This aminosulfonic acid is a zwitterion, like an α-aminocarboxylic acid.

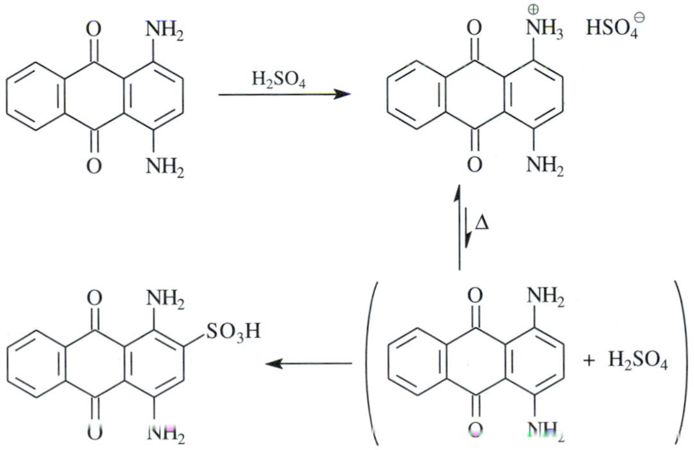

Fig. 5.15. Sulfonation of an aromatic amine.

Naphthalene can be sulfonated with concentrated sulfuric acid (Figure 5.16). At 80°C, this reaction proceeds under kinetic control and is therefore regioselective at the C1 center (cf. discussion of Figure 5.10). At 160°C the SO_3H group migrates into the

2-position under thermodynamic control. A mechanistic analysis of this reaction can be found in Figure 5.6.

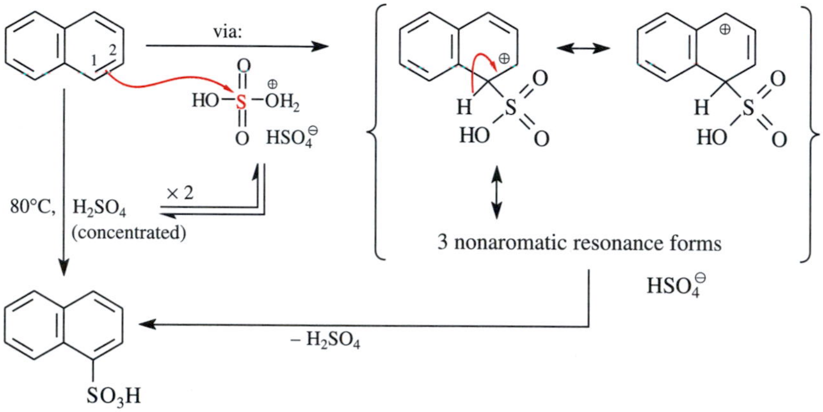

Fig. 5.16. Kinetically controlled sulfonation of naphthalene (thermodynamically controlled sulfonation of naphthalene: Figure 5.6).

Certain deactivated aromatic compounds can be sulfonated only with fuming rather than concentrated sulfuric acid:

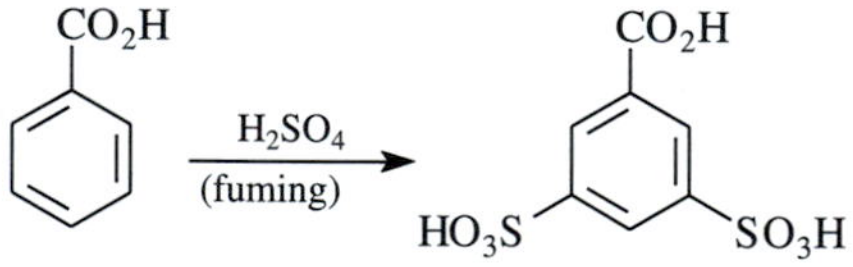

5.2.3 Ar—NO_2 Bond Formation by Ar-S_E Reaction

B

NO_2 groups are introduced into aromatic compounds through an Ar-S_E reaction with nitric acid **(nitration).** The HNO_3 molecule itself is a very weak electrophile, but it is the only electrophile present in dilute nitric acid. It is suitable for nitrating very strongly activated aromatic compounds only. (An NH_2 group would be protonated under these conditions, and the protonated group is deactivating. On the other hand, the AcNH group remains unprotonated; it therefore remains activating.)

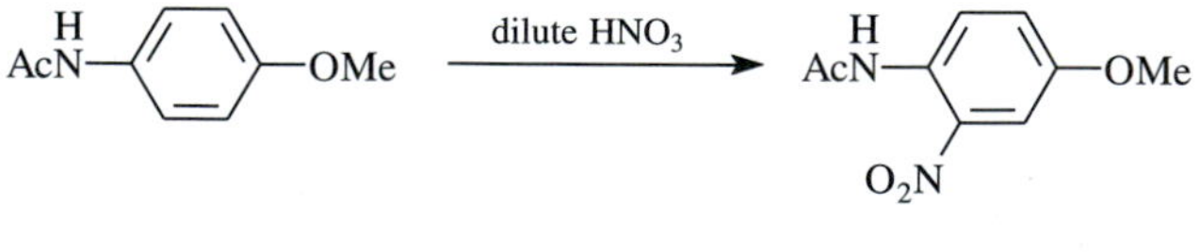

Moderately activated aromatic compounds can be nitrated only with concentrated nitric acid. Under these conditions NO_3^- and $H_2NO_3^+$ are produced in the autoprotolysis equilibrium (Figure 5.17). The NO_2^+ cation (nitronium ion), which is a considerably better electrophile than an HNO_3 molecule, is formed by eliminating a molecule of H_2O from $H_2NO_3^+$. Such a nitronium ion initiates, for example, the following reaction:

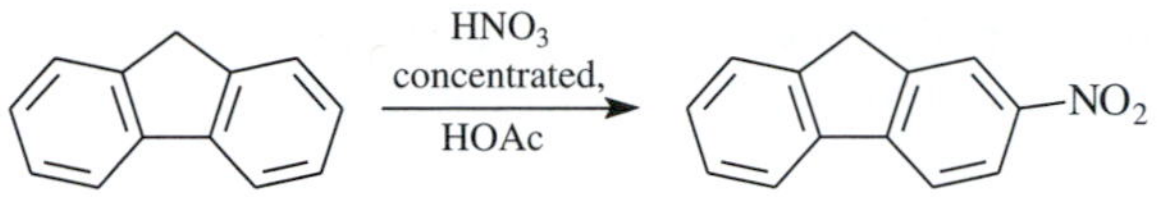

In this example the phenyl substituent with its +M effect determines the structure of the most stable Wheland complex intermediate and thus the regioselectivity. The competing +I effect of the alkyl substituent cannot prevail. This is understandable because it is not as effective at stabilizing an adjacent positive change as the phenyl group (cf. Table 5.3).

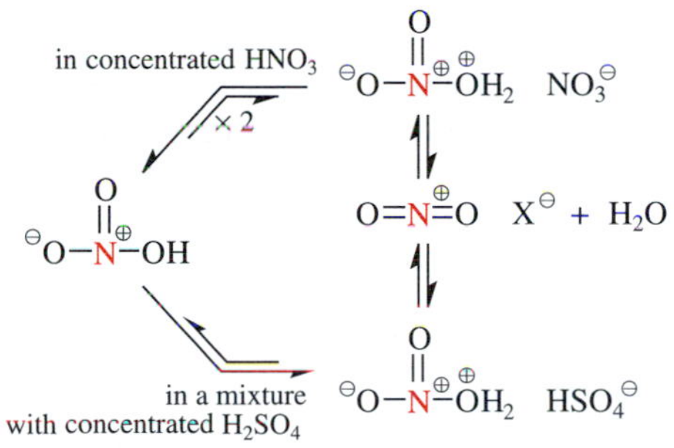

Fig. 5.17. Reactive electrophiles in the nitration of aromatic compounds with nitric acids with different concentrations and with H_2SO_4/HNO_3.

A higher fraction of nitronium ions is present in a mixture of concentrated sulfuric and concentrated nitric acid. In this medium the HNO_3 molecules are protonated to a larger extent than in nitric acid; this is essentially due to the sulfuric acid, which is a stronger acid than nitric acid (Figure 5.17).

An HNO_3/H_2SO_4 mixture is therefore also suitable for nitrating deactivated aromatic compounds. Aromatic amines are included in this category: In the very acidic reaction medium, they are protonated quantitatively. Thus, for example, the actual substrate of the nitration of *N,N*-dimethylaniline is an aromatic compound **A,** in which the ammonium substituent directs the attacking nitronium ion to the *meta* position because of its –I-effect:

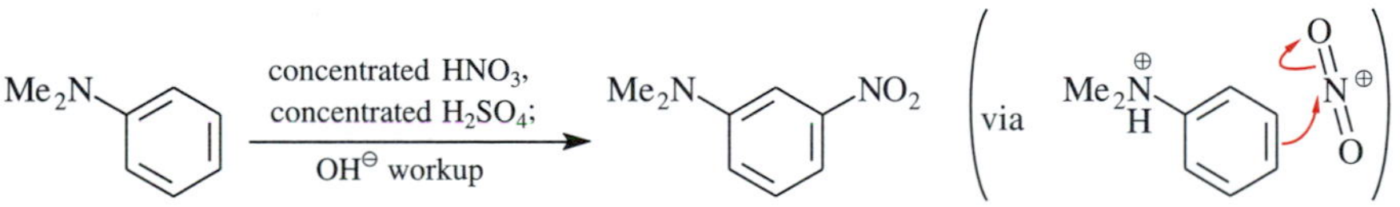

Still more strongly deactivated aromatic compounds are nitrated with *hot* HNO_3/H_2SO_4 or with a mixture of fuming nitric acid and concentrated sulfuric acid:

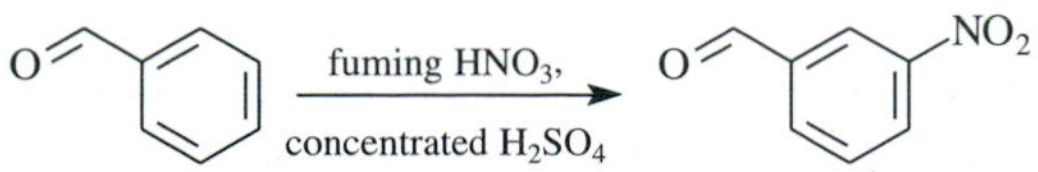

The third nitro group of the explosive 2,4,6-trinitrotoluene (TNT) is introduced under similarly drastic conditions (into 2,4- or 2,6-dinitrotoluene).

5.2.4 Ar—N=N Bond Formation by Ar-S_E Reaction

B

Aryldiazonium salts are weak electrophiles. Consequently, they undergo Ar-S_E reactions via Wheland complexes (azo couplings) only with the most strongly activated aromatic compounds. Phenolates and secondary as well as tertiary aromatic amines are therefore solely attacked. Primary aromatic amines react with diazonium salts, too, but via their N atom. Thus, triazenes, that is, compounds with the structure Ar—N=N—NH—Ar′ are produced. Phenol ethers or nondeprotonated phenols can be attacked by aryldiazonium salts only when the latter are especially good electrophiles, for example, when they are activated by nitro groups in the *ortho* or *para* position.

Fig. 5.18. Regioselective azo coupling between diazotized sulfanilic acid and sodium 1-naphtholate (synthesis of Orange I).

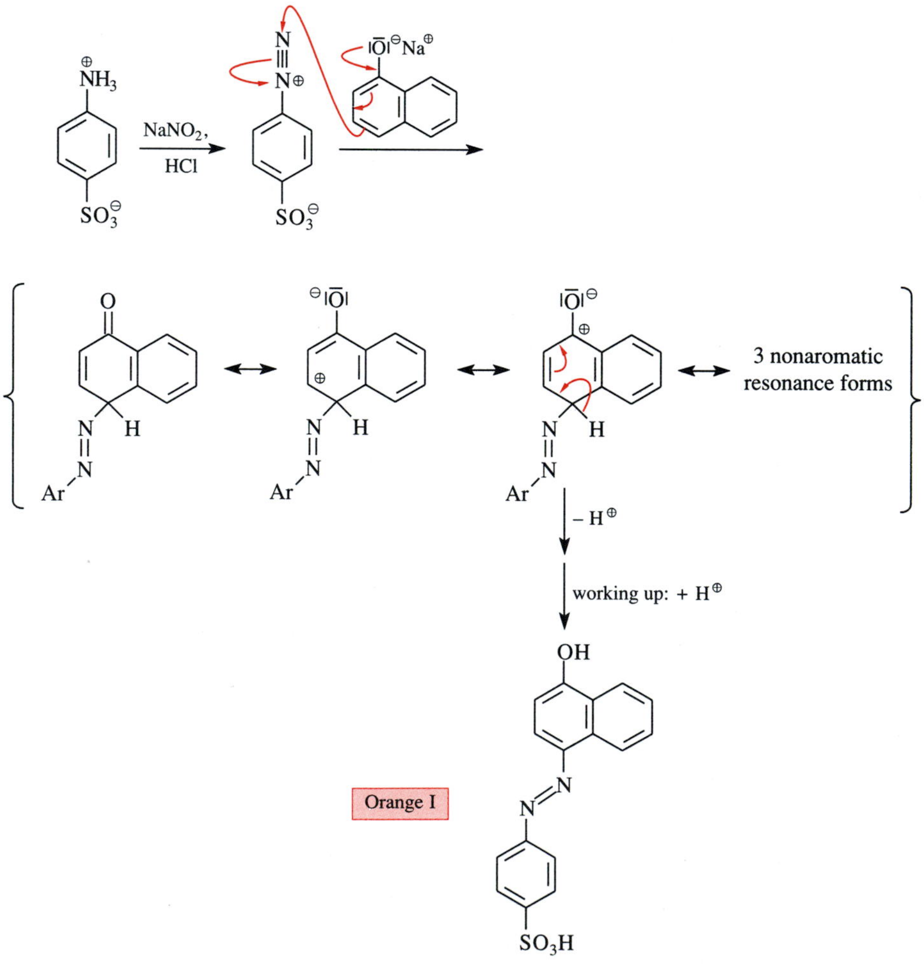

$NaNO_2$, HCl

3 nonaromatic resonance forms

$- H^{\oplus}$

working up: $+ H^{\oplus}$

Orange I

$NH_3^{\oplus}$ / $SO_3^{\ominus}$ —[$NaNO_2$, HCl; $Na^{\oplus}$ $^{\ominus}O$-naphthyl; $H_3O^{\oplus}$-working up]→ Orange II (HO, N=N, SO_3H)

Fig. 5.19. Regioselective azo coupling between diazotized sulfanilic acid and sodium 2-naphtholate (synthesis of Orange II).

On an industrial scale, azo couplings serve as a means of producing **azo dyes.** The first azo dyes included the azo compounds Orange I (Figure 5.18) and Orange II (Figure 5.19). They are obtained by azo couplings between diazotized *para*-aminobenzenesulfonic acid (sulfanilic acid) and sodium 1-naphtholate or sodium 2-naphtholate. The perfect regioselectivity of *both* couplings follows from the fact that in each case the most stable Wheland complex is produced as an intermediate. The most stable Wheland complex in both reactions is the only one of seven conceivable complexes for which there exists an especially stable *all-octet resonance form free of formal charges and with an aromatic sextet.* For both azo couplings this resonance form results from an Ar-S_E reaction on that ring of the naphtholate to which the O^- substituent is bound. Figure 5.18 shows as an example the most stable resonance form, which determines the regioselectivity of the formation of Orange I.

In the laboratory, azo couplings can also be carried out for another purpose. Azo compounds, such as those shown in Figures 5.18 and 5.19, can be reduced to aromatic amino compounds. The N=N unit is then replaced by two NH_2 groups. In this case, an azo coupling would be the first step of a two-step process with which naphthols or arylamines can be functionalized with an NH_2 group (i.e., aminated). For many other aromatic compounds, the same transformation can be achieved by nitrating with HNO_3/H_2SO_4 (Section 5.2.3) and subsequently reducing the nitro group to the amino group. However, this standard process cannot be used directly with naphthols, because they are easily oxidized by the reagent mixture to give 1,4- or 1,2-naphthoquinone.

5.2.5 Ar—Alkyl Bond Formations by Ar-S_E Reaction

Suitable Electrophiles and How They Are Attacked Nucleophilically by Aromatic Compounds

Activated aromatic compounds can be alkylated via Ar-S_E reactions (Friedel–Crafts alkylation). Suitable reagents are certain

- alkyl halides or alkyl sulfonates in the presence of catalytic amounts of a Lewis acid,
- alcohols in the presence of catalytic amounts of a Brønsted acid,
- olefins that can be converted by catalytic amounts of a Brønsted acid directly (i.e., by protonation) or indirectly (see the first reaction in Figure 3.45) into a carbenium ion.

The effective electrophiles in Friedel–Crafts alkylations are the species shown in Figure 5.20.

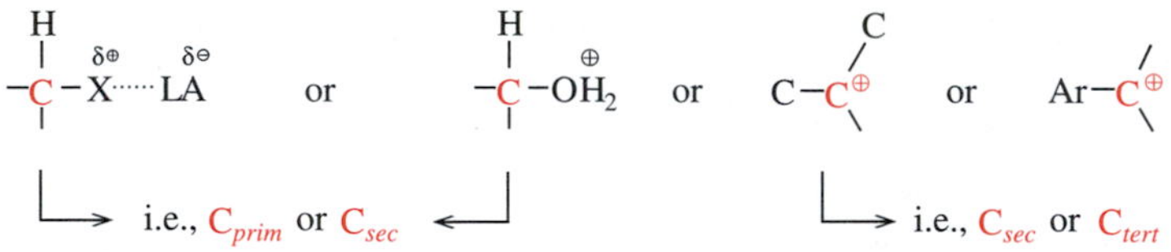

Fig. 5.20. Electrophiles that can initiate Friedel–Crafts alkylations without rearrangements; LA stands for Lewis acid.

The word "certain" was deliberately chosen at the beginning of this section. When the reactive electrophile is a carbenium ion, a constitutionally unique alkyl group can generally be introduced through an Ar-S_E reaction only when this carbenium ion does not isomerize competitively with its attack on the aromatic substrate. An isomerization is most reliably excluded for those Friedel–Crafts alkylations in which quite stable carbenium ions appear in the first step. These include *tert*-alkyl and benzyl cations (Figure 5.20, right). Suitable reagents are therefore all alkyl halides, alkyl sulfonates, alcohols, and olefins that give either *tert*-alkyl or benzyl cations with the additives enumerated in the above list. Accordingly, the aromatic compound displaces the leaving group from the first three of these four reagents, as in an S_N1 reaction.

Lewis acid complexes of alkyl halides and alkyl sulfonates (Figure 5.20, left) and protonated alcohols (Figure 5.20, middle) are additional reactive electrophiles in Friedel–Crafts alkylations. The aromatic compound displaces their respective leaving group in an S_N2 process. This is in principle possible (Section 2.4.4) for primary or secondary alkylating agents and alcohols.

Alternatively, namely in the presence of a Lewis acid, secondary alkyl halides and sulfonates can react with aromatic compounds via an S_N1 mechanism. The pair of reactions in Figure 5.21 shows that the balance between S_N2 and S_N1 mechanisms can be subject to subtle substituent effects. *S*-2-Chlorobutane and $AlCl_3$ alkylate benzene with complete racemization. This means that the chlorobutane is substituted by the aromatic substrate through an S_N1 mechanism, which takes place via a solvent-separated ion pair (cf. Section 2.5.2). On the other hand, the mesylate of *S*-methyl lactate and $AlCl_3$ alkylate benzene with complete inversion of the configuration of the stereocenter. This means that the mesylate group is displaced by the aromatic compound in an S_N2 reaction. There are two reasons why these substitution mechanisms are so different and so clearly preferred in each case. On the one hand, the lactic acid derivative cannot undergo an S_N1 reaction. The carbenium ion that would be gener-

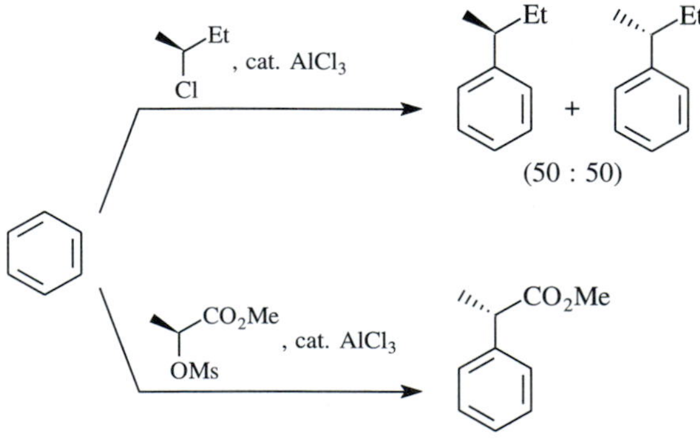

Fig. 5.21. Friedel–Crafts alkylations with secondary alkyl halides or sulfonates and $AlCl_3$: competition between S_N1 (top) and S_N2 (bottom) substitution of the electrophile.

ated, $H_3C—CH^+—CO_2Me$, would be strongly destabilized by the electron-withdrawing CO_2Me group (cf. discussion of Figure 2.24). On the other hand, S_N2 reactions in the α position to an ester group take place especially rapidly (cf. discussion of Figure 2.10).

Single or Multiple Alkylation by the Friedel–Crafts Reaction?

Friedel–Crafts alkylations differ from all other Ar-S_E reactions considered in Section 5.1 in that the reaction product is a better nucleophile than the starting material. This is because the alkyl group introduced is an *activating* substituent. Therefore, in a Friedel–Crafts alkylation we risk an overreaction of the primary product to further alkylation.

There Are Only Three Conditions under Which Multiple Alkylations Do Not Occur

First, *intermolecular* Friedel–Crafts alkylations take place as monofunctionalizations of the aromatic ring when the primary product cannot take up any additional alkyl group for steric reasons:

tert-BuCl, $FeCl_3$ → *tert*-Bu—biphenyl—*tert*-Bu (isomerically pure)

Second, Friedel–Crafts monoalkylations of aromatic compounds such as benzene or naphthalene, in which the introduction of a second alkyl group is not prevented by steric hindrance, can be carried out (only) with a trick: The reaction is performed with a large excess of the aromatic compound. There is almost no overalkylation at all. Simply for statistical reasons virtually the entire attack is directed at the excess starting material, not the primary product. For example, the following substitution succeeds in this way:

+ Cl, H_2SO_4 → Cl (via H ⊕ Cl)

This same situation occurred in the de-*tert*-butylation of Figure 5.3: mono- and not di-*tert*-butyl benzene was produced because the reaction was carried out *in* benzene.

The third possibility for a selective monoalkylation is provided by *intramolecular* Friedel–Crafts alkylations. There are no multiple alkylations simply because all electrophilic centers react most rapidly intramolecularly (i.e., only once). Friedel–Crafts alkylations of this type are ring closure reactions.

OH, MeO, H_2SO_4 → ⊕, MeO → MeO

Intramolecularly, even certain secondary carbenium ions can be introduced without isomerization in Friedel–Crafts alkylations. An example is the last step in the bicyclization of Figure 3.45.

In contrast to the intramolecular case just mentioned, in *intermolecular* Friedel–Crafts alkylations secondary carbenium ion intermediates often have sufficient time to undergo a Wagner–Meerwein rearrangement (cf. Section 11.3.1). This can lead to the formation of an unexpected alkylation product or product mixtures:

, cat. $AlCl_3$

Cl

+

Friedel–Crafts Alkylations with Multiply Chlorinated Methanes

B

If $CH_2Cl_2/AlCl_3$ is used as electrophile in the Friedel–Crafts alkylation, a benzyl chloride is first produced. However, this compound is itself a Friedel–Crafts electrophile, and, in the presence of $AlCl_3$, it benzylates unconsumed starting material immediately. As a result, one has linked two aromatic rings with a CH_2 group:

Br $\xrightarrow[\text{cat. } AlCl_3]{CH_2Cl_2,}$ (Br Cl $\xrightarrow{\text{cat. } AlCl_3}$) Br Br

Similarly, Friedel–Crafts alkylations with $CHCl_3$ and $AlCl_3$ lead to the linking of three aromatic rings through a CH group. A related alkylation occurs with CCl_4 and $AlCl_3$. At first, three aromatic rings are linked with a C—Cl group, so that a trityl chloride is obtained. This heterolyzes in the presence of the Lewis acid and gives a trityl cation. This step may seem to set the stage for yet another Friedel–Crafts alkylation. However, the unconsumed aromatic compound is too weak a nucleophile to be able to react with the well-stabilized trityl cation.

A Mechanistic Study: Substrate and Positional Selectivity in a Series of Analogous Friedel–Crafts Benzylations

A

Those who want to develop a flair for mechanistic reasoning should consider the four Friedel–Crafts benzylations in Table 5.4. The reagents are, in this order, *para*-nitro-, the unsubstituted, *para*-methyl-, and *para*-methoxybenzyl chloride. The substrate in each case is a large excess of a 1:1 mixture of toluene and benzene. The catalyst is always $TiCl_4$, a strong Lewis acid. As a consequence, the latter initiates Ar-S_E reactions, which represent S_N1 reactions with respect to the benzyl chlorides because the substrates toluene and benzene are weak nucleophiles. Each Ar-S_E reaction gives a mixture of four products: the *para-*, the *meta-*, and the *ortho*-benzylated toluene as well as the benzylated benzene. Table 5.4 sorts these mixtures according to the *para*-substituent

Table 5.4. Substrate and Positional Selectivity in Four Analogous Friedel–Crafts Benzylations*

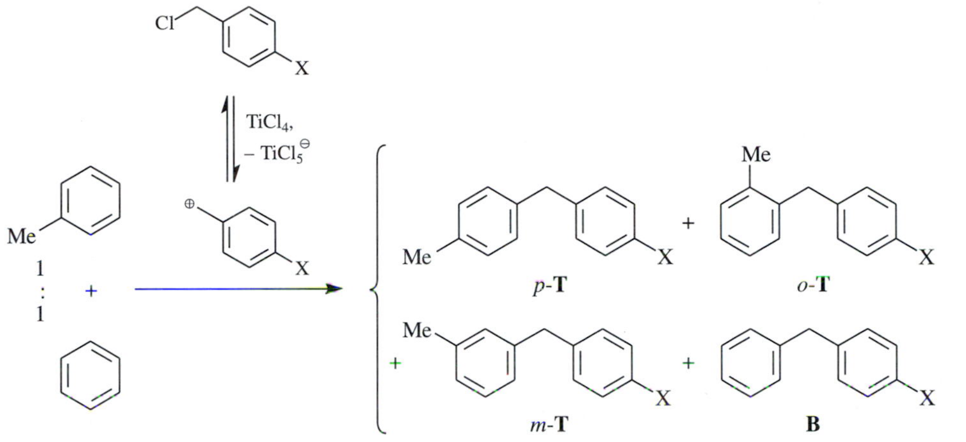

X	Amount in the product mixture in %			
	p-**T**	*o*-**T**	*m*-**T**	**B**
NO_2	24.4	42.6	4.4	28.6
H	47.6	35.0	3.7	13.7
Me	64.3	30.4	2.0	3.3
OMe	69.2	28.3	1.5	1.0

X	Rate constant ratios $k_{sub,\,para}$: $k_{sub,\,ortho}$: $k_{sub,\,meta}$: $k_{sub,\,benzene}$						
NO_2	11.1	:	9.68	:	1.0	:	2.17
H	25.7	:	9.46	:	1.0	:	1.23
Me	64.3	:	15.2	:	1.0	:	0.55
OMe	92.3	:	18.9	:	1.0	:	0.22

*See text for definitions of $k_{sub,para}$, $k_{sub,ortho}$, and $k_{sub,meta}$; $k_{sub,benzene}$ is the relative rate constant of the substitution in *one* position of benzene.

in the benzylating agent and depicts percentages of their compositions. Table 5.4 also defines the abbreviations *p*-**T**, *m*-**T**, *o*-**T**, and **B** used in Figures 5.22 and 5.23.

The *para*-substituents in the benzylating agent are varied from strongly electron withdrawing (NO_2) through electronically neutral (H) to weakly (Me) and strongly electron donating (MeO). The product palette and the rate constants in Table 5.4 are influenced by this variation, as follows:

1) The benzylation of toluene vs benzene is increasingly favored.
2) The regioselectivity of the benzylation of toluene shifts from $k_{sub,\,para} \geq k_{sub,\,ortho} > k_{sub,\,meta}$ to $k_{sub,\,para} > k_{sub,\,ortho} \gg k_{sub,\,meta}$. $k_{sub,\,para}$ is the relative rate constant of the substitution in the *para* position, $k_{sub,\,meta}$ is the relative rate constant of the substitution in a single *meta*-position, and $k_{sub,\,ortho}$ is the relative rate constant of substitution in a *single ortho* position. Thus, in calculating these rate constants from the product ratios of Table 5.4, it was taken into consideration that toluene has twice as many *ortho* and *meta* positions as *para* positions.

Energy order of the substrate/reagent pairs:

$\oplus$ / NO_2

is 1 kcal/mol more energy rich than

Me / $\oplus$ / NO_2

Energy order of the 4 Wheland complexes

Me, H, +0.30, NO_2, *p*-**T**$_{WC}$ > Me, +0.25, H, NO_2, *o*-**T**$_{WC}$

≫ +0.10, Me, H, NO_2, *m*-**T**$_{WC}$ > H, NO_2, **B**$_{WC}$

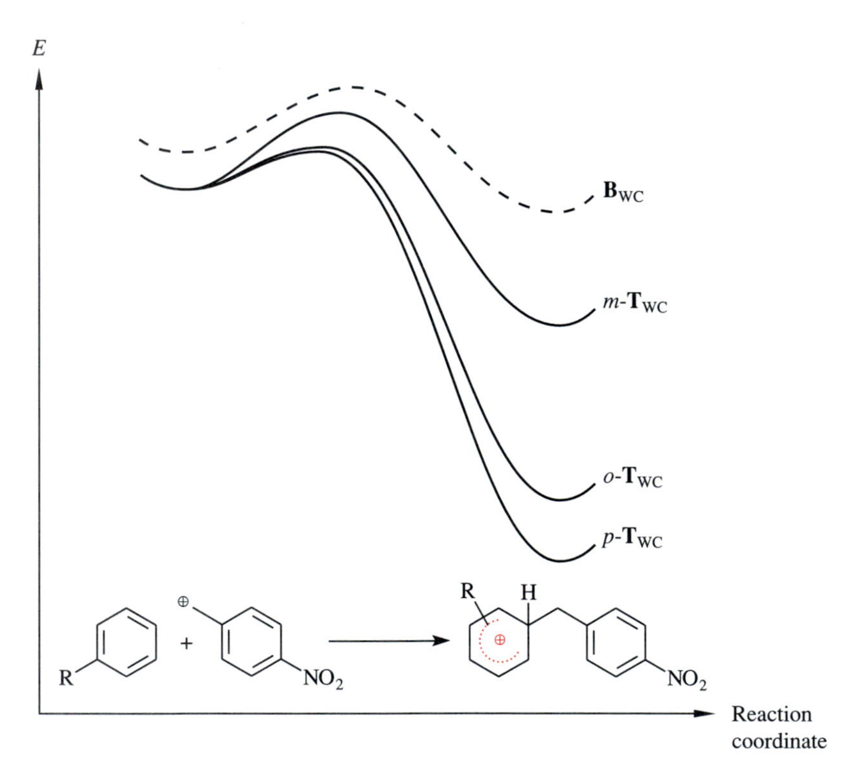

Fig. 5.22. Energy profile of the step of the benzylation of a 1 : 1 toluene:benzene mixture with $O_2N—C_6H_4—CH_2Cl/TiCl_4$ that determines the substrate selectivity and the regioselectivity. The benzyl cation involved in this step is destabilized electronically by the *para*-nitro group; WC stands for Wheland complex.

The reason for both trends is the same: the benzyl cations that attack the toluene or the benzene have different stabilities depending on the nature of their *para* substituent X. When X is the electron acceptor NO_2, we have the most electron-deficient cation, whereas when X is the electron donor MeO, we have the most electron-rich cation. The formation of Wheland complexes from *p*-$O_2N—C_6H_4—CH_2^+$ should therefore be exothermic and exergonic. Conversely, the formation of Wheland complexes

Energy order of the substrate/reagent pairs:

benzene / p-MeO—C_6H_4—CH_2^+

is 1 kcal/mol more energy rich than

toluene / p-MeO—C_6H_4—CH_2^+

Energy order of the 4 Wheland complexes

p-$\mathbf{T}_{WC}$ (+0.30) > o-$\mathbf{T}_{WC}$ (+0.25) ≫ m-$\mathbf{T}_{WC}$ (+0.10) > $\mathbf{B}_{WC}$

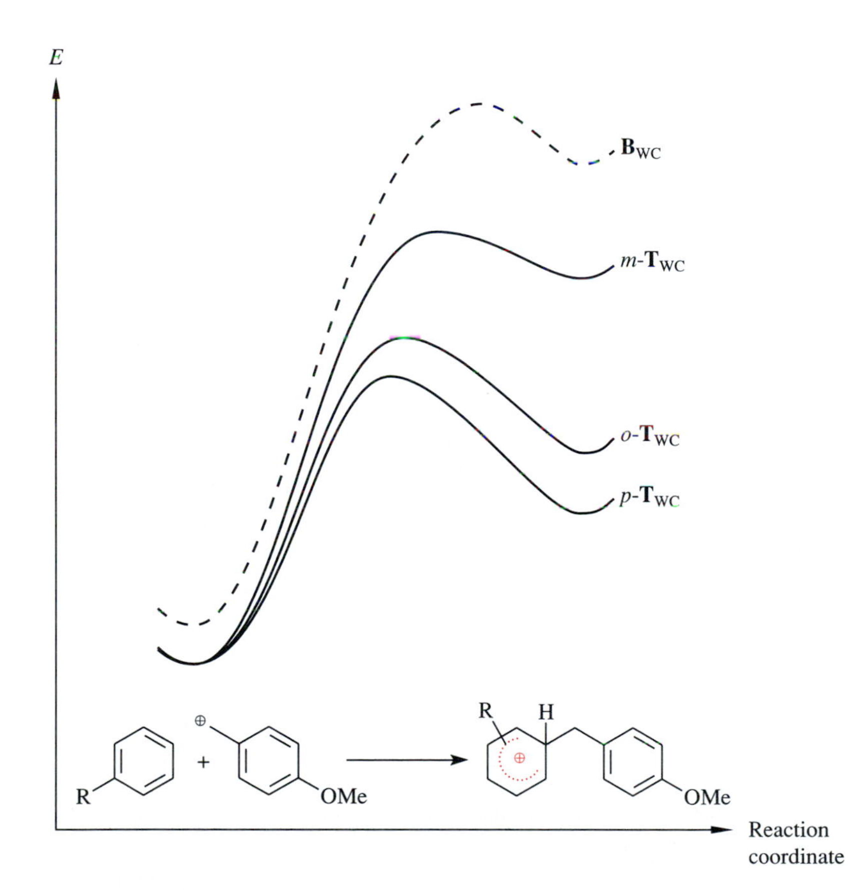

Fig. 5.23. Energy profile of that step of the benzylation of a 1:1 toluene:benzene mixture with pMeO—C_6H_4—$CH_2Cl/TiCl_4$, which determines the substrate selectivity and the regioselectivity. The benzyl cation involved in this step is destabilized electronically by the *para*-methoxy group; WC stands for Wheland complex.

from p-MeO—C_6H_4—CH_2^+ should take place endothermically and endergonically. According to the Hammond postulate, the Wheland complexes derived from the nitro-substituted cation should therefore be formed via early transition states (Figure 5.22), whereas the Wheland complexes derived from the methoxylated cation should be formed via late transition states (Figure 5.23).

Consequently, the early transition states of the O_2N—C_6H_4—CH_2^+ reactions are hardly influenced by the energy differences between the Wheland complexes produced. On the other hand, the late transition states of the reactions with MeO—C_6H_4—CH_2^+ almost fully reflect these energy differences. Therefore, only in Ar-S_E reactions of the *last* type, the substituent effect of the extra methyl group in the toluene- vs benzene-based Wheland complexes becomes sizable. This substituent effect pertains to the ability of the methyl group to stabilize Wheland complexes no matter where it is located, on the one hand, and in the order *para* > *ortho* > *meta,* on the other hand.

Let us turn to one last detail of the data compiled in Table 5.4: Why is a *meta* position of toluene attacked slightly more slowly by the two most reactive benzyl cations *p*-X—C_6H_4—CH_2^+ (X=NO_2, H) than a C atom in benzene? Does this not contradict the fact that a more stable Wheland complex (*m*-$\mathbf{T}_{WK}$ in Figure 5.22) is produced in the first attack than in the second ($\mathbf{B}_{WK}$ in Figure 5.22)? No! Because these reactions take place via *early* transition states and a ground-state effect predominates here: in the conversion into the Wheland complex, toluene loses a slightly greater resonance energy than benzene. *This* reduces its reactivity with electrophiles.

5.2.6 Ar—C(OH) Bond Formation by Ar-S_E Reactions and Associated Secondary Reactions

Aldehydes and ketones react with aromatic compounds in the presence of Brønsted or Lewis acids. The actual electrophile is the carboxonium ion formed in an equilibrium reaction by protonation or complexation, respectively. The primary product is a substituted benzyl alcohol, which, however, is not stable and easily forms a benzyl cation. The latter continues to react further, either via an S_N1 or an E1 reaction. Thereby, the following overall functionalizations are realized: Ar—H → Ar—C-Nu or Ar—H → Ar—C=C.

If the activated aromatic compound formalin, concentrated hydrochloric acid, and $ZnCl_2$ react with each other, the result is a so-called chloromethylation (Figure 5.24):

Fig. 5.24. Chloromethylation of aromatic compounds via Ar-S_E reaction.

MeS — Formalin, $ZnCl_2$, HCl (⟶ H_2ZnCl_4; via $H_2C{=}\overset{\oplus}{O}H$) → (MeS ... OH) — S_N1 reaction of the alcohol → MeS ... Cl

The stable reaction product is a primary benzyl chloride. This reaction is initiated by an electrophilic substitution by protonated formaldehyde; it is terminated by an S_N1 reaction in which a chloride ion acts as the nucleophile.

Phenols are such good nucleophiles that protonated carbonyl compounds functionalize two phenol molecules. The first phenol molecule is attacked in an Ar-S_E reaction by the carboxonium ion formed in an equilibrium reaction. Subsequently, the second equivalent of phenol becomes the substrate of a Friedel–Crafts alkylation. The electrophile is the benzyl cation that is formed from the initially obtained benzyl alcohol and the acid. Protonated acetone is only a weak electrophile for electronic and steric reasons: It contains two electron-donating and relatively large methyl groups on the electrophilic C atom. Therefore, it attacks phenol regioselectively in the *para* and not at all in the less favored *ortho* position (Figure 5.25). The benzyl cation formed there-

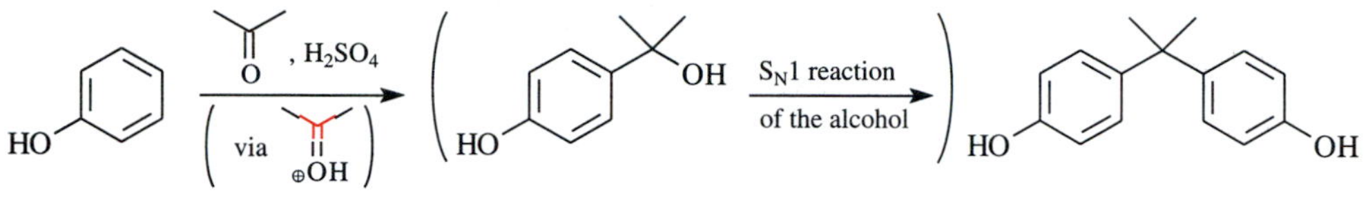

Fig. 5.25. Linking aromatic compounds with methylene units via Ar-S_E reactions. Preparation of bisphenol **A**, an industrially important component for the production of polycarbonates and epoxy resins.

after is a poor electrophile, too, and again for both electronic and steric reasons, it attacks the second phenol molecule with high *para* selectivity.

Under comparable reaction conditions the much more reactive formaldehyde and phenol give not only the *para-* but also the *ortho*-substituted phenol derivative. This reaction ultimately leads to the three-dimensional network of formaldehyde/phenol condensation resins such as Bakelite.

The *intramolecular* hydroxyalkylation of aromatic compounds is a ring closure reaction. In the reaction example in Figure 5.26 it is followed by an E1 elimination, which leads to a styrene derivative.

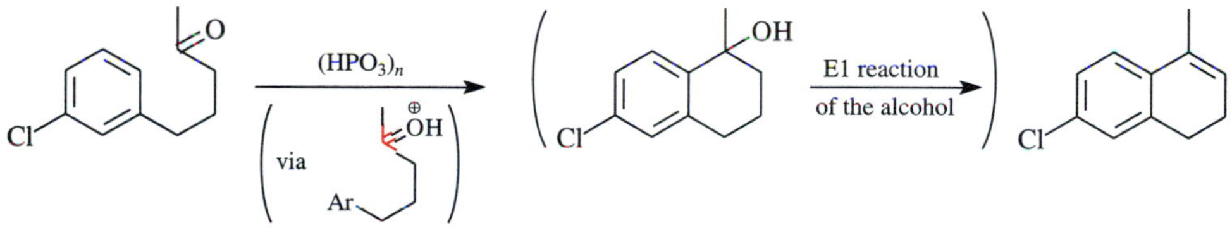

Fig. 5.26. Alkenylation of an aromatic compound by a combination of Ar-S_E and E1 reactions.

5.2.7 Ar—C(=O) Bond Formation by Ar-S_E Reaction

Bond formation between aromatic compounds and a C(=O)—$CR^1R^2R^3$ unit (R^n = H, alkyl, and/or aryl) is the domain of the Friedel–Crafts acylation. As reagents one uses:

B

- a carboxylic chloride with a stoichiometric amount of $AlCl_3$ (because the resulting ketone binds one equivalent of $AlCl_3$ in a Lewis acid complex),
- a carboxylic acid anhydride with a stoichiometric amount of $AlCl_3$
- a carboxylic acid anhydride together with a mineral acid or
- a carboxylic acid together with a mineral acid

In each case the reactive electrophile is produced from these reagents in an equilibrium reaction (Figure 5.27).

The first reagent combination, carboxylic acid chloride/$AlCl_3$, reacts via the $AlCl_3$ complex **A** of the acid chloride or via the acylium tetrachloroaluminate **B** formed from it by β-elimination. A carboxylic acid anhydride and $AlCl_3$ react via analogous electrophiles, namely via the $AlCl_3$ complex **D** of the anhydride or via the acylium salt **E** formed therefore by a β-elimination. The protonated anhydride **F** and the protonated carboxylic acid **C** are the reactive electrophiles of the Friedel–Crafts acylations catalyzed by Brønsted acids.

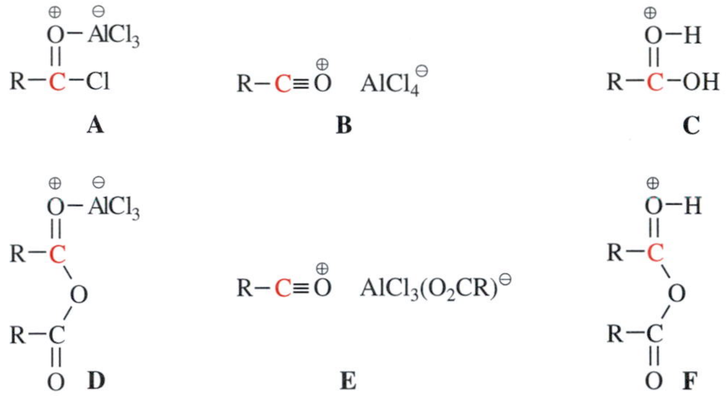

$$\underset{\textbf{A}}{\mathrm{R{-}C(=\overset{\oplus}{O}{-}\overset{\ominus}{Al}Cl_3){-}Cl}} \qquad \underset{\textbf{B}}{\mathrm{R{-}C{\equiv}\overset{\oplus}{O}\ \ AlCl_4^{\ominus}}} \qquad \underset{\textbf{C}}{\mathrm{R{-}C(=\overset{\oplus}{O}{-}H){-}OH}}$$

$$\underset{\textbf{D}}{\mathrm{R{-}C(=\overset{\oplus}{O}{-}\overset{\ominus}{Al}Cl_3){-}O{-}C(=O){-}R}} \qquad \underset{\textbf{E}}{\mathrm{R{-}C{\equiv}\overset{\oplus}{O}\ \ AlCl_3(O_2CR)^{\ominus}}} \qquad \underset{\textbf{F}}{\mathrm{R{-}C(=\overset{\oplus}{O}{-}H){-}O{-}C(=O){-}R}}$$

Fig. 5.27. Reactive electrophiles in the Friedel–Crafts acylation.

Importantly, a Friedel–Crafts *formylation* has not yet been successful. Formyl chloride and formic anhydride are not stable reagents. The mixed anhydride H—C(=O)—O—C(=O)CH_3 acts as a formylating reagent in reactions with many nucleophiles (cf. Section 6.3.3). However, in reactions with aromatic compounds under Friedel–Crafts conditions it acts as an acetylating agent rather than as a formylating agent. Last but not least, formic acid and mineral acids proceed to react via the acylium ion $H{-}C{\equiv}O^+$ to form carbon monoxide and water in an α-elimination.

Substrates of Friedel–Crafts acylations are benzene and naphthalene, as well as their halogen, alkyl, aryl, alkoxy, or acylamino derivatives. Acceptor-substituted aromatic compounds are inert. Because Friedel–Crafts acylations introduce an acceptor into the aromatic substrate, no multiple substitutions take place. This distinguishes them from Friedel–Crafts alkylations. Free OH and NH_2 groups in the aromatic compound prevent Friedel–Crafts acylations because they *themselves* are acylated. However, the *O*-acylphenols available in this way can later be rearranged with $AlCl_3$ into *ortho*-acylated isomers (**Fries rearrangement**).

The following reaction example demonstrates that Friedel–Crafts acylations take place more rapidly than Friedel–Crafts alkylations:

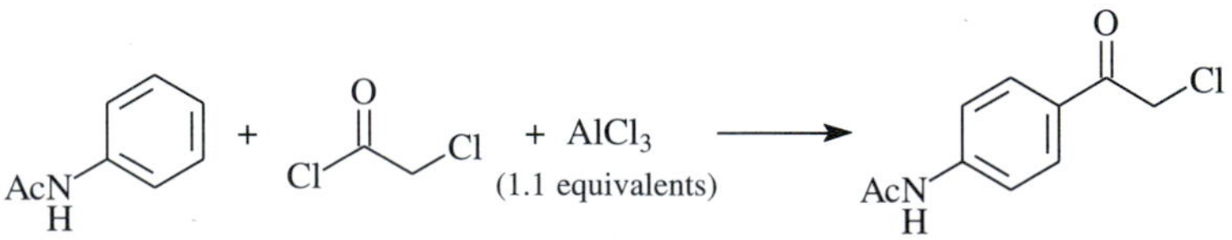

AcNH–C$_6$H$_5$ + Cl–C(=O)–CH$_2$–Cl + $AlCl_3$ (1.1 equivalents) ⟶ 4-AcNH–C$_6$H$_4$–C(=O)–CH$_2$–Cl

Friedel–Crafts acylations with anhydrides include the **Haworth synthesis** of naphthalenes from benzenes via the annulation of a C_4 unit (Figure 5.28). However, the two C—C(=O) bonds cannot be made in the same reaction: the acyl group that enters first deactivates the aromatic compound so thoroughly that it is protected from a second attack, even though in this case it would be an intramolecular attack. The second step of the Haworth synthesis is therefore a Wolff–Kishner reduction of the carbonyl to a methylene group (for the mechanism, see Section 14.4.6). In the aromatic compound, which now is activated again, a second Friedel–Crafts acylation is possible as step 3. It takes place through the carboxylic acid group in the presence of polyphos-

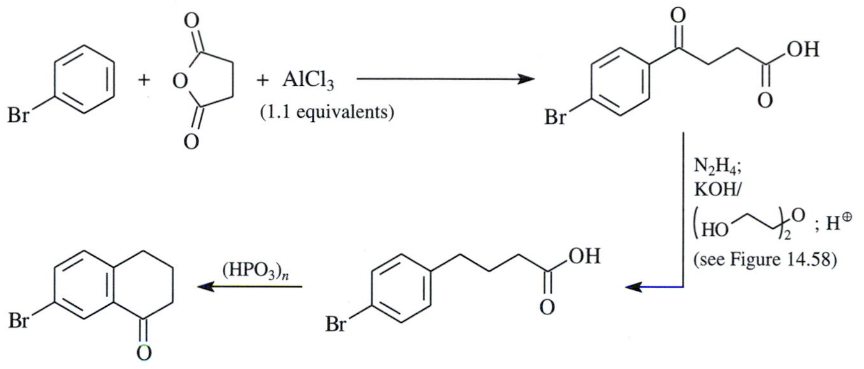

Fig. 5.28. Steps 1–3 of the five-step Haworth synthesis of substituted naphthalenes.

phoric acid. The attacking electrophile is either the protonated carboxylic acid or a mixed carboxylic acid/phosphoric acid anhydride.

Multiply hydroxylated benzene derivatives react with phthalic acid anhydride and a suitable activator when heated, in a *double* C acylation. Evidently, under these conditions, the activating influence of the additional OH groups is still sensible after the first acyl group has been introduced. The resulting doubly acylated product is a 9,10-anthraquinone. Compounds of this type are important as dyes:

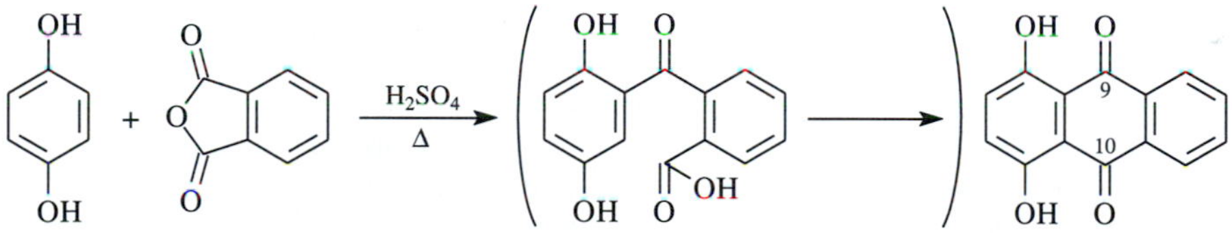

5.2.8 Ar—C(=O)H Bond Formation through Ar-S_E Reaction

As we have seen in Section 5.2.7, formylations by a Friedel–Crafts reaction are not possible. Therefore, the formation of Ar—C (=O)H bonds through Ar-S_E reactions is normally carried out in the form of the Vilsmeier–Haack formylation. The reagent in the Vilsmeier–Haack formylation is a 1:1 mixture of DMF and $POCl_3$. This forms the actual electrophile, an α-chlorinated iminium ion, according to the sequence of steps shown in Figure 5.29. This species is occasionally referred to as the Vilsmeier reagent. Being an iminium ion it is a weaker electrophile than, for example, the oxonium ion of the Friedel–Crafts acylation, because the +M effect of the NMe_2 group is greater than that of —O—$AlCl_3^-$. As one consequence, Vilsmeier–Haack formylations are possible only on especially nucleophilic aromatic compounds, i.e., on aniline, phenol, and their derivatives. After completion of the reaction, a benzylic iminium ion is present (Figure 5.29). It remains stable until it is worked up with water through an unstable *N,O*-acetal intermediate to finally give the desired aldehyde.

Because the iminium ion derived from *H*—C(=O)NMe_2 already is a poor electrophile, it is understandable why there are in general no analogous Vilsmeier–Haack

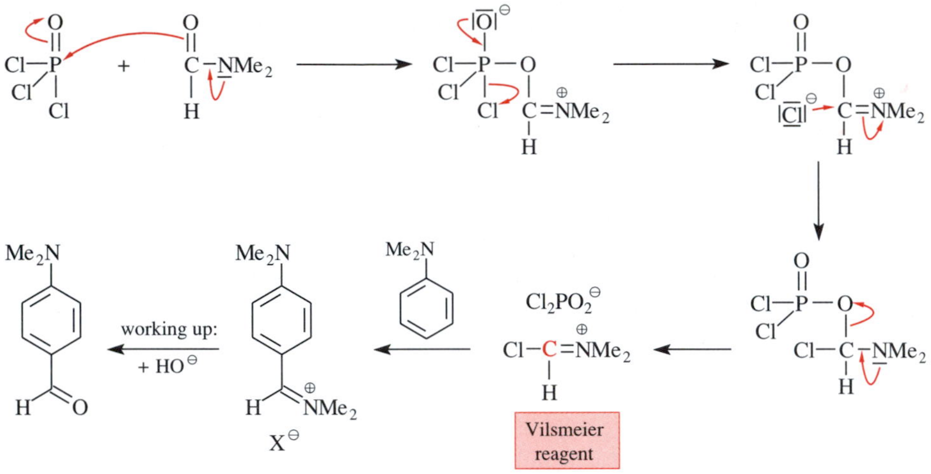

Fig. 5.29. Vilsmeier–Haack formylation ($X^- = Cl^-$ or $Cl_2PO_2^-$).

acylations using R—C(=O)NMe$_2$. The corresponding iminium ion R—C(Cl)=NMe$_2^+$ is a still poorer electrophile for steric and electronic reasons.

Basic aromatic compounds (i.e., aniline derivatives) *would* react with iminium ions of the latter substitution pattern (i.e., with R^1R^2 CH—C(Cl)=NMe$_2^+$), however, in a quite different way: they would deprotonate them to provide an α-chloroenamine. In this way one obtains from Me$_2$CH—C(=O)NMe$_2$ and $POCl_3$ via Me$_2$ CH—C(Cl)=NMe$_2^+$ "chloroenamine" [Me$_2$C=C(Cl)NMe$_2$], a reagent, which is used to convert carboxylic acids into acyl chlorides according to Figure 6.12.

5.3 Electrophilic Substitution Reactions on Metallated Aromatic Compounds

B

Aryl metal compounds Ar—M (M = Li, Mg—Hal) are much better nucleophiles than the corresponding metal-free aromatic compounds Ar—H. Consequently, they react with many electrophiles that could not attack the corresponding nonmetallated aromatic compounds through an Ar-S_E reaction (Sections 5.3.1 and 5.3.2). In addition, there are methods for preparing substituted aryl metal compounds that are isomerically pure. In this way one has an especially elegant possibility for carrying out *ortho*-selective Ar-S_E reactions (Section 5.3.1), which otherwise is practically impossible. Finally, aryl boron compounds are the best substrates for the electrophilic hydroxylation of aromatic compounds giving phenols. Moreover, aryl boron compounds make it possible to attach unsaturated hydrocarbon substituents to aromatic rings (Section 5.3.3).

5.3.1 Electrophilic Substitution Reactions of *ortho*-Lithiated Benzene and Naphthalene Derivatives

A

Certain derivatives of benzene and naphthalene can be lithiated with *sec*-butyllithium (*sec*-BuLi). If this reaction occurs at all it is regioselective: It takes place exclusively in the *ortho* position to a so-called metallation-directing group (MDG), whose presence, accordingly, is a prerequisite for such a metallation. Figure 5.30 gives examples of MDGs that are bound through a C, an O, or an N atom to the aromatic compound.

All MDGs shown contain an O atom that can loosely bind the electron-deficient Li atom of *sec*-BuLi by formation of complex **B** (Figure 5.30). It is believed that this Li atom remains complexed by the same O atom even during the H/Li exchange that follows. In the corresponding transition state (**C** in Figure 5.30) the complexation of the lithium takes place in such a way that the $C_{sec\text{-butyl}}$—Li bond exhibits considerable partial charges: a positive charge on the Li atom and a negative charge in the *sec*-butyl moiety. The *sec*-butyl moiety thus becomes so basic that it can remove one of the H atoms located in the immediate vicinity of the MDG, i.e., in the *ortho* position. Fortunately, this deprotonation does not create an energetically disfavored carbanion because, as indicated in the transition state structure **C** by the dotted bonds, concomitantly the fairly stable C—Li bond of the organolithium compound **A** has been established.

MDG is ...	... C-bound	... O-bound	... N-bound
and is ...	... an amide	... a carbamate	... a lithioamide

Fig. 5.30. *ortho*-Selective electrophilic functionalization of aromatic compounds via a substituent-controlled lithiation.

Any reaction Ar—H + *sec*-BuLi → Ar-Li + *sec*-BuH possesses a considerable driving force. In an MDG-controlled *ortho*-lithiation it is even greater than it would be normally. This is due to the stabilization of the *ortho*-lithiated aromatic compound **A** because of the intramolecular complexation of the Li atom by the donor oxygen of the neighboring MDG. The exclusive occurrence of *ortho*-lithiation is thus a consequence not only of the precoordination of *sec*-BuLi by the MDG but also of product-development control.

Figure 5.31 shows an *ortho*-selective bromination of an aromatic compound that contains a C-bonded MDG. Instead of the strong electrophile Br_2, the weak electrophile NBS is used. With respect to NBS, this reaction represents an S_N2 reaction of the organometallic on the bromine.

Fig. 5.31. Electrophilic functionalization *ortho* to a C-bound MDG (on the left for comparison: *meta*-selectivity of the analogous classic Ar-S_E reaction).

Br, *m*, O, NEt_2 ← Br_2, $FeBr_3$, Δ — O, NEt_2 — *sec*-BuLi; Br–N (succinimide, O, O) → *o*, Br, O, NEt_2

Figure 5.32 shows the preparation of an arylboronic acid as an example of the *ortho*-selective functionalization of an aromatic compound containing an O-bonded MDG. The $Li^+ArB(OMe)_3^-$ complex is generated and then hydrolyzed during workup.

Fig. 5.32. Electrophilic functionalization *ortho* to an O-bonded MDG; preparation of an arylboronic acid.

NEt_2, O, O — *sec*-BuLi; $B(OMe)_3$; $H_3O^{\oplus}$ → NEt_2, O, O, $B(OH)_2$

In Figure 5.33 you see how an aromatic compound can be functionalized with complete regioselectivity exploiting an *N*-lithioamide as an N-bonded MDG. The first reaction step is an *ortho*-lithiation. It allows the introduction of an *ortho*-methyl group by quenching with MeI. The second reaction arrow illustrates that an MDG can also activate a neighboring benzylic position toward lithiation by *sec*-BuLi. The benzyllithium obtained can then be combined with the entire palette of known electrophiles (e.g., the third reaction step in Figure 5.33).

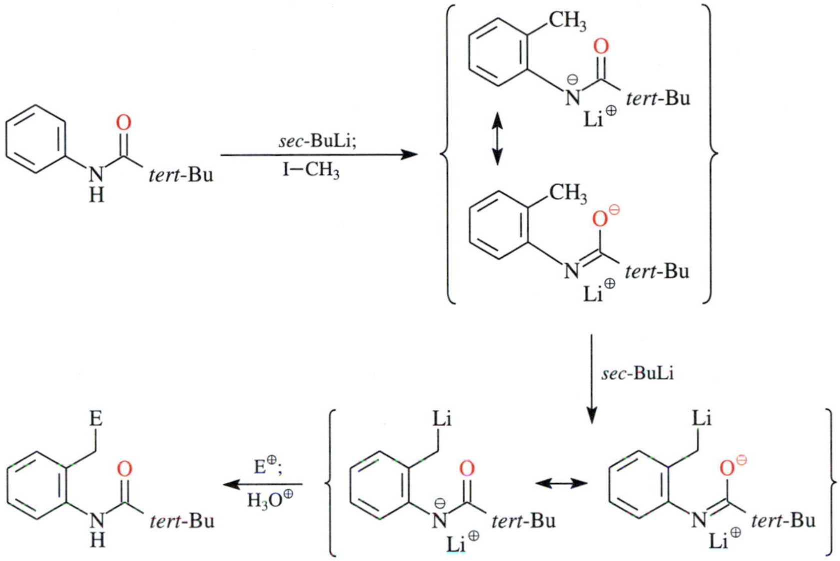

Fig. 5.33. Electrophilic functionalization of C_{sp^2}—H and CH_2—H groups in the position *ortho* to an N-bonded MDG. Here, the lithiation of the benzylic position is associated with a change from colorless to yellow, which can be used to determine the concentration of the alkyllithium solution used.

5.3.2 Electrophilic Substitution Reactions in Aryl Grignard and Aryllithium Compounds That Are Accessible from Aryl Halides

B

The title reactions offer a possibility for exchanging the halogen atom in aryl halides (Hal = Cl, Br, I) first with a metal (MgHal, Li) and then with an electrophile. It is generally easier to introduce bromine than chlorine or iodine into aromatic compounds. Accordingly, functionalizations of aryl bromides are the preparatively most important examples of the title reaction.

Aryl bromides can be converted to aryl Grignard or aryllithium compounds in three ways (Figure 5.34). In the first two methods the aryl bromide is reacted with Mg shavings or with Li wire, respectively (see Section 14.4.1 for the mechanism). In the third method—which is especially convenient for small-scale preparations—aryl bromides are converted into aryllithium compounds either with 1 equivalent of butyllithium (*n*-BuLi) or with 2 equivalents of *tert*-butyllithium (*tert*-BuLi) by the so-called Br/Li exchange reaction.

The mechanism of the Br/Li exchange reaction in aryl bromides is shown in Figure 5.34. It is the same as the mechanism of the Br/Li exchange reaction in bromo-olefins (cf. Figure 13.11). First, the aryl bromide and *n*- or *tert*-BuLi furnish the lithium salt **A.** Its anion part contains a negatively charged bromine atom with two bonds and a total of 10 valence electrons. This species is referred to as a bromate ion in *organic* nomenclature (in *inorganic* chemistry, "bromate" designates an anion, BrO_3^-). Accordingly, the lithium salt **A** is called an "–ate-complex." This –ate-complex decomposes when the Li^+ ion attacks one of its C—Br bonds as an electrophile. This attack occurs at ap-

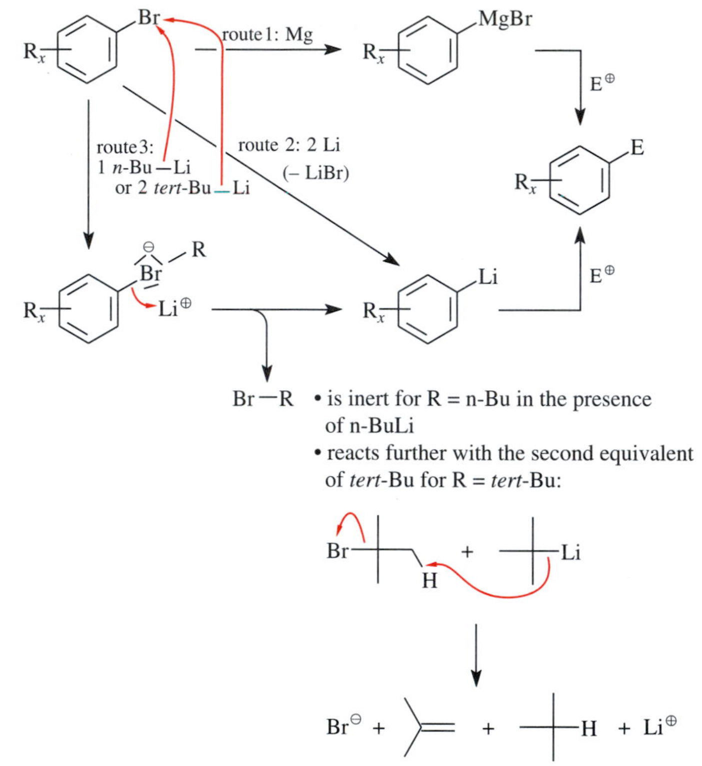

Fig. 5.34. Conversions of Ar—Br into Ar—MgBr or Ar—Li.

proximately the center of the respective C—Br bond, which is the site of the highest electron density. When the Li^+ attacks the C—Br bond that was established during the formation of **A,** the –ate-complex reverts to the starting materials. However, the preferred site of attack by the Li^+ is the C_{aryl}—Br bond of the –ate-complex **A.** In this way **A** proceeds to the desired lithio-aromatic compound as well as to *n*- or *tert*-butyl bromide, respectively.

Overall, these reactions give rise to the following equilibrium:

$$R_x\text{-}C_6H_4\text{-}Br + \begin{array}{c} Li-C_{sp^3}H_2Pr \\ \text{or} \\ Li-C_{sp^3}Me_3 \end{array} \rightleftharpoons R_x\text{-}C_6H_4\text{-}{}^{sp^2}Li + \begin{array}{c} Br-Bu \\ \text{or} \\ Br-tert\text{-}Bu \end{array}$$

It lies completely on the side of the lithio-aromatic compound and of *n*- or *tert*-butyl bromide. This is because the C atom of the C—Li bond is sp^2 hybridized in the lithio-aromatic compound and therefore more electronegative than the sp^3-hybridized C atom of the C—Li bond in *n*- or *tert*-BuLi. More electronegative C atoms stabilize C—Li bonds because these are very electron-rich.

If *n*-BuLi is used for the Br/Li exchange in Figure 5.34, the reaction is completed when the lithio-aromatic compound and *n*-butyl bromide have formed. It is different when this Br/Li exchange is carried out with *tert*-BuLi, where *tert*-BuLi and *tert*-butyl bromide necessarily would leave to coexist during the reaction. However, *tert*-BuLi reacts very fast with *tert*-butyl bromide: As a base *tert*-BuLi eliminates HBr from *t*-butyl bromide. Therefore, Br/Li exchange reactions with *tert*-BuLi can go to completion only when 2 equivalents of the reagent are used.

The reactions of electrophiles with metallated aromatic compounds obtained in this way provide access to products that are not accessible at all or not selectively through classic Ar-S_E reactions. Phenyllithium, for example, unlike benzene, reacts with ketones to form tertiary alcohols or with elemental selenium to form first phenylselenol and thereafter, through oxidation, diphenyl diselenide (Figure 5.35).

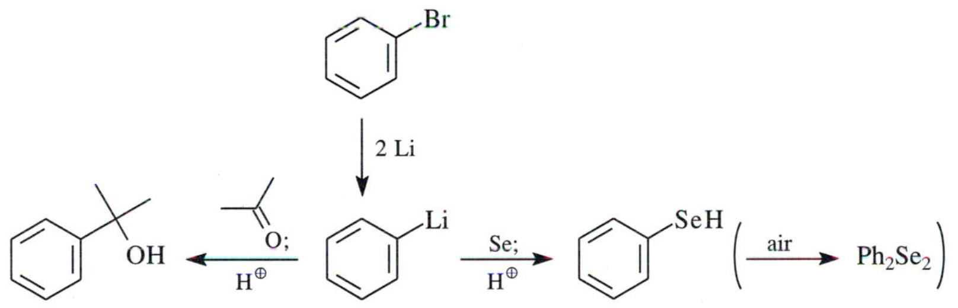

Fig. 5.35. Electrophilic functionalization of phenyllithium. Tertiary alcohols can be synthesized from ketones. Diphenyl diselenide can be prepared from elemental selenium.

There are no direct ways to introduce a CO_2H group, a single deuterium atom, or a single additional methyl group into mesitylene. However, the desired products can easily be prepared via metallated mesitylene as an intermediate. Mesitylene is first brominated on the aromatic ring, and the bromomesitylene formed is then subjected to a Li/Br exchange (Figure 5.36). The resulting mesityllithium is precisely the required nucleophile: with CO_2 it gives the first, with D_2O the second, and with MeI the third target molecule.

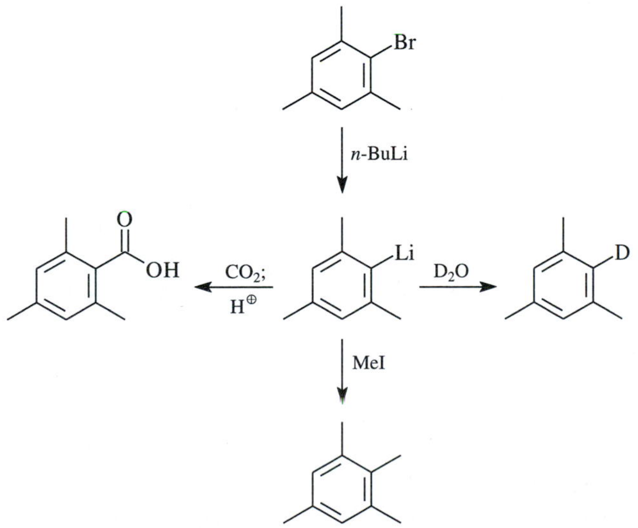

Fig. 5.36. Electrophilic functionalization of mesityllithium. A COOH group, a deuterium atom, or a methyl group can be introduced readily.

A

Interestingly, it is possible to obtain di-Grignard compounds from many dibromoaromatic compounds. These, naturally, react with 2 equivalents of an electrophile. $ClPPh_2$, the electrophile in the reaction example in Figure 5.37, would not react with metal-free 1-1-binaphthyl. However, with the di-Grignard compound shown it forms *R*-BINAP, a compound that has been used with great success as a ligand in transition metal complexes for the enantioselective hydrogenation of C═C and C═O bonds (Section 14.4.7).

Br Br 2 Mg MgBr MgBr $ClPPh_2$ PPh_2 PPh_2

(See Figure 5.50 for the preparation) *R*-BINAP

Fig. 5.37. Electrophilic functionalization of an aromatic di-Grignard compound; synthesis of *R*-BINAP.

The last electrophilic substitution reaction of a Grignard compound we want to consider is a transmetallation, namely one that leads to an arylboronic ester (Figure 5.38). Arylboronic esters or their hydrolysis products, the arylboronic acids, are valuable reagents in modern aromatic chemistry. They react with a series of electrophiles that would not react with Grignard or organolithium compounds (Sections 5.3.3 and 13.3.2).

O O Br Mg; $FB(OMe)_2$ O O $B(OMe)_2$

Fig. 5.38. Preparation of an arylboronic ester via a Grignard compound.

5.3.3 Electrophilic Substitutions of Arylboronic Acids and Arylboronic Esters

A

Arylboronic acids and their esters are nucleophiles in which the —$B(OH)_2$- or the —$B(OR)_2$- group can be substituted by many electrophiles. Representative examples are shown in Figures 5.39 and 5.40.

Arylboronic esters (Figure 5.39) can be oxidized with H_2O_2 in acetic acid to give aryl borates, which can then be hydrolyzed to provide phenols. The mechanism for this oxidation is similar to that for the OOH^- oxidation of trialkylboranes and is discussed in Section 11.4.3. In combination with the frequently simple synthesis of arylboronic esters one can thus achieve the overall conversions Ar—H → Ar—OH or Ar—Br → Ar—OH in two or three steps. There is no metal-free alternative for these hydroxylations even if occasionally there are readily applicable roundabout routes, referred to in Figures 11.32 and 11.34.

Arylboronic esters (Figure 5.39) and arylboronic acids (Figure 5.40) can also react with unsaturated electrophiles, which cannot be introduced in one step into metal-free aromatic compounds: these reactions are with alkenyl bromides, iodides, and triflates

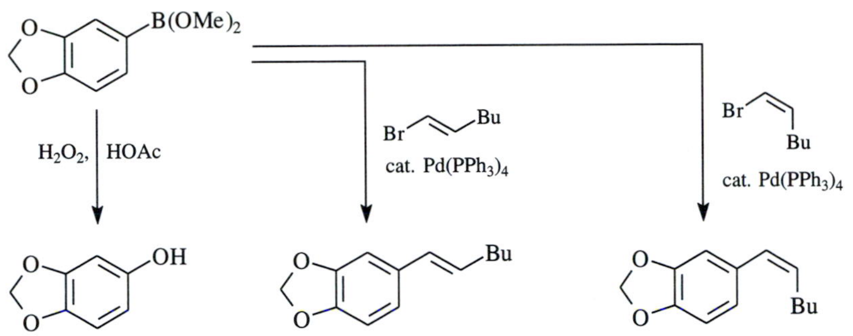

Fig. 5.39. Reactions of an arylboronic ester (for preparation, see Figure 5.38) with selected electrophiles.

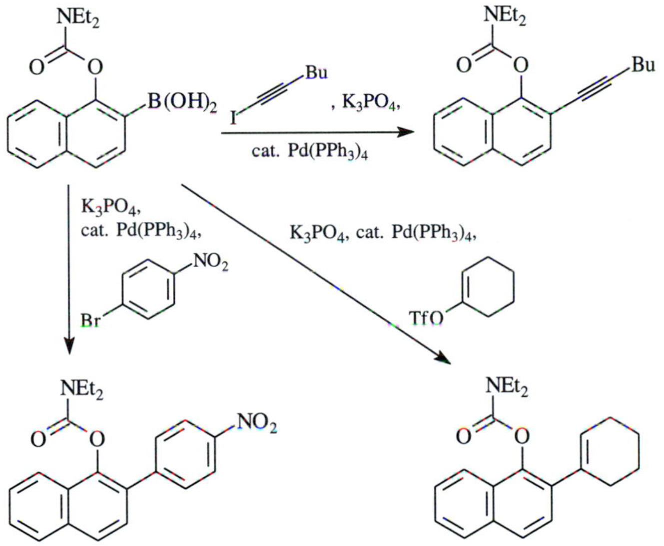

Fig. 5.40. Reactions of an arylboronic acid (for preparation, see Figure 5.32) with selected electrophiles; TfO stands for the triflate group.

(with retention of the double-bond geometry!); with aryl bromides, iodides, and triflates; and with iodoalkynes. All these compounds react (only) in the presence of Pd(0) catalysts according to the mechanisms presented in Section 13.3.

5.4 Nucleophilic Substitution Reactions of Aryldiazonium Salts

Nucleophilic substitution reactions via an S_N2 reaction with backside attack of the nucleophile are not possible in aromatic compounds because the configuration at the attacked C atom cannot be inverted.

B

Nucleophilic substitution reactions according to the S_N1 mechanism have to take place via aryl cations. This should be exceedingly difficult because the positive charge in aryl cations is located in a low-energy sp^2 AO (Figure 5.41, middle line), and is not delocalized as one might first think. In fact, aryl cations are even less stable than alkenyl cations, which are already quite unstable. However, in contrast to aryl cations, alkenyl cations can assume the favored linear geometry at the cationic center as shown in Figure 1.3. Consequently, the considerable energy expenditure required to generate an aryl cation in an S_N1 reaction can be supplied only by aryldiazonium salts. They do so successfully because the very stable N_2 molecule is produced as a second reaction product, which constitutes an "energy offset." Because this nitrogen continuously evades from the reaction mixture, the elimination of that entity is irreversible so that the entire diazonium salt is gradually converted into the aryl cation.

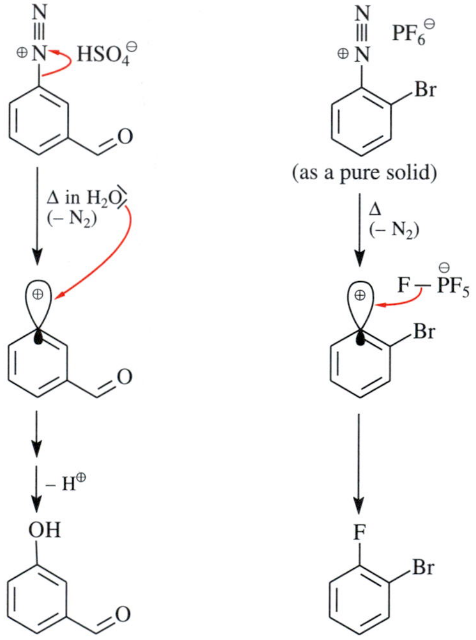

Fig. 5.41. S_N1 reactions of aryldiazonium salts: left, hydrolysis of a diazonium salt leading to phenol; right, Schiemann reaction.

Once the aryl cation is formed it reacts immediately with the best nucleophile in the reaction mixture. Being a very strong electrophile an aryl cation can even react with very weak nucleophiles. When the best nucleophile is H_2O—a weak one, indeed—a phenol is produced (Figure 5.41, left). Even when the nucleophile is a tetrafluoroborate or a hexafluorophosphate (which are *extremely* weak nucleophiles), S_N1 reactions run their course, and an aryl fluoride is obtained. As the so-called **Schiemann reaction,** this transformation is carried out by heating the dried aryldiazonium tetrafluoroborates or hexafluorophosphates (Figure 5.41, right).

With other nucleophiles (Figures 5.42–5.45) aryldiazonium salts react according to other mechanisms to form substitution products. These substitutions are possible because certain nucleophiles reduce aryldiazonium salts to form radicals Ar—N=N·. These radicals lose molecular nitrogen. A highly reactive aryl radical remains, which then reacts directly or indirectly with the nucleophile.

$$Ar{-}NO_2 \xleftarrow[\text{Cu(II)}]{NaNO_2,} Ar{-}\overset{\oplus}{N}{\equiv}N\ X^{\ominus} \xrightarrow{CuCl} Ar{-}Cl$$
$$\xrightarrow{CuBr} Ar{-}Br$$
$$Ar{-}H \xleftarrow[\text{Cu(II)}]{H_3PO_2,} \qquad \xrightarrow{CuCN} Ar{-}CN$$

Sandmeyer reactions

Fig. 5.42. S_N reactions of aryldiazonium salts via radicals.

On the one hand, it is possible to reduce aryldiazonium salts using copper salts. The crucial electron originates from Cu(I), which is either added as such or is formed *in situ* from Cu(II) by a redox reaction (Figure 5.42). The diazo radical Ar–N═N· is produced and fragments to Ar· + N≡N. Three of the ensuing reactions of these aryl radicals shown in Figure 5.42 are called **Sandmeyer reactions** and give aromatic chlorides, bromides, or cyanides. Aryldiazonium salts, sodium nitrite, and Cu(II) give aromatic nitro compounds by the same mechanism. Finally, through reduction with hypophosphoric acid, aryl radicals furnish defunctionalized aromatic compounds, following again the same mechanism.

$$Ar{-}\overset{\oplus}{N}{\equiv}N\ X^{\ominus} + \overset{+1}{Cu}Nu \longrightarrow Ar{-}N{=}N\cdot + \overset{+2}{Cu}NuX$$
$$Ar{-}N{=}N\cdot \longrightarrow Ar\cdot + N_2$$
$$Ar\cdot + \overset{+2}{Cu}NuX \longrightarrow Ar{-}\overset{+3}{Cu}NuX$$

mechanistic alternative: one-step conversion to final products

$$Ar{-}\overset{+3}{Cu}NuX \longrightarrow Ar{-}Nu + \overset{+1}{Cu}X$$

Fig. 5.43. Mechanism of the nucleophilic aromatic substitution reactions of Figure 5.42 carried out in the presence of Cu salts. Two alternative mechanisms are possible for the third step. Either the Cu(II) salt is bound to the aryl radical and the intermediate Ar—Cu(III)NuX decomposes to Cu(I)X and the substitution product Ar—Nu, or the aryl radical reacts with the Cu(II) salt of the nucleophile giving the substitution product Ar—Nu through ligand transfer and Cu(I) through concomitant reduction.

In the reaction of aryldiazonium salts with KI aryl iodides are formed (Figure 5.44). In the initiating step, the diazonium salt is reduced. Therefore, aryl radicals Ar· are obtained under these conditions, too. However, their fate presumably differs from that of the aryl radicals, which are faced with nucleophiles in the presence of Cu(I) (cf. Figure

$$O_2N{-}C_6H_2Br_2{-}\overset{\oplus}{N}{\equiv}N\ HSO_4^{\ominus} \xrightarrow[(-N_2)]{KI} O_2N{-}C_6H_2Br_2{-}I$$

$$X + \overset{+1}{Cu}Nu$$

Fig. 5.44. Reaction of aryldiazonium salts with KI.

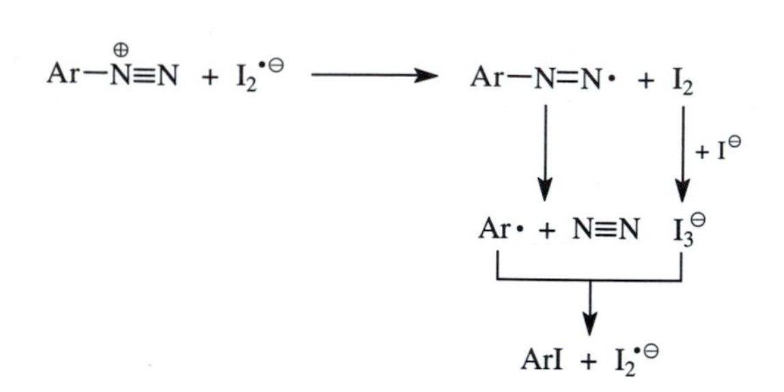

Fig. 5.45. Mechanism of the nucleophilic aromatic substitution reaction of Figure 5.44. The radical $I_2^{\cdot -}$ plays the role of the chain-carrying radical and also the important role of the initiating radical in this chain reaction. The scheme shows how this radical is regenerated. It remains to be added how it is presumably formed initially: (1) $Ar{-}N^+{\equiv}N + I^- \rightarrow Ar{-}N{=}N\cdot + I\cdot$; (2) $I\cdot + I^- \rightarrow I_2^{\cdot -}$.

5.43). The iodination mechanism of Figure 5.45 is a radical *chain* reaction, which consists of four propagation steps.

To conclude this section, Figure 5.46 shows an elegant possibility for carrying out substitution reactions on aryldiazonium salts without using them as the substrates proper. One advantage of this method is that it makes it possible to prepare fluoroaromatic compounds (Figure 5.46, top) without having to isolate the potentially explosive Schiemann diazonium salts (i.e., the starting material for the reaction on the

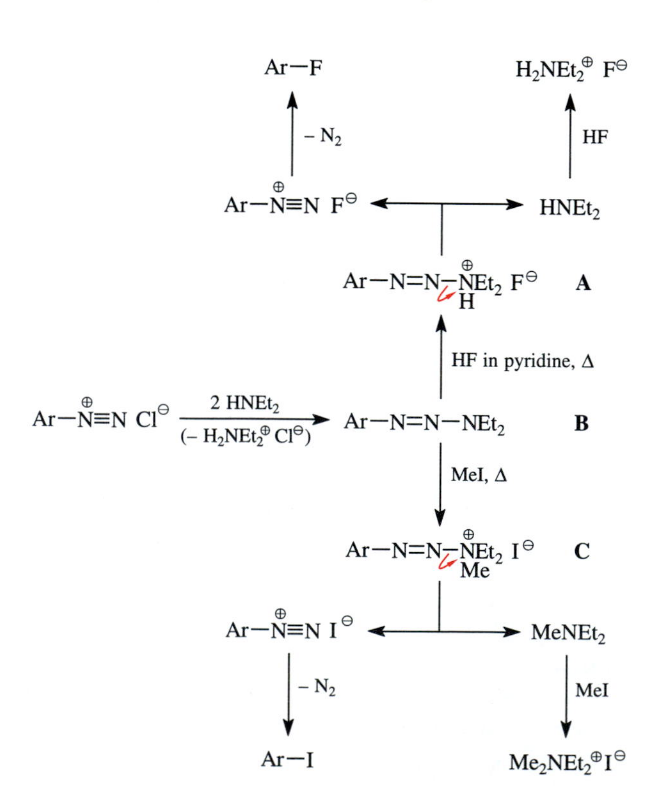

Fig. 5.46. Nucleophilic substitution reactions on a masked aryldiazonium salt.

right in Figure 5.41). In addition, by the same method it is possible to prepare aryl iodides (Figure 5.46, bottom) without the risk that a nucleophile other than iodide introduced during the *preparation* of the diazonium salt competes with the iodide and thereby gives rise to the formation of a second substitution product.

To conduct the substitution reactions of Figure 5.46, one neutralizes an acidic solution of the aromatic diazonium salt with diethylamine. This forms the diazoamino compound **B** (called a "triazene"). It is isolated and subjected to the substitution reactions in an organic solvent. In each case, at first a leaving group is generated from the NEt_2 moiety of the diazoamino group of compound **B.** This is achieved either by a protonation with HF (→ diazoammonium salt **A**) or by a methylation with MeI (→ diazoammonium salt **C**). The activated leaving groups are eliminated from the intermediates **A** and **C** as $HNEt_2$ or $MeNEt_2$, respectively. Aryldiazonium ions are formed, which are faced with only *one* nucleophile, namely, with F^- or I^-, respectively. These reactions give Ar—F or Ar—I, and they run so smoothly that the extra effort, which the isolation of intermediate product **B** requires in comparison to performing the corresponding direct substitutions (Schemes 5.41 and 5.44/5.45, respectively), is often warranted.

5.5 Nucleophilic Substitution Reactions via Meisenheimer Complexes

5.5.1 Mechanism

B

Let us once more view the mechanism of the "classic" Ar-S_E reaction of Figure 5.1: In the rate-determining step an electrophile attacks an aromatic compound. A carbenium ion is produced, which is called the Wheland complex. Therein a *positive* charge is delocalized over the five sp^2-hybridized centers of the former aromatic compound whose sixth center is sp^3-hybridized and linked to the electrophile (Figure 5.7). This sp^3 center also binds the substituent X, which is eliminated as a leaving group in the fast reaction step that follows. This substituent is almost always an H atom (→ elimination of H^+) and only in exceptional cases a *tert*-butyl group (→ elimination of *tert*-Bu^+) or a sulfonic acid group (→ elimination of SO_3H^+) (Section 5.1.2).

The counterpart to this mechanism in the area of nucleophilic aromatic substitution reaction exists as the so-called Ar-S_N reaction via a Meisenheimer complex (Figure 5.47). In this case, a *nucleophile* attacks the aromatic compound in the rate-determining step. A *carbanion* is produced, which is called the Meisenheimer complex (example: see Figure 5.48). A *negative* charge, which should be stabilized by electron-withdrawing substituents, is delocalized over the five sp^2-hybridized centers of the former aromatic compound. In the position *para* to the C atom, which has become linked to the nucleophile, a partial charge of -0.30 appears; in the *ortho* position, a partial charge of -0.25 appears, and in the *meta* position, a partial charge of -0.10 appears (see intermediate **A,** Figure 5.47). Corresponding to the magnitudes of these partial charges, the Meisenheimer complex is better stabilized by an electron acceptor in the *para* position than through one in the *ortho* position. Still, both of these provide con-

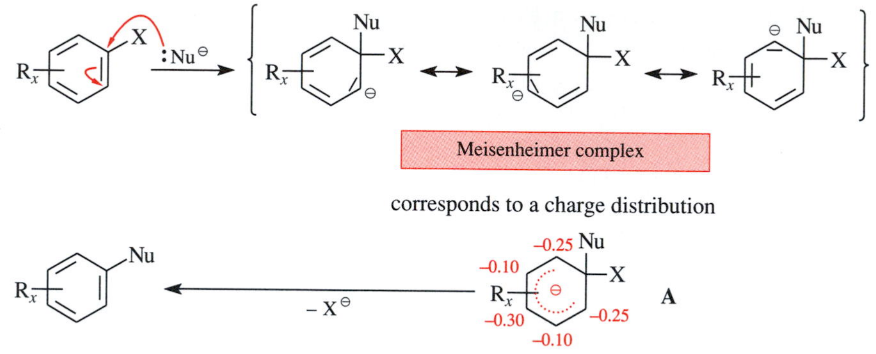

Fig. 5.47. Mechanism for the S_N reaction via Meisenheimer complexes.

Meisenheimer complex

corresponds to a charge distribution

siderably better stabilization than an electron acceptor in the *meta* position. Actually, only when such a stabilization is present can these anionic intermediates form with a preparatively useful reaction rate. The C atom that carries the former nucleophile is sp^3-hybridized in Meisenheimer intermediates and also linked to the substituent X, which is eliminated as X^- in the second, fast step. X is usually Cl (→ elimination of Cl^-) or, for example, in Sanger's reagent, F (→ elimination of F^-; see below).

The name "Meisenheimer complex" had originally been given to carbanions of structures related to **A** (Figure 5.47) *if they could be isolated or could be maintained in solution long enough so that they could be examined spectroscopically.* The best-known Meisenheimer complex is shown in Figure 5.48. It can be prepared from trinitroanisol and NaOMe. It can be isolated because its negative charge is very well stabilized by the nitro groups located in the two *ortho* positions and in the *para* position of the carbanion. As another stabilizing factor, the leaving group (MeO^-) is poor: as an alkoxide it is a high-energy species—so that it stays in the molecule instead of being expelled under rearomatization.

With a less comprehensive substitution by EWGs than in Figure 5.48 and/or with a better leaving group at the sp^3-C, the lifetimes of Meisenheimer complexes are considerably shorter. They then appear only as the short-lived intermediates of the Ar-S_N reactions of Figure 5.47.

Fig. 5.48. Formation of an isolable Meisenheimer complex.

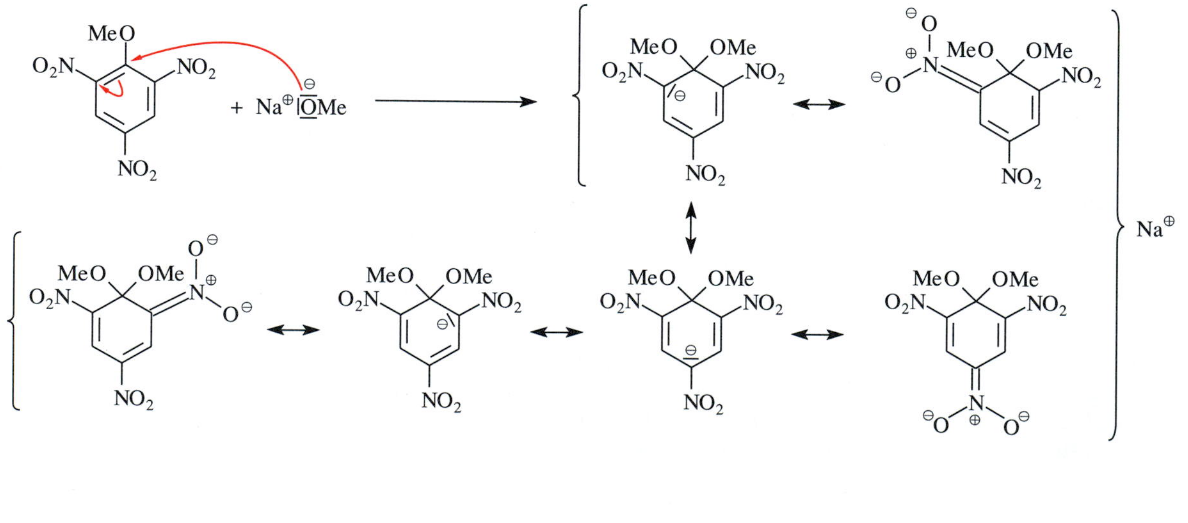

5.5.2 Examples of Reactions of Preparative Interest

Two suitably positioned nitro groups make the halogen-bearing carbon atom in 2,4-dinitrohalobenzenes a favored point of attack for nucleophilic substitution reactions. Thus, 2,4-dinitrophenyl hydrazine is produced from the reaction of 2,4-dinitrochlorobenzene with hydrazine:

$$\text{2,4-}(O_2N)_2C_6H_3Cl + NH_2{-}NH_2 \xrightarrow[-\ NH_2{-}\overset{\oplus}{N}H_3\ Cl^{\ominus}]{} \text{2,4-}(O_2N)_2C_6H_3{-}NH{-}NH_2$$

2,4-Dinitrofluorobenzene (Sanger's reagent) was used earlier in a different S_N reaction: for arylating the N atom of the N-terminal amino acid of oligopeptides. The F atom contained in this reagent strongly stabilizes the Meisenheimer complex because of its particularly great −I effect. A chlorine atom as the leaving group would not provide as much stabilization. Consequently, as inferred from the Hammond postulate, the F atom gives Sanger's reagent a higher reactivity toward nucleophiles compared to 2,4-dinitrochlorobenzene.

The negative charge, which is located essentially on three of the five sp^2-hybridized ring atoms of a Meisenheimer complex, can be stabilized extremely well not only by a nitro substituent at these atoms. Such a negative charge is comparably well stabilized when it is located on a ring nitrogen instead of a nitro-substituted ring carbon. Therefore, pyridines, pyrimidines, and 1,3,5-triazines containing a Cl atom in the 2, 4, and/or 6 position, likewise enter into Ar-S_N reactions via Meisenheimer complexes very readily.

Under forcing conditions chlorine can be displaced by nucleophiles according to

Side Note 5.2
Synthesis of a Dye and Its Binding to a Cotton Fiber

The chlorine atoms of 2,4,6-trichloro-1,3,5-triazine, for example, are rapidly substituted by nucleophiles. This is exploited in textile dyeing through the reaction sequence of Figure 5.49. Trichlorotriazine serves as the so-called triazine anchor, which means that it links the dye covalently to the cotton fiber making the fiber colorfast. First *one* Cl atom of the triazine is replaced, for example, by the nucleophilic amino group of an anthraquinone dye. Then, in basic solution, a deprotonated OH group of the cotton fiber substitutes the second Cl atom. The occurrence of these substitutions in a stepwise manner is due to two effects. First, each Meisenheimer complex intermediate of this substitution sequence is stabilized not only by the N atoms of the ring but also by the −I and −M effects of the chlorine substituents. If the first chlorine has been replaced by the amino group of the dye, the stability of the next Meisenheimer complex intermediate drops: Compared with Cl, an NH(Ar) group is a poorer −I acceptor and indeed no −M acceptor at all. Second, the NH(Ar) substituent stabilizes the triazine framework somewhat. This is because its +M effect gives rise to a type of amidine resonance. This stabilization must be overcome in the formation of the second Meisenheimer complex.

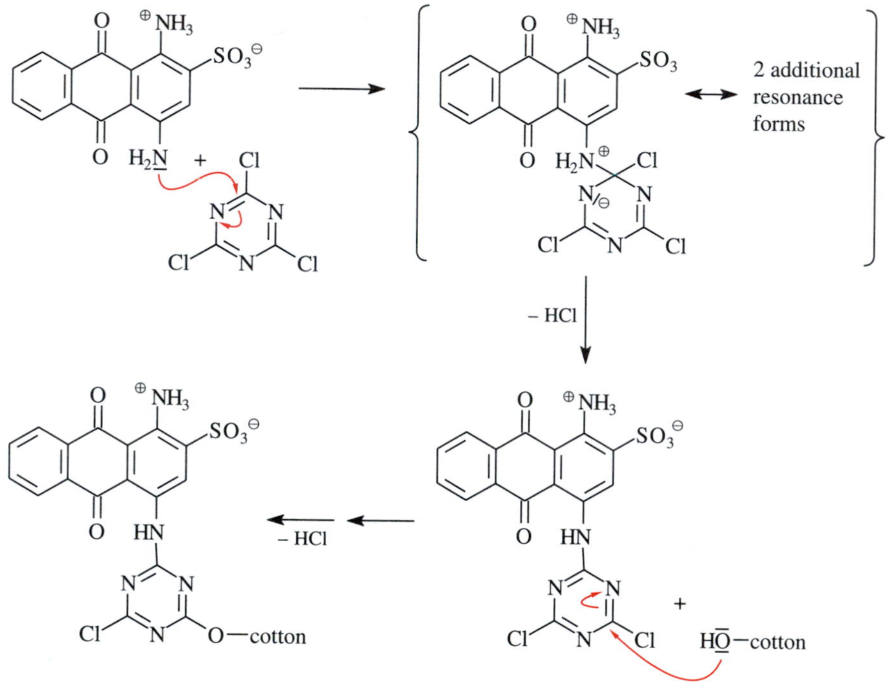

Fig. 5.49. Formation of a dye with subsequent binding to a cotton fiber by a sequence of two Ar-S_N reactions via Meisenheimer complexes.

the mechanism described in this chapter also in compounds with fewer or weaker electron accepting substituents than hitherto mentioned. This explains the formation of tetrachlorodibenzodioxin ("dioxin") from sodium trichlorophenolate in the well-known Seveso accident:

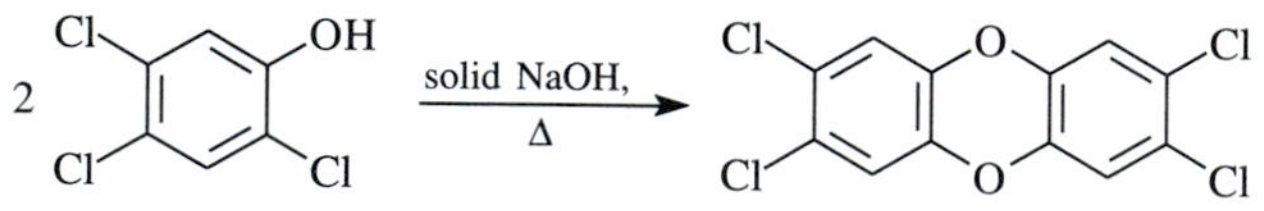

According to Section 5.14, naphthalene takes up an electrophile to form a Wheland complex more rapidly than benzene does. The reason was that in this step naphthalene gives up only ~30 kcal/mol of aromatic stabilization, whereas benzene gives up ~36 kcal/mol. For the very same reason, naphthalene derivatives react with nucleophiles faster to form Meisenheimer complexes than the analogous benzene derivatives do. In other words: Naphthalene derivatives undergo an Ar-S_N reaction more readily than analogous benzene derivatives. In fact, there are naphthalenes that undergo Ar-S_N reactions even when they do not contain *any* electron withdrawing substituent other than the leaving group.

Let us consider as an example the synthesis of a precursor of the hydrogenation catalyst *R*-2,2'-bis(diphenylphosphino)-1,1'-binaphthyl *R*-BINAP (Figure 5.50). The sub-

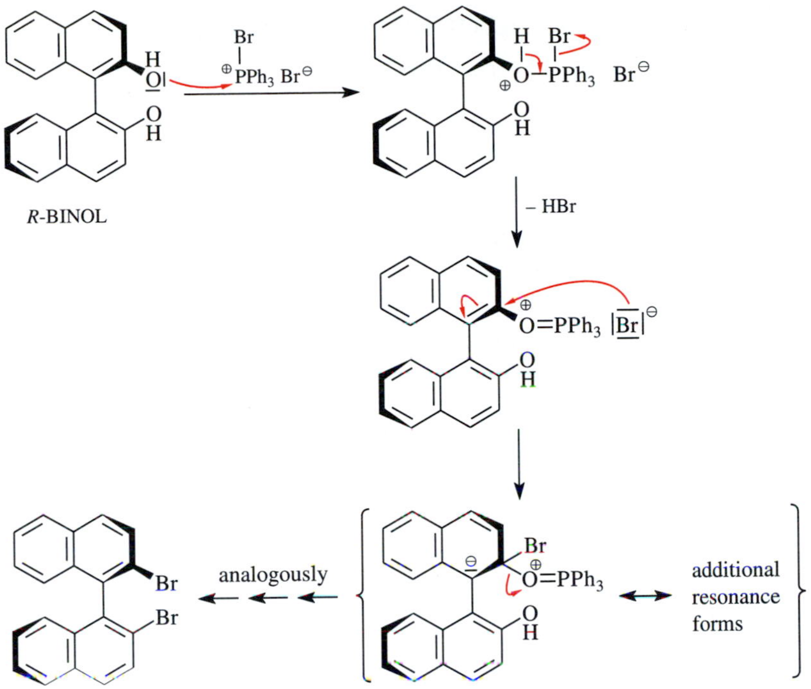

Fig. 5.50. Synthesis of a precursor of the hydrogenation catalyst *R*-BINAP (for completion of its synthesis, see Figure 5.37) from *R*-BINOL.

strate of this reaction is enantiomerically pure *R*-1,1'-bi-2-naphthol (*R*-BINOL). Its OH groups become leaving groups after activation with a phosphonium salt, which is prepared and which acts as discussed in the context of the redox condensations according to Mukaiyama (Figure 2.31). The bromide ion contained in the reagent then acts as the nucleophile.

5.5.3 A Special Mechanistic Case: Reactions of Aryl Sulfonates with NaOH/KOH in a Melt

B

It is unlikely that the reaction of aryl sulfonates with NaOH/KOH in a melt proceeds via a Meisenheimer complex. Such an intermediate would, of course, experience only a marginal stabilization—if any at all—through the SO_3^- substituent. It is not known what the actual mechanism is. Because of the preparative importance of this reaction, however, it is nonetheless presented here with one example:

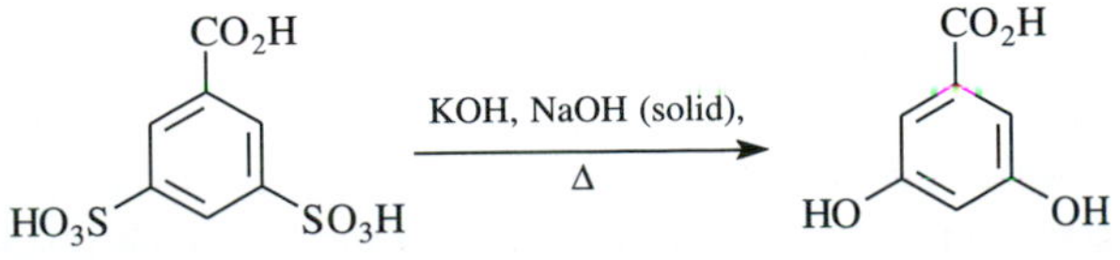

(preparation in Section 5.2.2)

5.6 Nucleophilic Aromatic Substitution via Arynes, *cine* Substitution

The significance of this type of reaction in preparative chemistry is limited. Therefore, we mention just the famous early phenol synthesis by Dow, which is no longer profitably carried out (Figure 5.51). The substrate of this substitution is chlorobenzene; the nucleophile is a hot aqueous solution of sodium hydroxide.

Cl
6% aqueous NaOH, 360°C, 200 bar; $H^{\oplus}$ workup
OH

NaOH
→ β-elimination
is identical with
+ $OH^{\ominus}$
+ $H^{\oplus}$, then − $H^{\oplus}$ (or O→C migration of $H^{\oplus}$)
sodium phenolate
$H_3O^{\oplus}$ workup
50 : 50

Fig. 5.51. The old phenol synthesis according to Dow: preparative (left) and mechanistic (right) aspects.

In a mechanistic investigation a ^{14}C label was introduced at the C1 center of the substrate and in the resulting phenol the OH group was located *50% at the location and 50% next to the location* of the label. Consequently, there had been 50% *ipso* and 50% so-called *cine* substitution. This finding can be understood when it is assumed that in the strongly basic medium chlorobenzene first eliminates HCl to give dehydrobenzene (or benzyne). This species is an alkyne suffering from an enormous angular strain. Thus, it is so reactive that Na^+OH^- can add to its C≡C bond. This addition, of course, takes place without regioselectivity with respect to the ^{14}C label.

The primary addition product formed is (2-hydroxyphenyl)sodium. However, this product undergoes immediately a proton transfer to give sodium phenolate, which is the conjugate base of the target molecule.

References

M. Sainsbury, "Aromatic Chemistry," Oxford University Press, Oxford, U.K., **1992.**

D. T. Davies, "Aromatic Heterocyclic Chemistry," Oxford University Press, New York, **1992.**

5.1

R. J. K. Taylor, "Electrophilic Aromatic Substitution," Wiley, Chichester, U.K., **1990.**

F. Effenberger, "1,3,5-Tris(dialkylamino)benzenes: Model compounds for the electrophilic substitution and oxidation of aromatic compounds," *Acc. Chem. Res.* **1989,** *22,* 27–35.

K. K. Laali, "Stable ion studies of protonation and oxidation of polycyclic arenes," *Chem. Rev.* **1996,** *96,* 1873–1906.

A. R. Katritzky and W. Q. Fan, "Mechanisms and rates of electrophilic substitution reactions of heterocycles," *Heterocycles* **1992,** *34,* 2179–2229.

5.2

M. R. Grimmett, "Halogenation of heterocycles: II. Six- and seven-membered rings," *Adv. Heterocycl. Chem.* **1993,** *58,* 271–345.

C. M. Suter and A. W. Weston, "Direct sulfonation of aromatic hydrocarbons and their halogen derivatives,"*Org. React.* **1946,** *3,* 141–197.

L. Eberson, M. P. Hartshorn, F. Radner, "Ingold's nitration mechanism lives!," *Acta Chem. Scand.* **1994,** *48,* 937–950.

B. P. Cho, "Recent progress in the synthesis of nitropolyarenes: A review," *Org. Prep. Proced. Int.* **1995,** *27,* 243–272.

J. H. Ridd, "Some unconventional pathways in aromatic nitration," *Acta Chem. Scand.* **1998,** *52,* 11–22.

H. Zollinger, "Diazo Chemistry I. Aromatic and Heteroaromatic Compounds," VCH Verlagsgesellschaft, Weinheim, Germany, **1994.**

C. C. Price, "The alkylation of aromatic compounds by the Friedel-Crafts method," *Org. React.* **1946,** *3,* 1–82.

G. A. Olah, R. Krishnamurti, G. K. S. Prakash, "Friedel-Crafts Alkylations," in *Comprehensive Organic Synthesis* (B. M. Trost, I. Fleming, Eds.), Vol. 3, 293, Pergamon Press, Oxford, **1991.**

H. Heaney, "The Bimolecular Aromatic Friedel-Crafts Reaction," in *Comprehensive Organic Synthesis* (B. M. Trost, I. Fleming, Eds.), Vol. 2, 733, Pergamon Press, Oxford, **1991.**

H. Heane, "The Intramolecular Aromatic Friedel-Crafts Reaction," in *Comprehensive Organic Synthesis* (B. M. Trost, I. Fleming, Eds.), Vol. 2, 753, Pergamon Press, Oxford, **1991.**

T. Ohwada, "Reactive carbon electrophiles in Friedel-Craft reactions," *Rev. on Heteroatom Chem.* **1995,** *12,* 179.

R. C. Fuson and C. H. McKeever, "Chloromethylation of aromatic compounds," *Org. React.* **1942,** *1,* 63–90.

E. Berliner, "The Friedel and Crafts reaction with aliphatic dibasic acid anhydrides," *Org. React.* **1949,** *5,* 229–289.

I. Hashimoto, T. Kawaji, F. D. Badea, T. Sawada, S. Mataka, M. Tashiro, G. Fukata, "Regioselectivity of Friedel-Crafts acylation of aromatic compounds with several cyclic anhydrides," *Res. Chem. Intermed.* **1996,** *22,* 855–869.

A. R. Martin, "Uses of the Fries rearrangement for the preparation of hydroxyaryl ketones," *Org. Prep. Proced. Int.* **1992,** *24,* 369.

O. Meth-Cohn and S. P. Stanforth, "The Vilsmeier-Haack Reaction," in *Comprehensive Organic Synthesis* (B. M. Trost, I. Fleming, Eds.), Vol. 2, 777, Pergamon Press, Oxford, **1991.**

G. Jones and S. P. Stanforth, "The Vilsmeier reaction of fully conjugated carbocycles and heterocycles," *Org. React.* **1997,** *49,* 1–330.

5.3

L. Brandsma, "Aryl and Hetaryl Alkali Metal Compounds," in *Methoden Org. Chem. (Houben-Weyl) 4th ed. 1952–, Carbanions* (M. Hanack, Ed.), Vol. E19d, 369, Georg Thieme Verlag, Stuttgart, **1993.**

H. Gilman and J. W. Morton, Jr., "The metalation reaction with organolithium compounds," *Org. React.* **1954,** *8,* 258–304.

V. Snieckus, "Regioselective synthetic processes based on the aromatic directed metalation strategy," *Pure Appl. Chem.* **1990,** *62,* 671.

V. Snieckus, "The directed ortho metalation reaction. Methodology, applications, synthetic links, and a nonaromatic ramification," *Pure Appl. Chem.* **1990,** *62,* 2047–2056.

V. Snieckus, "Directed ortho metalation. Tertiary amide and O-carbamate directors in synthetic strategies for polysubstituted aromatics," *Chem. Rev.* **1990,** *90,* 879–933.

V. Snieckus, "Combined directed ortho metalation-cross coupling strategies. Design for natural product synthesis," *Pure Appl. Chem.* **1994,** *66,* 2155–2158.

K. Undheim and T. Benneche, "Metalation and metal-assisted bond formation in π-electron deficient heterocycles," *Acta Chem. Scand.* **1993,** *47,* 102–121.

H. W. Gschwend and H. R. Rodriguez, "Heteroatom-facilitated lithiations," *Org. React.* **1979,** *26,* 1–360.

R. D. Clark and A. Jahangir, "Lateral lithiation reactions promoted by heteroatomic substituents," *Org. React.* **1995,** *47,* 1–314.

R. G. Jones and H. Gilman, "The halogen-metal interconversion reaction with organolithium compounds," *Org. React.* **1951,** *6,* 339–366.

W. E. Parham and C. K. Bradsher, "Aromatic organolithium reagents bearing electrophilic groups: Preparation by halogen-lithium exchange," *Acc. Chem. Res.* **1982,** *15,* 300.

A. R. Martin and Y. Yang, "Palladium catalyzed cross-coupling reactions of organoboronic acids with organic electrophiles," *Acta Chem. Scand.* **1993,** *47,* 221–230.

A. Suzuki, "New synthetic transformations via organoboron compounds," *Pure Appl. Chem.* **1994,** *66,* 213–222.

N. Miyaura and A Suzuki, "Palladium-catalyzed cross-coupling reactions of organoboron compounds," *Chem. Rev.* **1995,** *95,* 2457–2483.

N. Miyaura, "Synthesis of Biaryls via the Cross-Coupling Reaction of Arylboronic Acids" in *Advances in Metal-Organic Chemistry* (L. S. Liebeskind, Ed.), **1998,** *6,* JAI, Greenwich, CT.

5.4

R. K. Norris, "Nucleophilic Coupling with Aryl Radicals," in *Comprehensive Organic Synthesis* (B. M. Trost, I. Fleming, Eds.), Vol. 4, 451, Pergamon Press, Oxford, **1991.**

A. Roe, "Preparation of aromatic fluorine compounds from diazonium fluoroborates: The Schiemann reaction," *Org. React.* **1949,** *5,* 193–228.

N. Kornblum, "Replacement of the aromatic primary amino group by hydrogen," *Org. React.* **1944,** *2,* 262–340.

5.5

C. Paradisi, "Arene Substitution via Nucleophilic Addition to Electron Deficient Arenes," in *Comprehensive Organic Synthesis* (B. M. Trost, I. Fleming, Eds.), Vol. 4, 423, Pergamon Press, Oxford, **1991.**

F. Terrier, "Nucleophilic Aromatic Displacement. The Influence of the Nitro Group," VCH, New York, **1991.**

I. Gutman, (Ed.), "Nucleophilic Aromatic Displacement: The Influence of the Nitro Group," VCH, New York, **1991.**

N. V. Alekseeva and L. N. Yakhontov, "Reactions of pyridines, pyrimidines, and 1,3,5-triazines with nucleophilic reagents," *Russ. Chem. Rev.* **1990,** *59,* 514–530.

Further Reading

M. Carmack and M. A. Spielman, "The Willgerodt reaction," *Org. React.* **1946,** *3,* 83–107.

S. Sethna and R. Phadke, "The Pechmann reaction," *Org. React.* **1953,** *7,* 1–58.

H. Wynberg and E. W. Meijer, "The Reimer-Tiemann reaction," *Org. React.* **1982,** *28,* 1–36.

W. E. Truce, "The Gattermann synthesis of aldehydes," *Org. React.* **1957,** *9,* 37–72.

N. N. Crounse, "The Gattermann-Koch reaction," *Org. React.* **1949,** *5,* 290–300.

A. H. Blatt, "The Fries reaction," *Org. React.* **1942,** *1,* 342–369.

P. E. Spoerri and A. S. DuBois, "The Hoesch synthesis," *Org. React.* **1949,** *5,* 387–412.

W. E. Bachmann and R. A. Hoffman, "The preparation of unsymmetrical biaryls by the diazo reaction and the nitrosoacetylamine reaction," *Org. React.* **1944,** *2,* 224–261.

DeLos F. DeTar, "The Pschorr synthesis and related diazonium ring closure reactions," *Org. React.* **1957,** *9,* 409–462.

C. S. Rondestvedt, Jr., "Arylation of unsaturated compounds by diazonium salts," *Org. React.* **1960,** *11,* 189–260.

C. S. Rondestvedt, Jr., "Arylation of unsaturated compounds by diazonium salts (the Meerwein arylation reaction)," *Org. React.* **1976,** *24,* 225–259.

M. Braun, "New Aromatic Substitution Methods," in *Organic Synthesis Highlights* (J. Mulzer, H.-J. Altenbach, M. Braun, K. Krohn, H.-U. Reißig, Eds.), VCH, Weinheim, New York, etc., **1991,** 167–173.

J. F. Bunnett, "Some novel concepts in aromatic reactivity," *Tetrahedron* **1993,** *49,* 4477.

G. A. Artamkina, S. V. Kovalenko, I. P. Beletskaya, O. A. Reutov, "Carbon-carbon bond formation in electron-deficient aromatic compounds," *Russ. Chem. Rev. (Engl. Transl.)* **1990,** *59,* 750.

C. D. Hewitt and M. J. Silvester, "Fluoroaromatic compounds: Synthesis, reactions and commercial applications," *Aldrichimica Acta* **1988,** *21,* 3–10.

R. A. Abramovitch, D. H. R. Barton, J.-P. Finet, "New methods of arylation," *Tetrahedron* **1988,** *44,* 3039.

J. H. Clark, T. W. Bastock, D. Wails (Eds.), "Aromatic Fluorination," CRC Press, Boca Raton, FL, **1996.**

L. Delaude, P. Laszlo, K. Smith, "Heightened selectivity in aromatic nitrations and chlorinations by the use of solid supports and catalysts," *Acc. Chem. Res.* **1993,** *26,* 607–613.

J. K. Kochi, "Inner-sphere electron transfer in organic chemistry. Relevance to electrophilic aromatic nitration," *Acc. Chem. Res.* **1992,** *25,* 39–47.

L. Eberson, M. P. Hartshorn, F. Radner, "Electrophilic Aromatic Nitration Via Radical Cations: Feasible or Not?," in *Advances in Carbocation Chemistry* (J. M. Coxon, Ed.) **1995,** *2,* JAI, Greenwich, CT.

H. Ishibashi, M. Ikeda, "Recent progress in electrophilic aromatic substitution with α-Thiocarbocations," *Rev. Heteroatom Chem.* **1996,** *14,* 59–82.

H. Heaney, "The Bimolecular Aromatic Mannich Reaction," in *Comprehensive Organic Synthesis* (B. M. Trost, I. Fleming, Eds.), Vol. 2, 953, Pergamon Press, Oxford, **1991.**

P. E. Fanta, "The Ullmann synthesis of biaryls," *Synthesis* **1974,** 9.

J. A. Lindley, "Copper assisted nucleophilic substitution of aryl halogen," *Tetrahedron* **1984,** *40,* 1433.

G. P. Ellis and T. M. Romsey-Alexander, "Cyanation of aromatic halides," *Chem. Rev.* **1987,** *87,* 779.

I. A. Rybakova, E. N. Prilezhaeva, V. P. Litvinov, "Methods of replacing halogen in aromatic compounds by RS-functions," *Russ. Chem. Rev. (Engl. Transl.)* **1991,** *60,* 1331.

O. N. Chupakhin, V. N. Charushin, H. C. van der Plas, "Nucleophilic Aromatic Substitution of Hydrogen," Academic Press, San Diego, CA, 1994.

J. M. Saveant, "Mechanisms and reactivity in electron-transfer-induced aromatic nucleophilic-substitution—Recent advances," *Tetrahedron* **1994,** *50,* 10117.

A. J. Belfield, G. R. Brown, A. J. Foubister, "Recent synthetic advances in the nucleophilic amination of benzenes," *Tetrahedron* **1999,** *55,* 11399–11428.

Nucleophilic Substitution Reactions on the Carboxyl Carbon (Except through Enolates)

6

6.1 C=O-Containing Substrates and Their Reactions with Nucleophiles

B

C=O double bonds occur in a series of different classes of compounds:

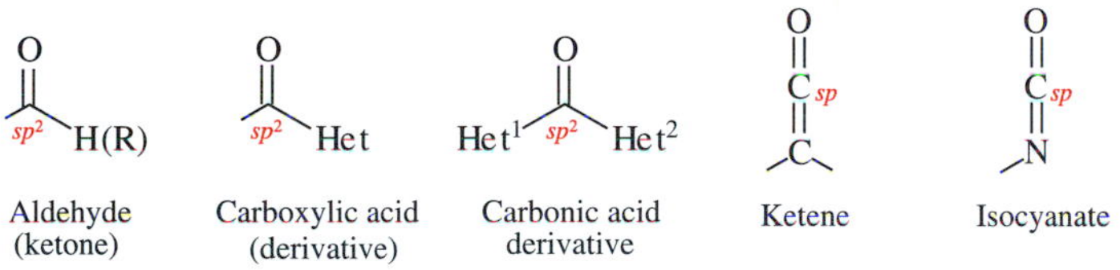

Aldehyde (ketone): O=C(sp^2)–H(R); Carboxylic acid (derivative): O=C(sp^2)–Het; Carbonic acid derivative: Het^1–C(sp^2)(=O)–Het^2; Ketene: O=C(sp)=C<; Isocyanate: O=C(sp)=N–

In aldehydes and ketones, which together are referred to as **carbonyl compounds,** C=O double bonds are part of a carbonyl group C_{sp^2}=O. In this group the C=O double bond—as formulated by organic chemists—connects a carbonyl carbon and a carbonyl oxygen. Carboxylic acids, carboxylic esters, and carboxylic amides, as well as all carboxylic acid derivatives used as acylating agents (see Section 6.3) are termed collectively as **carboxyl compounds** and are thereby distinguished from the carbonyl compounds. They contain a carboxyl group C_{sp^2}(=O)—Het. In the carboxyl group the C=O double bond—according to the nomenclature used throughout this chapter—connects a carboxyl carbon and a carboxyl oxygen. C=O double bonds are also part of carbonic acid derivatives Het^1—C_{sp^2}(=O)—Het^2. Carbonic acid derivatives contain a carboxyl carbon and a carboxyl oxygen, too. Thus there is no difference between the nomenclatures for the C=O groups of carbonic acid derivatives and carboxylic acid derivatives. Finally, there are C_{sp}=O double bonds; these occur in ketenes and isocyanates.

Each of the aforementioned C=O-containing compounds reacts with nucleophiles. Which kind of a reaction occurs depends almost exclusively on the nature of the substrate and hardly at all on the nucleophile:

- The typical reaction of carbonyl compounds with nucleophiles is the *addition* (Section 7.2); the C=O bond disappears:

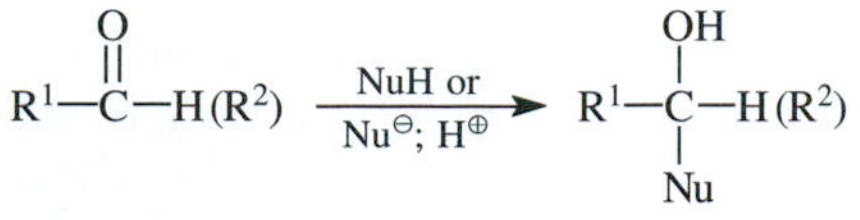

$$R^1-C(=O)-H(R^2) \xrightarrow[Nu^{\ominus};\ H^{\oplus}]{NuH\ or} R^1-C(OH)(Nu)-H(R^2)$$

- Ketenes and other C=O-containing heterocumulenes also react with nucleophiles in *addition* reactions; the C=O double bond, however, is nonetheless retained (Section 7.1):

$$R^1R^2C{=}C{=}O \xrightarrow[Nu^{\ominus};\ H^{\oplus}]{NuH\ or} R^1R^2CH{-}C({=}O){-}Nu \qquad Het{=}C{=}O \xrightarrow[Nu^{\ominus};\ H^{\oplus}]{NuH\ or} H{-}Het{-}C({=}O){-}Nu$$

(Het = NR, O, S)

- In contrast, C=O-containing carboxylic acid and carbonic acid derivatives react with nucleophiles in *substitution* reactions. The one group or one of the two groups bound through a heteroatom to the carboxyl carbon of these substrates is substituted so that compounds **A** or **B**, respectively, are obtained.

$$R{-}C({=}O){-}Het \xrightarrow[Nu^{\ominus};\ H^{\oplus}\ (-\ HetH)]{NuH\ or} \underset{\mathbf{A}}{R{-}C({=}O){-}Nu} \qquad Het^1{-}C({=}O){-}Het^2 \xrightarrow[Nu^{\ominus};\ H^{\oplus}\ (-\ Het^2{-}H)]{NuH\ or} \underset{\mathbf{B}}{Het^1{-}C({=}O){-}Nu}$$

These substitution products **A** and **B** need not yet be the final product of the reaction of nucleophiles with carboxylic acids, carboxylic acid derivatives, or carbonic acid derivatives. Sometimes they may be formed only as intermediates and continue to react with the nucleophile: Being carbonyl compounds (substitution products **A**) or carboxylic acid derivatives (substitution products **B**), they can in principle undergo, in addition, an addition reaction or another substitution reaction (see above). Thus carboxylic acid derivatives can react with as many as two equivalents of nucleophiles, and carbonic acid derivatives can react with as many as three.

It remains to be explained how these different chemoselectivities come about. Why do nucleophiles react . . .

a) with aldehydes and ketones by addition and not by substitution?
b) with ketenes and other heterocumulenes by addition and not by substitution?
c) with carboxylic acids (or their derivatives) and with carbonic acid derivatives by substitution and not by addition?

Regarding question (a): in substitution reactions, aldehydes and ketones would have to eliminate a hydride ion or a carbanion; both of which are extremely poor leaving groups (cf. Section 2.3).

Regarding question (b): ketenes and other heterocumulenes contain only double-bonded substituents at the attacked C atom. However, a leaving group must be single-

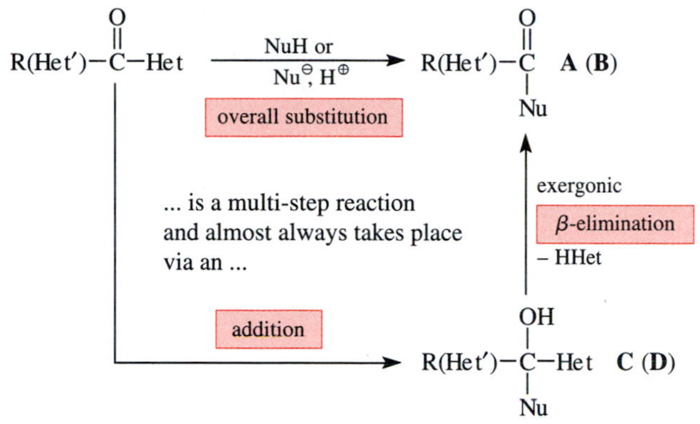

Fig. 6.1. Reactions of nucleophiles with C=O-containing carboxylic acid and carbonic acid derivatives. Substitution at the carboxyl carbon instead of addition to the acyl group.

bonded. Consequently, the structural prerequisite for the occurrence of a substitution reaction is absent.

Regarding question (c): the addition of a nucleophile to the C=O double bond of carboxylic or carbonic acid derivatives would give products of type **C** or **D** (Figure 6.1). However, these compounds are without exception thermodynamically less stable than the corresponding substitution products **A** or **B.** The reason for this is that the three bonds in the substructure C_{sp^3}(—O—H)—Het of the addition products **C** and **D** are together less stable than the double bond in the substructure C_{sp^2}(=O) of the substitution products **A** and **B** plus the highlighted single bond in the by-product H—Het. In fact, ordinarily (Section 6.2), substitution reactions on the carboxyl carbon leading ultimately to compounds of type **A** or **B** take place via neutral addition products of types **C** or **D** as *intermediates* (Figure 6.1). These addition products may be produced either continuously through attack of the nucleophile (cf. Figures 6.2, 6.5) or not until an aqueous workup has been carried out subsequent to the completion of the nucleophile's attack (cf. Figure 6.4). In both cases, once the neutral addition products **C** and **D** have formed, they decompose exergonically to furnish the substitution products **A** or **B** via rapid E1 eliminations, respectively.

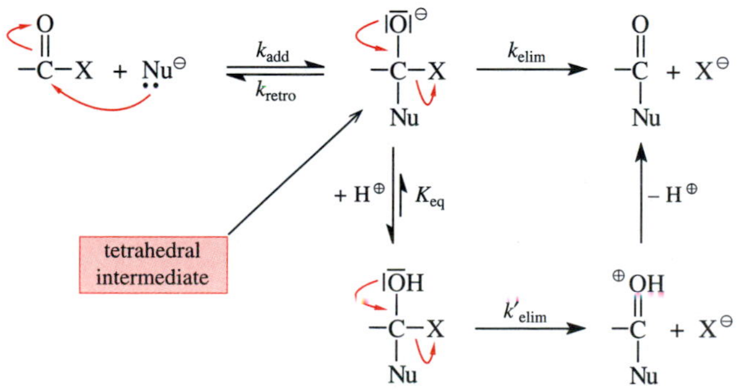

Fig. 6.2. Mechanism of S_N reactions of good nucleophiles at the carboxyl carbon: k_{add} is the rate constant of the addition of the nucleophile, k_{retro} is the rate constant of the back-reaction, and k_{elim} is the rate constant of the elimination of the leaving group; K_{eq} is the equilibrium constant for the protonation of the tetrahedral intermediate at the negatively charged oxygen atom.

6.2 Mechanisms, Rate Laws, and Rate of Nucleophilic Substitution Reactions on the Carboxyl Carbon

B

When a nucleophile containing a heteroatom attacks at a carboxyl carbon, S_N reactions occur that convert carboxylic acid derivatives to other carboxylic acid derivatives, or that convert carbonic acid derivatives to other carbonic acid derivatives. When an organometallic compound is used as the nucleophile, S_N reactions at the carboxyl carbon make it possible to synthesize aldehydes (from derivatives of formic acid), ketones (from derivatives of higher carboxylic acids), or—starting from carbonic acid derivatives—in some cases carboxylic acid derivatives. Similarly, when using a hydride transfer agent as the nucleophile, S_N reactions at a carboxyl carbon allow the conversion of carboxylic acid derivatives into aldehydes.

From the perspective of the nucleophiles, these S_N reactions constitute acylations. Section 6.2 describes which acid derivatives perform such acylations rapidly as "good acylating agents" and which ones undergo them only slowly as "poor acylating agents," and why this is so. Because of their greater synthetic importance, we will examine thoroughly only the acylations with *carboxylic acid derivatives*. Using the principles learned in this context, you can easily derive the acylating abilities of *carbonic acid derivatives*.

6.2.1 Mechanism and Rate Laws of S_N Reactions on the Carboxyl Carbon

B

Most S_N reactions of carboxylic acids and their derivatives follow one of three mechanisms (Figures 6.2, 6.4, and 6.5). A key intermediate common to all of them is the species in which the nucleophile is linked with the former carboxyl carbon for the first time. In this intermediate, the attacked carbon atom is tetrasubstituted and thus tetrahedrally coordinated. This species is therefore referred to as the **tetrahedral intermediate** (abbreviated as "Tet. Intermed." in the following equations). Depending on the nature of the reaction partners, the tetrahedral intermediate can be negatively charged, neutral, or even positively charged.

The tetrahedral intermediate is a high-energy intermediate. Therefore, independently of its charge and also independently of the detailed formation mechanism, it is formed via a *late* transition state; it also reacts further via an *early* transition state. Both properties follow from the Hammond postulate. Whether the transition state of the *formation* of the tetrahedral intermediate has a higher or a lower energy than the transition state of the subsequent *reaction* of the tetrahedral intermediate determines whether this intermediate is formed in an irreversible or in a reversible reaction, respectively. Yet, in any case, *the tetrahedral intermediate is a transition state model of the rate-determining step of the vast majority of S_N reactions at the carboxyl carbon*. In the following sections we will prove this statement by formal kinetic analyses of the most important substitution mechanisms.

S_N Reactions on the Carboxyl Carbon in Nonacidic Protic Media

B

Figure 6.2 shows the standard mechanism of substitution reactions carried out on carboxylic acid derivatives in neutral or basic solutions. The tetrahedral intermediate—formed in the rate-determining step—can be converted to the substitution product via two different routes. The shorter route consists of a single step: the leaving group X is eliminated with a rate constant k_{elim}. In this way the substitution product is formed in a total of two steps. The longer route to the same substitution product is realized when the tetrahedral intermediate is protonated. To what extent this occurs depends, according to Equation 6.1, on the pH value and on the equilibrium constant K_{eq} defined in Figure 6.2:

$$[\text{Protonated Tet. Intermed.}] = [\text{Tet. Intermed.}] \cdot K_{eq} \cdot 10^{-pH} \tag{6.1}$$

The protonated tetrahedral intermediate can eject the leaving group X with a rate constant k'_{elim}. Subsequently, a proton is eliminated. Thereby, the substitution product is formed in a total of four steps.

Which of the competing routes in Figure 6.2 preferentially converts the tetrahedral intermediate to the substitution product? An approximate answer to this question is possible if an equilibrium between the negatively charged and the neutral tetrahedral intermediate is established and if the rate constants of this equilibration are much greater than the rate constants k_{elim} and k'_{elim} of the reactions of the respective intermediates to give the substitution product. Under these conditions we have:

$$\frac{d[S_N\ \text{product}_{\text{two-step}}]}{dt} = k_{elim}\,[\text{Tet. Intermed.}] \tag{6.2}$$

$$\frac{d[S_N\ \text{product}_{\text{four-step}}]}{dt} = k'_{elim}\,[\text{Protonated Tet. Intermed.}] \tag{6.3}$$

The subscripts "two-step" and "four-step" in Equations 6.2 and 6.3, respectively, refer to the rates of product formation via the two- and four-step routes of the mechanism in Figure 6.2. If one divides Equation 6.2 by Equation 6.3 and integrates subsequently, one obtains:

$$\frac{\text{Yield } S_N\ \text{product}_{\text{two-step}}}{\text{Yield } S_N\ \text{product}_{\text{four-step}}} = \underbrace{\frac{k_{elim}}{k'_{elim}}}_{\gg 1} \cdot \frac{10^{pH}}{K_{eq}} \tag{6.4}$$

Equation 6.4 shows the following: in strongly basic solutions, S_N products can be produced from carboxylic acid derivatives in *two* steps. An example of such a reaction is the saponification $PhC({=}O)OEt + KOH \rightarrow PhC({=}O)O^-K^+ + EtOH$. However, in approximately neutral solutions the *four-step* path to the S_N product should predominate. An example of this type of reaction is the aminolysis $PhC({=}O)OEt + HNMe_2 \rightarrow PhC({=}O)NMe_2 + EtOH$.

Finally we want to examine the kinetics of two-step acylations according to the mechanism in Figure 6.2. We must distinguish between two cases. Provided that the tetrahedral intermediate forms *irreversibly* and consequently in the rate-determining step, a rate law for the formation of the S_N product (Equation 6.7) is obtained from Equation 6.5.

$$\frac{d[\text{Tet. Intermed.}]}{dt} = k_{add}[-C(=O)X][Nu^{\ominus}] - k_{elim}[\text{Tet. Intermed.}] \tag{6.5}$$

$$= 0 \text{ (because of the steady-state approximation)}$$

$$\Rightarrow k_{elim}[\text{Tet. Intermed.}] = k_{add}[-C(=O)X][Nu^{\ominus}] \tag{6.6}$$

Inserting Equation 6.6 in Equation 6.2 ⇒

$$\frac{d[S_N \text{ product}]}{dt} = k_{add}[-C(=O)X][Nu^{\ominus}] \tag{6.7}$$

In contrast, Equations 6.8–6.10 serve to derive the rate of product formation for an acylation by the two-step route in Figure 6.3 assuming that the tetrahedral intermediate is formed *reversibly*. Interestingly, it does not matter whether in the rate-determining step the tetrahedral intermediate is *formed* ($k_{retro} < k_{elim}$) or *reacts further* (then $k_{retro} > k_{elim}$).

$$\frac{d[\text{Tet. Intermed.}]}{dt} = k_{add}[-C(=O)X][Nu^{\ominus}] - k_{retro}[\text{Tet. Intermed.}] - k_{elim}[\text{Tet. Intermed.}]$$

$$= 0 \text{ (because of the steady-state approximation)} \tag{6.8}$$

$$\Rightarrow [\text{Tet. Intermed.}] = \frac{k_{add}}{k_{retro} + k_{elim}}[-C(=O)X][Nu^{\ominus}] \tag{6.9}$$

Inserting Equation 6.9 in Equation 6.2 ⇒

$$\frac{d[S_N \text{ product}]}{dt} = k_{add}\frac{1}{1 + k_{retro}/k_{elim}}[-C(=O)X][Nu^{\ominus}] \tag{6.10}$$

Equations 6.7 and 6.10 are indistinguishable in the kinetic experiment because the *experimental* rate law would have the following form in both cases:

$$\frac{d[S_N \text{ product}]}{dt} = \text{const.}[-C(=O)X][Nu^{\ominus}] \tag{6.11}$$

This means that from the kinetic analysis one could conclude only that the S_N product is produced in a *bimolecular* reaction. One has no information as to whether the tetrahedral intermediate forms irreversibly or reversibly.

What does one expect regarding the irreversibility or reversibility of the formation of the tetrahedral intermediate in Figure 6.2? Answering this question is tanta-

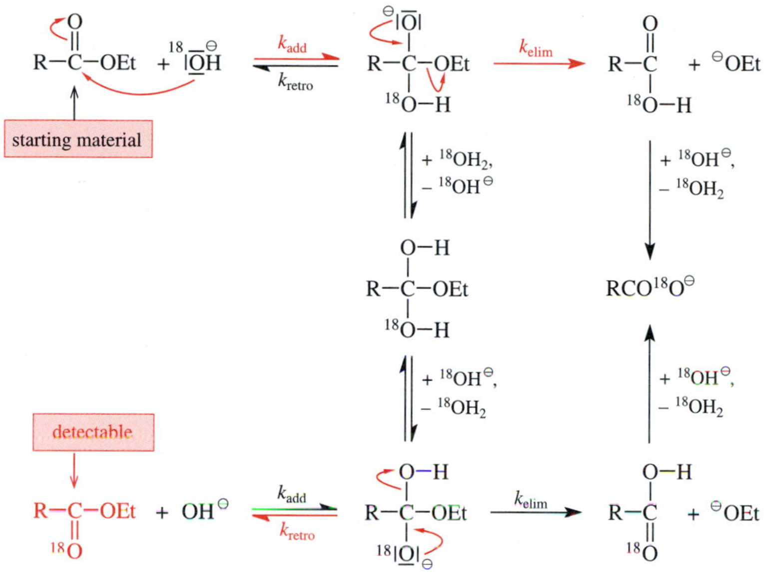

Fig. 6.3. Alkaline hydrolysis of carboxylic esters according to the mechanism of Figure 6.2: proof of the reversibility of the formation of the tetrahedral intermediate. In the alkaline hydrolysis of ethyl *para*-methylbenzoate in H_2O, for example, the ratio k_{retro}/k_{elim} is at least 0.13 (but certainly not much more).

mount to knowing the ratio in which the tetrahedral intermediate ejects the group X (with the rate constant k_{elim}) or the nucleophile (with the rate constant k_{retro}). The outcome of this competition depends on whether group X or the nucleophile is the better leaving group. When X is a good leaving group, the tetrahedral intermediate is therefore expected to form irreversibly. This should be the case for all *good* acylating agents. When X is a poor leaving group, the nucleophile may be a better one so that it reacts with the acylating agent to give the tetrahedral intermediate in a reversible reaction. The alkaline hydrolysis of amides is an excellent example of this kind of substitution. Finally, when X and the nucleophile have similar leaving group abilities, k_{retro} and k_{elim} are of comparable magnitudes. Then, the tetrahedral intermediate decomposes partly into the starting materials and partly into the products. A good example of this kind of mechanism is the alkaline hydrolysis of carboxylic acid esters. In that case, the reversibility of the formation of the tetrahedral intermediate was proven by performing the hydrolysis with ^{18}O-labeled NaOH (Figure 6.3): The unreacted ester that was recovered had incorporated ^{18}O atoms in its C=O double bond.

S_N Reactions at the Carboxyl Carbon via a Stable Tetrahedral Intermediate

B

A variant of the substitution mechanism of Figure 6.2 is shown in Figure 6.4. The tetrahedral intermediate is produced in an irreversible step and does not react further until it is worked up with aqueous acid. Overall, the substitution product is produced according to the four-step route of Figure 6.2. But in contrast to its standard course, two separate operations are required for the gross substitution to take place: first, the nucleophile must be added; second, H_3O^+ must be added.

The reactivity of carboxylic acid derivatives that react with nucleophiles according to the mechanism in Figure 6.4 cannot be measured via the rate of formation of the

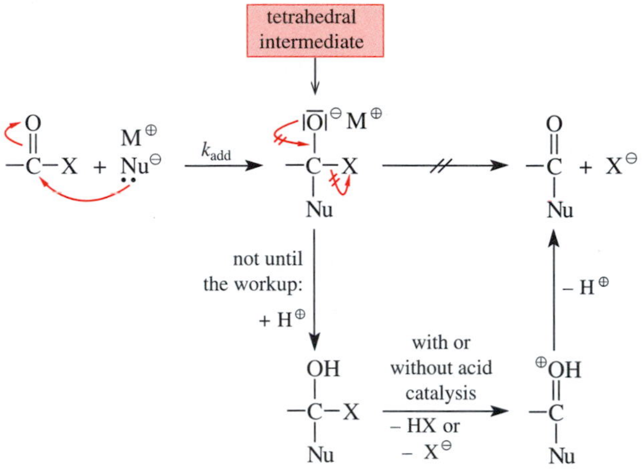

Fig. 6.4. Mechanism of S_N reactions at the carboxyl carbon via a stable tetrahedral intermediate.

substitution product. Instead, the decrease in the concentration of the starting material serves as a measure of the reactivity.

$$\frac{d[-C(=O)X]}{dt} = k_{add}[-C(=O)X][Nu^{\ominus}] \tag{6.12}$$

Almost all substitution reactions on the carboxyl carbon undertaken by hydride donors or organometallic reagents take place according to the mechanism of Figure 6.4 (cf. Section 6.5).

Proton-Catalyzed S_N Reactions on the Carboxyl Carbon

B

Figure 6.5 shows the third important mechanism of S_N reactions on the carboxyl carbon. It relates to proton-catalyzed substitution reactions of weak nucleophiles with weak acylating agents. When weak acylating agents are protonated in fast equilibrium reactions at the carboxyl oxygen, they turn into considerably more reactive acylating agents, namely, into carboxonium ions. Even catalytic amounts of acid can increase the reaction rate due to this effect because even a small equilibrium fraction of the highly reactive carboxonium ion can be attacked easily by the nucleophile. Because proto-

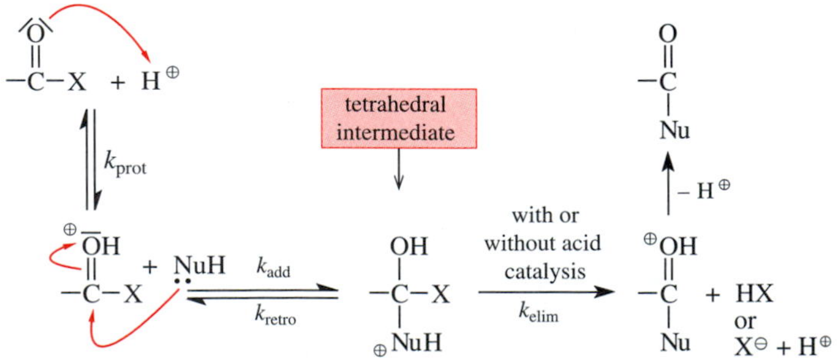

Fig. 6.5. Mechanism of proton-catalyzed S_N reactions at the carboxyl carbon; K_{prot} is the equilibrium constant of the protonation of the weak acylating agent used.

nation of the starting material continues to supply an equilibrium amount of the carboxonium ion, gradually the entire acylating agent reacts via this intermediate to give the S_N product. The proton-catalyzed esterifications of carboxylic acids or the acidic hydrolysis of amides exemplify the substitution mechanism of Figure 6.5.

The rate law for S_N reactions at the carboxyl carbon according to the mechanism shown in Figure 6.5 can be derived as follows:

$$\frac{d[S_N \text{ product}]}{dt} = k_{elim}[\text{Tet. Intermed.}] \tag{6.13}$$

$$\frac{d[\text{Tet. Intermed.}]}{dt} = k_{add}[-C(=\overset{\oplus}{O}H)X][Nu^{\ominus}] - k_{retro}[\text{Tet. Intermed.}] - k_{elim}[\text{Tet. Intermed.}]$$

$$= 0 \text{ (because of steady-state approximation)}$$

$$\Rightarrow [\text{Tet. Intermed.}] = \frac{k_{add}}{k_{retro} + k_{elim}}[-C(=\overset{\oplus}{O}H)X][Nu^{\ominus}] \tag{6.14}$$

For the initiating protonation equilibrium we have:

$$[-C(=\overset{\oplus}{O}H)X] = K_{prot}[-C(=O)X][H^{\oplus}] \tag{6.15}$$

Equation 6.14 and Equation 6.15 inserted in Equation 6.13 ⇒

$$\frac{d[S_N \text{ product}]}{dt} = k_{add}\frac{1}{1 + k_{retro}/k_{elim}} \cdot K_{prot}[-C(=O)X][H^{\oplus}][Nu^{\ominus}] \tag{6.16}$$

The initial Equation 6.13 reflects that the second-to-last reaction step in Figure 6.5 is substantially slower than the last step. Equation 6.13 is simplified using Equation 6.14. Equation 6.14 requires the knowledge of the concentration of the carboxonium ions, which are the acylating reagents in this mechanism. Equation 6.15 provides this concentration via the equilibrium constant K_{prot} of the reaction that forms the ion (Figure 6.5). Equations 6.13–6.15 allow for the derivation of the rate law of the S_N reaction according to this mechanism, which is given as Equation 6.16. This equation is the rate law of a trimolecular reaction: The rate of the reaction is proportional to the concentrations of the acylating agent, the nucleophile, and the protons.

The Rate-Determining Step of the Most Important S_N Reactions on the Carboxyl Carbon

B

Let us summarize: The rate laws for S_N reactions at the carboxyl carbon exhibit an important common feature regardless of whether the substitution mechanism is that of Figure 6.2 (rate laws: Equation 6.7 or 6.10), Figure 6.4 (rate law: Equation 6.12), or Figure 6.5 (rate law: Equation 6.16):

> The larger the rate constant k_{add} for the formation of the tetrahedral intermediate, the faster an acylating agent reacts with nucleophiles.

Therefore, *independent of the substitution mechanism,* the reactivity of a series of acylating agents with respect to a given nucleophile is characterized by the fact that the most reactive acylating agent exhibits the smallest energy difference between the acylating agent and the derived tetrahedral intermediate. This energy difference becomes small if the acylating agent R—C(=O)(—X) is energy rich and/or the derived tetrahedral intermediate R—C(—O^-)(—Nu)(—X) or R—C(—OH)(—Nu)(—X) is energy poor:

- The acylating agent R—C(=O)—X is generally higher in energy, the lower the resonance stabilization of its C=O double bond by the substituent X. This effect is examined in detail in Section 6.2.2.
- The tetrahedral intermediate is generally lower in energy the more it is stabilized by a –I effect of the leaving group X or by an anomeric effect. This will be discussed in detail in Section 6.2.3.

6.2.2 S_N Reactions on the Carboxyl Carbon: The Influence of Resonance Stabilization of the Attacked C=O Double Bond on the Reactivity of the Acylating Agent

B

Table 6.1 lists acylating agents that can react with nucleophiles *without prior protonation,* that is, according to the mechanism of Figure 6.2 or according to the mechanism of Figure 6.4. They are arranged from the top to the bottom in order of definitely decreasing (entries 1–3) or presumably decreasing (entries 4–12) resonance stabilization of the C=O double bond, which is attacked by the nucleophile. It was briefly indicated in the preceding section that carboxylic acid derivatives R—C(=O)—X lose that part of their resonance stabilization which the leaving group had provided to the C=O double bond of the substrate before it was attacked by the nucleophile, when they react with the nucleophile and thereby form the tetrahedral intermediate. This explains why the acylating agents of Table 6.1 are concomitantly arranged in the order of increasing reactivity.

At approximately 30 kcal/mol, the carboxylate ion (Table 6.1, entry 1) has the greatest resonance stabilization. As expected, it is the weakest acylating agent of all and can be attacked only by organolithium compounds as nucleophiles. Amides are also significantly resonance stabilized (stabilization ≈22 kcal/mol; entry 2). Accordingly, they are rather poor acylating agents, too. Nevertheless they react not only with organolithium compounds but also with Grignard reagents and hydride donors and, under harsher conditions, also with NaOH or amines. In carboxylic acids and carboxylic esters, the C=O double bond exhibits a resonance stabilization of ca. 14 kcal/mol (entry 3). Both compounds are therefore considerably more reactive than amides with respect to nucleophiles. However, it must be kept in mind that this is true for carboxylic acids only in the absence of a base. Bases, of course, deprotonate them to yield the almost inert carboxylate ions. The decrease in resonance stabilization of the carboxyl group in the acylating agents carboxylate > carboxylic amide > carboxylic (ester) specified earlier is caused by the decrease in the +M effect of the substituent on the carboxyl carbon in the order O^- > NR_2, NRH, NH_2 > OAlk, OH.

Table 6.1. Acylating Agents in the Order of Decreasing Resonance Stabilization of the Attacked C═O Double Bond*

(1)  { R–C(=O)–O⊖ ↔ R–C(–O⊖)=O } provides 30 kcal/mol stabilization

(2) { R–C(=O)–NR'_2 ↔ R–C(–O⊖)=NR'_2⊕ 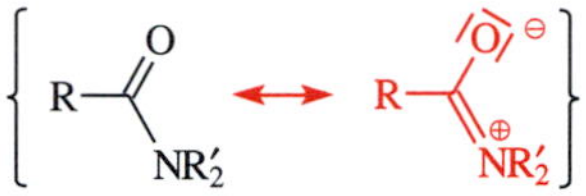 } provides 22 kcal/mol stabilization

R′ = Alk, H

(3) 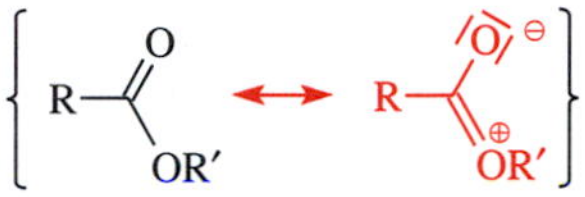 { R–C(=O)–OR′ ↔ R–C(–O⊖)=OR′⊕ } provides 14 kcal/mol stabilization

R′ = Alk, H

(4) 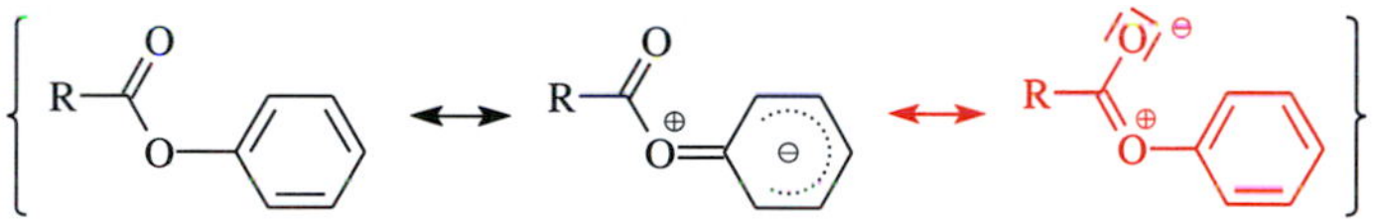 { R–C(=O)–O–Ph ↔ R–C(=O)–O⊕=C(ring ⊖) ↔ R–C(–O⊖)=O⊕–Ph }

(5) 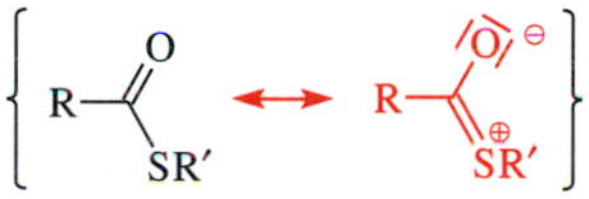 { R–C(=O)–SR′ ↔ R–C(–O⊖)=SR′⊕ }

{ R–C(=O)–O–C(=X)–Y ↔ R–C(=O)–O⊕=C(–X⊖)–Y ↔ R–C(=O)–O–C(–X⊖)=Y⊕ ↔ R–C(–O⊖)=O⊕–C(=X)–Y }

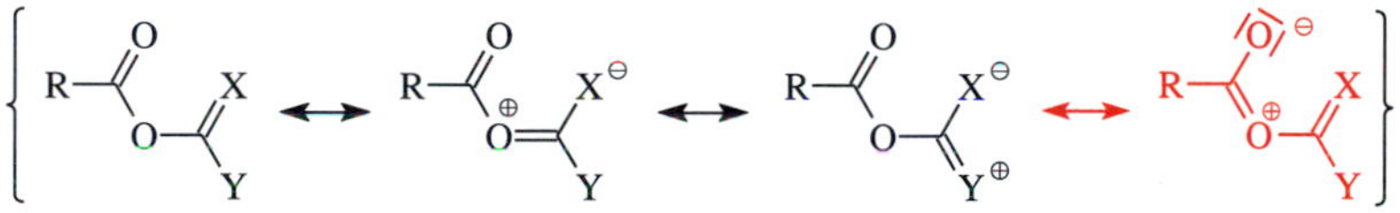

(6) • X = NAlk, Y = NHAlk: DCC activation (cf. Figure 6.15)

(7) • X = O, Y = OR: $ClCO_2iBu$ activation (cf. Figure 6.14)

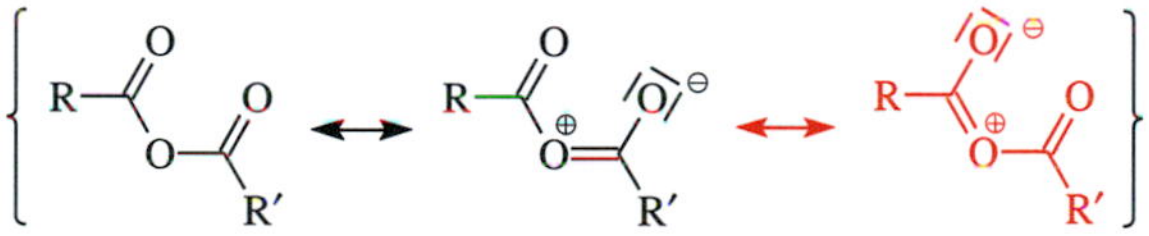

{ R–C(=O)–O–C(=O)–R′ ↔ R–C(=O)–O⊕=C(–O⊖)–R′ ↔ R–C(–O⊖)=O⊕–C(=O)–R′ }

(8) • R′ = R: inevitably 50% of the acylating agent does not go into the product but is lost as a leaving group

(9) • R′ ≠ R: as much as 100% of the contained RC(=O) can be incorporated into the product (cf. Figure 6.14)

(10) 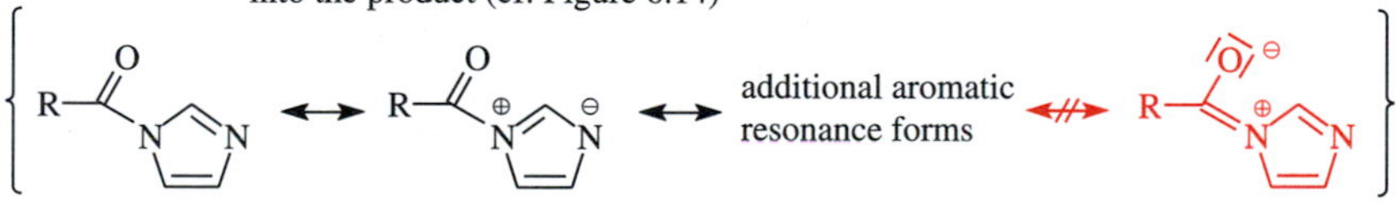 { R–C(=O)–N(imidazole)N ↔ R–C(=O)–N⊕(imidazole)N⊖ ↔ additional aromatic resonance forms ↮ R–C(–O⊖)=N⊕(imidazole)N }

(11)  { R–C(=O)–Cl ↮ R–C(–O⊖)=Cl⊕ }

(12) 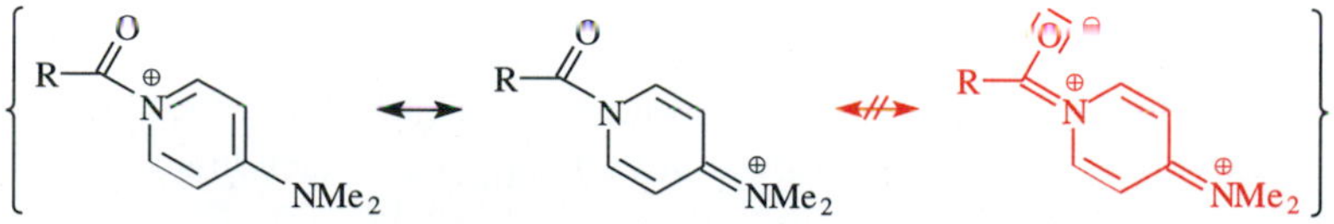 { R–C(=O)–N⊕(pyridine)–NMe_2 ↔ R–C(=O)–N(dihydropyridine)=NMe_2⊕ ↮ R–C(–O⊖)=N⊕(dihydropyridine)=NMe_2⊕ }

*Resonance forms drawn black contribute to the overall stabilization of the acylating agent but not to the stabilization of the C═O double bond, which is attacked by the nucleophile.

All other carboxylic acid derivatives in Table 6.1, in which the leaving group is bound to the carboxyl carbon through an O atom, are increasingly better acylating agents than carboxylic acid alkyl esters (entry 3) in the order carboxylic acid phenyl ester (entry 4) < acyl isourea (entry 6) < mixed carboxylic acid/carbonic acid anhydride (entry 7) < carboxylic acid anhydride (entry 8) ≤ mixed carboxylic acid anhydride (entry 9).

The reason for this increase in reactivity is the decreasing resonance stabilization of the attacked C═O double bond by the free electron pair on the *neighboring* O atom. In the mentioned series of compounds, this electron pair is available to an increasingly limited extent for stabilizing the attacked C═O double bond. This is because this lone pair also provides resonance stabilization to a second adjacent C═Het double bond. Note that the resonance stabilization of that second C═Het double bond is fully retained in the tetrahedral intermediate of the acylating agent. Consequently, the existence of *this* resonance does not lead to any decrease in the reactivity of the acylating agent. The demand on the lone pair of the single-bonded O atom by the second C═Het double bond of the acylating agents under consideration is naturally more pronounced, the greater the −M effect of the second C═Het group. The −M effect increases in the order C═C(as part of an aromatic compound) < —C(═NAlk)−NHAlk < —C(═O)—OR < —C(═O)—R. Consequently, in the corresponding acylating agents RC(═O)—X the +M effect of the carboxyl substituent X decreases in the order —O—Ar > —O—C(═NAlk)—NAlk > —O—C(═O)—OR > —O—C(═O)—R. The ease of acylation increases accordingly.

Thioesters (entry 5 of Table 6.1) are quite good acylating agents. Thus, they react with nucleophiles considerably faster than their oxa analogs, the carboxylic acid alkyl esters (entry 3). This difference in reactivity is essentially due to the fact that the +M effect of an —S—R group is smaller than that of an —O—R group. According to the-double- bond rule, sulfur as an element of the second long period of the periodic table is less capable of forming stable p_π, p_π double bonds than oxygen.

In carboxylic chlorides, the Cl atom, which likewise is an element of the second long period of the periodic table, is not able to stabilize the neighboring C═O group by resonance at all (Table 6.1, entry 11). The main reason for this is that according to the double-bond rule, chlorine has only a negligible ability to form stable p_π, p_π double bonds. Because of its greater electronegativity, the electron donor capability of chlorine is even lower than that of sulfur. Carboxylic chlorides are consequently among the strongest acylating agents.

The acylimidazoles (Table 6.1, entry 10) and the *N*-acylpyridinium salts (entry 12) occupy additional leading positions with respect to their acylation rates. In the acylimidazoles the "free" electron pair of the acylated N atom is essentially unavailable for stabilization of the C═O double bond by resonance because it is part of the π-electron sextet, which makes the imidazole ring an aromatic compound. For a similar reason there is no resonance stabilization of the C═O double bond in *N*-acylpyridinium salts: in the corresponding resonance form, the aromatic sextet of the pyridine would be destroyed in exchange for a much less stable quinoid structure.

Carboxylic amides, carboxylic esters, and carboxylic acids react with acid-stable heteroatom nucleophiles in a neutral solution much more slowly according to the mechanism of Figure 6.2 than in an acidic solution according to the mechanism of Figure 6.5. In an acidic solution their carboxonium ion derivatives, which result from the reversible protonation of the carboxyl oxygen, act as precursors of the tetrahedral intermediate. According to the discussion earlier in this section, this might at first surprise you: The resonance stabilization of these carboxonium ions is in fact greater than that of the nonprotonated C=O double bond in nonprotonated amides, esters, and acids, respectively (Table 6.2).

Table 6.2. Energy Gain through Resonance in Nonprotonated and Protonated Carboxylic Acid Derivatives

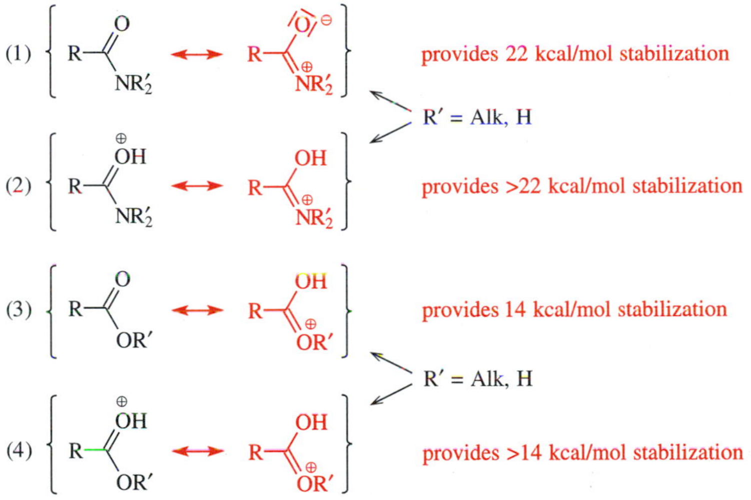

	Resonance structures	Stabilization	
(1)	$R-C(=O)NR'_2 \leftrightarrow R-C(-O^{\ominus})=N^{\oplus}R'_2$	provides 22 kcal/mol stabilization	R′ = Alk, H
(2)	$R-C(=O^{\oplus}H)NR'_2 \leftrightarrow R-C(OH)=N^{\oplus}R'_2$	provides >22 kcal/mol stabilization	
(3)	$R-C(=O)OR' \leftrightarrow R-C(-O^{\ominus})=O^{\oplus}R'$	provides 14 kcal/mol stabilization	R′ = Alk, H
(4)	$R-C(=O^{\oplus}H)OR' \leftrightarrow R-C(OH)=O^{\oplus}R'$	provides >14 kcal/mol stabilization	

The energy profiles of Figure 6.6 solve the apparent contradiction: the protonated forms of the acylating agents in question are in fact higher in energy than the corresponding nonprotonated forms, the higher resonance stabilization of the former not-withstanding. Consequently, in all the systems mentioned only a small fraction of the amide, ester, or acid present is protonated! Actually, the reason for the effectiveness of proton catalysis of these substitutions is quite different: In the presence of excess protons, the tetrahedral intermediate is similar to an alcohol (reaction **B** in Figure 6.6), while in the absence of excess protons (i.e., in basic or neutral solutions), it is similar to an alkoxide (reaction **A** in Figure 6.6). Being stronger bases, alkoxides are high-energy species in comparison to their conjugate acids, the alcohols. Accordingly, S_N reactions in acidic solutions make more stable tetrahedral intermediates **B** available than is the case in nonacidic solutions, where they would have the structure **A.**

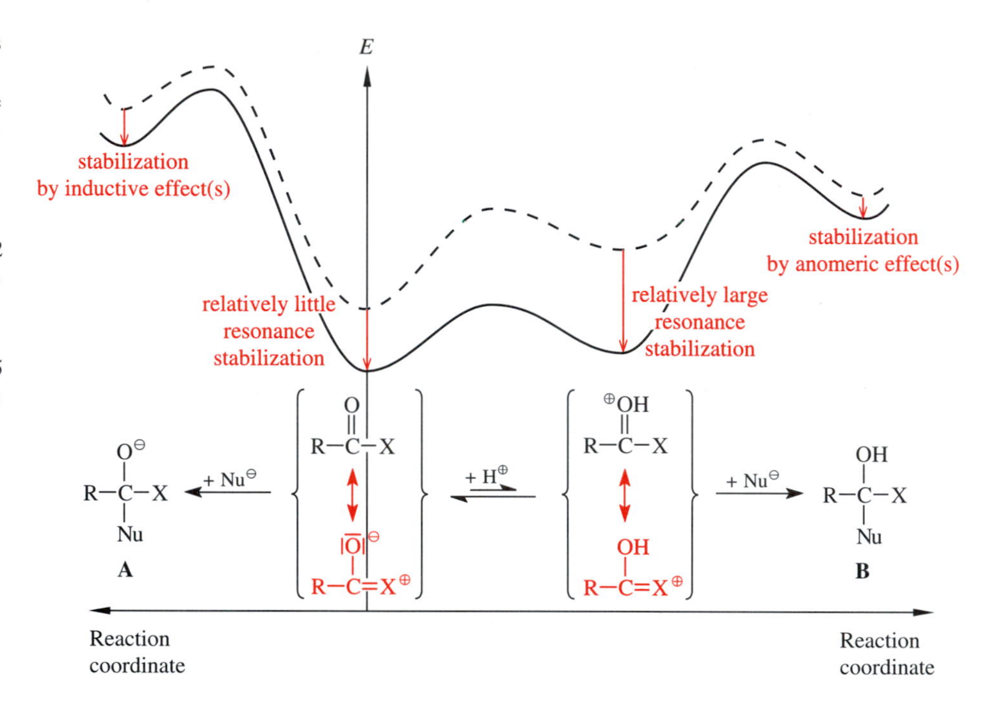

Fig. 6.6. Energy profiles for forming the tetrahedral intermediate from carboxylic amides, carboxylic esters, and carboxylic acids according to the mechanism of Figure 6.2 (the reaction coordinate goes to the left) or according to the mechanism of Figure 6.5 (the reaction coordinate goes to the right); solid curves, actual energy profiles taking stabilizing electronic effects into consideration; dashed curves, fictitious energy profiles for reactions that take place in the absence of stabilizing electronic effects.

6.2.3 S_N Reactions on the Carboxyl Carbon: The Influence of the Stabilization of the Tetrahedral Intermediate on the Reactivity

B

According to Section 6.2.1, the tetrahedral intermediate is the transition state model of the rate-determining step for any of the most important substitution mechanisms at the carboxyl carbon. In S_N reactions that take place according to the mechanisms of Figures 6.2 or 6.4, this tetrahedral intermediate is an alkoxide (**A** in Figure 6.7), and for those that take place according to Figure 6.5 it is an alcohol (**B** in Figure 6.7). According to the Hammond postulate, the formation of these intermediates should be favored kinetically when they experience a stabilizing substituent effect. If the reactivity of a variety of acylating agents toward a reference nucleophile were compared, any rate difference, which would be attributable to differential stabilization of the respective intermediates **A** or **B,** would be due only to a substituent

Fig. 6.7. Comparison of the substituent effects on the tetrahedral intermediates in S_N reactions at the carboxyl carbon.

Nu⊖, k_{add} — K_{prot}, H⊕ — Nu⊖, k_{add}

A **B**

effect of the leaving group X of the acylating agent. As it turns out, the nature of this substituent effect depends on whether the stabilization of alkoxide **A** or of alcohol **B** is involved.

In intermediate **A** of Figure 6.7, which is an alkoxide, the negative charge on the alkoxide oxygen must be stabilized. The leaving group X does this by its −I effect, and the greater this −I effect the better the leaving group's stabilities. The greatest −I effects are exerted by X = pyridinium and X = Cl. Therefore, *N*-acylpyridinium salts and carboxylic chlorides react with nucleophiles via especially well-stabilized tetrahedral intermediates **A.**

The stability of intermediate **B** of Figure 6.7, which is an alcohol, is less influenced by the leaving group X and in any event not primarily through its −I effect (because it is no longer important to delocalize the excess charge of an anionic center). Nonetheless, the substituent X may even stabilize a *neutral* tetrahedral intermediate **B.** It does so through a stereoelectronic effect. This effect is referred to as the **anomeric effect,** because it is very important in sugar chemistry. Anomeric effects can occur in compounds that contain the structural element :Het^1—C_{sp^3}—Het^2. The substituents "Het" either must be halogen atoms or groups bound to the central C atom through an O or an N atom. In addition, there is one more condition for the occurrence of an anomeric effect: The group :Het^1 must be oriented in such a way that the indicated free electron pair occupies an orbital that is oriented *parallel* to the C—Het^2 bond.

The stabilization of such a substructure :Het^1—C_{sp^3}—Het^2 **A** can be rationalized both with the VB theory and with the MO model (Figure 6.8). On the one hand, it is possible to formulate a no-bond resonance form **B** for this substructure **A.** In this substructure, a positive charge is localized on the substituent Het^1 and a negative charge on the substituent Het^2. The stability of this resonance form increases with an increasing +M effect of the substituent Het^1 and with increasing electronegativity of the substituent Het^2. The more stable the no-bond resonance form **B** is, the more resonance stabilization it contributes to intermediate **A.**

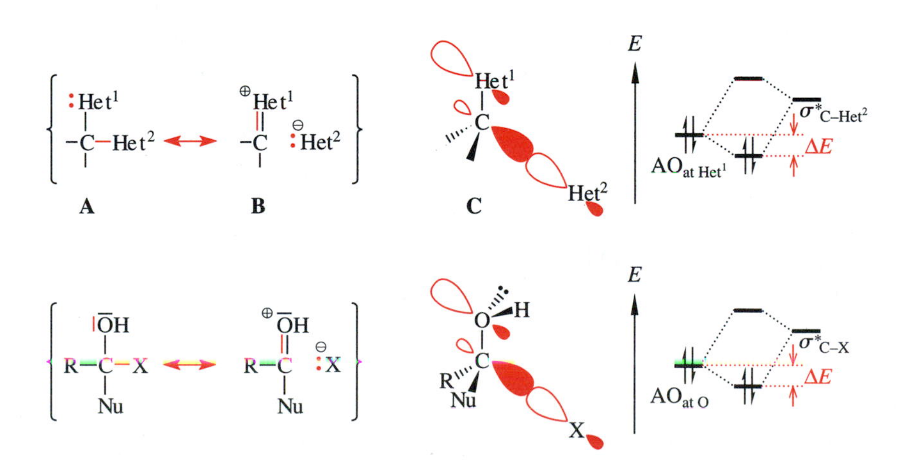

Fig. 6.8. VB explanation (left) and MO explanation (right) of the stabilization of a structure element :Het^1—C—Het^2 (upper line) and of a suitable conformer of intermediate **B** of Figure 6.7 (bottom line) through the "anomeric effect."

In MO theory, the mentioned conformer **A** of the :Het[1]—C—Het containing compound allows for overlap between the atomic orbital on the substituent Het[1], which accommodates the lone pair, and the $\sigma^*_{C—Het^2}$ orbital (see formula **C** in Figure 6.8). This overlap lowers the energy of the atomic orbital (diagram in Figure 6.8, top right). As we know, this is more effective the narrower the energy-gap between the overlapping orbitals. Nonbonding electron pairs have a higher energy at nitrogen than at oxygen, and at oxygen than at fluorine. Conversely, the energy of the $\sigma^*_{C—Het^2}$ decreases in the order $Het^2 = NR_2$, OR and F. Accordingly, for this group of compounds, calculations showed that the greatest anomeric effect occurs in the substructure $:NR_2—C_{sp^3}—F$ and the smallest in the substructure $:F—C_{sp^3}—NR_2$. (Only theory is able to separate the last effects from each other; in fact, these anomeric effects only occur together and can therefore only be observed as a sum effect.)

The lower part of Figure 6.8 shows the application of our considerations of the general substructure $:Het^1—C_{sp^3}—Het^2$ to the specific substructure $H\ddot{O}—C_{sp^3}—X$ of the tetrahedral intermediate **B** of the S_N reactions of Figure 6.7. This allows us to state the following: Suitable leaving groups X can stabilize intermediate **B** through an anomeric effect. This stabilization increases with increasing electronegativity of the leaving group X.

In other words: The higher the electronegativity of the leaving group X in the acylating agent R—C(=O)—X, the better stabilized is the tetrahedral intermediate of an S_N attack on the carboxyl carbon. Whether this tetrahedral intermediate happens to be an alkoxide and is stabilized inductively or whether it happens to be an alcohol and is stabilized through an anomeric effect plays a role only in the *magnitude* of the stabilization.

The observations in Sections 6.2.2 and 6.2.3 can be summarized as follows: Strongly electronegative leaving groups X make the acylating agent R—C(=O)—X reactive because they provide only little resonance stabilization to the C=O double bond of the acylating agent or no such stabilization at all, and because they stabilize the tetrahedral intermediate inductively or through an anomeric effect. We thus note that:

Carboxylic acid derivatives with a very electronegative leaving group X are good acylating agents, whereas carboxylic acid derivatives with a leaving group X of low electronegativity are poor acylating agents.

6.3 Activation of Carboxylic Acids and of Carboxylic Acid Derivatives

B

The conversion of a carboxylic acid into a carboxylic acid derivative, which is a more reactive acylating agent, is called "carboxylic acid activation." One can also convert an already existing carboxylic acid derivative into a more reactive one by "activating" it. Three methods are suitable for realizing such activations. One can activate carboxylic acids and some carboxylic acid derivatives through equilibrium reactions, in which, however, only part of the starting material is activated, namely, as much as is dictated by the respective equilibrium constant (Section 6.3.1). On the other hand, carboxylic acids can be converted into more reactive acylating agents *quantitatively*. One can distinguish between quantitative activations in which the acylating agent obtained must be isolated (Section 6.3.2) and quantitative activations that are effected *in situ* (Section 6.3.3).

6.3.1 Activation of Carboxylic Acids and Carboxylic Acid Derivatives in Equilibrium Reactions

Primary, secondary, and tertiary carboxylic amides, carboxylic esters, and carboxylic acids are protonated by mineral acids or sulfonic acids at the carboxyl oxygen to a small extent (Figure 6.9). This corresponds to an activation as discussed in Section 6.2.3. This activation is used in acid hydrolyses of amides and esters, in esterifications of carboxylic acids according to Fischer, and in Friedel–Crafts acylations of aromatic compounds with carboxylic acids.

In the Friedel–Crafts acylation, carboxylic acid chlorides and carboxylic acid anhydrides are activated with stoichiometric amounts of $AlCl_3$ (Section 5.2.7). However, this activation is only possible in the presence of very weak nucleophiles such as aromatic compounds. Stronger nucleophiles would attack the $AlCl_3$ instead of the carboxylic acid derivative. If one wants to acylate such stronger nucleophiles—for example, alcohols or amines—with carboxylic acid chlorides or with carboxylic acid anhydrides, and one wishes to speed up the reaction, these acylating agents can be activated by adding catalytic amounts of *para*-(dimethylamino)pyridine ("Steglich's catalyst"). Then, the acylpyridinium chlorides or carboxylates of Figure 6.9 form in equilibrium reactions (according to the mechanism shown in Figure 6.2); they are far more reactive acylating agents (cf. discussion of Table 6.1).

$$R-C(=O)-NR'_2\,(H_2) \xrightleftharpoons{HCl_{conc}} R-C(=\overset{\oplus}{O}H)-NR'_2\,(H_2)$$

$$R-C(=O)-OR'(H) \xrightleftharpoons{HCl \text{ or } p\text{-TsOH}} R-C(=\overset{\oplus}{O}H)-OR'(H)$$

$$R-C(=O)-Cl + \text{N}\langle\text{pyridine}\rangle-NMe_2 \rightleftharpoons R-C(=O)-\overset{\oplus}{N}\langle\text{pyridine}\rangle-NMe_2\ \ Cl^{\ominus}$$

$$\left(R-C(=O)\right)_2O + \text{N}\langle\text{pyridine}\rangle-NMe_2 \rightleftharpoons R-C(=O)-\overset{\oplus}{N}\langle\text{pyridine}\rangle-NMe_2\ \ {}^{\ominus}O_2CR$$

Fig. 6.9. Examples of the activation of carboxylic acid derivatives in equilibrium reactions.

6.3.2 Conversion of Carboxylic Acids into Isolable Acylating Agents

B

The most frequently used strong acylating agents are carboxylic chlorides. They can be prepared from carboxylic acids especially easily with $SOCl_2$ or with oxalyl chloride (Figure 6.10). In fact, if carboxylic acids are treated with one of these reagents, only gaseous by-products are produced: SO_2 and HCl from $SOCl_2$, and CO_2, CO, and HCl from oxalyl chloride. PCl_3, $POCl_3$, or PCl_5 do not provide this advantage, although they can also be used to convert carboxylic acids to carboxylic chlorides.

The carboxylic acid activation with $SOCl_2$ or with oxalyl chloride starts with the formation of the respective mixed anhydride (Figure 6.10). Carboxylic acids and $SOCl_2$ give a carboxylic acid/chlorosulfinic acid anhydride **A.** Carboxylic acids and oxalyl chloride furnish the carboxylic acid/chloro-oxalic acid anhydride **C,** presumably, following the mechanism of Figure 6.2, that is, by nucleophilic attack of the carboxylic acid on the carboxyl carbon of oxalyl chloride. The anhydride **A** is probably formed in an analogous S_N reaction, i.e., one which the carboxylic acid undertakes at the S atom of $SOCl_2$.

In the formation of anhydrides **A** and **C** one equivalent of HCl is released. It attacks the activated carboxylic carbon atom of these anhydrides in an S_N reaction. The carboxylic acid chloride is formed via the tetrahedral intermediates **B** or **D,** respec-

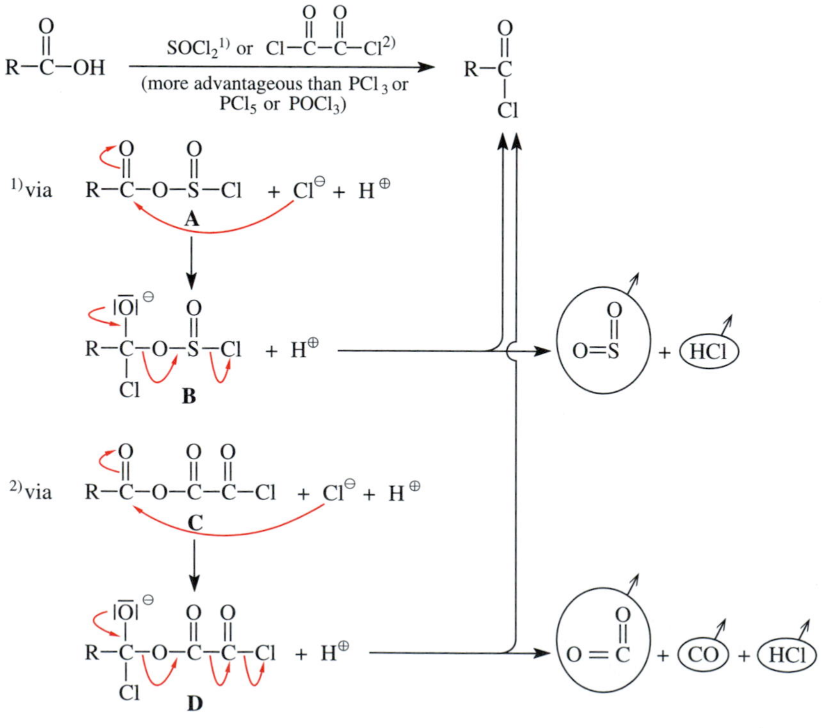

Fig. 6.10. Conversion of carboxylic acids to carboxylic chlorides with thionyl chloride or oxalyl chloride.

tively, according to the mechanism of Figure 6.2. At the same time the leaving group Cl—S(=O)—O^- or Cl—C(=O)—C(=O)—O^-, respectively, is liberated. Both leaving groups are extremely short-lived—if they are at all able to exist—and fragment immediately. After protonation the gaseous by-products SO_2 and HCl or CO_2, CO, and HCl, respectively, are produced.

The conversion of carboxylic acids and $SOCl_2$ into carboxylic chlorides is frequently catalyzed by DMF. The mechanism of this catalysis is shown in Figure 6.11. It is likely that $SOCl_2$ and DMF first react to give the Vilsmeier-Haack reagent **A.** It differs from the reactive intermediate of the Vilsmeier-Haack formylation (Figure 5.29) only insofar as the cation is associated here with a chloride ion, rather than a dichlorophosphate ion. Now, the carboxylic acid attacks the imminium carbon of intermediate **A** in an S_N reaction in which the Cl atom is displaced. This reaction takes place analogously to the mechanism shown in Figure 6.2. The substitution product is the *N*-methylated mixed anhydride **B** of a carboxylic acid and an imidoformic acid. This mixed anhydride **B** finally acylates the released chloride ion to yield the desired acid chloride. At the same time the catalyst DMF is regenerated.

Figure 6.12 shows that carboxylic acids can also be converted into carboxylic chlorides *without* releasing HCl. This is possible when carboxylic acids are treated with the chloro-enamine **A.** First the carboxylic acid adds to the C=C double bond of this reagent electrophilically (mechanism: Figure 3.40, see also Figure 3.42). Then, the addition product **B** dissociates completely to give the ion pair **C;** it constitutes the isopropyl analog of the Vilsmeier-Haack intermediate **B** of the DMF-catalyzed carboxylic chloride synthesis of Figure 6.11. The new Vilsmeier-Haack intermediate reacts exactly like the old one (cf. previous discussion): The chloride ion undertakes an S_N reaction at the carboxyl carbon. This produces the desired acid chloride and isobutyric *N*,*N*-dimethylamide.

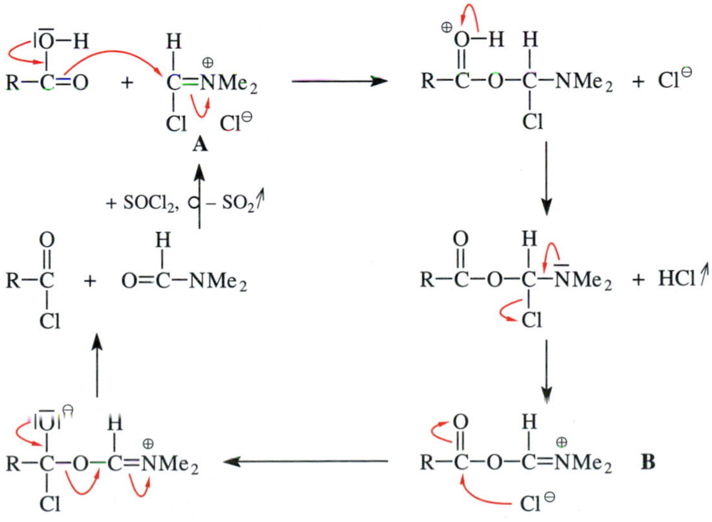

Fig. 6.11. Mechanism of the DMF-catalyzed conversion of carboxylic acids and $SOCl_2$ into carboxylic chlorides.

Fig. 6.12. Acid-free preparation of carboxylic chloride from carboxylic acids and a chloro-enamine.

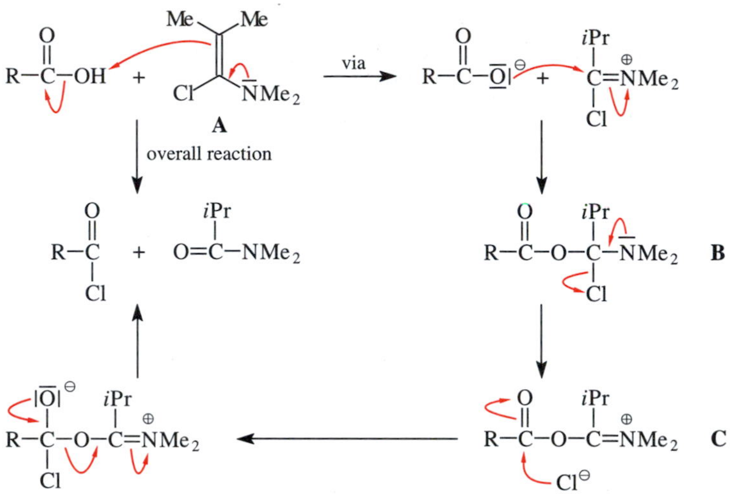

B

Another carboxylic acid activation in a neutral environment is shown in Figure 6.13 together with all mechanistic details: Carboxylic acids and carbonyldiimidazole (**A**) react to form the reactive carboxylic acid imidazolide **B.**

Carboxylic acids can also be activated by converting them to their anhydrides. For this purpose they are dehydrated with concentrated sulfuric acid, phosphorus pentoxide, or 0.5 equivalents of $SOCl_2$ (according to Figure 6.10, 1 equivalent of $SOCl_2$ reacts with carboxylic acids to form carboxylic chlorides rather than anhydrides). However, carboxylic anhydrides cannot transfer more than 50% of the original carboxylic acid to a nucleophile. The other 50% is released—depending on the pH value—either as the carboxylic acid or as a carboxylate ion; therefore, it is lost for the acylation. Consequently, in laboratory chemistry, the conversion of carboxylic acids into anhydrides is not as relevant as a carboxylic acid activation. Nonetheless, acetic anhydride is an important acetylating agent because it is commercially available and inexpensive.

6.3.3 Complete *in Situ* Activation of Carboxylic Acids

B

As can be seen from Table 6.1, a number of mixed anhydrides are good acylating agents. Still, only *one* mixed anhydride is commercially available as such: The formylating agent formyl acetate HC(=O)—OC(=O)—CH_3. All other acylating agents are prepared *in situ* from the carboxylic acid and a suitable reagent. Four of these mixed anhydrides will be discussed in more detail in the following.

The acylation of carboxylic acids with 2,4,6-trichlorobenzoyl chloride gives mixed anhydrides **A** (Figure 6.14). Triethylamine must be present in this reaction to scavenge the released HCl. Anhydrides **A** contain two different acyl groups. In principle, both of them could be attacked by a nucleophile. However, one observes the chemoselective attack on the acyl group that originates from the *acid* used. This is because the

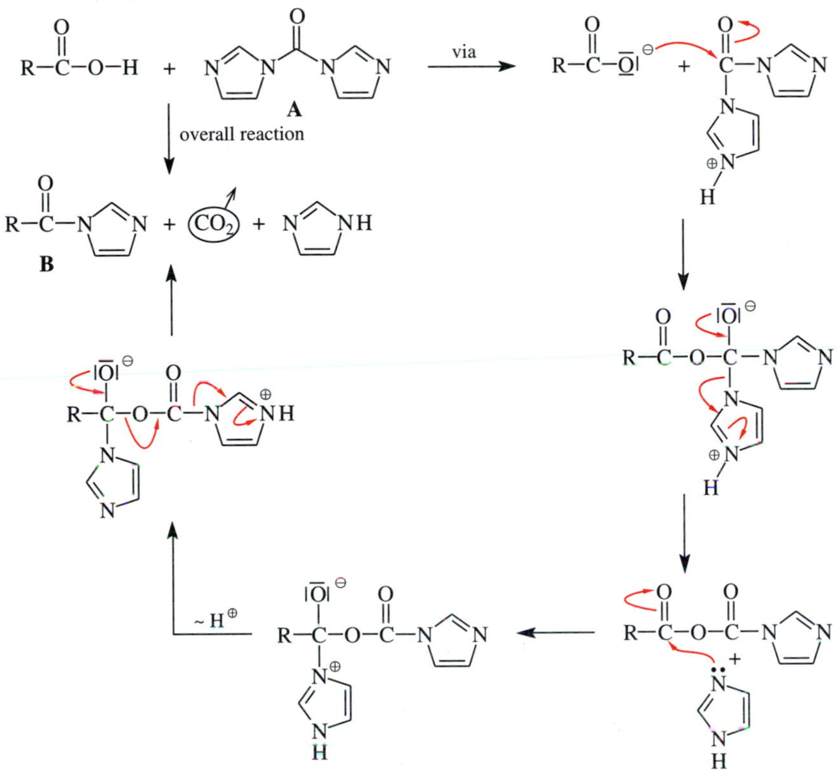

R–C(=O)–O–H + A (carbonyldiimidazole) —via→ R–C(=O)–O⁻ + C(=O)(imidazolyl)–N-imidazolium–H

overall reaction

R–C(=O)–imidazolide **B** + CO_2 + imidazole (NH)

~ $H^{\oplus}$

Fig. 6.13. Acid-free activation of carboxylic acids as carboxylic acid imidazolides.

carboxyl group located next to the aromatic ring is sterically hindered. In the most stable conformation the Cl atoms in the *ortho* positions lie in the halfspaces above and below the proximal C=O double bond and therefore block the approach of the nucleophile toward that part of the molecule.

Carboxylic acids can be activated *in situ* in a manner mechanistically analogous to chloroformic ester: as mixed anhydrides **B** (Figure 6.14), which are mixed anhydrides of a carboxylic acid and a carbonic acid halfester. As can be seen from Table 6.1, in

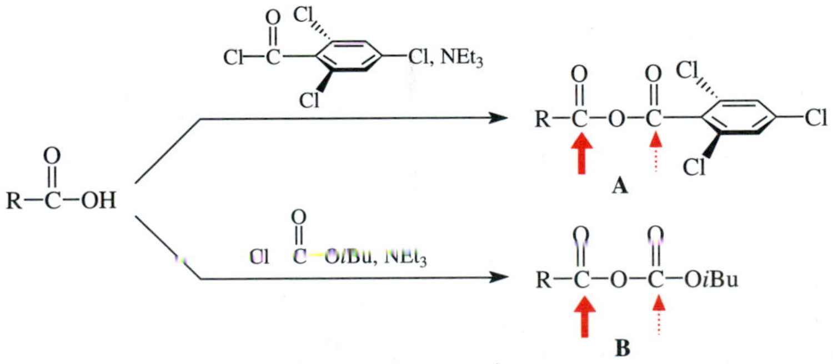

R–C(=O)–OH

Cl–C(=O)–$C_6H_2Cl_3$, NEt_3 → R–C(=O)–O–C(=O)–$C_6H_2Cl_3$ **A**

Cl–C(=O)–O*i*Bu, NEt_3 → R–C(=O)–O–C(=O)–O*i*Bu **B**

Fig. 6.14. *In situ* activation of carboxylic acids as mixed anhydrides.

anhydrides of this type the C═O double bond of the carboxylic acid moiety is stabilized less by resonance than the C═O double bond of the carbonic acid moiety. Therefore, a nucleophile attacks chemoselectively the carboxyl carbon of the carboxylic and not the carbonic acid ester moiety.

Whereas the mixed anhydrides **A** of Figure 6.14 acylate amines and alcohols, the mixed anhydrides **B** are suitable for acylating amines but unsuitable for acylating alcohols. The latter is true even if at the start of the reaction between anhydride **B** and an alcohol the desired acylation occurs. However, in its course, the leaving group *i*BuO—C(═O)—O^- is liberated. It is unstable and fragments to give CO_2 and isobutanol. This isobutanol, being an alcohol, too, competes more or less successfully—in any case, with *some* success—with the other alcohol for the remaining anhydride **B.** In contrast, an amine as the original reaction partner of the mixed anhydride **B** remains the best nucleophile even if in the course of its acylation isobutanol is released. Therefore, aminolyses of **B** succeed without problems.

In peptide synthesis, the *in situ* activation of carboxylic acids with dicyclohexylcarbodiimide (DCC; compound **A** in Figure 6.15) is very important. By adding the carboxylic acid to the C═N double bond of this reagent, one obtains compounds of type

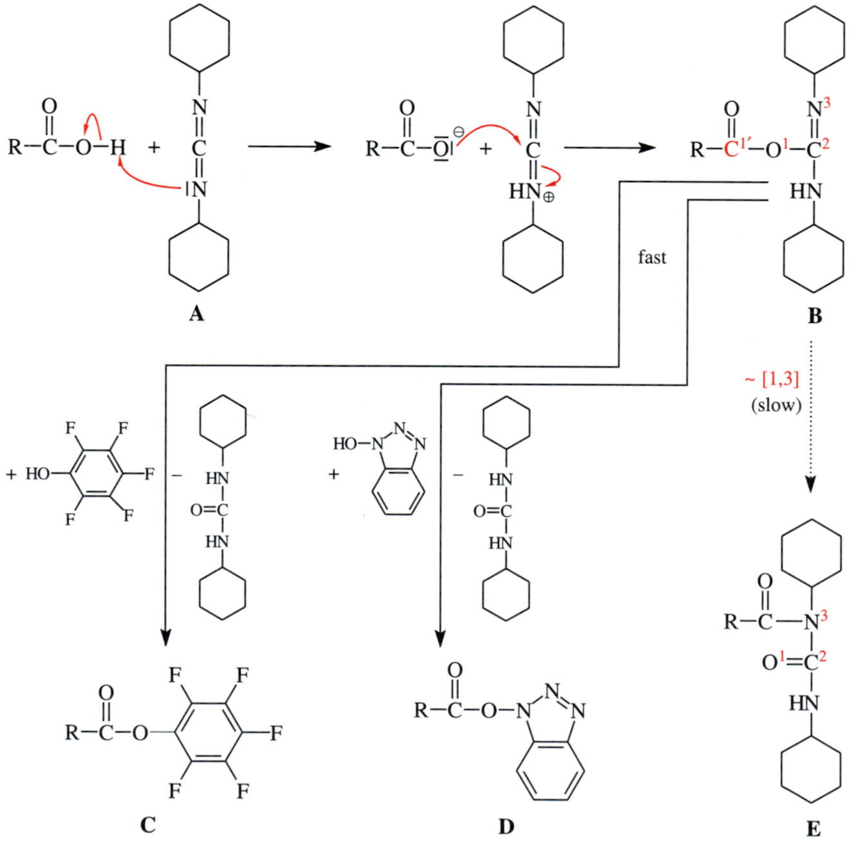

Fig. 6.15. Carboxylic acid activation with DCC. ~[1,3] means the intramolecular substitution of the oxygen atom O^1 by the N atom "3" via a cyclic four-membered tetrahedral intermediate. From the standpoint of the heteroatoms, this S_N reaction corresponds to a migration of the acyl group R-C═O from the oxygen to the nitrogen.

B, so-called O-acyl isoureas. To a certain extent, these constitute diaza analogs of the mixed anhydrides **B** of Figure 6.14. As can therefore be expected, O-acyl isoureas react with good nucleophiles with the same regioselectivity as their oxygen analogs: at the carboxyl carbon of the carboxylic acid moiety.

Poor nucleophiles react with acyl isoureas **B** so slowly that the latter start to decompose. They acylate themselves in a sense. The N atom designated with the positional number 3 intramolecularly substitutes the O-bound leaving group that is attached to the carboxyl carbon C1′. A four-membered cyclic tetrahedral intermediate is formed. When the C1′-O1 bond in this intermediate opens up, the *N*-acyl urea **E** is produced. Because compound **E** is an amide derivative it is no longer an acylating agent (cf. Section 6.2).

The "deactivation" of the *O*-acyl isoureas **B** in Figure 6.15 must be prevented when a poor nucleophile is to be acylated. In such a case, **B** is treated with a mixture of the poor nucleophile and an auxiliary nucleophile that must be a good nucleophile. The latter undergoes a substitution at the carboxyl carbon of the carboxylic acid moiety of **B.** However, in contrast to the acyl urea **E,** which is inert, the substitution product now obtained is still an acylating agent. Gratifyingly, it is a long-lived derivative of the originally used carboxylic acid, yet sufficiently reactive and indeed a so-called "active ester." An "active ester" is an ester that is a better acylating agent than an alkyl ester. As Table 6.1 shows, for example, phenyl esters are also "active esters." Compared with that species, Figure 6.15 shows two esters that are even more reactive, namely, the perfluorophenyl ester **C** and the hydroxybenzotriazole ester **D.** These active esters retain some of the reactivity of the acyl isourea **B.** In contrast to **B,** however, they are stable long enough even for a poor nucleophile to become acylated. The *in situ* activation of carboxylic acids to compounds of type **B, C,** or **D** is used in oligopeptide synthesis for activating *N*-protected α-amino acids (see Section 6.4.3).

A last *in situ* procedure for activating carboxylic acids is shown in Figure 6.16. There, the α-chlorinated *N*-methylpyridinium iodide **A** reacts with the carboxylic acid by an S_N reaction at a pyridine carbon. This leads to the pyridinium salt **C,** presumably via the Meisenheimer complex **B** and its deprotonation product **D** as intermediates. The

Fig. 6.16. *In situ* activation of carboxylic acids according to the procedure of Mukaiyama.

activated carboxylic acid **C** is not only an aryl ester but one in which the aryl group is positively charged. This charge keeps the single-bonded O atom of this species *completely* from providing any resonance stabilization by its +M-effect to the C=O double bond (cf. discussion of Table 6.1).

6.4 Selected S_N Reactions of Heteroatom Nucleophiles on the Carboxyl Carbon

B

Quite a few substitution reactions of heteroatom nucleophiles at the carboxyl carbon as well as their mechanisms are discussed in introductory organic chemistry courses. The left and the center columns of Table 6.3 summarize these reactions. Accordingly, we will save ourselves a detailed repetition of all these reactions and only consider the ester hydrolysis once more (Section 6.4.1). Beyond that, S_N reactions of this type will only be discussed using representative examples, namely:

- the formation of cyclic esters (lactones; Section 6.4.2),
- the formation of the amide bond of oligopeptides (Section 6.4.3), and
- acylations with carbonic acid derivatives (Section 6.4.4).

Table 6.3. Preparatively Important S_N Reactions of Heteroatom Nucleophiles on the Carboxyl Carbon of Carboxylic Acids and Their Derivatives

$Nu^{\ominus}$	Important acylations of this nucleophile, with which you are already familiar	Acylations of this nucleophile which are discussed *here* as prototypical examples
H_2O or $OH^{\ominus}$	Hydrolysis of esters and amides	Ester hydrolysis (Section 6.4.1)
ROH or $RO^{\ominus}$	Esterification of carboxylic acids; transesterification giving polyethyleneterephthalate (Dacron®)	Lactonization of hydroxy acids (Section 6.4.2)
RCO_2H or $RCO_2^{\ominus}$	Formation of anhydrides; carboxylic acid activation (Section 6.3)	–
NH_3 or RNH_2 or R_2NH	Amide formation from carboxylic acid derivatives (mild) or from carboxylic acids (Δ; technical synthesis of nylon-6,6); transamidation [caprolactame → nylon-6 (perlon)]	Peptide synthesis (Section 6.4.3)

In addition, Figures 6.17 to 6.19 briefly present a handful of other preparatively important S_N reactions on the carboxyl carbon. Figure 6.17 shows S_N reactions with H_2O_2. They are carried out in basic solution in order to utilize the higher reactivity of the HOO^- ion. All these reactions take place according to the mechanism of Figure 6.2.

excess H_2O_2/NaOH ($\rightarrow HOO^{\ominus}$); $H^{\oplus}$ workup → O–O–H **A**

excess H_2O_2/MgO → O–O–H, $O^{\ominus}$, $Mg^{2\oplus} \cdot 6\ H_2O$ **B**

Ph–C(=O)–Cl, H_2O_2/NaOH in deficit → Ph–C(=O)–O–O–C(=O)–Ph **C**

Fig. 6.17. S_N reactions with H_2O_2 at the carboxyl carbon. Syntheses of *meta*-chloroperbenzoic acid **(A),** magnesium monoperoxophthalate hexahydrate **(B),** and dibenzoyl peroxide **(C).**

By using these reactions it is possible—as a function of the structure of the acylating agent and the ratio of the reaction partners—to obtain the reagents *meta*-chloroperbenzoic acid (applications: Figures 3.14, 11.32, 14.26, 14.28, 14.29, 14.31, 14.32), magnesium monoperoxophthalate hexahydrate (applications: Figures 3.14, 11.31, 14.28, 14.29, 14.31), and dibenzoyl peroxide (applications: Figures 1.9, 1.27). Dibenzoyl peroxide is produced through two such substitution reactions: The HOO^- ion is the nucleophile in the first reaction, and the Ph-C(=O)—O—O^-ion is the nucleophile in the second reaction.

Two successive S_N reactions on two carboxyl carbons occur in the second step of the Gabriel synthesis of primary alkyl amines (Figure 6.18; first step: Figure 2.26). Hydrazine breaks both C(=O)—N bonds of the *N*-alkylphthalimide precursor **A.** The first bond cleavage is faster than the second because the first acylating agent **(A)** is an imide and the second acylating agent **(C)** is an amide, and according to Table 6.1 amides are comparatively inert toward nucleophiles. Still, under the conditions of Figure 6.18 even the amide **C** behaves as an acylating agent (giving the diacylhydrazide **B**). The reason for this relatively fast reaction is that it is *intramolecular*. Intramolecular reactions via a five- or a six-membered transition state are always much faster than analogous intermolecular reactions. Therefore, the *N*-alkylated phthalimide intermediates of the Gabriel synthesis are cleaved with (the carcinogen) hydrazine because the second acylation is intramolecular and favored and it can thus take place rapidly. If one were to take NH_3 instead of hydrazine, this would have to cleave the second C(=O)—N bond in an *intermolecular* S_N reaction, which would be impossible under the same conditions.

It remains to be clarified why the 1,2-diacylated hydrazine **B** and not the 1,1-diacylated hydrazine **D** is formed in the hydrazinolysis of Figure 6.18. In the intermediate **C** the only nucleophilic electron pair resides in the NH_2 group and not in the NH group because the electron pair of the NH group is involved in the hydrazide resonance **C** ↔ **C′** and is therefore not really available. The hydrazide resonance is about as important as the amide resonance (ca. 22 kcal/mol according to Table 6.2).

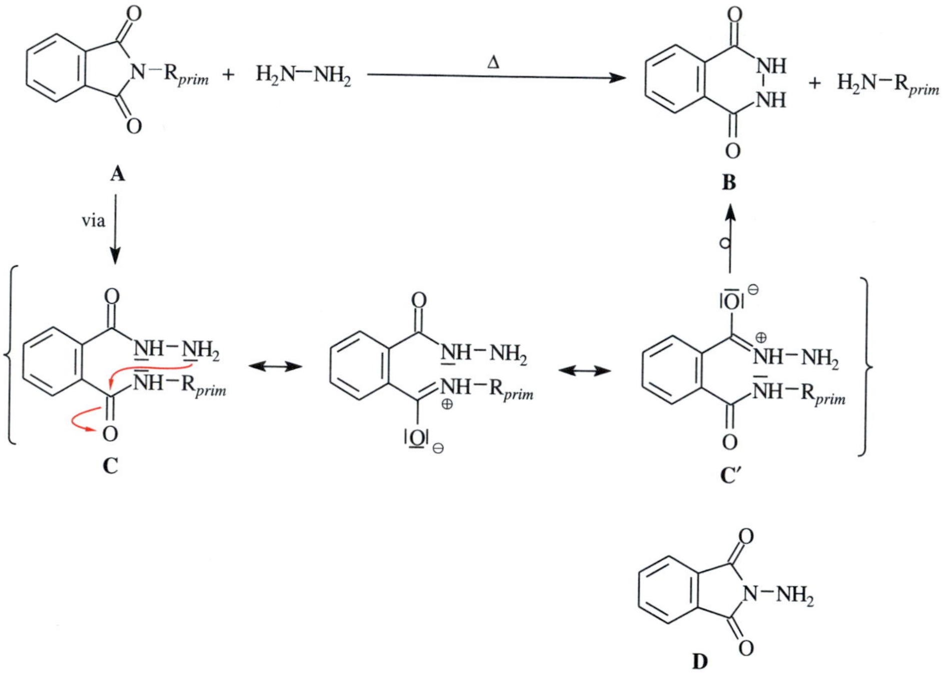

A + H_2N-NH_2 →(Δ) B + H_2N-R_{prim}

via C ↔ ↔ C′ → B

D

Fig. 6.18. Mechanism of the second step of the Gabriel synthesis of primary alkyl amines.

6.4.1 Hydrolysis of Esters

The hydrolysis of carboxylic esters can in principle take place either as **carboxyl-O cleavage**—i.e., as an S_N reaction at the carboxyl carbon—

$$R^1-C(=O)-O-R^2 \xrightarrow{+H_2O} R^1-COOH + HO-R^2$$

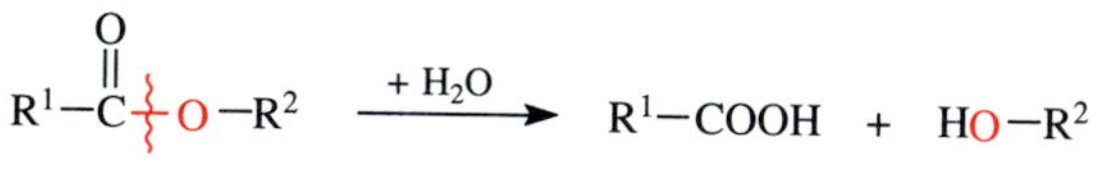

or as **alkyl-O cleavage** (this variant *does not* represent an S_N reaction at the carboxyl carbon):

$$R-C(=O)-O-Alk \xrightarrow{+H_2O} R-COOH + HO-Alk$$

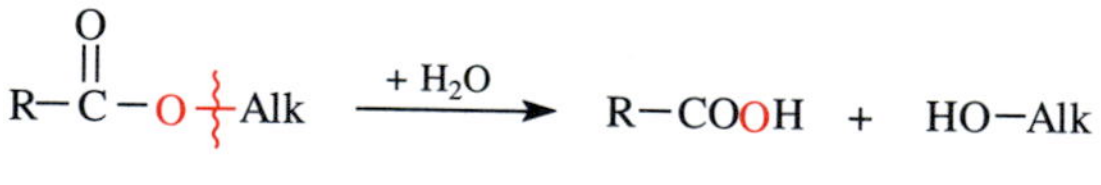

The hydrolysis is generally not carried out at pH 7 but either **acid-catalyzed** or **base-mediated** (i.e., as a so-called saponification). Base-*catalyzed* ester hydrolyses do not exist: The carboxylic acid produced protonates a full equivalent of base and thus consumes it.

Driving Force of Base-Mediated and Acid-Catalyzed Ester Hydrolyses

Base-mediated ester hydrolyses have a high driving force. This is because of the inevitably ensuing acid/base reaction between the carboxylic acid, which is first formed, and the base even when only 1 equivalent of OH^- is used. The resonance stabilization of the carboxylate is approximately 30 kcal/mol, which means a gain of about 16 kcal/mol compared to the starting material, the carboxylic ester (resonance stabilization 14 kcal/mol according to Table 6.1). Accordingly, the hydrolysis "equilibrium" lies completely on the side of the carboxylate.

Acid-catalyzed ester hydrolyses lack a comparable contribution to the driving force: The starting material and the product, the ester and the carboxylic acid, possess resonance stabilizations of the same magnitude of 14 kcal/mol each (Table 6.1). For this reason, acid-catalyzed ester hydrolyses can go to completion only when one starting material (H_2O) is used in great excess and the hydrolysis equilibrium is thereby shifted unilaterally to the product side. Five and six membered *cyclic* esters can be saponified only in basic media. In acidic solutions they are often spontaneously formed again (cf. Figure 6.22 and the explanation given).

The mechanisms of ester hydrolysis are distinguished with abbreviations of the type "medium$_{\text{designation as carboxyl-O or as alkyl-O cleavage}}$ reaction order." The medium of **a**cid-catalyzed hydrolyses is labeled **"A,"** and the medium of **b**ase-mediated hydrolyses is labeled **"B."** A carboxyl-O cleavage is labeled with **"AC"** (for **ac**yl-O cleavage), and an **al**kyl-O cleavage is labeled with **"AL."** The possible reaction orders of ester hydrolyses are 1 and 2. If all permutations of the cited characteristics were to occur, there would be eight hydrolysis mechanisms: The $A_{AC}1$, $A_{AC}2$, $A_{AL}1$, and $A_{AL}2$ mechanisms in acidic solutions and the $B_{AC}1$, $B_{AC}2$, $B_{AL}1$, and $B_{AL}2$ mechanisms in basic solutions. However, only three of these mechanisms are of importance: the $A_{AC}2$ mechanism (Figure 6.19), the $A_{AL}1$ mechanism (Figure 6.20), and the $B_{AC}2$ mechanism (Figure 6.21).

The $A_{AC}2$ mechanism (Figure 6.19) of ester hydrolysis represents an S_N reaction at the carboxyl carbon, which accurately follows the general mechanism of Figure 6.5. Acid-catalyzed hydrolyses of carboxylic esters that are derived from primary or from secondary alcohols take place according to the $A_{AC}2$ mechanism. The reverse reactions of these hydrolyses follow the same mechanism, namely, the acid-catalyzed esterifications of carboxylic acids with methanol, with primary or with secondary alcohols. In the esterifications, the same intermediates as during hydrolysis are formed but in the opposite order.

As you have already seen, in the system carboxylic ester + $H_2O \rightleftharpoons$ carboxylic acid + alcohol, the equilibrium constant is in general only slightly different from 1. *Com-*

Fig. 6.19. $A_{AC}2$ mechanism of the acid-catalyzed hydrolysis of carboxylic esters (read from left to right); $A_{AC}2$ mechanism of the Fischer esterification of carboxylic acids (read from right to left). $\sim H^+$ means migration of a proton.

hydrolysis of esters

esterification of carboxylic acids

$R^1-C(=O)-OR^2 + H_2O \xrightarrow{\text{cat. } H^\oplus} R^1-C(=O)-OH + HOR^2$

$R^1-C(=\overset{\oplus}{O}H)-OR^2 + OH_2 \rightleftharpoons R^1-C(OH)(\overset{\oplus}{O}H_2)-OR^2 \overset{\sim H^\oplus}{\rightleftharpoons} R^1-C(OH)(OH)-\overset{\oplus}{O}HR^2 \rightleftharpoons R^1-C(=\overset{\oplus}{O}H)-OH + HOR^2$

plete reactions in both directions are therefore only possible under suitably adjusted reaction conditions. Complete $A_{AC}2$ *hydrolyses* of carboxylic esters can be carried out with a large excess of water. Complete $A_{AC}2$ *esterifications* succeed when a large excess of the alcohol is used. For this purpose it is best to use the alcohol as the solvent. However, when the alcohol involved is difficult to obtain or expensive, this procedure cannot be used because the alcohol is affordable only in a stoichiometric amount. Its complete esterification by a carboxylic acid is then still possible, provided that the released water is removed. That can be done by continuously distilling it azeotropically off with a solvent such as cyclohexane. By removing one of the reaction products, the equilibrium is shifted toward this side, which is also the side of the desired ester.

In acidic media, carboxylic esters of tertiary alcohols are not cleaved according to the $A_{AC}2$ mechanism (Figure 6.19) but according to the $A_{AL}1$ mechanism (Figure 6.20). However, this cleavage would probably not be a "hydrolysis" even if the reaction mixture contained water. *This* mechanism for ester cleavage does not belong in Chapter 6 at all! It was already discussed in Section 4.5.3 (Figure 4.32) as the E1 elimination of carboxylic acids from *tert*-alkyl carboxylates.

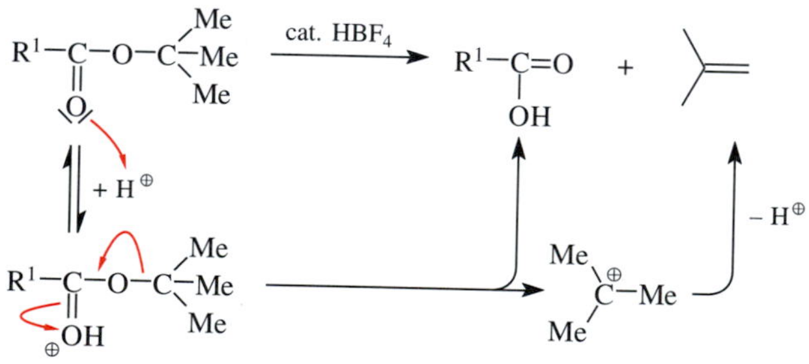

Fig. 6.20. $A_{AL}1$ mechanism of the acidic cleavage of *tert*-alkyl esters.

Carboxylic esters of any alcohol are saponified quantitatively (see above) in basic solution according to the $B_{AC}2$ mechanism (Figure 6.21). The $B_{AC}2$ mechanism is an S_N reaction at the carboxyl carbon that also proceeds according to the general mechanism of Figure 6.2. The reversibility of the formation of the tetrahedral intermediate in such hydrolyses was proven with the isotope labeling experiment of Figure 6.3.

In a $B_{AC}2$ saponification, the C—O bond of the released alcohol is not formed freshly, but it is already contained in the ester. Therefore, if the C atom of this C—O bond represents a stereocenter, its configuration is completely retained. This is used in the

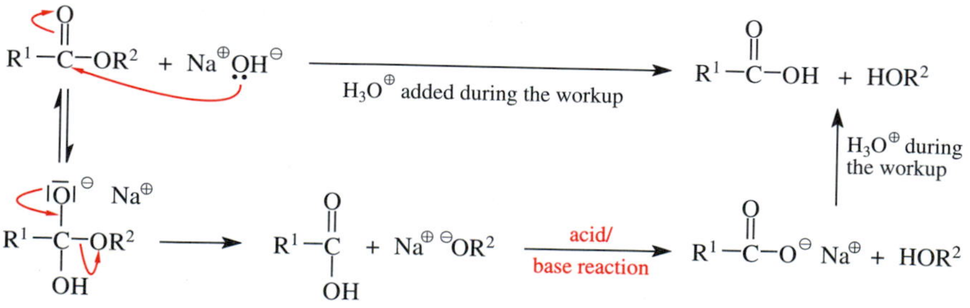

Fig. 6.21. $B_{AC}2$ mechanism of the basic hydrolysis of carboxylic esters.

$B_{AC}2$ hydrolysis of esters with the substructure —C(=O)—O—$CR^1R^2R^3$ to stereoselectively obtain the corresponding alcohols. An application thereof is the hydrolysis of the following lactone, whose preparation as a pure enantiomer can be found in Figure 11.31:

NaOH, H_2O → $O^{\ominus}$ $Na^{\oplus}$ … OH; acidify → OH … OH

Transesterifications in basic solutions can also follow the $B_{AC}2$ mechanism. The reactions also can release the corresponding alcohols with retention of configuration from sterically uniform esters with the substructure —C(=O)—O—$CR^1R^2R^3$. This kind of reaction is used, for example, in the second step of a Mitsunobu inversion, such as the following, which you have already seen in Figure 2.28:

OMe, OH → $EtO_2C-N=N-CO_2Et$, PPh_3, $Ph-CO_2H$ → OMe, Ph → K_2CO_3 in MeOH ($\rightarrow K^{\oplus}$ $^{\ominus}OMe$) → OMe, H + MeO_2CPh

According to what was generally discussed at the beginning of Section 6.2.1, the tetrahedral intermediate is also the best transition state model of the rate-determining step of the saponification of esters according to the $B_{AC}2$ mechanism. Knowing that, the substrate dependence of the saponification rate of esters is easily understood. As can be seen from Table 6.4, the saponification rate decreases sharply with increasing size of the acyl substituent because a bulky acyl substituent experiences more steric hindrance in the tetrahedral intermediate than in the starting material: In the tetrahedral intermediate, it has three vicinal O atoms compared with two in the starting material, and the three O atoms are no longer so far removed because the C—C—O bond angle has decreased from ~120° to ~109°.

Rate effects of the type listed in Table 6.4 make it possible to carry out chemoselective monohydrolyses of sterically differentiated diesters, for example:

$k_{rel} = 0.011$, *tert*-Bu, Me, $k_{rel} = 1$ → $OH^{\ominus}$ → HO, *tert*-Bu

Table 6.4. Substituent Effects on the Rate of the $B_{AC}2$ Saponification of Different Ethyl Esters

Increase in the steric interactions in the rate-determining step:

R	k_{rel}
Me	≡ 1.0
Et	0.47
*i*Pr	0.10
tert-Bu	0.011

2 × ca. 120° → 3 × ca. 109°

Table 6.5 shows that saponifications according to the $B_{AC}2$ mechanism are also slowed down when the esters are derived from sterically demanding alcohols. However, this structural variation takes place at a greater distance from the reaction center than the structural variation of Table 6.4. The substituent effects in Table 6.5 are therefore smaller.

Table 6.5. Substituent Effects on the Rate of the $B_{AC}2$ Saponification of Different Acetic Esters

Occurrence of *syn*-pentane interactions in the rate-determining step:

R	k_{rel}
Et	≡ 1.0
*i*Pr	0.70
tert-Bu	0.18
CEt_3	0.031

6.4.2 Lactone Formation from Hydroxycarboxylic Acids

γ- and δ-hydroxycarboxylic acids esterify very easily intramolecularly in the presence of catalytic amounts of acid; they are thereby converted to five-membered γ-lactones or six-membered δ-lactones (Figure 6.22). These lactonizations often take place so easily that they can hardly be avoided. In these cases it seems that the carboxylic acid moiety of the substrate effects acid catalysis by its own acidity.

The high *rate* of the lactonizations in Figure 6.22 is a consequence of the less negative than usual activation entropies, from which intramolecular reactions that proceed via three, five-, or six-membered transition states always profit. The high lactonization *tendency* stems from an increase in entropy, from which intermolecular esterifications do not profit: Only during lactonization does the number of molecules double (two

B

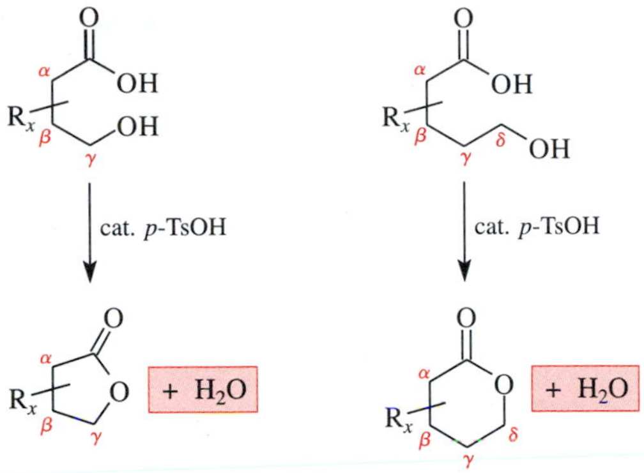

Fig. 6.22. Spontaneous lactonizations according to the $A_{AC}2$ mechanism of Figure 6.19.

molecules are produced from one). This entropy contribution to the driving force requires that the equilibrium constant of the reaction γ- or δ-hydroxycarboxylic acid $\rightleftharpoons \gamma$- or δ-lactone + H_2O be *greater than* 1 rather than *approximately* 1, as for intermolecular esterifications (Section 6.4.1).

Lactonizations of hydroxycarboxylic acids in which the OH and the CO_2H groups are separated from each other by six to ten C atoms have a reduced driving force or are even endergonic. These lactonizations lead to **"medium-sized ring lactones."** They are—similar to medium-sized ring hydrocarbons—destabilized by eclipsing and transannular interactions. The formation of **"large-ring lactones"** with 14 or more ring members is free of such disadvantages. Still, these lactones are not obtained from the corresponding hydroxycarboxylic acids by simple acidification for kinetic reasons: The two parts of the molecule that must react in these lactonizations are so far away from each other that they encounter each other less frequently, that is, much more slowly than the OH and the CO_2H group of γ- or δ-hydroxycarboxylic acids do (Figure 6.22). More accurately stated, the activation entropy of the formation of large-ring lactones from hydroxycarboxylic acids is quite negative. This is because only a few of the many conformers of the starting material are capable of forming the tetrahedral intermediate with nothing more than an attack of the OH group on the carboxyl group. All other conformers must first be converted into a cyclizable conformer through numerous rotations about the various C—C single bonds. Because of this difficulty, an attack of the OH group on the carboxyl group of a nearby but different molecule is more probable: almost every conformer of the hydroxycarboxylic acid is able to do that. Consequently, when long-chain ω-hydroxycarboxylic acids are heated up in the presence of an acid intermolecular esterifications occur instead of a lactonization. The former continue in an uncontrolled fashion beyond the stage of a monoesterification and give an ester/oligoester/polyester mixture.

Large-ring lactones are available in good yields from ω-hydroxycarboxylic acids only through a combination of two measures. First of all, the carboxylic acid moiety must be activated. This ensures that the highest possible percentage of the (still improbable) encounters between the alcoholic OH group and the carboxyl carbon of the same molecule lead to a successful reaction. In addition, one must make sure that the OH group is not acylated intermolecularly, that is, by an activated neighbor molecule. To this end, the hydroxycarboxylic acid is activated in a very dilute solution. This is based on the following consideration: The rate of formation of the tetrahedral intermediate that will deliver the lactone is as follows:

$$\frac{\mathrm{d}\,[\text{tetrahedral precursor of lactone}]}{\mathrm{d}t} = k_{\text{lactonization}}[\text{activated } \omega\text{-hydroxycarboxylic acid}] \quad (6.17)$$

(Here $k_{\text{lactonization}}$ is the rate constant of the lactonization.)

On the other hand, the rate of formation of the undesired acyclic (mono/oligo/poly) esters is:

$$\frac{\mathrm{d}\,[\text{tetrahedral precursor of acyclic ester}]}{\mathrm{d}t} = k_{\text{acyclic ester}}[\text{activated } \omega\text{-hydroxycarboxylic acid}]^2 \quad (6.18)$$

(Here $k_{\text{acyclic ester}}$ is the rate constant for the formation of acyclic esters.)

If one divides Equation 6.17 by Equation 6.18, the expression on the left side of the equals sign corresponds approximately to the yield ratio of the lactone and the mixture of the acyclic esters, oligo- and polyesters. Taking this into account, one obtains as a new equation the following:

$$\frac{\text{yield of lactone}}{\text{yield of acyclic ester}} = \frac{k_{\text{lactonization}}}{k_{\text{acyclic ester}}} \cdot \frac{1}{[\text{activated } \omega\text{-hydroxycarboxylic acid}]} \quad (6.19)$$

From Equation 6.19 it follows that the lower the concentration of the activated acid, the higher the selectivity with which the lactone is produced versus the acyclic ester, oligo- and polyesters. Macrolactones are therefore produced in very dilute (<1 μmol/L) solutions **(high dilution principle according to Ziegler and Ruggli).**

For work on a 1-mole scale one would thus have to use a 1000-liter flask to activate and then lactonize the entire ω-hydroxycarboxylic acid at once. Of course, it is much more practical to work in a smaller reaction vessel. However, there one must *also* not exceed the mentioned concentration limit of <1 μmol/L. Therefore, one can introduce only as much of the carboxylic acid in this smaller reaction vessel at a time so that its concentration does not exceed 1 μmol/L. Subsequently, one would have to activate this amount of acid and would then have to wait until it is lactonized. After that additional acid would have to be added and then activated, and so on. A more practical alternative is shown in Side Note 6.1.

Side Note 6.1
Continuous Process for Preparing Macrolactones

A

The discontinuous process for preparing macrolactones described in the text is impractical. Instead of this process one uses a continuous method: with a syringe pump one adds a solution of the hydroxycarboxylic acid very slowly—that is, in the course of hours or days—into a small flask, which contains ≥1 equivalent of the activator and, if necessary, just enough triethylamine to neutralize any released HCl. The rate at which the acid is added is regulated such that it is equal to or smaller than the lactonization rate. This procedure is called **"working under pseudo-high dilution."** At the end of the reaction the lactone solution can be relatively concentrated, e.g., 10 mmol/L, at least 10,000 times more concentrated than without the use of this trick.

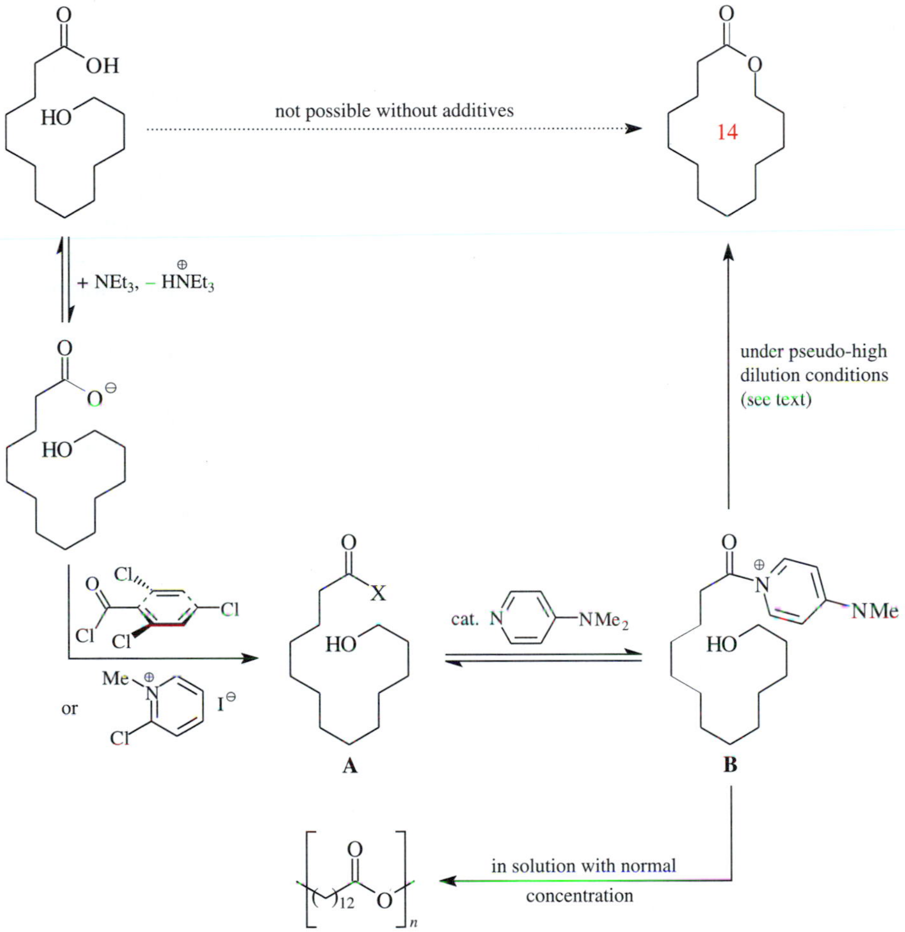

Fig. 6.23. Possibilities for macrolactonization.

Macrolactonizations are generally realized through an acyl activation (Figure 6.23; the alternative of an OH activation during macrolactonizations was shown in Figure 2.30). In the *in situ* activation process with trichlorobenzoyl chloride, the mixed anhydride **A** (Figure 6.23, X stands for O—C(=O)—$C_6H_2Cl_3$) is formed as shown in a more generalized manner in Figure 6.14. Alternatively, the *in situ* activation process from Figure 6.16 with the *N*-methylpyridinium halide gives the imminium analogue **A** (Figure 6.23, X stands for 1,2-dihydro-*N*-methylpyrid-2-yl) of a mixed anhydride.

A

Another level of refinement regarding *in situ* acyl group activations is reached when the activated hydroxycarboxylic acid **A** is converted with additionally added *para*-(dimethylamino)pyridine, Steglich's catalyst, in an equilibrium reaction into the corresponding *N*-acylpyridinium salt **B** (method: Figure 6.9). Under these conditions macrocyclizations routinely succeed in yields well above 50%.

6.4.3 Forming Peptide Bonds

An amino acid protected only at the N atom and a different amino acid in which only the CO_2H group is protected do not react with each other to form a peptide bond. On the contrary, they form an ammonium carboxylate in a fast acid/base reaction (Figure 6.24). Ammonium carboxylates can in principle be converted into amides by strong heating. Thus, for example, in the industrial synthesis of nylon-6,6 the diammonium dicarboxylate obtained from glutamic acid and hexamethylenediamine is converted to the polyamide at 300°C. However, this method is not suitable for peptide synthesis because there would be too many undesired side reactions.

To combine the amino acid from Figure 6.24, which is only protected at the N atom, and the other amino acid from Figure 6.24, in which only the CO_2H group is protected, into a dipeptide, the CO_2H group of the N-protected amino acid must be activated. As a general rule this is not possible via an acid chloride, the standard acylating agent according to Table 6.1. This is because if an amino acid chloride is treated with an amino acid ester with a free NH_2 group, the desired peptide formation is only part of what happens. At the same time, there is a disastrous side reaction (Figure 6.25).

The amino acid ester with the free NH_2 group can also react as a base. Hence, the amino acid chloride is deprotonated—reversibly—to the enolate. This is so readily pos-

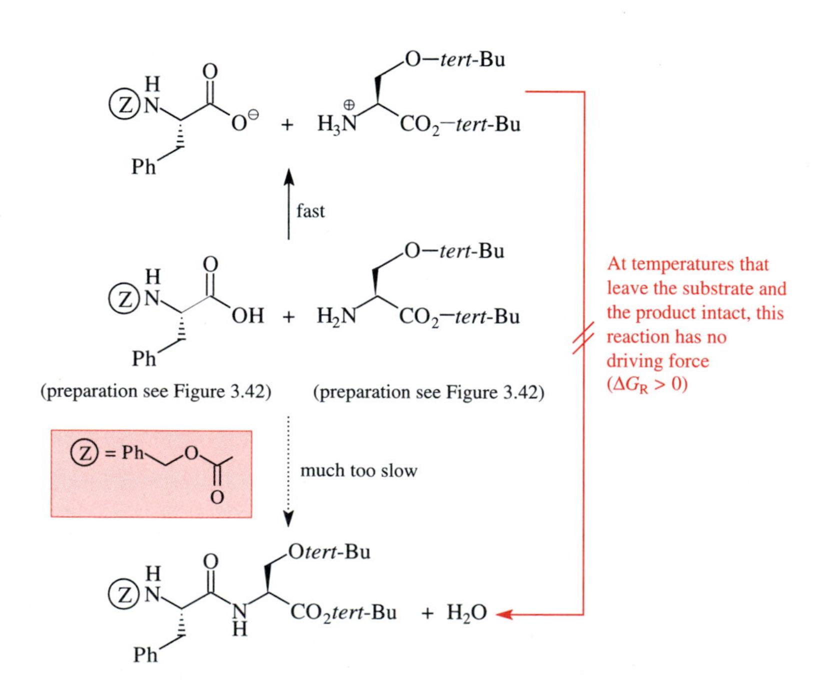

Fig. 6.24. Impossibility of preparing a dipeptide from an N-protected amino acid and an amino acid ester with a free NH_2 group.

Overall reaction (not going to completion)
partial enantiomerization,
thus partial racemization
via $-H^{\oplus}$ $+H^{\oplus}$ $+H^{\oplus}$ $+H^{\oplus}; -H^{\oplus}$ $+H^{\oplus}; -H^{\oplus}$ $+H^{\oplus}$

Fig. 6.25. Mechanism of the stereoisomerization of an N-protected amino acid chloride under the conditions of a peptide synthesis. The red reaction arrows stand for the steps of the racemization.

sible because this enolate is stabilized by the combined electron withdrawal by the Cl atom and by the protected amino group. The reprotonation of this enolate takes place without stereocontrol. It is not clear whether this protonation takes place at the C atom or at the O atom (whereupon the resulting enol would have to be protonated once more, now at the C atom). Be this as it may, what is important is that in this way the configurational integrity at the methine carbon of the activated amino acid is lost.

Let us analyze once more with a different emphasis what has just been said: An N-protected amino acid chloride can be deprotonated by an amino acid ester with a free NH_2 group because the enolate produced is stabilized, among other things, by the $-I$ effect of the Cl atom. This immediately suggests a solution to circumvent the described dilemma: Activating the N-protected amino acid requires a derivative in which the substituent at the carboxyl carbon is less electron-withdrawing than a Cl atom. In contrast to the Cl atom, a substituent of this type would not help enolate formation. Hence, one needs carboxyl substituents that activate the amino acid sufficiently but do not acidify it too much.

Acyl isoureas are extremely well suited for this purpose. Figure 6.26 shows a corresponding L-phenylalanine derivative as compound **A.** Substances such as these are prepared according to the mechanism of Figure 6.15 *in situ* by first adding dicyclohexylcarbodiimide to the N-protected L-phenylalanine. Subsequently, an amino acid ester with a free NH_2 group is added. This group is acylated without destroying the stereochemical integrity of the phenylalanine component **A** (upper part of Figure 6.26). The only drawback that remains is that the yield in this example leaves a great deal to be desired—unless the acyl isourea is used in large excess.

A

In its original form the DCC procedure for peptide synthesis has one basic disadvantage, which you have already learned about as side reaction **B** → **E** in Figure 6.14: When acyl isoureas are exposed to a poor nucleophile, they rearrange to form unreactive N-acylureas. These mismatched reactivities characterize precisely the situation in the reaction of Figure 6.26, in which O,O-di-*tert*-butylserine is used as nucleophile. The $-I$ effect of the ethereal *tert*-butoxy group of this compound makes its NH_2 group noticeably less nucleophilic than the NH_2 group in the esters of most other amino acids. In the discussion of Figure 6.15, we already presented, in general terms, a method that allows for the

Fig. 6.26. Stereoisomerization-free synthesis of a dipeptide according to the original DCC procedure (above) or a modified DCC procedure (below) (Z = benzyloxycarbonyl; preparation of compound **A:** Figure 6.29).

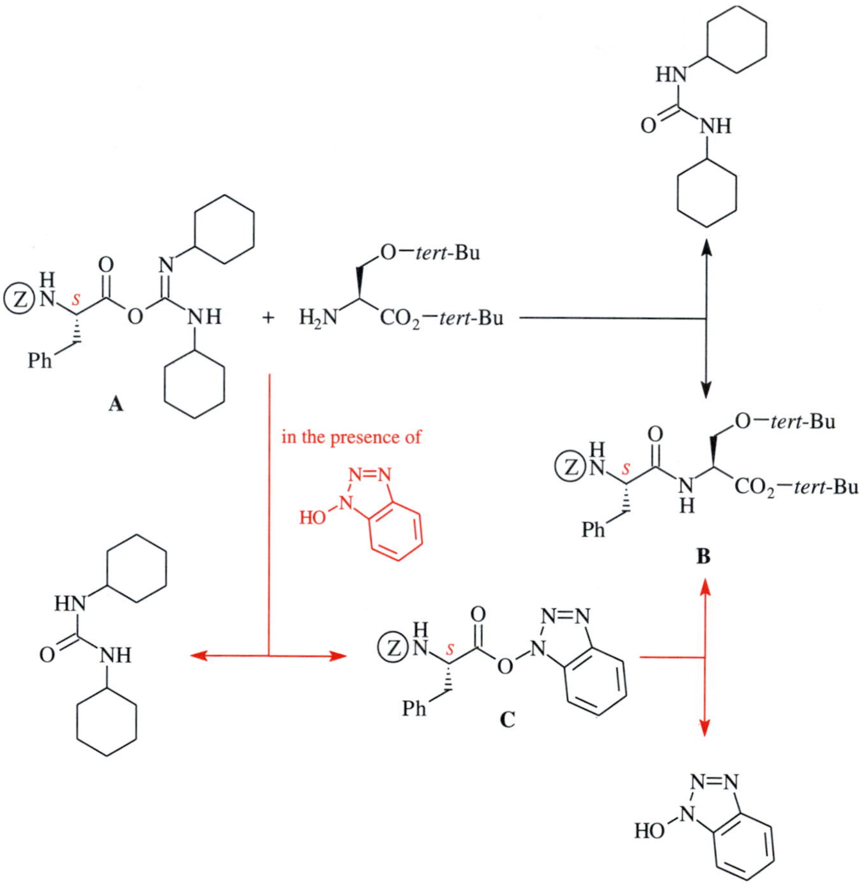

incorporation of this serine derivative into a dipeptide without losing part of the other, DCC-protected amino acid as an inert urea derivative. A specific solution can be found in the lower half of Figure 6.26. The unstable DCC adduct **A** is transacylated with *N*-hydroxybenzotriazole to give the stable amino acid derivative **C.** This is converted by the serine derivative into dipeptide **B** via an almost quantitative S_N reaction.

6.4.4 S_N Reactions of Heteroatom Nucleophiles with Carbonic Acid Derivatives

B

Many heteroatom nucleophiles can be added to heterocumulenes of the general structure $Het^1{=}C{=}Het^2$ to give carbonic acid derivatives of the type $Het^1{-}C{-}({=}O)\text{-}Het^{2\text{ or }3}$ (Section 7.1). Almost all of these carbonic acid derivatives can undergo S_N reactions with heteroatom nucleophiles at the carboxyl carbon. In this way other carbonic acid derivatives are produced. A short survey of the preparative possibilities for these reactions is given in Figure 6.27. Each of these reactions can be understood mechanistically in analogy to what you have learned in Sections 6.2–6.4 about S_N reactions at the carboxyl carbon of carboxylic acid derivatives. Moreover, the cross references placed in the figure refer mostly to explicitly presented mechanisms. With this information you yourself can think up suitable reagents and appropriate mechanisms for the reaction arrows that have not been labeled.

Figure 6.28 shows the S_N reaction of two equivalents of benzyl alcohol (formula **B**) with $Cl_3C—O—C(=O)—Cl$ (diphosgene, formula **C**; preparation: Section 1.7.1). It is preferable to work with diphosgene, which is a liquid, rather than with phosgene

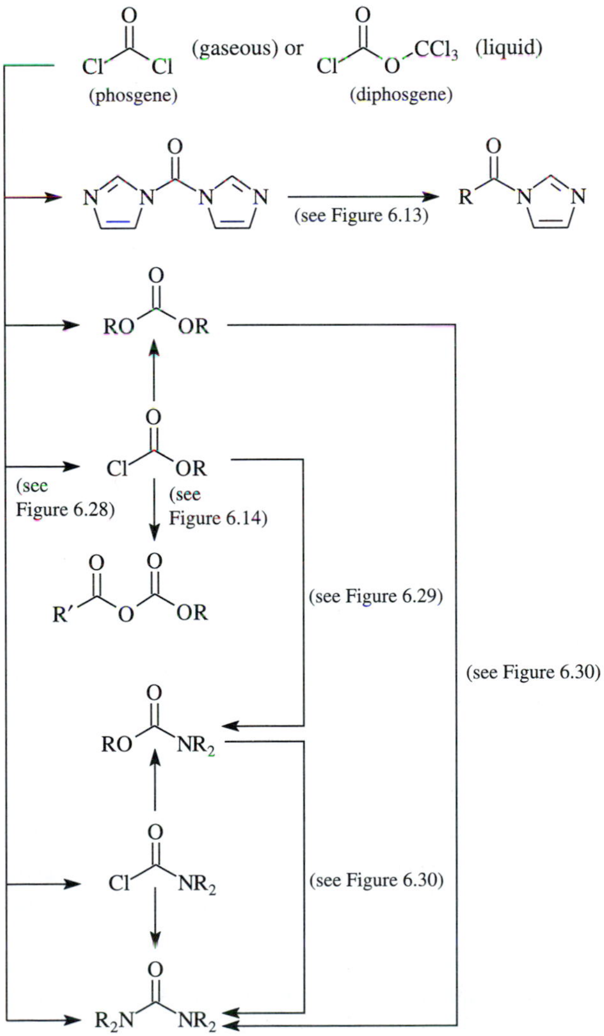

Fig. 6.27. Carbonic acid derivatives and their interconversions.

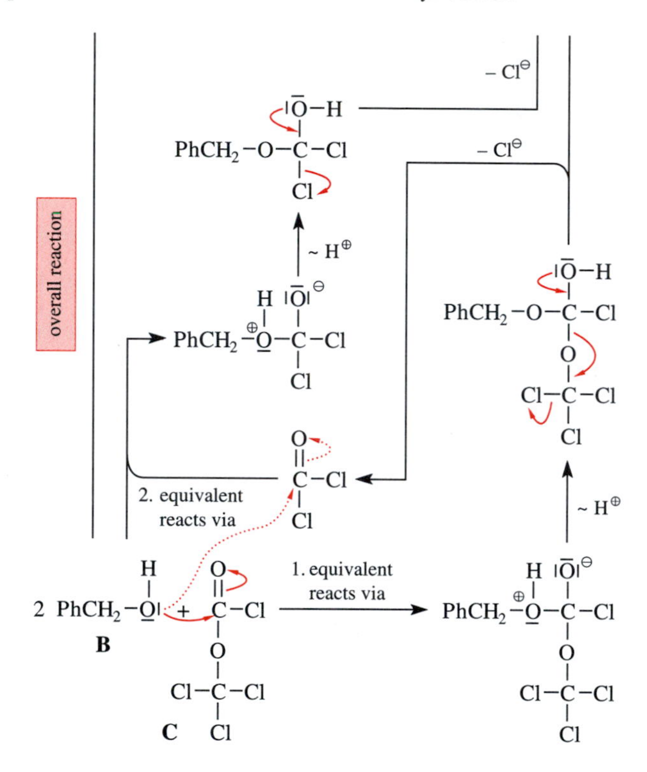

Fig. 6.28. Mechanism for obtaining a chlorocarbonate through an S_N reaction of an alcohol with diphosgene. The first equivalent of the alcohol reacts with diphosgene itself; the second equivalent reacts with the phosgene formed in the third step.

Cl—C(=O)—Cl, which is a gas, because both reagents are very poisonous. As the substitution product one obtains chloroformic acid benzyl ester **A,** which is also called "Z-chloride."

Various heteroatom nucleophiles carry out an S_N reaction at the carboxyl carbon of Z-chloride (whereby they displace a chloride ion). The most important nucleophiles of this type are amino acids. They give rise to benzyloxycarbonyl-protected amino acids (Z-protected amino acids; Figure 6.29). They are standard components for peptide synthesis (cf. Figure 6.26).

As the last example of an S_N reaction at the carboxyl carbon of a carbonic acid derivative, consider the synthesis of dicyclohexylurea in Figure 6.30. In this synthesis two equivalents of cyclohexylamine replace the two methoxy groups of dimethyl carbonate. Dicyclohexylurea can be converted into the carbodiimide dicyclohexylcarbodiimide (DCC) by treatment with tosyl chloride and triethylamine. Thereby, the urea is effectively dehydrated. The mechanism of this reaction is identical to the mechanism that is presented in Figure 7.5 for the similar preparation of a different carbodiimide.

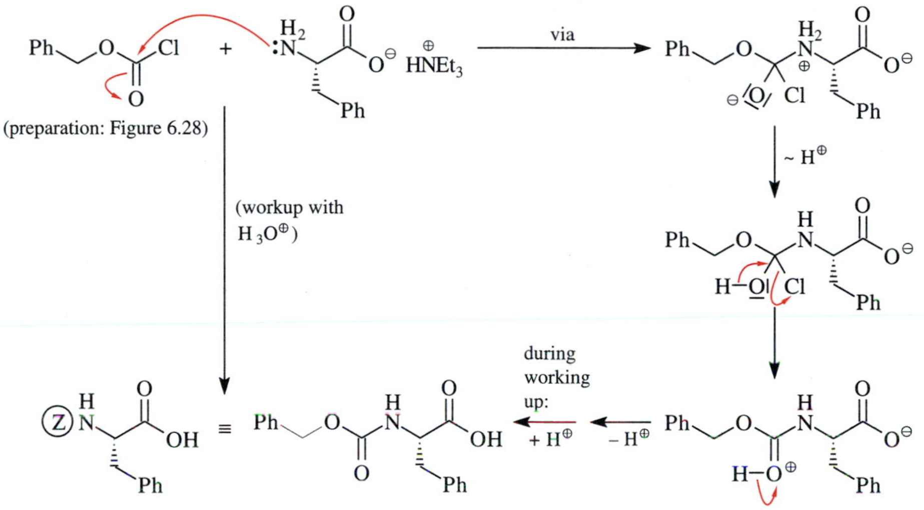

Fig. 6.29. Mechanism for the preparation of a Z-protected amino acid by acylation of an amino acid with Z-chloride.

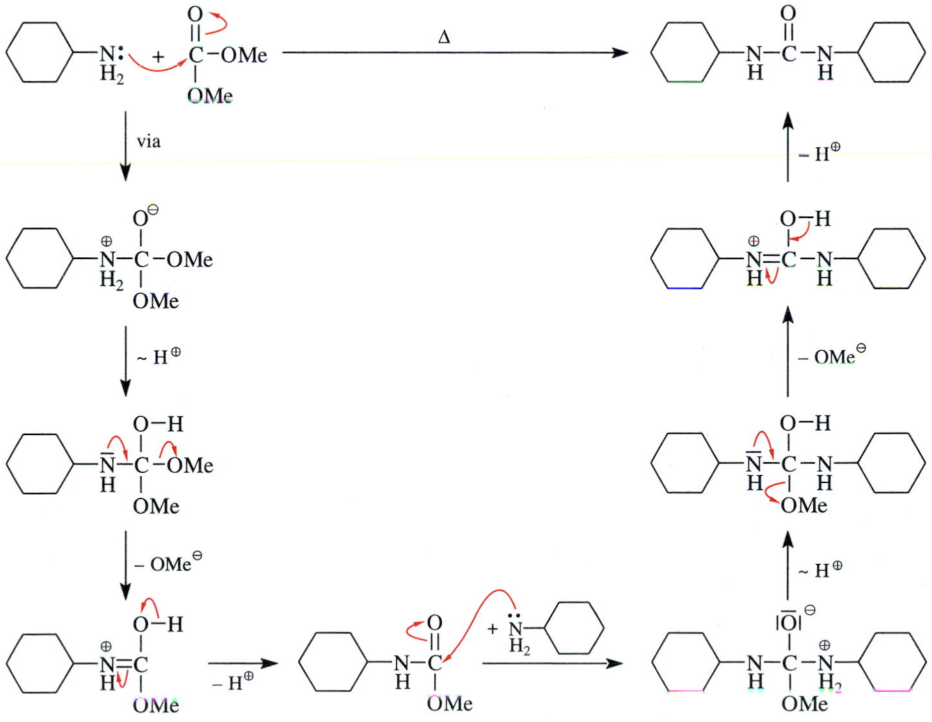

Fig. 6.30. Acylation of cyclohexylamine with dimethyl carbonate—a possibility for synthesizing the DCC precursor dicyclohexylurea.

6.5 S_N Reactions of Hydride Donors, Organometallics, and Heteroatom-Stabilized "Carbanions" on the Carboxyl Carbon

6.5.1 When Do Pure Acylations Succeed, and When Are Alcohols Produced?

B

Many hydride donors, organometallic compounds, and heteroatom-stabilized "carbanions" react with carboxylic acids and their derivatives. However, the corresponding substitution product, i.e., the acylation product (**C** in Figure 6.31) can only be *isolated* using very specific reagent/substrate combinations. In the other cases the result is an "overreaction" with the nucleophile. It occurs as soon as the acylation product **C** forms according to the mechanism of Figure 6.2 already at a time when the nucleophile is still present. Because this acylation product **C** is an aldehyde or a ketone, it is able to add any remaining nucleophile (cf. Chapter 8) furnishing alkoxides **D.**

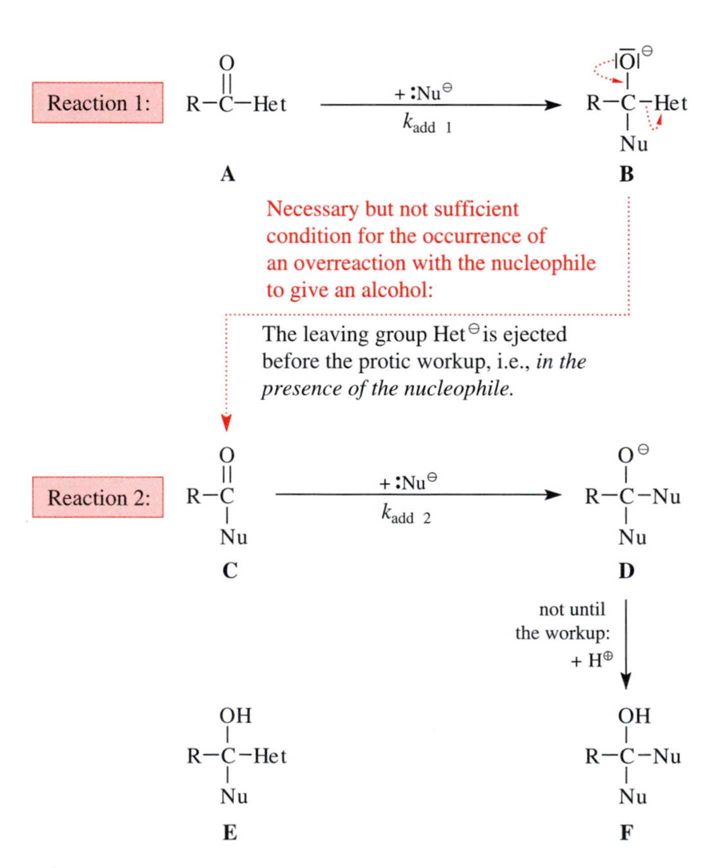

Fig. 6.31. On the chemoselectivity of the reactions of hydride donors, organometallic compounds, and heteroatom-stabilized "carbanions" with acylating agents ($k_{add\ 1}$ refers to the rate constant of the addition of the nucleophile to the carboxyl carbon, and $k_{add\ 2}$ refers to the rate constant of the addition of the nucleophile to the carbonyl carbon).

Upon aqueous workup, these provide secondary (R = H) or tertiary (R ≠ H) alcohols **F**. These compounds contain two identical substituents at the carbinol carbon both of which originate from the nucleophile.

Most S_N reactions of hydride donors, organometallic compounds, and heteroatom-stabilized "carbanions" at the carboxyl carbon follow the mechanism shown in Figure 6.2. Thus the substitution products, i.e., the aldehydes and ketones **C**, form *in the presence of the nucleophiles*. Thus, when the nucleophile and the acylating agent are used in a 2:1 ratio, alcohols **F** are *always* produced.

However, if one of the aforementioned nucleophiles and the acylating agent are reacted in a 1:1 ratio, it is under certain circumstances possible to stop the reaction chemoselectively at the stage of the carbonyl compound **C**.

From Figure 6.31 you can see that these particular conditions are fulfilled if the rate-determining step of the acylation—i.e., the formation of the tetrahedral intermediate **B**—is considerably faster than the further reaction of the carbonyl compound **C** giving the alkoxide **D**. In more quantitative terms, it would hence be required that the rate of formation $d[\mathbf{B}]/dt$ must be greater than the rate of formation $d[\mathbf{D}]/dt$. Thus, in order for acylations to occur chemoselectively, the following must hold:

$$k_{\text{add 1}}\,[\mathbf{A}]\,[\text{Nu}^-] \gg k_{\text{add 2}}\,[\mathbf{C}]\,[\text{Nu}^-] \tag{6.20}$$

This is ensured during the entire course of the reaction only if

$$k_{\text{add 1}} \gg k_{\text{add 2}} \tag{6.21}$$

For the reaction of hydride donors, organometallic compounds and heteroatom-stabilized "carbanions" with acylating agents or carbonyl compounds one encounters a universal reactivity order RC(=O)Cl > RC(=O)H > R_2C=O > RC(=O)OR > RC(=O)NR_2. It applies to both good and poor nucleophiles, but—in agreement with the reactivity/selectivity principle (Section 1.7.4)—for poor nucleophiles the reactivity differences are far larger.

Accordingly, if one wants to react a nucleophile and carboxylic acid derivative to produce a carbonyl compound in a chemoselective fashion according to the mechanism of Figure 6.2, then one best employs carboxylic acid chlorides or comparably strongly activated carboxylic acid derivatives. In addition, the respective reaction must be carried out with the weakest possible nucleophile because only such a nucleophile reacts *considerably* faster with the activated carboxylic acid derivative than with the product carbonyl compound (see above). The nucleophile must react "considerably" faster with the carboxylic acid derivative because at 95% conversion there is almost twenty times more carbonyl compound present than carboxylic acid derivative, but even at this stage the carboxylic acid derivative must be the preferred reaction partner of the nucleophile.

To make as *much* carboxylic acid derivative as possible available to the nucleophile at all stages of the reaction, the nucleophile is added dropwise to the carboxylic acid derivative and not the other way around. In Figure 6.32 the approach to chemoselective acylations of hydride donors and organometallic compounds, which we have just described, is labeled as "strategy 2" and compared to two other strategies, which we will discuss in a moment.

Chemoselective S_N reactions of nucleophiles with carboxylic acid derivatives are guaranteed to take place without the risk of an overreaction when the substitution mechanism of Figure 6.4 applies. This is because as long as the nucleophile is present, only one reaction step is possible: the formation of the negatively charged tetrahedral intermediate. Figure 6.31 summarizes this addition in the top line as "Reaction 1" (→ **B**).

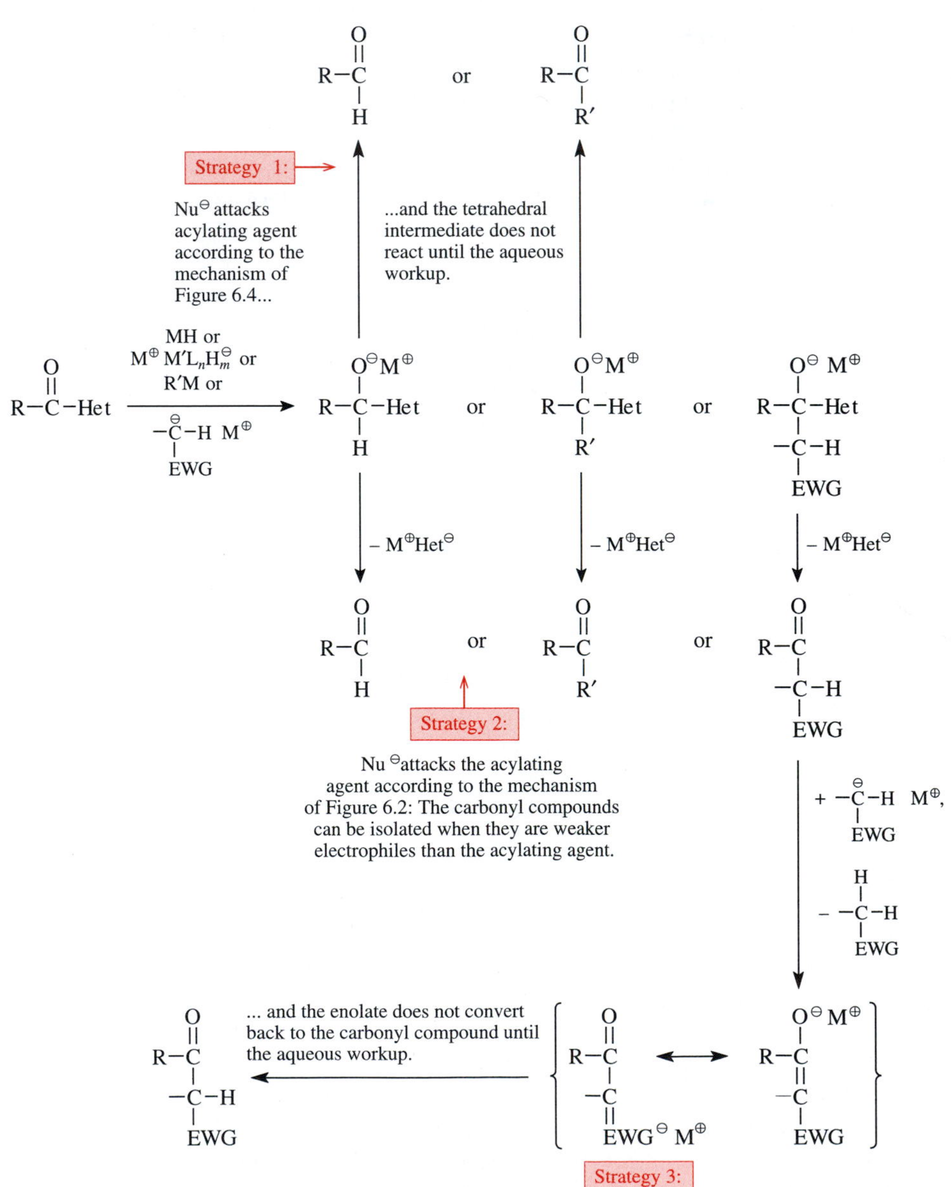

Fig. 6.32. Three strategies for the chemoselective acylation of hydride-donors, organometallics and heteroatom-stabilized "carbanions" with carboxylic acid derivatives.

There may be two reasons why such a tetrahedral intermediate **B** does not fragment into the carbonyl compound **C** before the aqueous workup. For one thing, the "Het" group in it may be too poor a leaving group. For another thing, the "Het" group may be bound through a suitable metal ion to the alkoxide oxygen of the tetrahedral intermediate and may thus be kept bound to it.

The S_N reaction under consideration is not terminated until water, a dilute acid, or a dilute base is added to the crude reaction mixture. The tetrahedral intermediate **B** is then protonated to give the compound **E.** Through an E1 elimination it liberates the carbonyl compound **C** (cf. discussion of Figure 6.4). Fortunately, at this point in time no overreaction of this aldehyde with the nucleophile can take place because the nucleophile has been destroyed during the aqueous workup by protonation or hydrolysis. In Figure 6.32 this process for chemoselective acylation of hydride donors, organometallic compounds, and heteroatom-stabilized "carbanions" has been included as "strategy 1."

As acylation "strategy 3," Figure 6.32 finally outlines a trick with which certain heteroatom-stabilized "carbanions" can be acylated chemoselectively. It is interesting to note that such a chemoselectivity exists although in this mechanism the tetrahedral intermediate does proceed to give the carbonyl compound while the nucleophile is still present. However, the heteroatom substituent, whose presence is required to make this trick possible, now exerts a significant electron withdrawal: It makes the carbonyl compound obtained C,H acidic (on α-C,H acidity of carbonyl compounds, especially those that carry an electron withdrawing group in the α position, cf. Section 10.1.2). Accordingly, this carbonyl compound is rapidly deprotonated by a second equivalent of the reagent (which is required for the success of this reaction) whereupon the enolate of the carbonyl compound is produced. This enolate can obviously not react further with any nucleophile still present. During the aqueous workup the desired carbonyl compound is formed again by protonation of the enolate. At the same time any remaining nucleophile is destroyed. Again, the carbonyl compound can no longer be attacked by the nucleophile at this point in time.

6.5.2 Acylation of Hydride Donors: Reduction of Carboxylic Acid Derivatives to Aldehydes

Chemoselective S_N reactions at the carboxyl carbon in which hydride-donors function as nucleophiles can be carried out using "strategy 1" or "strategy 2" from Figure 6.32. The substrates par excellence for "strategy 1" are tertiary amides and especially the so-called Weinreb amides. The latter are carboxylic amides (**A** in Figure 6.33) derived from *N*,*O*-dimethylhydroxylamine. Tertiary amides or Weinreb amides and **d**i**i**so**b**utyl**al**uminum hydride (DIBAL) furnish stable tetrahedral intermediates. In the tetrahedral intermediate, obtained from "normal" tertiary amides the $R^1R^2N^-$ group is a particularly poor leaving group. The same qualification applies to the (MeO)MeN$^-$ group in the tetrahedral intermediates (**B** in Figure 6.33) derived from Weinreb amides. In addition, the Weinreb intermediates **B** are stabilized by the chelation indicated in Figure 6.33 (this is why Weinreb amides contain the methoxy substituent!). During the aqueous workup these tetrahedral intermediates are hydrolyzed to give an aldehyde.

B

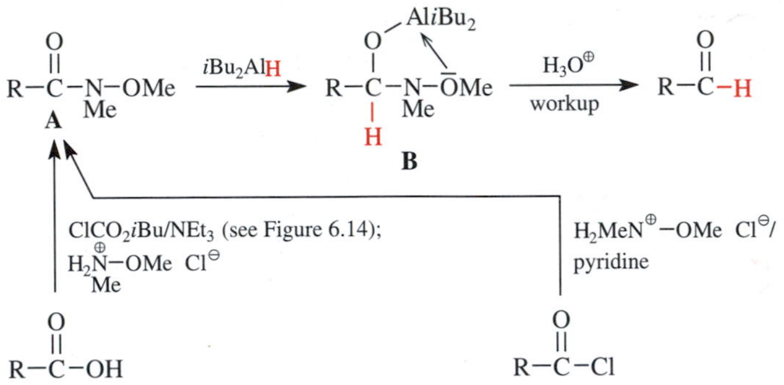

Fig. 6.33. Preparation of Weinreb amides through S_N reactions at the carboxyl carbon. Chemoselective reduction of Weinreb amides to aldehydes.

Weinreb amides can be reduced to aldehydes not only with DIBAL but also with $LiAlH_4$. In general this is not true for other amides.

A

tert-Hexylmonobromoborane reduces free carboxylic acids selectively to the aldehyde. Why this is so can be seen from Figure 6.34: First an acyloxyborane **A** is formed together with H_2. In **A,** the B atom strongly attracts the free electron pair of the single-bonded O atom to supplement its electron sextet. Thus the C=O double bond of the acyloxyborane **A** lacks any resonance stabilization whatsoever. Consequently, this compound contains an extremely electrophilic C=O double bond. A second equivalent of the borane adds to it in the second reaction step. The tetrahedral intermediate **B,** which resembles an acetal, is produced. Unlike a real acetal, compound **B** is stable under the reaction conditions, because **B** cannot heterolyze to give a carboxonium ion: If its C—O bonds were broken, an acceptor-substituted carbenium ion and not a carboxonium ion would be produced. Accordingly, this carbenium ion would bear an oxy-

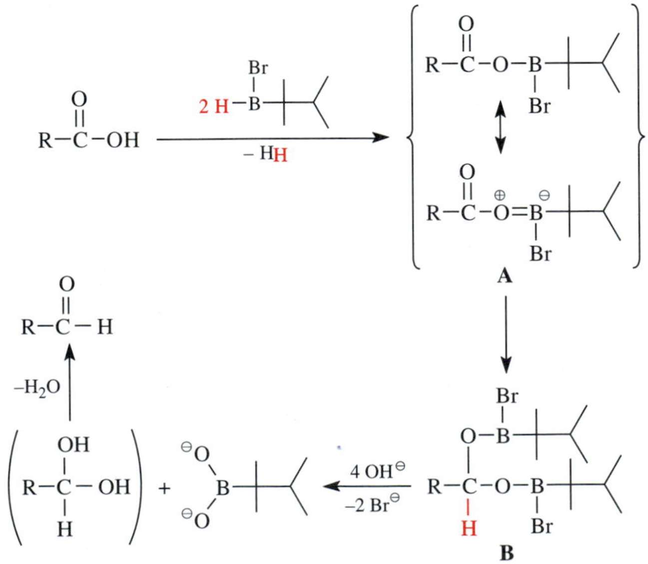

Fig. 6.34. Chemoselective reduction of free carboxylic acids to aldehydes. Intermediate **B** yields, upon hydrolysis, initially an aldehyde hydrate, which dehydrates to the aldehyde spontaneously (mechanism: Section 7.2.1).

gen substituent that acts as an electron acceptor and not as an electron donor. This is because this oxygen substituent would donate one of its free electron pairs completely to the boron atom, which thereby would fill up its valence electron sextet to give an octet. For this reason the intermediate **B** survives until the aqueous workup. There it hydrolyzes first to give the aldehyde hydrate, which then spontaneously decomposes to form the aldehyde (mechanism: Section 7.2.1).

With their alkoxy group, esters contain a better leaving group than amides or Weinreb amides. Also, esters lack a heteroatom that could be incorporated in a five-membered chelate, as starting from Weinreb amides. Therefore, tetrahedral intermediates formed from esters and hydride-donating reagents decompose considerably more readily than those resulting from amides or Weinreb amides. Only from DIBAL and esters—and even then only in noncoordinating solvents—is it possible to obtain tetrahedral intermediates that are stable until the aqueous workup and can then be hydrolyzed to an aldehyde (details: Figure 14.53). In contrast, in coordinating solvents DIBAL and esters give alcohols through an overreaction with the aldehydes that are now rapidly formed *in situ* (details: Figure 14.53). The reaction of $LiAlH_4$ and esters (Figure 14.52) always proceeds to alcohols through such an overreaction.

Following "strategy 2" from Figure 6.32, chemoselective S_N reactions of hydride-donors with carboxylic acid derivatives also succeed starting from carboxylic chlorides. For the reasons mentioned further above, *weakly* nucleophilic hydride donors are used for this purpose preferentially and should be added dropwise *to* the acylating agent in order to achieve success:

$$R-\overset{\overset{\displaystyle O}{\|}}{C}-Cl \xrightarrow[\text{or}\ Li^{\oplus}AlH(O-tert\text{-}Bu)_3^{\ominus},\ \text{low temperature}]{Na^{\oplus}BH_4^{\ominus},\ \text{low temperature}\ \text{or}\ Cu(PPh_3)_2^{\oplus}BH_4^{\ominus}} R-\overset{\overset{\displaystyle O}{\|}}{C}-H$$

6.5.3 Acylation of Organometallic Compounds and Heteroatom-Stabilized "Carbanions": Synthesis of Ketones

B

Tertiary amides in general and Weinreb amides in particular react according to "strategy 1" of Figure 6.32 not only with hydride-donors to give stable tetrahedral intermediates (Figure 6.33), but also with organolithium and Grignard compounds (reactions leading to **A** or to **B,** Figure 6.35). The aqueous workup of these intermediates **A** or **B** leads to pure acylation products. In this way, DMF or the Weinreb amide of formic acid and organometallic compounds give aldehydes. In the same way, tertiary amides or Weinreb amides of all higher monocarboxylic acids and organometallic compounds form ketones.

The same reaction mechanism—again corresponding to "strategy 1" of Figure 6.32—explains why carboxylic acids and two equivalents of an organolithium compound react selectively to form ketones (Figure 6.36). The first equivalent of the reagent de-

Fig. 6.35. Chemoselective acylations of organometallic compounds with tertiary amides or Weinreb amides giving aldehydes (R^1 = H) or ketones ($R^1 \neq$ H).

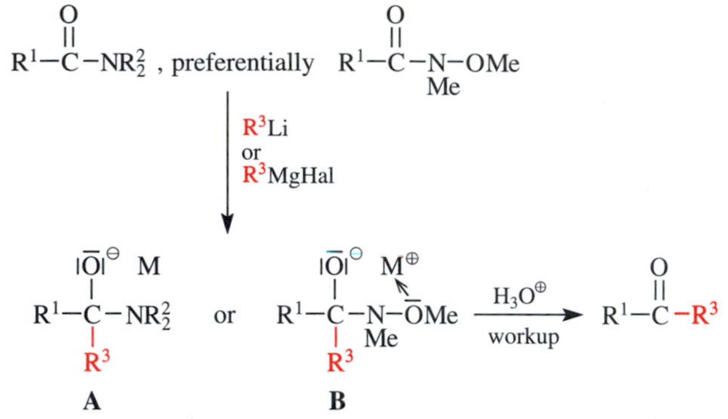

protonates the substrate to give a lithium carboxylate. According to Table 6.1, carboxylates are the weakest of all acylating agents. Actually, lithium carboxylates acylate only a single type of nucleophile, namely organolithium compounds. Therefore, the second equivalent of the organolithium compound reacts with the carboxylate forming the tetrahedral intermediate **A** in Figure 6.36. Because this species cannot possibly eliminate Li_2O, it is stable until the aqueous workup. It is then protonated to give the ketone hydrate from which the ketone is produced immediately.

Fig. 6.36. Chemoselective acylation of organolithium compounds with lithiumcarboxylates.

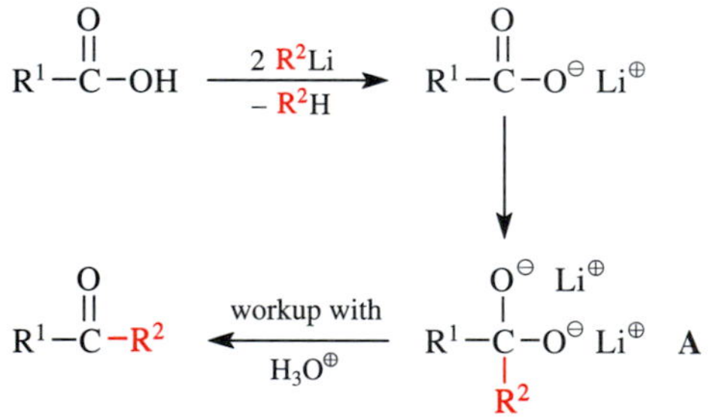

You are already familiar with "strategy 2" of Figure 6.32; this strategy also allows for the chemoselective acylation of weakly nucleophilic organometallic compounds, such as the ones shown in Figure 6.37, with carboxylic chlorides to afford various ketones. Suitable organometallic compounds are Gilman cuprates (R_2CuLi; preparation: Figure 6.37, left), Knochel cuprates [$R_{funct}Cu(CN)ZnX$; preparation: Figure 8.34, right; R_{funct} means a functionalized group], or lithium acetylides.

In Figure 6.32 the reaction of certain heteroatom-stabilized "carbanions" with carboxylic acid derivatives has been presented as "strategy 3" of Figure 6.32 for achieving chemoselective acylations. This strategy can be used to convert esters into β-ketophosphonic acid esters with sodium phosphonates (Figure 6.38) or to acylate the sodium salt of DMSO (dimethyl sulfoxide, Figure 6.39). The last reaction can, for example, be part of a two-step synthesis of methyl ketones from carboxylic esters, which does not require the use of organometallic compounds.

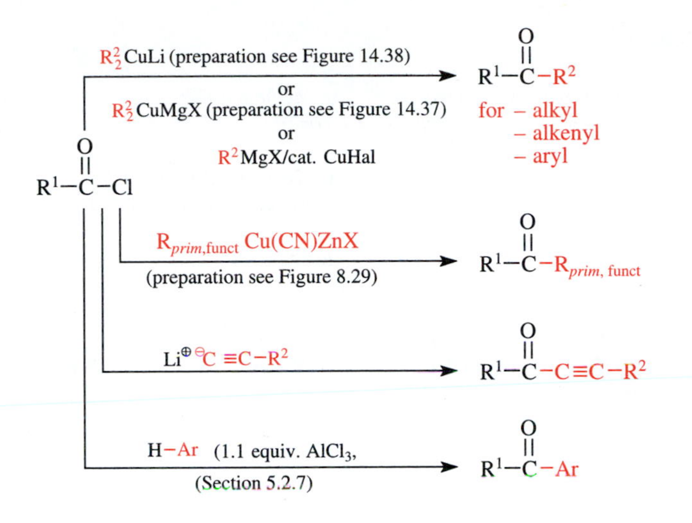

Fig. 6.37. Top three reactions: Chemoselective acylations of weakly nucleophilic organometallic compounds with carboxylic chlorides. Bottom reaction (as a reminder): Chemoselective acylation of an aromatic compound with an activated carboxylic chloride (Friedel–Crafts acylation, cf. Section 5.2.7).

$$R^1-\overset{O}{\overset{\|}{C}}-OR^2 + 2\,Na^{\oplus}\,\underset{R^3}{\overset{\ominus}{C}H}-\overset{O}{\overset{\|}{P}}(OEt)_2 \xrightarrow[\text{workup)}]{\text{(after } H_3O^{\oplus}} R^1-\overset{O}{\overset{\|}{C}}-\underset{R^3}{CH}-\overset{O}{\overset{\|}{P}}(OEt)_2$$

$$\underset{R^3}{H_2C}-\overset{O}{\overset{\|}{P}}(OEt)_2 \xrightarrow{\text{NaH}} Na^{\oplus}\,\underset{R^3}{\overset{\ominus}{C}H}-\overset{O}{\overset{\|}{P}}(OEt)_2$$

Fig. 6.38. Preparation of Horner–Wadsworth–Emmons reagents (synthesis applications: Section 9.4) by chemoselective acylation of a phosphonate-stabilized "carbanion" with an ester.

$$R^1-\overset{O}{\overset{\|}{C}}-OR^2 + 2\,Na^{\oplus}\,\overset{\ominus}{C}H_2-\overset{O}{\overset{\|}{S}}-CH_3 \xrightarrow[\text{workup)}]{\text{(after } H_3O^{\oplus}} R^1-\overset{O}{\overset{\|}{C}}-CH_2-\overset{O}{\overset{\|}{S}}-CH_3$$

$$H_3C-\overset{O}{\overset{\|}{S}}-CH_3 \xrightarrow{\text{NaH}} Na^{\oplus}\,\overset{\ominus}{C}H_2-\overset{O}{\overset{\|}{S}}-CH_3$$

$$R^1-\overset{O}{\overset{\|}{C}}-CH_2-\overset{O}{\overset{\|}{S}}-CH_3 \xrightarrow{\text{Al/Hg}} R^1-\overset{O}{\overset{\|}{C}}-CH_3$$

Fig. 6.39. Preparation of methyl ketones by (1) chemoselective acylation of a sulfinyl-stabilized "carbanion," (2) reduction (mechanism: analogous to Figure 14.41).

References

R. Sustmann and H.-G. Korth, "Carboxylic Acids," in *Methoden Org. Chem. (Houben-Weyl) 4th ed. 1952, Carboxylic Acids and Carboxylic Acid Derivatives* (J. Falbe, Ed.), Vol. E5, 193, Georg Thieme Verlag, Stuttgart, **1985.**

R. Sustmann and H.-G. Korth, "Carboxylic Acid Salts," in *Methoden Org. Chem. (Houben-Weyl) 4th ed. 1952, Carboxylic Acids and Carboxylic Acid Derivatives* (J. Falbe, Ed.), Vol. E5, 470, Georg Thieme Verlag, Stuttgart, **1985.**

R. Sustmann and H. G. Korth, "Carboxylic Acid Chlorides," in *Methoden Org. Chem. (Houben-Weyl) 4th ed. 1952, Carboxylic Acids and Carboxylic Acid Derivatives* (J. Falbe, Ed.), Vol. E5, 587, Georg Thieme Verlag, Stuttgart, **1985.**

M. A. Ogliaruso and J. F. Wolfe, "Carbocylic Acids," in *Comprehensive Organic Functional Group Transformations* (A. R. Katritzky, O. Meth-Cohn, C. W. Rees, Eds.), Vol. 5, 23, Elsevier Science, Oxford, U.K., **1995.**

6.2

K. B. Wiberg, "The interaction of carbonyl groups with substituents," *Acc. Chem. Res.* **1999,** *32,* 922–929.

R. S. Brown, A. J. Bennet, H. Slebocka-Tilk, "Recent perspectives concerning the mechanism of H_3O^+- and OH^-promoted amide hydrolysis," *Acc. Chem. Res.* **1992,** *25,* 481–488.

E. Juaristi and G. Cuevas, "Recent studies on the anomeric effect," *Tetrahedron* **1992,** *48,* 5019–5087.

A. J. Kirby (Ed.), "Stereoelectronic Effects," Oxford University Press, Oxford, U.K., *1996.*

6.3

K. B. Wiberg, "The interaction of carbonyl groups with substituents," *Acc. Chem. Res.* **1999,** *32,* 922–929.

R. Sustmann, "Synthesis of Acid Halides, Anhydrides and Related Compounds," in *Comprehensive Organic Synthesis* (B. M. Trost, I. Fleming, Eds.), Vol. 6, 301, Pergamon Press, Oxford, **1991.**

P. Strazzolini, A. G. Giumanini, S. Cauci, "Acetic formic anhydride," *Tetrahedron* **1990,** *46,* 1081–1118.

A. A. Bakibayev and V. V. Shtrykova, "Isoureas: Synthesis, properties, and applications," *Russ. Chem. Rev.* **1995,** *64,* 929–938.

A. R. Katritzky, X. Lan, J. Z. Yang, O. V. Denisko, "Properties and synthetic utility of N-substituted benzotriazoles," *Chem. Rev.* **1998,** *98,* 409–548.

U. Ragnarsson and L. Grehn, "Novel amine chemistry based on DMAP-catalyzed acylation," *Acc. Chem. Res.* **1998,** *31,* 494–501.

A. R. Katritzky and S. A. Belyakov, "Benzotriazole-based intermediates: Reagents for efficient organic synthesis," *Aldrichimica Acta,* **1998,** *31,* 35–45.

V. F. Pozdnev, "Activation of carboxylic acids by pyrocarbonates: Scope and limitations," *Org. Prep. Proced. Int.* **1998,** *30,* 631–655.

H. A. Staab, H. Bauer, K. M. Schneider, "Azolides in Organic Synthesis and Biochemistry," Wiley, New York, **1998.**

6.4

U. Ragnarsson and L. Grehn, "Novel Gabriel reagents," *Acc. Chem. Res.* **1991,** *24,* 285.

C. Salomon and E. G. Mata, "Recent developments in chemical deprotection of ester functional groups," *Tetrahedron* **1993,** *49,* 3691.

J. Mulzer, "Synthesis of Esters, Activated Esters and Lactones," in *Comprehensive Organic Synthesis* (B. M. Trost, I. Fleming, Eds.), Vol. 6, 323, Pergamon Press, Oxford, **1991.**

R. Sustmann and H. G. Korth, "Protecting Groups for Carboxylic Acids," in *Methoden Org. Chem. (Houben-Weyl) 4th ed. 1952, Carboxylic Acids and Carboxylic Acid Derivatives* (J. Falbe, Ed.), Vol. E5, 496, Georg Thieme Verlag, Stuttgart, **1985.**

E. Haslam, "Recent developments in methods for the esterification and protection of the carboxyl group," *Tetrahedron* **1980,** *36,* 2409.

J. Otera, "Transesterification," *Chem. Rev.* **1993,** *93,* 1449–1470.

N. F. Albertson, "Synthesis of peptides with mixed anhydrides, *Org. React.* **1962,** *12,* 157–355.

R. C. Sheppard, "Peptide Synthesis," in *Comprehensive Organic Chemistry* (E. Haslam, Ed.), **1979,** *5* (Biological Compounds), 321–366. Pergamon Press Ltd., England.

J. Jones, "The Chemical Synthesis of Peptides," Clarendon Press, Oxford, U.K., *1991.*

J. Jones, "Amino Acid and Peptide Synthesis (Oxford Chemistry Primers: 7)," Oxford University Press, Oxford, U.K., **1992.**

G. A. Grant, "Synthetic Peptides: A User's Guide," Freeman, New York, **1992.**

M. Bodanszky, "Peptide Chemistry: A Practical Textbook," 2nd ed., Springer Verlag, Berlin, **1993.**

M. Bodanszky, "Principles of Peptide Synthesis," 2nd ed., Springer Verlag, Berlin, **1993.**

C. Basava and G. M. Anantharamaiah (Eds.), "Peptides: Design, Synthesis and Biological Activity," Birkhaeuser, Boston, **1994.**

M. Bodanszky and A. Bodanszky, "The Practice of Peptide Synthesis," 2nd ed., Springer Verlag, Heidelberg, **1994.**

L. A. Carpino, M. Beyermann, H. Wenschuh, M. Bienert, "Peptide synthesis via amino acid halides," *Acc. Chem. Res.* **1996,** *29,* 268–274.

G. Jung, A. G. Beck-Sickinger, "Multiple peptide synthesis methods and their applications," *Angew. Chem., Int. Ed. Engl.* **1992,** *31,* 367.

P. Lloyd-Williams, F. Albericio, E. Giralt, "Convergent solid-phase peptide synthesis," *Tetrahedron* **1993,** *49,* 11065–11133.

T. A. Ryan, "Phosgene and Related Compounds," Elsevier Science, New York, **1996.**

L. Cotarca, P. Delogu, A. Nardelli, V. Sunjic, "Bis(trichloromethyl) carbonate in organic synthesis," *Synthesis* **1996,** 553–576.

6.5

M. P. Sibi, "Chemistry of N-methoxy-N-methylamides: Applications in synthesis," *Org. Prep. Proced. Int.* **1993,** *25,* 15–40.

J. L. Romine, "Bis-protected hydroxylamines as reagents in organic synthesis: A review," *Org. Prep. Proced. Int.* **1996,** *28,* 249–288.

G. Benz, K.-D. Gundermann, A. Ingendoh, L. Schwandt, "Preparation of Aldehydes by Reduction," in *Methoden Org. Chem. (Houben-Weyl) 4th ed. 1952, Aldehydes* (J. Falbe, Ed.), Vol. E3, 418, Georg Thieme Verlag, Stuttgart, **1983.**

E. Mosettig, "The synthesis of aldehydes from carboxylic acids," *Org. React.* **1954,** *8,* 218–257.

J. S. Cha, "Recent developments in the synthesis of aldehydes by reduction of carboxylic acids and their derivatives with metal hydrides," *Org. Prep. Proced. Int.* **1989,** *21,* 451–477.

R. A. W. Johnstone, "Reduction of Carboxylic Acids to Aldehydes by Metal Hydrides," in *Comprehensive Organic Synthesis* (B. M. Trost, I. Fleming, Eds.), Vol. 8, 259, Pergamon Press, Oxford, **1991.**

A. P. Davis, "Reduction of Carboxylic Acids to Aldehydes by Other Methods," in *Comprehensive Organic Synthesis* (B. M. Trost, I. Fleming, Eds.), Vol. 8, 283, Pergamon Press, Oxford, **1991.**

B. T. O'Neill, "Nucleophilic Addition to Carboxylic Acid Derivatives," in *Comprehensive Organic Synthesis* (B. M. Trost, I. Fleming, Eds.), Vol. 1, 397, Pergamon Press, Oxford, **1991.**

D. A. Shirley, "The synthesis of ketones from acid halides and organometallic compounds of magnesium, zinc, and cadmium," *Org. React.* **1954,** *8,* 28–58.

M. J. Jorgenson, "Preparation of ketones from the reaction of organolithium reagents with carboxylic acids," *Org. React.* **1970,** *18,* 1–9.

R. K. Dieter, "Reaction of acyl chlorides with organometallic reagents: A banquet table of metals for ketone synthesis," *Tetrahedron* **1999,** *55,* 4177–4236.

W. E. Bachmann and W. S. Struve, "The Arndt-Eistert reaction," *Org. React.* **1942,** *1,* 38–62.

Further Reading

M. Al-Talib and H. Tashtoush, "Recent advances in the use of acylium salts in organic synthesis," *Org. Prep. Proced. Int.* **1990,** *22,* 1–36.

S. Patai, (Ed.), "The Chemistry of Ketenes, Allenes, and Related Compounds" Wiley, New York, **1980.**

H. R. Seikaly and T. T. Tidwell, "Addition reactions of ketenes," *Tetrahedron* **1986,** *42,* 2587.

P. W. Raynolds, "Ketene," in *Acetic Acid and Its Derivatives* (V. H. Agreda, J. R. Zoeller, Eds.), 161, Marcel Dekker, New York, **1993.**

A.-A. G. Shaikh and S. Sivaram, "Organic carbonates," *Chem. Rev.* **1996,** *96,* 951–976.

Y. Ono, "Dimethyl carbonate for environmentally benign reactions," *Pure Appl. Chem.* **1996,** *68,* 367–376.

V. F. Pozdnev, "Activation of carboxylic acids by pyrocarbonates. Scope and limitations. A review," *Org. Prep. Proced. Int.* **1998,** *30,* 631–655.

Additions of Heteroatom Nucleophiles to Heterocumulenes. Additions of Heteroatom Nucleophiles to Carbonyl Compounds and Follow-up Reactions

7

7.1 Additions of Heteroatom Nucleophiles to Heterocumulenes

7.1.1 Mechanism of the Addition of Heteroatom Nucleophiles to Heterocumulenes

Heterocumulenes are compounds with a substructure X=C=Y, in which at least one of the groups X or Y is an O atom, an S atom, or an NR group. Heteroatom nucleophiles add to heterocumulenes, as was already discussed briefly with respect to the C=O-containing heterocumulenes in Section 6.1.1. Most of the heteroatom nucleophiles used are not charged.

B

The attack by the heteroatom nucleophile occurs at the *sp*-hybridized C atom in the middle of the functional group of the heterocumulene. Either the *neutral* heterocumulene or its protonated form **B** (Figure 7.1), which is formed from the neutral substrate and an acid in an equilibrium reaction, is attacked in this way. In each case the rate-determining step of the addition reaction is the formation of an intermediate, in which the attacked C atom is sp^2-hybridized and thus trigonal planar. When the nucleophile adds to the neutral heterocumulene, this intermediate is represented by the resonance forms **A.** If the addition involves the protonated heterocumulene **B,** the trigonal planar intermediate has the formula **C.** While intermediate **A** is then converted to a charge-free addition product through a proton *shift,* intermediate **C** provides the same product through a proton *loss.* The charge-free addition product may have an opportunity to stabilize further through tautomerism. Whether such tautomerism takes place and which isomer it favors depends on the heteroatoms in the primary addition product.

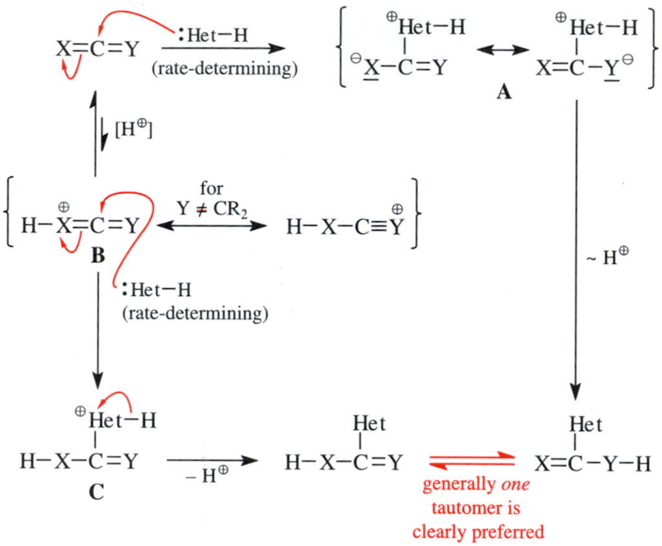

Fig. 7.1. Mechanism of the uncatalyzed addition (starting top left and proceeding clockwise) and the acid-catalyzed addition (starting top left and then proceeding counterclockwise) of heteroatom nucleophiles to heterocumulenes.

7.1.2 Examples of the Addition of Heteroatom Nucleophiles to Heterocumulenes

Additions to Ketenes

B

Ketenes are extremely powerful acylating agents for heteroatom nucleophiles. They react in each case according to the uncatalyzed mechanism of Figure 7.1. In this manner, ketenes can add

- H_2O to give carboxylic acids,
- ROH to give carboxylic esters,
- NH_3 to give primary carboxylic amides,
- RNH_2 to give secondary carboxylic amides, and
- R^1R^2NH to give tertiary carboxylic amides.

With very few exceptions ketenes cannot be isolated pure at room temperature (cf. Section 12.4). Consequently, they are prepared only *in situ* and in the presence of the heteroatom nucleophile. The Wolff rearrangement of α-diazoketones is often used for this purpose (Section 11.3.2). α-Diazoketones can be obtained, for example, by the reaction between a carboxylic chloride and diazomethane (Figure 7.2; see also Figure 11.25) or by treating the Na enolate of α-formylketones with tosyl azide (Figure 11.27).

Figure 7.2 shows a reaction sequence in which a carboxylic acid serves ultimately as the ketene precursor. For this purpose it is converted with $SOCl_2$ into the corresponding carboxylic chloride in a first reaction (mechanism without DMF catalysis, Figure 6.10; mechanism with DMF catalysis, Figure 6.11). As the second reac-

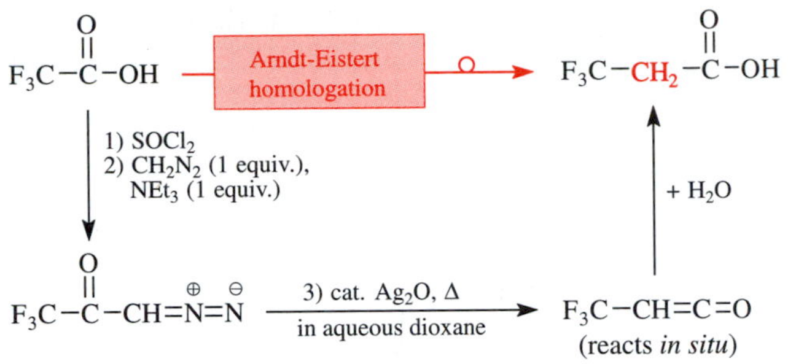

Fig. 7.2. Arndt–Eistert homologation of carboxylic acids—addition of H_2O to a ketene in part 2 of the third reaction of this three-step reaction sequence.

tion this carboxylic chloride is allowed to react with diazomethane according to the mechanism with which we are familiar from Figure 6.2. The reagent is acylated at the C atom whereby an α-diazoketone is produced. The desired ketene is finally obtained from this diazoketone through a Wolff rearrangment (mechanism: Figure 11.24) in part 1 of the third step of this sequence. In part 2 of the third step the heteroatom nucleophile H_2O, which must have been added before the rearrangement occurred adds to this ketene. In this way one obtains a carboxylic acid. The newly obtained carboxylic acid is one CH_2 group longer than the carboxylic acid one started with. Therefore, the three-step reaction sequence of Figure 7.2 is a method for preparing the next homolog of a carboxylic acid—the so-called Arndt–Eistert reaction.

Of course, one can also carry out Wolff rearrangements in the presence of nucleophiles other than H_2O. Ketenes are then produced in *their* presence, and the addition products of *these* nucleophiles are therefore isolated. Figure 11.26 shows how one can add an alcohol to a ketene in this way.

Additions to Symmetrical Heterocumulenes

B

The starting materials for the industrial synthesis of urea are the nucleophile NH_3 and the heterocumulene CO_2 (Figure 7.3). The reaction starts with the addition of NH_3 to

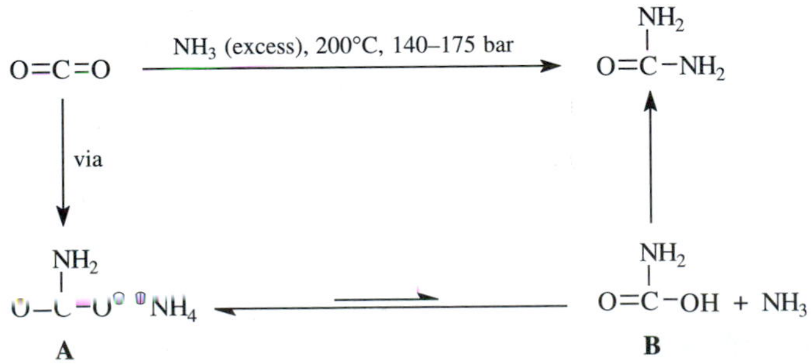

Fig. 7.3. Industrial urea synthesis.

one C=O double bond of the CO_2. It corresponds mechanistically to the uncatalyzed variant of the general addition mechanism of Figure 7.1. The zwitterionic primary addition product is deprotonated by a second equivalent of NH_3 to give the unsubstituted ammonium carbamate **A.** This is converted to a small extent into NH_3 and the unsubstituted carbamic acid **B** through a reversible acid/base reaction. Note that when one tries to isolate it unsubstituted carbamic acid decomposes into CO_2 and NH_3 by an exact reversal of the shown formation reaction. Under the conditions of the industrial urea synthesis an NH_3 molecule now undertakes an S_N reaction at the carboxyl carbon of the equilibrium fraction of this carbamic acid. H_2O is released as a leaving group and the resulting substitution product is urea.

Carbodiimides are diaza derivatives of CO_2. It is also possible to add heteroatom nucleophiles to them. The addition of carboxylic acids to dicyclohexyl carbodiimide was mentioned in the context of Figures 6.15 and 6.26, but there we looked at it only from the point of view of activating a carboxylic acid. This addition follows the proton-catalyzed mechanism of Figure 7.1.

Carbon disulfide is the dithio derivative of CO_2. It is only a weak electrophile. Actually, it is so unreactive that in many reactions it can be used as a solvent. Consequently, only good nucleophiles can add to the C=S double bond of carbon disulfide. For example, alkali metal alkoxides add to carbon disulfide forming alkali metal xanthates **A** (Figure 7.4). If one were to protonate this compound this would provide compound **B,** which is a derivative of free dithiocarbonic acid. It is unstable in the condensed phase in pure form, just as free carbonic acid and the unsubstituted carbamic acid (Formula **B** in Figure 7.3) are unstable. Compound **B** would therefore decompose spontaneously into ROH and CS_2. *Stable* derivatives of alkali metal xanthates **A** are their esters **C.** They are referred to as xanthic acid esters or xanthates. They are obtained by an alkylation (almost always by a methylation) of the alkali metal xanthates **A.** You have already learned about synthesis applications of xanthic acid esters in Figures 1.32, 4.13, and 4.14.

Fig. 7.4. Nucleophilic addition of alcoholates to carbon disulfide.

Additions to Isocyanic Acid and to Isocyanates

Wöhler's urea synthesis from the year 1828

$$O{=}C{=}N^{\ominus}\,NH_4^{\oplus} \xrightarrow{\Delta} \left(O{=}C{=}N{-}H + NH_3 \longrightarrow\right) O{=}C(NH_2){-}NH_2$$

and an analogous preparation of semicarbazide

$$O{=}C{=}N^{\ominus}\,K^{\oplus} + H_2N{-}\overset{\oplus}{N}H_3\,HSO_4^{\ominus} \longrightarrow \left(O{=}C{=}N{-}H + H_2N{-}NH_2\right) \longrightarrow O{=}C(NH{-}NH_2){-}NH_2$$

are addition reactions to the C=N double bond of isocyanic acid (H—N=C=O). These reactions belong to the very oldest organic chemical syntheses and take place according to the uncatalyzed addition mechanism of Figure 7.1.

At slightly elevated temperature, alcohols add to the C=N double bond of isocyanates according to the same mechanism. The addition products are called carbamic acid esters or urethanes:

$$O{=}C{=}NPh + HOR \longrightarrow O{=}C(OR){-}NHPh$$

In classical organic analysis liquid alcohols were in this way converted into the often easily crystallizable *N*-phenyl urethanes. These were characterized by their melting points and thus distinguished from each other. By comparing these melting points with the tabulated melting point of *N*-phenyl urethanes previously characterized, it was also possible to identify the alcohols used.

Side Note 7.1
Synthesis of Polyurethanes

Just as alcohols add to isocyanates, diols add to diisocyanates. In a polyaddition reaction polyurethanes are then produced, like the one shown here:

$$O{=}C{=}N{-}C_{10}H_6{-}N{=}C{=}O + HO{-}(CH_2)_4{-}OH \longrightarrow \left[{-}O{-}C(=O){-}NH{-}C_{10}H_6{-}NH{-}C(=O){-}O{-}(CH_2)_4{-}\right]_n$$

Polyurethanes are important synthetic macromolecules. They are manufactured, for example, in the form of foams. Such a foam is obtained when during the formation and solidification of the polyurethane a gas escapes from the reaction mixture and the material is thereby puffed up. An elegant possibility for generating such a gas uniformly distributed everywhere in the reaction medium is as follows: Besides the diol, one adds a small amount of H_2O to the diisocyanate. Then H_2O also adds to the C═N double bond of the diisocyanate. According to the uncatalyzed addition mechanism of Figure 7.1, this produces an *N*-arylated free carbamic acid Ar—NH—C(═O)—OH. However, such a compound decomposes just as easily as the *N*-unsubstituted carbamic acid **B** in Figure 7.3. The decomposition products are the strictly analogous ones: a primary aryl amine and *gaseous carbon dioxide.* The latter puffs up the polyurethane to give a foam.

The sensitivity of isocyanates toward hydrolysis to give primary amines and CO_2 achieved sad fame with the largest chemical accident in history. When in 1984 in Bhopal, India, water penetrated into or was conducted into a giant tank full of methyl isocyanate, the result was a humongous hydrolysis of the following type:

$$\mathrm{O{=}C{=}NMe} \xrightarrow{+\,H_2O} \left(\mathrm{O{=}C(OH){-}NHMe} \longrightarrow \right) \mathrm{O{=}C{=}O}\uparrow + \mathrm{H_2NMe}\uparrow$$

The released gases burst the tank, and about 40 tons of methylisocyanate was released. It poisoned thousands of people mortally and left tens of thousands with chronic ailments.

Both primary and secondary amines add to the C═N double bond of isocyanates (Figure 7.5) according to the uncatalyzed mechanism of Figure 7.1. These reactions produce ureas. These may contain the same or different substituents on the two N atoms. How an asymmetrical urea can be prepared in this way is shown in Figure 7.5 (compound **A**).

A

Urea **A** is the starting material for preparing the carbodiimide **C,** which activates carboxylic acids according to the same mechanism and for the same reason as DCC, with which you are already familiar (Figures 6.15 and 6.26). If the carbodiimide **C** from Figure 7.5 were not so much more expensive than DCC, everybody would use the former instead of the latter for carboxylic acid activation. There is a practical reason for this. When a heteroatom nucleophile is acylated with the DCC adduct of a carboxylic acid, besides the desired carboxylic acid derivative one obtains dicyclohexyl urea (formula **B** in Figure 7.5). This (stoichiometric) by-product must be separated from the acylation product, which is relatively laborious when realized by chromatography or by crystallization. When a carboxylic acid has been activated with the carbodiimide **C** and the subsequent acylation of a heteroatom nucleophile has been effected, one also obtains a urea as a stoichiometric by-product. It has the structure **D** and is therefore

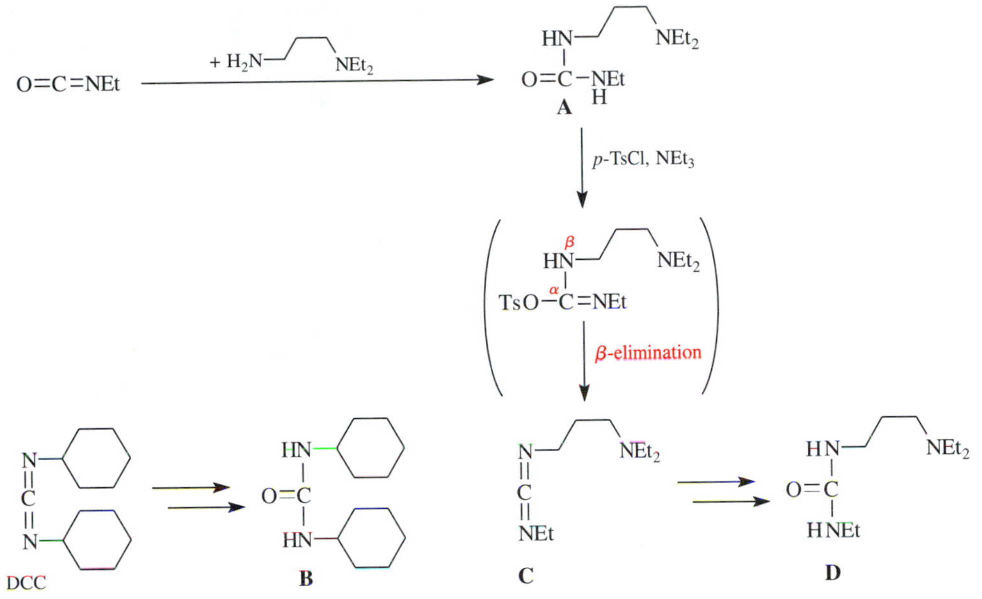

Fig. 7.5. Amine addition to an isocyanate as part of the synthesis of an asymmetrical carbodiimide.

also an amine. As such, this urea **D** can be separated from the acylation product much more readily than dicyclohexyl urea (**B**)—by an extraction with aqueous hydrochloric acid.

Additions to Isothiocyanates

A

Just as amines add to the C=N double bond of isocyanates giving ureas (Figure 7.5), they add to the C=N double bond of isothiocyanates, producing thioureas. An important example in this regard is the reaction **A** + Ph−N=C=S → **B** of the three-step Edman degradation of oligopeptides (Figure 7.6).

Side Note 7.2
Sequence Determination of Oligopeptides with the Edman Degradation

The Edman degradation is used to determine the amino acid sequence of oligopeptides starting from the N terminus. The oligopeptide to be sequenced is linked to a solid support in the form of a derivative **A** (formula in Figure 7.6). The advantage of working on such a solid support is the following. After the first cycle of the Edman degradation, one will have obtained a mixture of oligopeptide **D,** which is deprived of the originally present N-terminal amino acid, and of the heterocycle **G,** into which the amino acid removed has been incorporated. Because oligopeptide **D** is polymer-bound it is separable especially simply from heterocycle **G,** namely by filtration. This makes the isolation of **D** particularly easy and thus allows it to be subjected *quickly* to the next cycle of the Edman degradation.

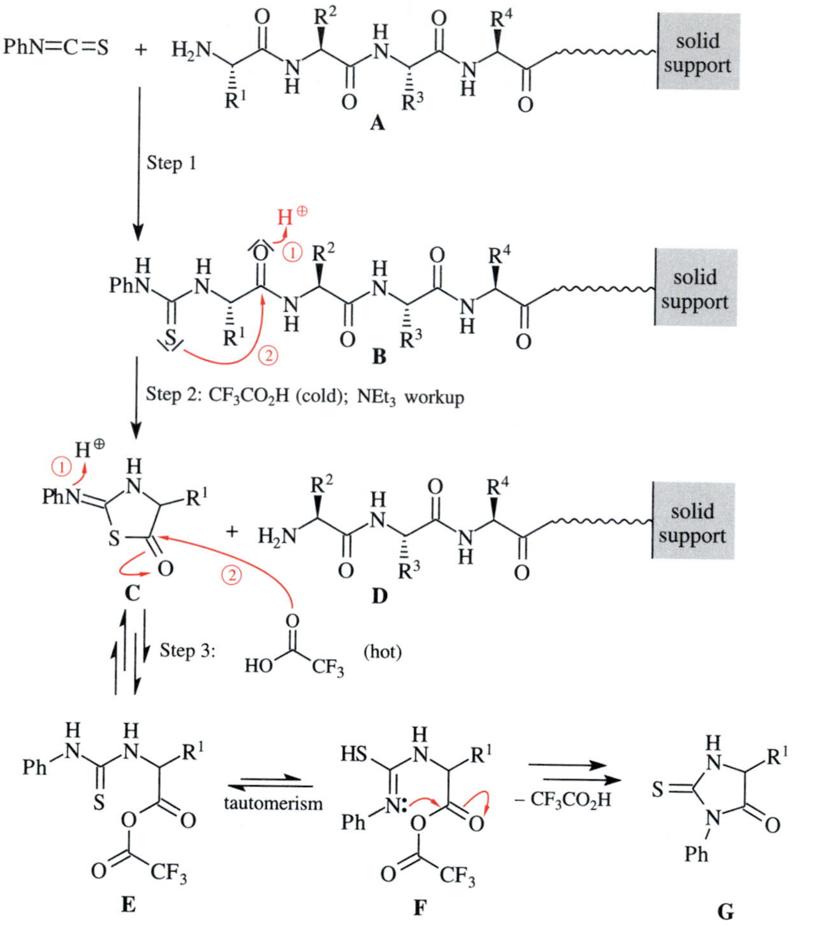

Fig. 7.6. Edman degradation of polymer-bound oligopeptides. One pass through the three-step reaction sequence is shown.

Step 1 of the Edman degradation is the addition of the NH_2 group of the N-terminal amino acid to the C=N double bond of phenyl isothiocyanate. Step 2 (**B** → **C**) is an intramolecular S_N reaction of an S nucleophile on the carboxyl carbon of a protonated amide. It follows the substitution mechanism shown in Figure 6.5. The substitution product **C** is a heterocyclic derivative of the N-terminal amino acid. The simultaneously formed second reaction product, the oligopeptide **D,** which has been shortened by one amino acid, is ejected as the leaving group.

Next, the new oligopeptide **D** is degraded according to Edman. It then releases *its* N-terminal amino acid—that is, the second amino acid of the original oligopeptide **A** counting from its N-terminus—in the form of an analogous heterocycle, and so on.

The analytical objective is to determine the side chain R^1—there are 20 possible side chains in proteinogenic amino acids—in each of the heterocycles of type **C,** one of them being released per pass through the Edman degradation. However, this determination is not easily possible, as it turns out. Therefore, the Edman degradation contains a third step, in which these difficult to analyze heterocycles **C** are isomerized. This takes place through the sequence acylation (formation of **E**), tautomerism (formation of **F**), and acylation (formation of **G**), shown in Figure 7.6 at the bottom. The reaction **C** → **E** is an S_N reaction on the carboxyl carbon, which follows the mechanism of Figure 6.5. The reaction **F** → **G** is also an S_N reaction on the carboxyl carbon, but it follows the mechanism of Figure 6.2. With the heterocycle **G**—a so-called thiohydantoin—one has obtained a compound in which the side chain R^1 can be identified very easily by a chromatographic comparison with authentic reference compounds.

7.2 Additions of Heteroatom Nucleophiles to Carbonyl Compounds

B

Only three heteroatom nucleophiles add to a significant extent to carbonyl compounds without being followed by secondary reactions such as S_N1 reactions (Section 7.3) or E1 reactions (Section 7.4): H_2O, alcohols, and, should the substitution pattern be suitable, the carbonyl compound itself.

H_2O or alcohols as nucleophiles give low-molecular-weight compounds when they add to the C=O double bond of carbonyl compounds. These addition products are called aldehyde or ketone hydrates (Section 7.2.1) and hemiacetals or hemiketals (Section 7.2.2), respectively, depending on whether they result fom the addition to an aldehyde or to a ketone. Today, one no longer distinguishes systematically between hemiacetals and hemiketals, but the expression "hemiacetal" is frequently used to cover both.

Aldehydes can add to themselves only occasionally (Section 7.2.3). If so, cyclic oligomers or acyclic polymers such as paraformaldehyde are formed.

7.2.1 On the Equilibrium Position of Addition Reactions of Heteroatom Nucleophiles to Carbonyl Compounds

B

The additions of H_2O or alcohols to the C=O double bond of carbonyl compounds as well as the oligomerizations or polymerizations of aldehydes are *reversible* reactions. Therefore, the extent of product formation is subject to thermodynamic control. The equilibrium constant of the formation of the respective addition product is influenced by steric and electronic effects.

For a given nucleophile the equilibrium lies farther on the product side the smaller the substituents R^1 and R^2 of the carbonyl compound are (Figure 7.7). Large substituents R^1 and R^2 prevent the formation of addition products. This is because they come closer to each other in the addition product, where the bonds to R^1 and R^2 enclose a tetrahedral angle, than in the carbonyl compound, where these substituents are separated

Fig. 7.7. Substituent effects on the equilibrium position of the addition reactions of H_2O (Het = OH), alcohols (Het = O-Alkyl) and the respective carbonyl compounds [Het = $O(—CR^1R^2—O)_n—H$] to aldehydes and ketones. EWG, electron-withdrawing group.

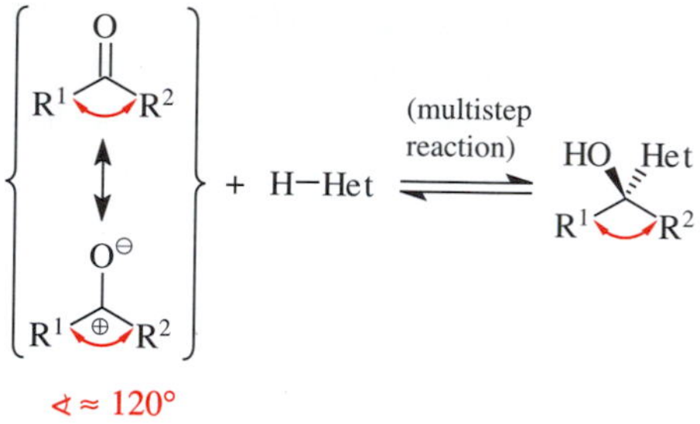

R^1	R^2	Substituent effect in starting material	Steric hindrance of the product	Consequence for equilibrium position
Alkyl	Alkyl	Stabilizes more	Significant	On starting material side
Alkyl	H	Stabilizes somewhat	Present but small	
H	H	= Reference	= Reference product	
EWG	H	Destabilizes clearly	Present but small	
EWG	EWG	Destabilizes much	Present but small	On product side

by a bond angle of about 120°. Formaldehyde is the sterically least hindered carbonyl compound. Thus, in H_2O this aldehyde is present completely as dihydroxymethane, and anhydrous formaldehyde is present completely as polymer. In contrast, acetone is already so sterically hindered that it does not hydrate, oligomerize, or polymerize at all.

The equilibrium position of addition reactions to carbonyl compounds is also influenced by electronic substituent effects, as was also shown in Figure 7.7 but not yet discussed:

Influence of Substituents on the Equilibrium Position of Addition Reactions to the Carbonyl Group

- Substituents with a +I effect (i.e., alkyl groups) stabilize the C═O double bond of aldehydes and ketones. They increase the importance of the zwitterionic resonance form by which carbonyl compounds are in part described. The driving force for the formation of addition products from carbonyl compounds therefore decreases in the order H—CH(═O) > R—CH(═O) > $R^1R^2C(═O)$.
- Alkenyl and aryl substituents stabilize the C═O double bond of carbonyl compounds even more than alkyl substituents. This is due to their +M effect, which allows one to formulate additional zwitterionic resonance forms for carbonyl compounds of this type. Thus, no hydrates, hemiacetals, oligomers, or polymers can be derived from unsaturated or aromatic aldehydes.
- Electron-withdrawing substituents at the carbonyl carbon destabilize the zwitterionic resonance form of aldehydes and ketones. Thus, they deprive these compounds of the resonance stabilization, which the alkyl substituents usually present would give them. Therefore, addition reactions to acceptor-substituted C═O double bonds have an increased driving force.

Table 7.1. Position of the Hydration Equilibrium in the System Carbonyl Compound/H_2O

Carbonyl compound	 acetone	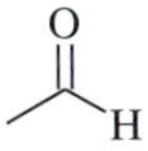 acetaldehyde	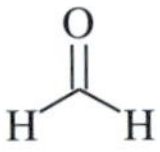 formaldehyde	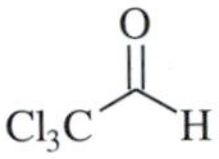 Cl_3C–CHO	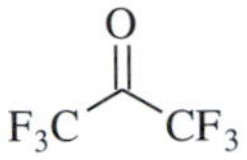 F_3C–CO–CF_3
% Hydrate at equilibrium	≪ 0.1	58	100[1]	100[2]	100[2]

[1] Not isolable. [2] Isolable.

The effects illustrated by Figure 7.7 are corroborated by the data in Table 7.1 for the hydration equilibria of differently substituted carbonyl compounds.

7.2.2 Hemiacetal Formation

Scope of the Reaction

In the presence of a base or an acid as a catalyst, alcohols add to aldehydes or ketones with the formation of hemiacetals. You should already be familiar with the corresponding formation mechanism and its intermediates from your introductory class. Therefore it is sufficient if we briefly review it graphically by means of Figure 7.8.

B

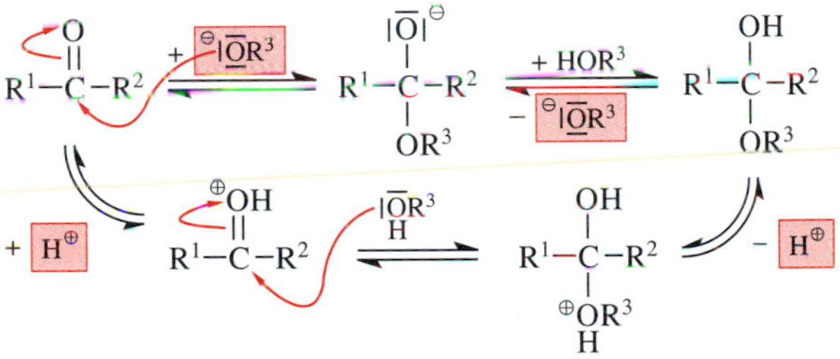

Fig. 7.8. Base-catalyzed (top) and acid-catalyzed (bottom) hemiacetal formation from carbonyl compounds and alcohols.

What you have learned in Section 7.2.1 about electronic substituent effects allows the understanding by analogy why intermolecular hemiacetal formations from the electron-deficient carbonyl compounds methyl glyoxalate **(A)** and ninhydrin **(B)** take place quantitatively:

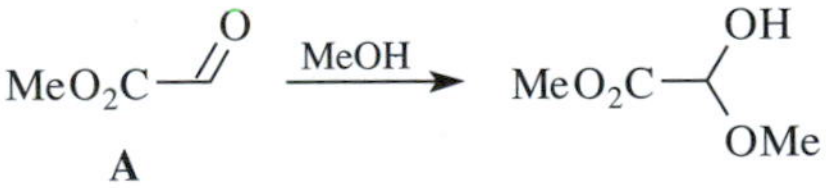

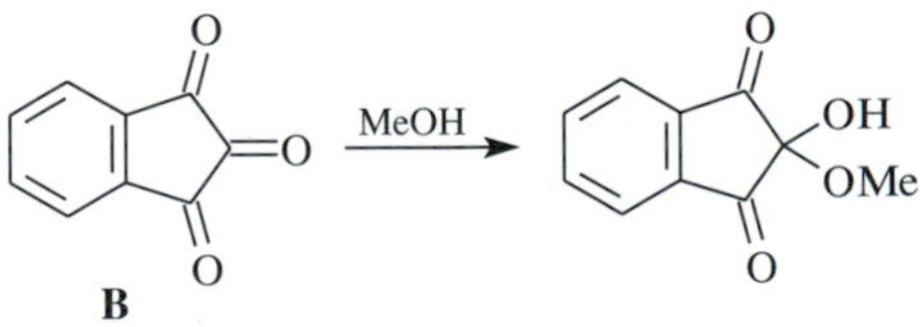

α-Hydroxylated aldehydes, too, are α-acceptor substituted carbonyl compounds and as such are capable of forming relatively stable hemiacetals. However, since these aldehydes contain an alcoholic OH group, no additional alcohol is required in order for hemiacetals to be produced from *them*. Rather, these aldehydes form a hemiacetal through the addition of the OH group of one of their own (i.e., the OH group *of a second molecule*). Figure 7.9 shows a reaction of this type using the example of the conversion of the α-hydroxyaldehyde **C** into the hemiacetal **D.**

Being a hemiacetal, **D** immediately reacts further, as is typical for this type of compound (Figure 7.10). The result is an addition of the δ-OH group (i.e., of the still available OH group of what was originally the first aldehyde) to the C=O double bond (i.e., to the still available C=O group of what was originally the second aldehyde). Thus, the bis(hemiacetal) **B** is produced. This second hemiacetal formation is more favorable entropically than the first one: The number of molecules that can move about independently remains constant in the second hemiacetal formation (**D** $\rightarrow$ **B**), whereas this number is reduced by a factor of one-half in the first hemiacetal formation (**C** $\rightarrow$ **D**). The second reaction of Figure 7.9 therefore "pulls" the entire reaction to the product side.

The quantitative formation of the bis(hemiacetal) **B** from the glyceraldehyde derivative **C,** as just discussed, led to the recent proposal to use the glyceraldehyde derivative **C** as a storage form of the glyceraldehyde derivative **A** (Figure 7.9; possible preparation, Figure 14.18). The latter compound cannot be stored because it is prone to racemization because its C=O group allows for ready enolate and enol formation (both of which destroy the stereochemical integrity of the stereocenter at C-α). Because compound **C** dimerizes quantitatively to give **B,** in contrast to **A, C** is not at all present as a carbonyl compound and is therefore not threatened by racemization through enolate or enol formation.

B

As can be seen from Figure 7.7, carbonyl compounds without electron-withdrawing α-substituents do not react *intermolecularly* with alcohols to form hemiacetals to any significant extent. However, while for such carbonyl compounds there is too little driving force for hemiacetalization to occur, the reaction is not drastically disfavored. This

at risk of racemization

A

PhCH$_2$O HO O OH O OCH$_2$Ph

B

(4 diastereomers)

O OH OCH$_2$Ph

C

not at risk of racemization because of . . .

× 2

O H O OCH$_2$Ph α β γ δ O OH OCH$_2$Ph

D

(2 diastereomers)

Fig. 7.9. Formation of a bis(hemiacetal) from an α-hydroxyaldehyde by dimerization.

explains why this type of compound undergoes almost complete hemiacetal formation provided it takes place intramolecularly and leads to a nearly strain-free five- or six-membered cyclic hemiacetal—a so-called **lactol** (Figure 7.10). What makes the difference is that only in the intramolecular hemiacetal formation is no translational entropy lost (because the number of molecules moving about independently of each other does not decrease).

R_x OH CH=O ⇌ [$H^{\oplus}$] or [$OH^{\ominus}$] ⇌ R_x OH CH=O

R_x 5 O OH R_x 6 O OH

Fig. 7.10. Hemiacetal formation from γ- or δ-hydroxyaldehydes.

Five- or six-membered cyclic hemi*ketals*, which are also referred to as lactols, can form from γ- or δ-hydroxy*ketones* in a strictly analogous manner. However, hemiketals of this type are not necessarily more stable than the acyclic hydroxyketone isomers because ketones are less thermodynamically suitable to take up nucleophiles than are aldehydes (Section 7.2.1).

In order to understand certain reactions of lactols with nucleophiles, you must now familiarize yourself with the **principle of microscopic reversibility:**

Principle of Microscopic Reversibility

When a molecule **A** is converted to a molecule **B** through a certain mechanism—and it does not matter whether this mechanism comprises one or more elementary reactions—a possible conversion of product **B** back into the starting material **A** takes place through the very same mechanism.

Lactols form, just like acyclic hemiacetals or hemiketals, according to one of the two mechanisms of Figure 7.8. From the principle of microscopic reversibility the following is therefore concluded: The ring opening of a lactol to give a hydroxycarbonyl compound proceeds exactly via the pathways shown in Figure 7.8. Hence, both in acidic or basic solutions, a lactol gives rise to an equilibrium amount of the hydroxycarbonyl compound. Through the latter, a lactol can often react with nucleophiles.

Stereochemistry

A

δ-Hydroxyvaleraldehyde is achiral. When it cyclizes to give the six-membered ring lactol, which, of course, takes place under thermodynamic control, the lactol is formed as a 50:50 mixture of the two enantiomers (Figure 7.11). Each of these enantiomers in turn is present as a 77:23 mixture of two chair conformers. The conformer with the axially oriented OH group is in each case more stable than the conformer with the equatorially oriented OH group. In the carba analog (i.e., in cyclohexanol) one finds almost ex-

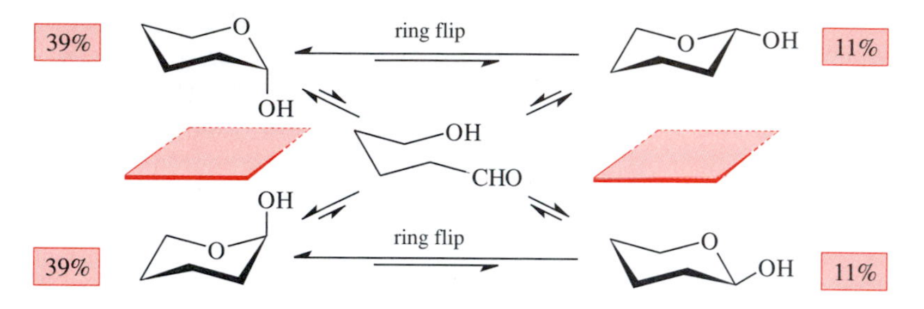

Fig. 7.11. Stereochemistry of the formation of δ-valerolactol (neat, at 38°C; ≤5% of the free hydroxyaldehyde is present). The top formulas represent the 77:23 mixture of the two chair conformers of one lactol enantiomer, whereas the bottom formulas represent the 77:23 mixture of the two chair conformers of the other lactol enantiomer.

actly the opposite conformational preference: At 25°C the axial and the equatorial alcohols are present in a ratio of 27:73. This corresponds to a $\Delta G°$ value of -0.60 kcal/mol for the conversion cyclohexanol$_{\text{OH,axial}} \rightarrow$ cyclohexanol$_{\text{OH,equatorial}}$. It is well-known that the equatorial conformer of a mono-substituted cyclohexane is more stable than the axial conformer due to a steric effect. This steric effect should be even somewhat greater in the axial conformer of the lactol of δ-hydroxyvaleraldehyde—the IUPAC name of which is 2-hydroxytetrahydropyran. It should be remembered that the $\Delta G°$ value for the conversion cyclohexane$_{\text{Me,axial}} \rightarrow$ cyclohexane$_{\text{Me,equatorial}}$ is -1.74 kcal/mol. On the other hand, $\Delta G°$ for the conversion tetrahydropyran$_{\text{2-Me,axial}} \rightarrow$ tetrahydropyran$_{\text{2-Me,equatorial}}$ is -2.86 kcal/mol. This means that in 2-substituted tetrahydropyrans, there is a greater preference for the substituent to be equatorial than in cyclohexane. This should hold for a hydroxy group as well as for the methyl substituent, possibly to the same extent. This would mean that the $\Delta G°$ value for the tetrahydropyran$_{\text{OH,axial}} \rightarrow$ tetrahydropyran$_{\text{OH,eq}}$ conversion is approximately $-0.6 \times (2.86/1.74) = -1.0$ kcal/mol. Since the more stable lactol is nonetheless the 2-hydroxytetrahydropyran with the axial OH group, an opposing effect must determine the position of the conformational equilibrium. This is the anomeric effect, which is a stereoelectronic effect (Figure 6.8). In a substructure :O—C_{sp^3}—O it contributes a stabilization energy of 1–2.5 kcal/mol. The higher stability of the axially vs equatorially hydroxylated δ-valerolactol results, in the MO picture, from an overlap between the axially oriented free electron pair on the ring O atom and the $\sigma^*_{\text{C—O}}$ orbital of the exocyclic C—OH bond.

The installment of another four hydroxy groups in the δ-hydroxyvaleraldehyde from Figure 7.11 generates the acyclic form of D-glucose (Fig. 7.12). Acyclic D-glucose is a *chiral* δ-hydroxyaldehyde. When it cyclizes to give the six-membered ring hemiacetal, a new stereocenter is formed so that two different hemiacetals can be produced. They are diastereomers and therefore not equally stable. Being formed under thermodynamic control, these hemiacetals therefore occur in a ratio different from 50:50, namely as a 63:37 mixture.

For each of these diastereomorphic hemiacetals there are in principle two different chair conformers, but in each case only one chair conformer is found. The other one would contain too many energetically unfavorable axial substituents. In an aqueous solution, the hemiacetal with an equatorial OH group at the newly produced stereo-

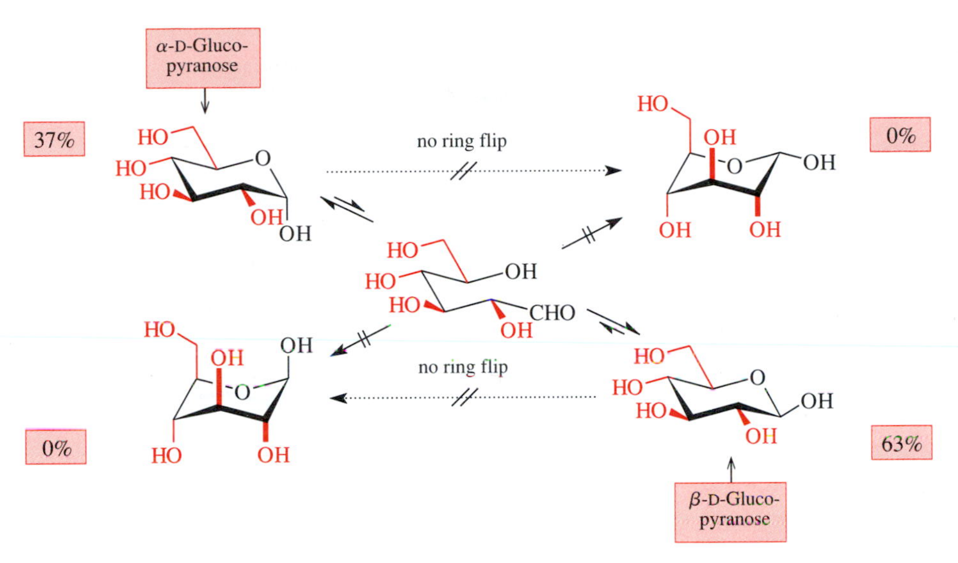

Fig. 7.12. Stereochemistry of the hemiacetal formation from D-glucose in an aqueous solution at 20°C (in the formulas, the carbon skeleton of δ-valerolactol is black—which emphasizes the relationship to the reaction from Figure 7.11—whereas the extra substituents as well as the bonds leading to them are red).

center—the so called **anomeric stereocenter**—predominates to the extent of 63%. It is the so-called β-D-glucopyranose. An axial OH group at the anomeric carbon is contained in the 37% fraction of the other hemiacetal, the α-D-glucopyranose.

The β-D-glucopyranose profits—as did the equatorially hydroxylated tetrahydropyran from Figure 7.11—exclusively from a steric effect but not from the anomeric effect. In contrast, a stabilizing anomeric effect and a destabilizing steric effect occur in the α-D-pyranose—as was the case in the axially hydroxylated tetrahydropyran of Figure 7.11. This means that although the stability order of the underlying core structure of Figure 7.11 suggests the opposite, the α-D-glucopyranose is nevertheless the minor and the β-D-glucopyranose the major hemiacetal diastereomer of Figure 7.12.

What is the reason for this apparent discrepancy? It is a solvent effect. In *aqueous* solution, the OH group at the anomeric C atom of the glucose becomes so voluminous due to hydration that it strives for the position in which the steric interactions are as weak as possible. Thus, it moves into the equatorial position—with a $\Delta G°$ value of approximately -1.6 kcal/mol—to avoid a *gauche* interaction with the six-membered ring skeleton. (Remember that axially oriented substituents on the chair conformer of cyclohexane are subject to two *gauche* interactions with the two next-to-nearest C_{ring}—C_{ring} bonds. They therefore have a higher energy than equatorial substituents, which are not exposed to any *gauche* interactions.) This steric benefit more than compensates the absence of an anomeric stabilization which only an axially oriented anomeric OH group would experience.

As already mentioned, δ-hydroxyketones may be present in the open-chain form or as an isomeric cyclic hemiketal, depending on their substitution pattern. For example, the polyhydroxylated δ-hydroxyketone, which D-fructose can be considered to be, is present exclusively in the form of hemiketals (Figure 7.13). Responsible for this equilibrium position is the fact that the carbonyl group involved contains an electron-withdrawing group in both α positions (cf. our discussion of the substituent effects on the equilibrium position of the hydration of carbonyl compounds; Section 7.2.1).

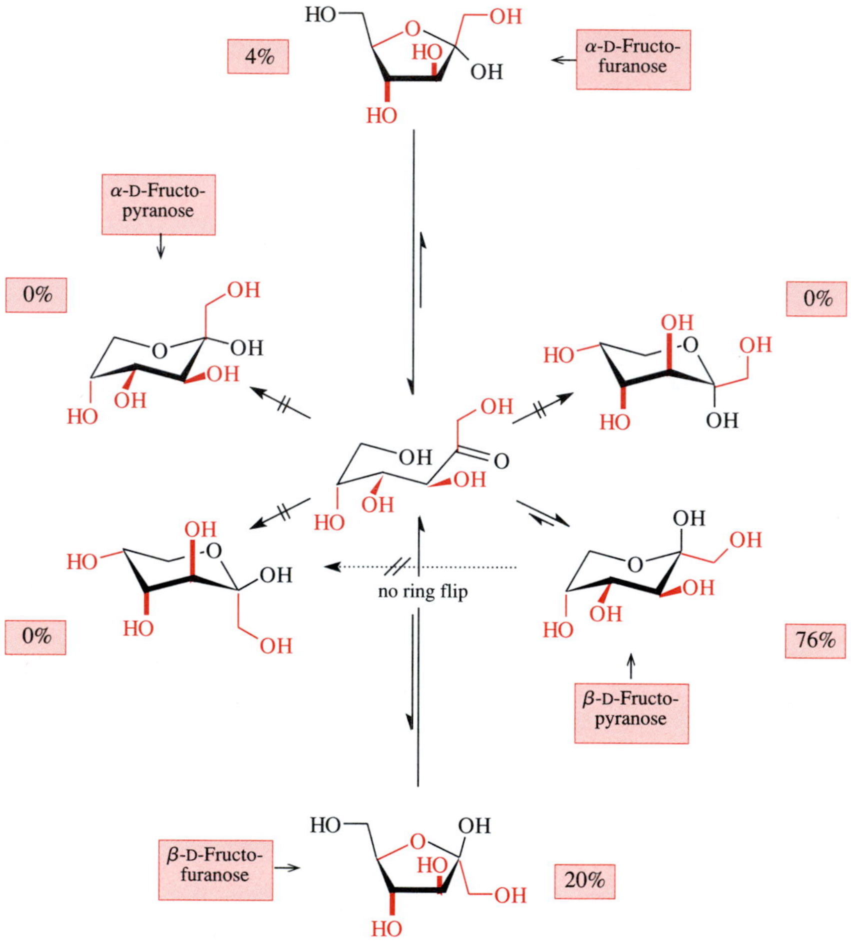

Fig. 7.13. Stereochemistry of the hemiketal formation from D-fructose in aqueous solution at 25°C (in the formulas for the six-membered ring hemiketal the carbon skeleton of δ-valerolactol is black, which emphasizes the relationship to the reaction from Figure 7.11, whereas the extra substituents as well as the bonds leading to them are red).

α-D-Fructofuranose 4%

α-D-Fructopyranose 0%

0%

no ring flip

0%

β-D-Fructopyranose 76%

β-D-Fructofuranose 20%

In contrast to the case of D-glucose (Figure 7.12), in D-fructose there are two five-membered hemiketals in addition to one six-membered hemiketal (Figure 7.13). The six-membered ring hemiketal of D-fructose occurs as a pure β-diastereomer. The substituents on the anomeric carbon of this hemiketal are arranged in such a way that the OH group is oriented axially and the CH_2OH group equatorially. The OH group thus profits from the anomeric effect and the larger CH_2OH group from the minimization of steric hindrance.

7.2.3 Oligomerization/Polymerization of Carbonyl Compounds

B

Some aldehydes oligomerize in the presence of acids. The polymerizations of formaldehyde (formation of **E**; see Figure 7.14) as well as that of a few α-acceptor-substituted aldehydes are even quite unavoidable reactions. The oligomerization of other unhin-

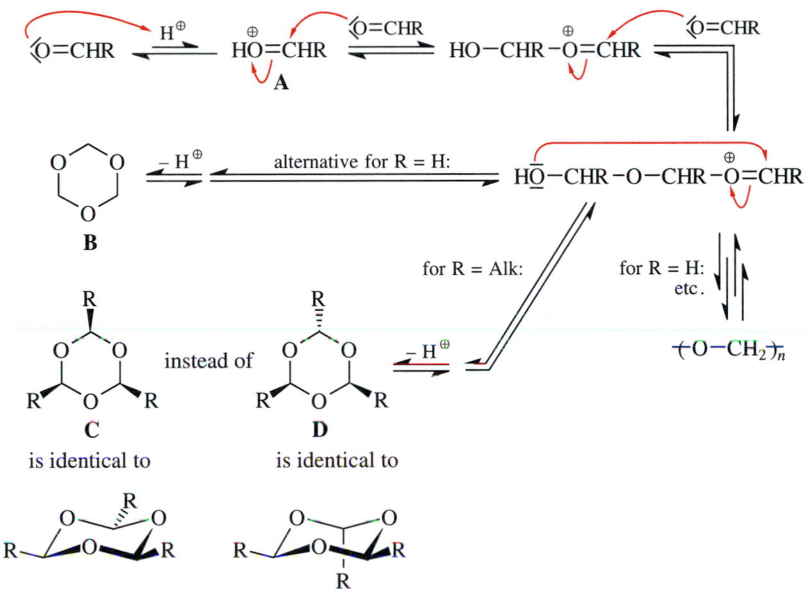

Fig. 7.14. Mechanism of the proton-catalyzed oligomerization or polymerization of aldehydes (in the absence of H_2O).

dered aliphatic aldehydes is also possible, but it takes place much more slowly. For example, in the presence of protons acetaldehyde and isobutyraldehyde trimerize easily or without much difficulty, respectively, to give the substituted trioxanes **C** (R = Me or *i*Pr, respectively) of Figure 7.14.

The polymer **F** of formaldehyde is called polyoxymethylene or paraformaldehyde. It contains two OH groups, one at each end of the molecule, but these are omitted for clarity in the structural formula shown in Figure 7.14. This shows that one molecule of water, which is present everywhere even in the most careful work, has been incorporated. If these terminal OH groups are functionalized, paraformaldehyde can be used as plastic. When paraformaldehyde is heated with acid, it is transformed (entropy gain) into the cyclic trimer **B** from Figure 7.14, which is called 1,3,5-trioxane.

The mechanistic analysis shows that the oligomerization or polymerization of aldehydes can be considered as an addition reaction of a heteroatom nucleophile to the C=O double bond of the (activated) aldehyde (Figure 7.14). The carbonyl oxygen of the (unprotonated) aldehyde functions as the nucleophilic center. The carboxonium ion **A** formed in an equilibrium reaction between the aldehyde and the acidic catalyst acts as the first electrophile.

As you can see from Figure 7.14, all steps of the oligomerization or polymerization of aldehydes in the presence of protons are reversible. The trimerizations of the mentioned aldehydes acetaldehyde and isobutyraldehyde are therefore thermodynamically controlled. This is the reason why they take place stereoselectively and the trimers **C** (R = Me or *i*Pr) are produced diastereoselectively with a *cis* configuration. In contrast to their *cis,trans,trans*-isomers **D,** the trimers **C** can occur as all equatorially substituted chair conformers.

7.3 Addition of Heteroatom Nucleophiles to Carbonyl Compounds in Combination with Subsequent S_N1 Reactions: Acetalizations

7.3.1 Mechanism

B

The addition of one equivalent of a heteroatom nucleophile to carbonyl compounds was described in Section 7.2. However, if two or more equivalents of the heteroatom nucleophile are present, in many cases the first obtained addition products (formula **A** in Figure 7.15) react further with the second equivalent of the nucleophile. This occurs in such a way that the OH group of **A** is replaced by that nucleophile. Thus, conversions of the type $R^1R^2C{=}O + 2HNu \rightarrow R^1R^2C(Nu)_2 + H_2O$ take place.

The product obtained (formula **B** in Figure 7.15) is, in the broadest sense of the word, an acetal. This is a collective term for compounds of the structure $R^1R^2C(Nu)_2$, where Nu can be OR^3 (so-called *O,O*-acetal), SR^3 (so-called *S,S*-acetal), or NR^3R^4 (so-called *N,N*-acetal). *O,O*- and *S,S*-acetals are isolable compounds. *N,N*-acetals can in general only be isolated when they are cyclic or substituted on both N atoms by an acceptor. There are also mixed acetals. By this one means analogous compounds

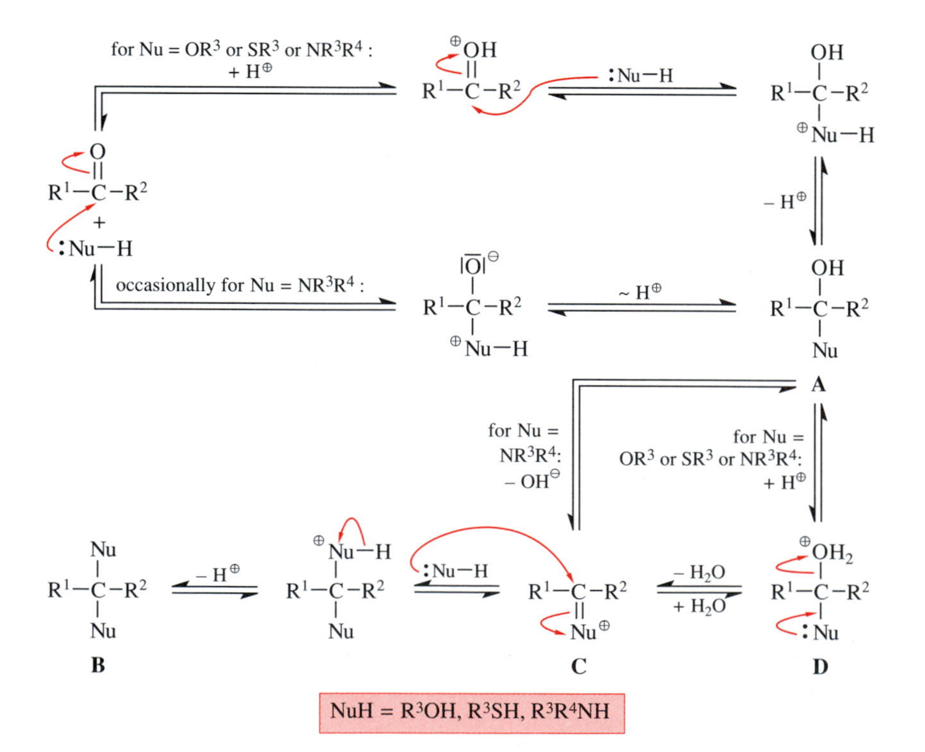

Fig. 7.15. Acetalization of carbonyl compounds.

$R^1R^2C(Nu)_2$ but with two different substituents Nu, such as $R^1R^2(OR^3)(OR^4)$ (so-called mixed *O,O*-acetal) or $R^1R^2(OR)(SR)$ (so-called *O,S*-acetal).

The mechanism for the acetalization is detailed in Figure 7.15. *O,O*-, *S,S*-, and *N,N*-acetals form via the addition products **A** of the respective nucleophiles to the carbonyl compounds. Under these conditions, the additions occur reversibly. For Nu = R^3S they can, in principle, proceed according to the same mechanism which Figure 7.8 showed only for Nu = R^3O. In order for these addition products **A** to be able to react further through an S_N1 reaction, they should be able to eject H_2O as a good leaving group rather than be forced to eject OH^- as a poor leaving group. This requires that **A** be protonated in the next step, which means that one should work in the presence of acid which in turn implies that **A** should also be generated in the presence of an acid. Interestingly, even N nucleophiles and carbonyl compounds react to give such addition products **A** in acidic solutions, including very acidic solutions. This is remarkable because under these conditions almost all N atoms are protonated. As a consequence, most of the N "nucleophile" added to the reaction mixture is no longer a nucleophile. The fact that nonetheless the overall reactivity is sufficiently high is due to two effects. First, under (strongly) acidic conditions there is a higher concentration of the carboxomium ions, which are particularly good electrophiles. Second, under (strongly) acidic conditions the addition products **A** (Nu = NR^3R^4) are protonated very efficiently to provide their conjugated acids **D** so that the iminium ions **C** ($Nu^+ = NR^3R^{4+}$) are formed very rapidly *en route* to the final products. It should be added that, in contrast to alcohols and thiols, ammonia and amines react with carbonyl compounds *also* under neutral or weakly basic conditions; this is possible because of their superior nucleophilicity.

The addition products **A** now undergo S_N1 reactions (Figure 7.15). A reversible protonation of the OH group of **A** leads to oxonium ions **D.** The leaving group H_2O is preformed in them. It is ejected from **D** without the nucleophile getting involved. Thereby the intermediate oxonium ions **C** (Nu = OR^3), thioxonium ions **C** (Nu = SR^3), or iminium ions **C** (Nu = NR^3R^4) are formed. These combine with the second equivalent of the nucleophile to form an *O,O*-acetal **B** (Nu = OR^3), an *S,S*-acetal **B** (Nu = SR^3), or, if possible, an *N,N*-acetal (Nu = NR^3R^4).

When the hemiaminals **A** (Figure 7.15, Nu = NR^3R^4) are formed in a neutral or weakly basic solution, they also have the possibility to react further by an S_N1 reaction albeit in a different manner than just described for the corresponding hemiacetals (Nu = OR^3) and hemithioacetals (Nu = SR^3): The OH group of hemiaminals **A** is then ejected without prior protonation (i.e., simply as an OH^- ion). This is possible because an especially well-stabilized carbocation is produced at the same time, that is, the iminium ion **C** (Nu = NR^3R^4). It finally reacts with the second equivalent of the N nucleophile. The excess proton is eliminated whereupon the *N,N*-acetal **B** (Nu = NR^3R^4) has been formed.

When one follows the reaction arrows in Figure 7.15 from the bottom upward, the following important information can be noted: In an acidic water-containing solution *O,O*-acetals are hydrolyzed to give carbonyl compounds and alcohols. Such a hydrolysis consists of seven elementary reactions. First, the hemiacetal **A** (Nu = OR^3) and one equivalent of alcohol are produced from the *O,O*-acetal and water in an exact reversal of the latter's formation reaction, i.e., through a proton-catalyzed S_N1 substitution (in four steps). What follows is a three-step decomposition of this hemiacetal to the carbonyl compound and a second equivalent of the alcohol.

Acetals are thus not only produced from carbonyl compounds in acidic solutions but also can be hydrolyzed in this medium.

7.3.2 Formation of *O*,*O*-Acetals

B

The formation of *O*,*O*-acetals from carbonyl compounds and alcohols takes place according to the mechanism of Figure 7.15. As acidic catalysts one uses a mineral acid (HCl or H_2SO_4) or a sulfonic acid (*p*-toluenesulfonic acid, camphorsulfonic acid, or a strongly acidic cation exchange resin):

O

+ 2 MeOH $\xrightleftharpoons{[H^{\oplus}]}$ MeO OMe + H_2O

The shown formation of the dimethylketal of cyclohexanone is an equilibrium reaction according to Figure 7.15. Complete conversion to the ketal is most simply achieved by working in methanol as a solvent and thus shifting the equilibrium to the ketal side.

A

One can also acetalize carbonyl compounds completely without using the alcohol in excess. This is the case when one prepares dimethyl or diethyl acetals from carbonyl compounds with the help of the orthoformic acid esters trimethyl orthoformate $HC(OCH_3)_3$ or triethyl orthoformate $HC(OC_2H_5)_3$, respectively. In order to understand these reactions, one must first clearly understand the mechanism for the hydrolysis of an orthoester to a normal ester (Figure 7.16). It corresponds step by step to the mechanism of hydrolysis of *O*,*O*-acetals, which was detailed in Figure 7.15. The fact that the individual steps are really strictly analogous becomes very clear when one takes successive looks at

- the indicated acetal substructure in the substrate,
- the hemiacetal substructure in the key intermediate, and
- the C=O double bond in the hydrolysis product (i.e., in the normal ester).

The driving force for a hydrolysis orthoester → normal ester + alcohol is greater than the driving force for a hydrolysis *O*,*O*-acetal → carbonyl compound + alcohol. This is because the newly formed C=O double bond receives approximately 14 kcal/mol resonance stabilization in the ester (Table 6.1)—that is, starting from an orthoester—but not in the carbonyl compound.

This difference in driving force is used in the already briefly mentioned dimethyl- or diethylacetalizations of carbonyl compounds with orthoformic acid esters: these acetalizations are simply *linked* with the hydrolysis of an orthoester into a normal ester.

According to variant 1 of this process, a 1:1 mixture of the carbonyl compound and

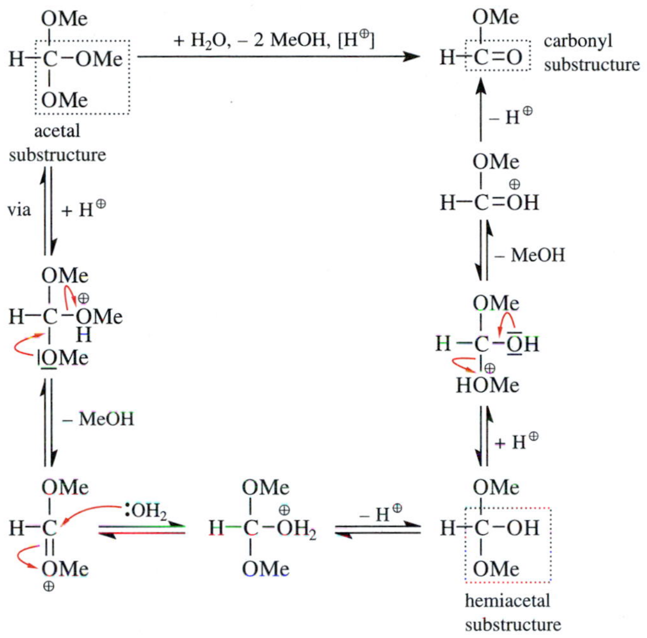

Fig. 7.16. Mechanism of the acid-catalyzed hydrolysis of orthoformic acid trimethyl ester.

the orthoester reacts in the presence of one of the previously mentioned acid catalysts in the alcohol from which the orthoester is derived. Under these conditions the acetalizations take place mechanistically as shown in Figure 7.15. The H_2O released then hydrolyzes the orthoester according to the mechanism shown in Figure 7.16.

Variant 2 of the coupled acetalizations/orthoester hydrolyses manages completely without alcohol. Here the carbonyl compound is acetalized only with the orthoester. Possibly, a trace of alcohol contaminating the orthoester would still make the acetalization via the mechanism of the preceding paragraph conceivable. However, perhaps the alternative acetalization mechanism of Figure 7.17 comes into play. The orthoester would then first be protonated so that thereafter MeOH would be eliminated, as if an E1 elimination started. This would provide the carboxonium ion **B,** which is a methylated ester and hence a good methylating agent. Because of this property, compound **B** could methylate the ketone reversibly to the carboxonium ion **A.** This step would ensure the success of the overall reaction if thereafter the methylated ketone **A** combined with MeOH. This would first produce the protonated dimethyl acetal, and the dimethyl acetal itself would be obtained by removal of a proton.

Acetalization equilibria of carbonyl compounds are shifted toward the side of the acetal when the reaction is carried out with one equivalent of a diol rather than with two equivalents of an alcohol (Figure 7.18). The reason for this is that now the reaction entropy is no longer negative: Using the diol, two product molecules are produced from two molecules of starting material, whereas during an acetalization with monofunctional alcohols, two product molecules are produced from three molecules of start-

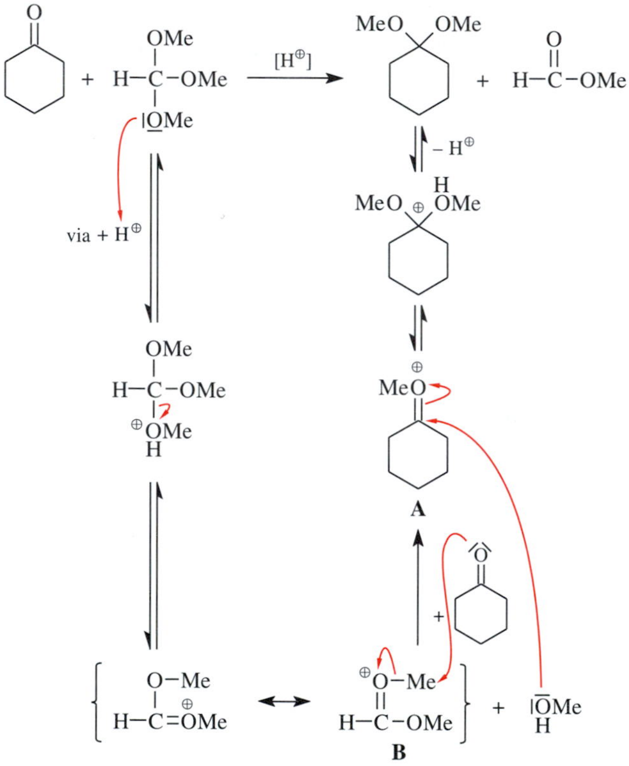

Fig. 7.17. Possible mechanism for the alcohol-free acetalization of a carbonyl compound with orthoformic acid trimethyl ester.

ing material. Ethylene glycol and 2,2-dimethyl-1,3-propanediol are frequently used acetalization reagents because of this enhanced driving force.

When an acetalization takes place intramolecularly, the reaction entropy can even be positive because the number of particles doubles. Therefore, in the presence of pro-

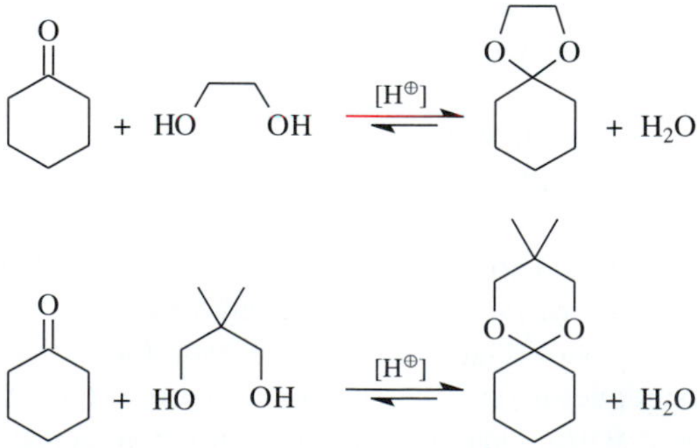

Fig. 7.18. Acetalization of a carbonyl compound with diols.

tons γ,γ'-, γ,δ'-, or δ,δ'-dihydroxyketones such as the following compound **A** react to give acetals of type **C** in rapid reactions and with high yields:

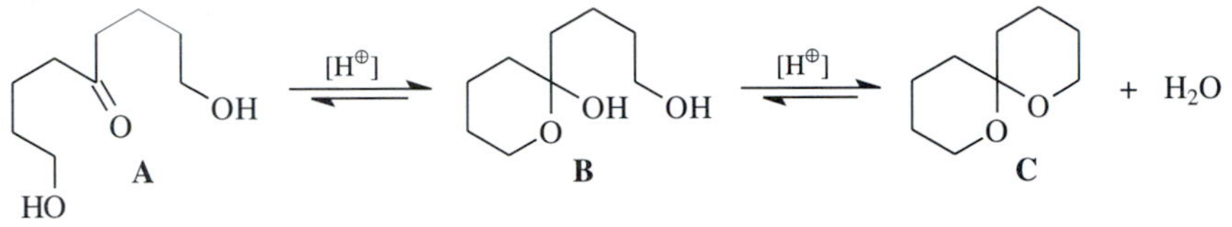

Acetals such as **C** are referred to as **spiroketals** because their acetal carbon is a spiro atom (in a "spiro compound" two rings are connected to each other via a single common atom that is called a "spiro atom"). The intermediates in this acetalization are lactols **B.** They resemble those lactols whose rapid formation from γ- or δ-hydroxyketones was shown in Figure 7.10. Note that because of the unfavorable reaction entropy, there is often no path back from spiroketals to the open-chain form: Usually, spiroketals cannot be hydrolyzed completely.

A

All acid-catalyzed acetalizations of carbonyl compounds with alcohols take place according to the mechanism of Figure 7.15—that is, through reversible steps. Therefore, they take place with thermodynamic control. Changing perspective, it also follows that acid-catalyzed acetalizations of triols (Figure 7.19) or of polyols (Figure 7.20) occur with thermodynamic control. This means that when several different acetals can in principle be produced from these reactants, the most stable acetal is produced preferentially. This circumstance can be exploited for achieving regioselective acetalizations of specific OH groups of polyalcohols.

An important polyalcohol for regioselective acetalizations is the triol of Figure 7.19 because it can be easily obtained from *S*-malic acid. This substrate contains both a 1,2-diol and a 1,3-diol. Each of these subunits can be incorporated into an acetal selectively—depending on the carbonyl compound with which the acetalization is carried out:

- With benzaldehyde the 1,3-diol moiety is acetalized exclusively. The resulting acetal **A** contains a strain-free, six-membered ring. It is present in the chair conformation

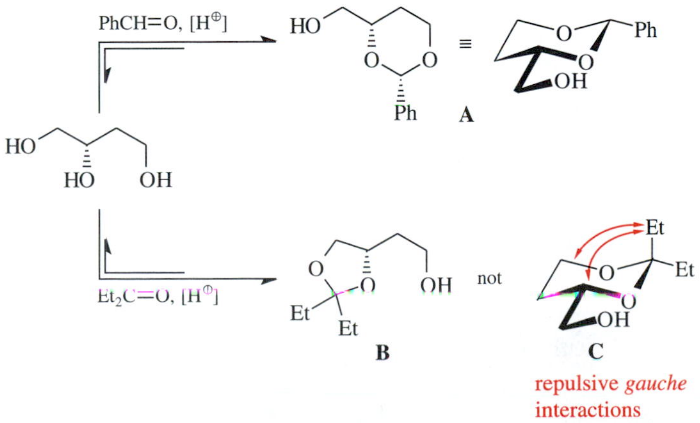

Fig. 7.19. Regioselective acetalizations of a 1,2,4-triol ("malic acid triol").

and accommodates both ring substituents in the energetically favorable equatorial orientation.

- With diethyl ketone it is possible to acetalize the 1,2-diol moiety of the malic acid triol selectively so that one obtains the five-membered acetal **B.** While, indeed, it exhibits more ring strain than the six-membered ring acetal **C,** this is overcompensated for because **B** suffers from considerably less Pitzer strain. The acetal isomer **C** in the most stable chair conformation would have to accommodate an axial ethyl group. However, because of their quite strong *gauche* interactions with the next-to-nearest O—C bonds of the ring skeleton, axial ethyl groups on C2 of six-membered ring acetals have almost as high a conformational energy and are therefore as unfavorable as an axial *tert*-butyl group on a cyclohexane.

Side Note 7.3 Regioselective Bisacetalization of a Pentaol

Figure 7.20 shows a biacetalization of a pentaol. There, the existence of thermodynamic control leads to the preferential production of one, namely **A,** of three possible bis(six-membered ring acetals)—**A, B,** and **C.** In the presence of catalytic amounts of *p*-toluenesulfonic acid the pentaol from Figure 7.20 is converted into the bisacetal *in* acetone as the solvent but by a *reaction* with the dimethylacetal of acetone. From the point of view of the dimethylacetal, this reaction is therefore a transacetalization. Each of the two transacetalizations involved—remember that a bisacetal is produced—takes place as a succession of two S_N1 reactions at the acetal carbon of the dimethyl-acetal. Each time the nucleophile is an OH group of the pentaol.

The equilibrium of the double transacetalization of Figure 7.20 lies completely on the side of the bisacetal. There are two reasons for this. On the one hand, the dimethylacetal is used in a large excess, which shifts the equilibrium to the product side. On the other hand, the transacetalization is favored by an increase in the translational

R 1 3 5 7 9 R
OH OH OH OH OH
O MeO OMe , , cat. *p*-TsOH
R 1 3 5 7 9 R
O O O O OH , not contaminated by
A
R 1 3 5 7 9 R O O OH O O or R 1 3 5 7 9 R OH O O O O
B **C**

Fig. 7.20. Regioselective bisacetalization of a 1,3,5,7,9-pentaol.

entropy: One molecule of pentaol and two molecules of the reagent give five product molecules.

Among the three conceivable bis(six-membered ring acetals) **A, B,** and **C** of the transacetalization of Figure 7.20, bisacetal **A** is the only one in which both six-membered rings can exist as chair conformers with three equatorial and only one axial substituent. In contrast, if the isomeric bis(six-membered ring acetals) **B** and **C** are also bis(chair conformers), they would contain on each six-membered ring not only two equatorial but also two axial substituents. This is energetically not as disadvantageous as usual. This is because one axial substituent would be located at the acetal carbon atom; this is especially unfavorable as we have a ready seen with respect to the instability of the acetal **C** from Figure 7.19. Also, there is a 1,3-diaxial interaction between the two axial substituents of each six-membered ring. Thus, if the two bisacetals **B** and **C** were in fact chair conformers they would not be able to cope with so much destabilization. Consequently, they prefer to be twist-boat conformers. However, the inherently higher energy of twist-boat vs chair conformers makes the acetals **B** and **C** still considerably less stable than acetal **A.** This is why acetal **A** is formed exclusively.

B

However, there is *one* type of carbonyl compound that cannot be converted into *O,O*-acetals of the type so far presented by treatment with alcohols and acid: γ- or δ-hydroxycarbonyl compounds. As you know from Figure 7.10, these compounds usually exist as lactols. In an acidic solution their OH group is exchanged very rapidly by an OR group according to the S_N1 mechanism of Figure 7.15. This produces an *O,O*-acetal, which is derived from two different alcohols, a so-called **mixed acetal.** However, there is no further reaction delivering an open-chain nonmixed *O,O*-acetal of the γ- or δ-hydroxycarbonyl compound. This is because such a reaction would be accompanied by and disfavored due to an entropy loss: A single molecule would have been produced from two.

Examples of the previously mentioned formation of mixed *O,O*-acetals from lactols are given by acid-catalyzed reactions of D-glucose with MeOH (Figure 7.21). They lead to mixed *O,O*-acetals, which in this case are named methyl glucosides. The six-membered ring mixed acetal **A** is formed using HCl as a catalyst (and under thermodynamic control). The less stable five-membered ring mixed acetal **C** is produced under $FeCl_3$ catalysis. The last reaction is remarkable in that this acetalization is achieved under kinetic control. The open-chain dimethyl acetal **B** never forms.

7.3.3 Formation of *S,S*-Acetals

B

In the presence of catalytic amounts of a sufficiently strong acid, thiols and carbonyl compounds form *S,S*-acetals according to the mechanism of Figure 7.15. The thermodynamic driving force for this type of *S,S*-acetal formation is greater than that for the analogous *O,O*-acetal formation (there is no clear reason for this difference). For ex-

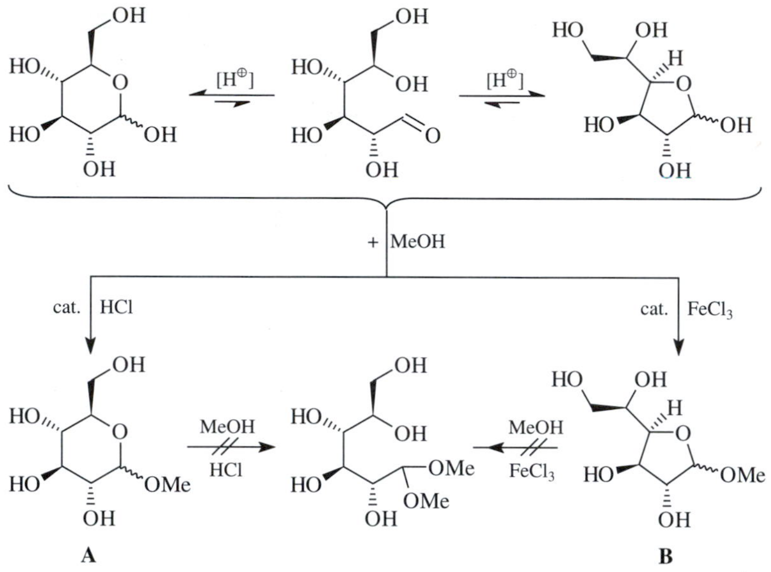

Fig. 7.21. Chemoselective formation of mixed *O*,*O*-acetals from D-glucose with MeOH.

ample, although D-glucose, as shown in Figure 7.21, cannot be converted into an open-chain *O*,*O*-acetal, it can react to give an open-chain *S*,*S*-acetal:

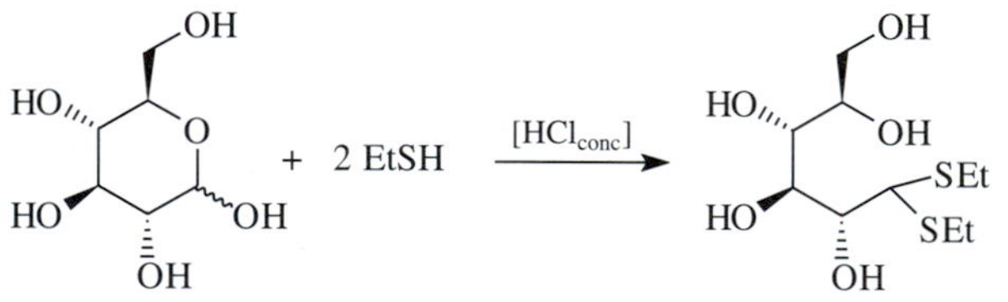

The most important *S*,*S*-acetals in organic chemistry are the six-membered ring *S*,*S*-acetals, the so-called **dithianes** (formulas **A**, **C**, and **F** in Figure 7.22). Dithianes are produced from carbonyl compounds and 1,3-propanedithiol usually in the presence of Lewis instead of Brønsted acids.

In the most simple dithiane **A** or in dithianes of type **C**, an H atom is bound to the C atom between the two S atoms. This H atom can be removed with LDA and thus be replaced by a Li atom. As an alternative, it can also be deprotonated with *n*-BuLi (Figure 7.22). In this way one obtains the lithiodithiane **B** or its substituted analog **D.** These compounds are good nucleophiles and can be alkylated, for example, with alkylating agents in S_N2 reactions (Figure 7.22; cf. Figure 2.26). The alkylated dithianes **C** and **E** can subsequently be hydrolyzed to carbonyl compounds. This hydrolysis is best done not simply in acids but in the presence of Hg(II) salts. Monoalkyldithianes **C** then give aldehydes, and dialkyldithianes **E** give ketones. The alkyl group resulting as the carbonyl substituent of these products has been incorporated as an electrophile in the syntheses of Figure 7.22. This distinguishes this so-called Corey–Seebach synthesis from most aldehyde and ketone syntheses that you know.

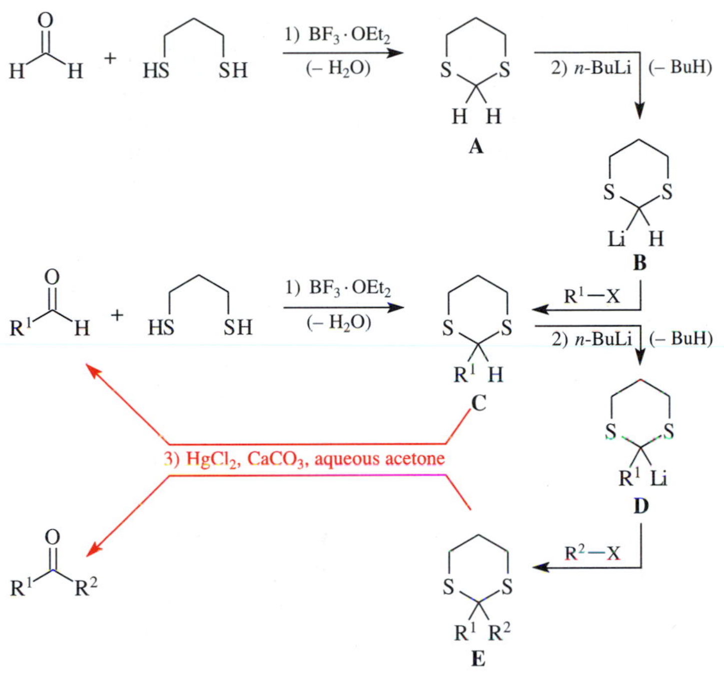

Fig. 7.22. Preparation, alkylation, and hydrolysis of dithianes.

7.3.4 Formation of *N*,*N*-Acetals

B

Ammonia, primary amines—and also many other compounds that contain an NH_2 group—as well as secondary amines add to many carbonyl compounds only to a certain extent; that is, in equilibrium reactions (formation of hemiaminals; formula **B** in Figure 7.23). This addition is almost always followed by the elimination of an OH^- ion. As the result, one obtains an iminium ion (formula **C** in Figure 7.23).

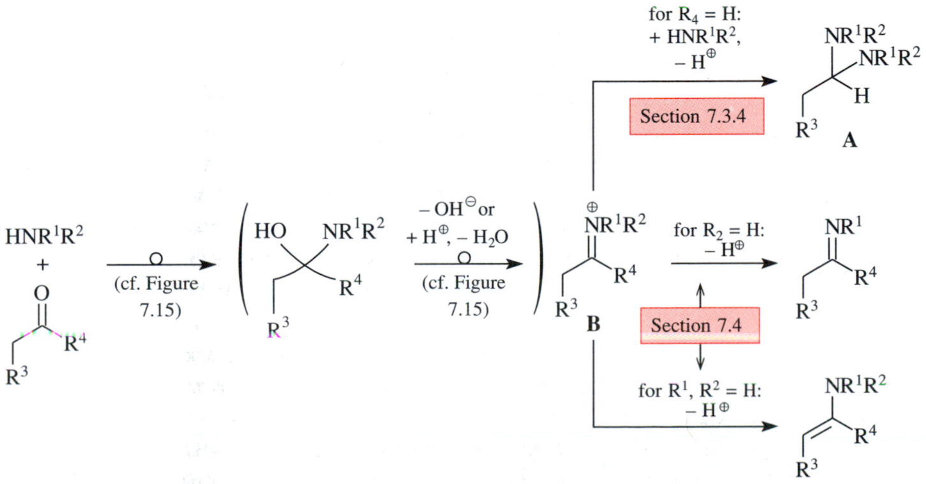

Fig. 7.23. Survey of the chemistry of iminium ions produced *in situ*.

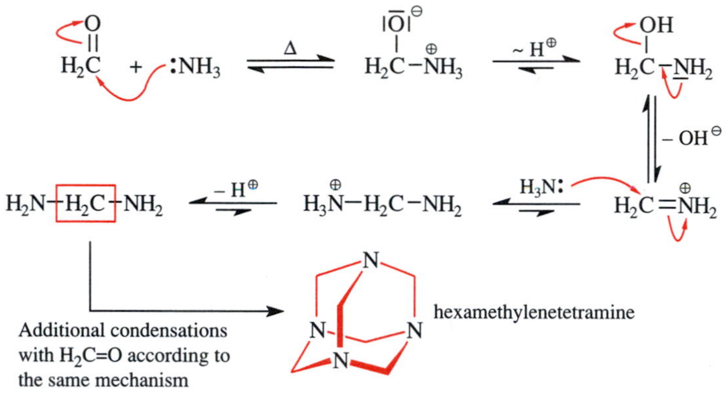

Fig. 7.24. Mechanism for the formation of hexamethylenetetramine from ammonia and formalin.

Additional condensations with $H_2C{=}O$ according to the same mechanism

hexamethylenetetramine

This iminium ion **C** can be stabilized by combining with a nucleophile. As Figure 7.15 shows, when Nu = R^1R^2N, this step completes an S_N1 reaction in the initially formed addition product **B.** Thereby, an *N,N*-acetal (formula **A** in Figure 7.23) or a derivative thereof (see Figure 7.23) is produced. We will treat this kind of reaction in more detail in Section 7.4. The other important reaction mode of the iminium ions **C** in Figure 7.23 is the elimination of a proton. This step would complete an E1 elimination of H_2O from the primary adduct **B.** This kind of reaction also occurs frequently and will be discussed in Section 7.4, as well.

The most important reaction examples for the formation of *N,N*-acetals involve formaldehyde because it tends more than most other carbonyl compounds to undergo additions (Section 7.2.1). With ammonia formaldehyde gives hexamethylenetetramine (Figure 7.24). At its six C atoms this compound contains six *N,N*-acetal subunits.

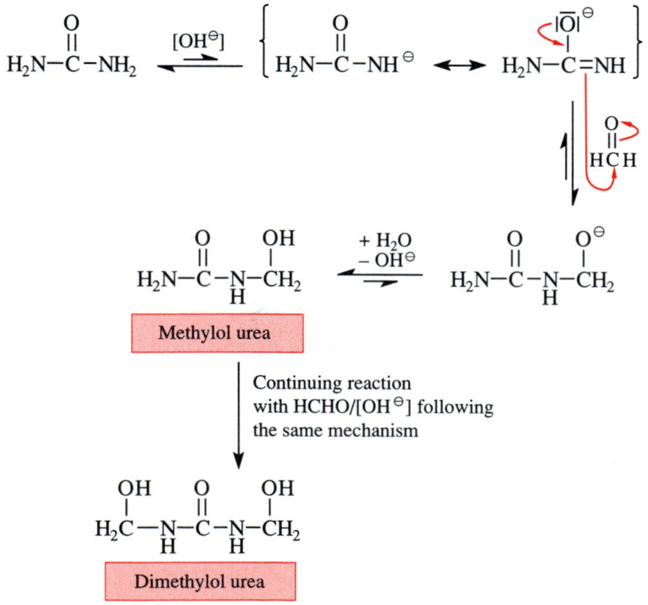

Methylol urea

Continuing reaction with HCHO/[$OH^{\ominus}$] following the same mechanism

Dimethylol urea

Fig. 7.25. Reaction of urea with formaldehyde at room temperature.

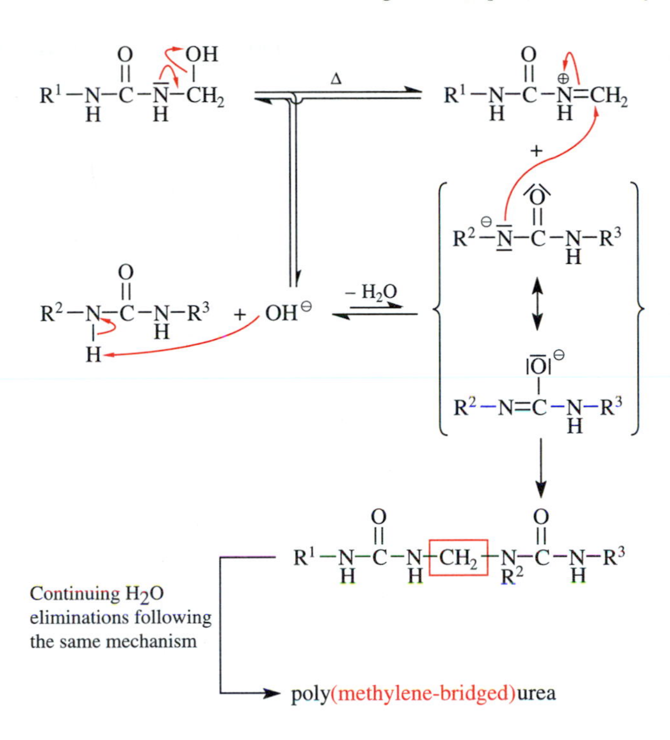

Fig. 7.26. Mechanism for the formation of a urea/formaldehyde resin from methylol urea (R^1 = H in formula **A;** possible preparation: Figure 7.25) or dimethylol urea (R^1 = HO—CH_2 in formula **A;** possible preparation: Figure 7.25). The substituents R^1, R^2, and R^3 represent the growing ···—CH_2—NH—C(=O)—NH—CH_2—··· chains as well as derivatives thereof with two CH_3 substituents on the N atoms.

A

With urea, formaldehyde forms two stable *N,O*-hemiacetals (Figure 7.25): a 1:1 adduct ("methylol urea") and a 1:2 adduct ("dimethylol urea"). When they are heated, both compounds are converted to *macromolecular N,N*-acetals (Figure 7.26). A three-dimensionally cross linked **urea/formaldehyde resin** is produced; it is an important plastic.

A structurally related macromolecular *N,N*-acetal is obtained from melamine (2,4,6-triamino-1,3,5-triazine) and formaldehyde. It is called **melamine/formaldehyde resin** and is likewise used as a plastic.

7.4 Addition of Nitrogen Nucleophiles to Carbonyl Compounds in Combination with Subsequent E1 Eliminations: Condensation Reactions of Nitrogen Nucleophiles with Carbonyl Compounds

B

Iminium ions which stem from nitrogen nucleophiles and formaldehyde are more electrophilic and sterically less hindered than iminium ions obtained from higher aldehydes let alone from ketones. This explains why the first type of iminium ion is likely to combine with an extra molecule of the nitrogen nucleophile and in this way reacts further to give *N,N*-acetals (Section 7.3.4). The more highly substituted iminium ions are less electrophilic because of their electron-donating nonhydrogen substituent and because they are also more sterically hindered. Therefore any excess N nucleophile would add only slowly to iminium ions of this type. Instead such an N nucleophile uses a different reaction mode: It reacts with these iminium ions as a base, i.e., deprotonates them. This takes place regioselectively: Secondary iminium ions are deprotonated at the nitrogen, whereas tertiary iminium ions are deprotonated at the β carbon. In this way, the initially obtained addition products of N nucleophiles to the C=O double bond of higher aldehydes or ketones are converted to unsaturated products through an E1 elimination of H_2O. Preferably, these products contain a C=N double bond. Only if this is impossible is a C=C double bond formed.

Reactions such as those just described, i.e., ones in which two molecules of starting material react to form one product molecule plus a by-product of considerably lower molecular weight are referred to as **condensation reactions.** *Here,* as in many other condensations this by-product is H_2O. Table 7.2 summarizes the most important condensation reactions of nitrogen nucleophiles with carbonyl compounds, in which C=N double bonds are produced.

You should already be familiar with approximately half of the reactions listed in Table 7.2 from your introductory class. Moreover, if you have taken or are taking a classical analytical laboratory course, you have possibly tried or are trying to prepare an oxime, a phenylhydrazone, a 2,4-dinitrophenylhydrazone, or a semicarbazone. These compounds serve as crystalline derivatives with sharp and characteristic melting points for identifying aldehydes and ketones and for distinguishing them. When spectroscopic methods for structure elucidation are not available, carbonyl compounds are more difficult to characterize unambiguously because of their volatility.

Most of the other product types in Table 7.2 are used in preparative chemistry. Aldimines, especially those that are derived from cyclohexylamine, serve as precursors to azaenolates (Figure 10.30). The same holds for the SAMP hydrazones, except that enantiomerically pure azaenolates are accessible through them (Figures 10.31 and 14.31; also see Figure 10.32). Starting from phenylhydrazones it is possible to synthesize indoles according to the Fischer procedure. Tosylhydrazones are used in two-step reductions of carbonyl compounds to hydrocarbons (Figures 14.60 and 14.61). Occasionally, semicarbazones are used similarly (Figure 14.59).

Secondary amines react with ketones that contain an H atom in the α position through an addition and subsequent E1 elimination to form **enamines** (Figure 7.27). In order for enamines to be formed *at all* in the way indicated, one must add an acid catalyst. In order for them to be formed *completely,* the released water must be distilled off azeotropically. The method of choice for preparing enamines is therefore to heat a solution of the carbonyl compound, the amine, and a catalytic amount of tolu-enesulfonic acid in cyclohexane to reflux in an apparatus connected to a Dean–Stark trap.

A

The deprotonation of an iminium ion (formula **A** in Figure 7.27) to give an enamine is reversible under the usual reaction conditions. Therefore, the most stable enamine possible is produced preferentially. Figure 7.28 emphasizes this using the example of an enamine formation from α-methylcyclohexanone (i.e., from an asymmetrical ketone). The enamine with the trisubstituted double bond is produced regioselectively and not the enamine with the tetrasubstituted double bond. Since the stability of olefins usually increases with an increasing degree of alkylation, this result is at first

Table 7.2. Condensation Reactions of Nitrogen Nucleophiles with Carbonyl Compounds through Which C═N Double Bonds Are Established—Mechanism and Scope

via
$+ H^{\oplus}$
$\sim H^{\oplus}$
$- H_2O$
$- H^{\oplus}$ (at the latest in the basic workup)
$H^{\oplus}X^{\ominus}$ (cat. or stoichiometric amounts);
$OH^{\ominus}$-workup
fast
at room temperature
overall reaction

R^3	Name of the N nucleophile	Name of the condensation product
H	Ammonia	Product unstable
Alk	Primary alkylamine	For $R^1 = R^2 = H$; product unstable For $R^1 \neq H$, $R^2 = H$: aldimine For $R^1, R^2 \neq H$: ketimine
Ph	Aniline	Anil
OH	Hydroxylamine	Oxime
(pyrrolidin-1-yl with 2-CH₂OMe)	(S)-N-**Amino**-**p**rolinol methylether (SAMP)	SAMP hydrazone
NHPh	Phenylhydrazine	Phenylhydrazone
NH–C₆H₃(NO₂)₂ (2,4-dinitrophenyl)	2,4-Dinitrophenylhydrazine	2,4-Dinitrophenylhydrazone
NHTs	Tosylhydrazide	Tosylhydrazone
NH–C(=O)–NH₂	Semicarbazide	Semicarbazone

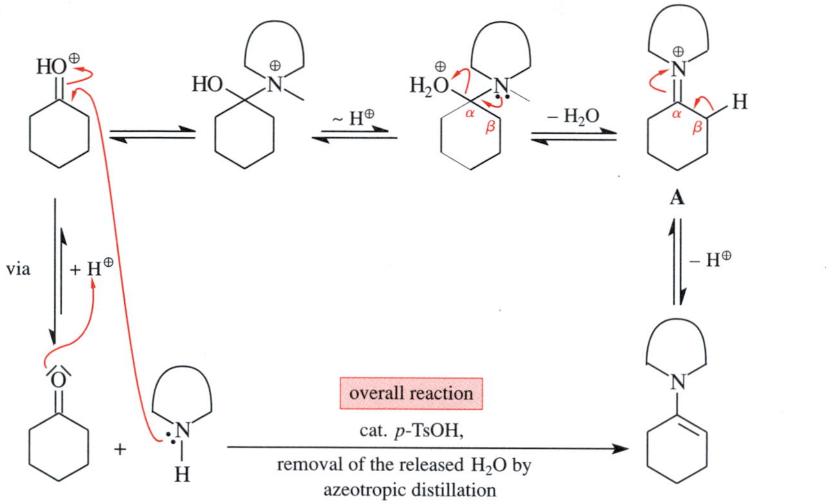

Fig. 7.27. Mechanism of the formation of enamines from secondary amines and cyclohexanone. Overall reaction in the bottom line.

surprising. However, the apparent contradiction disappears when one recalls the following: The C═C double bond of enamines is stabilized considerably by the enamine resonance $C{=}C{-}NR_2 \leftrightarrow {}^{-}C{-}C{=}NR_2^{+}$. The feasability of this resonance requires an sp^2 hybridization of the nitrogen atom. Furthermore, the $2p_z$ AO at the N atom and the π-orbital of the C═C double bond must be parallel. In other words, for op-

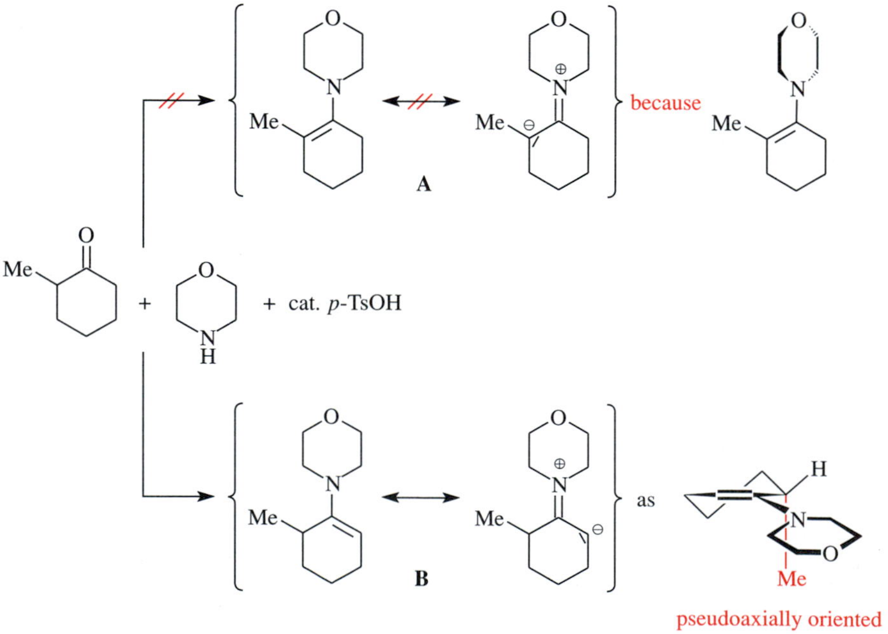

Fig. 7.28. Regioselective formation of an enamine from an asymmetrical ketone.

timumen amine resonance to be possible the C=C double bond, the attached nitrogen and the five other atoms bound to these C and N atoms all must lie in a single plane.

This is precisely what is impossible in the more highly alkylated enamine **A** of Figure 7.28. A considerable repulsion (1,3-allylic strain) between the methyl group and the heterocycle would be introduced; it would be so great that the nitrogen-containing ring would be rotated out of the plane of the C=C double bond. As a consequence, there would be no enamine resonance at all. On the other hand, the observed enamine **B** would benefit from enamine resonance. However, even this molecule must avoid the occurrence of strain, namely of *syn*-pentane strain between the methyl group and the heterocycle. This strain is avoided when the methyl group adopts an unusual pseudoaxial orientation.

References

J. K. Whitesell, "Carbonyl Group Derivatization," in *Comprehensive Organic Synthesis* (B. M. Trost, I. Fleming, Eds.), Vol. 6, 703, Pergamon Press, Oxford, **1991.**

7.1

H. R. Seikaly and T. T. Tidwell, "Addition reactions of ketenes," *Tetrahedron* **1986,** *42,* 2587.

W. E. Hanford and J. C. Sauer, "Preparation of ketenes and ketene dimers," *Org. React.* **1946,** *3,* 108–140.

7.2

A. Klausener, "O/O-Acetals: Special (OH/OH-Acetals, Lactols, 1,3,5-Trioxanes, OR/OEne-Acetals, OEne-Acetals etc.)," in *Methoden Org. Chem. (Houben-Weyl) 4th ed. 1952, O/O- and O/S-Acdetals* (H. Hagemann, D. Klamann, Eds.), Vol. E14a/1, 591, Georg Thieme Verlag, Stuttgart, **1991.**

M. M. Joullie and T. R. Thompson, "Ninhydrin and ninhydrin analogs: Syntheses and applications," *Tetrahedron* **1991,** *47,* 8791–8830

M. B. Rubin and R. Gleiter, "The chemistry of vicinal polycarbonyl compounds," *Chem. Rev.* **2000,** *100,* 1121–1164.

7.3

A. Klausener, "O/O-Acetals: Introduction," in *Methoden Org. Chem. (Houben-weyl) 4th ed. 1952, O/O-Acetals* (H. Hagemann, D. Klamann, Eds.), Vol. E14a/1, XIX, Georg Thieme Verlag, Stuttgart, **1991.**

O. Lockhoff, "Hal/O-, O/O-Acetals as Anomeric Centers of Carbohydrates," in *Methoden Org. Chem. (Houben-Weyl) 4th ed. 1952, Hal/O(S,N)-, S/S(N)-, N,N-Acetals and Hal/O-, O/O-Acetals as Anomeric Centers of Carbohydrates* (H. Hagemann, D. Klamann, Eds.), Vol. E14a/3, 621, Georg Thieme Verlag, Stuttgart, **1992.**

F. A. J. Meskens, "Methods for the preparation of acetals from alcohols or oxiranes and carbonyl compounds," *Synthesis* **1981,** 501.

L. Hough and A. C. Richardson, "Monosaccharide Chemistry," in *Comprehensive Organic Chemistry* (E. Haslam, Ed.), **1979,** *5* (Biological Compounds), 687–748, Pergamon Press Ltd., England.

F. Perron and K. F. Albizati, "Chemistry of spiroketals," *Chem. Rev.* **1989,** *89,* 1617–1661.

S. Pawlenko and S. Lang-Fugmann, "S/S-Acetals as Anomeric Centers of Carbohydrates," in *Methoden Org. Chem. (Houben-Weyl) 4th ed. 1952, Hal/O(S,N)-, S/S(N)-, N,N-Acetals and Hal/O-, O/O-Acetals as Anomeric Centers of Carbohydrates* (H. Hagemann, D. Klamann, Eds.), Vol. E14a/3, 403, Georg Thieme Verlag, Stuttgart, **1992.**

P. C. B. Page, M. B. Van Niel, J. C. Prodger, "Synthetic uses of the 1,3-dithiane grouping from 1977 to 1988," *Tetrahedron* **1989,** *45,* 7643–7677.

D. P. N. Satchell and R. S. Satchell, "Mechanisms of hydrolysis of thioacetals," *Chem. Soc. Rev.* **1990,** *19,* 55–81.

7.4

A. G. Cook (Ed.), "Enamines: Synthesis, Structure, and Reactions," Marcel Dekker, New York, **1988.**

Z. Rappoport (Ed.), "The Chemistry of Enamines," Wiley, Chichester, UK, **1994.**

Further Reading

A. J. Kirby, "Stereoelectronic effects on acetal hydrolysis," *Acc. Chem. Res.* **1984,** *17,* 305.

E. Juaristi and G. Cuevas, "Recent studies on the anomeric effect," *Tetrahedron* **1992,** *48,* 5019–5087.

A. J. Kirby (ed.), "Stereoelectronic Effects," Oxford University Press, Oxford, U.K., **1996.**

T. Y. Luh, "Regioselective C-O bond cleavage reactions of acetals," *Pure Appl. Chem.* **1996,** *68,* 635.

P. Wimmer, "O/S-Acetals," in *Methoden Org. Chem.* (*Houben-Weyl*) *4th ed. 1952, O/O- and O/S-Acetals* (H. Hagemann, D. Klamman, Eds.), Vol. E14a/1, 785, Georg Thieme Verlag, Stuttgart, **1991.**

S. Pawlenko, S. Lang-Fugmann, "N/N-Acetals as Anomeric Centers of Carbohydrates," in *Methoden Org. Chem. (Houben-Weyl) 4th ed. 1952, Hal/O(S,N)-, S/S(N)-, N,N-Acetals and Hal/O-, O/O-Acetals as Anomeric Centers of Carbohydrates* (H. Hagemann, D. Klamann, Eds.), Vol. E14a/3, 545, Georg Thieme Verlag, Stuttgart, **1992.**

S. Pawlenko and S. Lang-Fugmann, "S/N-Acetals as Anomeric Centers of Carbohydrates," in *Methoden Org. Chem. (Houben-Weyl) 4th ed. 1952, Hal/O(S,N)-, S/S(N)-, N,N-Acetals and Hal/O-, O/O-Acetals as Anomeric Centers of Carbohydrates* (H. Hagemann, D. Klamann, Eds.), Vol. E14a/3, 483, Georg Thieme Verlag, Stuttgart, **1992.**

Addition of Hydride Donors and Organometallic Compounds to Carbonyl Compounds

8

In Section 6.5 you learned that the acylations of hydride donors or of organometallic compounds, which give aldehydes or ketones, often are followed by an unavoidable second reaction: the addition of the hydride or organometallic compound to the aldehyde or the ketone. In this chapter, we will study the intentional execution of such addition reactions. They do not start from carbonyl compounds produced *in situ* but from carbonyl compounds used as such.

8.1 Suitable Hydride Donors and Organometallic Compounds and a Survey of the Structure of Organometallic Compounds

B

The addition of a hydride donor to an aldehyde or to a ketone gives an alcohol. This addition is therefore also a redox reaction, namely, the reduction of a carbonyl compound to an alcohol. Nevertheless, this type of reaction is discussed here and not in the redox chapter (Chapter 14).

Hydride donors can be subdivided into three classes. They are

- ionic, soluble hydrido complexes of B or Al ("complex hydrides"),
- covalent compounds with at least one B—H or one Al—H bond, or
- organometallic compounds which contain no M—H bond but do contain a transferable H atom at the C atom in the position β to the metal.

The first group of H nucleophiles includes $NaBH_4$ (which can be used in MeOH, EtOH, or HOAc), the considerably more reactive $LiAlH_4$ (with which one works in THF or ether), and alcoholysis products of these reagents such as Red-Al [$NaAlH_2(O—CH_2—CH_2—OMe)_2$] or the sterically very hindered LiAlH(O-*tert*-Bu)$_3$. The last important hydride donor in this group is LiBH(*sec*-Bu)$_3$ (L-Selectride), a sterically very hindered derivative of the rarely used $LiBH_4$.

A

The hydride donor with a covalent M—H bond which is most frequently used for reducing carbonyl groups is $i\text{Bu}_2\text{AlH}$ (DIBAL stands for **di**iso**b**utyl**al**uminum hydride; it can be used in ether, THF, toluene, saturated hydrocarbons, or CH_2Cl_2).

B

The most important organometallic compound which transfers an H atom along with the electron pair from the β position to the carbonyl carbon is Alpine-Borane (cf. Figure 8.19). It should be mentioned that certain Grignard reagents with C—H bonds in the β position (e.g., *i*BuMgBr) can act as H nucleophiles rather than C nucleophiles (cf. Figure 8.22) with respect to the carbonyl carbon of sterically hindered ketones.

In this chapter, reagents which transfer a carbanion (in contrast to an enolate ion) to the C atom of a C═O double bond are referred to as **C nucleophiles**. The most important nucleophiles of this kind are organolithium compounds and Grignard reagents. On the other hand, organocopper compounds transfer their carbanion moieties to the carbonyl carbon far less easily and usually not at all.

In the majority of cases, organolithium compounds and Grignard reagents contain polarized but covalent carbon—metal bonds. Lithioalkanes, -alkenes, and -aromatics, on the one hand, and alkyl, alkenyl, and aryl magnesium halides, on the other hand, are therefore formulated with a hyphen between the metal and the neighboring C atom. Only lithiated alkynes and alkynyl Grignard reagents are considered to be ionic—that is, species with carbon, metal bonds similar to those in LiCN or $Mg(CN)_2$.

In covalent organolithium compounds and covalent Grignard reagents neither the lithium nor the magnesium possesses a valence electron octet. This is energetically disadvantageous. In principle, the same mechanism can be used to stabilize these metals, which monomeric boranes $BH_{3-n}R_n$ use to attain a valence electron octet at the boron atom (Section 3.3.3): the formation either of oligomers or, with suitable electron pair donors, of Lewis acid/Lewis base complexes.

Alkyllithium compounds occur in hydrocarbon solutions as hexamers, tetramers, or dimers depending on the alkyl substituent. For a given substituent, the degree of association drops in diethyl ether or especially in THF solutions because the O atoms of these solvents can occupy vacant coordination sites around the electron-deficient lithium. The monomeric form of organolithium compounds can sometimes be stabilized in hydrocarbons and always in ether or THF by adding TMEDA (**t**etra**m**ethyl**e**thylene**dia**mine; $Me_2N—CH_2—CH_2—NMe_2$). The N atoms of this additive then occupy two of the three vacant coordination sites of the lithium. The monomerization of oligomeric organolithium compounds takes place most reliably in the presence of up to three equivalents of HMPA (structural formula in Figure 2.16). The basic oxygen of this additive is an excellent electron pair donor.

That the valence electron shell of magnesium can be filled up to some extent through the interaction with the O atoms of suitable ethers is even a prerequisite for obtaining Grignard reagents from halides and magnesium. In general, this Grignard formation can be done successfully only in diethyl ether, in THF, or in the seldom used dimethoxyethane (DME) (mechanism: Figure 14.37). Ethyl magnesium bromide is present in diethyl ether essentially as dimer **A** (Figure 8.1, X = Br, R = Et). On the one hand, this species dissociates reversibly to yield a small amount of the monomer. It also participates in the Schlenk equilibrium, i.e., in the equilibration with Et_2Mg and $MgBr_2$ (Figure 8.1).

Neither the mechanism for all addition reactions of hydride donors to the carbonyl carbon nor the mechanism for all addition reactions of organometallic compounds to the carbonyl carbon is known in detail. It is even doubtful whether only ionic intermediates occur. For instance, for some $LiAlH_4$ additions an electron transfer mecha-

A [Et(R₂O)Mg(μ-X)₂Mg(OR₂)Et] ⇌ $Et_2Mg + MgX_2$

Fig. 8.1. Schlenk equilibrium of ethyl magnesium halides.

nism might apply. The same might be true for the addition of a number of organometallic compounds to aromatic ketones. Often it is not even clear whether oligomeric and/or monomeric organometallic compounds are the species involved in the additions. Also, the counterion X of Grignard reagents RMgX can exert an influence on their reactivity (compare Tables 8.1 and 8.3) but it is not known how.

Therefore, we will proceed pragmatically in this chapter: For the additions discussed, the substrates, the reagents formulated as monomers, and the tetrahedral intermediates considered established will be shown by structural formulas. How exactly the reagents are converted into these intermediates remains unknown. However, this conversion should occur in the rate-determining step.

8.2 Chemoselectivity of the Addition of Hydride Donors to Carbonyl Compounds

B

In the addition of hydride donors to aldehydes (other than formaldehyde) the tetrahedral intermediate is a primary alkoxide. In the addition to ketones it is a secondary alkoxide. When a primary alkoxide is formed, the steric hindrance is smaller. Also, when the C=O double bond of an aldehyde is broken due to the formation of the CH(OM) group of an alkoxide, less resonance stabilization of the C=O double bond by the flanking alkyl group is lost than when the analogous transformation occurs in a ketone (c.f. Table 7.1). For these two reasons aldehydes react faster with hydride donors than ketones. With a moderately reactive hydride donor such as $NaBH_4$ at low temperature one can even chemoselectively reduce an aldehyde in the presence of a ketone (Figure 8.2, left).

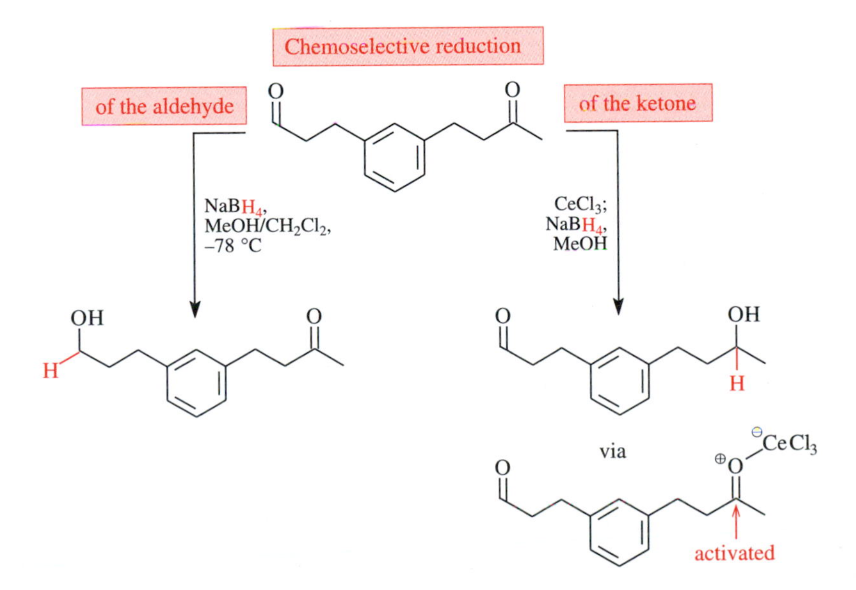

Fig. 8.2. Chemoselective carbonyl group reductions I. On the left side a chemoselective reduction of the aldehyde takes place, whereas on the right side a chemoselective reduction of the ketone is shown.

Because of the +I effect of two flanking alkyl groups, the carbonyl oxygen of ketones is more basic than that of aldehydes. Therefore, ketones form more stable Lewis acid/Lewis base complexes with electrophilic metal salts than aldehydes. This is exploited in the **Luche reduction** for the chemoselective reduction of a ketone in the presence of an aldehyde (Figure 8.2, right). As the initiating step, the keto group is complexed selectively with one equivalent of $CeCl_3$. This increases its electrophilicity so that it exceeds that of the aldehyde. Therefore, the $NaBH_4$ which is now added reduces the ketonic C=O double bond preferentially.

Of two ketonic C=O double bonds, the sterically less hindered one reacts preferentially with a hydride donor. The higher the selectivity, the bulkier the hydride donor is. This makes **L-Selectride** the reagent of choice for reactions of this type (Figure 8.3, left).

A

Conversely, the more easily accessible C=O group of a diketone can be selectively complexed with a Lewis acid, which responds sensitively to the differential steric hindrance due to the substituents at the two C=O groups. A suitable reagent for such a se-

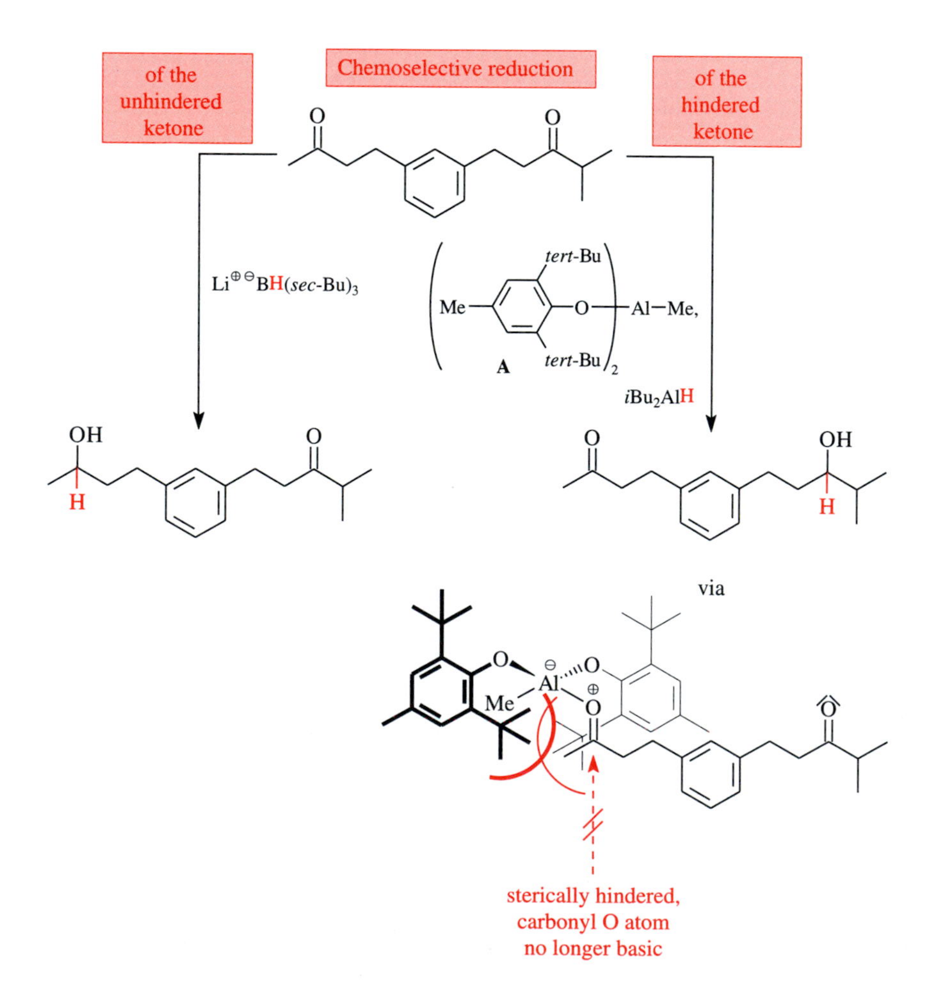

Fig. 8.3. Chemoselective carbonyl group reductions II. A chemoselective reduction of the less hindered ketone takes place on the left side, and a chemoselective reduction of the more strongly hindered ketone takes place on the right side.

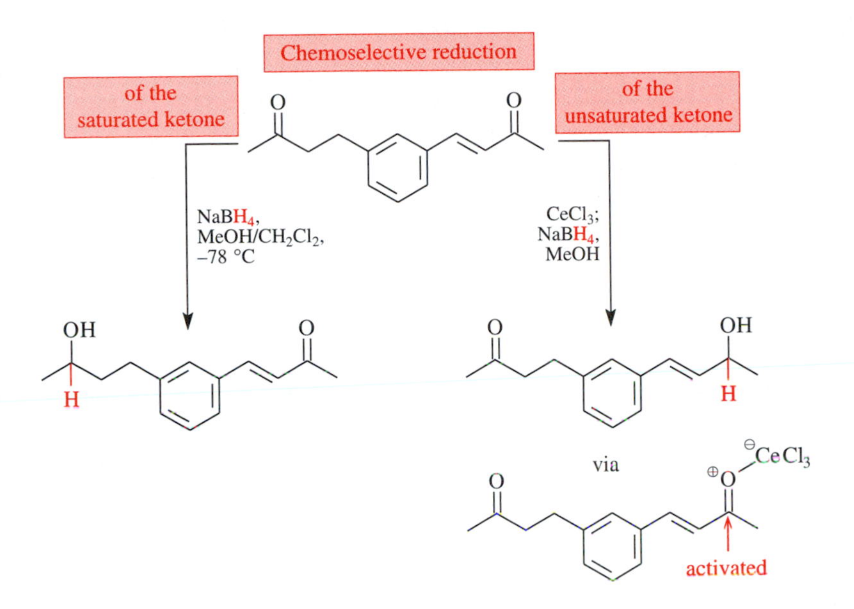

Fig. 8.4. Chemoselective carbonyl group reductions III. Reduction of a saturated ketone in the presence of an unsaturated ketone (left) and reduction of an unsaturated ketone in the presence of a saturated ketone (right).

lective complexation is the aluminoxane **A** (Figure 8.3, right). After the complexation has taken place, DIBAL is added as a reducing agent. In contrast to $NaBH_4$, this reagent exhibits a certain electrophilicity because of the electron-deficient aluminum atom. Therefore, it preferentially attacks the C=O double bond which is not complexed by the aluminoxane **A.** This is because that is the (only) C=O double bond which offers the aluminum the free electron pair of the O atom as a nucleophilic point of attack. The result is that DIBAL reacts with the more strongly hindered ketonic C=O double bond.

B

An α,β-unsaturated ketone and a saturated ketone have different reactivities with respect to hydride donors (Figure 8.4). $NaBH_4$ reacts to form the tetrahedral intermediate preferentially with the nonconjugated C=O double bond (Figure 8.4, left). Product-development control is responsible for this: In this addition no conjugation between the C=O and the C=C double bond is lost. This would be the case if the addition of $NaBH_4$ occurred at the C=O double bond of the unsaturated ketone. Conversely, the unsaturated substituent, because of its +M effect, increases the basicity of the conjugated ketonic carbonyl group. Therefore, this group can be complexed preferentially with one equivalent of $CeCl_3$. Thereby, however, it becomes more electrophilic and can be reduced chemoselectively with subsequently added $NaBH_4$ in a Luche reduction (Figure 8.4, right).

8.3 Diastereoselectivity of the Addition of Hydride Donors to Carbonyl Compounds

B

When the plane of the double bond of a carbonyl compound is flanked by diastereotopic half-spaces, a stereogenic addition of a hydride can take place diastereoselectively (cf. Section 3.4.1). In Section 8.3.1 we will investigate which diastereomer is preferentially produced in such additions to the C=O double bond of cyclic ketones. In Sections 8.3.2 and 8.3.3 we will discuss which diastereomer is preferentially formed in stereogenic additions of hydride donors and acyclic chiral ketones or acyclic chiral aldehydes (in such reactions with acyclic chiral aldehydes, deuterium labeling is required either in the reagent or in the substrate because otherwise no stereogenic reaction would take place).

8.3.1 Diastereoselectivity of the Addition of Hydride Donors to Cyclic Ketones

B

In the reacting moiety of many cyclic or bicyclic molecules a stereostructure is present in which one can identify a convex and a concave side. Because reactions usually take place in such a way that the attacking reagent is exposed to the least possible steric hindrance, convex/concave substrates are generally attacked on their convex side.

Figure 8.5 shows an application of this principle in the diastereoselective addition of a hydride donor to a bicyclic ketone: With L-Selectride [$= Li^+ \ ^-BH(sec\text{-}Bu)_3$] the *endo*-alcohol is produced exclusively.

endo* and *exo

The stereodescriptors "*endo*" and "*exo*" are used to distinguish between the positions of a substituent beneath or outside the concave area of a bent molecule. For example, in the major product of the reduction of norbornanone with $Li^+ \ ^-BH(sec\text{-}Bu)_3$ (Figure 8.6) the newly entered H atom is oriented *exo* and the OH group obtained *endo*.

Other cyclic or bicyclic ketones do not have a convex side at all in the reacting moiety of the molecule but only a less concave and a more concave side. Thus, a hydride donor can add to such a carbonyl group only from a concave side. Because of the steric hindrance, this normally results in a decrease in the reactivity. However, the addition of this hydride donor is still less disfavored when it takes place from the less concave (i.e., the less hindered) side. As shown in Figure 8.6 by means of the comparison of two reductions of norbornanone, this effect is more noticeable for a bulky hydride donor such as L-Selectride than for a small hydride donor such as $NaBH_4$.

Fig. 8.5. Diastereoselective addition of a bulky hydride donor (L-Selectride) to a bicyclic ketone. The *endo*-alcohol is formed exclusively.

$NaBH_4$:	86	:	14
$Li^{\oplus}\,{}^{\ominus}BH(sec\text{-}Bu)_3$:	99.6	:	0.4

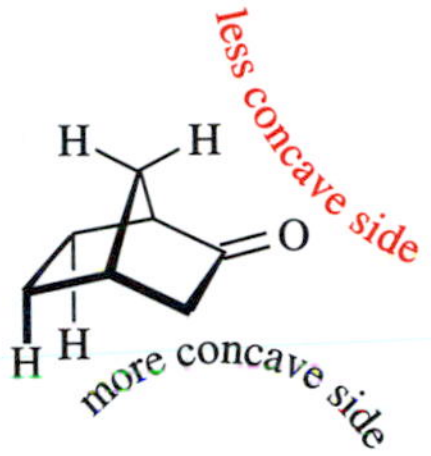

Fig. 8.6. Addition of a bulky hydride donor (L-Selectride) to the less concave side of the C=O double bond of norbornanone.

The preferred direction of addition of hydride donors to norbornanone (Figure 8.6) is shown once more in Figure 8.7 (substrate **A**). As can also be seen from this figure, the additions of the same hydride donors to the norbornanone derivatives **B** (norbornenone) and **C** (camphor) take place with the opposite diastereoselectivity. As indicated for each substrate, the *common* selectivity-determining factor remains the principle that the hydride attack takes place preferentially from the less concave side of the molecule.

In the hydride additions of Figures 8.5–8.7, the C=O groups are part of a bent moiety of the molecule because of the *configuration* of the substrate. *Conformational* preferences can also bring C=O groups into positions in which they have a convex side that is easier to attack and a concave side that is more difficult to attack. In 4-*tert*-butylcyclohexanone, for example, the equatorially fixed *tert*-butyl group enforces such a geometry. Again, addition of the hydride donor from the convex side of this molecule is

A

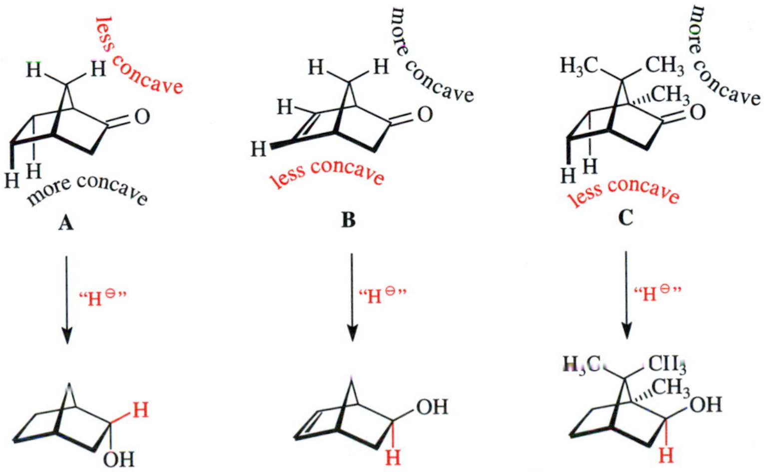

Fig. 8.7. Reaction of hydride donors with norbornanone (**A**), norbornenone (**B**), and camphor (**C**).

sterically favored; it corresponds to an equatorial attack (Figure 8.8). Therefore, the sterically demanding hydride donor L-Selectride converts 4-*tert*-butylcyclohexanone diastereoselectively into the cyclohexanol with the axial OH group (Figure 8.8, top reaction).

In contrast, sterically undemanding hydride donors such as $NaBH_4$ or $LiAlH_4$ reduce 4-*tert*-butylcyclohexanone preferentially through an axial attack. This produces mainly the cyclohexanol with the equatorial OH group (Figure 8.8, middle and bottom reactions). This difference results from the fact that there is also a stereoelectronic effect which influences the diastereoselectivity of the reduction of cyclohexanones.

In the explanation favored today, the reason for this stereoelectronic effect is as follows: The electronically preferred direction of attack of a hydride donor on the C=O double bond of cyclohexanone is the direction in which two of the C—H bonds at the neighboring α positions are exactly opposite the trajectory of the approaching nucleophile. Only the axial C—H bonds in the α positions can be in such an antiperiplanar position while the equatorial C—H bonds cannot. Moreover, these axial C—H bonds are antiperiplanar with regard to the trajectory of the H nucleophile only if the nucleophile attacks via a transition state **B,** that is, axially (what was to be shown). The "antiperiplanarity" of the two axial C—H bonds in the α positions is reminiscent of the "antiperiplanarity" of the electron-withdrawing group in the α position relative to the nucleophile in the Felkin–Anh transition state (formula **C** in Fig. 8.8; cf. Fig. 8.11, middle row).

In the attack by sterically undemanding reducing agents, this stereoelectronic effect is *fully* effective (for a completely different but perfectly diastereoselective reduction of 4-*tert*-butylcyclohexanone to the equatorial alcohol, see Figure 14.46). However, in the attack of such a bulky hydride donor as L-Selectride the stereoelectronic effect is *overcompensated* by the opposing steric effect discussed above.

Fig. 8.8. Addition of different hydride donors to 4-*tert*-butylcyclohexanone. For L-Selectride the equatorial attack is preferred (formula **A**), whereas for sterically undemanding hydride donors the axial attack via transition state **B** is preferred. For easier comparison: the Felkin–Anh transition state **C** (from Fig. 8.11; EWG, electron-withdrawing group).

Reagent	axial OH		equatorial OH
$Li^{\oplus}\,{}^{\ominus}BH(sec\text{-}Bu)_3$:	93	:	7
$NaBH_4$:	20	:	80
$LiAlH_4$:	8	:	92

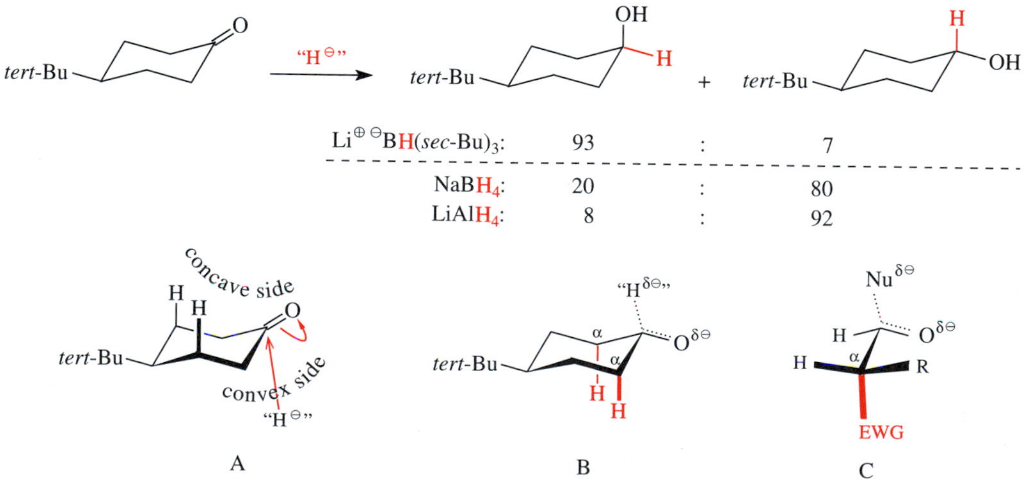

8.3.2 Diastereoselectivity of the Addition of Hydride Donors to α-Chiral Acyclic Carbonyl Compounds

B

In this section as well as in Section 8.5.3, carbonyl compounds that contain a stereocenter in the position α to the C=O group are referred to succinctly with the term "α-chiral carbonyl compound." Because of the presence of the stereocenter, the half-spaces on both sides of the plane of the C=O double bond of these compounds are diastereotopic. In this section we will study in detail stereogenic addition reactions of hydride donors to the C=O double bond of α-chiral carbonyl compounds. Additions of this type can take place faster from one half-space than from the other—that is, they can be diastereoselective.

Introduction: Representative Experimental Findings

B

If one wants to study a stereogenic addition reaction of a hydride donor to an α-chiral *aldehyde,* one must use, for instance, $LiAlD_4$ as the reducing agent (thus, a deuteride donor). In contrast, stereogenic addition reactions of hydride donors to α-chiral *ketones* require no deuterium labeling: They can be observed in "ordinary" reductions, for instance, with $LiAlH_4$. Which alcohol is preferentially produced in each case depends on, among other things, which atoms are connected to the stereocenter at C-α.

A

Additions of hydride donors to α-chiral carbonyl compounds that bear only hydrocarbon groups or hydrogen at C-α typically take place with the diastereoselectivities of Figure 8.9. One of the resulting diastereomers and the relative configuration of its stereocenters are referred to as the Cram product. The other diastereomer that results and its stereochemistry are referred to with the term anti-Cram product.

$LiAlD_4$ preferentially OH, D , little OH, D

racemic mixture or pure enantiomer

Ph, *tert*-Bu, O $LiAlH_4$ preferentially Ph, H, HO *tert*-Bu , little Ph, H, HO *tert*-Bu

R_{medium}, R_{large}, R, O $M^{\oplus} Nu^{\ominus}$ preferentially R_{large}, R_{medium}, Nu, HO R , little R_{large}, R_{medium}, Nu, HO R

Cram product

anti-Cram product

Fig. 8.9. Examples and structural requirements for the occurrence of Cram-selective additions of hydride donors to α-chiral carbonyl compounds. In the three compounds at the bottom R_{large} refers to the large and R_{medium} refers to the medium-sized substituent.

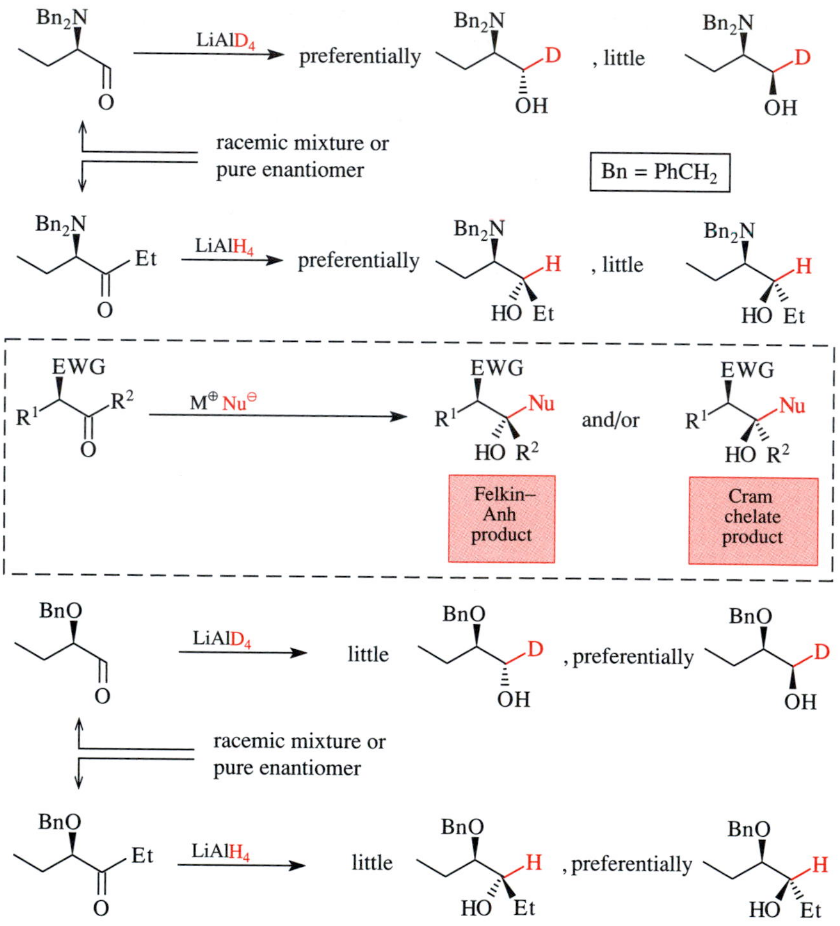

Fig. 8.10. Examples and structural requirements for the occurrence of Felkin–Anh selective (top) or chelation-controlled (bottom) additions of hydride donors to α-chiral carbonyl compounds. EWG, electron-withdrawing group.

In contrast, the diastereoselectivities of Figure 8.10 can be observed for many additions of hydride donors to carbonyl compounds which contain a stereocenter in the α position with an O or N atom bound to it. One of the product diastereomers and the relative configuration of its stereocenters is called the Felkin–Anh product. The other diastereomer and its stereochemistry are referred to as the so-called Cram chelate product. If the latter is produced preferentially, one also talks about the "occurrence of chelation control" or—only in laboratory jargon—of the predominance of the "chelation-controlled product."

The reason for the diastereoselectivities presented in Figures 8.9 and 8.10 will be explained in the following section.

The Reason for Cram and Anti-Cram Selectivity and for Felkin–Anh and Cram Chelate Selectivity; Transition State Models

B

Whether, in additions of hydride donors to α-chiral carbonyl compounds, Cram or anti-Cram selectivity, on the one hand, or Felkin–Anh or Cram chelate selectivity, on the other hand, occurs is the result of kinetic control. The rate-determining step in either of these additions is the formation of a tetrahedral intermediate. It takes place irreversibly. The tetrahedral intermediate that is accessible via the most stable transition state is produced most rapidly (as may be inferred from Hammond's postulate, irrespective of the enthalpy or free energy of this step). However, in contrast to what is found in many other considerations in this book, this intermediate does not represent a good transition state model for its formation reaction. The reason for this "deviation" is that it is produced in an exothermic and exergonic step. Thus, according to the Hammond postulate, the tetrahedral intermediate here is formed in an early transition state. Consequently, the latter does not resemble the tetrahedral intermediate but rather the carbonyl compound and the reducing agent. Calculations confirmed this for the Cram transition state (Figure 8.11, left) and for the Felkin–Anh transition state (Figure 8.11, middle). Therefore, the existence of an early transition state is plausible also in the case of chelation-controlled addition reactions (Figure 8.11, right).

From the transition state structures in Figure 8.11 it is seen that

1. The additions of hydride donors to the C=O double bonds of α-chiral carbonyl compounds take place via transition states whose stereostructures do not reflect the preferred conformations of the substrates.
2. The addition of a hydride donor to an α-chiral carbonyl compound, *without* an O or N atom in the α position, regardless of whether it is racemic or enantiomerically pure, takes place through the so-called Cram transition state. It is shown in Figure 8.11 (left; $Nu^- = H^-$) both as a Newman projection and in the sawhorse representation. In the Cram transition state the H atom (or the smallest hydrocarbon substituent) at C-α is aligned approximately antiperiplanar to the attacked C=O double bond. The hydride donor attacks the C=O double bond from the half-space that does *not* contain the largest hydrocarbon substituent at C-α. Incidentally, the nucleophile attacks the C=O double bond at an angle of ca. 103° (i.e., slightly from the rear). In other words, in the Cram transition state the hydride donor attacks the carbonyl carbon in a trajectory which is almost *anti* to the bond that connects C-α to the largest α substituent.
3. The addition of a hydride donor to an α-chiral aldehyde *with* an O or an N atom in the α position or to an analogous ketone takes place through the so-called Felkin–Anh transition state provided that the heteroatom at C-α is *not* incorporated in a five-membered chelate ring together with the O atom of the carbonyl group. This transition state is also shown in Figure 8.11 (center; $Nu^-{=}H^-$), both as a Newman projection and in the sawhorse representation. The H atom (or the second largest hydrocarbon substituent) at C-α is aligned approximately antiperiplanar to the attacked C=O double bond. The hydride donor approaches the substrate from the half-space that does *not* contain the heteroatom in the α position. Again, the nucleophile attacks the C=O double bond at an angle of ca. 103° (i.e., slightly from the rear). In other words, in the Felkin–Anh transition state the C—H bond being formed and the C_α—heteroatom bond are arranged almost *anti*.
4. The addition of a hydride donor to an α-chiral aldehyde with an OR or NR_2 substituent at C-α or to an analogous ketone takes place via the so-called Cram chelate

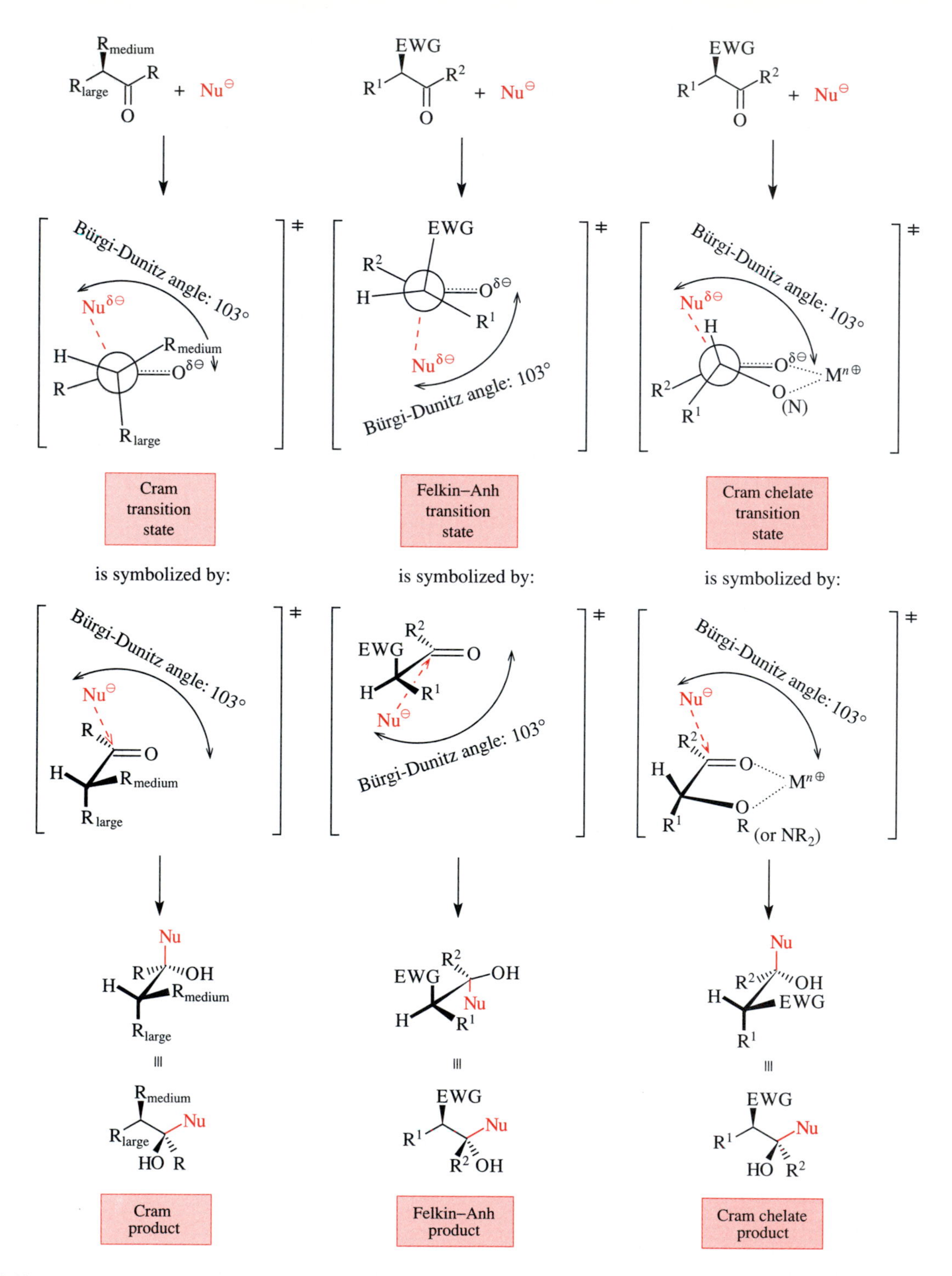

Fig. 8.11. The three transition state models for the occurrence of diastereoselectivity in the addition of hydride donors ($Nu^- = H^-$) or organometallic compounds ($Nu^- = R^-$) to α-chiral carbonyl compounds (R_{large} refers to the large substituent, R_{medium} refers to the medium-sized substituent, and EWG refers to an electron-withdrawing group).

transition state provided that the heteroatom at C-α and the O atom of the carbonyl group are incorporated in a five-membered chelate ring. This transition state is also shown in Figure 8.11 (right; $Nu^- = H^-$) in two different perspectives. The hydride donor approaches the carbonyl carbon from the less hindered half-space. This is the one that contains the H atom (or the second largest hydrocarbon group) in the α position.

Side Note 8.1 Historical Explanation for the Selectivities Named after Cram

B

After Cram had discovered the selectivities now named after him, he proposed the transition state model for the formation of *Cram chelate* products which is still valid today. However, his explanation for the preferred formation of *Cram* products was different from current views. Cram assumed that the transition state for the addition of nucleophiles to α-alkylated carbonyl compounds was so early that he could model it with the carbonyl compound alone. His reasoning was that the preferred conformation of the free α-chiral carbonyl compound defines two sterically differently encumbered half-spaces on both sides of the plane of the C=O double bond. The nucleophile was believed to attack from the less hindered half-space.

Today, it is known that Cram was wrong about this preferred conformation (he assumed that a C_α—hydrogen bond was oriented *syn* to the C=O double bond; however, a C_α—carbon bond actually occupies this position). If Cram had known this conformation at that time, he would certainly not have based his explanation of Cram selectivity upon an attack on the most stable conformer of the aldehyde. This is because, in order to establish the experimentally found product geometry, the nucleophile would have been required to approach the C=O double bond of the really favored conformer from the sterically more hindered side. However, Cram *should not* have based his argument on *any* preferred conformation of the aldehyde. This is because in doing that he committed an error—an error which even at that time could have been avoided (see below).

Curtin–Hammett Principle

B

Many organic chemists have overlooked the fundamental error in Cram's old argument and are continuously inclined to commit all sorts of analogous errors in proposing "explanations" for other stereoselectivities. This type of error in reasoning is unfortunately very tempting. However, it is as easy as this: In more than 99.9% of all stereogenic reactions converting a given substrate into two diastereomers the activation barriers of the selectivity-determining steps are distinctly higher than the rotational barriers which separate the different conformers of the substrate from each other. Figure 8.12 illustrates these relationships using stereogenic addition reactions to α-methylisovaleraldehyde as an example.

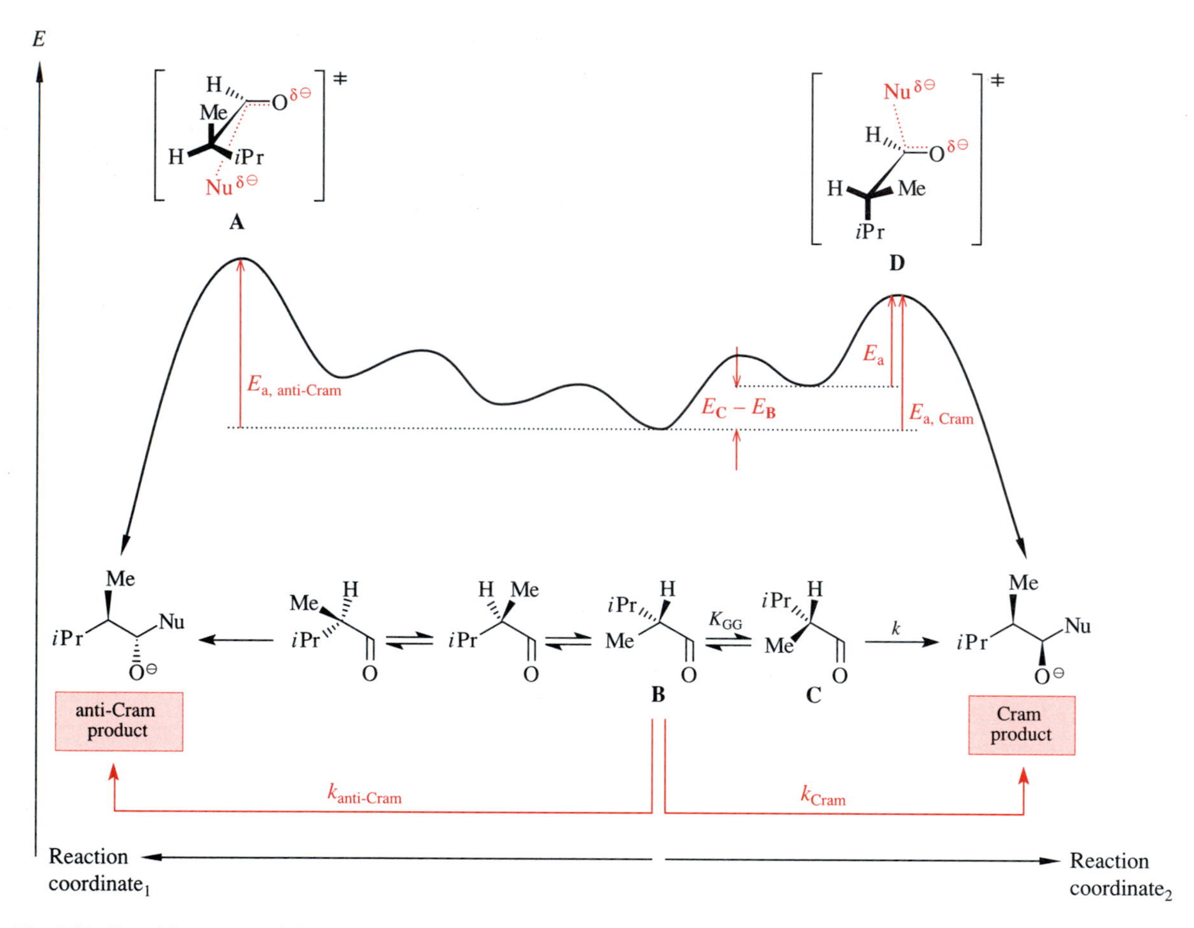

Fig. 8.12. Transition states of the selectivity-determining step of a stereogenic addition of a hydride donor to an α-chiral carbonyl compound (the energy profile would be allowed to contain additional local energy maxima provided that they do not have a higher energy than the two highest maxima shown in the figure).

Curtin–Hammett Principle

In stereogenic reactions leading from a substrate to two diastereomers, for determining the favored reaction it is irrelevant which conformer of the substrate is preferred. The favored reaction is the one that takes place via the transition state with the lowest energy. This is also true when the transition structure is not derived from the preferred conformation of the substrate. The extent of diastereoselectivity encountered in the stereogenic reaction under scrutiny results from the competition between the favored and the disfavored reaction paths; it depends exclusively on the free enthalpy difference between the competing, diastereomorphic transition states.

Here, the Curtin–Hammett principle will be proven using the example of the reaction pair from Figure 8.12. The Cram product forms from a conformer **C** of the α-chiral aldehyde via the Cram transition state **D** with the rate constant *k*. The corresponding rate law is obtained as Equation 8.1. However, the conformer **C** of the aldehyde also equilibrates rapidly with the more stable conformer **B.** Therefore, only a small fraction of the total aldehyde is present as **C.** The concentration of conformer **C** is to a very good approximation described by Equation 8.2 as a function of the concentration of conformer **B.** Equation 8.3 follows from the fact that virtually all of the aldehyde present adopts the conformation **B.** (If substantial amounts of other conformers were also present, one would have to introduce a proportionality factor <1 on the right side of Equation 8.3. However, this proportionality factor would have no influence on the following derivation.)

$$\frac{d[\text{Cram product}]}{dt} = k[\mathbf{C}][\mathrm{Nu}^{\ominus}] \tag{8.1}$$

$$[\mathbf{C}] \approx K_{eq} \cdot [\mathbf{B}] \tag{8.2}$$

$$[\mathbf{B}] \approx [\text{aldehyde}] \tag{8.3}$$

Equation 8.4 is obtained by successive insertions of Equations 8.2 and 8.3 into Equation 8.1. The rate constant k of Equation 8.4 can also be written in the form of Equation 8.5 which is the Arrhenius law. In addition, one can rewrite the equilibrium constant K_{eq} of the conformer equilibrium $\mathbf{B} \rightleftharpoons \mathbf{C}$ as Equation 8.6, which relates K_{eq} to the energy difference between **C** and **B.**

$$\frac{d[\text{Cram product}]}{dt} = k \cdot K_{eq} \cdot [\text{aldehyde}][\mathrm{Nu}^{\ominus}] \tag{8.4}$$

$$k = A \cdot \exp\left(-\frac{E_a}{RT}\right) \tag{8.5}$$

$$K_{eq} = \exp\left(-\frac{E_{\mathbf{C}} - E_{\mathbf{B}}}{RT}\right) \tag{8.6}$$

We obtain Equation 8.7 by multiplying Equation 8.5 with Equation 8.6. When we now take Figure 8.12 into consideration, we can summarize the energy terms in the numerator of Equation 8.7, as shown in Equation 8.8. By using the Arrhenius relationship, we can abbreviate the exponent on the right side of Equation 8.8 as rate constant k_{Cram}. The result is Equation 8.9.

$$k \cdot K_{eq} = A \cdot \exp\left[-\frac{E_a + (E_{\mathbf{C}} - E_{\mathbf{B}})}{RT}\right] \tag{8.7}$$

$$= A \cdot \exp\left(-\frac{E_{a,\text{Cram}}}{RT}\right) \tag{8.8}$$

$$= k_{\text{Cram}} \tag{8.9}$$

By inserting Equation 8.9 into Equation 8.4, we finally obtain

$$\frac{d[\text{Cram product}]}{dt} = k_{\text{Cram}} \cdot [\text{aldehyde}][\mathrm{Nu}^{\ominus}] \tag{8.10}$$

Equation 8.11 for the rate of formation of the anti-Cram product is derived in a similar way:

$$\frac{\mathrm{d[anti\text{-}Cram\ product]}}{\mathrm{d}t} = k_{\text{anti-Cram}} \cdot [\text{aldehyde}][\text{Nu}^{\ominus}] \tag{8.11}$$

One now divides Equation 8.10 by Equation 8.11, reduces and integrates from the beginning of the reaction to the time of the work up.

In this way, and by using the Arrhenius relationship one more time, one obtains for the ratio of the yields:

$$\frac{\text{Yield of Cram product}}{\text{Yield of anti-Cram product}} = \frac{k_{\text{Cram}}}{k_{\text{anti-Cram}}} = \exp\left\{-\frac{E_{\text{a,Cram}} - E_{\text{a,anti-Cram}}}{RT}\right\} \tag{8.12}$$

Equation 8.12 states that for the diastereoselectivity of stereogenic additions of nucleophiles to α-chiral aldehydes it is neither important which conformation the substrate prefers nor is it important of how many conformers of the substrate exist.

Felkin–Anh or Cram Chelate Selectivity in the Addition of Hydride Donors to Carbonyl Compounds with an O or N Atom in the α-Position?

A

In order for the Cram chelate product to predominate after the addition of a hydride donor to a chiral carbonyl compound, which contains a heteroatom in the α position, this heteroatom and part of the reagent must be able to form a five-membered ring chelate. If this is not possible, one observes Felkin–Anh selectivity (provided one observes selectivity at all). This has the following interesting consequences for synthesis.

The α-chiral ketone from Figure 8.13—the α substituent is a benzyloxy group—is reduced to the Cram chelate product by $Zn(BH_4)_2$, a Lewis acidic reducing agent. The Zn^{2+} ion first links the benzyl and the carbonyl oxygen to a chelate. Only this species is subsequently reduced by the BH_4^- ion because a Zn^{2+}-complexed C=O group is a better electrophile than a Zn^{2+}-free C=O group. On the other hand, the Felkin–Anh product (Figure 8.13) is produced from the same ketone and $KBH(sec\text{-}Bu)_3$ (K-Selectride, i.e., the potassium analog of L-Selectride). Neither the K^+ ion nor the tetra-coordinated boron atom of this reagent acts as a Lewis acid. Consequently, no chelate intermediate can be formed.

If, for example, the mild Lewis acid $LiAlH_4$—it is mildly acidic because its acidic moiety is Li^+—is used as a hydride donor, the diastereoselectivity of the reduction of α-chiral carbonyl compounds with a heteroatom in the α position can be controlled by

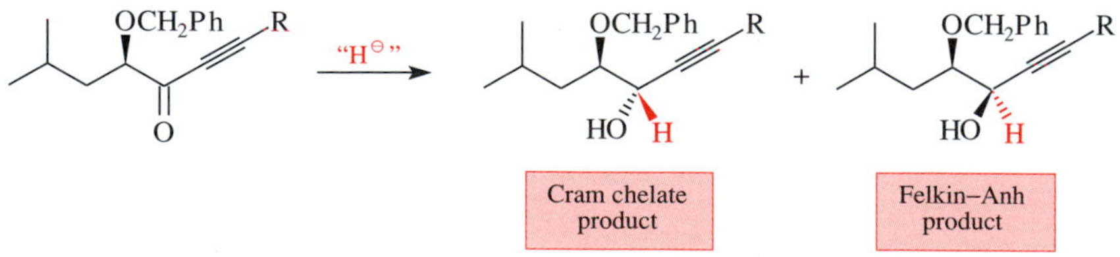

	Cram chelate product		Felkin–Anh product
$Zn(BH_4)_2$:	95	:	5
$K^{\oplus}\,{}^{\ominus}BH(sec\text{-}Bu)_3$:	10	:	90

Fig. 8.13. Ensuring Felkin–Anh versus Cram chelate selectivity by varying the Lewis acidity of the hydride donor.

	Cram chelate product		Felkin–Anh product
R = $PhCH_2$:	98	:	2
R = Me_2*tert*-BuSi :	5	:	95

Fig. 8.14. Ensuring Felkin–Anh or Cram chelate selectivity in the addition of hydride donors to α-chiral carbonyl compounds by varying the protecting group on the stereo-directing heteroatom.

a judicious choice of the protecting group on the heteroatom. Figure 8.14 shows this using the example of two differently protected α-oxygenated ketones. In the first ketone, a benzyl group is attached to the O atom at C-α. This protecting group is not bulky because it is not branched. Accordingly, it allows the incorporation of the protected α oxygen in a five-membered ring chelate. Therefore, the reduction of this substrate with $LiAlH_4$ leads preferentially to the Cram chelate product. On the other hand, if one employs a trialkylsilyl group with branched alkyl substituents as the protecting group at the α oxygen, it is bulky enough to prevent this formation of a chelate ring. Therefore, the hydride addition to the silylether from Figure 8.14 takes place via a Felkin–Anh transition state and delivers the Felkin–Anh product.

Of course, the highest diastereoselectivities in the reduction of α-chiral α-oxygenated carbonyl compounds are expected when the reducing agent and the protecting group direct in one and the same direction; that is, when both make possible or both prevent

anti-1,2-Diol

syn-1,2-Diol

Fig. 8.15. Ensuring Felkin–Anh or Cram chelate selectivity in the addition of hydride donors to α-chiral carbonyl compounds by a combined variation of the hydride donor and the protecting group.

chelation control. In the reaction examples depicted in Figure 8.15, it was possible in this way to achieve a diastereoselectivity of 98:2 in the sense of Cram chelation control or optionally of 95:5 in the opposite Felkin–Anh direction. Figure 8.15 puts these reductions into the context of how diastereopure glycols can be synthesized. Stereochemically uniform glycols of this type are used, for example, as the starting materials of Corey–Winter eliminations (cf. Section 4.7.5).

8.3.3 Diastereoselectivity of the Addition of Hydride Donors to β-Chiral Acyclic Carbonyl Compounds

A

In this section and in Section 8.5.3, the term "β-chiral carbonyl compound" will be used as an abbreviation for carbonyl compounds that contain a stereogenic center in the position β to the C=O group. Stereogenic addition reactions of hydride donors to β-chiral carbonyl compounds have been described especially for substrates in which the stereocenter in the β position is connected to an O atom.

β-Hydroxyketones can be reduced diastereoselectively with a hydride donor in an analogous way to that shown in Figure 8.15 for α-hydroxyketones by using chelation control. This succeeds best when the β-hydroxyketone is fixed in a chelate complex of structure **A** through a reaction with diethylborinic acid methyl ester (Figure 8.16). In chelate **A** the diastereotopic sides of the ketonic C=O group are not equally well accessible. Thus, $NaBH_4$ added after formation of this chelate leads to the transfer of a hydride ion from that side of the molecule which lies opposite to the pseudo-

Fig. 8.16. Diastereoselective reduction of β-hydroxyketones to *syn*-configured 1,3-diols.

Fig. 8.17. Diastereoselective reduction of β-hydroxyketones to *anti*-configured 1,3-diols.

equatorially oriented substituent R′ in formula **B** of chelate **A.** This selectivity is not so much caused by the avoidance of the still quite distant R′ substituent by the $NaBH_4$. Rather, a transition state is preferred, in which the hydride ion to be transferred assumes an axial orientation in the forming saturated six-membered ring and that six-membered ring assumes a chair rather than boat conformation. Under these conditions the borane-containing part of the reagent resides on the convex, that is, the less hindered side of the least-strained conformer of the transition state.

The addition of a hydride donor to a β-hydroxyketone can also be conducted such that the opposite diastereoselectivity is observed. However, the possibility previously discussed for additions to α-chiral carbonyl compounds is not applicable here. One must therefore use a different strategy as is shown in Figure 8.17, in which the OH group at the stereocenter C-β of the substrate is used *to bind the hydride donor before it attacks the C=O double bond.* Thus, the hydridoborate **A** reacts intramolecularly. This species transfers a hydride ion to the carbonyl carbon after the latter has been protonated and thereby made more electrophilic. The hydride transfer takes place via a six-membered chair-like transition state, which preferentially assumes the stereostructure **B.** It is characterized by the energetically favorable pseudo-equatorial arrangement of the ring substituents R^1 and R^2. This lets the double bonded oxygen atom, as the least voluminous substituent, assume the pseudo-axial orientation shown in the figure.

8.4 Enantioselective Addition of Hydride Donors to Carbonyl Compounds

B

Stereogenic additions of hydride donors to achiral deuterated aldehydes R—C(=O)D or to achiral ketones $R^1R^2C(=O)$ take place without stereocontrol using the reagents which you learned about in Section 8.3. Thus, *racemic* deuterated alcohols R—C(OH)D or *racemic* secondary alcohols $R^1R^2C(OH)H$ are produced. The reason for this is that the C=O double bond of the mentioned substrates is flanked by enantiotopic half-spaces. From these spaces achiral hydride donors *must* attack with the same reaction rates (cf. Section 3.4.1). On the other hand, as you can extrapolate from Section 3.4.1, chiral and enantiomerically pure reducing agents might attack from each of the half-spaces with different rates. Therefore, reducing agents of this type can in principle effect enantioselective reductions of achiral deuterated aldehydes or achiral ketones. The most important among these reagents are the Noyori reagent (Figure 8.18), Alpine-Borane (Figure 8.19), Brown's chloroborane (Figure 8.20), and the oxazaborolidines by Corey and Itsuno (Figure 8.21). The first three of these reagents allow enantioselective carbonyl group reductions if they are used in stoichiometric amounts, while the fourth class of reagents make *catalytic* enantioselective carbonyl group reductions possible.

A

The hydride donor of the **Noyori reduction** of ketones is the hydrido aluminate *R*-BINAL-H shown in Figure 8.18 or its enantiomer *S*-BINAL-H. The new C—H bond is presumably formed via a cyclic six-membered transition state of stereostructure **A.** Unfortunately, there is no easy way to rationalize why enantioselectivity in this kind of addition is limited to substrates in which the carbonyl group is flanked by one conjugated substituent (C≡C, aryl, C=C).

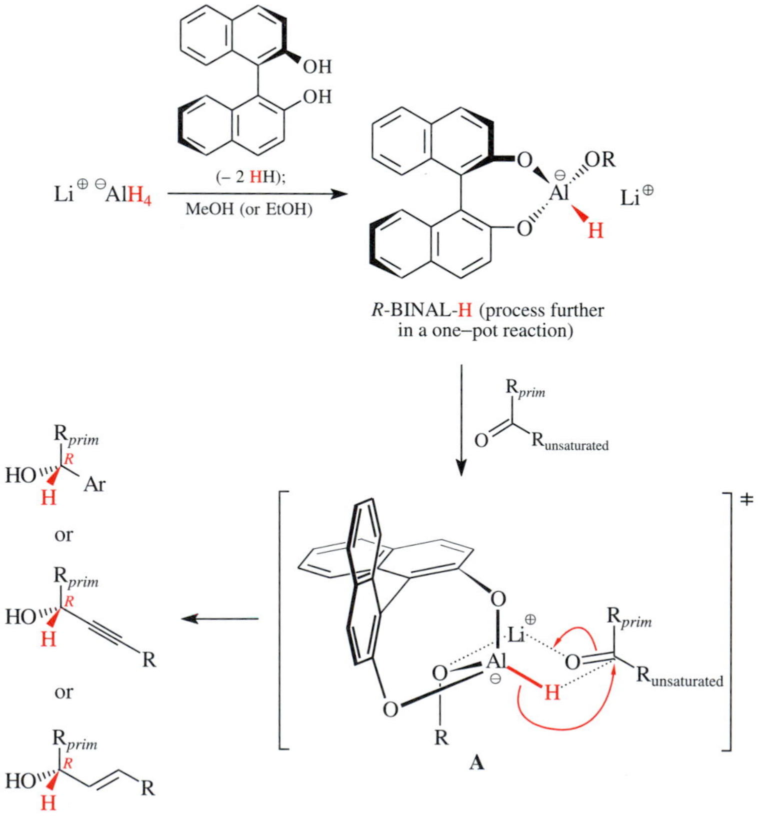

Fig. 8.18. Asymmetric carbonyl group reduction with the Noyori reagent.

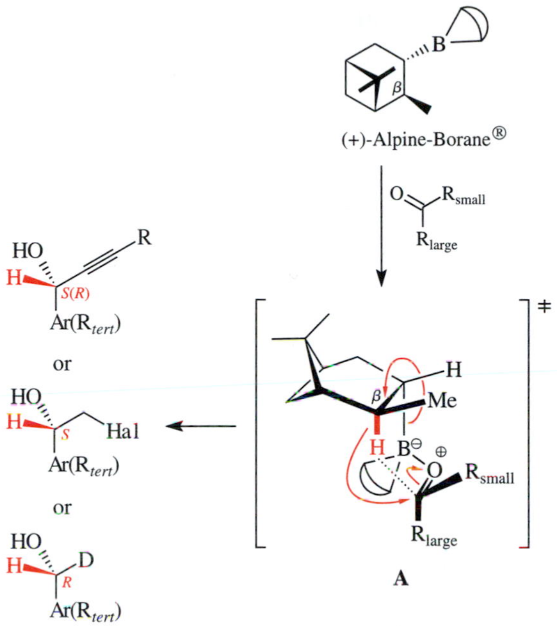

Fig. 8.19. Asymmetric carbonyl group reduction with Alpine-Borane (preparation: Figure 3.21).

B

Another set of carbonyl compounds can be reduced enantioselectively with **Alpine-Borane** (Figure 8.19). Alpine-Borane, just like **diisopinocampheylchloroborane (Brown's chloroborane)** (Figure 8.20), is a reducing reagent of the type $H{-}^{\beta}C{-}^{\alpha}C{-}ML_n$. Each of these reagents transfers the H atom which is bound to the C atom in the position β to the boron atom to the carbonyl carbon of the substrate. The metal and its substituents L_n are transferred concomitantly to the carbonyl oxygen. At the same time, the atoms C-β and C-α become part of an olefin, which is formed as a stoichiometric by-product. Thus, Alpine-Borane as well as diisopinocampheylchloroborane release α-pinene in these reductions. The carbonyl compounds whose C═O double bonds are reduced by these boranes in highly enantioselective hydride transfer reactions are the same for both reagents: aryl conjugated alkynones, phenacyl halides, and deuterated aromatic aldehydes.

The stereoselectivities of the carbonyl group reductions with Alpine-Borane (Figure 8.19) or with diisopinocampheylchloroborane (Figure 8.20) are explained as shown in the formula schemes. Being Lewis acids, these reagents first bind to the carbonyl oxygen of the substrate. Then the H atom located in the position β to the boron atom is transferred to the carbonyl carbon. This takes place via six-membered transition states **A,** which are necessarily in the boat form. The preferred orientation in **A** of the substituents on the carbonyl carbon relative to the reagent follows from the requirement to minimize 1,3-diaxial interactions with the methyl group on the β C atom of the reagent. This interaction is smallest when this methyl group lies opposite the smaller substituent of the carbonyl carbon and not opposite the larger substituent.

Reductions with diisopinocampheylchloroborane [$(IPC)_2BCl$] often afford higher *ee* values than reductions with Alpine-Borane because $(IPC)_2BCl$ is the stronger Lewis

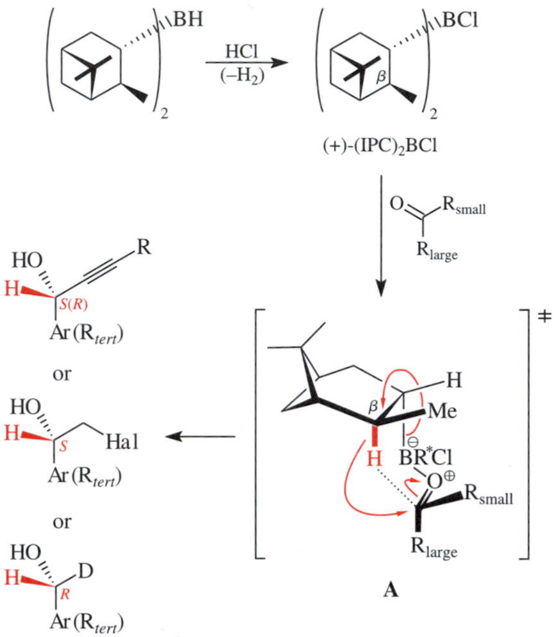

Fig. 8.20. Asymmetric carbonyl group reduction with diisopinocampheylchloroborane [Brown's chloroborane, $(IPC)_2BCl$].

acid. Therefore, in the transition state **A** (Figure 8.20) the O—B bond is stronger than in the Alpine-Borane analog and accordingly also shorter. This enhances the 1,3-diaxial interactions between the carbonyl substituents and the methyl group at C-β of the reagent so that the need to minimize them is also enhanced.

There is a second effect. In the transition state in which the stronger Lewis acid complexes the carbonyl oxygen, the carbonyl group is a better electrophile. Therefore, it becomes a better hydride acceptor for Brown's chloroborane than in the hydride transfer from Alpine-Borane. Reductions with Alpine-Borane can actually be so slow that decomposition of this reagent into α-pinene and 9-BBN takes place as a competing side reaction. The presence of this 9-BBN is problematic because it reduces the carbonyl compound competitively and of course without enantiocontrol.

A

The enantioselective carbonyl group reduction by the Corey–Itsuno process is particularly elegant (Figure 8.21). There, in contrast to the processes of Figures 8.18–8.20, one requires a stoichiometric amount only of an achiral reducing agent (BH_3), which is much cheaper than the stoichiometrically required reducing reagents of Figures 8.18–8.20. Enantiocontrol in the Corey-Itsuno reduction stems from the enantiomerically pure auxiliary **A.** While this *is* comparatively expensive, fortunately, one needs only a few mole percent thereof. That with this reagent combination enantioselective reductions are achieved at all is due to the fact that, in the absence of the chiral auxiliary **A,** BH_3 reacts only quite slowly with ketones of the type R_{large}—C(=O)—R_{small}. However, the chiral auxiliary **A** forms a Lewis acid/Lewis base complex **B** with such a ketone. This increases the electrophilicity of the carbonyl group such that it can now rapidly be reduced by BH_3. Consequently, we encounter here a new example of a *ligand-accelerated reaction*—that is, a class of reactions which also includes the Sharpless epoxidations discussed in Section 3.4.6. Additional reactions of this type are the organozinc reactions of Figures 8.30 and 8.31.

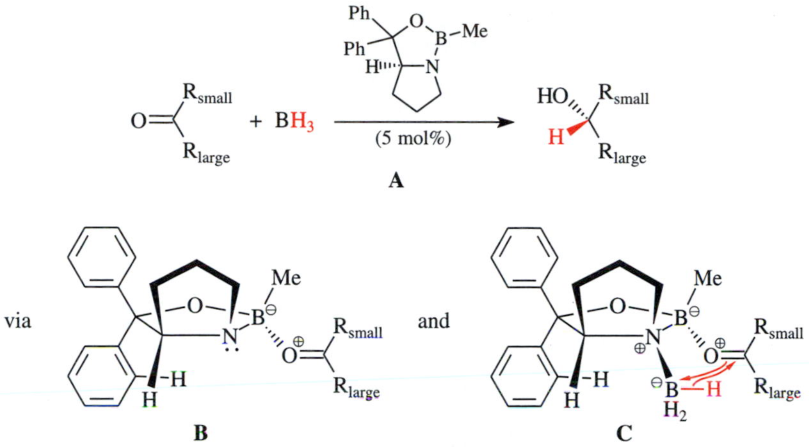

Fig. 8.21. Catalytic asymmetric carbonyl group reduction according to Corey and Itsuno.

The N atom of the ketone-associated heterocycle **B** of the Corey–Itsuno reduction (Figure 8.21) is a Lewis base: Other than in the ketone-free heterocycle **A,** its lone electron pair is not shared with the boron atom because the latter has already acquired a valence electron octet. Therefore, the electron pair is available for binding a molecule of the Lewis acid BH_3, whereby the ternary complex **C** is produced.

Looking at all the reactions that took place for the formation of the complex **C,** the heterocycle **A** plays the role of a "molecular glue," which is possible because it represents both a Lewis acid and a Lewis base. As a result of this dual role, **A** fixes the electrophile (the ketone) and the hydride donor (the BH_3) in close proximity. In this way the complex **C** makes possible a quasi-intramolecular reduction of the ketone. It takes place stereoselectively in such a way as the preferred, special arrangement of the reaction partners in **C** readily suggests (Figure 8.21): being a bicyclic compound with a convex and a concave side, the heterocycle binds both the ketone and the BH_3 on its less hindered convex side. The presence of the phenyl group on the convex side of the molecule ensures that the axis of the O=C bond of the coordinated ketone points away from it. Given this orientation of the ketone axis, the methyl group on the heterocyclic B atom forces the large ketone substituent R_{large} out of its way. This suffices to define *unambiguously* the face selectivity of the BH_3 addition to the C=O double bond. Consequently, Corey–Itsuno reductions exhibit high *ee* values.

8.5 Addition of Organometallic Compounds to Carbonyl Compounds

B

Alkyllithium and alkyl Grignard reagents, alkenyllithium and alkenyl Grignard reagents, aryllithium and aryl Grignard reagents, as well as alkynyllithium and alkynyl Grignard reagents, can be added to the carbonyl group of aldehydes and ketones. The addition of alkynyllithium compounds to sterically more demanding ketones, however, is sometimes only possible after adding one equivalent of $CeCl_3$. The addition then takes place via an organocerium compound of the composition R—C≡C—$CeCl_2$, which is formed *in situ.* Diorganozinc compounds do not add to aldehydes as such. Yet, they do so after reacting with CuCN and in the presence of a Lewis acid. Alkyl zinc iodides behave similarly under the same conditions, but the Lewis acid can be replaced by Me_3SiCl. On the other hand, organocopper compounds which do not contain an additional metal besides copper and lithium are, as a rule, not able to add to the carbonyl carbon.

8.5.1 Simple Addition Reactions of Organometallic Compounds

Similarities and Differences in the Reactions of Organolithium vs Grignard Reagents with Carbonyl Compounds

B

Often, the addition of organolithium compounds to carbonyl compounds gives the same alcohols that one would also get from the analogous Grignard reagents and the same carbonyl compounds. For example, one could prepare all the products in Figures 8.25–8.28, which result from the addition of Grignard reagents as well as from the addition of the analogous organolithium compounds.

In certain additions of aryl nucleophiles to C=O double bonds, however, it is advantageous to use organolithium instead of organomagnesium compounds:

- When one is working on a very small scale, one can prepare aryllithium compounds through a Br/Li exchange reaction (path 3 in Figure 5.34) in a much simpler way than aryl Grignard reagents.
- By means of an *ortho*-lithiation of suitably functionalized aromatics it is possible to obtain aryllithium compounds from halogen-free aromatics in a way and with a regioselectivity (Section 5.3.1) for which there is no analogy in the preparation of aryl Grignard reagents.

Some addition reactions to C=O double bonds can be carried out at all only with organolithium and not with Grignard reagents. For instance, bulky Grignard reagents do not react as C nucleophiles with sterically hindered ketones (Figure 8.22), but the analogous lithium organometallics do since they are stronger nucleophiles (Figure 8.23).

When bulky Grignard reagents contain an H atom in the position β to the magnesium, they can transfer this hydrogen to the C=O double bond in a way that is similar to that of the reduction of carbonyl compounds with Alpine-Borane (Figure 8.19) or with diisopinocampheylchloroborane (Figure 8.20)—via a six-membered transition state. Here, it is derived from a Lewis acid/Lewis base complex of structure **C** (Figure 8.22). A magnesium alkoxide is produced, which has the structure **E.** In the aqueous work-up compound **E** gives the reduction product of the ketone used and not the addition product of the Grignard reagent. This unexpected reaction is referred to as a **Grignard reduction.**

Being a *better* nucleophile, the lithium analog of the Grignard reagent in Figure 8.22 adds to the same ketone without problems (Figure 8.23, left). At first this furnishes

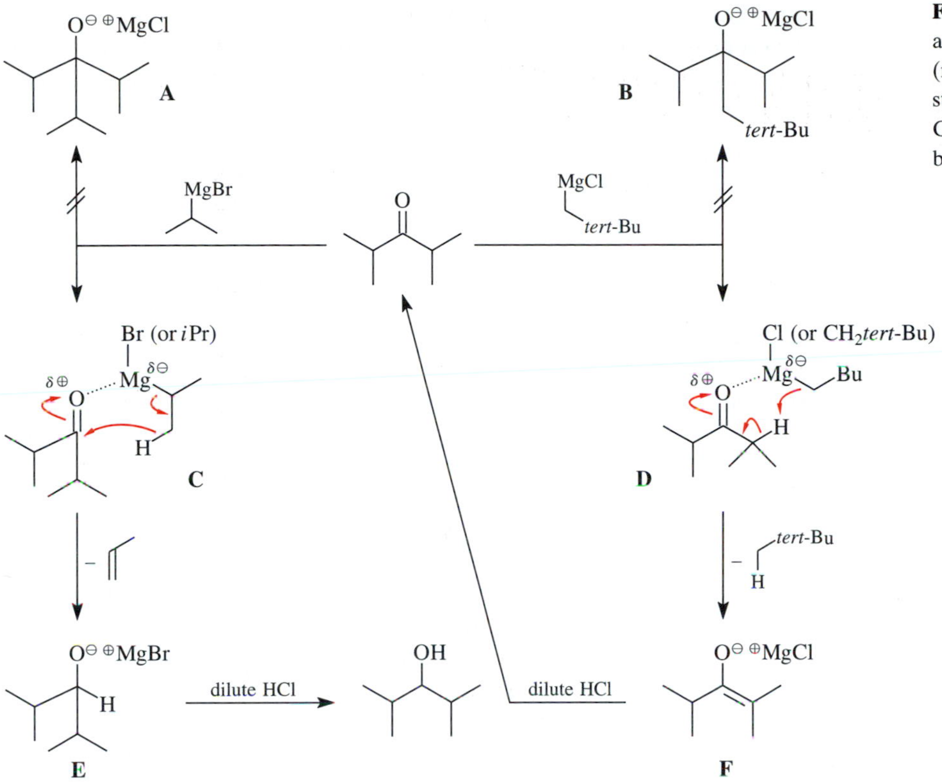

Fig. 8.22. Reduction (left) and enolate formation (right) in reactions of sterically demanding Grignard reagents with bulky ketones.

the lithium analog of the inaccessible magnesium alkoxide **A** of Figure 8.22. Its protonation under the weakly acidic workup conditions furnishes the sterically very hindered triisopropyl carbinol.

Neopentyl magnesium chloride also does not react with a hindered ketone such as diisopropyl ketone to form a magnesium alkoxide (formula **B** in Figure 8.22). Instead

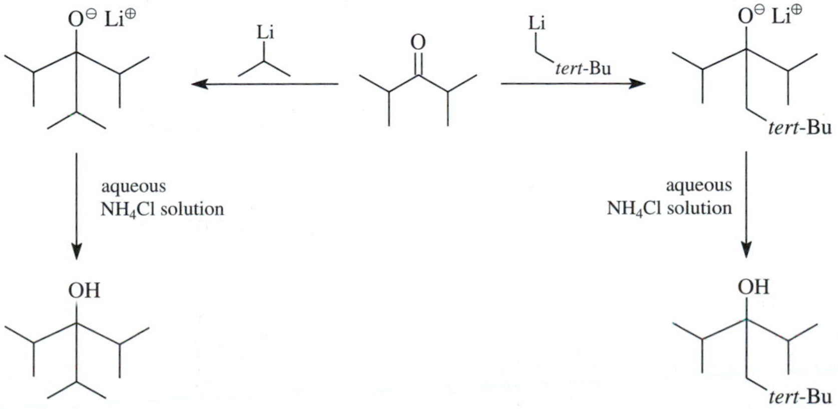

Fig. 8.23. Synthesis of sterically hindered alcohols by the reaction of sterically demanding organolithium compounds with bulky ketones. Preparation of triisopropyl carbinol (left) and diisopropylneopentyl carbinol (right).

the reactants first form a Lewis acid/Lewis base complex. This time it has the structure **D**. The neopentyl group then deprotonates the complexed ketone to give the magnesium enolate **F** via a six-membered cyclic transition state. In the aqueous workup one thus obtains only the unchanged starting ketone. Whereas diisopropylneopentyl carbinol is not accessible in *this* way, the use of the analogous lithium reagent as a C nucleophile provides help: It adds to diisopropyl ketone as desired (Figure 8.23, right).

There is one more type of addition reaction in which the use of organolithium compounds instead of Grignard reagents is a must: when the goal is to prepare an allyl alcohol through the addition of an organometallic compound to the C=O double bond of an α,β-unsaturated ketone (Figure 8.24). Grignard reagents often attack α,β-unsaturated ketones both at the carbonyl carbon and at the center C-β. However, only in the first case is the desired addition initiated. With respect to the distance between the added organic group and the metal ion in the initially obtained alkoxide, this addition mode is called a **1,2-addition.** On the other hand, the attack of a Grignard reagent at the center C-β of an α,β-unsaturated ketone leads to an enolate. Therein, the newly added organic group and the MgX^+ ion are located in positions 1 and 4. Consequently, this additon mode is called a **1,4-addition.**

A smooth 1,2-addition is almost always observed in the reaction of α,β-unsaturated ketones with organolithium compounds (Figure 8.24, bottom). Only in extreme cases such as additions to α,β-unsaturated trityl ketones do organolithium compounds also undertake 1,4-additions.

The intermediate formed in a 1,2-addition to an α,β-unsaturated ketone is an alkoxide, whereas the intermediate of the analogous 1,4-addition is an enolate (Fig. 8.24). Alkoxides are stronger bases than enolates because alcohols are weaker acids compared to enols. This is due to the localization of the negative charge in the alkoxide, whereas the charge is delocalized in the enolate. Hence, the alkoxide intermediate of a 1,2 addition is higher in energy compared to the enolate intermediate of a 1,4-addition.

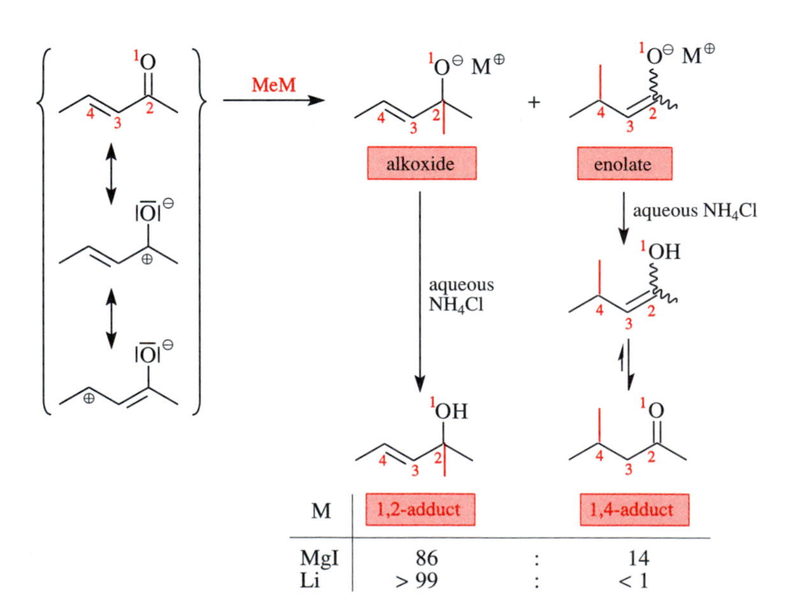

Fig. 8.24. Competition of 1,2- and 1,4-addition in the reaction of organolithium compounds and Grignard reagents with α,β-unsaturated ketones.

In the reaction of Grignard reagents, which are weaker C nucleophiles than organolithium compounds, some product-development control could occur in the transition state of the addition to α,β-unsaturated ketones. This would explain why the more stable intermediate (i.e., the enolate) is obtained starting from a Grignard reagent rather than starting from an organolithium compound.

Selective 1,4-additions of organometallic compounds to α,β-unsaturated ketones can also be achieved, namely, starting from organocopper and organozinc reagents. We will treat this in Section 8.6.

Addition of Grignard Reagents to Carbonyl Compounds: The Range of Products

B

Grignard reagents add to the C=O double bond of formaldehyde, higher aldehydes, and sterically unhindered or moderately hindered ketones (remember, though, that combined with sterically very hindered ketones, Grignard reagents undergo side reactions, as shown in Figure 8.22). In each case the primary product is a magnesium alkoxide. Its protonation with dilute HCl or dilute H_2SO_4 then gives an alcohol as the final product. In the preparation of acid-sensitive alcohols—that is, of tertiary alcohols or allyl alcohols—an aqueous NH_4Cl solution or a weakly acidic phosphate buffer is used to effect the protonation (Figures 8.26 and 8.27). If in these cases one were to also use the dilute mineral acids mentioned previously, one would induce acid-catalyzed secondary reactions of these acid-sensitive alcohols, for example, E1 reactions (Figure 8.28). This is because such mineral acids would protonate not only the alkoxide but also the resulting alcohol, and the latter would lead to an irreversible elimination.

tert-BuBr $\xrightarrow{\text{Mg}}$ *tert*-BuMgBr $\xrightarrow[\text{dilute HCl}]{\text{HCH=O;}}$ *tert*-Bu$-CH_2-$OH

i.e., formaldehyde $\longrightarrow$ $R_{prim}-OH$

Fig. 8.25. Preparation of a primary alcohol from formaldehyde and a Grignard reagent.

The addition of Grignard reagents to formaldehyde gives primary alcohols (Figure 8.25), and the addition to higher aldehydes gives secondary alcohols (Figure 8.26).

Fig. 8.26. Preparation of a secondary alcohol from a higher aldehyde and a Grignard reagent.

Finally, the addition of Grignard reagents to ketones gives tertiary alcohols when one works up with a weak acid (Figure 8.27). Adding a strong acid to the magnesium alkoxide intermediate leads to further reactions of these tertiary alcohols giving, for example, E1 products instead (Figure 8.28).

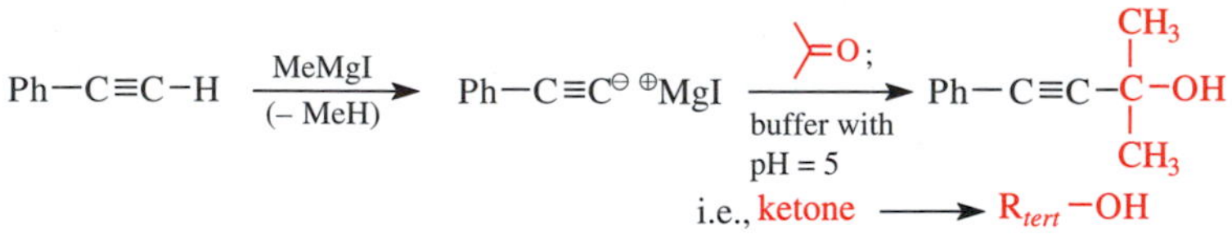

Fig. 8.27. Preparation of a tertiary alcohol by the addition of Grignard reagents to ketones and subsequent workup with weak acids.

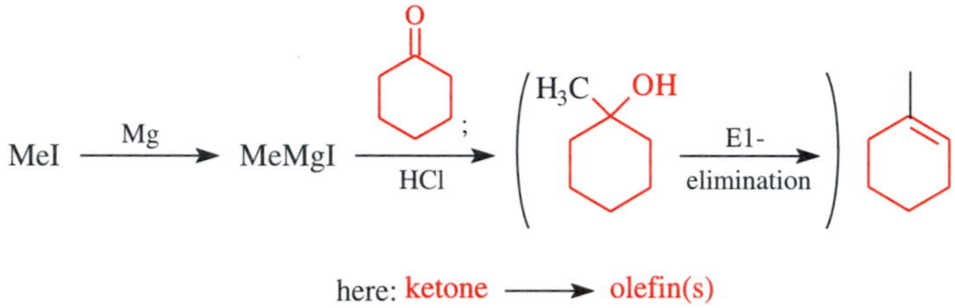

Fig. 8.28. Preparation of an olefin by adding a Grignard reagent to a ketone and subsequent workup with a strong acid.

Addition of Knochel Cuprates to Aldehydes

B

The preparation of organolithium compounds and of Grignard reagents from halides is successful only when these reagents contain virtually no functional groups. Furthermore, these C nucleophiles can be added to the C═O double bond of a carbonyl compound only when this carbonyl compound also contains practically no additional electrophilic group(s).

Organozinc compounds exhibit a much greater compatibility with functional groups both in the reagent and in the substrate of an addition reaction. The standard preparation of organozinc compounds starts with primary alkyl iodides (see also Figure 14.38). These iodides are reduced with metallic zinc (mechanism: analogous to Figure 14.37) to give zinc analogs R_{FG}—Zn—I (FG, contains a functional group) of the almost always unfunctionalized Grignard reagents R—Mg—I. These alkyl zinc iodides R_{FG}—Zn—I can contain, for example, the following functional groups: CO_2R, R^1R^2C═O, C≡N, Cl, RNH, NH_2RC(═O)NH, sulfoxide, sulfone, and an internal or a terminal C≡C triple bond.

Alkyl zinc iodides R_{FG}—Zn—I are poor nucleophiles. However, they are turned into good nucleophiles when they are converted into the so-called **Knochel cuprates** R_{FG}—Cu(CN)ZnHal with solubilized CuCN—that is, CuCN containing LiHal. Knochel cuprates add to aldehydes in the presence of Lewis acid, as shown in Figure 8.29.

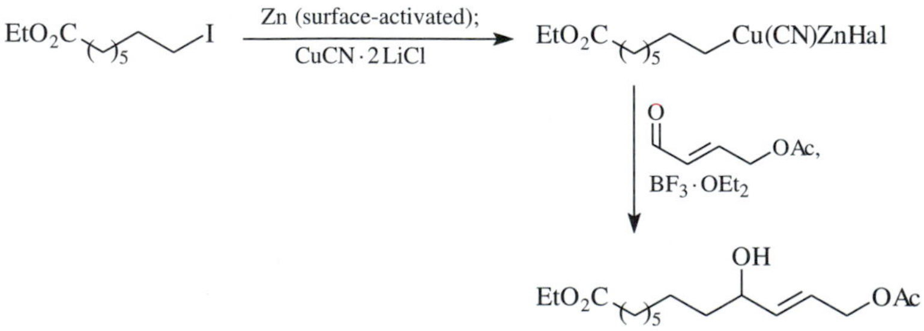

Fig. 8.29. Addition of a Knochel cuprate to an aldehyde.

Another type of organozinc compound also adds to aldehydes in the presence of Lewis acids. The significance of this reaction is that it can also be achieved enantioselectively (cf. Section 8.5.2).

8.5.2 Enantioselective Addition of Organozinc Compounds to Carbonyl Compounds: Chiral Amplification

Zinc-containing C nucleophiles, which tolerate the presence of diverse functional groups, are not limited to the mentioned Knochel cuprates obtained from alkyl iodides (Figures 8.29 and 8.34 right). The other zinc-based C nucleophiles, which can contain a comparably broad spectrum of functional groups, are dialkylzinc compounds. They are best prepared from terminal olefins. Figure 8.30 shows in the first two reactions how this is done.

One starts with a hydroboration with $HBEt_2$. The substrate of Figure 8.30 being a diene, this borane adds exclusively to the electron-rich and not at all to the acceptor-substituted C=C double bond. The reactive C=C double bond takes up the reagent regioselectively in such a way that the terminal C atom forms the C—B bond. This regioselectivity corresponds to the anti-Markovnikov selectivity one should expect according to Table 3.1. The trialkylborane **A** thereby formed is then subjected to a B/Zn exchange reaction with $ZnEt_2$, an equilibrium reaction that leads to the dialkylzinc compound **B** and BEt_3. The latter is removed continuously from the equilibrium by distillation. Thereby, the equilibrium is shifted completely toward the side of the desired dialkylzinc compound **B**.

Fig. 8.30. Preparation of a dialkylzinc compound **B** and its subsequent enantioselective addition to an aldehyde.

Reagents of this type are suitable for performing catalytic asymmetric additions to aldehydes. To that end, an enantiomerically pure Lewis acid is generated *in situ* from $Ti(OiPr)_4$ and the enantiomerically pure bis(sulfonamide) **C.** It catalyzes the enantioselective addition of functionalized (or unfunctionalized) dialkylzinc compounds to widely variable aldehydes. There is no detailed, substantiated rationalization of the underlying addition mechanism.

A different method for the catalytic asymmetric addition of a dialkylzinc compound—Et_2Zn and aromatic aldehydes have almost always been used—is shown in Figure 8.31. With regard to stereoselective synthesis, this method has an importance

Fig. 8.31. Catalytic asymmetric addition of Et_2Zn to Ph—C(=O)H. Chiral amplification through a mutual kinetic resolution of the $(auxiliary/ZnEt)_2$ complex.

which goes beyond the reaction shown there. This is because of the surprising finding that the addition product may have an *ee* of 95% even when the chiral auxiliary (−)-**A** is used with a relatively low enantiomeric purity of 15% *ee.* The *ee* value of the product would thus exceed the *ee* value of the chiral auxiliary quite considerably. This phenomenon is referred to as **chiral amplification.**

The occurrence of chiral amplification in this case is explained by the mechanism of Figure 8.31. When chiral amplification is observed one always finds—as is the case here—that two molecules of the chiral auxiliary become linked to each other, albeit indirectly (i.e., via another component of the reaction mixture). In other words, a "derivative of the dimer" of the chiral auxiliary is formed. When chiral amplification occurs, this derivative of the dimer exists in the form of only two of its three conceivable stereoisomers. The reason for this is a mutual kinetic resolution (cf. Section 3.4.3).

In the reaction of Figure 8.31, the chiral auxiliary and Et_2Zn at first form as much "dimer" **C** as possible by a combination of the two enantiomers with each other. Therefore, the entire fraction of *racemic* chiral auxiliary **A** is used up. The remaining chiral auxiliary (−)-**A** is *enantiomerically 100% pure.* It reacts with additional Et_2Zn to form the "dimer" **B.** This species is less stable than "dimer" **C. B,** therefore dissociates—in contrast to **C**—reversibly to a small equilibrium fraction of the monomer **D. D** is *enantiomerically 100% pure,* just like **B,** *and represents the effective catalyst of the addition reaction which is now initiated.*

The trivalent Zn atom of the complex **D** coordinates as a Lewis acid to the carbonyl carbon of the reaction partner benzaldehyde. This occurs on the convex face of the heterocyclic moiety such that the Zn-bound ethyl group pushes the phenyl group as far away as possible. At this stage a second molecule of $ZnEt_2$ gets involved. It binds to the O atom of the substructure C_{sp^3}—O—Zn—on the least hindered, i.e., the convex face of the heterocycle—and thereby creates an oxonium ion next to both the first and the second Zn atom. In the resulting complex **E** the electrophilicity of the first Zn atom is therefore greatly increased. As a consequence, the electrophilicity of the benzaldehyde bound to this Zn atom is also greatly increased. Thus, it can finally bind to an ethyl group. This ethyl group is transferred from the second, negatively charged Zn atom with very high enantioselectivity.

8.5.3 Diastereoselective Addition of Organometallic Compounds to Carbonyl Compounds

As observed in Section 8.3.1, stereogenic addition reactions to cyclic ketones, in which one side of the carbonyl group lies in a concave region of the molecule and the other group lies in a convex region, take place preferentially from the convex side. With respect to the addition of sterically demanding hydride donors to the conformationally fixed ketone 4-*tert*-butyl cyclohexanone, this means that according to Figure 8.8 an equatorial attack is favored. However, as can also be seen from Figure 8.8, this kind of selectivity does not apply to the addition of sterically undemanding hydride donors. Because of a stereoelectronic factor *they* preferentially attacked axially.

A

Table 8.1 shows analogous additions of methylmagnesium bromide to the same *tert*-butylcyclohexanone. As can be seen from the complete absence of diastereoselectiv-

Table 8.1. Equatorial and Axial Addition of Classic Grignard Reagents and Reetz–Grignard Reagents to a Conformationally Fixed Cyclohexanone

MeMgX + *tert*-Bu–(cyclohexanone) ⟶ *tert*-Bu–(cyclohexanol, axial OH, equatorial Me) + *tert*-Bu–(cyclohexanol, axial Me, equatorial OH)

MeLi → MeMgX (for X ≠ Br: + MgX_2, (– LiX))

X =	OH axial / Me equatorial		Me axial / OH equatorial
Br:	49	:	51
OTf:	73	:	27
OTs:	85	:	15
O_3S–(2,4,6-trimethylphenyl):	90	:	10

ity, in this addition, the mentioned steric and stereoelectronic effects on diastereocontrol cancel each other out.

Grignard prepared the alkylmagnesium halides named after him in the 19th century. The analogous alkylmagnesium *sulfonates* were not described until 90 years later by Reetz. They were prepared by transmetallation of alkyllithium compounds with magnesium sulfonates (Table 8.1, left). Reetz–Grignard reagents add with higher diastereoselectivities to 4-*tert*-butylcyclohexanone than the corresponding classic Grignard reagent (Table 8.1). The larger the sulfonate group, the more it favors equatorial attack.

Cram selectivity occurs (Table 8.2) in the addition of methylmagnesium bromide to α-chiral aldehydes and ketones that do not contain a heteroatom at the stereocenter C-α. In agreement with the Cram transition state model of Figure 8.11 ($Nu^- = Me^-$), the Cram product is here produced via the transition state **A.** Whether the corresponding anti-Cram product is formed via the transition state **B** or the transition state **C** has not yet been determined (by model calculations). As one can see from Table 8.2, for a given substitution pattern at the stereocenter the fraction of anti-Cram product decreases with increasing size of the other, achiral substituent of the carbonyl group. In a transition state **B** this effect would be understandable: A destabilizing interaction (1,2-allyl strain) would occur between such a large substituent R at the carbonyl group and the neighboring methyl group, which is oriented almost *syn*-periplanar.

Table 8.3 shows how the Cram selectivity of the least Cram-selective addition reaction in Table 8.2, namely, the addition to the α-chiral aldehyde, can be increased considerably by using Reetz–Grignard compounds.

Grignard reagents also add diastereoselectively to α-chiral, α-oxygenated aldehydes and ketones. The chelation-controlled product constitutes the major product (Figure 8.32). It is the diastereomer that is produced via the Cram chelate transition state of Figure 8.11 ($Nu^- = R^-$). In fact, the Lewis acidity of the magnesium is so high that even a bulky protecting group at the O atom in the α position of the carbonyl group does not completely prevent chelate formation. Therefore, an addition of Grignard reagents to α-oxygenated aldehydes, which takes place exclusively via the Felkin–Anh transition state, is unknown.

Table 8.2. Diastereoselectivity of the Addition of MeMgBr to α-Chiral Carbonyl Compounds

MeMgBr + R(C=O)CH(Me)Ph ⟶ Cram product + anti-Cram product

R	Cram product	anti-Cram product
H	71	29
Et	86	14
*i*Pr	90	10
tert-Bu	96	4

A → Cram product

B → anti-Cram product ← C

Table 8.3. Diastereoselectivity of the Addition of Classic Grignard and Reetz–Grignard Reagents to an α-Chiral Aldehyde

MeMgX + OHC–CH(Me)Ph ⟶ Cram product + anti-Cram product

X	Cram product	anti-Cram product
Cl, Br, I	ca. 70	30
OAc	91	9
OTs	92	8
O_3S–(2,4,6-trimethylphenyl)	94	6
O_2C–*tert*-Bu	94	6

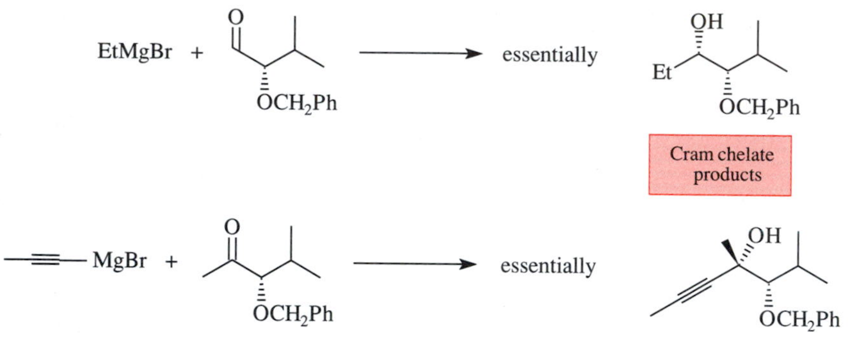

Fig. 8.32. Diastereoselective addition of Grignard reagents to α-chiral, α-oxygenated carbonyl compounds.

EtMgBr + [aldehyde, OCH_2Ph] → essentially [product: Et, OH, OCH_2Ph]

Cram chelate products

MgBr + [ketone, OCH_2Ph] → essentially [product: OH, OCH_2Ph]

Grignard reagents and α-chiral aldehydes which contain a dialkylamino group at their stereocenter, as does compound **A** of Figure 8.33 react selectively via the Felkin–Anh transition state of Figure 8.11 ($Nu^- = Ph^-$). The N atom in substrate **A** belongs to a tertiary amino group, which is, accordingly, attached to three non-hydrogen substituents. It is so severely hindered by the two benzyl substituents that it cannot be incorporated into a chelate.

A sterically less hindered N atom is found, for example, in the α-aminated α-chiral aldehyde **B** in Figure 8.33. This is because this N atom belongs to a secondary amino group and, accordingly, bears only two non-hydrogen substituents. Because of its single benzylidene substituent, the N atom of aldehyde **B** is so unhindered that it can be bound firmly in a chelate by the magnesium. Consequently, the addition of a Grignard reagent to aldehyde **B** affords exclusively the chelation-controlled product.

Grignard reagents add to the C═O double bond of β-chiral, β-oxygenated carbonyl compounds with only little diastereoselectivity. Highly diastereoselective additions to such substrates would be expected to take place under chelation control (cf. Section 8.3.3). However, in order to enforce a sufficient amount of chelate formation with these substrates, one must use organometallic compounds, which are stronger Lewis acids than Grignard reagents.

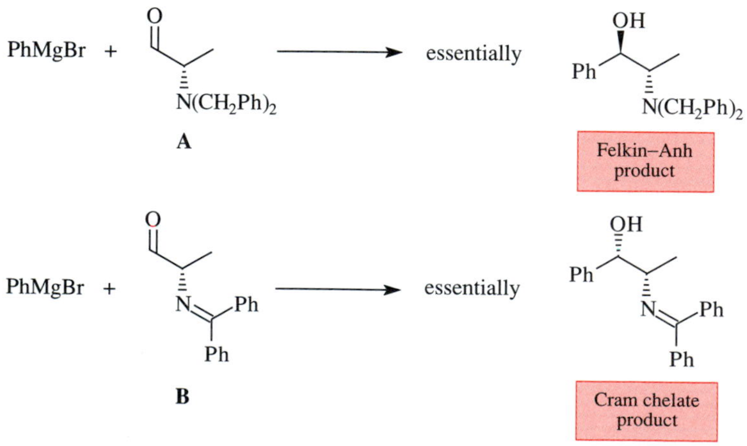

PhMgBr + **A** [aldehyde, $N(CH_2Ph)_2$] → essentially [product: Ph, OH, $N(CH_2Ph)_2$]

Felkin–Anh product

PhMgBr + **B** [aldehyde, N=C(Ph)Ph] → essentially [product: Ph, OH, N=C(Ph)Ph]

Cram chelate product

Fig. 8.33. Diastereoselective addition of Grignard reagents to α-chiral, α-aminated aldehydes.

8.6 1,4-Additions of Organometallic Compounds to α,β-Unsaturated Ketones

B

Many copper-containing organometallic compounds add to α,β-unsaturated ketones in smooth 1,4-additions (for the term "1,4-addition," see Figure 8.24). The most important ones are **Gilman cuprates** (Figure 8.34, left). For sterically hindered substrates, their rate of addition can be increased by adding Me_3SiCl to the reactants.

Grignard reagents are converted into C nucleophiles capable of undergoing 1,4-additions selectively when they are transmetallated with Cu(I) compounds to give so-called **Normant cuprates** (Figure 8.34, middle). Usually, this transmetallation is carried out with only catalytic amounts of CuI or $CuBr{\cdot}SMe_2$. This means that Normant cuprates undergo the 1,4-addition to α,β-unsaturated ketones considerably faster than the Cu-free Grignard reagents undergo the nonselective 1,2-/1,4-addition (cf. Figure 8.24). This is another example of a ligand-accelerated reaction (see also the comments to Figure 8.21).

Knochel cuprates (preparation: Figure 8.29) likewise add 1,4-selectively to α,β-unsaturated ketones but only in the presence of Me_3SiCl or $BF_3{\cdot}OEt_2$ (Figure 8.34, right). Under the same conditions, however, they add to α,β-unsaturated aldehydes with 1,2-selectivity (Figure 8.29).

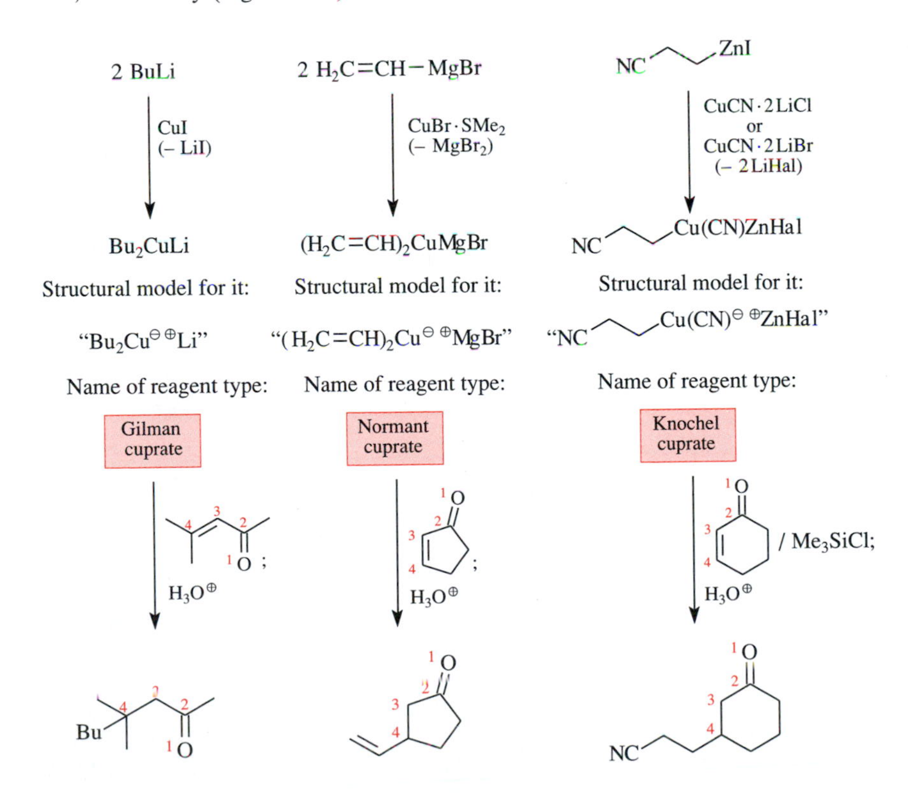

Fig. 8.34. 1,4-Addition of various organometallic compounds to α,β-unsaturated ketones.

These 1,4-additions have not been completely elucidated mechanistically. It is possible that they may not even take place uniformly by a common mechanism. Figure 8.35 shows two possibilities just for the mechanism of the 1,4-addition of Gilman cuprates to cyclohexenone.

If the addition mechanism were *polar,* a π-complex **A** would probably arise first. This makes it possible to understand why Gilman cuprates do not react with saturated ketones: They cannot form a π-complex. Next, there is an oxidative addition, which leads to the Cu(III)-containing lithium enolate **C.**

The further transformations of the enolate **C** start with a reductive elimination (additional examples of this type of reaction can be found in Chapter 13), which gives the enolate **D.** This compound is not a normal lithium enolate because it is associated with one equivalent of CuR. The CuR-containing enolate **D** remains inert until the aqueous workup. As you can see from Figure 8.35, 50% of the groups R contained in the Gilman cuprate are lost through formation of the stoichiometric by-product CuR. This disadvantage does not occur in the 1,4-additions of Normant and Knochel cuprates.

Enolate **D** gives a saturated ketone upon protonation (Figure 8.35). However, **D** can also be functionalized with other electrophiles, provided that they are highly reactive. Thus, with methyl iodide or allyl bromide, CuR-containing enolates **D** often form the

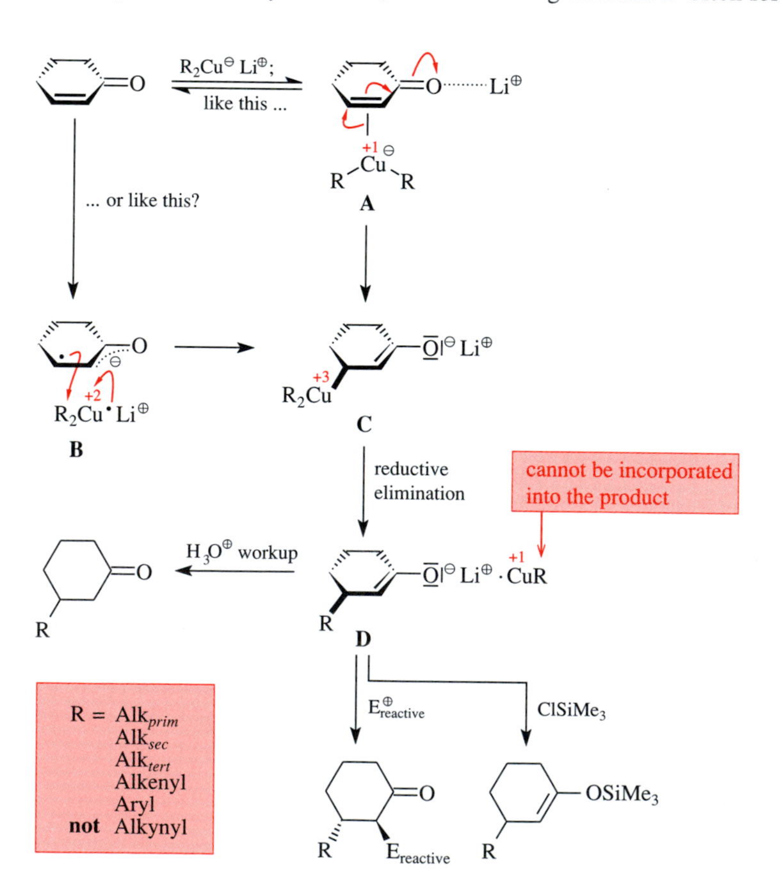

Fig. 8.35. Mechanistic possibilities for the 1,4-addition of a Gilman cuprate to an α,β-unsaturated ketone.

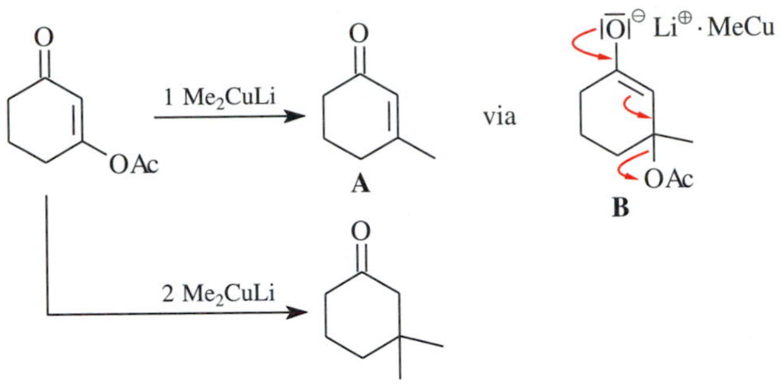

Fig. 8.36. Nucleophilic substitution I on an acceptor-substituted olefin with a leaving group in the β position.

expected alkylation products (Figure 10.27). With Me_3SiCl enolates **D** reliably give silylenol ethers (Figure 10.16).

The enol acetate from Figure 8.36 and one equivalent of Me_2CuLi react to give the methylcyclohexenone **A.** Presumably, in this reaction the enolate **B** is formed first—according to one of the two mechanisms shown in Figure 8.35. An acetate ion is cleaved off from this enolate as in the second step of an $E1_{cb}$ elimination. As an alternative, the enol acetate of Figure 8.36 can also react with two equivalents of Me_2CuLi. The second equivalent adds to the enone **A,** which is formed as before, and converts it into 3,3-dimethylcyclohexanone.

A

Other reactions of copper-containing organometallic compounds with α,β-unsaturated carbonyl compounds or with α,β-unsaturated esters (they do not always react), both containing a leaving group at C-β, are shown in Figures 8.37 and 8.38. It is not clear whether in these reactions as in Figure 8.36, a 1,4-addition takes place first and is then followed by the second part of an $E1_{cb}$ elimination or whether the organometallics substitute C_{sp^2}-bound leaving groups in the respective β-positions via a Cu-catalyzed mechanism, which is similar to the mechanisms in Chapter 13.

There is a serious limitation with respect to the structure of ketones that are accessible by 1,4-addition of Gilman or Normant cuprates to α,β-unsaturated ketones. As a

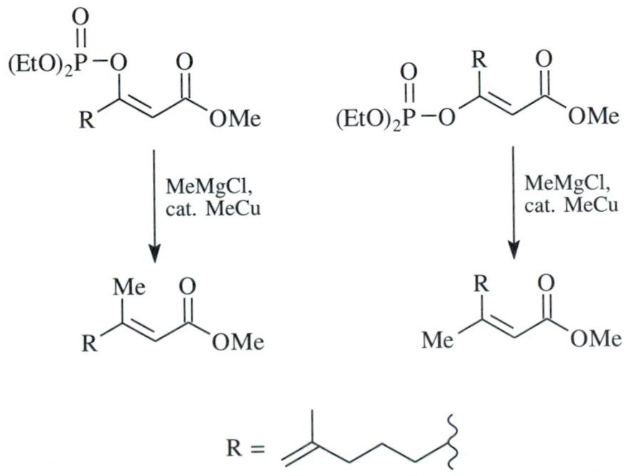

Fig. 8.37. Nucleophilic substitution II on an acceptor-substituted olefin with a leaving group in the β position (possibility for preparing the starting materials; Figure 10.22).

Fig. 8.38. Nucleophilic substitution III on acceptor-substituted olefins with a leaving group in the β position.

result of preparing these reagents from organolithium or from Grignard reagents, only functional groups can be incorporated into these cuprates that are compatible with forming the Li- or Mg-containing precursors. This essentially limits the substrates to contain only ethers, acetals, and internal C≡C triple bonds as functional groups.

B

As mentioned in the discussion of Figure 8.29, in Knochel cuprates a large variety of functional groups can be present: CO_2R, $R^1R^2C{=}O$, C≡N, Cl, RNH, NH_2, RC(=O)NH, sulfoxide, sulfone, and an internal or a terminal C≡C triple bond. This compatibility with functional groups makes Knochel cuprates the reagents of choice for 1,4-additions of organometallic compounds that contain *functional groups.* An example of such a reaction is shown in Figure 8.34 (right).

A

A second type of organometallic compound, the dialkylzinc compounds, is compatible with the presence of the most varied functional groups, too. However, they only react with those α,β-unsaturated ketones that contain a leaving group in the β position. The preparation of the required dialkylzinc compounds from olefins was described in Figure 8.30. Figure 8.38 shows how such a species, through the intermediacy of a Cu-containing derivative of unknown composition, can be reacted with a suitable substrate.

References

8.2

N. Greeves, "Reduction of C=O to CHOH by Metal Hydrides," in *Comprehensive Organic Synthesis* (B. M. Trost, I. Fleming, Eds.), Vol. 8, 1, Pergamon Press, Oxford, **1991.**

A. P. Davis, M. M. Midland, L. A. Morell, "Formation of C-H Bonds by Reduction of Carbonyl Groups (C=H), Reduction of Carbonyl Groups with Metal Hydrides," in *Methoden Org. Chem. (Houben-Weyl) 4th ed. 1952, Stereoselective Synthesis* (G. Helmchen, R. W. Hoffmann, J. Mulzer, E. Schaumann, Eds.), Vol. E21d, 3988, Georg Thieme Verlag, Stuttgart, **1995.**

M. M. Midland, L. A. Morell, K. Krohn, "Formation of C-H Bonds by Reduction of Carbonyl Groups (C=O)—Reduction with Hydride Donors," in *Methoden Org. Chem.* (*Houben-Weyl*) *4th ed. 1952, Stereoselective Synthesis* (G. Helmchen, R. W. Hoffmann, J. Mulzer, E. Schaumann, Eds.), Vol. E21d, 4082, Georg Thieme Verlag, Stuttgart, **1995.**

T. Ooi and K. Maruoka, "Exceptionally bulky lewis acidic reagent, MAD," *Rev. Heteroatom Chem.* **1998,** *18,* 61–85.

8.3

J. Mulzer, "Cram's Rule: Theme and Variations," in *Organic Synthesis Highlights* (J. Mulzer, H.-J. Altenbach, M. Braun, K. Krohn, H.-U. Reißig, Eds.), VCH, Weinheim, New York, **1991,** 3–8.

E. L. Eliel, "Application of Cram's Rule: Addition of Achiral Nucleophiles to Chiral Substrates," in *Asymmetric Synthesis* (J. D. Morrison, Ed.), Vol. 2, 125, Academic Press, New York, **1983.**

E. L. Eliel, S. V. Frye, E. R. Hortelano, X. B. Chen, "Asymmetric synthesis and Cram's (chelate) rule," *Pure Appl. Chem.* **1991,** *63,* 1591–1598.

A. Mengel and O. Reiser, "Around and beyond Cram's rule," *Chem. Rev.* **1999,** *99,* 1191–1223.

A. P. Davis, "Diastereoselective Reductions," in *Stereoselective Synthesis* (Houben-Weyl) 4th ed. 1996, (G. Helmchen, R. W. Hoffmann, J. Mulzer, E. Schaumann, Eds.), **1996,** Vol. E 21 (Workbench Edition), 7, 3988–4048, Georg Thieme Verlag, Stuttgart.

M. T. Reetz, "New approaches to the use of amino acids as chiral building blocks in organic synthesis," *Angew. Chem.* **1991,** *103,* 1559–1573; *Angew. Chem. Int. Ed. Engl.* **1991,** *30,* 1531–1546.

M. T. Reetz, "Structural, mechanistic, and theoretical aspects of chelation-controlled carbonyl addition reactions," *Acc. Chem. Res.* **1993,** *26,* 462–468.

8.4

M. M. Midland and L. A. Morell, "Enantioselective Reductions," in *Stereoselective Synthesis* (Houben-Weyl) 4th ed. 1995, (G. Helmchen, R. W. Hoffmann, J. Mulzer, E. Schaumann, Eds.), **1996,** Vol. E21 (Workbench Edition), 7, 4049–4066, Georg Thieme Verlag, Stuttgart.

M. Nishizawa and R. Noyori, "Reduction of C=X to CHXH by Chirally Modified Hydride Reagents," in *Comprehensive Organic Synthesis* (B. M. Trost, I. Fleming, Eds.), Vol. 8, 159, Pergamon Press, Oxford, **1991.**

E. R. Grandbois, S. I. Howard, J. D. Morrison, "Reductions with Chiral Modifications of Lithium Aluminium Hydride," in *Asymmetric Synthesis* (J. D. Morrison, Ed.), Vol. 2, 71, Academic Press, New York, **1983.**

H. Haubenstock, "Asymmetric reductions with chiral complex aluminium hydrides and tricoordinate aluminium reagents," *Top. Stereochem.* **1983,** *14,* 213.

M. M. Midland, "Asymmetric reductions with organoborane reagents," *Chem. Rev.* **1989,** *89,* 1553–1561.

H. C. Brown and P. V. Ramachandran, "Asymmetric reduction with chiral organoboranes based on a-pinene," *Acc. Chem. Res.* **1992,** *25,* 16–24.

R. K. Dhar, "Diisopinocamphenylchloroborane, (DIP-chloride), an excellent chiral reducing reagent for the synthesis of secondary alcohols of high enantiomeric purity," *Aldrichimica Acta* **1994,** *27,* 43–51.

B. B. Lohray and V. Bhushan, "Oxazaborolidines and dioxaborolidines in enantioselective catalysis," *Angew. Chem.* **1992,** *104,* 740–741; *Angew. Chem. Int. Ed. Engl.* **1992,** *31,* 729–730.

S. Wallbaum and J. Martens, "Asymmetric syntheses with chiral oxazaborolidines," *Tetrahedron: Asymmetry* **1992,** *3,* 1475–1504.

S. Itsuno, "Enantioselective reduction of ketones," *Org. React.* **1998,** *52,* 395–576.

E. J. Corey and C. J. Helal, "Reduction of carbonyl compounds with chiral oxazaborolidine catalysts: A new paradigm for enantioselective catalysis and a powerful new synthetic method," *Angew. Chem.* **1998,** *110,* 2092–2118; *Angew. Chem. Int. Ed. Engl.* **1998,** *37,* 1987–2012.

V. K. Singh, "Practical and useful methods for the enantioselective reduction of unsymmetrical ketones," *Synthesis* **1992,** 607–617.

M. Wills and J. R. Studley, "The asymmetric reduction of ketones", *Chem. Ind.* **1994,** 552–555.

8.5

G. Salem and C. L. Raston, "Preparation and Use of Grignard Reagents and Group II Organometallics in Organic Synthesis," in *The Use of Organometallic Compounds in Organic Synthesis* (F. R. Hartley, Ed.), Vol. 4, 159, Wiley, Chichester, **1987.**

R. M. Kellogg, "Reduction of C=X to CHXH by Hydride Delivery from Carbon," in *Comprehensive Organic Synthesis* (B. M. Trost, I. Fleming, Eds.), Vol. 8, 79, Pergamon Press, Oxford, **1991.**

P. Knochel, M. J. Rozema, C. E. Tucker, C. Retherford, M. Furlong, S. A. Rao, "The chemistry of polyfunctional organozinc and copper reagents," *Pure Appl. Chem.* **1992,** *64,* 361.

P. Knochel and R. D. Singer, "Preparation and reactions of polyfunctional organozinc reagents in organic synthesis," *Chem. Rev.* **1993,** *93,* 2117–2188.

P. Knochel, "New preparations of polyfunctional dialkylzincs and their application in asymmetric synthesis," *Chemtracts: Organic Chemistry* **1995,** *8,* 205.

R. M. Devant and H.-E. Radunz, "Formation of C-C Bonds by Addition to Carbonyl Groups (C=O)—σ-Type Organometallic Compounds," in *Methoden Org. Chem. (Houben-Weyl) 4th ed. 1952, Stereoselective Synthesis* (G. Helmchen, R. W. Hoffmann, J. Mulzer, E. Schaumann, Eds.), Vol. E21b, 1151, Georg Thieme Verlag, Stuttgart, **1995.**

D. Hoppe, "Formation of C-C Bonds by Addition to Carbonyl Groups (C=O)—Benzyl-Type Organometallic Compounds," in *Methoden Org. Chem. (Houben-Weyl) 4th ed. 1952, Stereoselective Synthesis* (G. Helmchen, R. W. Hoffmann, J. Mulzer, E. Schaumann, Eds.), Vol. E21b, 1335, Georg Thieme Verlag, Stuttgart, **1995.**

D. Hoppe, W. R. Roush, E. J. Thomas, "Formation of C-C Bonds by Addition to Carbonyl Groups (C=O)—Allyl-Type Organometallic Compounds," in *Methoden Org. Chem. (Houben-Weyl) 4th ed. 1952, Stereoselective Synthesis* (G. Helmchen, R. W. Hoffmann, J. Mulzer, E. Schaumann, Eds.), Vol. E21b, 1357, Georg Thieme Verlag, Stuttgart, **1995.**

G. Solladié, "Formation of C-C Bonds by Addition to Carbonyl Groups (C=O)—Metalated Sulfoxides or Sulfoximides," in *Methoden Org. Chem. (Houben-Weyl) 4th ed. 1952, Stereoselective Synthesis* (G. Helmchen, R. W. Hoffmann, J. Mulzer, E. Schaumann, Eds.), Vol. E21b, 1793, Georg Thieme Verlag, Stuttgart, **1995.**

A. B. Sannigrahi, T. Kar, B. G. Niyogi, P. Hobza, P. v. R. Schleyer, "The lithium bond reexamined," *Chem. Rev.* **1990,** *90,* 1061–1076.

W. Bauer and P. von Rague Schleyer, "Recent Results in NMR Spectroscopy of Organolithium Compounds," in *Advances in Carbanion Chemistry* (V. Snieckus, Ed.), **1992,** *1,* Jai Press, Greenwich, CT.

K. Maruyama and T. Katagiri, "Mechanism of the Grignard reaction," *J. Phys. Org. Chem.* **1989,** *2,* 205–213.

D. M. Huryn, "Carbanions of Alkali and Alkaline Earth Cations: (ii) Selectivity of Carbonyl Addition Reactions," in *Comprehensive Organic Synthesis* (B. M. Trost, I. Fleming, Eds.), Vol. 1, 49, Pergamon Press, Oxford, **1991.**

G. S. Silverman and P. E. Rakita (Eds.), "Handbook of Grignard Reagents," [in *Chem. Ind.* 1996; 64] Dekker, New York, **1996.**

R. Noyori and M. Kitamura, "Enantioselective addition of organometallic reagents to carbonyl compounds: Chirality transfer, multiplication and amplification," *Angew. Chem., Int. Ed. Engl.* **1991,** *30,* 49.

K. Soai and S. Niwa, "Enantioselective addition of organozinc reagents to aldehydes," *Chem. Rev.* 1992, *92,* 833–856.

P. Knochel, "Stereoselective reactions mediated by functionalized diorganozincs," *Synlett* **1995,** 393–403.

Y. L. Bennani and S. Hanessian, "Trans-1,2-diaminocyclohexane derivatives as chiral reagents, scaffolds, and ligands for catalysis: Applications in asymmetric synthesis and molecular recognition," *Chem. Rev.* **1997,** *97,* 3161–3195.

C. Girard and H. B. Kagan, "Nonlinear effects in asymmetric synthesis and stereoselective reactions: Ten years of investigation," *Angew. Chem.* **1998,** *110,* 3088–2127; *Angew. Chem. Int. Ed. Engl.* **1998,** *37,* 2923–2959.

K. Soai, "Rational design of chiral catalysis for the enantioselective addition reaction of dialkylzincs," *Enantiomer* **1999,** *4,* 591–598.

D. R. Fenwick and H. B. Kagan, "Asymmetric amplification," *Top. Stereochem.* **1999,** *22,* 257–296.

D. G. Blackmond, "Kinetic aspects of nonlinear effects in asymmetric catalysis," *Acc. Chem. Res.* **2000,** *33,* 402–411.

M. T. Reetz, "Chelation or non-chelation control in addition reactions of chiral α- or β-alkoxy carbonyl compounds," *Angew. Chem., Int. Ed. Engl.* **1984,** *23,* 556.

M. T. Reetz, "New approaches to the use of amino acids as chiral building blocks in organic synthesis," *Angew. Chem.* **1991,** *103,* 1559–1573; *Angew. Chem. Int. Ed. Engl.* **1991,** *30,* 1531–1546.

M. T. Reetz, "Synthesis and diastereoselective reactions of N,N-dibenzylamino aldehydes and related compounds," *Chem. Rev.* **1999,** *99,* 1121–1162.

M. T. Reetz, "Structural, mechanistic, and theoretical aspects of chelation-controlled carbonyl addition reactions," *Acc. Chem. Res.* **1993,** *26,* 462–468.

J. Jurczak and A. Golebiowski, "Optically active N-protected α-amino aldehydes in organic synthesis," *Chem. Rev.* **1989,** *89,* 149–164.

E. L. Eliel, "Application of Cram's Rule: Addition of Achiral Nucleophiles to Chiral Substrates," in *Asymmetric Synthesis* (J. D. Morrison, Ed.), Vol. 2, 125, Academic Press, New York, **1983.**

E. L. Eliel, S. V. Frye, E. R. Hortelano, X. B. Chen, "Asymmetric synthesis and Cram's (chelate) rule," *Pure Appl. Chem.* **1991,** *63,* 1591–1598.

A. Mengel and O. Reiser, "Around and beyond Cram's rule," *Chem. Rev.* **1999,** *99,* 1191–1223.

R. M. Pollack, "Stereoelectronic control in the reactions of ketones and their enolates," *Tetrahedron* **1989,** *45,* 4913.

8.6

Y. Yamamoto, "Addition of Organometallic Compounds to α,β-Unsaturated Carbonyl Compounds," in *Stereoselective Synthesis* (Houben-Weyl) 4th ed. 1996, (G. Helmchen, R. W. Hoffmann, J. Mulzer, E. Schaumann, Eds.), **1996,** Vol. E21 (Workbench Edition), 4, 2041–2067, Georg Thieme Verlag, Stuttgart.

G. H. Posner, "Conjugate addition reactions of organocopper reagents," *Org. React.* **1972,** *19,* 1–114.

E. Erdik, "Copper(I)-catalyzed reactions of organolithiums and grignard reagents," *Tetrahedron* **1984,** *40,* 641.

R. J. K. Taylor, "Organocopper conjugate addition-enolate trapping reactions," *Synthesis* **1985,** 364.

M. J. Chapdelaine and M. Hulce, "Tandem vicinal difunctionalization: β-Addition to α,β-unsaturated carbonyl substrates followed by α-functionalization," *Org. React.* **1990,** *38,* 225–653.

J. A. Kozlowski, "Organocuprates in the Conjugate Addition Reaction," in *Comprehensive Organic Synthesis* (B. M. Trost, I. Fleming, Eds.), Vol. 4, 169, Pergamon Press, Oxford, **1991.**

B. H. Lipshutz and S. Sengupta, "Organocopper reagents: Substitution, conjugate addition, carbo/metallocupration, and other reactions," *Org. React.* (*N.Y.*) **1992,** *41,* 136.

R. A. J. Smith and A. S. Vellekoop, "1,4-Addition Reactions of Organocuprates with α,β-Unsaturated Ketones," in *Advances in Detailed Reaction Mechanisms* (J. M. Coxon, Ed.), Vol. 3, Jai Press, Greenwich, CT, **1994.**

Further Reading

D. A. Hunt, "Michael addition of organolithium compounds: A review," *Org. Prep. Proced. Int.* **1989,** *21,* 705–749.

G. Boche, "The structure of lithium compounds of sulfones, sulfoximides, sulfoxides, thioethers and 1,3-dithianes, nitriles, nitro compounds and hydrazones," *Angew. Chem.* **1989,** *101,* 286–306, *Angew. Chem. Int. Ed. Engl.* **1989,** *28,* 277–297.

Y. Yamamoto, "Selective synthesis by use of lewis acids in the presence of organocopper and related reagents," *Angew. Chem., Int. Ed. Engl.* **1986,** *25,* 947.

E. Nakamura, "New tools in synthetic organocopper chemistry," *Synlett* **1991,** 539.

B. H. Lipshutz, "The evolution of higher order cyanocuprates," *Synlett* **1990,** *3,* 119.

B. H. Lipshutz, "Organocopper Reagents," in *Comprehensive Organic Synthesis* (B. M. Trost, I. Fleming, Eds.), Vol. 1, 107, Pergamon Press, Oxford, **1991.**

B. H. Lipshutz, "Synthetic Procedures Involving Organocopper Reagents," in *Organometallics: A Manual* (M. Schlosser, Ed.), 283, Wiley, Chichester, **1994.**

R. J. K. Taylor, "Organocopper Chemistry: An Overview," in *Organocopper Reagents: A Practical Approach* (R. J. K. Taylor, Ed.), 1, Oxford University Press, Oxford, U.K., **1994.**

R. J. K. Taylor, J. M. Herbert, "Compilation of Organocopper Preparations," in *Organocopper Reagents: A Practical Approach* (R. J. K. Taylor, Ed.), 307, Oxford University Press, Oxford, U.K., **1994.**

R. J. K. Taylor, "Organocopper Reagents: A Practical Approach," Oxford University Press, Oxford, U.K., **1995.**

N. Krause and A. Gerold, "Regioselective and stereoselective syntheses with organocopper reagents," *Angew. Chem.* **1997,** *109,* 194–213; *Angew. Chem. Int. Ed. Engl.* **1997,** *36,* 187–204.

E. Erdik, "Use of activation methods for organozinc reagents," *Tetrahedron* **1987,** *43,* 2203.

E. Erdik, "Transition metal catalyzed reactions of organozinc reagents," *Tetrahedron* **1992,** *48,* 9577.

E. Erdik (Ed.), "Organozinc Reagents in Organic Synthesis," CRC Press, Boca Raton, FL, **1996.**

Y. Tamaru, "Unique Reactivity of Functionalized Organozincs," in *Advances in Detailed Reaction Mechanism. Synthetically Useful Reactions* (J. M. Coxon, Ed.), **1995,** *4,* JAI Press, Greenwich, CT.

P. Knochel, "Organozinc, Organocadmium and Organomercury Reagents," in *Comprehensive Organic Synthesis* (B. M. Trost, I. Fleming, Eds.), Vol. 1, 211, Pergamon Press, Oxford, **1991.**

P. Knochel, M. J. Rozema, C. E. Tucker, C. Retherford, M. Furlong, S. A. Rao, "The chemistry of polyfunctional organozinc and copper reagents," *Pure Appl. Chem.* **1992,** *64,* 361–369.

P. Knochel, "Zinc and Cadmium: A Review of the Literature 1982–1994," in *Comprehensive Organometallic Chemistry II* (E. W. Abel, F. G. A. Stone, G. Wilkinson, Eds.), Vol. 11, 159, Pergamon, Oxford, UK, **1995.**

P. Knochel, "Preparation and Application of Functionalized Organozinc Reagents," in *Active Metals* (A. Fürstner, Ed.), 191, VCH, Weinheim, Germany, **1996.**

P. Knochel, J. J. A. Perea, P. Jones, "Organozinc mediated reactions," *Tetrahedron,* **1998,** *54,* 8275–8319.

M. Melnik, J. Skorsepa, K. Gyoryova, C. E. Holloway, "Structural analyses of organozinc compounds," *J. Organomet. Chem.* **1995,** *503,* 1.

P. Knochel, "Carbon-Carbon Bond Formation Reactions Mediated by Organozinc Reagents," in *Metal-Catalyzed Cross-coupling Reactions,* Hrsg.: F. Diederich, P. J. Stang, Wiley-VCH, Weinheim, **1998,** 387–416.

P. Knochel and P. Joned, (Eds.) "Organozinc Reagents: A Practical Approach," Oxford University Press, Oxford, U.K., **1999.**

R. Bloch, "Additions of organometallic reagents to C:N bonds: Reactivity and selectivity," *Chem. Rev.* **1998,** *98,* 1407–1438.

A. Alexakis, "Asymmetric Conjugate Addition," in *Organocopper Reagents: A Practical Approach* (R. J. K. Taylor, Ed.), 159, Oxford University Press, Oxford, U.K., **1994.**

T. Ooi and K. Maruoka, "Carbonyl-Lewis Acid Complexes," in *Modern Carbonyl Chemistry,* (J. Otera, Ed.), Wiley-VCH, Weinheim, **2000,** 1–32.

S. Saito and H. Yamamoto, "Carbonyl Recognition," in *Modern Carbonyl Chemistry,* (J. Otera, Ed.), Wiley-VCH, Weinheim, **2000,** 33–67.

J. M. Coxon and R. T. Luibrand, "π-Facial Selectivity in Reaction in Carbonyls: A Computational Approach," in *Modern Carbonyl Chemistry,* (J. Otera, Ed.), Wiley-VCH, Weinheim, **2000,** 155–184.

F. Effenberger, "Cyanohydrin Formation," in *Stereoselective Synthesis* (Houben-Weyl) 4th ed. 1996, (G. Helmchen, R. W. Hoffmann, J. Mulzer, E. Schaumann, Eds.), **1996,** Vol. E21 (Workbench Edition), 3, 1817–1821, Georg Thieme Verlag, Stuttgart.

W. Nagata and M. Yoshioka, "Hydrocyanation of conjugated carbonyl compounds," *Org. React.* **1977,** *25,* 255–476.

A. S. Cieplak, "Inductive and resonance effects of substituents on π-face selection," *Chem. Rev.* **1999,** *99,* 1265–1336.

W. Adcock and N. A. Trout, "Nature of the electronic factor governing diastereofacial selectivity in some reactions of rigid saturated model substrates," *Chem. Rev.* **1999,** *99,* 1415–1435.

Reaction of Ylides with Saturated or α,β-Unsaturated Carbonyl Compounds

9

9.1 Ylides/Ylenes

B

It is possible to remove a proton from the methyl group of a trialkylmethylammonium halide with strong bases. Thereby, a betaine (see Section 4.7.4) is produced with the structure $R_3N^+—CH_2^-$. A betaine in which the positive and the negative charges are located on adjacent atoms as in $R_3N^+—CH_2^-$ is called an **ylide.** The "yl" part of the name ylide refers to the *covalent bond* in the substructure $N^+—CH_2^-$. The "ide" part indicates that it also contains an *ionic bond.* When one wants to distinguish the ylide $R_3N^+—CH_2^-$ from other ylides, it is called an **ammonium ylide** or an **N ylide.**

With strong bases it is also possible to remove a proton from the methyl group of triarylmethylphosphonium halides, trimethylsulfonium iodide, or trimethylsulfoxonium iodide (Figure 9.1). Here, too, species are produced that are betaines of the ylide type.

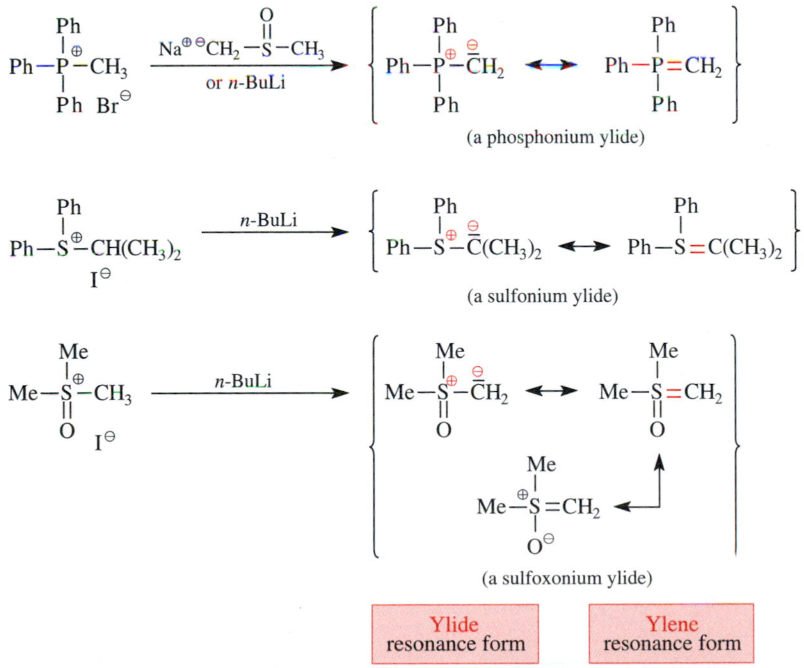

$$Ph_3P^{\oplus}-CH_3\ Br^{\ominus} \xrightarrow[\text{or } n\text{-BuLi}]{Na^{\oplus}\ {}^{\ominus}CH_2-S(=O)-CH_3} \left\{ Ph_3P^{\oplus}-\overset{\ominus}{C}H_2 \longleftrightarrow Ph_3P=CH_2 \right\}$$

(a phosphonium ylide)

$$Ph_2S^{\oplus}-CH(CH_3)_2\ I^{\ominus} \xrightarrow{n\text{-BuLi}} \left\{ Ph_2S^{\oplus}-\overset{\ominus}{C}(CH_3)_2 \longleftrightarrow Ph_2S=C(CH_3)_2 \right\}$$

(a sulfonium ylide)

$$Me_2S^{\oplus}(=O)-CH_3\ I^{\ominus} \xrightarrow{n\text{-BuLi}} \left\{ Me_2S^{\oplus}(=O)-\overset{\ominus}{C}H_2 \longleftrightarrow Me_2S(=O)=CH_2 \leftarrow Me_2S^{\oplus}(-O^{\ominus})=CH_2 \right\}$$

(a sulfoxonium ylide)

Ylide resonance form

Ylene resonance form

Fig. 9.1. Representative phosphonium, sulfonium, and sulfoxonium ylides—formation reactions and valence bond formulas.

In analogy with the nomenclature introduced previously, they are referred to as phosphonium, sulfonium, and sulfoxonium ylides, respectively, or as P or S ylides.

The ionic representation of the ylides in Figure 9.1 shows only one of two conceivable resonance forms of such species. In contrast to the N atom in the center of the N ylides, the P or S atoms in the centers of the P and S ylides (Figure 9.1) may exceed their valence electron octets and share a fifth electron pair. For P and S ylides one can therefore also write a resonance form with a C=P or C=S double bond, respectively; these are resonance forms free of formal charges (Figure 9.1). For the sulfoxonium ylide there is a second resonance form in which the S atom exceeds its valence electron octet; however, this does contain formal charges. Resonance forms of ylides in which the heteroatom exceeds its valence electron octet are called ylene resonance forms. The "ene" part of the designation "ylene" refers to the double bond between the heteroatom and the deprotonated alkyl group.

It should be remembered that the heteroatom under scrutiny belongs to the second full period of the periodic table of the elements. Therefore, the double bond rule also applies to its double bonds. Thus, the double bond of an ylene form is not a p_π, p_π, but a d_π, p_π double bond. Because of the presence of this (albeit not very stable) d_π, p_π double bond, one would a priori attribute to the ylene resonance form a certain importance for describing the electron distribution in the ylides from Figure 9.1. This should be all the more true because the ylide resonance form must cope with the energetically unfavorable charge separation (opposite charges attract each other and, if possible, annihilate each other), whereas the ylene resonance form does not have this disadvantage.

Nevertheless, it is doubtful as to whether the negative charge on the carbanionic C atom of P and S ylides is indeed stabilized by partial double bond formation to the heteroatom, as is expressed by the ylene resonance form. Probably, the formal charge at the carbanionic center of these ylides is rather stabilized to a considerable extent by two other effects:

1. Electron withdrawal due to the $-I$ effect of the heteroatom, and
2. Stabilization of the electron pair on the carbanionic carbon atom by an anomeric effect.

The occurrence of an anomeric effect was discussed with the help of Figure 6.8 using the example of a structural element $:\text{Het}^1\text{—C—Het}^2$. In the ylides of Figure 9.1 the anomeric effect arises in an analogous substructure, namely, in $:\text{CH}_2^-\text{—HetL}_{n-1}^+\text{—L}$. In the MO picture this is due to an interaction between the electron pair in the nonbonding carbanionic sp^3-AO and an appropriately oriented vacant σ^* orbital of the bond between the heteroatom and its substituent L. Such an anomeric stabilization of an ylide is explained in Figure 9.2 using the example of the P-ylide $:\text{CH}_2^-\text{—PPh}_2^+\text{—Ph}$.

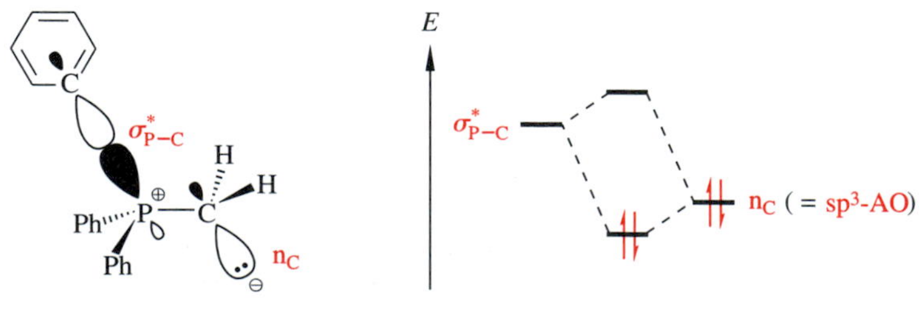

Fig. 9.2. The $Ph_3P^+\text{—}CH_2^-$ conformer, which is capable of maximal anomeric stabilization, and the associated MO diagram.

In order for this stabilization to be optimal, the crucial P^+—$C_{phenyl\ ring}$ bond must be oriented *anti* with respect to the nonbonding sp^3-AO on the (pyramidalized) carbanionic center. This arrangement is in agreement with the findings of crystal structure analyses.

P and S ylides are useful reagents in organic synthesis:

Applications of Ylides in Synthesis: A Survey

- P ylides condense with saturated and unsaturated carbonyl compounds to give olefins (Section 9.3).
- Both S ylides from Figure 9.1 react with α,β-unsaturated *esters* to give cyclopropanes (Section 9.2). Sulfoxonium ylides also react with α,β-unsaturated *carbonyl compounds* to give cyclopropanes (Section 9.2). Sulfonium ylides cannot do this because they react to form epoxides.
- Both S ylides from Figure 9.1 convert *saturated* carbonyl compounds into epoxides (Section 9.2). Sulfonium ylides also convert α,β-*unsaturated* carbonyl compounds into epoxides (Section 9.2). Sulfoxonium ylides cannot do this because they react to form cyclopropanes.

9.2 Reactions of S Ylides with Saturated Carbonyl Compounds or with Michael Acceptors: Three-Membered Ring Formation

9.2.1 Mechanism for the Formation of Cyclopropanes and Epoxides

B

With *trans*-crotonic acid methyl ester, dimethylsulfoxonium methylide forms a cyclopropane stereoselectively (Figure 9.3, bottom). Therein, the methyl group and the CO_2Me group retain the *trans* orientation they had in the crotonic ester. Thus, the *trans*-crotonic ester is cyclopropanated with complete retention of the configuration.

The isomeric *cis*-crotonic acid methyl ester also reacts with the same dimethylsulfoxonium methylide stereoselectively to form a cyclopropane (Figure 9.3, top). However, in this reaction there is a complete inversion of the configuration: The methyl group and the CO_2Me group are arranged *cis* in the starting material and *trans* in the product. This inversion of configuration proves that the cyclopropanation of the *cis*-crotonic acid ester takes place in several steps. The start is a 1,4-addition of the S ylide, which leads to the ester enolate **A.** In this enolate a rapid rotation takes place about the C—C single bond, which has evolved from the C=C double bond. In this way, conformer **B** of the ester enolate is produced. In an intramolecular S_N reaction the enolate carbon finally displaces the (somewhat unusual) leaving group, namely, a DMSO molecule. In this way, the cyclopropane ring closes.

The *trans* selectivity of the last cyclopropanation is due to the fact that the enolate conformer **B** cyclizes considerably faster to give the *trans*-cyclopropane than does the

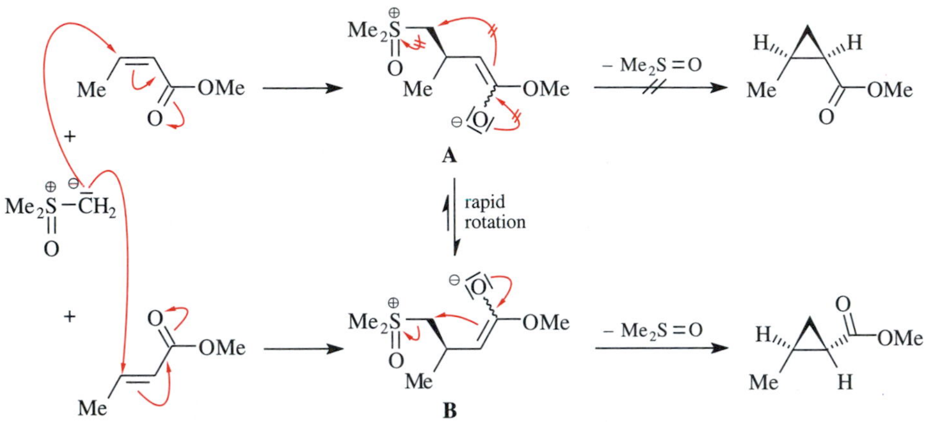

Fig. 9.3. The pair of reactions illustrates the occurrence of stereoselectivity and the lack of stereospecificity in two cyclopropanations with an S ylide. The *trans*-crotonic acid methyl ester reacts with complete retention of the configuration, whereas the *cis*-crotonic acid methyl ester reacts with complete inversion.

enolate conformer **A** to give the *cis*-cyclopropane. This again is a consequence of product-development control: The *trans*-isomer of the cyclopropane is more stable.

It seems obvious that the same mechanism could also apply to the cyclopropanation of the *trans*-crotonic ester (Figure 9.3, bottom). A Michael addition of the S ylide would then convert it directly into the ester enolate **B.** This species would cyclize exclusively to the *trans*-disubstituted cyclopropane exactly as when it is produced from the *cis*-crotonic ester in two steps via the conformer **A.**

All other three-membered ring formations from S ylides, which you will learn about in Section 9.2, are also in agreement with a two-step mechanism. It makes no difference whether in these cases a C=C or a C=O double bond is attacked—that is, whether a cyclopropane or an epoxide is produced. Probably, the first and rate-determining step is always the nucleophilic attack of the carbanionic C atom of the respective ylide on an electrophilic C atom to produce a zwitterion. Its negatively charged moiety subsequently acts as a nucleophile in an intramolecular S_N reaction in which it displaces the positively charged sulfur-containing group. The resulting substitution product is the three-membered ring, and the sulfur-containing by-product is Me_2S or DMSO.

If an acceptor-substituted C=C double bond is attacked in the first step of these ylide reactions and a zwitterion is formed whose anionic moiety is an enolate, then a cyclopropane is produced in the ensuing intramolecular S_N reaction. On the other hand, if a C=O double bond is attacked in the first step and an alkoxide results, then the subsequent S_N reaction gives an epoxide.

S ylides can react with α,β-unsaturated esters only by the first of the mentioned reaction modes (i.e., by converting the C=C double bond into a cyclopropane). Conversely, S ylides can react with saturated aldehydes or ketones only by the second reaction mode, namely, by the conversion of the C=O double bond into an epoxide. However, α,β-unsaturated aldehydes and ketones possess both C=C and C=O double bonds. Which of these double bonds is attacked by the respective S ylide and whether a cyclopropane or an epoxide is produced from these substrates (see Section 9.2.2) depend on the ylide structure.

9.2.2 Stereoselectivity and Regioselectivity of Three-Membered Ring Formation from S Ylides

A

Figure 9.4 shows stereogenic epoxide formations with S ylides and a ketone. The substrate is a conformationally fixed—because it represents a *trans*-decalin—cyclohexanone. Both the dimethylsulfoxonium methylide and the dimethylsulfonium methylide convert this cyclohexanone into an epoxide diastereoselectively. As Figure 9.4 shows, the observed diastereoselectivities are complementary. The sulfoxonium methylide attacks the carbonyl carbon equatorially, whereas the attack by the sulfonium ylide takes place axially.

The reason for this complementarity is that the CH_2 group of the sulfoxonium ylide is less nucleophilic than that of the sulfonium ylide. In the first case the CH_2 group is adjacent to the substituent $Me_2S^+{=}O$ and in the second case it is adjacent to the substituent Me_2S^+. The extra oxygen in the first substituent reduces the nucleophilicity of the sulfoxonium ylide because it stabilizes the negative charge of its carbanionic moiety especially well.

Fig. 9.4. Comparison of sulfonium and sulfoxonium ylides I—diastereoselectivity in the formation of epoxides. The sulfoxonium ylide attacks the carbonyl group equatorially and the sulfonium ylide attacks axially.

Because of the Hammond postulate, the sulfoxonium ylide should thus add to the cyclohexanone from Figure 9.4 through a later transition state than the more nucleophilic sulfonium ylide. Therefore, one would expect that product-development control plays a certain role in the sulfoxonium ylide reaction but not in the sulfonium ylide reaction. In Section 9.2.1, it was mentioned that the rate-determining step of epoxide formations from S ylides is probably the conversion of the reagents into an alkoxide. If, therefore, product-development control occurs in the attack of the sulfoxonium ylide on the C=O group of the cyclohexanone of Figure 9.4, then the configuration at the newly produced stereocenter of the alkoxide **A** is preferentially the one that allows the larger substituent to be equatorial. This substituent is the $Me_2S^+(=O)—CH_2$ and not the O^- group.

The more nucleophilic dimethylsulfonium methylide reacts with the same cyclohexanone from Figure 9.4 evidently via such an early transition state that no significant product-development control occurs. What was stated quite generally in the discussion of the reactions of Figure 8.8 and Table 8.1 regarding the diastereoselectivity of the addition of sterically unhindered nucleophiles to the C=O double bond of conformationally fixed cyclohexanones thus applies here: For stereoelectronic reasons, small nucleophiles preferentially attack such cyclohexanones axially. Accordingly, in the rate-determining step here the alkoxide **B** is preferentially formed, in which the newly introduced $Me_2S^+—CH_2$ group is axial.

B

There may be another explanation for why the sulfoxonium ylide and the cyclohexanone from Figure 9.4 preferentially give the epoxide with the equatorial $C_{spiro}—C_{epoxide}$ bond. One can argue that initially both the alkoxide **A** with the equatorial $Me_2S^+(=O)—CH_2$ group and its diastereomer with an axial $Me_2S^+(=O)—CH_2$ group are formed, but in reversible reactions. Thus, these alkoxides could revert to the starting materials and thereby equilibrate so that the more stable alkoxide **A** accumulates until it alone proceeds to the epoxide.

The greater stability of sulfoxonium compared to sulfonium ylides and the associated consequence that product-development control influences, which of several conceivable zwitterionic intermediates is formed preferentially, is also noticeable in reactions of S ylides with α,β-unsaturated carbonyl compounds. This can be seen in Figure 9.5 using the example of two reactions with cyclohexenone. Because of product-

Fig. 9.5. Comparison of sulfonium and sulfoxonium ylides II—chemoselectivity in the reaction with an α,β-unsaturated carbonyl compound. The sulfoxonium ylide reacts through the enolate intermediate **A**, whereas the sulfonium ylide reacts through the alkoxide intermediate **B**.

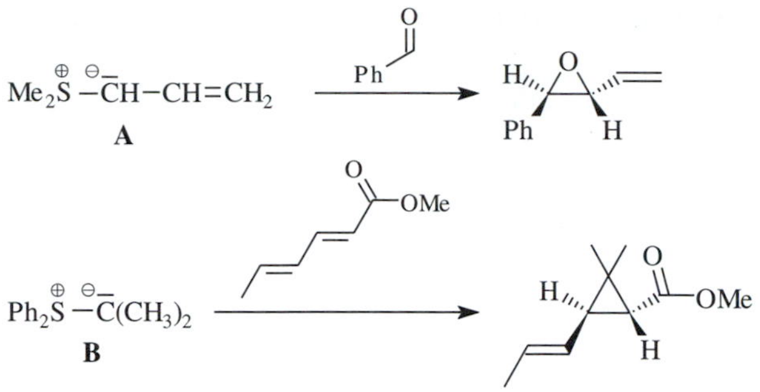

Fig. 9.6. Formation of functionalized three-membered rings from functionalized S ylides.

development control, the more stable sulfoxonium ylide reacts with this substrate through the more stable zwitterionic intermediate, namely, the *enolate* **A** with its *delocalized* charge. In contrast, the sulfonium ylide reacts through the less stable intermediate—that is, the alkoxide **B** with its *localized* charge. Consequently, the sulfoxonium ylide *cyclopropanates* the cyclohexenone while the sulfonium ylide converts it into an *epoxide*.

The simplest S ylides shown in Figures 9.3 and 9.4 transfer CH_2 groups to C=C or C=O double bonds. Monosubstituted or disubstituted S ylides can transfer CHR or CR_2 groups to C=C or C=O double bonds according to the same mechanism (Figure 9.6). However, the other substituents at the sulfonium center must be selected in such a way that they do not protonate the carbanionic center. Independently of the nature of these other heteroatom-bonded substituents, this condition is satisfied as soon as the carbanionic center of the ylide carries a *conjugated* substituent. In that case the negative formal charge of the ylide can be stabilized by resonance. This explains, for example, the stability of the vinyl-substituted ylide **A** (Figure 9.6). In addition, one can prepare substituted S ylides also independently of the nature of the substituent on their carbanionic center: This is always possible when the other substituents located at the heteroatom do not contain a proton in the α position—that is, when the other substituents are not acidic. This is the case, for example, in the S ylide **B** in Figure 9.6, with phenyl groups as nonacidic substituents.

9.3 Condensation of P Ylides with Carbonyl Compounds: Wittig Reaction

B

The Wittig reaction is a C,C-forming olefin synthesis from phosphonium ylides and carbonyl compounds (see also Section 4.7.4). In more than 99% of all Wittig reactions, ylides of the structure Ph_3P^+—CH^-—X (i.e., triphenylphosphonium ylides) are used. Therein, X usually stands for H, alkyl, aryl, or CO_2-alkyl and seldom for other substituents.

9.3.1 Nomenclature and Preparation of P Ylides

B

Most P ylides for Wittig reactions are prepared *in situ* and not isolated. Actually, they are always prepared *in situ* when the ylide $Ph_3P^+—CH^-—X$ contains a substituent X which is not at all able to stabilize the negative formal charge of the carbanionic center. This type of P ylide is called a **nonstabilized ylide.** The so-called **semi-stabilized ylides** $Ph_3P^+—CH^-—X$, on the other hand, contain a substituent X which slightly stabilizes the carbanionic center. This type of ylide is also prepared *in situ*. The **stabilized ylides** are the third and last P ylide type. They carry a *strongly* electron-withdrawing substituent on the carbanionic carbon atom and are the only triphenylphosphonium ylides which are shelf-stable and can be stored. Therefore, they can be (and generally are) added as neat compounds to Wittig reactions.

All P ylides for Wittig reactions are obtained by deprotonation of phosphonium salts. Depending on whether one wants to prepare a nonstabilized, a semi-stabilized, or a stabilized ylide, certain bases are especially suitable (Table 9.1). In stereogenic Wittig reactions with aldehydes, P ylides exhibit typical stereoselectivities. These depend mainly on whether the ylide involved is nonstabilized, semi-stabilized, or stabilized. This can also be seen in Table 9.1.

Table 9.1. Triphenylphosphonium Ylides: Nomenclature, Preparation, and Stereoselectivity of Their Wittig Reactions

P-Ylide	$Ph_3\overset{\oplus}{P}—\overset{\ominus}{C}HAlkyl$	$Ph_3\overset{\oplus}{P}—\overset{\ominus}{C}HAryl$	$Ph_3\overset{\oplus}{P}—\overset{\ominus}{C}H—CO_2R$
Ylide type	nonstabilized ylide	semi-stabilized ylide	stabilized ylide
Ylide is prepared...	... *in situ* ...	... *in situ* ...	... in prior reactions ...
... from $Ph_3\overset{\oplus}{P}—CHR\ Hal^{\ominus}$ and ...	*n*-BuLi or $Na^{\oplus\ominus}CH_2S(=O)CH_3$ [1] or $Na^{\oplus}\ NH_2^{\ominus}$ [1] or $K^{\oplus\ominus}O$*tert*-Bu [2]	NaOEt or aqueous NaOH	aqueous NaOH
1,2 disubstituted olefins typically result [3] ...	...with ≥ 90% *cis*-selectivity	... as *cis*-, *trans*-mixture	... with > 90% *trans*-selectivity

1) So-called salt-free Wittig reaction.
2) So-called high-temperature Wittig reaction, which takes place via the equilibrium fraction of the ylide.
3) For $Ph_3\overset{\oplus}{P}—\overset{\ominus}{C}HAlkyl$ under conditions 1 or 2.

9.3.2 Mechanism of the Wittig Reaction

cis-Selective Wittig Reactions

One has a detailed conception of the mechanism of the Wittig reaction (Figure 9.7). It starts with a one-step [2+2]-cycloaddition of the ylide to the aldehyde. This leads to a heterocycle called an oxaphosphetane. The oxaphosphetane decomposes in the second step—which is a one-step [2+2]-cycloreversion—to give triphenylphosphine oxide and an olefin. This decomposition takes place stereoselectively (cf. Figure 4.39): A *cis*-disubstituted oxaphosphetane reacts exclusively to give a *cis*-olefin, whereas a *trans*-disubstituted oxaphosphetane gives only a *trans*-olefin. In other words, stereospecificity occurs in a pair of decomposition reactions of this type.

The [2+2]-cycloaddition between P ylides and carbonyl compounds to give oxaphosphetanes can be stereogenic. It *is* stereogenic when the carbanionic C atom of the ylide bears—besides the P atom—two different substituents and when this holds true for the carbonyl group, too. The most important stereogenic oxaphosphetane formations of this type start from monosubstituted ylides Ph_3P^+—CH^-—X and from substituted aldehydes R—CH=O rather than formaldehyde. We will therefore study this case in Figure 9.7.

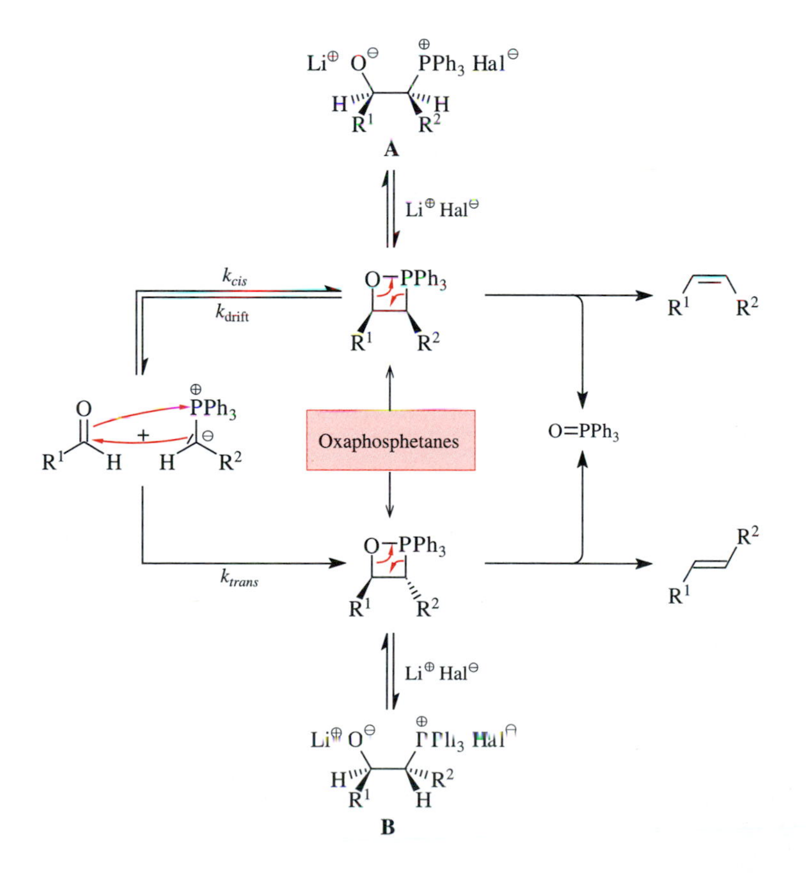

Fig. 9.7. Mechanism of the Wittig reaction. k_{cis} is the rate constant for the formation of the *cis*-oxaphosphetane, k_{trans} is the rate constant for the formation of the *trans*-oxaphosphetane, and k_{drift} is the rate constant for the isomerization of *cis*- to *trans*-configured oxaphosphetane, which is called "stereochemical drift."

The [2+2]-cycloaddition between the mentioned reaction partners to form an oxaphosphetane is not only stereogenic but frequently also exhibits a considerable degree of stereoselectivity. The latter is more precisely called **simple diastereoselectivity** or synonymously referred to as **noninduced diastereoselectivity.**

Simple (Noninduced) Diastereoselectivity

By simple diastereoselectivity we mean the occurrence of diastereoselectivity when bond formation between the C atom of a C═X double bond and the C atom of a C═Y double bond establishes a stereocenter at each of these C atoms.

The simple diastereoselectivity of the formation of oxaphosphetanes depends mainly on which ylide type from Table 9.1 is used.

Stereogenic Wittig reactions of nonstabilized ylides of the structure Ph_3P^+—CH^-—R^2 have been studied in-depth in many instances. They give the *cis*-configured oxaphosphetane rapidly, with the rate constant k_{cis}, and reversibly (Figure 9.7). On the other hand, the same nonstabilized ylide produces the *trans*-oxaphosphetane slowly, with the rate constant k_{trans}, and irreversibly. The primary product of the [2+2]-cycloaddition of a nonstabilized P ylide to a substituted aldehyde is therefore a *cis*-oxaphosphetane. Why this is so has not been ascertained despite the numerous suggestions about details of the mechanism which have been made.

The initially obtained *cis*-oxaphosphetane can subsequently isomerize irreversibly to a *trans*-oxaphosphetane with the rate constant $k_{\text{drift}} \times k_{trans}/(k_{trans} + k_{cis})$ (Figure 9.7). This isomerization is referred to as **stereochemical drift.**

The stereochemical drift in Wittig reactions of nonstabilized ylides can be suppressed if these reactions are carried out in the absence of Li salts ("salt-free"). However, Li salts are necessarily present in the reaction mixture when the respective nonstabilized ylide is prepared from a phosphonium halide and a lithium-containing base such as *n*-BuLi or LDA. In order to avoid a stereochemical drift in this case, the respective ylide must be prepared from a phosphonium halide and Na- or K-containing bases, for example, by using $NaNH_2$, KO-*tert*-Bu, KHMDS (K-hexamethyldisilazide, the potassium analog of LiHMDS, whose structural formula is shown in Figure 4.17). The latter deprotonation conditions initiate Wittig reactions, which in the literature are referred to as **Wittig reactions under salt-free conditions** or briefly as **"salt-free" Wittig reactions.**

Under salt-free conditions, the *cis*-oxaphosphetanes formed from nonstabilized ylides can be kept from participating in the stereochemical drift and left intact until they decompose to give the olefin in the terminating step. This olefin is then a pure *cis*-isomer. In other words, salt-free Wittig reactions of nonstabilized ylides represent a stereoselective synthesis of *cis*-olefins.

A

However, according to the data in Figure 9.8, the previous remark applies without limitation only to Wittig olefinations of *saturated* aldehydes. Indeed, α,β-unsaturated, α,β—C≡C-containing, or aromatic aldehydes have a tendency to react similarly, but the achievable *cis*-selectivities are lower. However, almost optimum *cis*-selectivities are also obtained with these unsaturated aldehydes when they are treated not with the common triphenyl-substituted P ylides but instead with its unusual tris(*ortho*-tolyl) analogs (Figure 9.9).

$Ar_3\overset{\oplus}{P}-CH_2-Me\ Br^{\ominus} \xrightarrow[\text{THF}]{NaNH_2,} Ar_3\overset{\oplus}{P}-\overset{\ominus}{C}H-Me \xrightarrow{R-CHO} R-CH=CH-Me$

% *cis*-Olefin for ...	... R = Pent	Pr–CH=CH–	Ph	Pr–C≡C–
Ar = Ph	96	90	92	84
Ar = *o*-Tolyl	98	97	96	95

Fig. 9.8. Optimum *cis*-selectivities of Wittig olefinations of different aldehydes with nonstabilized ylides under "salt-free" conditions.

The P ylides react with C=O double bonds faster the more electrophilic these C=O double bonds are. One can therefore occasionally olefinate aldehydes even in the presence of ketones. If one works salt-free, this is also accomplished with *cis*-selectivity (Figure 9.9).

B

$Ph_3\overset{\oplus}{P}-CH_2-Hex\ Br^{\ominus} \underset{\text{THF}}{\overset{K^{\oplus}\ {}^{\ominus}Otert\text{-}Bu}{\rightleftharpoons}} Ph_3\overset{\oplus}{P}-\overset{\ominus}{C}H-Hex$

Oct–CO–CH₂CH₂–CHO → Oct–CO–CH₂CH₂–CH=CH–Hex (98.5% *cis*)

Fig. 9.9. Chemoselective and stereoselective Wittig olefination with a nonstabilized ylide.

The C=O double bond of esters is usually not electrophilic enough to be olefinated by P ylides. Only formic acid esters can react with $Ph_3P{=}CH_2$ and then they give enol ethers of the structure H—C(=CH_2)—OR. α,β-Unsaturated esters can sometimes react with P ylides but this then results in a cyclopropanation similar to that which occurs with S ylides (Figure 9.3), not in an olefination giving an enol ether.

Wittig Reactions without Stereoselectivity

If nonstabilized P ylides react with carbonyl compounds in the presence of Li salts, i.e., not under salt-free conditions, several changes take place compared to the previously discussed mechanism of Figure 9.7. First, the *cis*-oxaphosphetanes are produced with a lower selectivity; that is, in contrast to the salt-free case, here the rate constant k_{cis} is not much greater than k_{trans}. Secondly, some of the initially formed *cis*-oxaphosphetane has an opportunity to isomerize to give the *trans*-oxaphosphetane—that is, to undergo the stereochemical drift—before an olefin is produced; however, *cis* → *trans* isomerization is not complete. For these reasons, the olefin ultimately obtained is a *cis/trans* mixture. Because such a mixture is normally useless, aldehydes are usually not treated with nonstabilized P ylides under non-salt-free conditions. On the other hand, non-salt-free Wittig reactions of nonstabilized ylides can be used without disadvantage when no stereogenic double bond is produced:

$$\text{cyclohexanone} \xrightarrow[\substack{\text{(prepared from}\\ Ph_3\overset{\oplus}{P}-CH_3\ Br^{\ominus}\\ +\ n\text{-BuLi)}}]{Ph_3\overset{\oplus}{P}-\overset{\ominus}{C}H_2} \text{methylenecyclohexane}$$

$$R\text{-CHO} \xrightarrow[\substack{\text{(prepared from}\\ Ph_3\overset{\oplus}{P}-CHMe_2\ I^{\ominus}\\ +\ n\text{-BuLi)}}]{Ph_3\overset{\oplus}{P}-\overset{\ominus}{C}Me_2} R\text{-CH=CMe}_2$$

There is a third Li effect under non-salt-free conditions. At first sight, it is only relevant mechanistically. Li salts are able to induce a heterolysis of the O–P bond of oxaphosphetanes. They thereby convert oxaphosphetanes into the so-called lithiobetaines (formulas **A** and **B** in Figure 9.7). The disappearance of the ring strain and the gain in Li^+O^- bond energy more than compensate for the energy required to break the P–O bond and to generate a cationic and an anionic charge. Until fairly recently, lithium-*free* betaines were incorrectly considered intermediates in the Wittig reaction. Today, it is known that lithium-*containing* betaines are formed in a dead-end side reaction. They must revert back to an oxaphosphetane—which occurs with retention of the configuration—before the actual Wittig reaction can continue.

trans-Selective Wittig Reactions

A

A different reaction mode of lithiobetaines is used in the **Schlosser variant** of the Wittig reaction. Here, too, one starts from a nonstabilized ylide and works under non-salt-free conditions. However, the Schlosser variant is an olefination which gives a pure *trans*-olefin rather than a *trans,cis* mixture. The experimental procedure looks like magic at first:

$$Ph_3\overset{\oplus}{P}-CH_2-Me\ \ Br^{\ominus} \xrightarrow[\substack{PhLi/LiBr;\\ HCl;\ KO\mathit{tert}\text{-Bu}}]{\substack{PhLi/LiBr;\\ R\text{-CHO};}} R\text{-CH=CH-Me}$$

	% *trans*-Olefin
for R = Pent	99
R = Ph	97

What is going on? First, the nonstabilized ylide is prepared by deprotonating the phosphonium halide with PhLi. For this purpose the deprotonating agent is best prepared according to the equation PhBr + 2 Li → PhLi + LiBr as a 1:1 mixture with LiBr. When the ylide, which then contains *two* equivalents of LiHal, is added to an aldehyde, a *cis,trans* mixture of the oxaphosphetanes is produced. However, because of the considerable excess of lithium ions, these oxaphosphetanes are immediately and completely converted to the corresponding lithiobetaines. The plural form is adequate since these lithiobetaines are obtained as a mixture of the diastereomers **A** and **C** (Figure 9.10). For kinetic and thermodynamic reasons, this ring opening of the oxaphosphetanes takes place faster and more completely, respectively, when two equivalents of LiHal are present than if only one equivalent were present. In fact, the first objective of the Schlosser olefination is to convert the starting materials into lithiobetaines as fast and completely as possible: Thereby, one keeps the initially formed

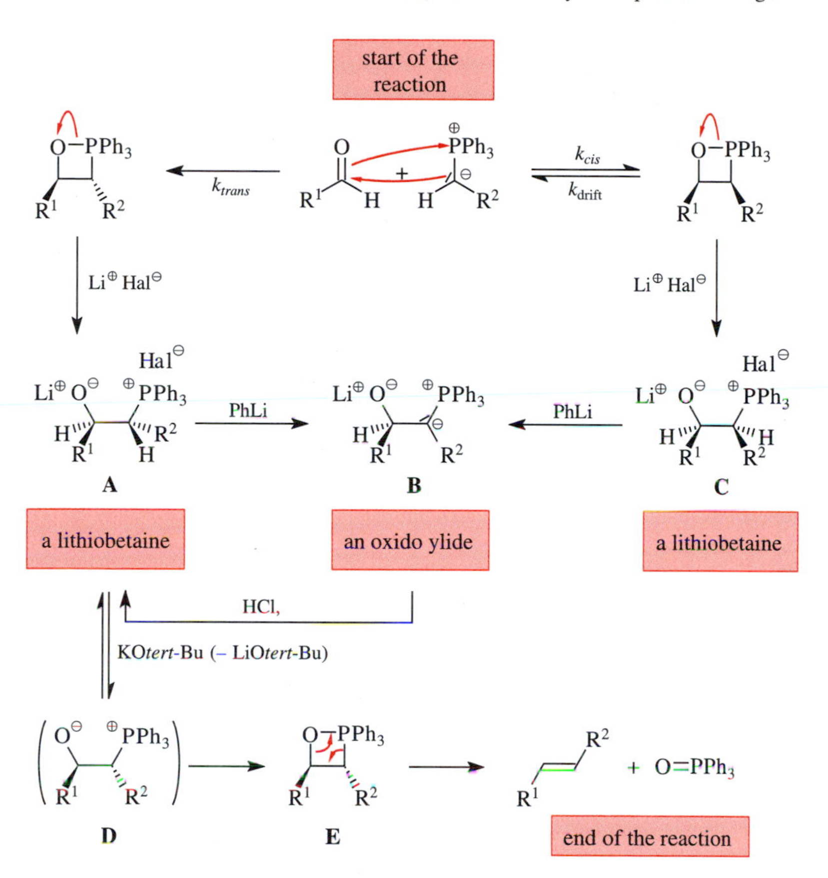

Fig. 9.10. Mechanism of the Schlosser variant of the Wittig reaction of nonstabilized ylides.

oxaphosphetane mixture from fragmenting prematurely to some extent to give an olefin fraction, which would be an undesired *cis,trans* mixture.

The lithiobetaines **A** and **C** in Figure 9.10 are phosphonium salts. As such, they contain an acidic H atom in the position α to the P atom. By adding a second equivalent of the previously mentioned PhLi/LiBr reagent to the reaction mixture, this H atom is now removed as a proton. In this way, a single so-called *oxido ylide* **B** is produced from each of the diastereomorphic lithiobetaines **A** and **C.** At this point, exactly 1.0 equivalent of HCl is added. The HCl protonates the oxido ylide. This produces a lithiobetaine again, but this time it is obtained diastereomerically pure as compound **A.** This species reacts very cleanly to give the olefin when it is treated with KO*tert*-Bu. This treatment establishes an equilibrium reaction in which the Li^+ ion migrates from the lithiobetaine **A** to the *tert*-BuO$^-$ ion. This creates a betaine **D,** which lacks the stabilizing Li^+ ion. However, as you have learned, lithium-free betaines are relatively unstable. Consequently, **D** collapses exergonically to the oxaphosphetane **E.** Of course, the stereocenters in **E** have the same configuration as the stereocenters in its precursor molecule **D** and the same configuration as in the highly diastereoselectively obtained lithiobetaine **A.** The oxaphosphetane **E** is therefore uniformly *trans*-configured.

$Ph_3P=CH-CO_2Me$

Fig. 9.11. *trans*-Selective Wittig olefination of aldehydes I—Preparation of a *trans*-configured α,β-unsaturated ester (preparation of the starting material: Figure 14.18).

Accordingly, the decomposition of **E** gives the pure *trans*-olefin in the last step of the Schlosser variant of the Wittig olefination.

B

Semistabilized ylides generally react with aldehydes to form mixtures of *cis*- and *trans*-oxaphosphetanes before the decomposition to the olefin starts. Therefore, stereogenic reactions of ylides of this type usually give olefin mixtures regardless of whether the work is carried out salt-free or not.

On the other hand, stabilized ylides react with aldehydes almost exclusively via *trans*-oxaphosphetanes. Initially, a small portion of the *cis*-isomer may still be produced. However, all the heterocyclic material isomerizes very rapidly to the *trans*-configured, four-membered ring through an especially pronounced stereochemical drift. Only after this point does the [2+2]-cycloreversion start. It leads to triphenylphosphine oxide and an acceptor-substituted *trans*-configured olefin. This *trans*-selectivity can be used, for example, in the C_2 extension of aldehydes to *trans*-configured α,β-unsaturated esters (Figure 9.11) or in the *trans*-selective synthesis of polyenes such as β-carotene (Figure 9.12).

Fig. 9.12. *trans*-Selective Wittig olefination of aldehydes II—Synthesis of β-carotene from a dialdehyde.

9.4 Horner–Wadsworth–Emmons Reaction

Although the Horner–Wadsworth–Emmons reaction is not an ylide reaction, it represents methodologically such an important supplement to the Wittig reaction discussed in Figure 9.3 that we include its discussion here.

9.4.1 Horner–Wadsworth–Emmons Reactions with Achiral Substrates

B

Horner–Wadsworth–Emmons reactions are C=C-forming condensation reactions between the Li, Na, or K salt of a β-keto- or an α-(alkoxycarbonyl)phosphonic acid dialkyl ester and a carbonyl compound (cf. Figure 4.41). These reactions furnish α,β-unsaturated ketones or α,β-unsaturated esters, respectively, as the desired products and a phosphoric acid diester anion as a water-soluble by-product. In general, starting from aldehydes, the desired compounds are produced *trans*-selectively or in the case of olefins with trisubstituted C=C double bonds *E*-selectively.

The precursors for these Horner–Wadsworth–Emmons reagents are β-ketophosphonic acid dialkyl esters or α-(alkoxycarbonyl)phosphonic acid dialkyl esters. The first type of compound, i.e., a β-ketophosphonic acid dialkyl ester is available, for example, by acylation of a metallated phosphonic acid ester (Figure 6.38). The second type of compound, i.e., an α-(alkoxycarbonyl)phosphonic acid dialkyl ester, can be conveniently obtained via the Arbusov reaction (Figure 9.13; cf. Figure 2.26).

A

Condensations between aldehydes and metallated phosphonic acid dialkyl esters other than those mentioned previously are also referred to as Horner–Wadsworth–Emmons reactions. Nevertheless, in these esters, too, the carbanionic center carries a substituent with a $-$M effect, for example, an alkenyl group, a polyene or a C≡N group. The Horner–Wadsworth–Emmons reactions of these reagents are also stereoselective and form the new C=C double bond *trans*-selectively.

B

The mechanism of Horner–Wadsworth–Emmons reactions has not been definitively established. A contemporary rationalization is shown in Figure 9.14 for the reaction between an aldehyde and a metallated phosphonic acid diester, whose carbanionic center is conjugated with a keto group or an ester function. The phosphonate ions from Figure 9.14 have a delocalized negative change. It is not known whether they react as conformers, in which the O=P—C$^-$—C=O substructure is U-shaped and bridged through the metal ion in a six-membered chelate. In principle, this substructure could also have the shape of a sickle or a W. It is also not known for certain whether this attack leads to the oxaphosphetane **B** in two steps through the alkoxide **A** or directly in a one-step cycloaddition. It is not even known whether the oxaphosphetane is formed in a reversible or an irreversible reaction.

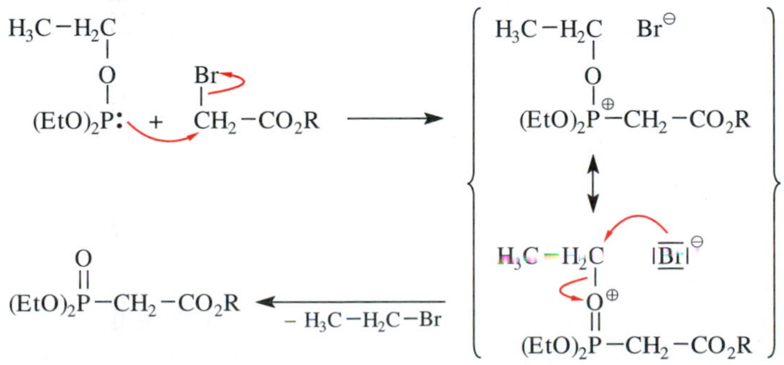

Fig. 9.13. The Arbusov reaction, which provides the most important access to Horner–Wadsworth–Emmons reagents.

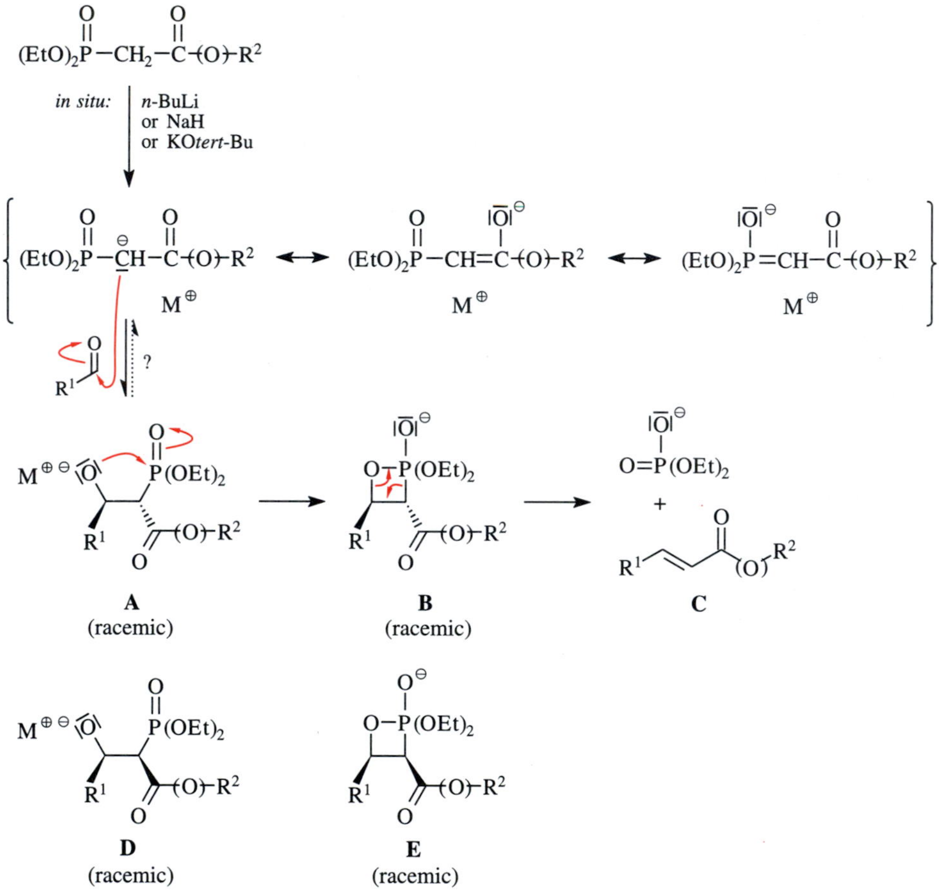

Fig. 9.14. The currently assumed mechanism for the Horner–Wadsworth–Emmons reaction.

The formal analogy to the Wittig reaction (Figure 9.7) would perhaps suggest a one-step formation of the oxaphosphetane **B** in Figure 9.14. On the other hand, the presumably closer analogy to the Wittig–Horner reaction would argue for a two-step formation of this oxaphosphetane. The latter reaction refers to the olefin-producing condensation reaction of a metallated diphenylalkylphosphine oxide [$Ph_2P(=O)—CHR^- M^+$] and an aldehyde. There, the primary product is definitely an alkoxide. If the Horner–Wadsworth–Emmons reaction takes place analogously, the phosphonate ion from Figure 9.14 would first react with the aldehyde to form the alkoxide **A,** which would then cyclize to give the oxaphosphetane **B.** The decomposition of this heterocycle by a one-step [2 + 2]-cycloreversion (cf. Figure 4.41) would finally lead to the *trans*-olefin **C.**

Why *trans*-selectivity occurs is not known because of the lack of detailed knowledge about the mechanism. Perhaps the reason is that only the alkoxide **A** is cyclized to the more stable *trans*-oxaphosphetane shown. This is conceivable because the diastereomorphic alkoxide (**D** in Figure 9.14) should cyclize comparatively slowly to the less stable *cis*-oxaphosphetane **E** if product-development control were to occur in this step.

It would thus be possible that first both the alkoxide **A** and its diastereomer **D** form unselectively but reversibly from the phosphonate ion and the aldehyde. Then an irreversible cyclization of the alkoxide **A** would give the *trans*-oxaphosphetane **B**. The alkoxide **D** would also gradually be converted into the *trans*-oxaphosphetane **B** through the equilibrium **D** $\rightleftharpoons$ starting materials $\rightleftharpoons$ **A.** In summary, in this explanation one would deal with the "productive" formation of an alkoxide intermediate **A** finally leading to the olefin and with the "vain" formation of an alkoxide intermediate **D** which cannot react further to an olefin.

According to Figure 9.14, Horner–Wadsworth–Emmons reactions always produce α,β-unsaturated *ketones* that are *trans-* or *E*-configured. In general, this also applies for similarly prepared α,β-unsaturated *esters* (Figure 9.15, left). However, an apparently innocent structural variation in the phosphonic ester moiety of metallated α-(alkoxycarbonyl)phosphonates completely reverses the stereochemistry of the formation of α,β-unsaturated esters: the replacement of the $H_3C—CH_2—O$ groups by $F_3C—CH_2—O$ groups. This structural variation is the main feature of the **Still–Gennari** variant of the Horner–Wadsworth–Emmons reaction. It makes *cis*-substituted acrylic esters and *Z*-substituted methacrylic esters accessible (Figure 9.15).

The stereostructure of the alkoxide intermediate of a Horner–Wadsworth–Emmons reaction which finally leads to the *trans*-olefin was recorded in Figure 9.14 (as formula **A**). The Still–Gennari variant of this reaction (Figure 9.15) must proceed via an alkoxide with the inverse stereostructure because an olefin with the opposite configuration is produced. According to Figure 9.16, this alkoxide is a 50:50 mixture of the enantiomers **C** and *ent*-**C.** Each of these enantiomers contributes equally to the formation of the finally obtained *cis*-configured acrylic ester **D.**

Fig. 9.15. Preparation of *trans-* or *E*-configured α,β-unsaturated esters by the Horner–Wadsworth–Emmons reaction (left) or preparation of their *cis-* or *Z*-isomers by the Still–Gennari variant of it (right). 18-Crown-6 is a so-called crown ether and contains a saturated 18-membered ring containing six successive $—CH_2—O—CH_2$-units. 18-Crown-6 dissociates the K^+ ions of the Homer-Wadsworth-Emmons reagent by way of complexation.

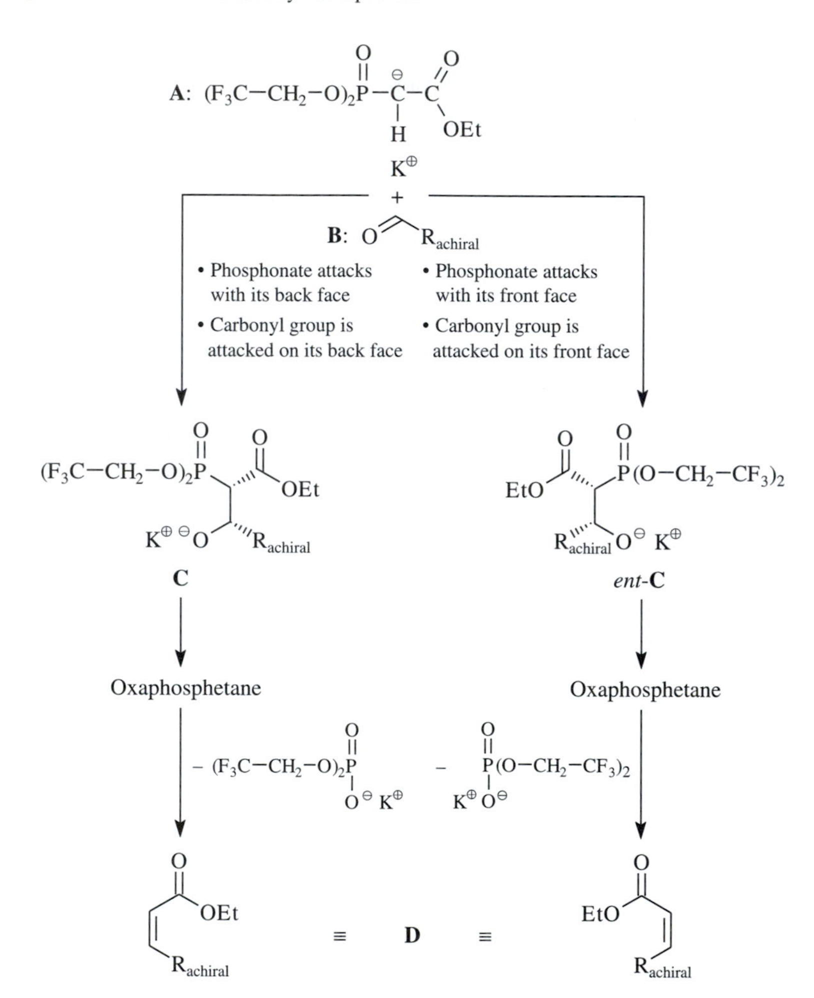

Fig. 9.16. Analysis of the overall stereoselectivity of a Still–Gennari olefination such as the one in Figure 9.15. Simple diastereoselectivity of the formation of the alkoxide intermediate from the achiral phosphonate **A** and the achiral aldehyde **B.** For both reagents the terms "back face" and "front face" refer to the selected projection. Because of their clarity, these terms were given preference over the projection-independent designations *si*- and *re*-face (in the case of **A**) or *re*- and *si*-face (in the case of **B**; explanation of the *re/si* terminology: Figure 10.41), respectively.

9.4.2 Horner–Wittig–Emmons Reactions between Chiral Substrates: A Potpourri of Stereochemical Specialties

A

Let us now consider the stereostructures **C**/*ent*-**C** of the two enantiomorphic Still–Gennari intermediates of Figure 9.16 from another point of view: One can unambiguously derive with which simple diastereoselectivity (see Section 9.3.2) the phosphonate **A** and the aldehyde **B** must combine in order for the alkoxides **C** and *ent*-**C** to be produced. If we use the formulas as written in the figure, this simple diastereoselectivity can be described as follows: The phosphonate ion **A** and the aldehyde **B** react with each other preferentially in such a way that a back face$_{\text{phosphonate}}$/back face$_{\text{aldehyde}}$ linkage (formation of alkoxide **C**) and a front face$_{\text{phosphonate}}$/front face$_{\text{aldehyde}}$ linkage (formation of alkoxide *ent*-**C**) take place *concurrently*.

The Still–Gennari olefinations of Figures 9.17–9.22 start from similar substrates as those shown in Figure 9.16. However, at least one of them is chiral. Since each of these

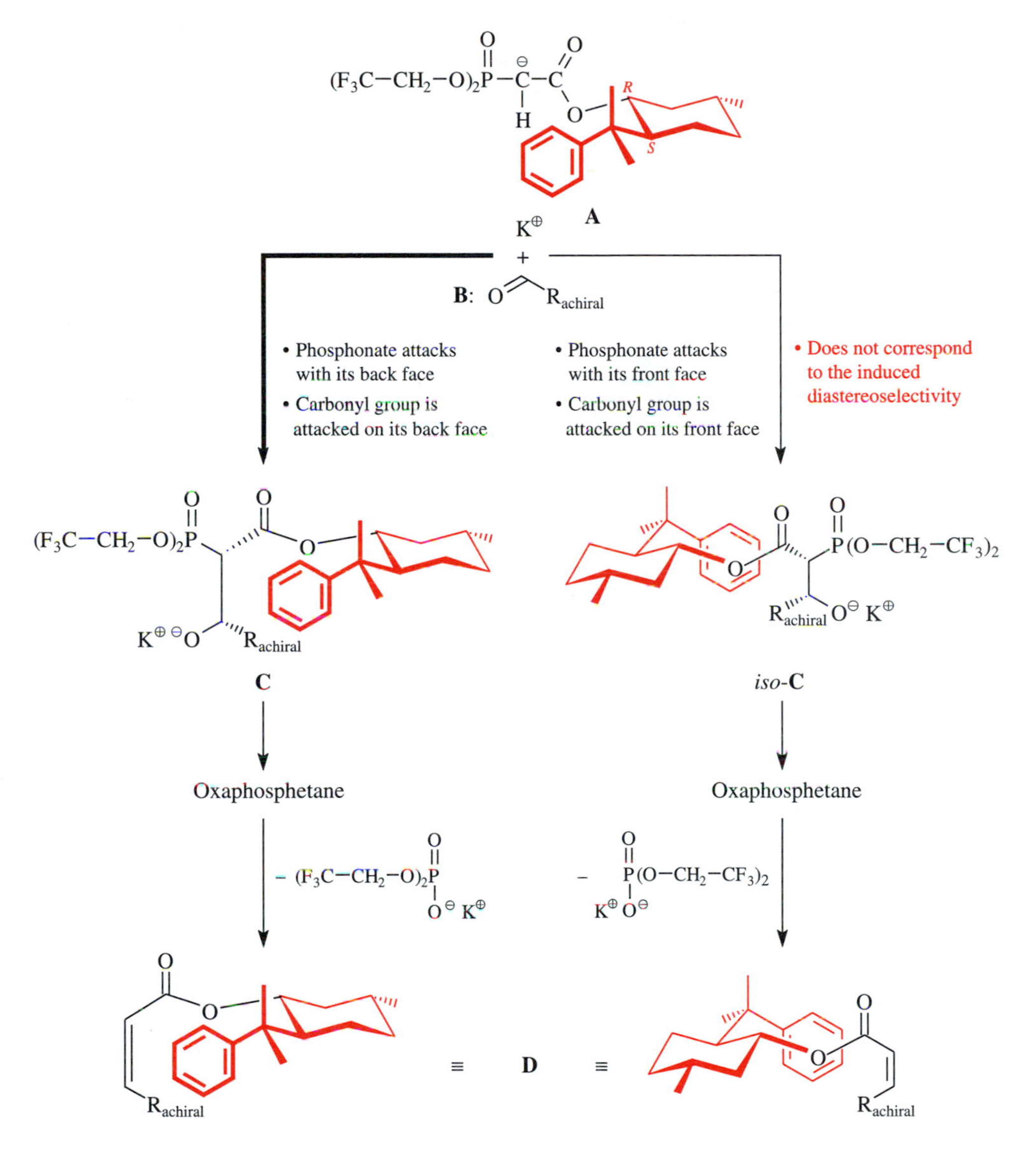

Fig. 9.17. Analysis of the simple diastereoselectivity of a Still–Gennari olefination which starts from the enantiomerically pure phosphonate **A** and the achiral aldehyde **B.**

Still-Gennari olefinations is *cis*-selective, it is clear that, when their reactants combine to form an alkoxide intermediate, they must do so with the same type of simple diastereoselectivity which we have just unraveled for the Still–Gennari reaction of Figure 9.16: Either the back faces of both reactants are linked or the front faces are linked. However, with regard to how important the back face/back face linkage is in comparison to the front face/front face linkage, these various Still–Gennari reactions will differ from one another. In fact, we will study in each individual olefination of Figures 9.17–9.22 how each of these two possible linkages is energetically influenced by the steric or electronic peculiarities of the substrates involved. This influence stems

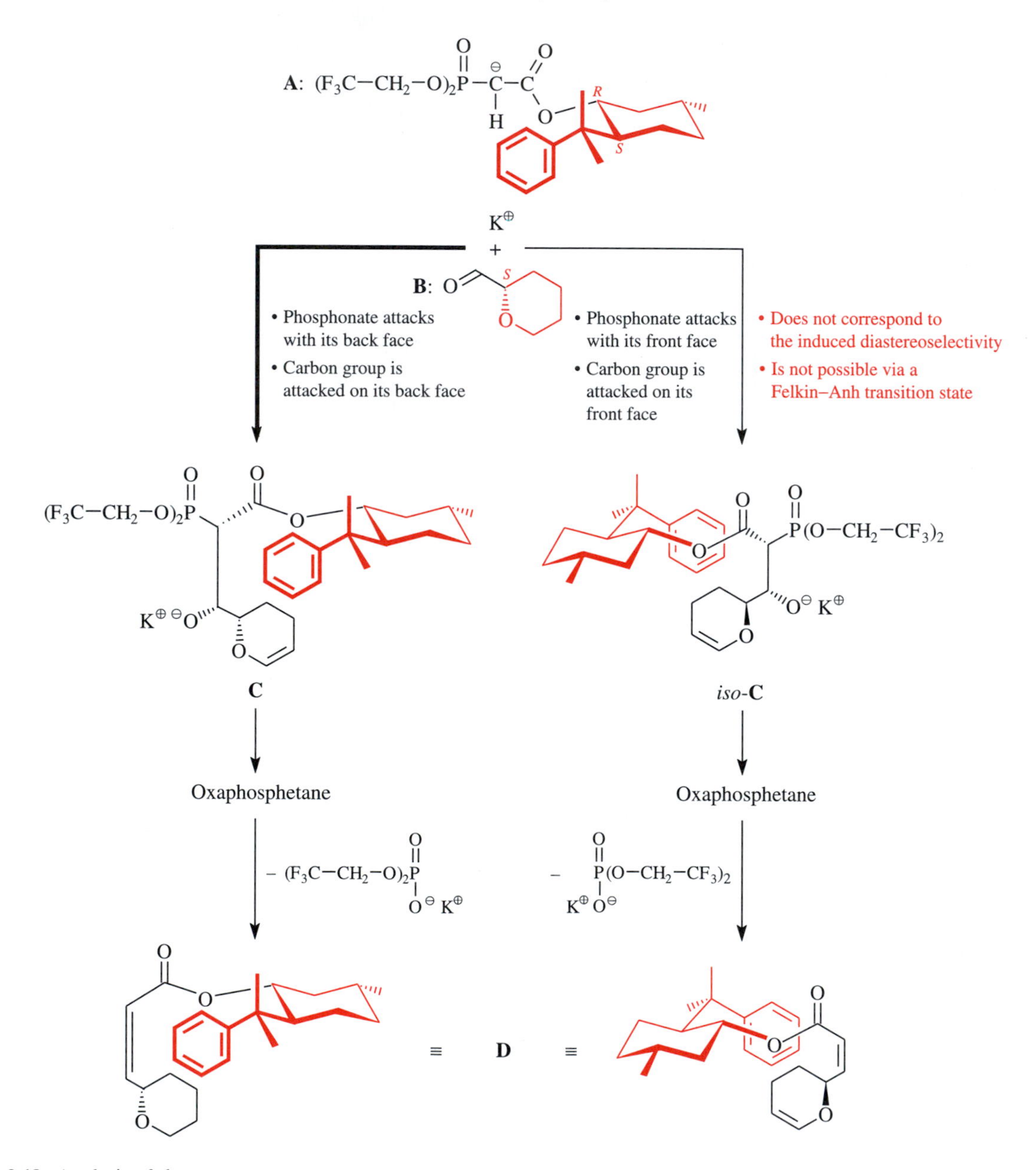

Fig. 9.18. Analysis of the simple diastereoselectivity of a Still–Gennari olefination which starts from the enantiomerically pure phosphonate **A** and the enantiomerically pure α-chiral aldehyde **B.**

from the stereocenters in the *chiral* substrates of Figures 9.17–9.22. They, of course, may exert reagent control or substrate control of the diastereoselectivity.

Figure 9.17 shows what happens when the achiral aldehyde **B** from Figure 9.16 is olefinated with an enantiomerically pure chiral phosphonate ion **A.** Of course, a *cis*-olefin (formula **D**) is again formed preferentially, and in principle it can again be produced in two ways: via the alkoxide **C** or via its diastereomer *iso*-**C.** If (only) a simple

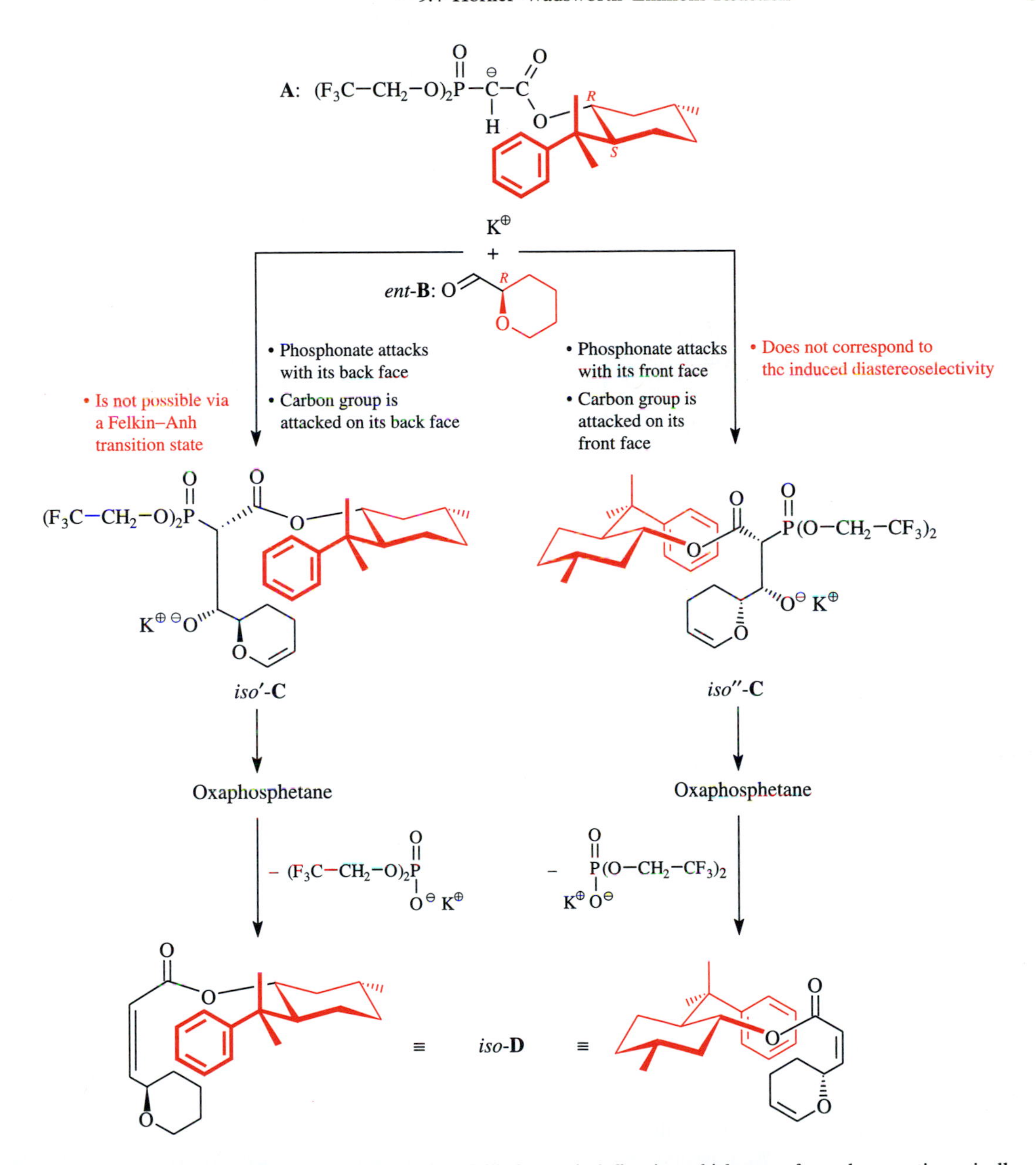

Fig. 9.19. Analysis of the simple diastereoselectivity of a Still–Gennari olefination which starts from the enantiomerically pure phosphonate **A** and the enantiomerically pure α-chiral aldehyde *ent*-**B** (the naming of the compounds in this figure is in agreement with the nomenclature in Figure 9.18).

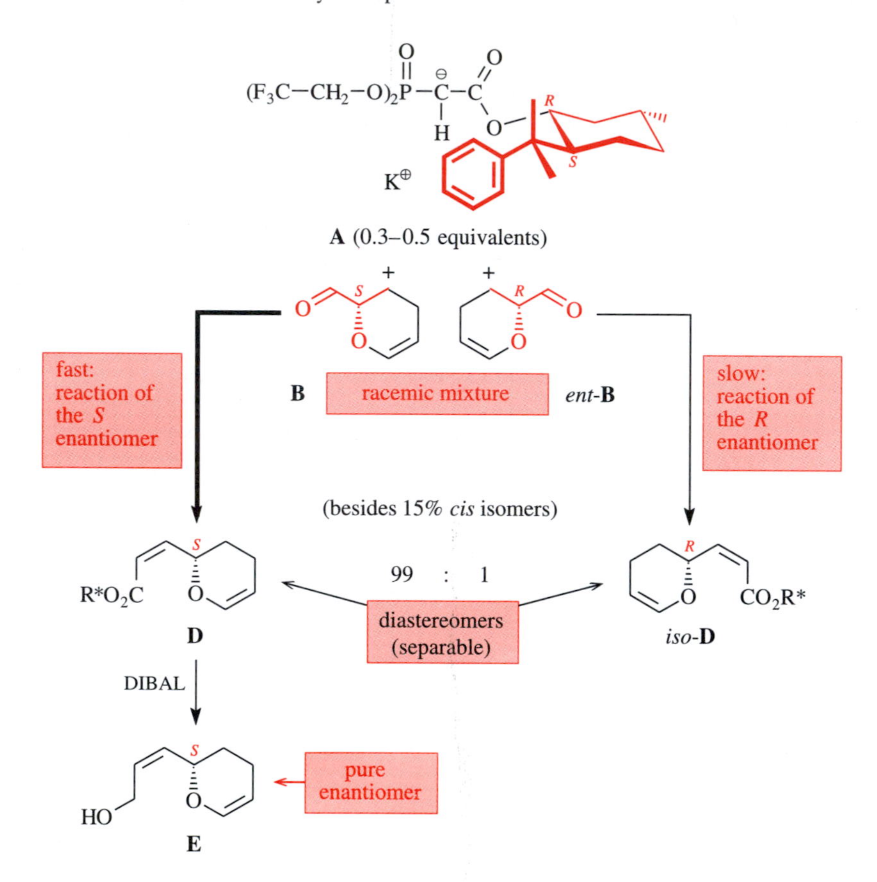

Fig. 9.20. Still–Gennari olefination of a racemic α-chiral aldehyde with an enantiomerically pure phosphonate as kinetic resolution I—Loss of the unreactive enantiomer *ent*-**B** of the aldehyde (R* stands for the phenylmenthyl group in the Horner–Wadsworth–Emmons products; the naming of the products in this figure is in agreement with the nomenclature of Figures 9.18 and 9.19).

diastereoselectivity as in Figure 9.16 were effective, this would mean that these alkoxides would be produced in a 50:50 ratio. However, there is, moreover, a reagent control of the diastereoselectivity since the phosphonate ion **A** of Figure 9.17 is chiral. The reagent control stems from the shielding of the front face of this phosphonate in its presumed reactive conformation **A** by the protruding phenyl ring. This shielding makes a front face$_{\text{phosphonate}}$ attack virtually impossible. Therefore, this phosphonate ion can combine with aldehyde **B** only at the back face of the phosphonate. For this to occur, the simple diastereoselectivity of the Horner–Wadsworth–Emmons reaction requires a back face$_{\text{phosphonate}}$/back face$_{\text{aldehyde}}$ linkage. The alkoxide intermediate of the olefination of Figure 9.17 is thus mainly compound **C.**

We now analyze the Still–Gennari olefination of Figure 9.18. The reagents there are the enantiomerically pure chiral phosphonate **A,** with which you are familiar from Figure 9.17, and an enantiomerically pure α-chiral aldehyde **B.** The diastereoselectivity of the formation of the crucial alkoxide intermediate(s) is in this case determined by the interplay of three factors:

1) The simple diastereoselectivity (as it already occurred in the olefinations of Figures 9.16 and 9.17).

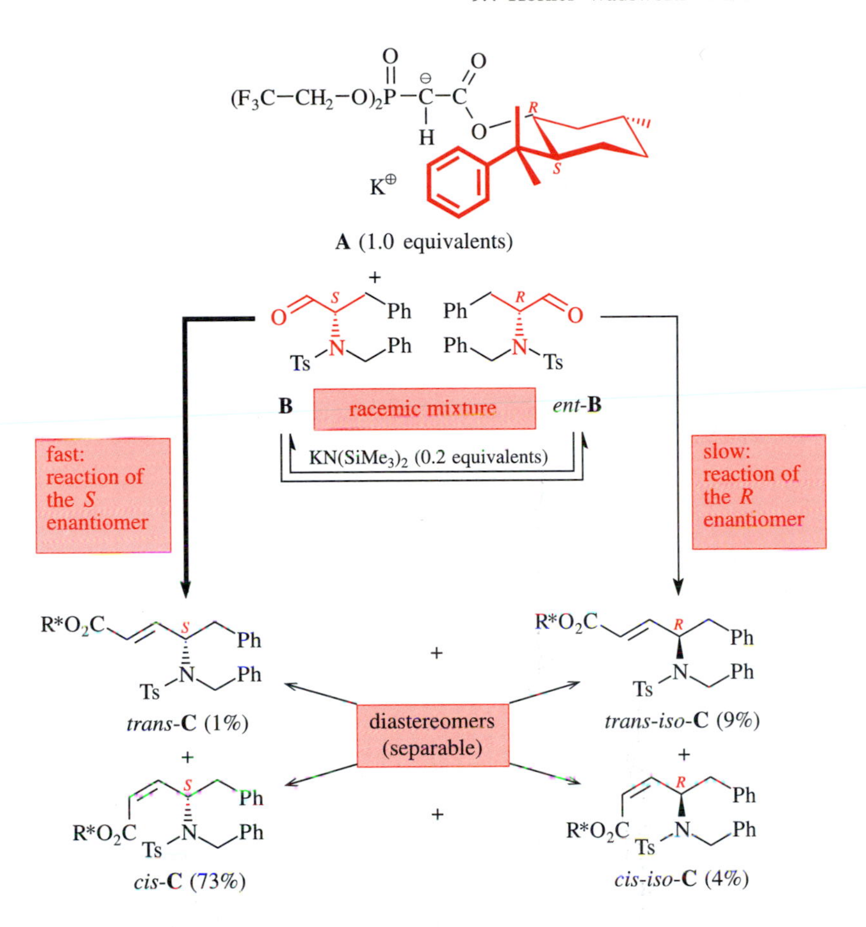

Fig. 9.21. Still–Gennari olefination of a racemic α-chiral aldehyde with an enantiomerically pure phosphonate as kinetic resolution II—Use of the unreactive enantiomer *ent*-**B** of the aldehyde (R* stands for the phenylmenthyl group in the Horner–Wadsworth–Emmons products).

2) The reagent control of the diastereoselectivity originating from phosphonate **A** (which you know from the olefination of Figure 9.17).
3) The substrate control of the diastereoselectivity, which originates from aldehyde **B** and makes a Felkin–Anh transition state preferred in the attack of the phosphonate ion on the C═O double bond (compare Figure 8.11, middle).

When reagent control and substrate control occur in one reaction *at the same time*, according to Section 3.4.4 this is a case of double stereodifferentiation. In a matched substrate/reagent pair both effects act in the same direction, whereas in a mismatched substrate/reagent pair they act against each other. Accordingly, in the Still–Gennari olefination of Figure 9.18 we have a case of double stereodifferentiation, and a matched substrate/reagent pair is realized. Reagent *and* substrate control cooperate and, in combination with the simple diastereoselectivity, cause the reaction to proceed practically exclusively via the alkoxide **C.**

The mismatched case of a Still–Gennari olefination, which corresponds to the matched case of Figure 9.18, can be found in Figure 9.19. There, reagent control and substrate control act against each other. The substrate in Figure 9.19 is the enantiomer

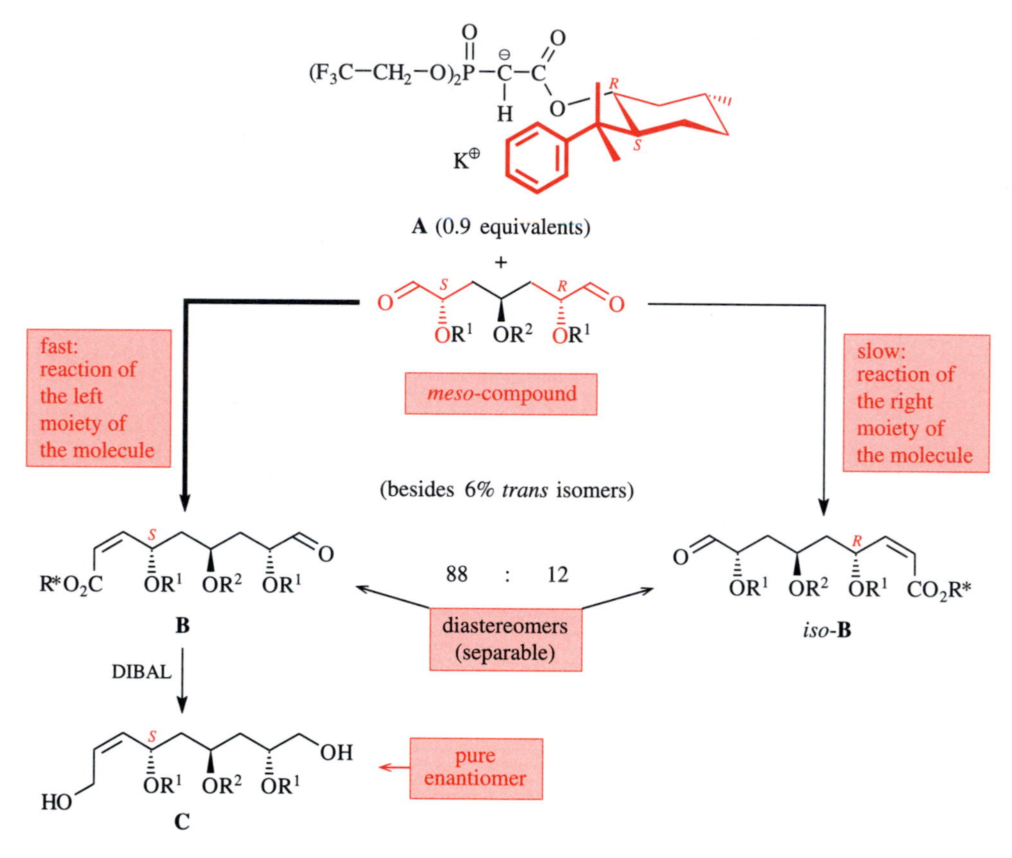

Fig. 9.22. Still–Gennari olefination of a *meso*-aldehyde with an enantiomerically pure phosphonate—Conversion of a *meso*- into an enantiomerically pure compound (R* stands for the phenylmenthyl group in the Horner–Wadsworth–Emmons products).

ent-**B** of the aldehyde used in Figure 9.18. Although the olefination product of Figure 9.19 must be *iso*-**D,** it is impossible to predict by which of the two conceivable routes it is formed: olefin *iso*-**D** can only be produced if either the substrate control (through formation of the alkoxide intermediate *iso′*-**C**) or the reagent control of the diastereoselectivity (through formation of the alkoxide intermediate *iso″*-**C**) is violated. It is not known which of these alkoxide intermediates is less disfavored and is therefore formed less slowly. One thing is certain: The olefin *iso*-**D** is produced in a considerably slower reaction than the olefin **D** of Figure 9.18.

Let us express the outcomes of Figures 9.18 and 9.19 once more in a different way: The *S*-configured chiral aldehyde **B** is olefinated by the enantiomerically pure phosphonate **A** considerably faster than its *R* enantiomer *ent*-**B.** This can be exploited when 0.5 equivalents of the enantiomerically pure phosphonate **A** are treated with a 1:1 mixture of the aldehydes **B** and *ent*-**B**—that is, treated with the racemic aldehyde (Figure 9.20). The *S* enantiomer **B** reacts with the phosphonate **A** in a matched pair (according to Figure 9.18), hence fast and consequently virtually completely. On the other hand, the *R* enantiomer *ent*-**B** of the aldehyde is almost not touched because it forms a mismatched pair with the phosphonate (according to Figure 9.19). Thus, the result is a kinetic resolution, and the *cis*-esters formed in Figure 9.20 are formed in a 99:1 ratio from the *S* or *R* aldehyde, respectively. Since these esters represent a 99:1 mixture of diastereomers—**D** and *iso*-**D**—they can be purified chromatographically to give the pure *S* ester **D.** The chiral auxiliary can be removed by a reduction with DIBAL (method: Figure 14.54). In this way one obtains the allyl alcohol **E** *in 100% enantiomeric purity.*

A completely analogous kinetic resolution succeeds with the Still–Gennari olefination of Figure 9.21. Here the racemic substrate is a different α-chiral aldehyde. It carries a sulfonamido group at the stereocenter in the α-position. Nucleophiles also add to the carbonyl group of such an aldehyde preferentially (cf. Figure 8.33) via a Felkin–Anh transition state (Figure 8.11, middle). Thus, in the Horner–Wadsworth–Emmons reaction of this aldehyde, exactly as in the analogous one from Figure 9.20, the *S* enantiomer is olefinated faster than the *R* enantiomer. However, the α-chiral aldehyde in Figure 9.21 differs from the one in Figure 9.20 in that it racemizes under the reaction conditions. It is therefore not only the initially available 50% of the reactive *S* enantiomer that can be olefinated. A major fraction of the *R* aldehyde, which is not olefinated to a significant extent (only 13% of the *R* olefins *cis-iso-* and *trans-iso*-**C** are produced), isomerizes to the *S* aldehyde and is subsequently *also* converted to the *S* olefin *cis*-**C.** The latter's yield is 73% based on the racemate but is 146% based on the *S* aldehyde used. Here we have the case of an especially efficient kinetic resolution—one which includes the enantiomerization of the undesired antipode. Such a kinetic resolution is termed **dynamic kinetic resolution.**

The Still–Gennari olefination in Figure 9.22 is recommended to anyone who wants to enjoy a third stereochemical delicacy. The substrate is a dialdehyde which contains oxygenated stereocenters in both α positions. Nevertheless, this aldehyde is achiral because it has a mirror plane and thus represents a *meso*-compound. *Meso*-compounds can sometimes be converted into enantiomerically pure chiral compounds by reacting them with enantiomerically pure reagents. Figure 9.22 shows how this can be done using a Still–Gennari olefination with the phosphonate ion **A.** The outcome of this olefination is readily understood when it is compared with the olefination in Figure 9.20. The left moiety of the *meso*-dialdehyde in Figure 9.22 corresponds structurally to the *reactive* enantiomer **B** of the aldehyde in Figure 9.20. Similarly, the right moiety of the *meso*-dialdehyde in Figure 9.22 corresponds structurally to the *unreactive* enantiomer *ent*-**B** of the aldehyde in Figure 9.20. We already discussed in detail why in Figure 9.20 the aldehyde enantiomer **B** is olefinated and the aldehyde enantiomer *iso*-**B** is not. These analogies rationalize perfectly why in the *meso*-dialdehyde of Figure 9.22 the carbonyl group on the left is olefinated faster than the one on the right.

The *cis*-olefin **B** is consequently the major product of the olefination of Figure 9.22. It can be separated chromatographically from the minor *cis*-olefin *iso*-**B** and also from the *trans*-olefins formed in small amounts because these species are diastereomers. Finally, the chiral auxiliary can be removed reductively from the obtained *pure* olefin **B** (cf. the analogous reaction at the bottom of Figure 9.20). The resulting allyl alcohol **C** is 100% enantiomerically pure. Figure 9.22 thereby gives an example of the conversion of a *meso*-compound into an *enantiomerically pure* substance; a 100% yield is possible in principle in such reactions.

References

J. M. J. Williams (Ed.), "Preparation of Alkenes: A Practical Approach," Oxford University Press, Oxford, U.K., **1996.**

9.1

H. J. Bestmann and R. Zimmerman, "Synthesis of Phosphonium Ylides," in *Comprehensive Organic Synthesis* (B. M. Trost, I. Fleming, Eds), Vol. 6, 171, Pergamon Press, Oxford, **1991.**

A. W. Johnson, "Ylides and Imines of Phosphorus," Wiley, New York, **1993**

O. I. Kolodiazhnyi, "C-Element-substituted phosphorus ylids," *Tetrahedron* **1996,** *52,* 1855–1929.

9.3

A. Maercker, "The Wittig reaction," *Org. React.* **1965,** *14,* 270–490.

H. J. Bestmann and O. Vostrowski, "Selected topics of the Wittig reaction in the synthesis of natural products," *Top. Curr. Chem.* **1983,** *109,* 105.

H. Pommer and P. C. Thieme, "Industrial applications of the Wittig reaction," *Top. Curr. Chem.* **1983,** *109,* 165.

B. E. Maryanoff and A. B. Reitz, "The Wittig olefination reaction and modifications involving phosphorylstabilized carbanions: Stereochemistry, mechanism, and selected synthetic aspects," *Chem. Rev.* **1989,** *89,* 863–927.

E. Vedejs and M. J. Peterson, "Stereochemistry and mechanism in the Wittig reaction," *Top. Stereochem.* **1994,** *21,* 1.

K. Becker, "Cycloalkenes by intramolecular Wittig reaction," *Tetrahedron* **1980,** *36,* 1717.

B. M. Heron, "Heterocycles from intramolecular Wittig, Horner and Wadsworth-Emmons reactions," *Heterocycles* **1995,** *41,* 2357.

9.4

W. S. Wadsworth, Jr., "Synthetic applications of phosphoryl-stabilized anions," *Org. React.* **1977,** *25,* 73–253.

D. F. Wiemer, "Synthesis of nonracemic phosphonates," *Tetrahedron* **1997,** *53,* 16609–16644.

H. B. Kagan and J. C. Fiaud, "Kinetic resolution," *Top. Stereochem.* **1988,** *18,* 249–330.

Further Reading

P. J. Murphy and J. Brennan, "The Wittig olefination reaction with carbonyl compounds other than aldehydes and ketones," *Chem. Soc. Rev.* **1988,** *17,* 1–30.

H. J. Cristau, "Synthetic applications of metalated phosphonium ylides," *Chem. Rev.* **1994,** *94,* 1299.

J. Seyden-Penne, "Lithium coordination by Wittig-Horner reagents formed by a-carbonyl substituted phosphonates and phosphine oxide," *Bull. Soc. Chim. Fr.* **1988,** *238.*

C. Yuan, S. Li, C. Li, S. Chen, W. Huang, G. Wang, C. Pan, Y. Zhang, "New strategy for the synthesis of functionalized phosphonic acids," *Heteroatom Chem.* **1997,** *8,* 103–122.

F. R. Harley (Ed.), "The chemistry of organophosphorus compounds: phosphine strategy for the synthesis of functionalized phosphonic acids," *Heteroatom Chem.* **1997,** *8,* 103–122.

F. R. Hartley (Ed.), "The Chemistry of Organophosphorus Compounds: Phosphine Oxides, Sulfides, Selenides, and Tellurides: The Chemistry of Functional Groups," Wiley, Chichester, U.K., **1992.**

J. Clayden and S. Warren, "Stereocontrol in organic synthesis using the diphenylphosphoryl group," *Angew. Chem.* **1996,** *108,* 261–291; *Angew. Chem. Int. Ed. Engl.* **1996,** 35, 241–270.

Chemistry of the Alkaline Earth Metal Enolates

10

B

Aldehydes, ketones, carboxylic esters, carboxylic amides, imines and *N,N*-disubstituted hydrazones react as electrophiles at their sp^2-hybridized carbon atoms. These compounds also become nucleophiles, if they contain an H atom in the α position relative to their C═O or C═N bonds. This is because they are C,H-acidic at that position, that is, the H atom in the α position can be removed with a base (Figure 10.1). The deprotonation forms the conjugate bases of these substrates, which are called **enolates.** Depending on the origins of these enolates, they may be called aldehyde enolates, ketone enolates, ester enolates, or amide enolates. The conjugate bases of imines and hydrazones are called **aza-enolates.** The reactions discussed in this chapter all proceed via enolates.

$$M^{\oplus}\,ba:^{\ominus} + H{-}\overset{|}{\underset{|}{C}}{-}C(=O){-}X \xrightarrow{-baH} {>}C{=}C(O^{\ominus}\,M^{\oplus}){-}X$$

$(M^{\oplus} = Li^{\oplus}, Na^{\oplus}, K^{\oplus})$

X	
H	Aldehyde enolate[1]
alkyl, aryl	Ketone enolate[1]
Oalkyl, Oaryl	Ester enolate
NR^1R^2	Amide enolate

[1]Also called simply enolate.

$$Li^{\oplus}\,iPr_2N:^{\ominus} + H{-}\overset{|}{\underset{|}{C}}{-}C(=NR){-}X \xrightarrow{-\,iPr_2NH} {>}C{=}C({}^{\ominus}NR\;Li^{\oplus}){-}X$$

X = H, alkyl; R = alkyl, $N(alkyl)_2$ } aza-enolate

Fig. 10.1. Formation of enolates from different C,H acids.

10.1 Basic Considerations

10.1.1 Notation and Structure of Enolates

B

In valence bond theory, every enolate can be described by two resonance forms. The negative formal charge is located at a C atom in one of these resonance forms and at an O or an N atom in the other resonance form. In the following, we refer to these resonance forms as the *carbanion* and the *enolate* resonance forms, respectively. Only the enolate resonance form is shown in Figure 10.1 because this resonance form has the higher weight according to resonance theory. The enolate resonance form places negative charge on the more electronegative heteroatom O or N. These heteroatoms stabilize the negative charge better than the less electronegative C atom in the carbanion resonance form.

In Figure 10.1 the enolate structures are shown with the charge on the heteroatom and with the heteroatom in association with a metal ion. The metal ion stems from the reagent used in the enolate formation. In the majority of the reactions in Chapter 10, the enolate is generated by deprotonation of C,H acids. The commonly employed bases contain the metal ions Li^+, Na^+, or K^+. Therefore, in Chapter 10, we will essentially consider the chemistry of lithium, sodium, and potassium enolates.

A

It is known that the chemistry of enolates depends on the nature of the metal. Moreover, the metals are an integral part of the structures of enolates. Lithium enolates are most frequently employed, and in the solid state the lithium cations definitely are associated with the heteroatoms rather than with the carbanionic C atoms. Presumably the same is true in solution. The bonding between the heteroatom and the lithium may be regarded as ionic or polar covalent. However, the heteroatom is not the only bonding partner of the lithium cation irrespective of the nature of the bond between lithium and the heteroatom:

- Assuming ionic Li^+O^- or Li^+NR^- interactions, it may be appropriate to draw a parallel between the structures of enolates and ionic crystals of the Li^+Cl^- or Li^+H^- types. In the latter structures, every lithium is coordinated by six neighboring anions.
- From the viewpoint of polar, yet covalent Li—O and Li—N bonds, lithium would be unable to reach a valence electron octet in the absence of bonding partners in addition to the heteroatom. The lithium thus has to surround itself by other donors in much the same way as has been seen in the case of the organolithium compounds (cf. Section 8.1).

Be this as it may, lithium attempts to bind to several bonding partners, and the structural consequences for the enolates of a ketone, an ester, and an amide are shown in Figure 10.2: In contrast to the usual notation, these enolates are not monomers at all! The heteroatom that carries the negative charge in the enolate resonance form is an excellent bonding partner, such that several such heteroatoms are connected to every lithium atom. Lithium enolates often result in "tetramers" if they are crystallized in the absence of other lithium salts and in the absence of other suitable neutral donors. The lithium enolate of *tert*-butyl methyl ketone, for example, crystallizes from THF in the form shown in Figure 10.3.

"Tetramers" like the one in Figure 10.3 contain cube skeletons in which the corners are occupied in an alternating fashion by lithium atoms and enolate oxygens. Every lithium atom is surrounded by three enolate oxygen atoms, and vice versa. Every lithium binds a molecule of THF as its fourth ligand. It is for this reason that the term "tetramer" was used in quotation marks; the overall structure is a THF complex of the tetramer.

Figure 10.2 shows structures that contain two lithium enolates each. But again, these structures are not *pure* dimers. Both lithium atoms employ two of their coordination

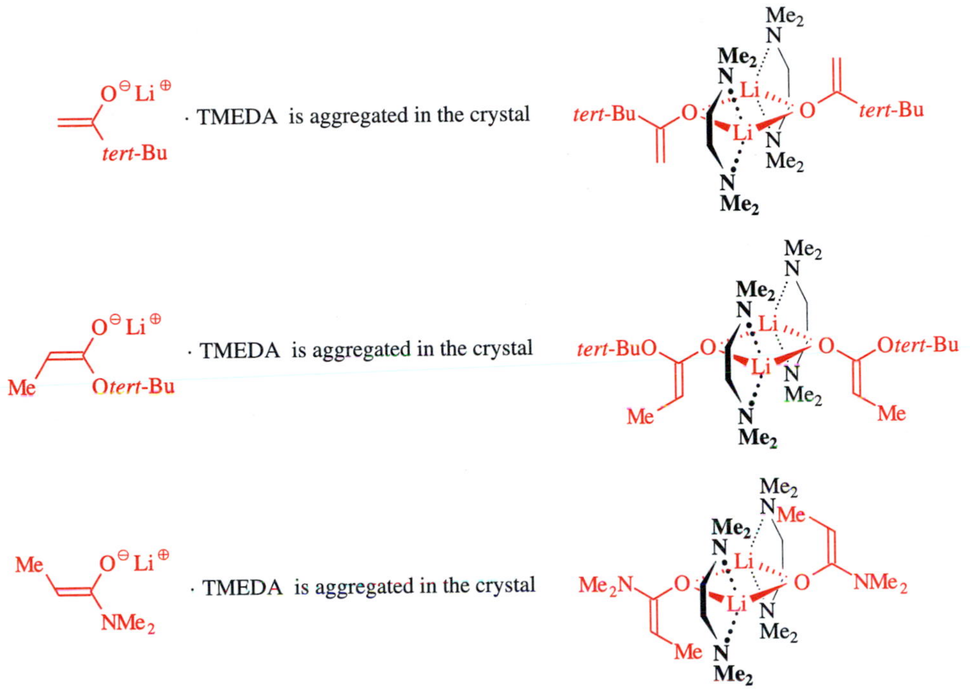

· TMEDA is aggregated in the crystal

· TMEDA is aggregated in the crystal

· TMEDA is aggregated in the crystal

Fig. 10.2. X-ray single crystal structures of lithium enolates. TMEDA, tetramethylethylenediamine.

sides to bind to an N atom of the bidentate ligand TMEDA (see Figure 10.2 for name and structure).

"Oligomeric" enolates along with the associated neutral ligands also are called **aggregates.** Lithium enolates are likely to exist as aggregates also in solution. The neutral ligands in these aggregates can be TMEDA, DMPU (structure in Figure 2.16), HMPA (structure in Figure 2.16), THF and/or $HN(iPr)_2$. Lithium enolates may occur in such **homoaggregates,** but they also may be part of so-called **mixed aggregates.** The latter are aggregates that also contain other lithium compounds (e.g., LiHal, LiOR or LDA).

B

It is not known whether lithium enolates exist in solution as homoaggregates or as mixed aggregates, nor is it known whether lithium enolates react as aggregates or via other species that might be present in low concentration. But it is certain that

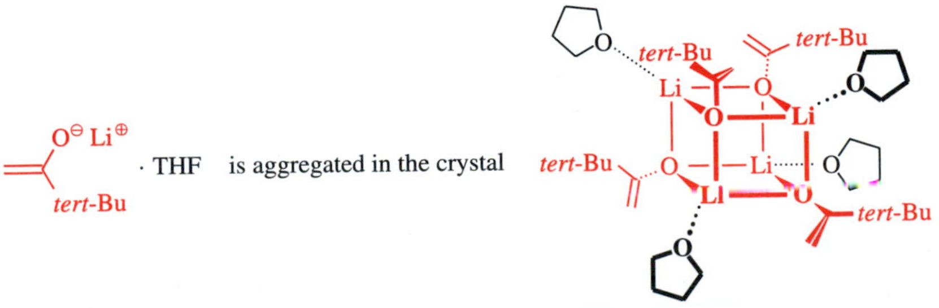

· THF is aggregated in the crystal

Fig. 10.3. X-ray crystal structure of $H_2C{=}C(O^-Li^+)(tert\text{-}Bu)$ · THF.

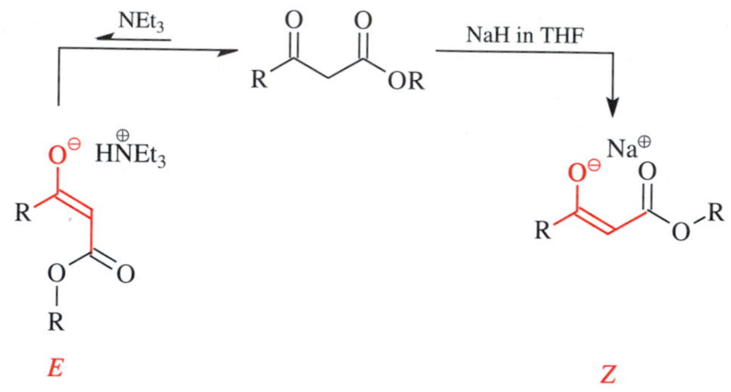

Fig. 10.4. Stereoselective deprotonation of a β-ketoester to trialkylammonium or sodium enolates. The *E*- and *Z*-enolates are formed when NEt_3 and NaH, respectively, are employed.

the reactivity of lithium enolates is affected by the presence or absence of molecules that are capable of forming aggregates. However, all these insights about aggregation do not preclude focusing on the enolate monomers in discussions of the elemental aspects of enolate reactivity. Hence, in Chapter 10, all reactions are formulated in a simplified and unified format considering monomeric enolates. There also are enolates of metals other than Li, Na, or K, but these are not considered in this book. In addition, there are some metal-free enolates, the ammonium enolates, which can be generated in equilibrium reactions between amines and so-called active-methylene compounds. These are compounds that contain two geminal acceptor substituents with strong −M effects.

Starting from β-ketoesters (Figure 10.4, see Figure 10.22 for a synthetic application), or β-ketoaldehydes (Figure 10.5), ammonium enolates are formed with a stereostructure (*E*-enolates) that differs from that of the corresponding alkaline earth metal enolates (*Z*-enolates). In the latter, the lithium atom forms a bridge between the two neg-

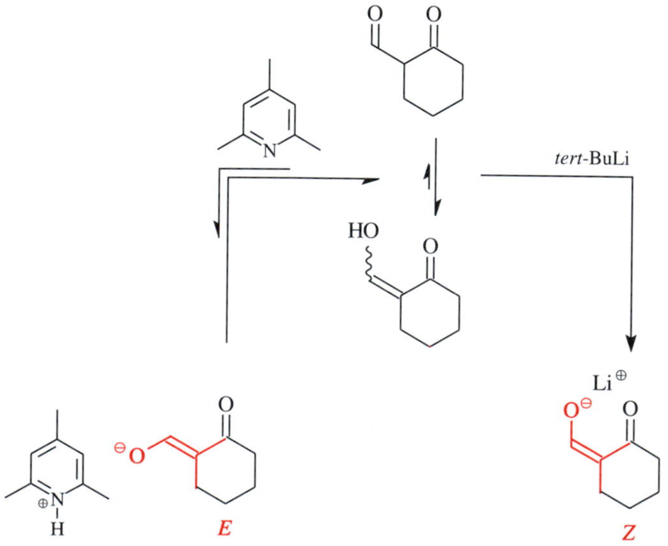

Fig. 10.5. Stereoselective deprotonation of a β-ketoaldehyde and its enol tautomer to substituted pyridinium or lithium enolates, respectively. Similar to the deprotonation of Figure 10.4, the *E*-enolate is formed when the amine is used and the *Z*-enolate is formed when the metal-containing base is employed.

atively charged O atoms such that a six-membered chelate results. In contrast, the ammonium ion cannot play such a role in the ammonium enolate, and the more stable *E*-enolate is formed as the result of product-development or thermodynamic control. The negatively charged O atoms are at a greater distance from each other in the *E*-enolate than in the *Z*-enolate. The greater distance between the O atoms in the *E*-enolate reduces charge–charge and dipole–dipole repulsions.

10.1.2 Preparation of Enolates by Deprotonation

Suitable Bases

B

According to Figure 10.1 carbon-bound H atoms are acidic if they are bound to carbon atoms that are in the α position with respect to an electron acceptor that exerts a strong −M-effect. Carbon-bound H atoms are even more acidic if they are located in the α position of *two* such electron acceptors which is the case in the so-called **active-methylene compounds.** Enolates derived from active-methylene compounds require three resonance forms for their description, and resonance forms **A** and **B** (Figure 10.6) are the more important ones. Compounds that contain an H atom in the α position with respect to three electron acceptors are even more acidic than active-methylene compounds. However, such compounds do not play a significant role in organic chemistry.

Table 10.1 lists the pK_a values of C,H-acidic compounds with a variety of electron acceptors. It shows that multiple substitution by a given acceptor enhances the acidity of the α-H-atom more than monosubstitution. Table 10.1 also shows that the nitro group is the most activating substituent. One nitro group causes the same acidity of an α-H-atom as do two carbonyl or two ester groups. Apparently, the nitro group with its three heteroatoms is best at stabilizing the negative charge in the conjugate base of a C,H acid.

The acidifying effect of the remaining acceptor substituents of Table 10.1 decreases in the order —C(=O)—H > —C(=O)—alkyl > —C(=O)—O-alkyl, and the amide group —C(=O)—NR_2 is even less effective. This ordering essentially reflects substituent effects on the stability of the C=O double bond in the respective C,H-acidic compound. The resonance stabilization of these C=O double bonds drastically increases in the order R—C(=O)—H < R—C(=O)—alkyl < R—C(=O)—O—alkyl < R—C(=O)— NR_2 (cf. Table 6.1; see Section 7.2.1 for a comparison between the C=O double bonds in aldehydes and ketones). This resonance stabilization is lost completely once the α-H-atom has been removed by way of deprotonation and the respective enolate has formed.

$$M^{\oplus}\ ba:^{\ominus} + H{-}C(EWG^2){-}EWG^1 \xrightarrow{-\ baH} \left\{ {}^{-}C(EWG^1)^{\ominus}(EWG^2)\ (\mathbf{A}) \longleftrightarrow {}^{-}C(EWG^1)(EWG^2)^{\ominus}\ (\mathbf{B}) \right\} M^{\oplus}$$

($M^{\oplus} = Na^{\oplus}$, $K^{\oplus}$; rarely used, but possible: $Bu_4N^{\oplus}$)

Fig. 10.6. Enolate formation of active-methylene compounds.

Table 10.1. Effects of Substituents on C,H-Acidity*

pK_a of ... for EWG =	H–C–EWG	H–C(EWG)–EWG
$-NO_2$	10.2	3.6
–C(=O)–H	16	5
–C(=O)–Me	19.2	9.0
–C(=O)–OMe	24.5	13.3

* EWG, electron-withdrawing group.

The equilibrium constant K_{eq} of the respective deprotonation equilibrium shows whether a base can deprotonate a C,H-acidic compound quantitatively, in part, or not at all:

$$(EWG)_{1 \text{ or } 2}C{-}H + \text{base}^{\ominus} \underset{}{\overset{K_{eq}}{\rightleftharpoons}} (EWG)_{1 \text{ or } 2}C^{\ominus} + \text{H-base}$$

$$K_{eq} = \frac{K_{a,\text{C,H acid}}}{K_{a,\text{H-base}}} \tag{10.1}$$

$$= 10^{pK_{a,\text{H-base}} - pK_{a,\text{C,H acid}}} \tag{10.2}$$

Equation 10.1 shows that these equilibrium constants in turn depend on the acidity constants of the two weak acids involved, that is, the acidity constant $K_{a,\text{C,H acid}}$ of the C,H acid and the acidity constant $K_{a,\text{H-base}}$ of the conjugate acid (H-base) of the base (base^-) employed. Equation 10.2 makes the same statement in terms of the corresponding pK_a values. From this equation it follows:

Rules of Thumb Regarding the Position of the Equilibria of C,H-Acidic Compounds

1) A C,H acid is deprotonated quantitatively (or nearly so) by an equimolar amount of base if the pK_a value of the conjugate acid of the base employed is higher than the pK_a value of the C,H acid.
2) A C,H acid is deprotonated by an equimolar amount of base only to the extent of 10, 1, or 0.1%, and so on, if the pK_a value of the conjugate acid of the base employed is 1, 2, 3, . . . units lower than the pK_a value of the C,H acid.
3) In cases of incomplete C,H acid deprotonation, an excess of base can be employed to increase the enolate fraction. According to the principle of Le Chatelier, the base excess increases the enolate fraction by a factor that equals the square root of the number of mole equivalents of the base employed.

Table 10.2. Basicity of Typical Reagents Employed in the Generation of Enolates via Deprotonation

Reagent	pK_a-value of the conjugate acid
$Li^{\oplus}NR_2^{\ominus}$	35 bis 40
$K^{\oplus}O$*tert*-$Bu^{\ominus}$	19
$Na^{\oplus}OEt^{\ominus}$	15.7
$Na^{\oplus}OH^{\ominus}$	15.5
NEt_3	10.2

With the pK_a values of the conjugate acids of the most commonly used organic bases (Table 10.2) and the pK_a values of the C,H acids compiled in Table 10.1, the foregoing statements lead to the following deductions:

- All aldehydes, ketones, and carboxylic esters can be deprotonated quantitatively to enolates with lithium amides. The same substrates and alkoxides give only small amounts of enolate in equilibrium reactions, but even these small amounts of enolate may be large enough to allow for enolate reactions.
- For the quantitative deprotonation of nitroalkanes and methylene-active compounds, there is no need to employ the "heavy artillery" of lithium amides; rather, it suffices to employ alkaline earth metal alkoxides or alkaline earth metal hydroxides. In addition, equilibrium reactions between these C,H acids and amines form enough enolate to initiate enolate reactions.

The foregoing classification is of fundamental significance for the understanding of enolate chemistry. For every pair of C,H acid and base, one needs to know whether the combination effects quantitative or partial enolate formation. If deprotonation is only partial, then the unreacted substrate may represent an electrophile which can react with the enolate nucleophile. In such a case, it depends on the specific circumstances whether an enolate reacts with any remaining substrate or whether it reacts only with an added *different* electrophile. The occurrence of a reaction between enolate and unreacted substrate is avoided if the C,H acid is deprotonated completely with a stoichiometric amount of a sufficiently strong base.

There is only one exception to the last statement in that aldehydes cannot be converted quantitatively into aldehyde enolates. Any attempt to achieve a quantitative deprotonation of an aldehyde—with a lithium amide, for example—necessarily leads to a situation in which some aldehyde enolate is already formed while some aldehyde substrate is still present, and these species cannot coexist even at temperatures as low as that of dry ice. The aldehyde is such an excellent electrophile that it reacts much faster with the enolate than it is deprotonated by the base.

Table 10.3 allows for a comparison of the basicities of the strongest lithium-containing bases. The basicities are measured by the heats of deprotonation liberated upon mixing the reference acid isopropanol with these bases. These heats of deproto-

Table 10.3. Thermochemistry of Selected Acid/Base Reactions: Deprotonation Enthalpies (kcal/mol) for Deprotonations of *i*PrOH with Various Organolithium Compounds and Lithium Amides

tert-BuLi	–56.2	LTMP	–30.4
sek-BuLi	–52.8	LDA	–28.6
n-BuLi	–50.0	LiHMDS	–12.1
PhLi	–42.3		

nation reveal that organolithium compounds are even stronger bases than lithium amides. Their basicities decrease from *tert*-BuLi via *sec*-BuLi and *n*-BuLi to PhLi.

Considering these heats of deprotonation, one wonders whether organolithium compounds should not be at least as suitable as lithium amides for effecting the deprotonation of carbonyl and carboxyl compounds. However, this is usually not the case, since organolithium compounds react almost always as nucleophiles rather than as bases. Organolithium compounds thus would add to the carbonyl carbon (Section 8.5) or engage in a substitution reaction at the carboxyl carbon (Section 6.5).

Obviously, only **nonnucleophilic bases** can be employed for the formation of enolates from carbonyl and carboxyl compounds. A base is nonnucleophilic if it is very bulky. The only nonnucleophilic organolithium compounds that *deprotonate* carbonyl and carboxyl compounds are mesityllithium (2,4,6-trimethylphenyllithium) and trityllithium (triphenylmethyllithium). However, these bases do not have any significance for the generation of enolates because of the difficulties associated with their preparation and with the separation of their conjugate acid hydrocarbons.

Alkaline earth metal amides have a unique place in enolate chemistry in light of the preceding discussion. Yet, amides without steric demand—from $NaNH_2$ to $LiNEt_2$—also are not suitable for enolate formations, since their nucleophilicities exceed their basicities. On the other hand, the amides LTMP, LDA, and LiHMDS (structures in Figure 4.17) are so bulky that they can never act as nucleophiles and always deprotonate C,H acids to the respective enolates.

A

Table 10.3 also shows that the deprotonation of isopropanol with LiHMDS is less than half as exothermic as the deprotonations with LDA or LTMP. Hence, LiHMDS is a much weaker base than the other two amides. This is due to the ability of the $SiMe_3$ groups of LiHMDS to stabilize the negative (partial) charge in the α position at the N atom. The mechanism of this stabilization might be the same as in the case of the isoelectronic triphenylphosphonium center in P-ylides (Figure 9.2), that is, a combination of −I-effect and anomeric effect. On the other hand, an SiR_3 group can possibly also stabilize a negative charge in the α position with its −M-effects, that is, via a d_π/p_π interaction. Because of its relatively low basicity, LiHMDS is employed for the preparation of enolates primarily when it is important to achieve high chemoselectivity.

The basicity of LDA is so high that it is even possible to generate bisenolates from β-diketones and β-ketoesters (Figure 10.7). Even carboxylates can be deprotonated at the α carbon if the strongest organic bases are employed (Figure 10.8). In contrast, the twofold deprotonation of phenylacetic acid by ethylmagnesium bromide is not com-

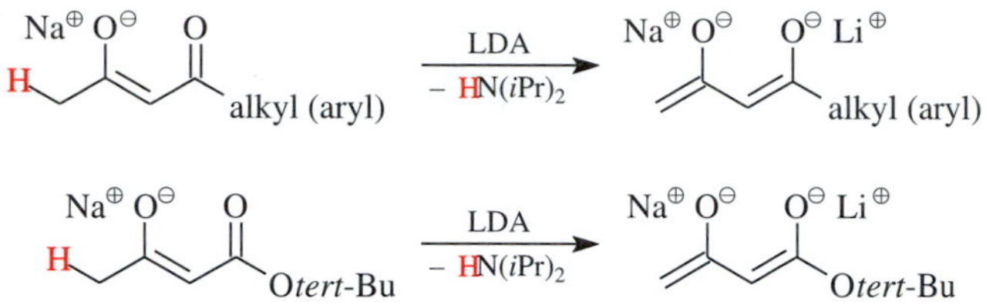

Fig. 10.7. Formation of bisenolates.

monly used. This reaction is mentioned in Figure 10.8 only because the resulting enolate **A** acts as the nucleophile in the Ivanov reaction in Figure 10.40.

Regiocontrol in the Formation of Lithium Enolates

B

Only one enolate can be generated from aldehydes or their aza analogs, from symmetric ketones or their aza analogs, or from carboxylic esters or carboxylic amides. For the moment we are ignoring the possibility that two stereoisomers, *E*- and *Z*-enolates, may occur for each of these enolates. On the other hand, constitutionally isomeric (regioisomeric) enolates may be derived from asymmetric ketones and from their aza analogs if they contain acidic H atoms at the C_α *and* $C_{\alpha'}$ centers. From certain asymmetric ketones or their aza analogs one or sometimes even both of these enolates can be generated regioselectively (see also Figure 10.32).

For example, 2-phenylcyclohexanone can be deprotonated regioselectively with LDA (Figure 10.9). This reaction is most successful at −78°C in THF because the reaction is irreversible under these conditions as long as a small excess of LDA is employed. Hence, the reaction is kinetically controlled and proceeds via the most stable transition state. The standard transition state of *all* enolate formations from C,H acids with LDA is thought to be cyclic, six-membered, and preferentially in the chair conformation (**A** and **B** in Figure 10.9). To be as stable as possible, this transition state should not feature any steric hindrance that can be avoided. In particular, the transition state should not contain any substituent in the six-membered ring that is parallel with the quasi–axially oriented amide *N*-isopropyl group if that can be avoided. Such a substituent would suffer from 1,3-diaxial repulsion because of its interaction with the isopropyl group. It follows that transition state **A** of Figure 10.9 is less stable than transition state **B.** The enolate formation thus proceeds selectively *via* transition state **B** and results in the **kinetic enolate D.**

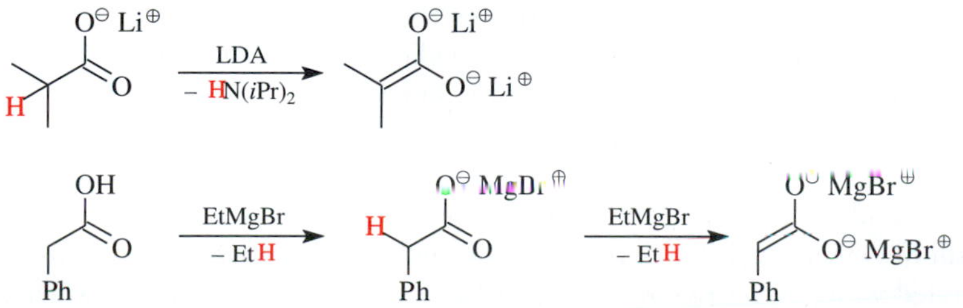

Fig. 10.8. Formation of carboxylate enolates ("carboxylic acid dianions") by reaction of carboxylic acid salts with strong bases.

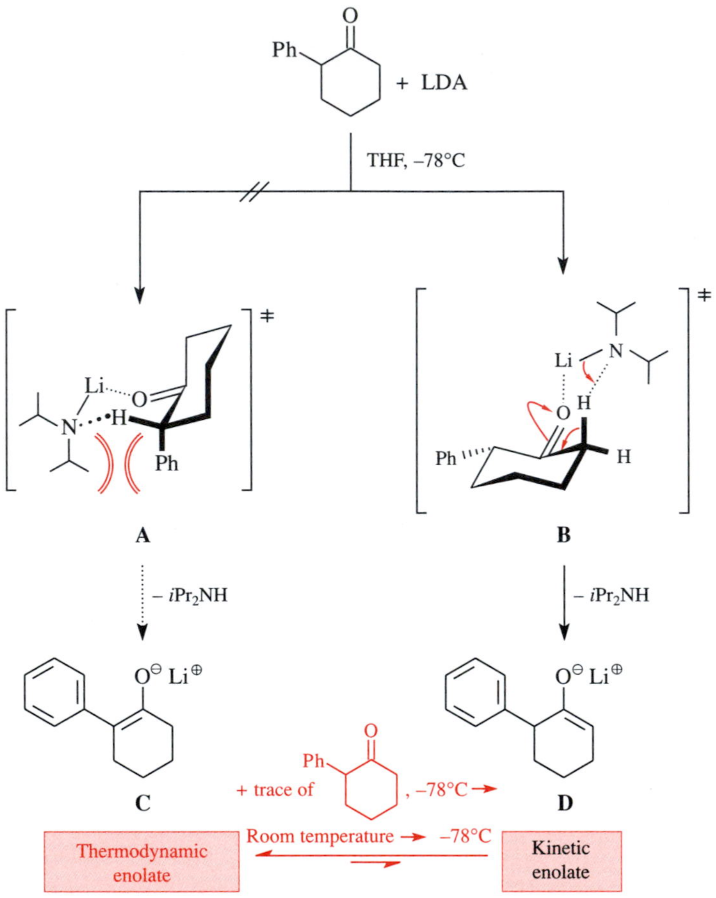

Fig. 10.9. Regioselective generation of ketone enolates I: the effects of different substituents in the α and α' positions. Enolate **D** is formed in THF at −78°C with LDA irrespective of whether a substoichiometric amount or an excess of LDA is used. However, if one employs slightly less than the stoichiometric amount of LDA (so that a trace of the neutral ketone is present), then, upon warming, the initially formed enolate **D** isomerizes quantitatively to enolate **C** with its more highly substituted C═C double bond. It should be noted that LDA removes an axially oriented α-H from the cyclohexanone; this is because only then the lone pair resulting in its place receives optimum stabilization by the adjacent C═O bond.

The C═C double bond of **D** is not conjugated to the phenyl ring. It is therefore less stable than the regioisomeric enolate **C,** which benefits from such a conjugation. In this context **C** is called the **thermodynamic enolate.** Because **C** is more stable than **D, C** can be generated from the kinetic enolate **D** as soon as the opportunity for isomerization is provided. The opportunity for isomerization arises if a weak acid is present that allows for the protonation of the enolate to the ketone. Even diisopropylamine (product of LDA) can serve as such an acid, provided there is not a stronger one present. However, it is more practical to allow for the presence of a trace of unreacted substrate ketone. The latter is ensured if one treats the substrate ketone with a slightly less than stoichiometric amount of LDA. Under these conditions, at temperatures above −78°C, the remaining substrate ketone reacts with the kinetic enolate **D** to yield the thermodynamic enolate **C** and newly formed substrate ketone. This occurs fast enough to effect a quantitative isomerization of **D** into **C.**

We thus reach the following interesting result. Depending on the reaction conditions, both the kinetic and the thermodynamic enolates of 2-phenylcyclohexanone can be generated with perfect regiocontrol. The same is true for many ketones that carry a different number of *alkyl* groups at the C_α and $C_{\alpha'}$ centers.

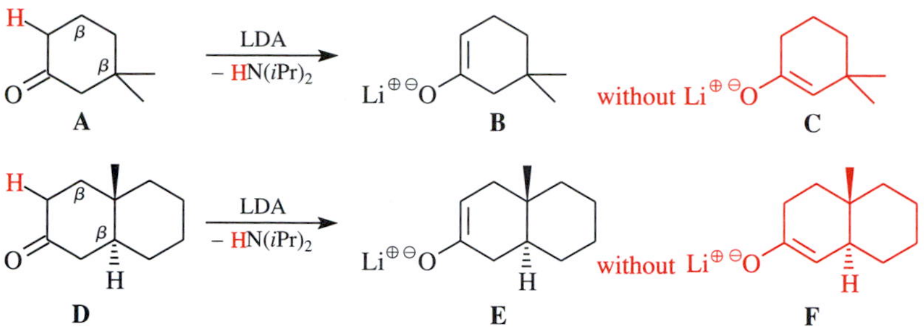

Fig. 10.10. Regioselective generation of ketone enolates II: the effects of different substituents in the β and β' positions. (For the regioselective preparation of enolates **C** and **F**, see Figures 10.16 and 10.17, respectively.)

An asymmetric ketone may even lead to the generation of one enolate in a selective fashion if the asymmetry is caused by the substitution patterns in the β rather than α positions. The difference in the β positions may be due to the number or the kind of substituents there. This point is emphasized in Figure 10.10 with cyclohexanones that contain one **(D)** or two **(A)** β substituents. In this case, deprotonation occurs preferentially on the side opposite to the location of the extra substituent; that is, the sterically less hindered acidic H atom is attacked.

A

In the discussion of Figure 10.9, we mentioned that LDA abstracts acidic H atoms from the α position of C,H-acidic compounds via a cyclic and six-membered transition state and that *the carbonyl group is an integral part of this six-membered ring.* This emphasis explains why the deprotonation of conjugated ketones with LDA yields the kinetic enolate (**A** in Figure 10.11) in a regioselective fashion instead of the thermodynamic enolate (**B** in Figure 10.11).

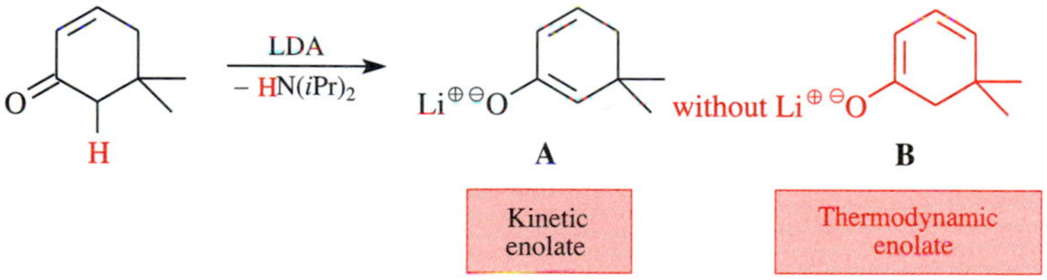

B

Fig. 10.11. Regioselective generation of ketone enolates III: the effects of conjugated versus nonconjugated substituents. (For the regioselective preparation of enolate **B,** see Figure 10.18.)

Stereocontrol in the Formation of Lithium Enolates

We have seen that LDA forms enolates of carbonyl compounds, carboxylic esters, and carboxylic amides via cyclic and six-membered transition states with the chair conformation. This geometry of the transition state for enolate formation has consequences if a stereogenic C=C double bond is generated.

A

The reaction of unhindered aliphatic ketones with LDA yields more *E*- than *Z*-enolates. On the other hand, the *Z*-enolate is formed selectively if either a bulky aliphatic or a conjugating aromatic group is the inert group attached to the carbonyl carbon. Figure 10.12 shows how this *Z*-selectivity results from the differing steric demands in the potentially available transition states for deprotonation. The point is made using the example of an ethyl ketone with a sterically demanding substituent. Transition state **A** is so strongly destabilized by the 1,2-interaction indicated that deproto-

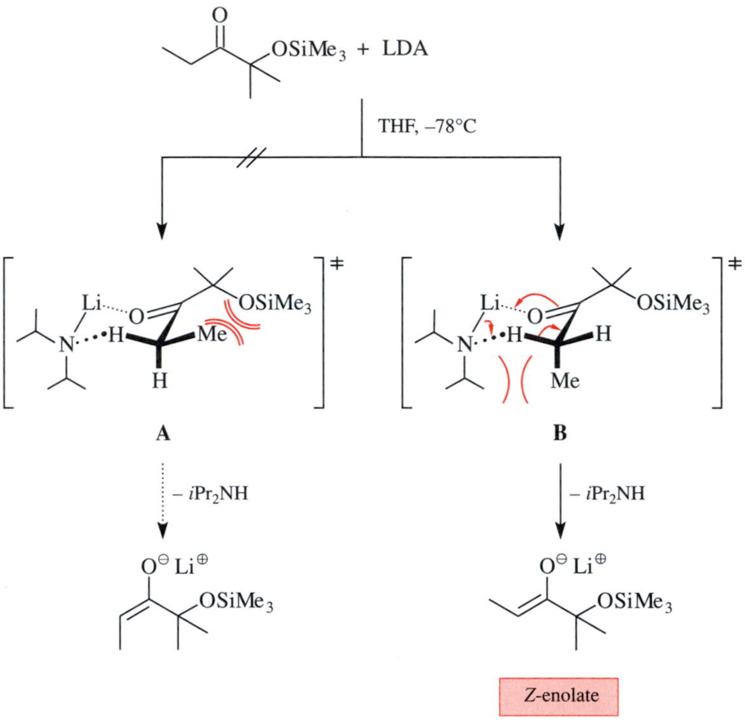

Fig. 10.12. Highly *Z*-selective generation of a ketone enolate. The transition state **A** is destabilized so strongly by 1,2-interactions that deprotonation occurs exclusively via the transition state **B.**

nation occurs exclusively via transition state **B** in spite of the repulsive 1,3-interaction in the latter.

Unhindered aliphatic ketones selectively yield *E*-enolates if they are deprotonated by a lithium amide via a transition state structure of type **A** of Figure 10.12. This occurs, for example, when the **B**-type transition state is destabilized because of the use of a base that is even more sterically demanding than LDA such as, for example, LTMP (for structure, see Figure 4.17). For example, diethyl ketone and LTMP form the *E*-enolate with $ds = 87{:}13$.

Deprotonation of α-alkylated acetic acid esters (e.g., the propionic acid ester of Figure 10.13) with LDA at -78°C selectively yields the "*E*"-enolates. The quotation marks indicate that this application of the term is based on an extension of the *E/Z*-nomenclature: here, the Cahn–Ingold–Prelog priority of the O^-Li^+ substituent is considered to be higher than the priority of the OR group. The deprotonation of the ester shown in Figure 10.13 occurs via the strain-free transition state **A.** The alternative transition state **B** is destabilized by a 1,3-diaxial interaction.

"*Z*"-Enolates also are accessible from the same propionic acid esters with the same high stereoselectivity (Figure 10.14). This is a complete reversal of stereoselectivity in comparison to the deprotonation shown in Figure 10.13. This reversal is solely due to the change in solvent from THF to a mixture of THF and DMPU. Probably, stereocontrol is again the result of kinetic control. However, no model of the respective transition state has been generally accepted. It is likely that the transition state struc-

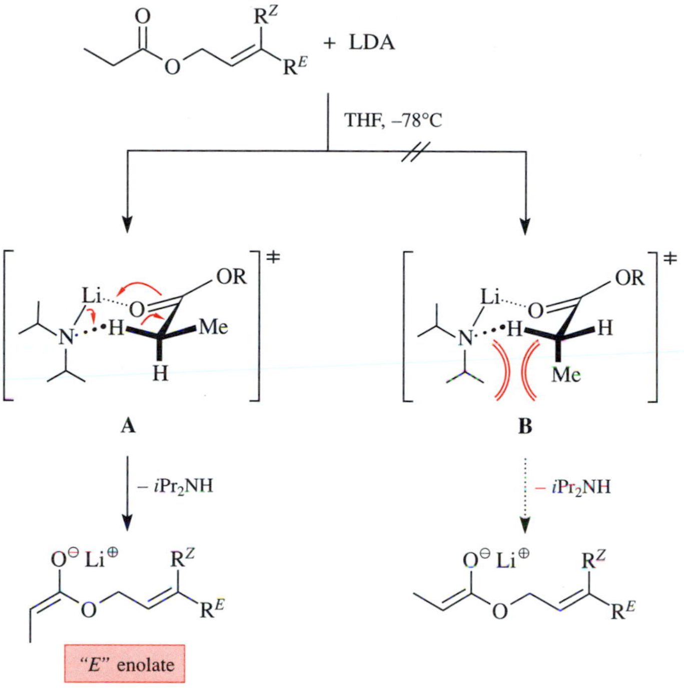

Fig. 10.13. Highly "*E*"-selective generation of ester enolates. The deprotonation of the ester occurs preferentially via the strain-free transition state **A.**

ture is acyclic. Moreover, it might well be that the lithium of LDA coordinates with the O-atom of DMPU in this transition state rather than with the ester oxygen of the substrate (as would be the case in the absence of DMPU).

The generation of amide enolates by the reaction of carboxylic amides and LDA at −78°C occurs with complete stereoselectivity and yields the "*Z*"-enolate (Figure 10.15). This is just the opposite selectivity than in the case of the generation of ester enolates under the same conditions (Figure 10.13). The NR_2 group of an amide is branched and therefore sterically more demanding than the OR group of an ester. Hence, in the transition state for deprotonation, the NR_2 group of an amide requires more space than the OR group of an ester. Consequently, the propionic acid amide of Figure 10.15 as well as all other carboxylic amides cannot react via transition states **A:** the 1,2-interaction would be *prohibitively* high. Therefore, carboxylic amides are deprotonated via **B**-type transition state structures in spite of the repulsive 1,3-inter-

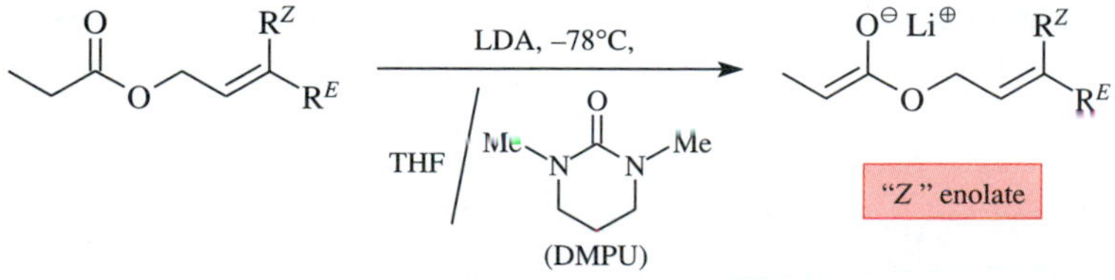

Fig. 10.14. Highly "*Z*"-selective generation of ester enolates in a THF/DMPU solvent mixture. DMPU, *N,N′*-dimethylpropyleneurea.

Fig. 10.15. Highly "*Z*"-selective generation of amide enolate.

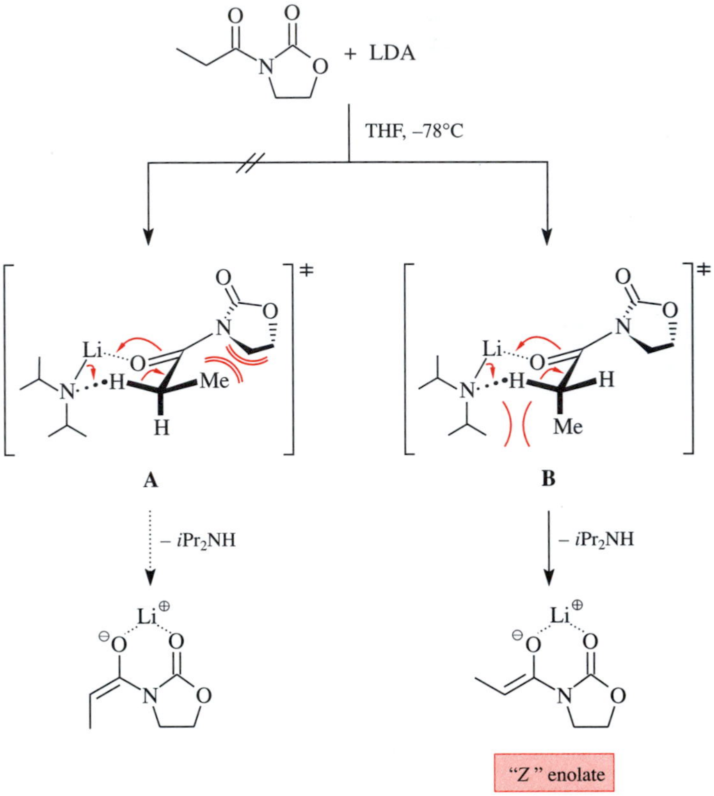

action. Remember: it is owing to this repulsive 1,3-interaction that these kinds of transition state structures (**B** in Figure 10.13) are *not involved* in the deprotonations of esters.

10.1.3 Other Methods for the Generation of Enolates

B

Figures 10.10 and 10.11 demonstrate that deprotonation might afford certain enolates with only *one* regioselectivity. However, there might be other reaction paths, which lead to the other regioisomer (Figures 10.16–10.18).

One of these alternative synthetic paths consists of the addition of Gilman cuprates (for preparation, see Figure 8.34) to α,β-unsaturated ketones. In the discussion of this addition mechanism (Figure 8.35) it was pointed out that the enolates formed are associated with CuR. It is assumed that mixed aggregates are formed. As it turns out, the enolate fragments contained in such aggregates are significantly less reactive than CuR-free enolates. It is possible, however, to convert these CuR-containing enolates

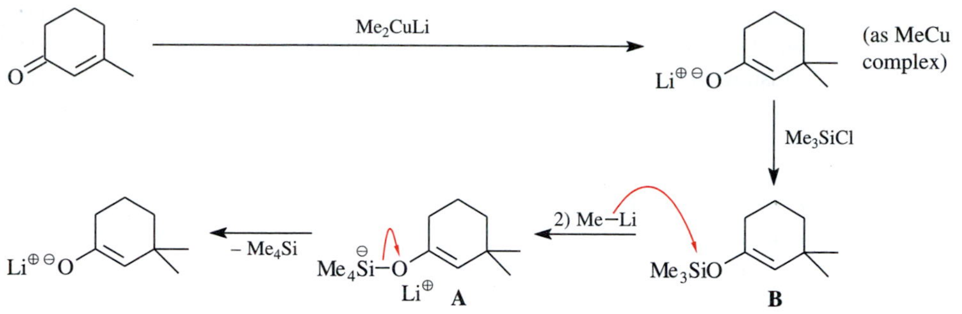

Fig. 10.16. Generation of Cu-containing enolates from Gilman cuprates and their use as substrates for the preparation of Cu-free enolates via silyl enol ethers.

into the CuR-free lithium enolates. This conversion typically starts with the reaction between the CuR-containing enolate and Me_3SiCl to form a silyl enol ether such as **B** (Figure 10.16). The silyl enol ether reacts with MeLi via the silicate complex **A** (pentacoordinated Si center) to provide the Cu-free enolate.

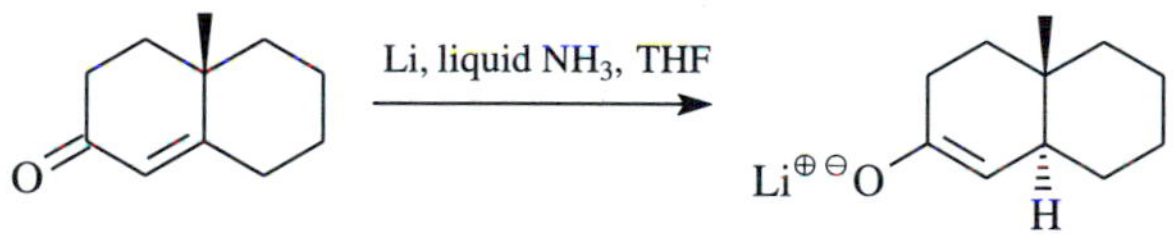

Fig. 10.17. Generation of an enolate via Birch reduction.

Accordingly, trimethylsilyl enol ethers are enolate precursors (Figure 10.16). Fortunately, they can be prepared in many ways. For instance, silyl enol ethers are produced in the silylation of ammonium enolates. Such ammonium enolates can be generated at higher temperature by partial deprotonation of ketones with triethylamine (Figure 10.18). The incompleteness of this reaction makes this deprotonation reversible. Therefore, the regioselectivity of such deprotonations is subject to thermodynamic control and assures the preferential formation of the more stable enolate. Consequently, upon

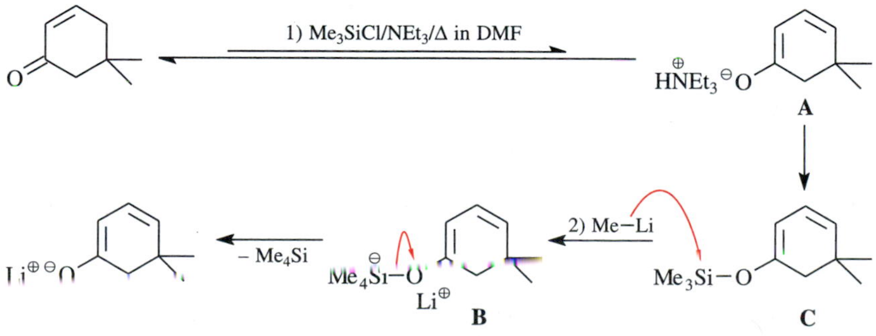

Fig. 10.18. Generation of enolates from silyl enol ethers.

heating with Me_3SiCl and NEt_3, α,β-unsaturated ketones are deprotonated to give 1,3-dienolates **A** with the O^- substituent in the 1-position. This enolate is more stable than the isomeric 1,3-enolate in which the O^- substituent is in the 2-position; three resonance forms can be written for the former enolate while only two can be written for the latter. Me_3SiCl then attacks the dienolate **A** at the oxygen atom. The dienol silyl ether **C** is obtained in this way. Silyl ether **C** reacts with MeLi—in analogy to the reaction **B** → **A** shown in Figure 10.16—via the silicate complex **B** to give the desired enolate. Note that the product enolate is not accessible by treatment of cyclohexanone with LDA (Figure 10.11).

10.1.4 Survey of Reactions between Electrophiles and Enolates and the Issue of Ambidoselectivity

B

Enolates and aza-enolates are so-called **ambident** or **ambifunctional nucleophiles.** This term describes nucleophiles with two nucleophilic centers that are in conjugation with each other. In principle, enolates and aza-enolates can react with electrophiles either at the heteroatom or at the carbanionic C atom. Ambidoselectivity occurs if one of these alternative modes of reaction dominates.

Most enolates exhibit an ambidoselectivity toward electrophiles that depends on the electrophile and not on the substrate. The extent of the ambidoselectivity almost always is complete:

1) Only very few electrophiles attack the enolate *oxygen* of the enolates of aldehydes, ketones, esters, and amides, and these few electrophiles are
   - Silyl chlorides (examples in Figures 10.16, 10.18, and 10.19).
   - Derivatives of sulfonic acids such as the *N*-phenylbisimide of trifluoromethanesulfonic acid. Examples are given in Figures 10.20 and 13.1. Note that alkenyl triflates are obtained in this way and these are the substrates of a variety of Pd-mediated C,C-coupling reactions (Chapter 13).
   - Derivatives of chlorophosphonic acid such as chlorophosphonic acid diamide (for example, see Figure 10.21) or chlorophosphonic acid esters (Figure 10.22).
2) Essentially all other electrophiles attack the enolate *carbon* of the enolates of aldehydes, ketones, esters, and amides (important examples are listed in Table 10.4).

Fig. 10.19. O-Silylation of an ester enolate to give a silyl ketene acetal. (The formation of the silicate complex in step 1 of the reaction is plausible but has not yet been proven.)

O

LDA

$Li^{\oplus}$ $\overline{|O|}^{\ominus}$ ① ② SO_2-CF_3 NPh SO_2-CF_3

$-Li^{\oplus\ominus}N-Ph$ SO_2-CF_3

OSO_2-CF_3

Fig. 10.20. O-Sulfonylation of a ketone enolate to give an enol triflate. (See Figure 10.9 regarding the regioselectivity of the enolate formation.)

O*tert*-Bu

H H H H O

LDA

O*tert*-Bu

H H H H

$Li^{\oplus\ominus}O$

① ②

$(Me_2N)_2P(=O)-Cl$

– LiCl

O*tert*-Bu

H H H H

$(Me_2N)_2P(=O)-O$

Fig. 10.21. O-Phosphorylation of a ketone enolate to afford an enol phosphonamide. (See Figure 10.10, bottom row, regarding the regioselectivity of the enolate formation.)

NaH in THF

$Na^{\oplus}$ $^{\ominus}O$ O

R_{prim} OMe

$(EtO)_2P(=O)-Cl$

– NaCl

$(EtO)_2P(=O)-O$ O

R_{prim} OMe

O O

R_{prim} OMe

NEt_3

$O^{\ominus}$ $HNEt_3^{\oplus}$

R_{prim}

O OMe

$(EtO)_2P(=O)-Cl$

$-HNEt_3^{\oplus}$ $Cl^{\ominus}$

$(EtO)_2P(=O)-O$

R_{prim}

O OMe

Fig. 10.22. O-Phosphorylation of a ketone enolate to afford an enol phosphate. (See Figure 10.4 regarding the stereochemistry of the enolate formation.)

Table 10.4. Electrophiles That Attack Ambidoselectively at the C Atom of Enolates Derived from Aldehydes, Ketones, Esters, and Amides

Electrophile	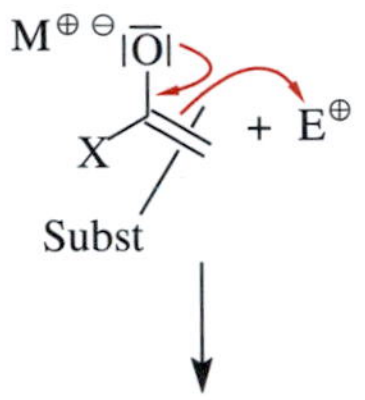 $M^{\oplus}$ $^{\ominus}$O–C(X)=C(Subst) + $E^{\oplus}$ ↓	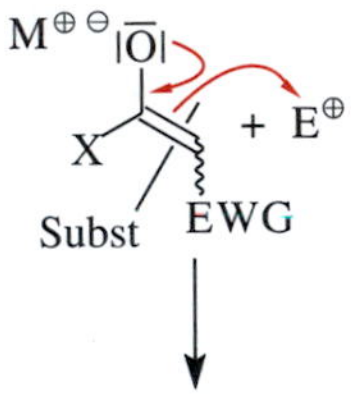 $M^{\oplus}$ $^{\ominus}$O–C(X)=C(Subst)(EWG) + $E^{\oplus}$ ↓	For details or applications in synthesis, see ...
PhSe–SePh	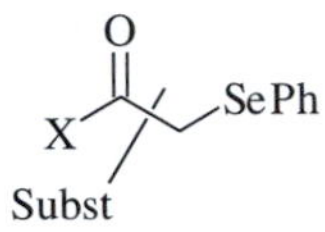 X–C(=O)–CH(Subst)–SePh	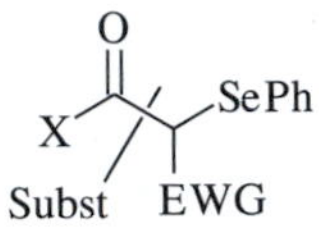 X–C(=O)–C(Subst)(EWG)–SePh	Fig. 4.12
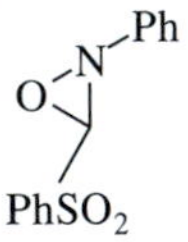 Oxaziridine (N–Ph, $PhSO_2$)	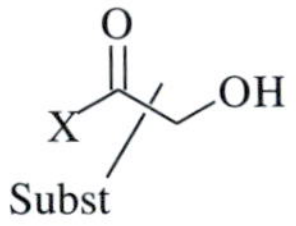 X–C(=O)–CH(Subst)–OH	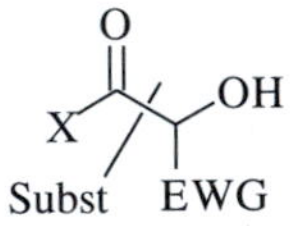 X–C(=O)–C(Subst)(EWG)–OH	–
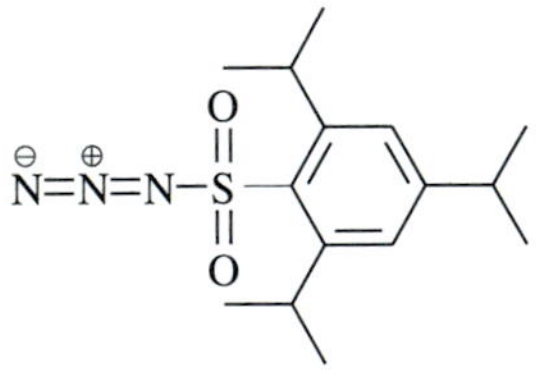 $N^{\ominus}=N^{\oplus}=N$–SO_2–(2,4,6-triisopropylphenyl)	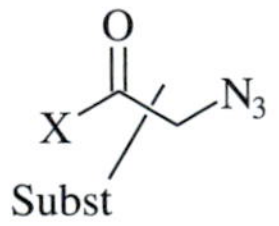 X–C(=O)–CH(Subst)–N_3		–
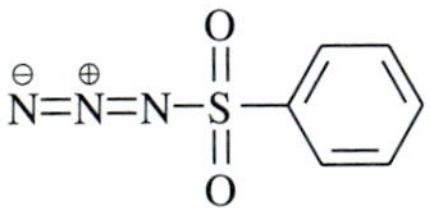 $N^{\ominus}=N^{\oplus}=N$–SO_2–Ph	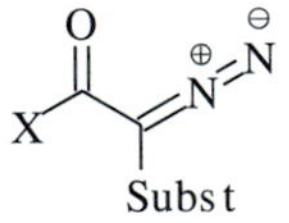 X–C(=O)–C(Subst)=$N^{\oplus}=N^{\ominus}$	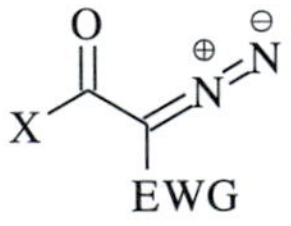 X–C(=O)–C(EWG)=$N^{\oplus}=N^{\ominus}$	Fig. 11.26, Fig. 11.29
RX	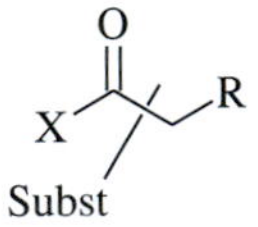 X–C(=O)–CH(Subst)–R	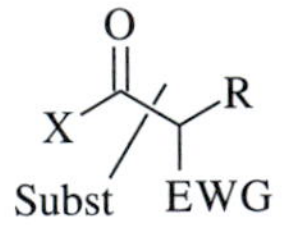 X–C(=O)–C(Subst)(EWG)–R	Section 10.2
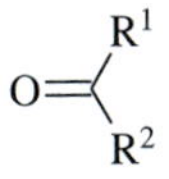 O=C(R^1)(R^2)	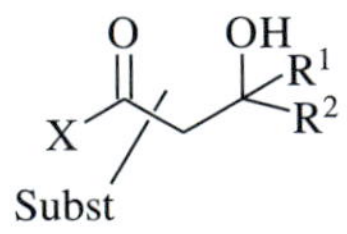 X–C(=O)–CH(Subst)–C(OH)(R^1)(R^2)	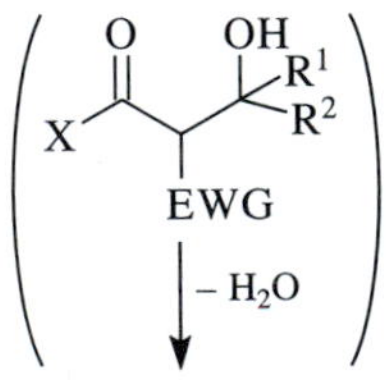 (X–C(=O)–CH(EWG)–C(OH)(R^1)(R^2) ↓ – H_2O)	Section 10.3
	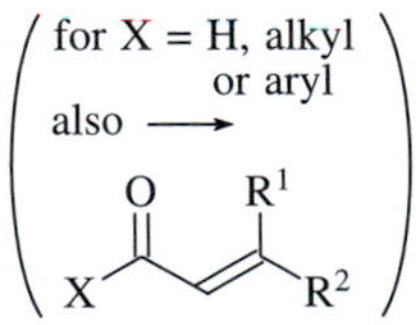 (for X = H, alkyl or aryl also → X–C(=O)–CH=C(R^1)(R^2))	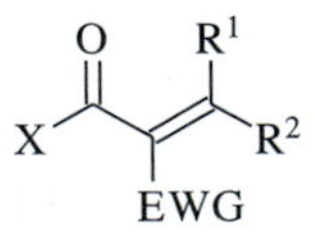 X–C(=O)–C(EWG)=C(R^1)(R^2)	Section 10.4

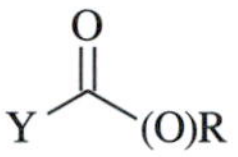 O, Y, (O)R	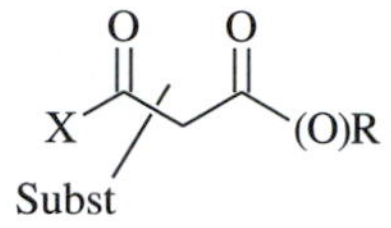 O, O, X, (O)R, Subst		
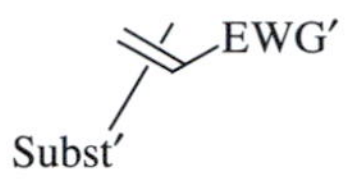 EWG′, Subst′	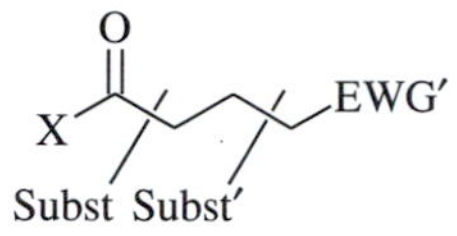 O, X, EWG′, Subst, Subst′	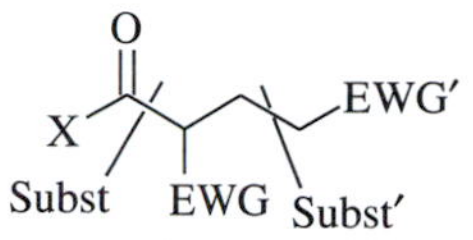 O, X, EWG′, Subst, EWG, Subst′	Section 10.6

10.2 Alkylation of Quantitatively Prepared Enolates and Aza-Enolates; Chain-Elongating Syntheses of Carbonyl Compounds and Carboxylic Acid Derivatives

B

All the reactions discussed in this section are S_N2 reactions with respect to the alkylating reagent. The most suitable alkylating reagents for enolates and aza-enolates are therefore the most reactive alkylating reagents (Section 2.4.4), that is, MeI, R_{prim}—X, and especially $H_2C{=}CH{-}CH_2{-}X$ and $Ar{-}CH_2{-}X$ (X = Hal, OTs, OMs). Isopropyl bromide and iodide also can alkylate enolates in some instances. Analogous compounds R_{sec}—X and R_{tert}—X either do not react with enolates at all or react via E2 eliminations to afford alkenes.

10.2.1 Chain-Elongating Syntheses of Carbonyl Compounds

Acetoacetic Ester Synthesis of Methyl Ketones

B

Acetoacetic ester is an active-methylene compound and it can be deprotonated (cf. Table 10.1) with one equivalent of NaOEt in EtOH to the sodium enolate **A** (Figure 10.23). As is depicted in Figure 10.23, **A** is *monoalkylated* by butyl bromide. This is possible even though the butylated sodium enolate **B** is already present while the reaction is still under way. The sodium enolate **B** is formed in an equilibrium reaction between not-yet-butylated enolate **A** and the already formed butylation product **C. B** represents a nucleophilic alternative to unreacted enolate **A.** However, the butylated enolate **B** is sterically more demanding than the nonbutylated (or not-yet-butylated) enolate **A.** The first butylation of **A** thus is faster than the second butylation reaction, that is, the butylation of **B.** This reactivity difference is not large enough to cause 100% monobutylation and 0% dibutylation. Still, the main product is the product of monobutylation **C.** Distillation is required to separate the monobutylation product from the dibutylation product and from unreacted substrate.

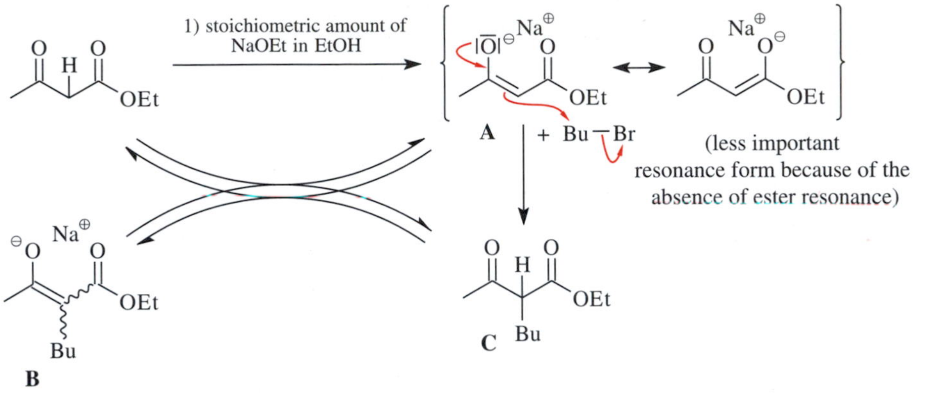

1) stoichiometric amount of NaOEt in EtOH

(less important resonance form because of the absence of ester resonance)

Fig. 10.23. Acetoacetic ester synthesis of methyl ketones I: preparation of an alkylated acetoacetic ester.

The butylated β-ketoester **C** of Figure 10.23 is not the final synthetic target of the *acetoacetic ester synthesis of methyl ketones.* In that context the β-ketoester **C** is converted into the corresponding β-ketocarboxylic acid via acid-catalyzed hydrolysis (Figure 10.24; for the mechanism, see Figure 6.19). This β-ketocarboxylic acid is then heated either in the same pot or after isolation to effect decarboxylation. The β-ketocarboxylic acid decarboxylates via a cyclic six-membered transition state in which three valence electron pairs are shifted at the same time. The reaction product is an enol, which isomerizes immediately to a ketone in general and to phenyl methyl ketone in the specific example shown. In general, alkyl methyl ketones are obtained by such acetoacetic ester syntheses.

Other β-ketoesters can be converted into other ketones under the reaction conditions of the acetoacetic ester synthesis and with the same kinds of reactions as are shown in Figures 10.23 and 10.24. Two examples are provided in Figures 10.25 and 10.26. These β-ketoesters also are first converted into sodium enolates by reaction with NaOEt in EtOH, and the enolates are then reacted with alkylating reagents. The reaction shown in Figure 10.25 employs a β-ketoester that can be synthesized by a Claisen condensation (see later, Figure 10.51). The alkylating reagent employed in Figure 10.25 is bifunctional and reacts at both its reactive centers, thereby cross-linking two originally separate β-ketoester molecules. The new bis(β-ketoester) is hydrolyzed to afford

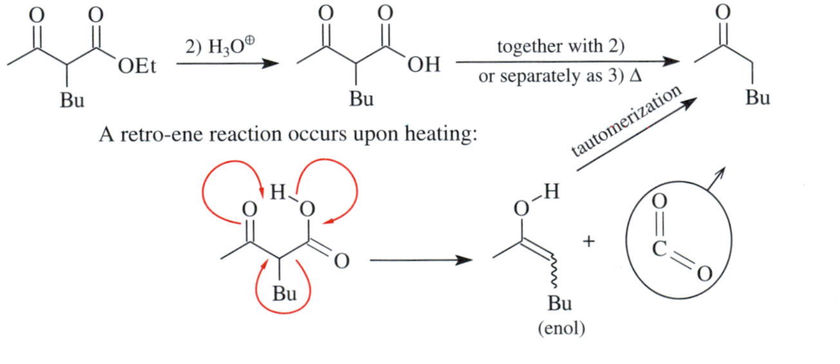

2) $H_3O^{\oplus}$

together with 2) or separately as 3) Δ

A retro-ene reaction occurs upon heating:

tautomerization

(enol)

Fig. 10.24. Acetoacetic ester synthesis of methyl ketones II: hydrolysis and decarboxylation of the alkylated acetoacetic ester.

a bis(β-ketocarboxylic acid), a twofold decarboxylation of which occurs subsequently according to the mechanism depicted in Figure 10.24. The reaction sequence of Figure 10.25 represents the synthesis of a diketone and illustrates the value of the acetoacetic ester synthesis to access a variety of *prim*-alkyl and *sec*-alkyl ketones.

A

β-Ketoesters derived from cyclic five- or six-membered ketones are conveniently accessible via the Dieckmann condensation (for example, see Figure 10.52). Such β-ketoesters can be converted into cyclic ketones under the reaction conditions of the acetoacetic ester synthesis. Step 1 in Figure 10.26 shows how such a β-ketoester is allylated at its activated position. The allylation product **A** *could* be converted into the alkylated ketone **B** as steps 2 and 3 of this sequence with the chemistry depicted in Figures 10.24 and 10.25, that is, a sequence comprising hydrolysis and decarboxylation. There exists, however, an alternative to achieve the special transformation of β-keto *methyl* esters into CO_2Me-free ketones. It is shown in Figure 10.26. This alternative is based on the knowledge that good nucleophiles like the iodide ion (from Li^+I^-; see Figure 10.26) or the phenylthiolate ion (from Na^+PhS^-) undergo an S_N2 reaction at the methyl group of the β-ketoester. The reaction is carried out at temperatures above 100°C. The β-ketocarboxylate leaving group decarboxylates immediately under these conditions, producing the enolate of the desired ketone. This eno-

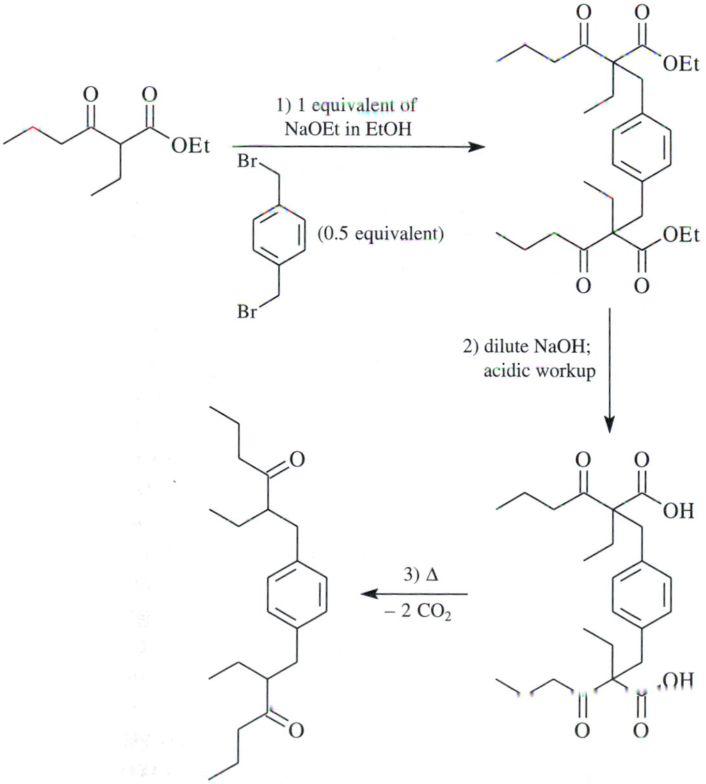

Fig. 10.25. Synthesis of complicated ketones in analogy to the acetoacetic ester synthesis I: generation of a diketone.

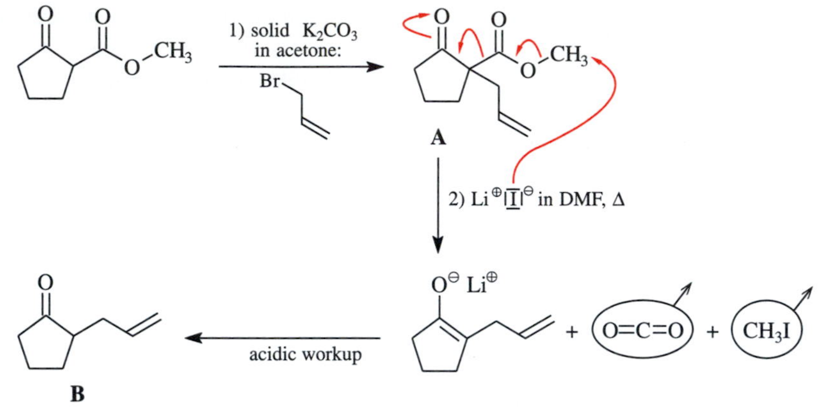

Fig. 10.26. Synthesis of complicated ketones in analogy to the acetoacetic ester synthesis II: generation of a cyclic ketone. In the first step, the β-ketoester is alkylated at its activated position. In the second step, the β-ketoester is treated with Li^+I^-. S_N2 attack of the iodide at the methyl group generates the β-ketocarboxylate ion as the leaving group. The β-ketocarboxylate decarboxylates immediately under the reaction conditions (temperature above 100°C) and yields the enolate of a ketone.

late is protonated to the ketone either by acidic contaminants of the solvent or later, during aqueous workup.

Alkylation of Ketone Enolates

B

Ester-substituted ketone enolates are stabilized, and these enolates can be alkylated (acetoacetic ester synthesis). Alkylation is, however, also possible for enolates that are not stabilized. In the case of the stabilized enolates, the alkylated ketones are formed in two or three steps, while the nonstabilized enolates afford the alkylated ketones in one step. However, the preparation of nonstabilized ketone enolates requires more aggressive reagents than the ones employed in the acetoacetic ester synthesis.

Figure 10.27 shows that even sterically hindered ketone enolates can be alkylated and that this reaction works even when the reactivity of these enolates is reduced because of being associated with CuR groups. The carbon atom in the β position relative to the carbonyl carbon of an α,β-dialkylated α,β-unsaturated ketone can be converted

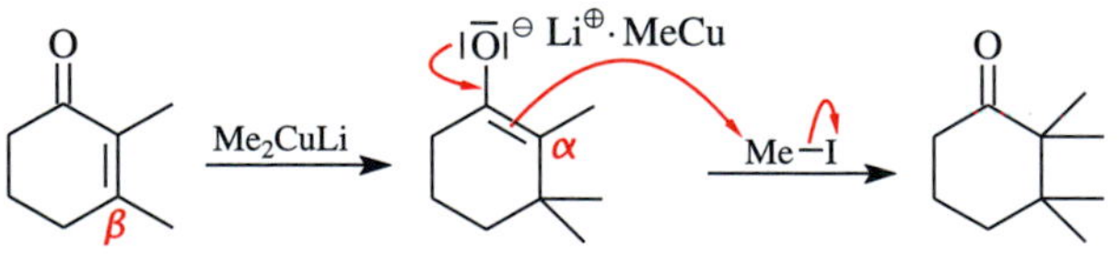

Fig. 10.27. 1,4-Addition plus enolate alkylation—a one-pot process for the α- and β-functionalization of α,β-unsaturated ketones. Two quaternary C atoms can be constructed via addition of a Gilman reagent and subsequent alkylation with MeI.

Fig. 10.28. Alkylation of a chiral ketone enolate. The attack of methyl iodide occurs preferentially from the side opposite to the side of the axially oriented methyl group at the bridgehead carbon.

into a quaternary C atom via 1,4-addition of a Gilman cuprate (for conveivable mechanisms, see Figure 8.35). As can be seen, a subsequent alkylation allows for the construction of another quaternary C atom in the α position even though it is immediately adjacent to the quaternary center generated initially in the β position.

Diastereoselective alkylations of enolates may occur if the enolate is chiral, i.e., surrounded by diastereotopic half-spaces. This was discussed in Section 3.4.1. In general, it is difficult to predict the preferred side of attack of the alkylating reagent on such enolates. For cyclic enolates the situation is relatively simple, because these enolates always are attacked from the less-hindered side. Hence, for the methylation of the enolate in Figure 10.28, the attack of methyl iodide occurs preferentially equatorially, that is, from the side that is opposite to the axially oriented methyl group at the bridgehead.

A

Bisenolates (for preparation, see Figure 10.7) such as compound **A** derived from acetoacetic ester in Figure 10.29 react with one equivalent of alkylating reagent in a regioselective fashion to give the ketone enolate **C,** which is conjugated to the acceptor. This could be the result of product-development control, since the isomeric alkylation product would be less stable. The acceptor-substituted enolate **C** resembles the nucleophile of the acetoacetic ester synthesis (Figure 10.23) and can be alkylated likewise. One can employ different alkylating reagents in the first and second alkylations to obtain a β-ketoester **B** with two new substituents. This product may feature a substitution pattern that could not be constructed via a Claisen condensation (Figures 10.51 and 10.53), as is true for the example presented in Figure 10.29.

Alkylation of Lithiated Aldimines and Lithiated Hydrazones

B

The quantitative conversion of aldehydes into enolates with lithium amides hardly ever succeeds because an aldol reaction (cf. Section 10.1.2) would occur while the deprotonation with LDA was in progress. Aldol additions also occur upon conversion of a small fraction of the aldehyde into the enolate with a weak base (Section 10.3.1). Hence, it is generally impossible to alkylate an aldehyde without simultaneous occurrence of an aldol addition. There is only one exception: certain α-branched aldehydes can be deprotonated to their enolates in equilibrium reactions, and these enolates can be reacted with alkylating reagents to obtain tertiary aldehydes.

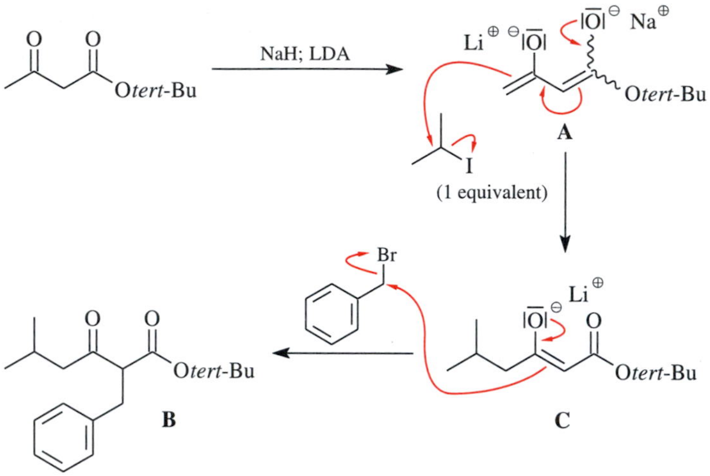

Fig. 10.29. Regiocontrolled bisalkylation of a bisenolate. The conjugated ketone enolate **C** is formed with one equivalent of the alkylating reagent. **C** can be alkylated again, and **B** is formed in that way.

While, accordingly, only very special aldehydes can be converted into an α-alkylated aldehyde directly, there are some "detours" available. Figure 10.30 shows such a detour for the conversion of an aldehyde (without α-branching) into an α-alkylated aldehyde. At first, the aldehyde is reacted with a primary amine—cyclohexyl amine is frequently used—to form the corresponding aldimine (for the mechanism, see Table 7.2). Aldimines can be deprotonated with LDA or *sec*-BuLi to give aza-enolates. The success of the deprotonation with *sec*-BuLi shows that aldimines are much weaker electrophiles than aldehydes: *sec*-BuLi would immediately add to an aldehyde.

The obviously low electrophilicity of the C=N double bonds of aldimines precludes the addition of already formed aza-enolate to still undeprotonated aldimine in the

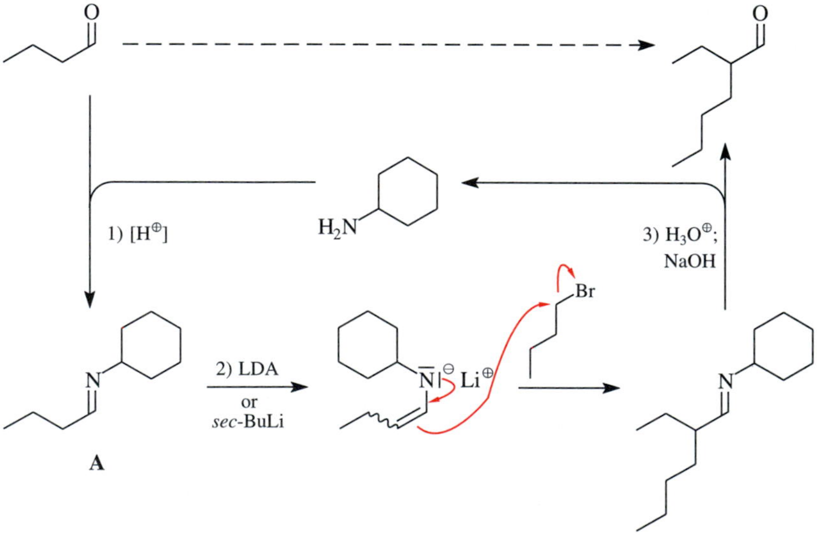

Fig. 10.30. α-Alkylation of an aldehyde via an imine derivative.

course of aldimine deprotonation. The aldimine enolate is obtained quantitatively and then reacted with the alkylating reagent. This step results cleanly in the desired product, again because of the low electrophilicity of imines: as the alkylation progresses, aza-enolate and already alkylated aldimine coexist without reacting with each other. *All* the aza-enolate is thus converted into the alkylated aldimine. In step 3 of the sequence of Figure 10.30, the imine is subjected to an acid-catalyzed hydrolysis, and the alkylated aldehyde results.

The formation of the alkylated aldimine in Figure 10.30 involves the generation of a stereocenter, yet without stereocontrol. The aldehyde derived from this aldimine consequently is obtained as a racemate. Figure 10.31 shows how a variation of this procedure allows for the enantioselective generation of the same aldehyde.

A

The "aldimine" of Figure 10.31 is a chiral and enantiomerically pure aldehydrazone **C.** This hydrazone is obtained by condensation of the aldehyde, which shall be alkylated, and an enantiomerically pure hydrazine **A** (see Table 7.2 for the mechanism), the *S*-proline derivative ***S*-aminoprolinol methyl ether (SAMP).** The hydrazone **C** derived from aldehyde **A** is called the SAMP hydrazone, and the entire reaction sequence of Figure 10.31 is the **Enders SAMP procedure.** The reaction of the aldehydrazone

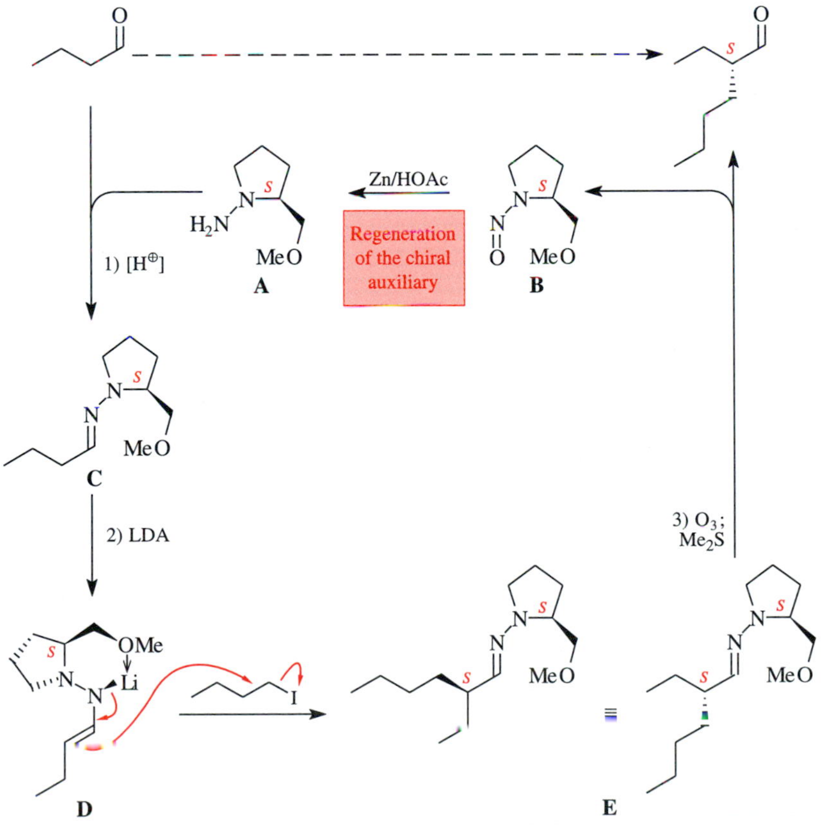

Fig. 10.31. Enders' SAMP method for the generation of enantiomerically pure α-alkylated carbonyl compounds; SAMP, *S*-aminoprolinol methyl ether = *S*-2-methoxymethyl-1-pyrrolidinamine.

C with LDA results in the chemoselective formation of an aza-enolate **D,** as in the case of the analogous aldimine **A** of Figure 10.30. The C=C double bond of the aza-enolate **D** is *trans*-configured. This selectivity is reminiscent of the *E*-preference in the deprotonation of sterically unhindered aliphatic ketones to ketone enolates (Section 10.1.2, paragraph "Stereocontrol in the Formation of Lithium Enolates") and, in fact, the origin is the same: both deprotonations occur via six-membered ring transition states with chair conformations. The transition state structure with the least steric interactions is preferred in both cases, which is the one that features the C atom in the β position of the C,H acid in the pseudoequatorial orientation.

The N—Li bond of aza-enolate **D** lies outside the plane of the enolate. This Li—N bond is covalent but highly polarized, with the lithium carrying a partial positive charge. This partial positive charge is large enough to allow the lithium atom to coordinate to the halogen atom of an added alkyl halide. It is by way of this coordination that the lithium directs the approach of an alkyl halide to the enolate carbon *from one face.* The alkylation product **E** is formed preferentially with the *S*-configuration shown. Only traces of the *R*-configured product are formed. The main and trace products are diastereoisomers, which can be completely separated by using chromatography. The separation affords a diastereomerically and enantiomerically pure SAMP hydrazone **E.**

In the third step of the reaction sequence depicted in Figure 10.31, the hydrazone **E** is converted into the desired sterically homogeneous aldehyde. This transformation can be achieved, for example, by ozonolysis of the C=N double bond. One of the products of ozonolysis is the desired enantiomerically pure α-butylated butanal. The other product of the ozonolysis also is valuable, since it is the nitroso derivative **B** of reagent **A.** The N=O group of **B** can be reduced to give an amino group whereby **A** would be regenerated from **B.** The possibility of recycling valuable chiral auxiliaries greatly enhances the attractiveness of any method for asymmetric synthesis.

The strategy of Figure 10.31 also is suitable for the synthesis of enantiomerically pure α-alkylated *ketones.* Figure 10.32 shows a procedure for the synthesis of the *S*-configured 6-methyl-2-cyclohexenone. The desired *S*-configuration is achieved with the help of a so-called RAMP hydrazone **C,** which is a derivative of the ***R*-amino**pro**l**inol methyl ether **A.** The reaction sequence of Figure 10.32 is the **Enders RAMP procedure.** In step 2 of the RAMP procedure, hydrazone **C** is deprotonated with LDA to give the aza-enolate **D.** This deprotonation occurs with the same regioselectivity as the formation of the kinetic enolate **A** in the reaction of cyclohexenone with LDA. The common regioselectivities have the same origin. Deprotonations with LDA prefer cyclic transition state structures that are six-membered rings and include the heteroatom of the acidifying C=X double bond (X = O, N).

As with the aza-enolate of Figure 10.31, the aza-enolate **D** in Figure 10.32 contains a polar, covalent N—Li bond that is twisted out of the plane of the enolate. And again as with Figure 10.31, the lithium of this N—Li bond directs the added alkyl halide *from the side of the lithium* to the enolate carbon. The kethydrazone **E** is formed with high diastereoselectivity and, after chromatographic separation, it is obtained in 100% stereochemically pure form.

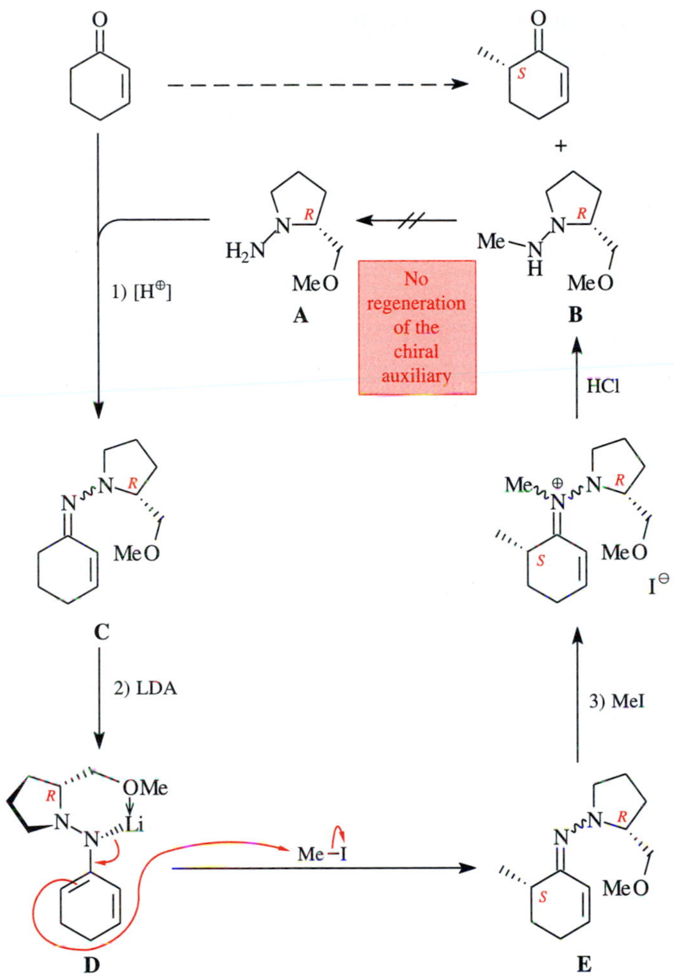

Fig. 10.32. Enders' RAMP method for the generation of enantiomerically pure α-alkylated carbonyl compounds; RAMP, *R*-aminoprolinol methyl ether.

To complete the reaction sequence of Figure 10.32, the desired alkylated ketone needs to be released from the kethydrazone. Ozonolysis cannot be used in the present case in contrast to the example of Figure 10.31. Ozonolysis would cleave not only the C═N double bond but also the C═C double bond and, of course, one wants to keep the latter. Another method must therefore be chosen to cleave the C═N double bond of compound **E** of Figure 10.32. The kethydrazone is alkylated to give a iminium ion. The iminium ion is much more easily hydrolyzed than the hydrazone itself, and mild hydrolysis yields the desired *S*-enantiomer of 6-methyl-2-cyclohexenone. The other product of hydrolysis is a RAMP derivative. This RAMP derivative carries a methyl group at the N atom and cannot be recycled to the enantiomerically pure chiral auxiliary **A** that was employed initially.

10.2.2 Chain-Elongating Syntheses of Carboxylic Acid Derivatives

Malonic Ester Synthesis of Substituted Acetic Acids

B

Malonic ester syntheses are the classical analog of acetoacetic ester syntheses of methyl ketones. Neither case requires the use of an amide base for the enolate formation, and in both cases alkoxides suffice to deprotonate the substrate completely. Malonic esters are active-methylene compounds just like acetoacetic ester and its derivatives.

One equivalent of NaOEt in EtOH deprotonates malonic acid diethyl ester completely to give the sodium enolate **A** (Figure 10.33). This enolate is monoalkylated upon addition of an alkylating reagent such as BuBr, and a substituted malonic ester **C** is formed. *During* the alkylation reaction, the substituted malonic ester **C** reacts to a certain extent with some of the enolate **A,** resulting in the butylated enolate **B** and unsubstituted neutral malonic ester. It is for this reason that the reaction mixture contains *two* nucleophiles—the original enolate **A** and the butylated enolate **B.** The alkylation of **A** with butyl bromide is much faster than that of **B,** since **A** is less sterically hindered than **B.** The main product is therefore the product of monoalkylation. Distillation can be used to separate the main product from small amounts of the product of dialkylation.

The butylated malonic ester **C** of Figure 10.33 is not the actual synthetic target of the *malonic ester synthesis of substituted acetic acids.* Instead, **C** is subjected to further transformations as shown in Figure 10.34. Ester **C** first is hydrolyzed with acid-catalysis to afford the corresponding alkylated malonic acid (for the mechanism, see Figure 6.19). The alkylated malonic acid then is heated either directly in the hydrolysis mixture or after it has been isolated. This heating leads to decarboxylation. The mechanism of this decarboxylation resembles the mechanism of the decarboxylation of β-ketocarboxylic acids (cf. Figure 10.24), and it involves a cyclic, six-membered ring transition state in which three valence electron pairs are shifted at the same time. The primary products of this decomposition are carbon dioxide and the enol of the carboxylic acid. The enol immediately tautomerizes to give the carboxylic acid. This carboxylic acid—an alkylated acetic acid—represents the typical final product of a malonic ester synthesis.

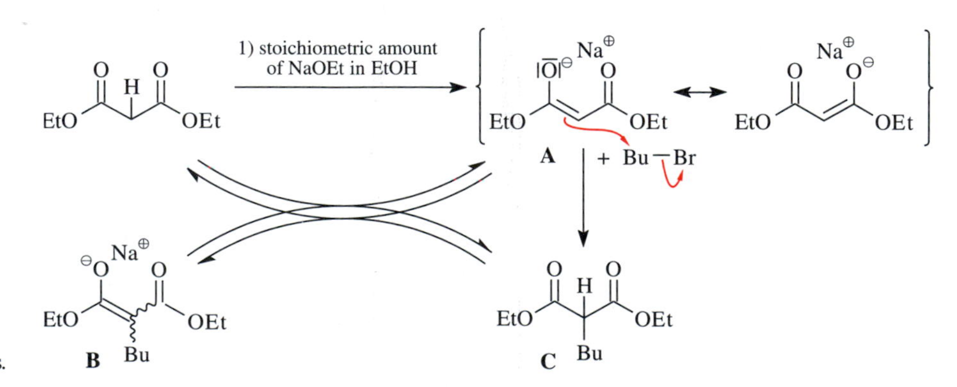

Fig. 10.33. Malonic ester synthesis of alkylated acetic acids I: preparation of alkylated malonic esters.

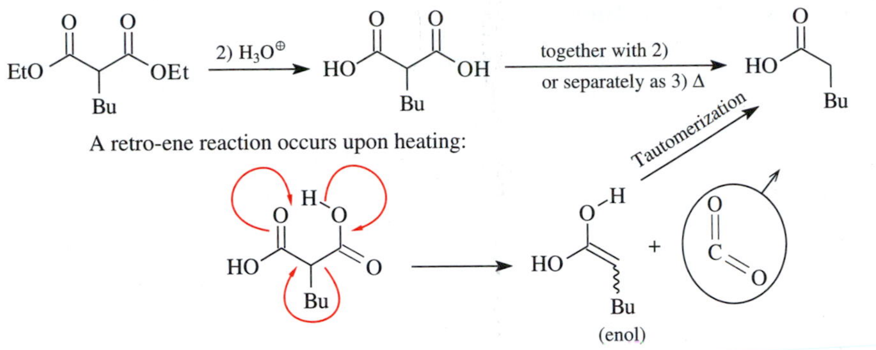

A retro-ene reaction occurs upon heating:

Fig. 10.34. Malonic ester synthesis of alkylated acetic acids II: hydrolysis and decarboxylation of the alkylated malonic ester.

Both acidic H atoms of a malonic ester can be replaced by alkyl groups. These dialkylated malonic esters are formed by successively removing the acidic protons with sodium alkoxide and treatment of the enolates with an alkylating reagent. The subsequent hydrolysis and decarboxylation of these dialkylated malonic esters affords α,α-dialkylated acetic acids as another class of products accessible via the malonic ester synthesis.

If one employs *monofunctional* alkylating reagents in the alkylation of malonic esters, one obtains dialkylated acetic acids in which the two α-alkyl groups are not connected with each other. On the other hand, if one employs a *difunctional* alkylating reagent, the dialkylated acetic acid synthesized is a cycloalkane carboxylic acid. This is the case when the second alkylation occurs in an intramolecular instead of an intermolecular fashion. Over 100 years ago, Perkin employed this principle and succeeded at the synthesis of cyclopropane carboxylic acid (Figure 10.35), the first cyclopropane derivative ever made. Until that time, the synthesis of a cyclopropane was thought to be impossible because of its high Baeyer strain ("angular strain").

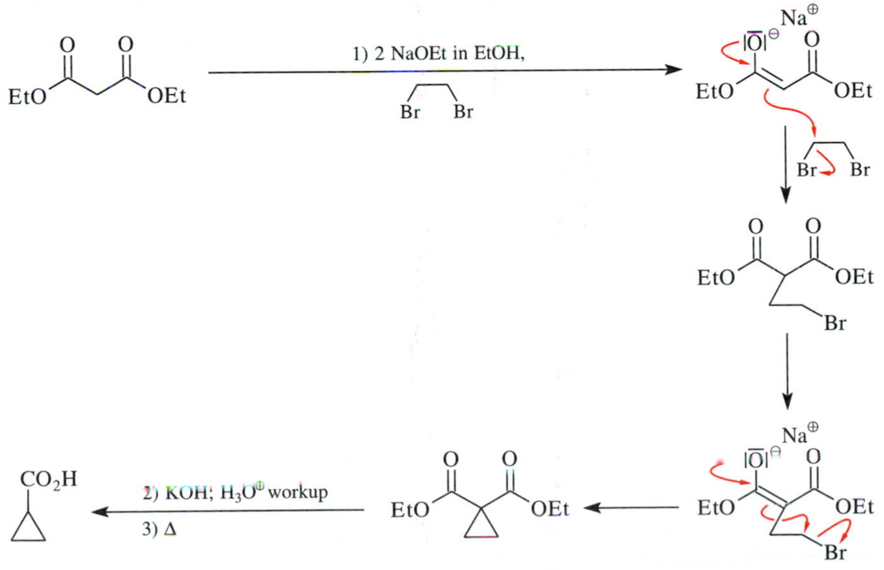

Fig. 10.35. Perkin's first cyclopropane synthesis via a malonic ester synthesis.

Alkylation of Ester Enolates

B

Ester enolates are generated by the reaction between an ester and LDA at −78°C in THF, enolate formation usually being "*E*"-selective (Figure 10.13). The "*Z*"-enolate is obtained in analogous deprotonations of esters that carry an alkoxy group in the *α* position relative to the C=O double bond (Figure 10.36). Product-development control is the reason for the latter stereochemical outcome: the "*Z*"-enolate and the lithium form an energetically favored five-membered chelate ring.

Fig. 10.36. Alkylation of an ester enolate for the preparation of an *α*-hydroxycarboxylic acid. The initially formed benzyl ester **B** contains two benzylic C—O bonds, which can be cleaved by means of hydrogenolysis.

Many ester enolates can be alkylated, and this is irrespective of whether they are "*E*"- or "*Z*"-configured. The example of Figure 10.36 shows the butylation of a "*Z*"-configured *α*-oxygenated ester enolate. The butylated ester **B** is both a benzyl ester and a benzyl ether. The two benzylic C—O bonds in this compound can be removed subsequently by way of hydrogenolysis (cf. Figure 14.44). Overall, this reaction sequence represents a method that allows for the elongation of alkylating agents to *α*-hydroxycarboxylic acids **A.**

Diastereoselective Alkylation of Chiral Ester and Amide Enolates: Generation of Enantiomerically Pure Carboxylic Acids with Chiral Centers in the *α* Position

The alkylation of an *achiral* ester enolate to give an *α*-alkylated carboxylic ester can generate a new stereocenter. If so, this stereocenter is formed without stereocontrol. For such a substrate, the two half-spaces above and below the enolate plane are enantiotopic. Consequently, the attack of an achiral alkylating reagent occurs from both faces with the same rate constant (cf. discussion of Section 3.4.1). Thus, one obtains the alkylated ester as a racemic, i.e., 50:50 *R/S* mixture. In the alkylation of an achiral amide enolate, the outcome is entirely analogous: the resulting *α*-alkylated amide either is achiral or a racemate.

The situation changes when chiral ester enolates or chiral amide enolates are alkylated. There, the half-spaces on the two sides of the enolate planes of the substrates are diastereotopic, and alkylating reagents can attack from one of the sides selectively (cf. discussion in Section 3.4.1). Stereogenic alkylations of such enolates therefore may occur diastereoselectively. Especially important examples of such diastereoselective alkylations are shown in Figures 10.37 and 10.38. Figure 10.37 shows the alkylation of a chiral propionic acid ester—the ester is derived from an enantiomerically pure alcohol—via the "*E*"- and "*Z*"-enolates. Figure 10.38 shows alkylations of two propi-

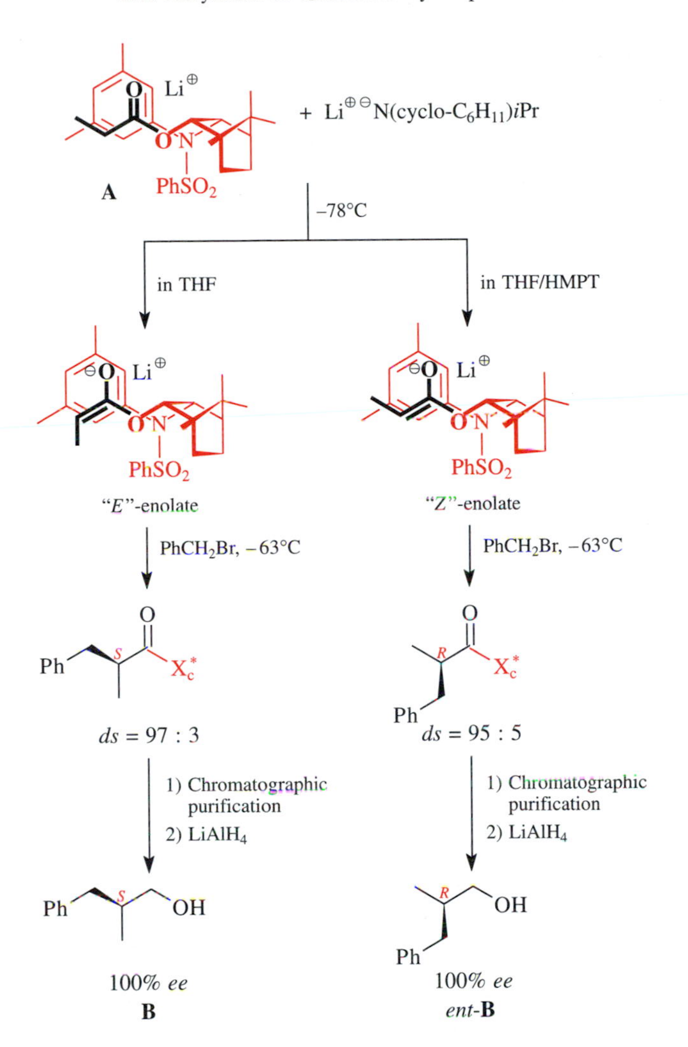

Fig. 10.37. Helmchen synthesis of enantiomerically pure α-alkylated carboxylic acids. The deprotonation of the propionic acid ester results in the "*E*"-enolate in the solvent THF and in the "*Z*"-enolate in the solvent mixture THF/HMPA. In these projections, both enolates are attacked preferentially from the front. The "*E*"-enolate results in a 97 : 3 mixture of *S*- and *R*-configured α-benzylpropionic acid esters (X_c^* marks the chiral alkoxide group), while the "*Z*"-enolate results in a 5:95 mixture. Chromatographic separation and reduction of the C(=O)—X_c^* groups afford alcohol **B** with 100% *ee* from the "*E*"-enolate and alcohol *ent*-**B** with 100% *ee* from the "*Z*"-enolate.

onic acid amides in which the N atom is part of an enantiomerically pure heterocycle; these alkylations occur via the "*Z*"-configured amide enolates.

Both enantiomers of camphor are commercially available. One of the camphor enantiomers can be converted in five steps into the enantiomerically pure, chiral alcohol contained in the propionic acid ester **A.** Ester **A** is employed in the **Helmchen synthesis** of Figure 10.37. The enantiomer of this ester can be obtained from the other camphor isomer. Each of these esters can be alkylated with high diastereoselectivity, as shown in Figure 10.37 for two alkylations of ester **A.** The highest selectivities are achieved if the ester is deprotonated at −78°C with lithium cyclohexyl isopropyl amide. This reagent is an amide base with a somewhat higher steric demand than LDA. The deprotonation with this reagent is a stereogenic reaction just like the LDA deprotonation of the propionic acid esters of Figures 10.13 and 10.14, and the same stereoselectivities result: in pure THF propionic acid ester **A** yields the "*E*"-enolate with high diastereoselectivity (Figure 10.37, left). In THF/HMPA mixtures, on the other hand, the reaction of the same propionic acid ester **A** with the same base occurs with complete reversal of stereochemistry, i.e., yields the "*Z*"-enolate (Figure 10.37, right). Accordingly, THF/HMPA mixtures have the same effect on the stereoselectivity of ester enolate formation as we discussed for THF/DMPU mixtures in the context of Figure 10.14. Fortunately, in *this* case the HMPA (carcinogenic) can be replaced by DMPU (not carcinogenic). This option was not known at the time when the investigations described in Figure 10.37 were carried out.

The chiral alcohol group in Figure 10.37 was chosen to differentiate as much as possible between the half-spaces on both sides of the enolate plane. One half-space should be left entirely unhindered while the other should be blocked as completely as possible. The attack of the alkylating reagent then occurs preferentially, and in the ideal case exclusively, from the unhindered half-space. The stereostructures of the two ester enolates of Figure 10.37 therefore model the enolate moieties of the (early!) transition states of these alkylations. The part of the transition state structure that contains the alkylating reagent is not shown.

It is assumed that the preferred conformation of the substructure C═C—O—C of the "*E*"-configured ester enolate in the preferred transition state of the alkylation is that depicted in the center of the left-hand column of Figure 10.37. In the projection shown, the alkylating reagent attacks the enolate from the front side for the reasons just stated. The reaction occurs with a diastereoselectivity of 97:3. Chromatography allows for the complete separation of the main diastereoisomer (97 rel-%) from the minor diastereoisomer. Reduction of the main diastereoisomer (for the mechanism, see Section 14.4.3) affords the alcohol **B,** a derivative of *S*-α-benzyl propionic acid, with an optical purity of 100% *ee. Hydrolysis* of the benzylated esters without isomerization is impossible, however, so that optically active α-benzylpropionic *acids* cannot be obtained in this way.

The center of the right-hand column of Figure 10.37 shows the assumed stereostructure of the substructure C═C—O—C of the "*Z*"-configured ester enolate in the preferred transition state of the alkylation. In the chosen projection, the alkylating reagent again attacks from the front side. With a diastereoselectivity of 95 : 5, the benzylated ester that was the minor product in the alkylation of the "*E*"-configured ester enolate is now formed as the main product. Again, chromatography allows one to separate the minor from the major diastereoisomer. The main product is then reduced without any isomerization to afford the α-benzylated propionic acid *ent*-**B** of the *R* series with an *ee* value of 100%.

Why are the benzylated esters of Figure 10.37 not obtained with higher diastereoselectivities than 95 or 97%, respectively? One of the reasons, and perhaps the only reason, lies in the failure of both the "*E*"- and the "*Z*"-enolate to form with perfect stereocontrol. Small contaminations of these enolates by just 5 or 3% of the corresponding enolate with the opposite configuration would explain the observed amounts of the minor diastereoisomers, even if every enolate were alkylated with 100% diastereoselectivity.

Only chiral propionic acid amides can be alkylated with still higher diastereoselectivity than chiral propionic acid esters. This is because according to Figure 10.15, the selectivity for the formation of a "*Z*"-configured amide enolate is higher than the selectivities that can be achieved in the conversion of esters to the "*E*"- and "*Z*"-enolates. The alkylation of the "*Z*"-configured lithium enolates of the two enantiomerically pure propionic acid amides in the **Evans synthesis** of Figure 10.38 proceeds with particularly high diastereoselectivity. These amide enolates contain an oxazolidinone ring, and the presence of this ring causes conformational rigidity of the enolates: lithium bridges between the enolate oxygen and the carbonyl O atom of the heterocycle to form a six-membered ring.

Both oxazolidinones in Figure 10.38 are selected such that the substituent marked by a red circle occupies one of the two half-spaces of the enolate. The oxazolidinone to the left in Figure 10.38 can be prepared from *S*-valine in two steps. The isopropyl group ensures that the most stable transition state for the alkylation of the "*Z*-enolate

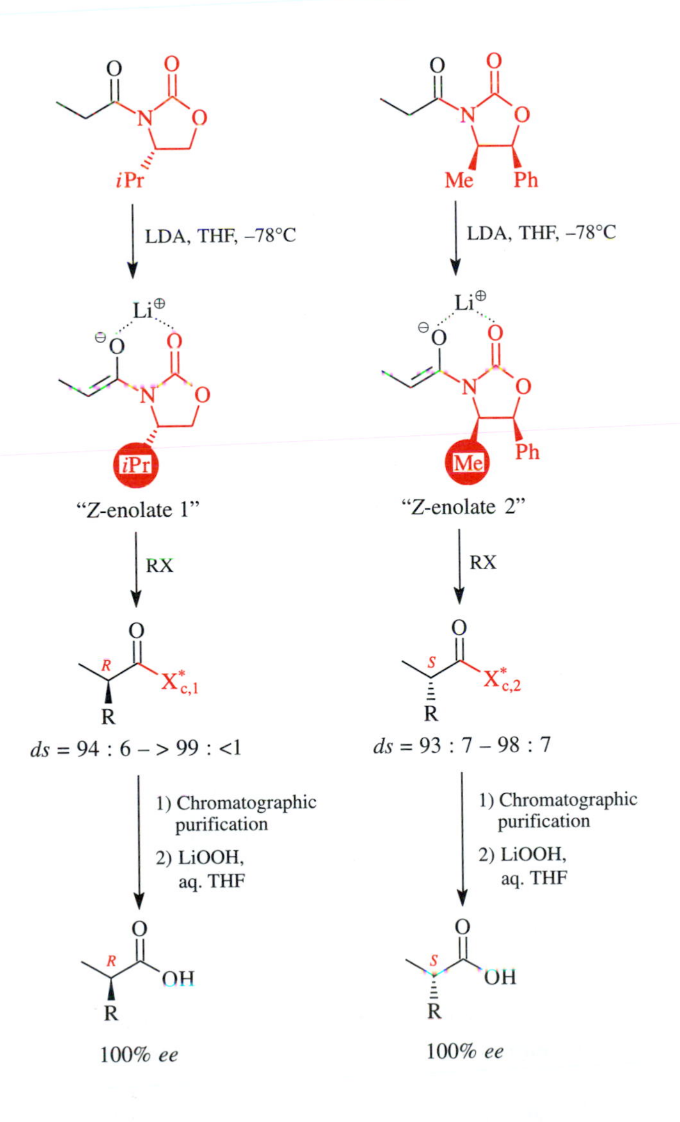

Fig. 10.38. Evans synthesis of enantiomerically pure α-alkylated carboxylic acids. The amides are derived from oxazolidinones and yield "*Z*"-enolates with high stereoselectivity. The alkylating agent attacks in both cases from the side that is opposite to the side of the substituent highlighted in red. Alkaline hydrolysis accelerated by hydrogen peroxide proceeds with retention of configuration and yields enantiomerically pure α-alkylated carboxylic acids; $X^*_{c,1}$ and $X^*_{c,2}$ are the chiral amide groups.

1" involves attack of the alkylating reagent from the front side (with respect to the selected projection). The alkylating agent thus attacks preferentially from the side opposite to the isopropyl group. Similar considerations apply to the most stable transition state structure of the alkylation of the oxazolidinone "*Z*-enolate 2" in Figure 10.38. This transition state results from the backside attack of the alkylating agent (again with regard to the projection drawn). But again, this is an attack from that side of the molecule that is opposite to the substituent marked by the red circle.

The alkylations of the oxazolidinone-containing amide enolate of Figure 10.38 occur with diastereoselectivities of 93:7 and >99:1, respectively. Therefore, a chromatographic purification often is not even warranted. The hydrogen peroxide–accelerated alkaline hydrolysis of these compounds occurs with complete retention of the previously established configuration at the α stereocenter. To date, the Evans synthesis offers the most versatile access to enantiomerically pure α-alkylated carboxylic acids.

10.3 Hydroxyalkylation of Enolates with Carbonyl Compounds ("Aldol Addition"): Synthesis of β-Hydroxyketones and β-Hydroxyesters

B

An "aldol addition" in present-day terminology involves the addition of the α-C atom of a carbonyl compound, a carboxylic acid, a carboxylic ester, or a carboxylic amide to the C=O double bond of an aldehyde or a ketone. In the past, the term "aldol addition" was used in a more restricted sense, referring to the addition of the α-C atom of nothing but carbonyl compounds to the C=O double bond of aldehydes and ketones. The products of aldol additions in today's usage of the term are β-hydroxylcarbonyl compounds (aldols), β-hydroxycarboxylic acids, β-hydroxycarboxylic esters, or β-hydroxycarboxylic amides.

10.3.1 Driving Force of Aldol Additions and Survey of Reaction Products

B

The addition of an alkaline earth metal enolate **A** to a carbonyl compound is always an exergonic process irrespective of whether the enolate is derived from a ketone, an ester, or an amide and whether the carbonyl compound is an aldehyde or a ketone (Figure 10.39, top). One of the reasons for this exergonicity lies in the fact that the alkaline earth metal ion is part of a chelate in the alkoxide **B** of the aldol addition product. The driving forces for the additions of alkaline earth metal enolates of esters and amides to carbonyl compounds are further increased because the aldol adducts **B** are resonance-stabilized, whereas the enolates are not.

Table 10.5 shows the various aldol adducts that can be obtained if one reacts a quantitatively formed ester enolate or a quantitatively formed (kinetic) ketone enolate with

Table 10.5. Representative Aldol Adducts Formed by Addition of Ketone or Ester Enolates to Selected Carbonyl Compounds

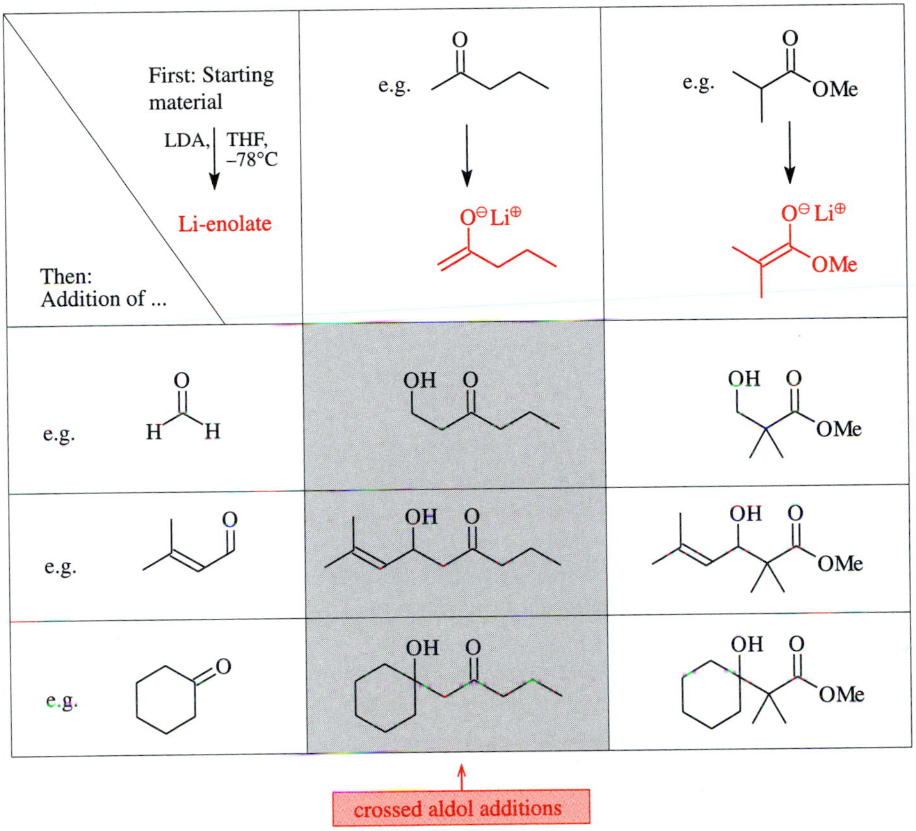

First: Starting material; LDA, THF, −78°C ↓ Li-enolate / Then: Addition of ...	e.g. O → O⊖Li⊕	e.g. O, OMe → O⊖Li⊕, OMe
e.g. O, H H	OH O	OH O, OMe
e.g. O	OH O	OH O, OMe
e.g. O	OH O	OH O, OMe

crossed aldol additions

three representative carbonyl compounds. **Crossed aldol adducts** are adducts that result from the addition of the enolate of *one* carbonyl compound to the C=O double bond of a *second* carbonyl compound (center column in Table 10.5).

In principle it also is possible to obtain the β-hydroxycarbonyl compounds directly in neutral form rather than in form of their alkoxides (Figure 10.39, bottom). This is accomplished by the reaction of one carbonyl compound or of a mixture of two carbonyl compounds with a catalytic amount of MOH or MOR. Aldehyde enolates and ketone enolates are then formed in small amounts (cf. the rules of thumb at the beginning of Section 10.1.2). These enolates add to the C=O double bond of nondeprotonated substrate molecules or, if a mixture of carbonyl compounds is employed, they add to the C=O double bond of the more reactive of the carbonyl compounds. The alkoxides **B** of the aldol adducts are formed initially but are converted immediately and quantitatively into the aldols by way of protonation.

This base-catalyzed aldol addition is an equilibrium reaction, and all steps of this reaction are reversible. The free enthalpy of reaction ΔG_r° of such aldol reactions is close

Fig. 10.39. Different driving forces of aldol additions depending on whether they involve quantitatively prepared enolates (top) or only small equilibrium concentrations of enolates (bottom).

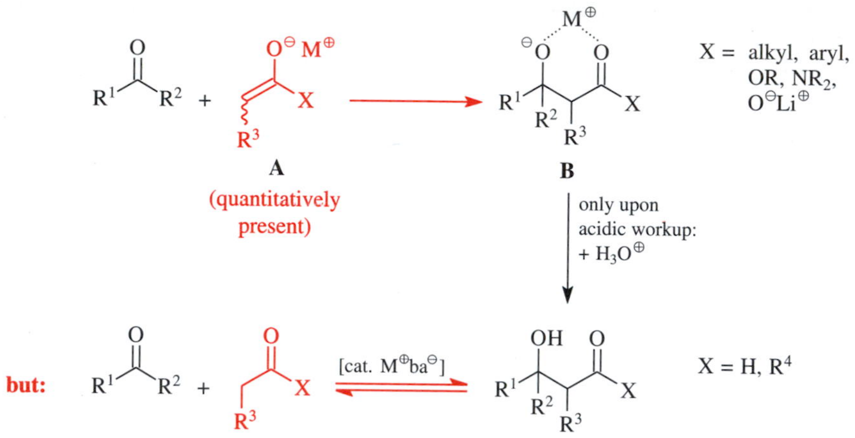

to zero. In fact, ΔG_r° is negative only if there are "many H atoms" among the substituents R^1, R^2, and R^3 of the two reacting components (structures in Figure 10.39, bottom). Otherwise, the formation of the aldol adduct is endergonic because of the destabilization due to the van der Waals repulsions between these substituents. A base-catalyzed aldol addition between two *ketones,* therefore, is never observed.

Esters and amides are much weaker C,H acids than aldehydes and ketones. Neither the ester nor the amide is deprotonated to any significant extent if a base such as MOH or MOR is added to a mixture of these esters or amides with a carbonyl compound. Hence, neither esters nor amides afford aldol adducts in base-*catalyzed* reactions.

10.3.2 Stereocontrol

A

The preparation of aldol adducts may occur with simple diastereoselectivity. A definition of the term was given in Section 9.3.2. In a slightly different formulation, simple diastereoselectivity means that a single relative configuration is established at two neighboring C atoms that become stereocenters for the following reasons. (1) Both C atoms were sp^2-hybridized in the reactants; one was part of a nonhomotopic C=X double bond and the other was part of a nonhomotopic C=Y double bond. (2) The formation of a σ bond between these C atoms causes them to be sp^3-hybridized in the reaction product.

The simple diastereoselectivity of aldol reactions was first studied in detail for the **Ivanov reaction** (Figure 10.40). The Ivanov reaction consists of the addition of a carboxylate enolate to an aldehyde. In the example of Figure 10.40, the diastereomer of the β-hydroxycarboxylic acid product that is referred to as the *anti*-diastereomer is formed in a threefold excess in comparison to the *syn*-diastereoisomer. Zimmerman and Traxler suggested a transition state model to explain this selectivity, and their transition state model now is referred to as the **Zimmerman–Traxler model** (Figure 10.41). This model has been applied ever since with good success to explain the simple diastereoselectivities of a great variety of aldol reactions.

upon workup + $H_3O^{\oplus}$

anti (racemic) 76 : 24 syn (racemic)

Fig. 10.40. The Ivanov reaction. For the generation of the carboxylate enolate, see Figure 10.8.

The key idea of the Zimmerman–Traxler model is that aldol additions proceed via six-membered ring transition state structures. In these transition states, the metal (a magnesium cation in the case of the Ivanov reaction) coordinates both to the enolate oxygen and to the O atom of the carbonyl compound. By way of this coordination, the metal ion guides the approach of the electrophilic carbonyl carbon to the nucleophilic enolate carbon. The approach of the carbonyl and enolate carbons occurs in a transition state structure with chair conformation. C—C bond formation is fastest in the transition state with the maximum number of quasi–equatorially oriented and therefore sterically unhindered substituents.

The application of the Zimmerman–Traxler model to the specific case of the Ivanov reaction of Figure 10.40 is illustrated in Figure 10.41. The reaction proceeds preferentially through transition state **B** and its mirror image *ent*-**B** and results in the formation of a racemic mixture of the enantiomers **A** and *ent*-**A** of the main diastereoisomer. Both phenyl groups are quasi-equatorial in these transition states. All other transition state stuctures are less stable because they contain at least one phenyl group in a quasi-axial orientation. For example, the phenyl group of the enolate is in a quasi-axial position in transition state **C** and its mirror image *ent*-**C.** In transition state **D** and its mirror image *ent*-**D,** the phenyl groups of the benzaldehyde occupy quasi-axial positions. The latter transition state structures are therefore just as disfavored as the pair **C** and *ent*-**C.** In fact, the *syn*-configured minor product of the Ivanov reaction, a racemic mixture of enantiomers **E** and *ent*-**E,** must be formed via the transition states **C** and *ent*-**C** or via **D** and *ent*-**D,** but it is not known which path is actually taken.

The *anti* adducts **A**/*ent*-**A** also could, in principle, result via the pair of transition states that contain *both* phenyl groups in quasi-axial positions. This reaction path might contribute a small amount of the *anti* adduct, but this is rather improbable.

Lithium enolates of ketones and esters also add to aldehydes by way of Zimmerman–Traxler transition states. However, the Li—O bond is weaker and longer than those—O bond. The lithium-containing transition state structures thus are less compact than those containing Mg. Therefore, in a Li-containing Zimmerman–Traxler transition state, a quasi-axial and thus unfavorably positioned aldehyde substituent suffers only a weak *gauche* interaction with the skeleton of the six-membered ring. In fact, this destabilization generally is too small to render such a transition state structure inaccessible. Hence, the addition of lithium enolates to aldehydes only occurs with high diastereoselectivity if the aldehyde substituent does not assume a quasi-axial position as the consequence of *another* destabilizing interaction.

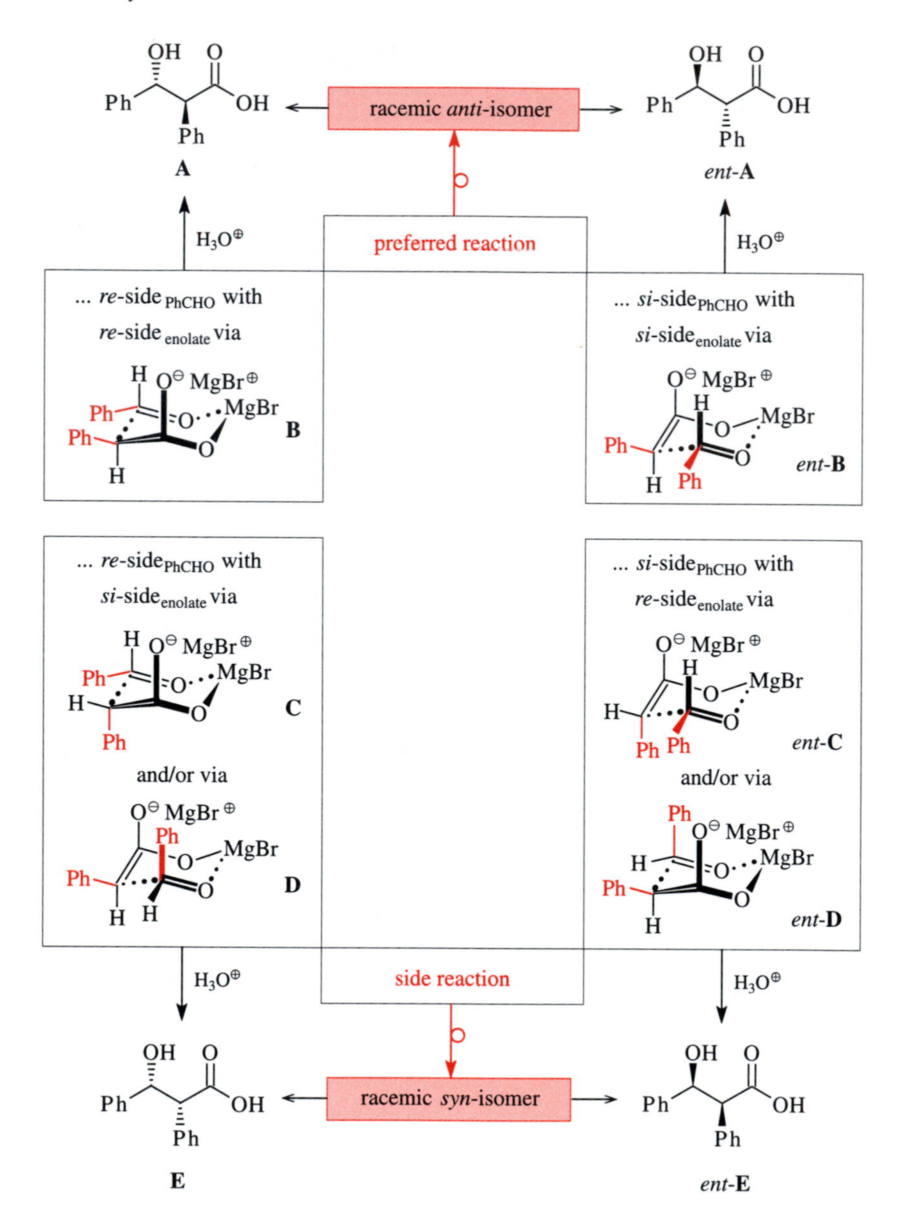

Fig. 10.41. Explanation of the *anti*-selectivity of the Ivanov reaction of Figure 10.40 by means of the Zimmerman–Traxler model. The stereodescriptors *re* and *si* are defined as follows. Suppose you are looking down on the plane of an alkene, in which an sp^2-hybridized C atom is connected to three different substituents. You are on the *re* side of the double bond if the Cahn–Ingold–Prelog priorities of these substituents decrease going clockwise, and on the *si* side otherwise.

Heathcock identified such a destabilizing interaction in the 1,3-diaxial interaction of the aldehyde substituent with a substituent at the C atom to which the enolate oxygen is attached. In spite of the relatively long Li—O distance, this 1,3-interaction can be sufficiently strong if the substituent of the aldehyde is extremely bulky. In that case, and only in that case, the aldehyde group is forced into the quasi-equatorial orientation also in a lithium-containing Zimmerman–Traxler transition state. If, in addition, the lithium enolate of the aldehyde contains a homogeneously configured C=C(—O$^-$) double bond, a highly diastereoselective aldol addition of a lithium enolate occurs. Two configurationally homogeneous lithium enolates with suitably bulky substituents are the *Z*-configured ketone enolate of Figure 10.42 with its $Me_2C(OSiMe_3)$ group and the "*E*"-configured ester enolate of Figure 10.43 with its (2,6-di-*tert*-butyl-4-methoxyphenyl)oxy group.

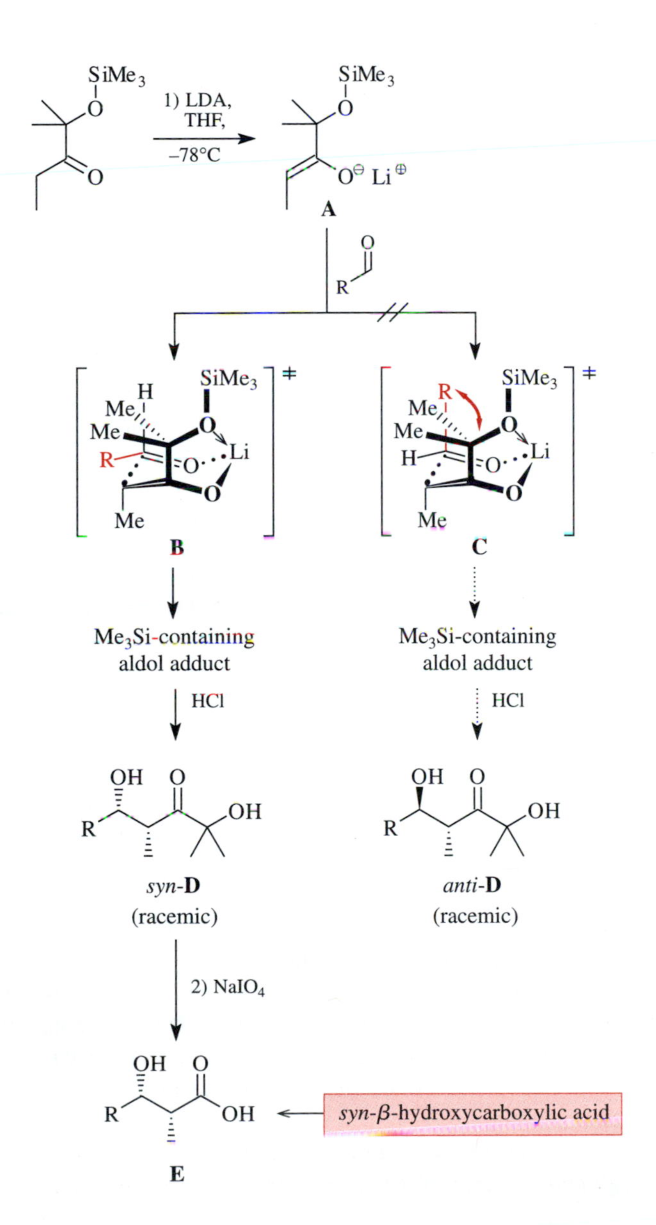

Fig. 10.42. *syn*-Selectivity of the aldol addition with a Heathcock lithium enolate and mechanistic explanation. The Zimmerman–Traxler transition state **C** is destabilized by a 1,3-diaxial interaction, while the Zimmerman–Traxler transition state **B** does not suffer from such a disadvantage. The reaction thus occurs exclusively via transition state **B.**

The ketone enolate **A** of Figure 10.42 is generated in a *Z*-selective fashion (as we saw in Figure 10.12). The bulky and branched enolate substituent destabilizes the Zimmerman–Traxler transition state **C** by way of the discussed 1,3-diaxial interaction, while the transition state structure **B** is not affected. Hence, the aldol addition of enolate **A** occurs almost exclusively via transition state **B,** and the *syn*-configured aldol adducts **D** (Figure 10.42) are formed with a near-perfect simple diastereoselectivity. The acidic workup converts the initially formed trimethysilyloxy-substituted aldol adducts into the *hydroxylated* aldol adducts.

It is for good reason that the bulky ketone substituent of enolate **A** contains a Me_3SiO group, which is carried on into the *syn*-configured aldol adduct: its acid-catalyzed hydrolysis affords an OH group in the α position of the C=O double bond. Such an α-hydroxycarbonyl compound can be oxidatively cleaved with sodium periodate to afford a carboxylic acid. The mechanism of this oxidation is described later in connection with Figure 14.17. It involves the oxidation of the corresponding hydrate of the carbonyl compound. The hydrate of the α-hydroxycarbonyl compound is formed in an equilibrium reaction. This oxidation converts the *syn*-configured aldol adduct **D**—which contains a synthetically less useful ketone substituent—into a synthetically more valuable *syn*-configured β-hydroxycarboxylic acid, **E.**

The aldol addition of Figure 10.42 can also be carried out in such a fashion that the crude silyl ether-containing aldol adducts are treated directly with periodic acid without prior aqueous workup. In that case, the silyl ether and α-hydroxyketone cleavages both occur in one and the same operation.

Anti-configured β-hydroxycarboxylic acids are accessible via the reaction sequence depicted in Figure 10.43. The ester enolate **A** is generated by using LDA in THF with the usual "*E*"-selectivity (cf. Figure 10.13). The enolate contains a phenyl group with two *ortho*-attached *tert*-butyl groups. A phenyl substituent with such a substitution pattern must be twisted out of the plane of the enolate. This is true also in the Zimmerman–Traxler transition states **B** and **C.** One of the *tert*-butyl groups ends up directly on top of the chair structure. Being forced into this position above the ring, the *tert*-butyl group necessarily repels the non-hydrogen substituent of the aldehyde in transition state structure **C.** The associated destabilization of **C** does not occur in the diastereomeric transition state **B.** The aldol addition of Figure 10.43 thus proceeds exclusively via **B,** and perfect simple diastereoselectivity results.

The reaction mixture of Figure 10.43 is worked up such that it yields the β-*acetoxy*carboxylic acid esters **D** instead of the β-*hydroxy*carboxylic acid esters. These products are obtained with diastereoselectivities $ds > 98{:}2$, and this is independent of the nature of the aldehyde employed. The acetoxyesters **D** are prepared instead of the hydroxyesters because the latter would not survive the ester cleavage still to ensue. The problem is that this ester group cannot be taken off by way of hydrolysis because it is so bulky. However, it can be removed via oxidation with ceric ammonium nitrate, because then, a quinone is the leaving group. The resulting β-hydroxycarboxylic acids **E** retain the high *anti* stereochemistry established in **C.**

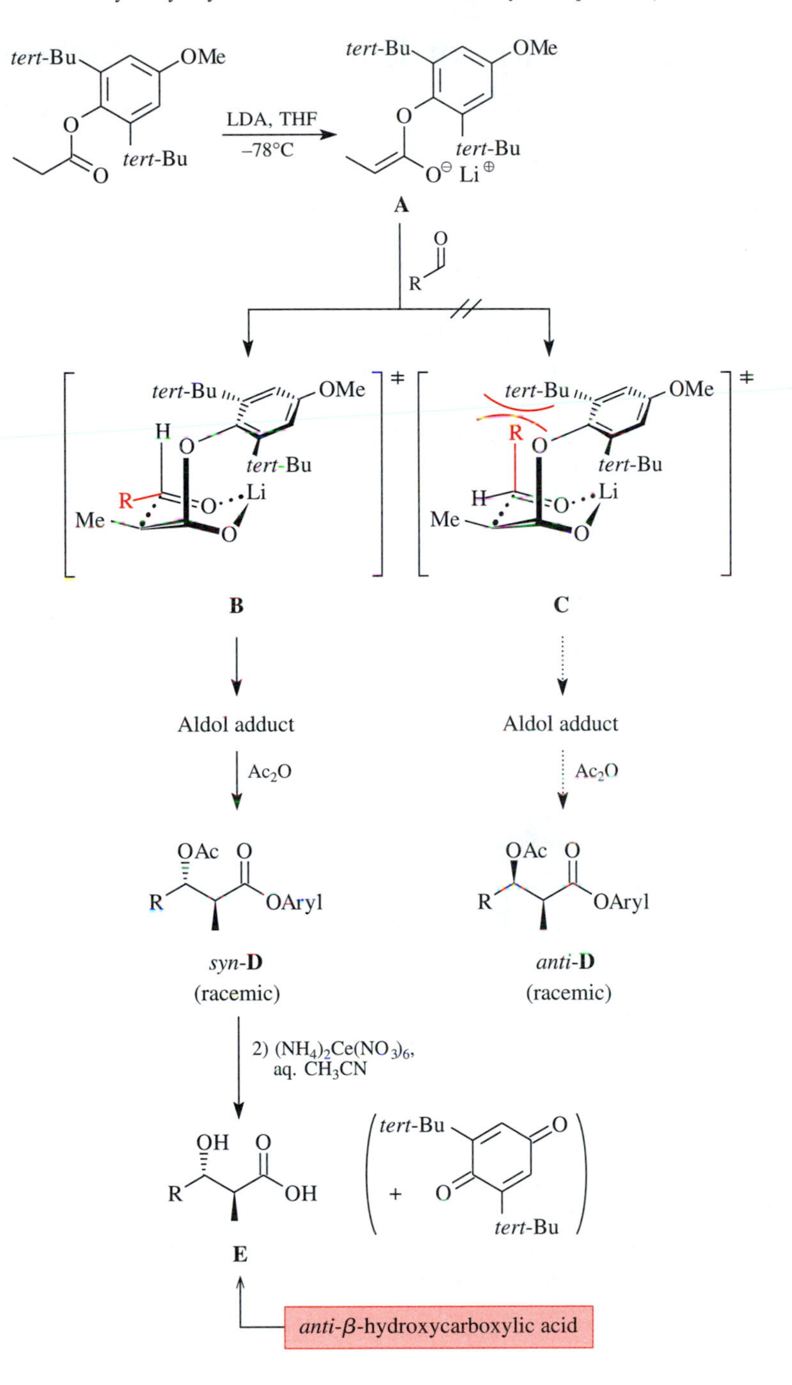

Fig. 10.43. *anti*-Selectivity of the aldol addition with a Heathcock lithium enolate and mechanistic explanation. The Zimmerman–Traxler transition state **C** is destabilized by a 1,3-diaxial interaction, while the Zimmerman–Traxler transition state **B** does not suffer from such a disadvantage. The reaction thus occurs exclusively via transition state **B.**

10.4 Condensation of Enolates with Carbonyl Compounds: Synthesis of Michael Acceptors

10.4.1 Aldol Condensations

B

An **aldol reaction** is a reaction between two carbonyl compounds in which (a derivative of) one carbonyl compound plays the role of a nucleophile while the other carbonyl compound acts as an electrophile. The term "aldol reaction" covers two types of reactions, namely, aldol additions in the older and more restrictive sense (cf. Section 10.3) as well as aldol condensations. The aldol reactions that lead to β-hydroxycarbonyl compounds belong to the class of aldol *additions*. Aldol *condensations* start from the same substrates but result in α,β-unsaturated carbonyl compounds (Figure 10.44).

Aldol reactions often proceed as aldol condensations if the participating aldehyde or ketone enolates **C** are formed only in equilibrium reactions, i.e., incompletely (Figure 10.44). Under these reaction conditions an aldol addition occurs first: it leads to the formation of **D** proceeding by way of the mechanism shown in Figure 10.39 (bottom). Then an $E1_{cb}$ elimination takes place: in an equilibrium reaction, aldol **D** forms a small amount of enolate **E** which eliminates NaOH or KOH.

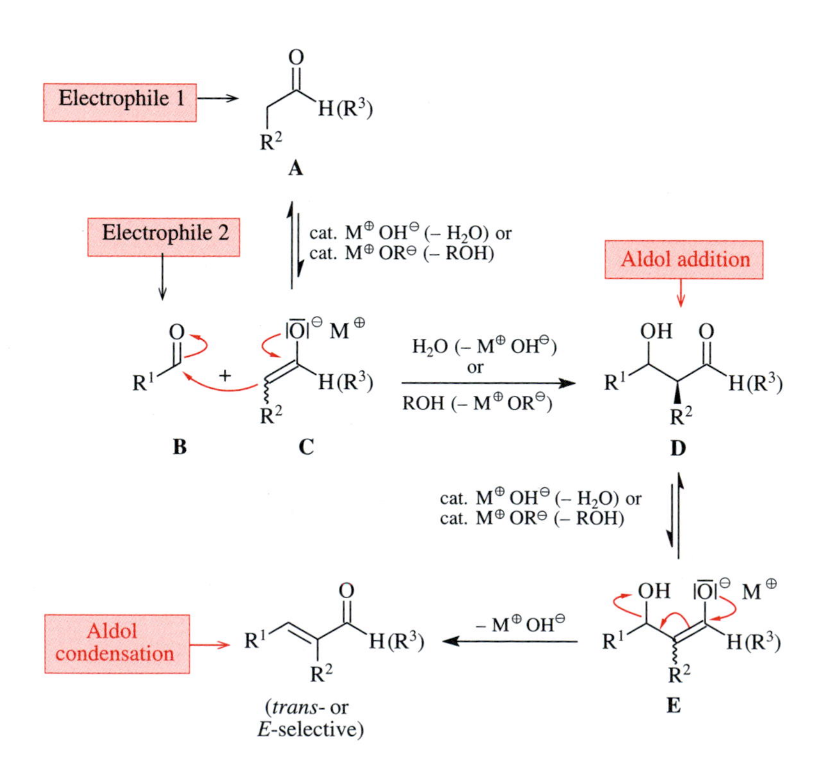

Fig. 10.44. Mechanisms of base-catalyzed aldol reactions: aldol addition (steps 1 and 2) and aldol condensation (up to and including step 4).

If a stereogenic double bond is established by this $E1_{cb}$ elimination, one usually observes a *trans*- or an *E*-selectivity. This experimental finding could have two origins: (1) product-development control (cf. Section 4.1.3), if the stereoselectivity occurs under kinetic control, or (2) thermodynamic control. Thermodynamic control comes into play as soon as the *cis,trans*- or *E,Z*-isomeric condensation products can be interconverted via a reversible 1,4-addition of NaOH or KOH. In the *trans*- or *E*-isomer of an α,β-unsaturated carbonyl compound the formyl or acyl group may lie *unimpeded* in the plane of the C═C double bond. This geometry allows one to take *full advantage* of the resonance stabilization {C═C—C═O ↔ $^+$C—C═C—O$^-$}. On the other hand, in the *cis*- or *Z*-isomer of an α,β-unsaturated carbonyl compound the formyl or acyl group interferes to such an extent with the substituent at the other end of the C═C double bond that a planar geometry is no longer possible and the resonance stabilization consequently is reduced.

As with MOH- or MOH-*catalyzed* aldol additions, MOH- or MOH-*catalyzed* aldol condensations can be carried out only with aldehyde or ketone enolates, not with ester or amide enolates. The reason for this is the same as discussed before, namely, that ester and amides are less acidic than carbonyl compounds and the amounts of enolate they form with the bases mentioned are much too small.

According to Figure 10.39 ketones often do not engage in base-catalyzed aldol *additions* because of a lack of driving force. Hence, ketones also are less suitable electrophiles than aldehydes in aldol *condensations.* However, for ketones, too, the condensation step is irreversible and they *can* therefore form α,β-unsaturated carbonyl compounds. It is not always possible to isolate the α,β-unsaturated carbonyl compounds thus formed. If the C-β atom is not sterically hindered, these products can act as electrophiles and add any residual ketone enolate; the α,β-unsaturated carbonyl compound acts as a Michael acceptor in this case (Section 10.6.1).

A broad spectrum of α,β-unsaturated carbonyl compounds becomes accessible in this way because a great variety of aldehydes is suited for aldol condensation. Table 10.6 exemplifies the broad scope by way of the reactions of an aldehyde enolate (center) and of the enolate of an asymmetrical ketone (right). The right column of Table 10.6 also shows that the regioselectivity of the aldol condensation of the ketone is not easy to predict. Subtle substituent effects decide whether in a given case the thermodynamic enolate **A** or the kinetic enolate **B** (present in much smaller amounts) is responsible for the reaction. The thermodynamic enolate enters the aldol addition, while the kinetic enolate enters the aldol condensation.

Only one of the aldol condensations of Table 10.6 (top right) concerns the reaction of a carbonyl compound *with itself.* In all other reactions of Table 10.6, the α,β-unsaturated carbonyl compounds are formed by two different carbonyl compounds. Such aldol condensations are referred to as **crossed aldol condensations** (cf. the discussion of crossed aldol additions in Section 10.3.1).

Of course, it is the goal of a crossed aldol condensation to produce a single α,β-unsaturated carbonyl compound. One has to keep in mind that crossed aldol condensations may result in up to four constitutionally isomeric condensation products (starting from two aldehydes or an aldehyde and a symmetric ketone) or even in eight

Table 10.6. Representative α,β-Unsaturated Carbonyl Compounds Generated by Aldol Condensations of Carbonyl Compounds with Selected Aldehydes

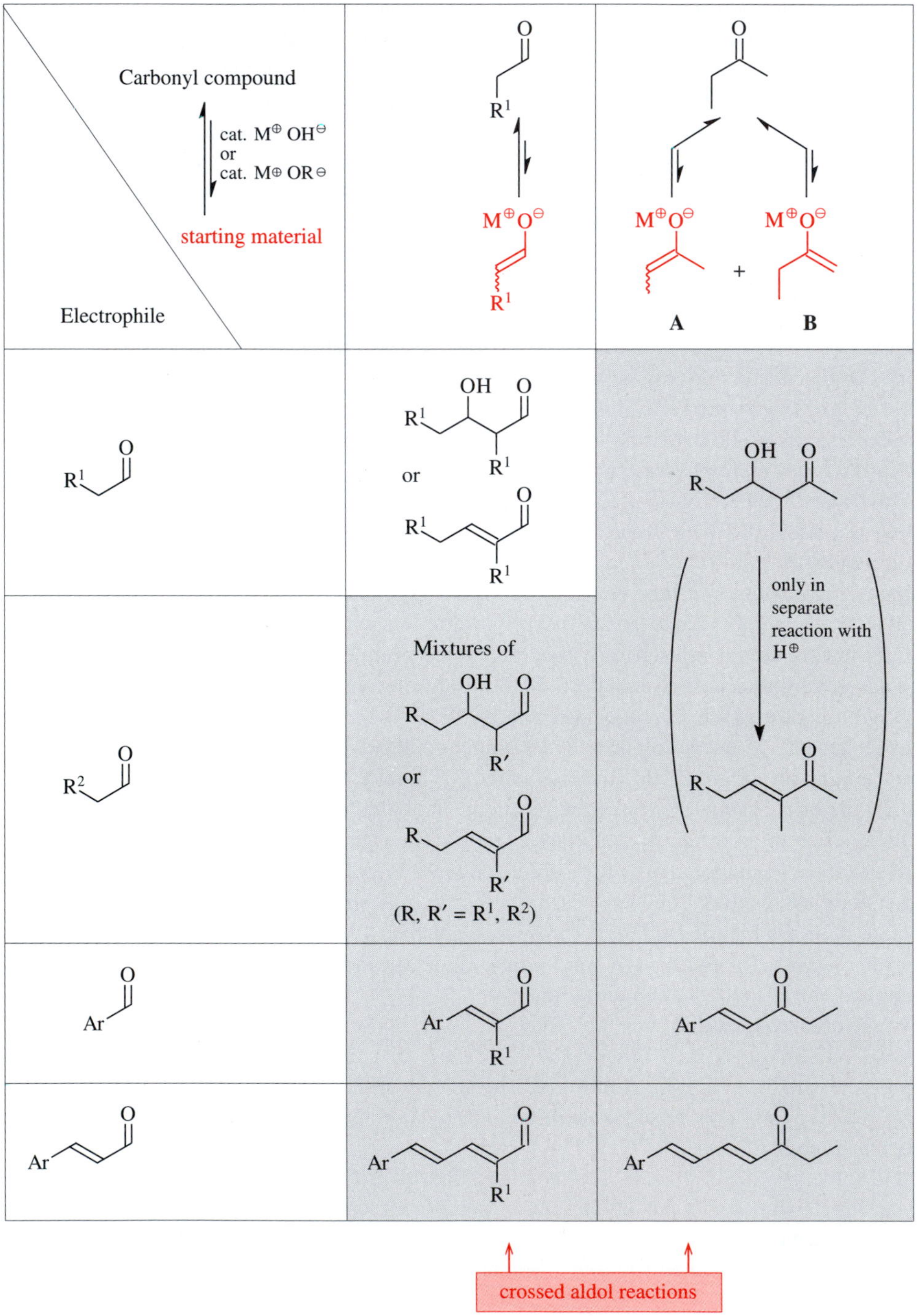

Carbonyl compound (cat. $M^{\oplus} OH^{\ominus}$ or cat. $M^{\oplus} OR^{\ominus}$; starting material) / Electrophile	R^1CH_2CHO ⇌ $M^{\oplus}O^{\ominus}$ enolate (R^1)	butanone ⇌ $M^{\oplus}O^{\ominus}$ enolates **A** + **B**
R^1CH_2CHO	aldol (R^1, OH, O, R^1) or enal (R^1, O, R^1)	aldol (R, OH, O); only in separate reaction with $H^{\oplus}$ → enone (R, O)
R^2CH_2CHO	Mixtures of aldol (R, OH, O, R′) or enal (R, O, R′) (R, R′ = R^1, R^2)	
ArCHO	Ar, O, R^1	Ar, O
ArCH=CHCHO	Ar, O, R^1	Ar, O

crossed aldol reactions

constitutional isomers (starting with an aldehyde and an asymmetric ketone). These maximum numbers of structural isomers result if both starting materials

- can react as electrophiles and as nucleophiles
- can react with molecules of their own kind as well as with other molecules. The product variety is further increased if
- asymmetric ketones can react via two regioisomeric enolates.

Crossed aldol condensations occur with chemoselectivity only if some of the foregoing options cannot be realized. The following possibilities exist:

Chemoselectivity of Crossed Aldol Condensations

1) Ketones generally react only as nucleophiles in crossed aldol additions because the addition of an enolate to their C=O double bond is thermodynamically disadvantageous (Figure 10.39).
2) *Benzaldehyde, cinnamic aldehyde, and their derivatives* do not contain any α-H-atoms; therefore, they can participate in crossed aldol additions only as electrophiles.
3) *Formaldehyde* also does not contain an α-H-atom. However, formaldehyde is such a reactive electrophile that it tends to undergo multiple aldol additions instead of a simple aldol condensation. This type of reaction is exploited in the pentaerythritol synthesis.

These guidelines allow one to understand the following observations concerning crossed aldol condensations that proceed via the mechanism shown in Figure 10.44.

1) Crossed aldol condensations between *benzaldehyde or cinnamic aldehyde or their derivatives* on the one hand and *ketones* on the other hand pose no chemoselectivity problems. The least sterically hindered ketone, acetone, may condense with benzaldehyde, cinnamic aldehyde, and their derivatives with both enolizable positions if an excess of the aldehyde is employed.

2) Crossed aldol condensations between *aliphatic aldehydes and ketones* succeed only in two steps via the corresponding crossed aldol adducts. The latter can be obtained by adding the aldehyde dropwise to an inert mixture of the ketone and base. The aldol adducts subsequently must be dehydrated with acid catalysis.

3) Crossed aldol condensations between *aliphatic aldehydes* on the one hand and *benzaldehyde* or *cinnamic aldehyde* or their derivatives on the other also are possible. The reaction components can even be mixed together. The aldol adducts are formed without chemoselectivity, as a mixture of isomers, but their formations are reversible. The $E1_{cb}$ elimination to an α,β-unsaturated carbonyl compound is fast only if the newly created C=C double bond is conjugated to an aromatic system or to another C=C double bond already present in the substrate. This effect is due to product-development control. All the starting materials thus react in this way via the most reactive aldol adduct

4) Chemoselective crossed aldol condensations between *two different C,H-acidic aldehydes* are impossible. There is only a single exception, and that is the intramolecular aldol condensation of an asymmetric dialdehyde.

10.4.2 Knoevenagel Reaction

B

A Knoevenagel reaction is a condensation reaction between an active-methylene compound or the comparably C,H-acidic nitromethane and a carbonyl compound. The product of a Knoevenagel reaction is an alkene that contains two geminal acceptor groups (**B** in Figure 10.45) or one nitro group (**B** in Figure 10.46).

Knoevenagel reactions are carried out in mildly basic media—in the presence of piperidine, for example—or in neutral solution—catalyzed by piperidinium acetate, for example. The basicity of piperidine or of acetate ions, respectively, suffices to generate a sufficiently high equilibrium concentration of the ammonium enolate of the active-methylene compound (**A** in Figure 10.45) or to generate a sufficiently high equilibrium concentration of the ammonium nitronate of the nitroalkane (**A** in Figure 10.46). The rather high acidity of the nitroalkanes (Table 10.1) alternatively allows the formation of nitroalkenes by way of reaction of nitroalkanes with aldehydes in the presence of basic aluminum oxide powder as base.

The enolate **A** or the nitronate **A,** respectively, initially adds to the C=O double bond of the aldehyde or the ketone. The primary product in both cases is an alkoxide, **D,** which again contains the structural motif of a fairly strong C,H-acid, namely, of an active-methylene compound or of a nitroalkane, respectively. Hence, intermediate **D** is protonated at the alkoxide oxygen and the C-β atom is deprotonated to about the same extent as in the case of the starting materials. An OH-substituted enolate **C** is formed (Figures 10.45 and 10.46), which then undergoes an $E1_{cb}$ elimination, lead-

Fig. 10.45. Mechanism of the Knoevenagel reaction of active-methylene compounds; $\sim H^+$ indicates the migration of a proton.

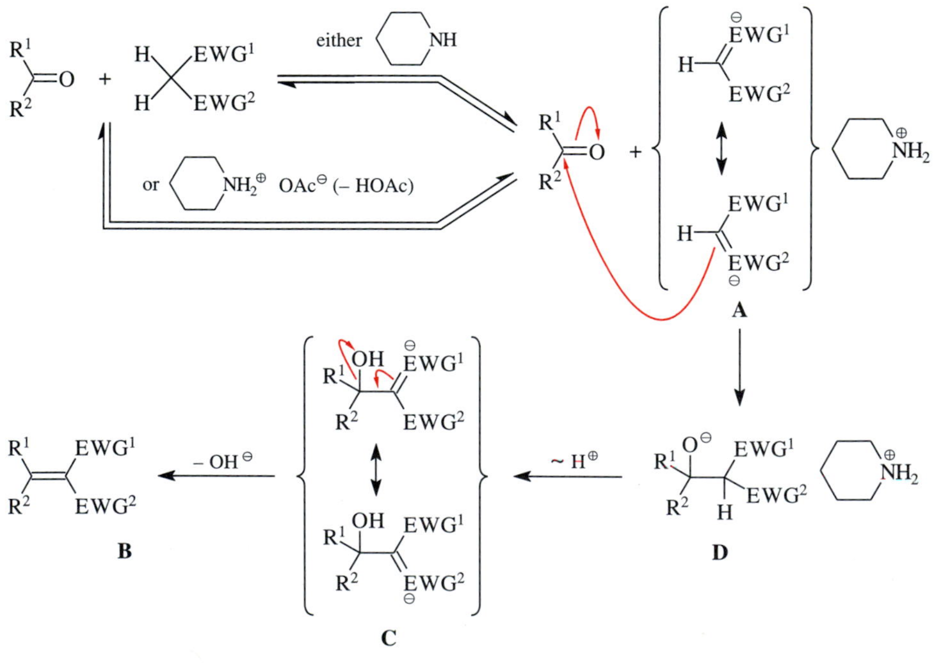

Fig. 10.46. Mechanism of a Knoevenagel reaction with nitromethane. Alkaline aluminum oxide powder is sufficiently basic to deprotonate nitromethane. The small amount of the anion generated from nitromethane suffices for the addition to aldehydes to proceed. The elimination of water via an $E1_{cb}$ mechanism follows quickly if a conjugated C=C double bond is formed, as in the present case.

ing to the condensation product **B.** The Knoevenagel condensation and the aldol condensation have in common that both reactions consist of a sequence of an enolate hydroxyalkylation and an $E1_{cb}$ elimination.

The Knoevenagel reaction can be employed for the synthesis of a wide variety of condensation products because the carbonyl component as well as the active-methylene component can be varied.

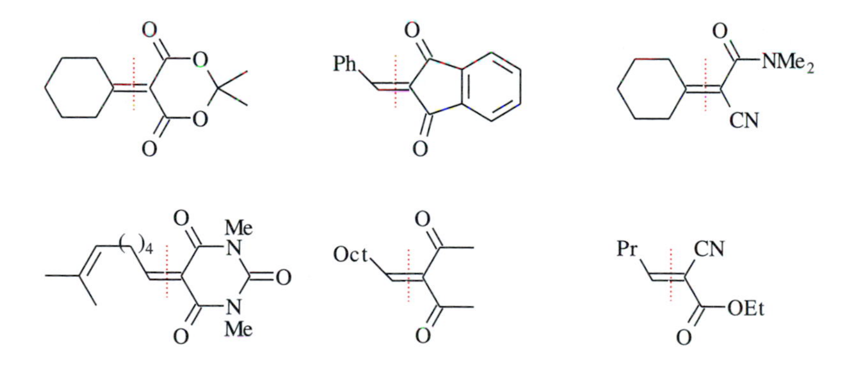

Fig. 10.47. Products of Knoevenagel condensations and indication of their potential synthetic origin. The left molecule halves stem from the carbonyl compounds and the right fragments come from the active-methylene compounds.

10.4.3 A Knoevenagel Reaction "with a Twist"

Malonic acid itself can react with aldehydes in the presence of piperidine by way of a Knoevenagel condensation. A decarboxylation occurs after the condensation, and this decarboxylation cannot be avoided. Figure 10.48 shows how the overall reaction can be employed for the synthesis of cinnamic acids and of their derivatives. This reaction

B

Fig. 10.48. A Knoevenagel reaction "with a twist": preparation of cinnamic acid.

sequence occurs under much milder conditions than the Perkin synthesis of cinnamic acids. (The Perkin synthesis consists of the condensation of aromatic aldehydes with acetic acid anhydride in the presence of sodium acetate.)

Benzaldehyde itself could be the electrophile in this Knoevenagel reaction. However, it is also possible that the piperidinium salt derived from benzaldehyde acts as the electrophile. Similarly, several plausible mechanisms can be formulated for the decarboxylation step—they depend, among other things, on the stage at which it may occur. Because of these ambiguities, we do not want to discuss any of the details of *these* reaction steps. Instead, we want to focus on a mechanistic detail of the first step and that is the question of which species acts as the nucleophile and initiates the C—C bond formation.

Certainly, the reactive species has to be an enolate. The conceivable candidates are the enolate **D** (malonic acid "monoanion," Figure 10.49), the enolate **E** (malonic acid "dianion"), and the enolate **F** (malonic acid "trianion"). The nucleophilicity increases greatly in this order, but the concentrations of these species also drastically decrease. The *combined* effect of nucleophilicity and abundance determines which nucleophile initiates the Knoevenagel condensation. It is therefore important to know the concentrations of the various species.

It can be assumed that the small amount of piperidine in the reaction mixture is completely protonated by malonic acid because piperidine is more basic than pyridine. Hence, only the pyridine is available for the formation of the enolate of malonic acid (formation of enolate **D,** Figure 10.49), of the enolate of the monocarboxylate **B** (formation of enolate **E**), or of the enolate of the dicarboxylate **C** (formation of **F**). Pyridine reacts with any of these C,H acids in an equilibrium reaction:

$$\text{baH} + \text{pyridine} \rightleftharpoons \text{ba}^{\ominus} + \text{H-pyridinium}^{\oplus}$$

The extent to which this equilibrium reaction proceeds follows from the law of mass action and can be expressed as Equation 10.3.

$$\frac{[\text{ba}^{\ominus}][\text{pyrH}^{\oplus}]}{[\text{baH}][\text{pyr}]} = \frac{10^{-\text{p}K_{\text{a,baH}}}}{10^{-\text{p}K_{\text{a,pyrH}^+}}} = 10^{5.2-\text{p}K_{\text{a,baH}}} \tag{10.3}$$

$$\frac{[\text{ba}^{\ominus}]}{[\text{baH}]} \approx 10^{5.2-\text{p}K_{\text{a,baH}}} \tag{10.4}$$

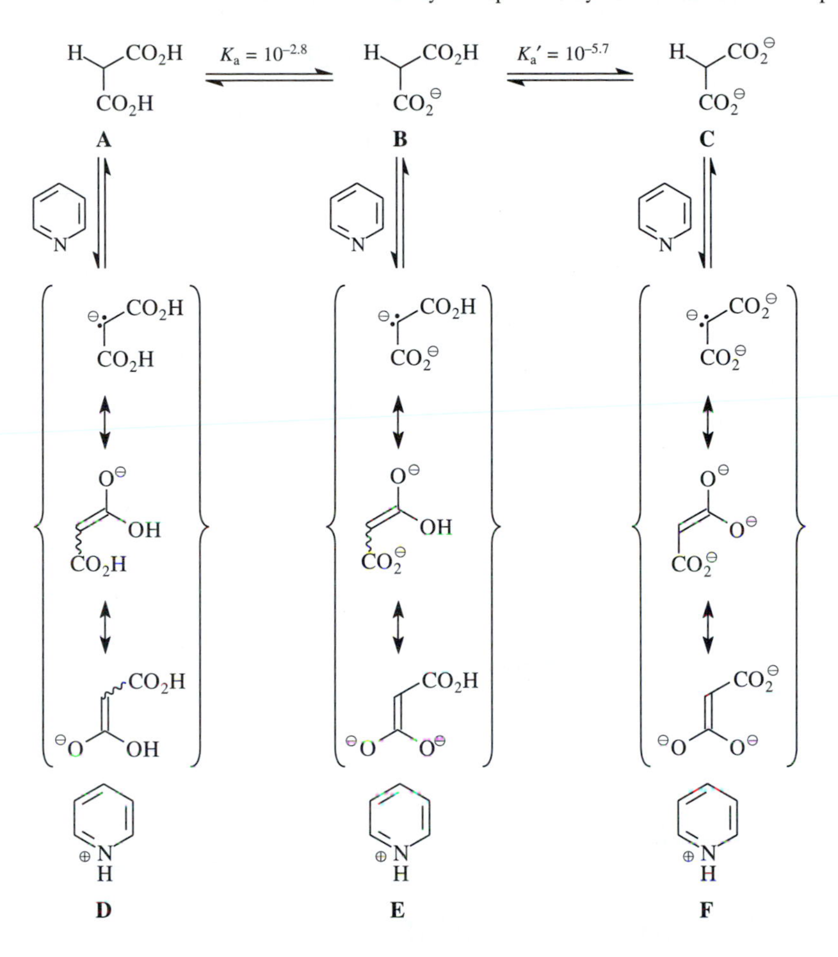

Fig. 10.49. Malonic acid "monoanion" **D,** malonic acid "dianion" **E,** and malonic acid "trianion" **F.** Nucleophiles that might participate in the Knoevenagel condensation of Figure 10.48 and their formation by deprotonation of the corresponding C,H acids. Note that of two methylene hydrogens, only the one that is being removed is shown explicitly in **A–C.**

The composition of the reaction mixture of Figure 10.48 shows that the concentrations of pyridine and piperidinium ions are about equal. One is thus justified in setting the ratio of these two concentrations to ≈1 in Equation 10.3. This approximation leads to Equation 10.4. Equation 10.4 could be used to calculate the equilibrium concentrations of all acids and bases of Figure 10.49, if their pK_a values were known. However, estimates will have to do since, unfortunately, these values are indeed unknown.

The pK_a value of malonic acid *with regard to its C,H-acidity* should be close to the pK_a value of malonic acid diethyl ester (pK_a = 13.3). The pK_a value of malonic acid monocarboxylate (**B** in Figure 10.49) *with regard to its C,H-acidity* could be as high as the pK_a value of Ph—CH*H*—CO_2^- $MgBr^+$ or of MeO_2C—CH*H*—CO_2^- $MgBr^+$. Both compounds can be deprotonated with EtMgBr (cf. Figure 10.8 for Ph—CH*H*—CO_2^- $MgBr^+$). The C,H-acidity of malonic acid monocarboxylate **B** therefore should be higher than the acidity of CH_2*H*—CO_2^- Li^+. Since the latter can be enolized with LDA, and since the pK_a value of $(i\text{Pr})_2$NH (the conjugate acid of LDA) is about 35, it follows that the pK_a value

$$\left[\begin{array}{c}CO_2H\\ CO_2^{\ominus}\end{array}\right] : \left[\begin{array}{c}CO_2^{\ominus}\\ CO_2^{\ominus}\end{array}\right] : \left[\begin{array}{c}{}^{\ominus}\ CO_2H\\ CO_2H\end{array}\right] : \left[\begin{array}{c}{}^{\ominus}\ CO_2H\\ CO_2^{\ominus}\end{array}\right] : \left[\begin{array}{c}{}^{\ominus}\ CO_2^{\ominus}\\ CO_2^{\ominus}\end{array}\right]$$

B : **C** : **D** : **E** : **F**

$$= \ {}^3/_4 \ : \ {}^1/_4 \ : \ 10^{-5} \ : \ < 10^{-20} \ : \ < 10^{-35} \ \ \frac{\text{moles}}{\text{moles of malonic acid}}$$

Fig. 10.50. Estimates of the mole ratios of the anionic species that could be generated from one mole of malonic acid and pyridine under the conditions of Figure 10.48.

of malonic acid monocarboxylate **B** must fall between the limits of "no less than 25" and "about 35." The pK_a value of malonic acid dicarboxylate (**C** in Figure 10.49) is at least 35—if not higher—because this substrate cannot be deprotonated by LDA.

With these estimates for the pK_a values of the CH-groups of malonic acid and the corresponding carboxylates on the one side and the known pK_a values of the two carboxyl groups of malonic acid (Figure 10.49, top row) on the other, all equilibrium constants are available that are required to estimate the relative amounts of all conceivable products **B–F** of the deprotonation of malonic acid with pyridine using Equation 10.4. Figure 10.50 gives the mole ratios of products **B–F**, relative to one mole of malonic acid dissolved in pyridine.

As one sees, not even a single molecule of the most reactive enolate **F** is present in the reaction flask, and it can be concluded safely that the Knoevenagel reaction does not proceed via enolate **F.** The second most nucleophilic species is enolate **E,** and its concentration is 10^{15} times smaller than the concentration of the least nucleophilic enolate **D.** The enolate **D** also does not occur in "high" concentration, but at least there is 10^{-5} mole of this species for every mole of malonic acid employed. Based on these numbers, it would seem reasonable to assume that the malonic acid monoenolate **D** is the most effective nucleophile in the Knoevenagel condensation under consideration.

10.5 Acylation of Enolates

10.5.1 Acylation of Ester Enolates

B

A **Claisen condensation** is the acylation of an ester enolate by the corresponding ester. By deprotonating an ester with MOR, only a small concentration of the ester enolate is generated and this enolate is in an equilibrium with the ester (cf. Table 10.1). The mechanism of the Claisen condensation is illustrated in detail in Figure 10.51 for the example of the condensation of butyric acid ethyl ester. Both the deprotonation of the ester to give enolate **A** and the subsequent acylation of the latter are reversible. This acylation occurs via a tetrahedral intermediate (**B** in Figure 10.51) just like the acylations of other nucleophiles (Chapter 6). The equilibrium between two molecules of butyric acid ester and one molecule each of the condensation product **C** and ethanol does not lie completely on the side of the products. In fact, Claisen condensations go to completion only

- if a stoichiometric amount of alkoxide is present, or

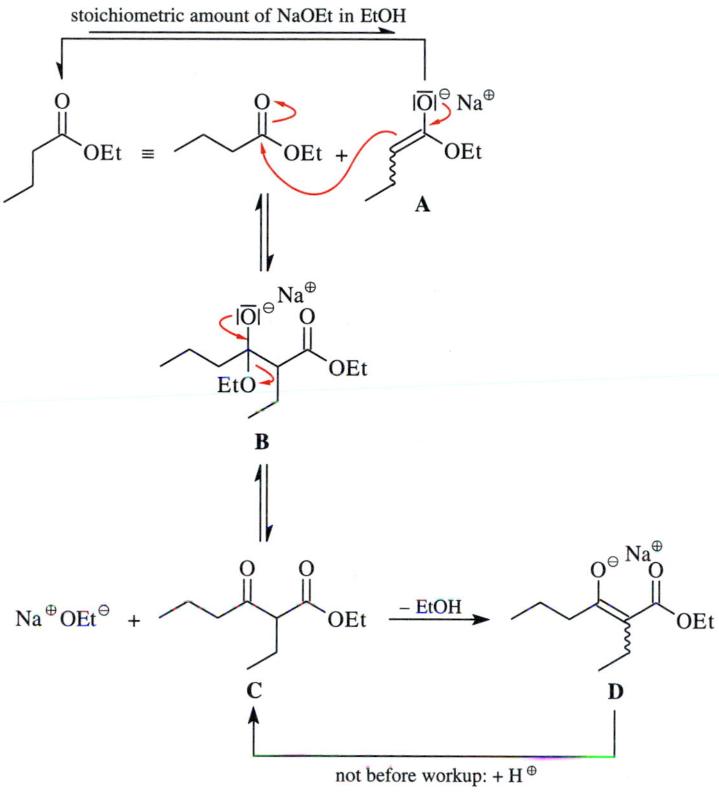

Fig. 10.51. Mechanism of a Claisen condensation. The deprotonation step $Na^+OEt^- + \mathbf{C} \rightarrow \mathbf{D}$ + EtOH is irreversible, and it is for this reason that eventually all the starting material will be converted into the enolate **D.**

- if a stoichiometric amount of alkoxide can be generated from the one equivalent of alcohol liberated in the course of the Claisen condensation and a stoichiometric amount of Na or NaH.

What is the effect of the stoichiometric amount of strong base that allows the Claisen condensation to proceed to completion? The β-ketoester **C,** which occurs in the equilibrium, is an active-methylene compound and rather C,H-acidic. Therefore, its reaction with the alkoxide to form the ester-substituted enolate **D** occurs *with considerable driving force.* This driving force is strong enough to render the deprotonation step $\mathbf{C} \rightarrow \mathbf{D}$ essentially irreversible. Consequently, the overall condensation also becomes irreversible. In this way, all the substrate is eventually converted into enolate **D.** The neutral β-ketoester can be isolated after addition of one equivalent of aqueous acid during workup.

Intramolecular Claisen condensations, called **Dieckmann condensations,** are ring-closing reactions that yield 2-cyclopentanone carboxylic esters (Figure 10.52) or 2-cyclohexanone carboxylic esters. The mechanism of the Dieckmann condensation is, of course, identical to the mechanism of the Claisen condensation (Figure 10.51). To ensure that the Dieckmann condensation goes to completion, the presence of a stoichiometric amount of base is required. As before, the neutral β-ketoester (**B** in Figure

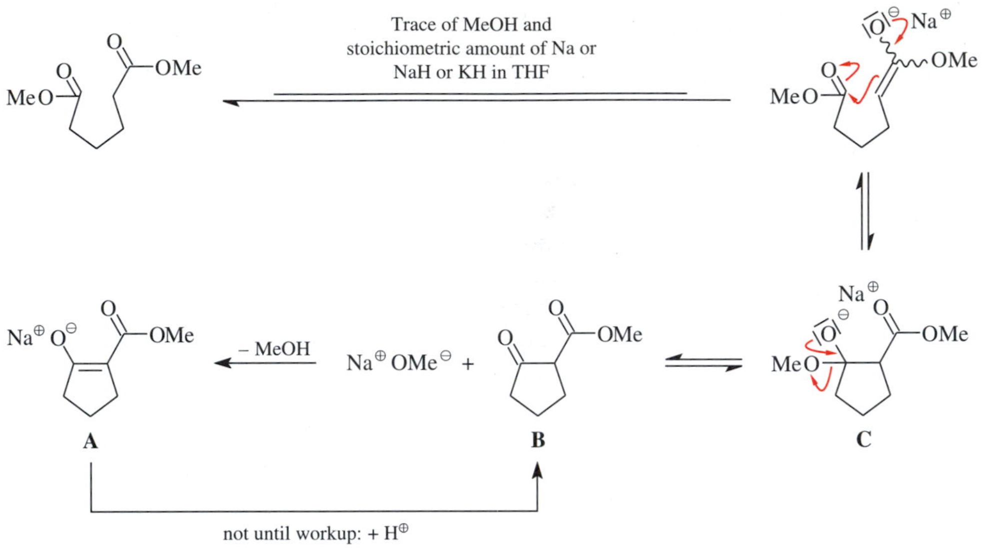

Trace of MeOH and stoichiometric amount of Na or NaH or KH in THF

– MeOH

not until workup: + $H^{\oplus}$

Fig. 10.52. Mechanism of a Dieckmann condensation. The Dieckmann condensation is an intramolecular Claisen condensation.

10.52) is formed in a reversible reaction under basic conditions. However, the back-reaction of the β-ketoester **B** to the diester is avoided by deprotonation to the substituted enolate **A.** This enolate is the thermodynamic sink to which all the substrate eventually is converted. The β-ketoester **B** is regenerated in neutral form again later, namely, during workup with aqueous acid.

Acylations of ester enolates with other esters are called **crossed Claisen condensations** if they are carried out—just like normal Claisen condensations—in the presence of a stoichiometric amount of alkoxide, Na, or NaH. Crossed Claisen condensations can in principle lead to four products. In order that only a single product is formed in a crossed Claisen condensation, the esters employed need to be suitably differentiated: one of the esters must be prone to enolate formation, while the other must possess a high propensity to form a tetrahedral intermediate (see example in Figure 10.53).

The use of an ester without acidic α-H-atoms ensures that this ester can act only as the electrophile in a crossed Claisen condensation. Moreover, this nonenolizable ester should be no less electrophilic than the other ester. This is because the larger fraction of the latter is present in its nondeprotonated form; that is, it represents a possible electrophile, too, capable of forming a tetrahedral intermediate when attacked by an enolate.

Accordingly, crossed Claisen condensations occur without any problems if the acylating agent is a better electrophile than the other, nondeprotonated ester. This is the case, for example, if the acylating agent is an oxalic ester (with an electronically activated carboxyl carbon) or a formic ester (the least sterically hindered carboxyl carbon).

Crossed Claisen condensations can be chemoselective even when the nonenolizable ester is *not* a better electrophile than the enolizable ester. This can be accomplished by a suitable choice of reaction conditions. The nonenolizable ester is mixed with the base and the enolizable ester is added slowly to that mixture. The enolate of the enolizable ester then reacts mostly with the nonenolizable ester for statistical reasons; it reacts much less with the nonenolized form of the enolizable ester, which is present

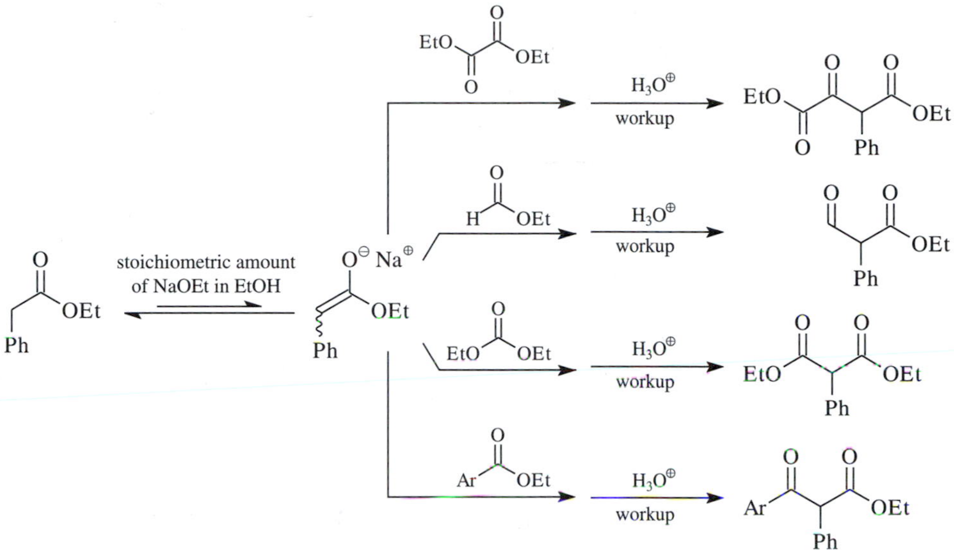

Fig. 10.53. Crossed Claisen condensation. Although the tautomers of the acylation products shown are not the major tautomer except for the third case from the top, they are presented because they show best the molecules from which these products were derived.

only in rather small concentration. Carbonic acid esters and benzoic acid esters are nonenolizable esters of the kind just described.

A

Under different reaction conditions, esters still other than the ones shown in Figure 10.53 can be employed for the acylation of ester enolates. In such a case, one *completely* deprotonates *two* equivalents of an ester with LDA or a comparable amide base and then adds one equivalent of the ester that serves as the acylating agent. The acylation product is a β-ketoester, and thus a stronger C,H acid than the conjugate acid of the ester enolate employed. Therefore, the initially formed β-ketoester reacts immediately in an acid/base reaction with the second equivalent of the ester enolate: The β-ketoester protonates this ester enolate and thereby consumes it completely.

In some acylations it may even be necessary to employ three equivalents of the ester enolate. The example of Figure 10.54 is such a case. The acylating ester contains an alcohol group and, of course, the H atom of the hydroxyl group is acidic. Thus, it destroys the first equivalent of the ester enolate through transfer of the proton to form the neutral ester. The second equivalent of the ester enolate is consumed in building up the β-ketoester intermediate, whereas the third equivalent of the ester enolate deprotonates this intermediate quantitatively.

10.5.2 Acylation of Ketone Enolates

B

Remember what we discussed in the context of Figure 10.39: *ketones* usually do not undergo aldol additions if they are deprotonated to only a small extent by an alkaline earth metal alkoxide or hydroxide. The driving force behind that reaction simply is

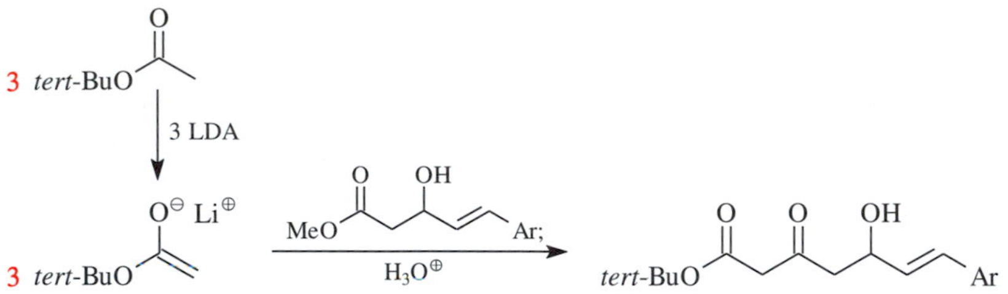

Fig. 10.54. Crossed ester condensation via acylation of a quantitatively prepared ester enolate. Three equivalents of ester enolate must be employed because the acylating ester contains a free OH group with an acidic H atom: one for the deprotonation of the OH group of the substrate, one for the substitution of the MeO group, and one for the transformation of the C,H-acidic substitution product into an enolate.

too weak. In fact, only a very few ketones can react with themselves in the presence of alkaline earth metal alkoxides or alkaline earth metal hydroxides. And if they do, they rather engage in an aldol condensation. Cyclopentanone and acetophenone, for example, show this reactivity.

The relative inertness of ketone enolates toward ketones makes it possible to react nonquantitatively obtained ketone enolates with esters instead of with ketones. These esters—and reactive esters in particular—then act as acylating reagents.

In contrast to ketones, aldehydes easily undergo a base-catalyzed aldol reaction (Figure 10.39), and this reaction may even progress to an aldol condensation (Section 10.4.1). It is therefore not possible to acylate aldehyde enolates that are present only in equilibrium concentrations. Any such enolates would be lost completely to an aldol reaction.

Oxalic esters (for electronic reasons) and formic esters (because of their low steric hindrance) are reactive esters that can acylate ketone enolates formed with NaOR in equilibrium reactions. Formic esters acylate ketones to provide formyl ketones (for example, see Figure 10.55). It should be noted that under the reaction conditions the conjugate base of the active-methylene formyl ketone is formed. The neutral formyl ketone is regenerated upon acidic workup.

Most of the other carboxylic acid derivatives can acylate only ketone enolates that occur without the presence of ketones. In these reactions, the acylation product is a β-diketone, i.e., another active-methylene compound. (An exception is the case of complete substitution of the methylene carbon, that is, for a methylene-carbon that does not carry any H atoms.) As a consequence it is so acidic that it will

Fig. 10.55. Acylation of a ketone enolate with a formic ester to generate a formyl ketone. The ketone enolate intermediate (not shown) is formed in an equilibrium reaction.

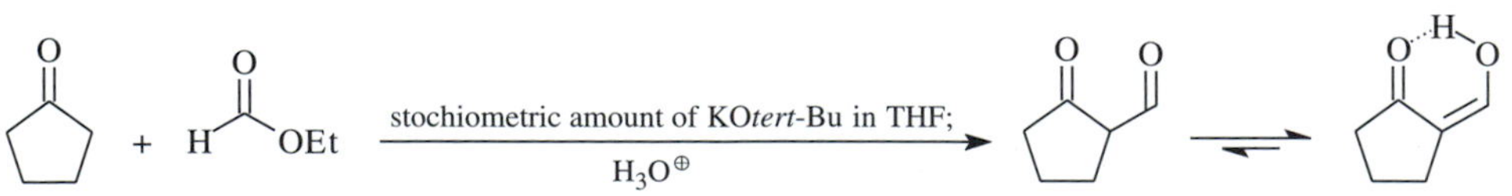

be deprotonated quantitatively. This deprotonation will be effected by the ketone enolate. Therefore, a complete acylation of this type can be achieved only if two equivalents of the ketone enolate are reacted with one equivalent of the acylating agent. Of course, proceeding in that manner would mean an unacceptable waste in the case of a valuable ketone.

Against Wastefulness: A Practical Hint Regarding the Acylation of Ketone Enolates

The following protocol requires no more than the stoichiometric amount of a ketone enolate to achieve a complete acylation. An ester is added dropwise to a 1:1 mixture of one equivalent each of the ketone enolate and LDA. The acidic proton of the β-diketone, which is formed, then is abstracted by the excess equivalent of LDA rather than by the ketone enolate.

A

The protocol described also can be used for the acylation of ketone enolates with carbonic acid derivatives (Figure 10.56). Especially good acylating agents are cyanocarbonic acid methyl ester (Figure 10.56, top) and dialkyl pyrocarbonates (bottom). Usually it is not possible to use dimethyl carbonate for the acylation of ketone enolates generated in an equilibrium reaction because dimethyl carbonate is a weaker electrophile than cyanocarbonic acid methyl ester or dialkyl pyrocarbonates.

Fig. 10.56. Acylation of ketone enolates with carbonic acid derivatives. Especially good acylation reagents are cyanocarbonic acid methyl ester (top) and dialkyl pyrocarbonates (bottom).

Weinreb amides are acylating agents that react according to the general mechanism outlined in Figure 6.4. Thus, the acylation product is not released from the tetrahedral intermediate as long as nucleophile is still present. Accordingly, the acylation of a ketone enolate by a Weinreb amide does not immediately result in the formation of the β-ketocarbonyl compound. Instead, the reaction proceeds just as an addition reaction and a tetrahedral intermediate is formed stoichiometrically (e.g., **C** in Figure 10.57). This tetrahedral intermediate is not an active-methylene compound but a donor-substituted (O^- substituent!) ketone. This intermediate therefore cannot act as a C,H acid with the ketone enolate or, in the present case, not even with the bis(ketone enolate) **B**. For reasons discussed in the context of Figure 6.33, the tetrahedral intermediate **C** is stable until it is protonated upon aqueous workup. Only then is the acylation product formed.

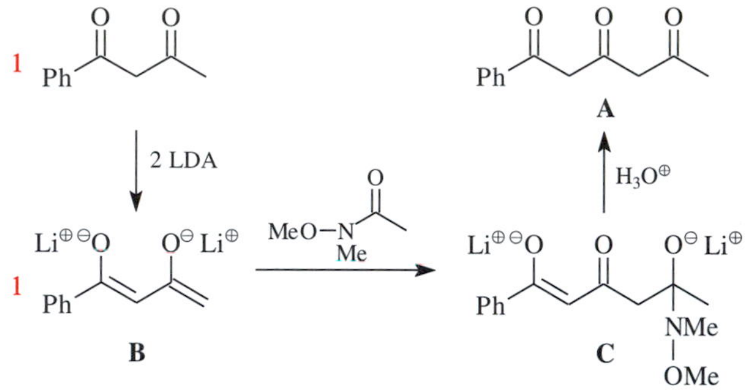

Fig. 10.57. Acylation of a bis(ketone enolate) with one equivalent of a Weinreb amide.

10.6 Michael Additions of Enolates

10.6.1 Simple Michael Additions

B

A Michael addition consists of the addition of the enolate of an active-methylene compound, the anion of a nitroalkane, or a ketone enolate to an acceptor-substituted alkene. Such Michael additions can occur in the presence of catalytic amounts of hydroxide or alkoxide. The mechanism of the Michael addition is shown in Figure 10.58. The ad-

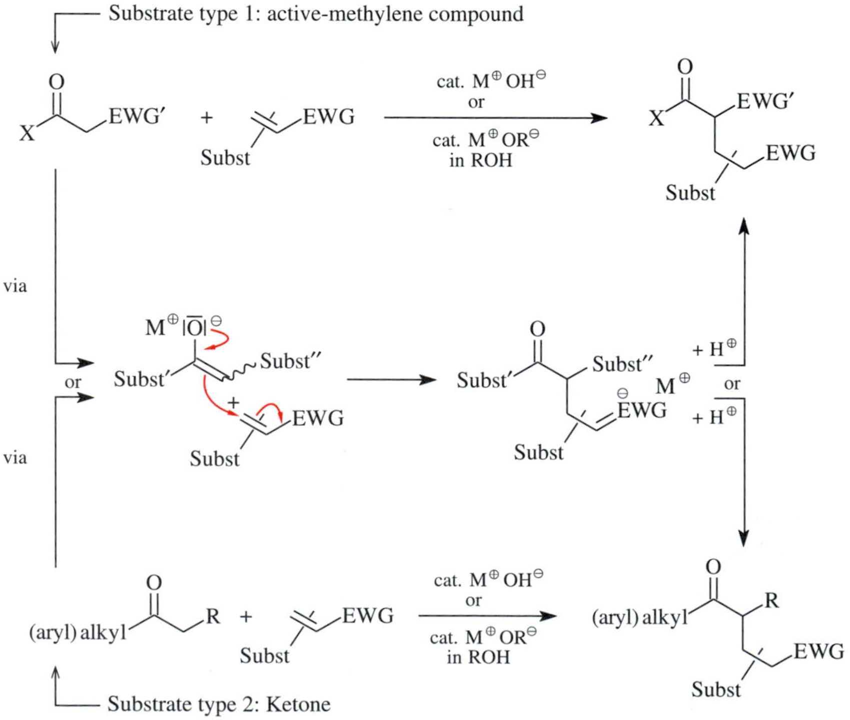

Fig. 10.58. Mechanism of the base-catalyzed Michael addition of active-methylene compounds (top) and of ketones (bottom), respectively; Subst indicates a substituent, and EWG an electron-withdrawing group.

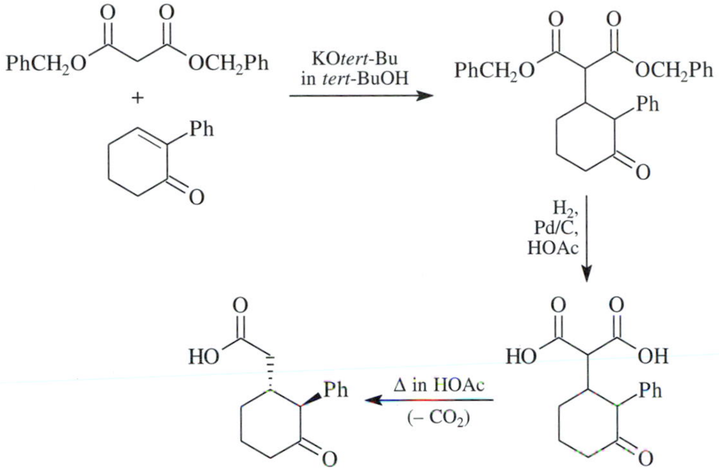

Fig. 10.59. Michael addition to an α,β-unsaturated ketone. A sequence of reactions is shown that effects the 1,4-addition of acetic acid to the unsaturated ketone. See Figure 14.44 regarding step 2 and Figure 10.24 for the mechanism of step 3. The stereochemistry of reaction steps 1 and 2 has not been discussed. The third step consists of a decarboxylation as well as an acid-catalyzed epimerization of the carbon in the α position relative to the carbonyl group. This epimerization allows for an equilibration between the *cis,trans*-isomeric cyclohexanones and causes the *trans* configuration of the major product.

dition step of the reaction initially leads to the conjugate base of the reaction product. Protonation subsequently gives the product in its neutral and more stable form. The Michael addition is named after the American chemist Arthur Michael.

Acceptor-substituted alkenes that are employed as substrates in Michael additions include α,β-unsaturated ketones (for example, see Figure 10.59), α,β-unsaturated esters (Figure 10.60), and α,β-unsaturated nitriles (Figure 10.61). The corresponding reaction products are bifunctional compounds with C=O and/or C≡N bonds in positions 1 and 5. Analogous reaction conditions allow Michael additions to vinyl sulfones or nitroalkenes. These reactions lead to sulfones and nitro compounds that carry a C=O and/or a C≡N bond at the C4 carbon.

Beyond the scope discussed so far, Michael additions also include additions of stoichiometrically generated enolates of ketones, SAMP or RAMP hydrazones, or esters to the C=C double bond of α,β-unsaturated ketones and α,β-unsaturated esters. These Michael additions convert one kind of enolate into another. The driving force stems from the C—C bond formation, not from differential stabilities of the enolates. It is important that the addition of the preformed enolate to the Michael acceptor is faster than the addition of the resulting enolate to another

Fig. 10.60. Michael addition to an α,β-unsaturated ester.

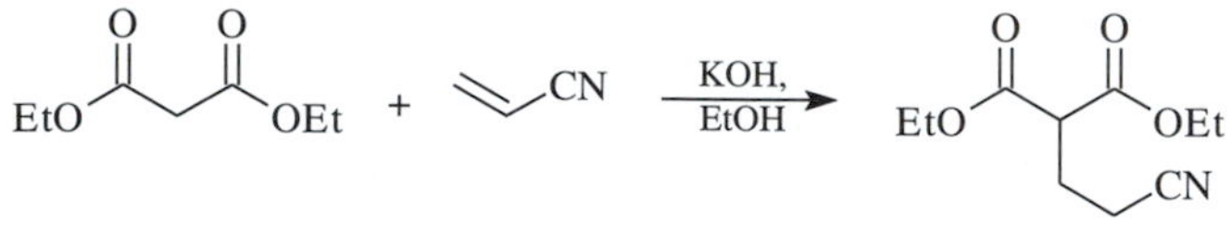

Fig. 10.61. Michael addition to an α,β-unsaturated nitrile.

molecule of the Michael acceptor. If that reactivity order were not true, an anionic polymerization of the Michael acceptor would occur. In many Michael additions, however, the enolate created is more hindered sterically than the enolate employed as the starting material, and in these cases Michael additions are possible without polymerization.

10.6.2 Tandem Reactions Consisting of Michael Addition and Consecutive Reactions

B

If a Michael addition of an enolate forms a *ketone enolate* as the primary reaction product, this enolate will be almost completely protonated to give the respective ketone. The reaction medium is of course still basic, since it still contains OH^- or RO^- ions. The Michael adduct, a ketone, is therefore reversibly deprotonated to a small extent.

This deprotonation may reform the ketone enolate that was the intermediate en route to the Michael adduct. However, the regioisomeric ketone enolate also can be formed. Figures 10.62–10.64 show such enolate isomerizations **B** $\rightarrow$ **D,** which proceed via the intermediacy of a neutral Michael adduct **C.** This neutral adduct is a 1,5-diketone in Figure 10.62, a δ-ketoaldehyde in Figure 10.63, and a δ-ketoester in Figure 10.64.

The new enolate carbon is located in position 6 of intermediate **D.** In this numbering scheme, position 1 is the C═O double bond of the keto group (Figure 10.62), the aldehyde group (Figure 10.63), or the ester group (Figure 10.64). Because of the dis-

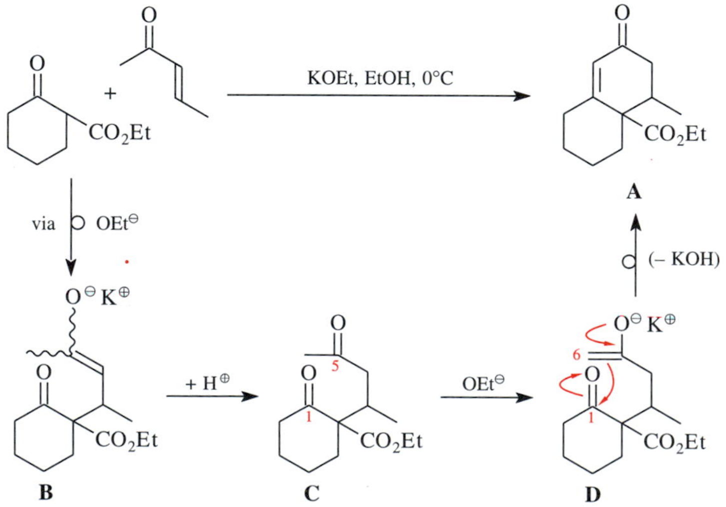

Fig. 10.62. Tandem reaction I, consisting of a Michael addition and an aldol condensation: Robinson annulation reaction for the synthesis of six-membered rings that are condensed to an existing ring.

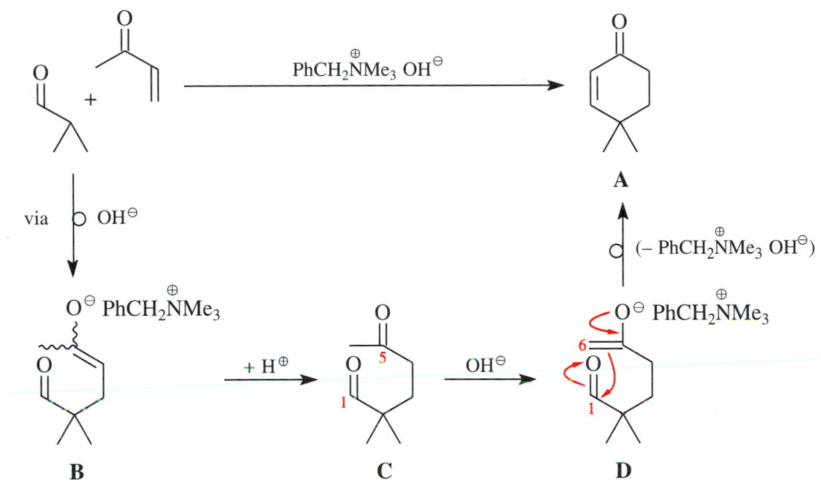

Fig. 10.63. Tandem reaction II, consisting of a Michael addition and an aldol condensation.

tance between the enolate position and the C=O double bond, a bond might form between C1 and C6:

- Enolate **D** of Figure 10.62 can undergo an aldol reaction with the C=O double bond of the ketone. The bicyclic compound **A** is formed as the condensation product. It is often possible to combine the formation and the consecutive reaction of a Michael adduct in a one-pot reaction. The overall reaction then is an annulation of a cyclohexanone to an enolizable ketone. The reaction sequence of Figure 10.62 is the **Robinson annulation,** an extraordinarily important synthesis of six-membered rings.

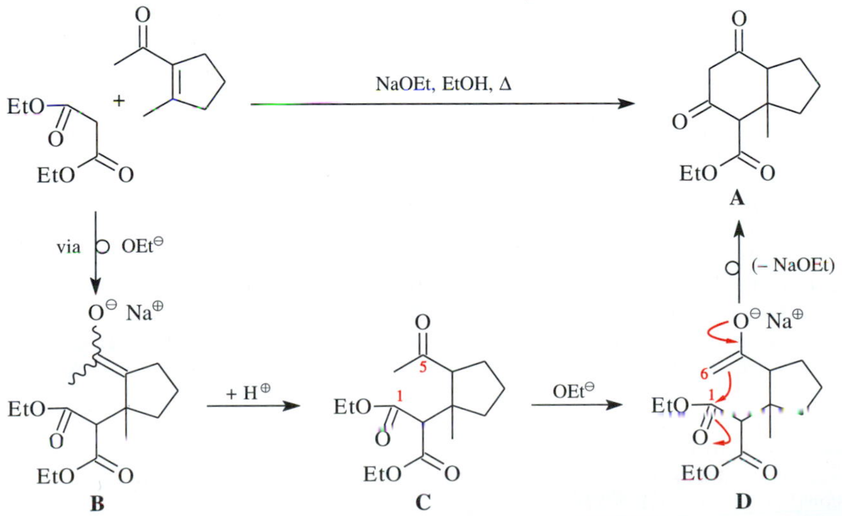

Fig. 10.64. Tandem reaction, consisting of a Michael addition and an enolate acylation. The major tautomer of the reaction product is not shown.

- Enolate **D** of Figure 10.63 undergoes an aldol condensation with the C═O double bond. The bicyclic compound **A** is the condensation product. This reaction represents a six-membered ring synthesis even though it is not a six-ring annulation.
- Enolate **D** of Figure 10.64 is acylated by the ester following the usual mechanism. The bicyclic compound **A** is a product, which contains a new six-membered ring that has been annulated to an existing ring.

References

10.1

H. B. Mekelburger and C. S. Wilcox, "Formation of Enolates," in *Comprehensive Organic Synthesis* (B. M. Trost, I. Fleming, Eds.), Vol. 2, 99, Pergamon Press, Oxford, **1991.**

C. H. Heathcock, "Modern Enolate Chemistry: Regio- and Stereoselective Formation of Enolates and the Consequence of Enolate Configuration on Subsequent Reactions," in *Modern Synthetic Methods* (R. Scheffold, Ed.), Vol. 6, 1, Verlag Helvetica Chimica Acta, Basel, Switzerland, **1992.**

I. Kuwajima and E. Nakamura, "Reactive enolates from enol silyl ethers," *Acc. Chem. Res.* **1985,** 18, 181.

D. Seebach, "Structure and reactivity of lithium enolates: From pinacolone to selective C-alkylations of peptides. Difficulties and opportunities afforded by complex structures," *Angew. Chem., Int. Ed. Engl.* **1988,** 27, 1624.

L. M. Jackman and J. Bortiatynski, "Structures of Lithium Enolates and Phenolates in Solution," in *Advances in Carbanion Chemistry* (V. Snieckus, Ed.), Vol. 1, 45, Jai Press Inc., Greenwich, **1992.**

10.2

A. C. Cope, H. L. Holmes, H. O. House, "The alkylation of esters and nitriles," *Org. React.* **1957,** *9,* 107–331.

D. Caine, "Alkylations of Enols and Enolates," in *Comprehensive Organic Synthesis* (B. M. Trost, I. Fleming, Eds.), Vol. 3, 1, Pergamon Press, Oxford, **1991.**

G. Frater, "Alkylation of Ester Enolates," in *Stereoselective Synthesis* (Houben-Weyl) 4th ed. 1996, (G. Helmchen, R. W. Hoffmann, J. Mulzer, E. Schaumann, Eds.), **1996,** Vol. E21 (Workbench Edition), *2,* 723–790, Georg Thieme Verlag, Stuttgart.

H.-E. Högberg, "Alkylation of Amide Enolates," in *Stereoselective Synthesis* (Houben-Weyl) 4th ed. 1996, (G. Helmchen, R. W. Hoffmann, J. Mulzer, E. Schaumann, Eds.), **1996,** Vol. E21 (Workbench Edition), 2, 791–915, Georg Thieme Verlag, Stuttgart.

T. Norin, "Alkylation of Ketone Enolates," in *Stereoselective Synthesis* (Houben-Weyl) 4th ed. 1996, (G. Helmchen, R. W. Hoffmann, J. Mulzer, E. Schaumann, Eds.), **1996,** Vol. E21 (Workbench Edition), *2,* 697–722, Georg Thieme Verlag, Stuttgart.

P. Fey, "Alkylation of Azaenolates from Imines," in *Stereoselective Synthesis* (Houben-Weyl) 4th ed. 1996, (G. Helmchen, R. W. Hoffmann, J. Mulzer, E. Schaumann, Eds.), **1996,** Vol. E21 (Workbench Edition), *2,* 973–993, Georg Thieme Verlag, Stuttgart.

P. Fey, "Alkylation of Azaenolates from Hydrazones," in *Stereoselective Synthesis* (Houben-Weyl) 4th ed. 1996, (G. Helmchen, R. W. Hoffmann, J. Mulzer, E. Schaumann, Eds.), **1996,** Vol. E21 (Workbench Edition), *2,* 994–1015, Georg Thieme Verlag, Stuttgart.

N. Petragnani and M. Yonashiro, "The reactions of dianions of carboxylic acids and ester enolates," *Synthesis* **1982,** 521.

D. A. Evans, "Studies in asymmetric synthesis—The development of practical chiral enolate synthons," *Aldrichimica Acta* **1982,** 15, 23.

D. A. Evans, "Stereoselective Alkylation Reactions of Chiral Metal Enolates," in *Asymmetric Syn-*

thesis—Stereodifferentiating Reactions, Part B (J. D. Morrison, Ed.), Vol. 3, 1, AP, New York, **1984.**

D. Enders, L. Wortmann, R. Peters, "Recovery of carbonyl compounds from *N,N*-dialkylhydrazones," *Acc. Chem. Res.* **2000,** *33,* 157–169.

J. E. McMurry, "Ester cleavages via S_N2-type dealkylation," *Org. React.* **1976,** *24,* 187–224.

A. J. Kresge, "Ingold lecture: Reactive intermediates: Carboxylic acid enols and other unstable species," *Chem. Soc. Rev.* **1996,** 25, 275–280.

10.3

D. A. Evans, J. V. Nelson, T. R. Taber, "Stereoselective aldol condensations," *Top. Stereochem.* **1982,** 13, 1.

C. H. Heathcock, "The Aldol Addition Reaction," in *Asymmetric Synthesis—Stereodifferentiating Reactions, Part B* (J. D. Morrison, Ed.), Vol. 3, 111, Academic Press, New York, **1984.**

M. Braun, "Recent Developments in Stereoselective Aldol Reactions," in *Advances in Carbanion Chemistry* (V. Snieckus, Ed.), Vol. 1, 177, Jai Press Inc., Greenwich, CT, **1992.**

M. Braun, L. S. Liebeskind, J. S. McCallum, W.-D. Fessner, "Formation of C—C Bonds by Addition to Carbonyl Groups (C=O)—Enolates," in *Methoden Org. Chem. (Houben-Weyl) 4th ed. 1952, Stereoselective Synthesis* (G. Helmchen, R. W. Hoffmann, J. Mulzer, E. Schaumann, Eds.), Vol. E21b, 1603, Georg Thieme Verlag, Stuttgart, **1995.**

C. H. Heathcock, "The Aldol Reaction: Group I and Group II Enolates," in *Comprehensive Organic Synthesis* (B. M. Trost, I. Fleming, Eds.), Vol. 2, 181, Pergamon Press, Oxford, **1991.**

C. H. Heathcock, "Modern Enolate Chemistry: Regio- and Stereoselective Formation of Enolates and the Consequence of Enolate Configuration on Subsequent Reactions," in *Modern Synthetic Methods* (R. Scheffold, Ed.), Vol. 6, 1, Verlag Helvetica Chimica Acta, Basel, Switzerland, **1992.**

10.4

G. Jones, "The Knoevenagel condensation," *Org. React.* **1967,** *15,* 204–599.

A. T. Nielsen and W. J. Houlihan, "The Aldol condensation," *Org. React.* **1968,** *15,* 1–438.

L. F. Tietze and U. Beifuss, "The Knoevenagel Reaction," in *Comprehensive Organic Synthesis* (B. M. Trost, I. Fleming, Eds.), Vol. 2, 341, Pergamon Press, Oxford, **1991.**

M. Braun, "Syntheses with Aliphatic Nitro Compounds," in *Organic Synthesis Highlights* (J. Mulzer, H.-J. Altenbach, M. Braun, K. Krohn, H.-U. Reißig, Eds.), VCH, Weinheim, New York, **1991,** 25–32.

10.5

B. R. Davis and P. J. Garratt, "Acylation of Esters, Ketones and Nitriles," in *Comprehensive Organic Synthesis* (B. M. Trost, I. Fleming, Eds.), Vol. 2, 795, Pergamon Press, Oxford, **1991.**

T. H. Black, "Recent progress in the control of carbon versus oxygen acylation of enolate anions," *Org. Prep. Proced. Int.* **1988,** 21, 179–217.

S. Benetti, R. Romagnoli, C. De Risi, G. Spalluto, V. Zanirato, "Mastering ß-keto esters," *Chem. Rev.* **1995,** 95, 1065-1115.

10.6

Y. Yamamoto, S. G. Pyne, D. Schinzer, B. L. Feringa, J. F. G. A. Jansen, "Formation of C–C Bonds by Reactions Involving Olefinic Double Bonds—Addition to α,β-Unsaturated Carbonyl Compounds (Michael-Type Additions)," in *Methoden Org. Chem. (Houben-Weyl) 4th ed.* **1952,** *Stereoselective Synthesis* (G. Helmchen, R. W. Hoffmann, J. Mulzer, E. Schaumann, Eds.), Vol. E21b, 2041, Georg Thieme Verlag, Stuttgart, **1995.**

P. Perlmutter, "Conjugate Addition Reactions in Organic Synthesis," Pergamon Press, Oxford, U.K., **1992.**

A. Bernardi, "Stereoselective conjugate addition of enolates to α,β-unsaturated carbonyl compounds," Gazz. *Chim. Ital.* **1995,** 125, 539–547.

R. D. Little, M. R. Masjedizadeh, O. Wallquist, J. I. McLoughlin, "The intramolecular Michael reaction," *Org. React.* **1995,** 47, 315–552.

D. A. Oare and C. H. Heathcock, "Stereochemistry of the base-promoted Michael addition reaction," *Top. Stereochem.* **1989,** 19, 227–407.

J. A. Bacigaluppo, M. I. Colombo, M. D. Preite, J. Zinczuk, E. A. Ruveda, "The Michael-Aldol condensation approach to the construction of key intermediates in the synthesis of terpenoid natural products," *Pure Appl. Chem.* **1996,** 68, 683.

Further Reading

D. B. Collum, "Solution structures of lithium dialkylamides and related *N*-lithiated species: Results from ^{6}Li-^{15}N double labeling experiments," *Acc. Chem. Res.* **1993,** 26, 227–234.

P. Brownbridge, "Silyl enol ethers in synthesis," *Synthesis* **1983,** 85.

J. M. Poirier, "Synthesis and reactions of functionalized silyl enol ethers," *Org. Prep. Proced. Int.* **1988,** 20, 317–369.

H. E. Zimmerman, "Kinetic protonation of enols, enolates, analogues": The stereochemistry of ketonisation," *Acc. Chem. Res.* **1987,** 20, 263.

D. Seebach and A. R. Sting, M. Hoffmann, "Self-regeneration of stereocenters (SRS)—Applications, limitations, and abandonment of a synthetic principle," *Angew. Chem.* **1996,** 108, 2880–2921; *Angew. Chem. Int. Ed. Engl.* **1997,** 35, 2708–2748.

F. A. Davis and B. C. Chen, "Formation of C—O Bonds by Oxygenation of Enolates," in *Methoden Org. Chem. (Houben-Weyl) 4th ed. 1952, Stereoselective Synthesis* (G. Helmchen, R. W. Hoffmann, J. Mulzer, E. Schaumann, Eds.), Vol. E21e, 4497, Georg Thieme Verlag, Stuttgart, **1995.**

P. Fey and W. Hartwig, "Formation of C—C Bonds by Addition to Carbonyl Groups (C=O)—Azaenolates or Nitronates," in *Methoden Org. Chem. (Houben-Weyl) 4th ed. 1952, Stereoselective Synthesis* (G. Helmchen, R. W. Hoffmann, J. Mulzer, E. Schaumann, Eds.), Vol. E21b, 1749, Georg Thieme Verlag, Stuttgart, **1995.**

K. Krohn, "Stereoselective Reactions of Cyclic Enolates," in *Organic Synthesis Highlights* (J. Mulzer, H.-J. Altenbach, M. Braun, K. Krohn, H.-U. Reißig, Eds.), VCH, Weinheim, New York, **1991,** 9–13.

K. F. Podraza, "Regiospecific alkylation of cyclohexenones. A review," *Org. Prep. Proced. Int.* **1991,** 23, 217–235.

C. M. Thompson and D. L. C. Green, "Recent advances in dianion chemistry," *Tetrahedron* **1991,** 47, 4223–4285.

B. M. Kim, S. F. Williams, S. Masamune, "The Aldol Reaction: Group III Enolates," in *Comprehensive Organic Synthesis* (B. M. Trost, I. Fleming, Eds.), Vol. 2, 239, Pergamon Press, Oxford, **1991.**

M. Sawamura and Y. Ito, "Asymmetric Carbon-Carbon Bond Forming Reactions: Asymmetric Aldol Reactions," in *Catalytic Asymmetric Synthesis* (I. Ojima, Ed.), 367, VCH, New York, **1993.**

A. S. Franklin and I. Paterson, "Recent developments in asymmetric aldol methodology," *Contemporary Organic Synthesis* **1994,** 1, 317.

C. J. Cowden and I. Paterson, "Asymmetric aldol reactions using boron enolates," *Org. Prep. Proced. Int.* **1997,** 51, 1–200.

A. Fürstner, "Recent advancements in the Reformatsky reaction," *Synthesis* **1989,** 8, 571–590.

S. M. Luk'yanov and A. V. Koblik, "Acid-catalyzed acylation of carbonyl compounds compounds," *Russ. Chem. Rev.* **1996,** 65, 1–26.

B. C. Chen, "Meldrum's acid in organic synthesis," *Heterocycles* **1991,** 32, 529–597.

G. W. Kabalka and R. M. Pagni, "Organic reactions on alumina," *Tetrahedron* **1997,** 53, 7999–8064.

C. F. Bernasconi, "Nucleophilic addition to olefins. Kinetics and mechanism," *Tetrahedron* **1989,** 45, 4017–4090.

Rearrangements

11

The term "rearrangement" is used to describe two different types of organic chemical reactions. A rearrangement may involve the *one-step* migration of an H atom or of a larger molecular fragment within a relatively short-lived intermediate. On the other hand, a rearrangement may be a *multistep reaction* that includes the migration of an H atom or of a larger molecular fragment as one of its steps. The Wagner–Meerwein rearrangement of a carbenium ion (Section 11.3.1) exemplifies a rearrangement of the first type. Carbenium ions are so short-lived that neither the starting material nor the primary rearrangement product can be isolated. The Claisen rearrangement of allyl alkenyl ethers also is a one-step rearrangement (Section 11.5). In contrast to the Wagner–Meerwein rearrangement, however, both the starting material and the product of the Claisen rearrangement are molecules that can be isolated. The ring expansion reaction shown in Figure 11.23 and the alkyne synthesis depicted in Figure 11.29 are examples of multistep rearrangement reactions.

B

11.1 Nomenclature of Sigmatropic Shifts

B

In many rearrangements, the migrating group connects to one of the direct neighbors of the atom to which it was originally attached. Rearrangements of this type are the so-called **[1,2]-rearrangements** or **[1,2]-shifts.** These rearrangements can be considered as **sigmatropic processes,** the numbers "1" and "2" characterizing the subclass to which they belong. The adjective "sigmatropic" emphasizes that a σ-bond migrates in these reactions. How far it migrates is described by specifying the positions of the atoms between which the bond is shifted. The atoms that are initially bonded are assigned positions 1 and $1'$. The subsequent atoms in the direction of the σ-bond migration are labeled 2, 3, and so forth, on the side of center 1 and labeled $2'$, $3'$, and so forth, on the side of center $1'$. After the rearrangement, the σ bond connects two atoms in positions n and m'. The rearrangement can now be characterized by the positional numbers n and m' in the following way: the numbers are written between brackets, separated by a comma, and the primed number is given without the prime. Hence, an $[n,m]$-rearrangement is the most general description of a sigmatropic process. A [1,2]-rearrangement is the special case with $n = 1$ and $m' = 2$ (Figure 11.1). [3,3]-Rearrangements occur when $n = m' = 3$ (Figure 11.2). Many other types of rearrangements are known, including [1,3]-, [1,4]-, [1,5]-, [1,7]-, [2,3]-, and [5,5]-rearrangements.

In this chapter we will be dealing primarily with [1,2]-rearrangements. In addition, the most important [3,3]-rearrangements, namely, the **Claisen** and the **Claisen–Ireland rearrangements,** will be discussed.

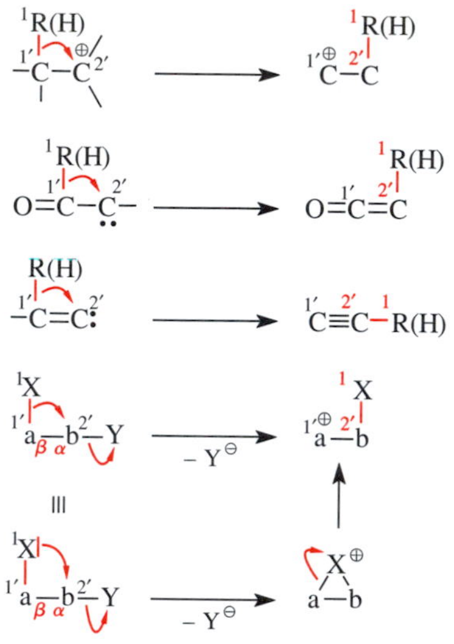

Fig. 11.1. The three reactions on top show [1,2]-rearrangements to a sextet carbon. The two reactions at the bottom show [1,2]-rearrangements to a neighboring atom that is coordinatively saturated but in the process of losing a leaving group.

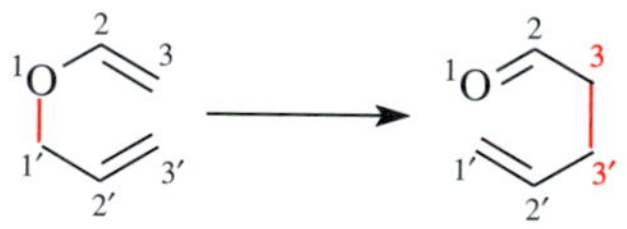

Fig. 11.2. A Claisen rearrangement as an example of a [3,3]-rearrangement.

11.2 Molecular Origins for the Occurrence of [1,2]-Rearrangements

B

Figure 11.1 shows the structures of the immediate precursors of one-step [1,2]-sigmatropic rearrangements. These formulas reveal two different reasons for the occurrence of rearrangements in organic chemistry. Rows 1–3 of Figure 11.1 reveal the first reason for [1,2]-rearrangements to take place, namely, the occurrence of a valence electron sextet at one of the C atoms of the substrate. This sextet may be located at the C^+ of a carbenium ion or at the C: of a carbene. Carbenium ions are extremely reactive species. If there exists no good opportunity for an intermolecular reaction (i.e., no good possibility for stabilization), carbenium ions often undergo an intramolecular reaction. This intramolecular reaction in many cases is a [1,2]-rearrangement.

Suppose a valence electron sextet occurs at a carbon atom and the possibility exists for a [1,2]-rearrangement to occur. The thermodynamic driving force for the potential [1,2]-rearrangement will be *significant* if the rearrangement leads to a structure with octets on all atoms. It is for this reason that acylcarbenes rearrange quantitatively to give ketenes (row 2 in Figure 11.1) and that vinylcarbenes rearrange quantitatively to give acetylenes (row 3 in Figure 11.1). In contrast, another valence electron sextet species is formed if the [1,2]-rearrangement of a carbenium ion leads to another carbenium ion. Accordingly, the driving force of a [1,2]-rearrangement of a carbenium ion is much smaller than the driving force of a [1,2]-rearrangement of a carbene. The following rules of thumb summarize all cases for which nonetheless quantitative carbenium ion rearrangements are possible.

[1,2]-Rearrangements of carbenium ions occur quantitatively only

- if the new carbenium ion is substantially better stabilized electronically by its substituents than the old carbenium ion.
- if the new carbenium ion is substantially more stable than the old carbenium ion because of other effects such as reduced ring strain.
- or if the new carbenium ion is captured in a subsequent, irreversible reaction.

Notwithstanding these cases, many [1,2]-rearrangements of carbenium ions occur reversibly because of the small differences in the free enthalpies.

Consideration of rows 4 and 5 of Figure 11.1 suggests a second possible cause for the occurrence of [1,2]-rearrangements. In those cases the substrates contain a b—Y bond. The heterolysis of this bond would lead to a reasonably stable leaving group Y^- but would also produce a cation b^+ with a sextet. Such a heterolysis would be possible only—even if assisting substituents were present—if the b—Y bond was a C—Y bond and the product of heterolysis was a carbenium ion. If the b—Y bond were an N—Y or an O—Y bond, the heterolyses would generate nitrenium ions $R^1R^2N^+$ and oxenium ions RO^+, respectively. Neither heterolysis has ever been observed. Nitrenium and oxenium ions have much higher heats of formation than carbenium ions because the central atoms nitrogen and oxygen are substantially more electronegative than carbon.

Heterolyses of N—Y and O—Y bonds are, however, not *entirely* unknown. This is because these heterolyses can occur concomitantly with a [1,2]-rearrangement. In a way, such a [1,2]-rearrangement presents a "preventive measure to avoid the formation of an unstable valence electron sextet." The b—Y bond thus undergoes heterolysis and releases Y^-. However, the formation of an electron sextet at center b (= NR or O) is avoided because center b shares another electron pair by way of binding to another center in the molecule. This new electron pair may be in either the β or the γ position relative to the position of the leaving group Y. If the b center binds to an electron pair in the β position relative to the leaving group Y (row 4 in Figure 11.1), this electron pair is the bonding electron pair of the σ_{a-X} bond in the substrate. On the other hand, if the b center binds to an electron pair in the γ position relative to the leaving group Y (row 5 in Figure 11.1), then this electron pair is a nonbonding lone pair of the X group attached to the β position. The engagement of this free electron pair leads to the formation of a positively charged three-membered ring. The substituent X exerts a so-called neighboring group effect if subsequently this b—X bond is again broken (Section 2.7). On the other hand, a [1,2]-rearrangement occurs if the a—X bond of the three-membered ring is broken. Such a "neighboring-group-effect initiated [1,2]-rearrangement" was discussed in connection with the deuteration experiment of Figure 2.22.

11.3 [1,2]-Rearrangements in Species with a Valence Electron Sextet

11.3.1 [1,2]-Rearrangements of Carbenium Ions

Wagner–Meerwein Rearrangements

B

Wagner–Meerwein rearrangements are [1,2]-rearrangements of H atoms or alkyl groups in carbenium ions that do not contain any heteroatoms attached to the valence-unsaturated center C1 or to the valence-saturated center C2. The actual rearrangement step consists of a reaction that cannot be carried out separately because both the starting material and the product are extremely short-lived carbenium ions that cannot be isolated. Wagner–Meerwein rearrangements therefore occur only as part of a reaction sequence in which a carbenium ion is generated in one or more steps, and the rearranged carbenium ion reacts further in one or several steps to give a valence-saturated compound. The sigmatropic shift of the Wagner–Meerwein rearrangement therefore can be embedded between a great variety of carbenium-ion-generating and carbenium-ion-annihilating reactions (Figures 11.3–11.11).

In Section 5.2.5, we discussed the Friedel–Crafts alkylation of benzene with 2-chloropentane. This reaction includes a Wagner–Meerwein reaction in conjunction with other elementary reactions. The Lewis acid catalyst $AlCl_3$ first converts the chloride into the 2-pentyl cation **A** (Figure 11.3). Cation **A** then rearranges into the isomeric 3-pentyl cation **B,** in part or perhaps to the extent that the equilibrium ratio is reached. The new carbenium ion **B** is not significantly more stable than the original one (**A**),

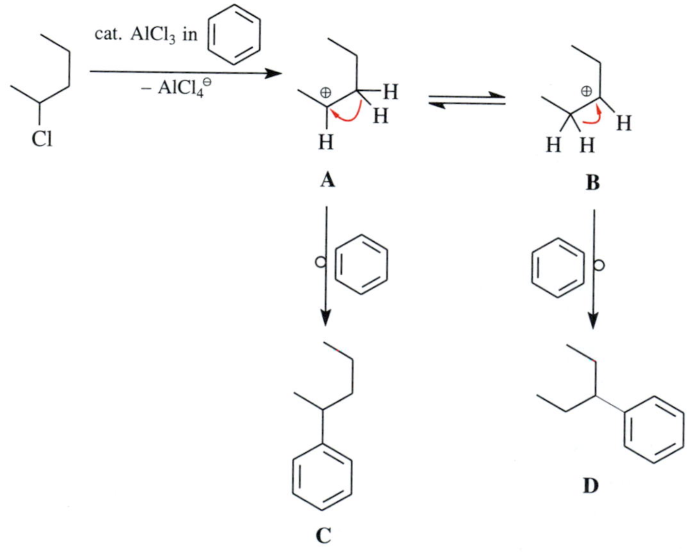

cat. $AlCl_3$ in

$- AlCl_4^{\ominus}$

Fig. 11.3. Mechanism of an Ar-S_E reaction (details: Section 5.2.5), which includes a reversible Wagner–Meerwein rearrangement.

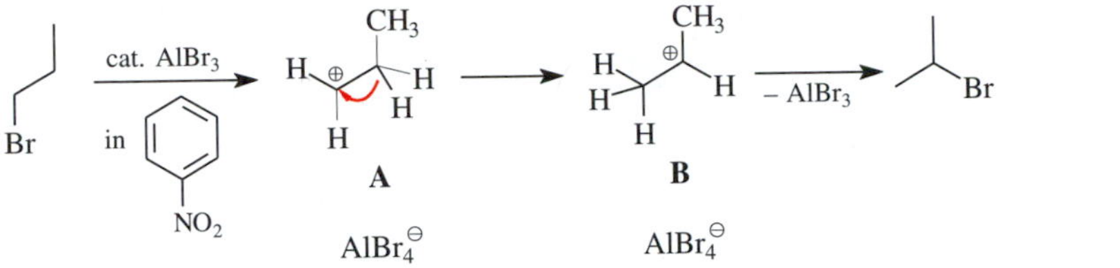

Fig. 11.4. Wagner–Meerwein rearrangement in the isomerization of an alkyl halide.

but it also is not significantly less stable. In addition, both the cations **A** and **B** are relatively unhindered sterically, and each can engage in an Ar-S_E reaction with a comparable rate of reaction. Thus, aside from the alkylation product **C** with its unaltered alkyl group, the isomer **D** with the isomerized alkyl group also is formed.

A Wagner–Meerwein rearrangement can be part of the isomerization of an alkyl halide (Figure 11.4). For example, 1-bromopropane isomerizes quantitatively to 2-bromopropane under Friedel–Crafts conditions. The [1,2]-shift **A** → **B** involved in this reaction again is an H-atom shift. In contrast to the thermoneutral isomerization between carbenium ions **A** and **B** of Figure 11.3, in the present case an energy gain is associated with the formation of a secondary carbenium ion from a primary carbenium ion. Note, however, that the different stabilities of the carbenium ions are not responsible for the complete isomerization of 1-bromopropane into 2-bromopropane. The position of this isomerization equilibrium is determined by thermodynamic control at the level of the alkyl halides. 2-Bromopropane is more stable than 1-bromopropane and therefore formed exclusively.

There also are Wagner–Meerwein reactions in which alkyl groups migrate rather than H atoms (Figures 11.3 and 11.4). Of course, these reactions, too, are initiated by carbenium-ion-generating reactions, as exemplified in Figure 11.5 for the case of an E1 elimination from an alcohol (Section 4.5). The initially formed neopentyl cation—a primary carbenium ion—rearranges into a tertiary carbenium ion, thereby gaining considerable stabilization. An elimination of a β-H-atom is possible only after the rearrangement has occurred. It terminates the overall reaction and provides an alkene (Saytzeff product).

The sulfuric acid catalyzed transformation of pinanic acid into abietic acid shown in Figure 11.6 includes a Wagner–Meerwein shift of an alkyl group. The initially formed carbenium ion, the secondary carbenium ion **A,** which is a localized carbenium ion, is generated by protonation of one of the C=C double bonds. A [1,2]-sigmatropic shift

A

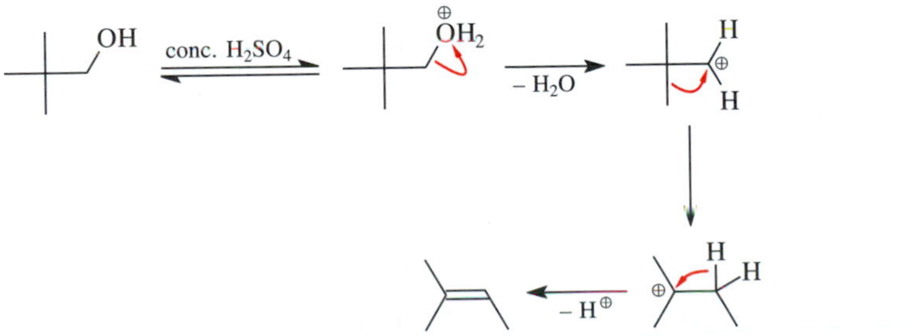

Fig. 11.5. Wagner–Meerwein rearrangement as part of an isomerizing E1 elimination.

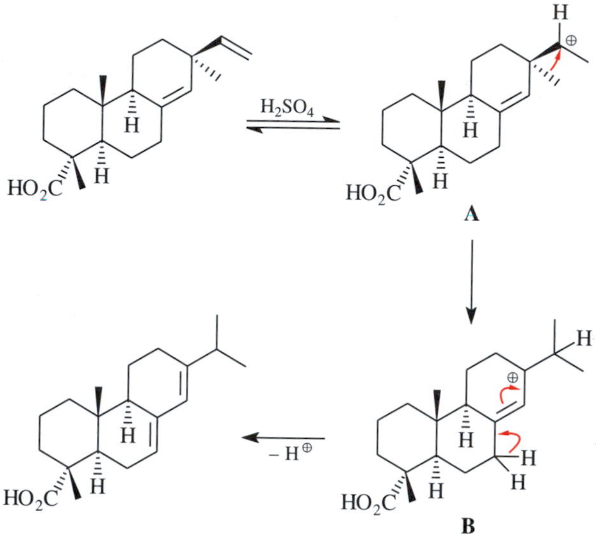

Fig. 11.6. Wagner–Meerwein rearrangement as part of an alkene isomerization.

of a methyl group occurs in **A,** and the much more stable, delocalized, and tetraalkyl-substituted allyl cation **B** is formed. Cation **B** is subsequently deprotonated and a 1,3-diene is obtained. Overall, Figure 11.6 shows the isomerization of a less stable diene into a more stable diene. The direction of this isomerization is determined by thermodynamic control. The product 1,3-diene is conjugated and therefore more stable than the unconjugated substrate diene.

In the carbenium ion **A** of Figure 11.6, there are three different alkyl groups in α positions with respect to the carbenium ion center, and each one could in principle undergo the [1,2]-rearrangement. Yet, only the migration of the methyl group is observed. Presumably, this is the consequence of product-development control. The migration of either one of the other two alkyl groups would have resulted in the formation of a seven-membered and therefore strained ring. Only the observed methyl shift **A** → **B** retains the energetically advantageous six-membered ring skeleton.

B

Even in rearrangements of carbenium ions that show no preference for the migration of a particular group which would be based on thermodynamic control or on product-development control one can observe chemoselectivity. This is because certain potentially migrating groups exhibit different intrinsic tendencies toward such a migration: in Wagner–Meerwein rearrange- ments, and in many other [1,2]-migrations, tertiary alkyl groups migrate faster than secondary, secondary alkyl groups migrate faster than primary, and primary alkyl groups in turn migrate faster than methyl groups. It is for this ordering that cation **A** in Figure 11.7 rearranges into cation **B** by way of a C_{tert} migration rather than into the cation **C** via a C_{prim} migration. Both cations **B** and **C** are secondary carbenium ions, and both are bicyclo[2.2.1]heptyl cations; thus they can be expected to be comparable in stability. If there were no intrinsic migratory preference of the type $C_{tert} > C_{prim}$, one would have expected the formation of comparable amounts of **B** and **C.**

The [1,2]-alkyl migration **A** → **B** of Figure 11.7 converts a cation with a well-stabilized tertiary carbenium ion center into a cation with a less stable secondary carbenium ion

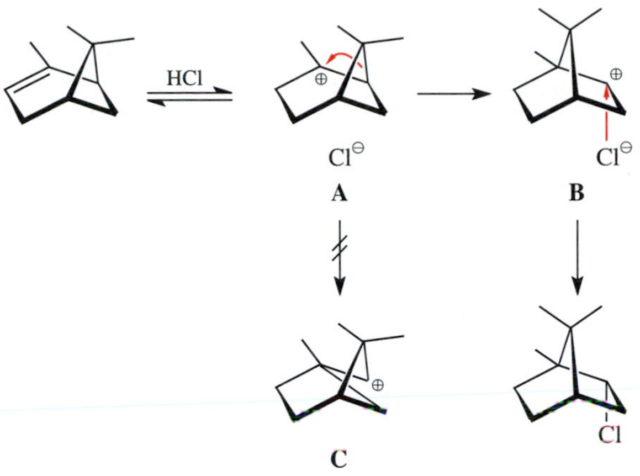

Fig. 11.7. Wagner–Meerwein rearrangement as part of an HCl addition to a C=C double bond.

center. This is possible only because of the driving force that is associated with the reduction of ring strain: a cyclobutyl derivative **A** is converted into a cyclopentyl derivative **B**.

Wagner–Meerwein Rearrangements in the Context of Tandem and Cascade Rearrangements

A

A carboxonium ion (an all-octet species) may become less stable than a carbenium ion (a sextet species) only when ring-strain effects dominate. In such cases carbenium ions can be generated from carboxonium ions by way of a Wagner–Meerwein rearrangement. Thus, the decrease of ring strain can provide a driving force strong enough to overcompensate for the conversion of a more stable into a less stable cationic center. In Figure 11.8, for example, the carboxonium ion **A** rearranges into the carbenium ion **B** because of the release of cyclobutane strain (about 26 kcal/mol) in the formation of the cyclopentane (ring strain of about 5 kcal/mol). Cation **B** stabilizes itself by way of another [1,2]-rearrangement. The resulting cation **C** has comparably little ring strain but is an

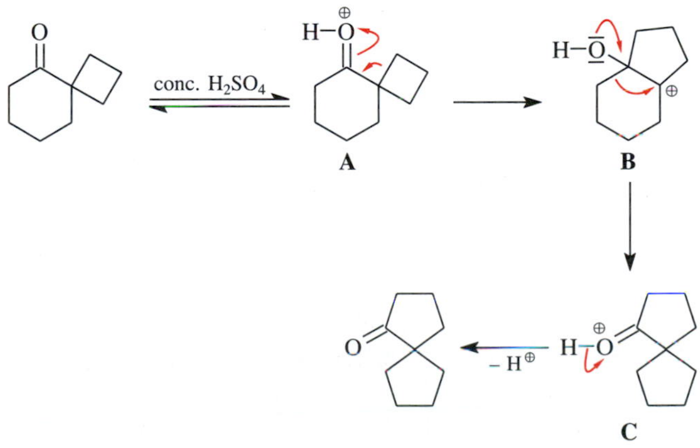

Fig. 11.8. Tandem rearrangement comprising a Wagner–Meerwein rearrangement and a semipinacol rearrangement.

electronically favorable carboxonium ion. The pinacol and semipinacol rearrangements (see below) include [1,2]-shifts that are just like the second [1,2]-shift of Figure 11.8, the only difference being that the β-hydroxylated carbenium ion intermediates analogous to **B** are generated in a different manner.

Camphorsulfonic acid is generated by treatment of camphor with concentrated sulfuric acid in acetic anhydride (Figure 11.9). Protonation of the carbonyl oxygen leads, in an equilibrium reaction, to the formation of a small amount of the carboxonium ion **A.** **A** undergoes a Wagner–Meerwein rearrangement into cation **B.** However, this rearrangement occurs only to a small extent, since an all-octet species is converted into an intermediate with a valence electron sextet and there is no supporting release of ring strain. Hence, the rearrangement **A** → **B** is an endothermic process. Nevertheless, the reaction ultimately goes *to completion* in this energy-consuming direction because the carbenium ion **B** engages in irreversible consecutive reactions.

Cation **B** is first deprotonated to give the hydroxycamphene derivative **C.** **C** is then electrophilically attacked by a reactive intermediate of unknown structure that is generated from sulfuric acid under these conditions. In the discussion of the sulfonylation of aromatic compounds (Figure 5.14), we have mentioned protonated sulfuric acid $H_3SO_4^+$ and its dehydrated derivative HSO_3^+ as potential electrophiles, which, accordingly, might assume the same role here, too. In any case, the attack of whatever H_2SO_4-based electrophile on the alkene **C** results in the formation of carbenium ion **E.**

A carbenium ion with a β-hydroxy group, **E** stabilizes itself by way of a carbenium ion → carboxonium ion rearrangement. Such a rearrangement occurs in the third step

Fig. 11.9. Preparation of optically active camphorsulfonic acid via a path involving a Wagner–Meerwein rearrangement (**A** → **B**) and a semipinacol rearrangement (**E** → **D**).

of the pinacol rearrangement (Figure 11.12) and also in many semipinacol rearrangements (Figures 11.17 and 11.19). The carbenium ion **E** therefore is converted into the carboxonium ion **D.** In the very last step, **D** is deprotonated and a ketone is formed. The final product of the rearrangement is camphorsulfonic acid.

While the sulfonation of C10 of camphor involves two [1,2]-rearrangements (Figure 11.9), the bromination of dibromocamphor involves even four of these shifts, namely, **A** → **B**, **B** → **D, I** → **H,** and **H** → **G** (Figure 11.10). Comparison of the mechanisms of sulfonylation (Figure 11.9) and bromination (Figure 11.10) reveals that the cations marked **B** in the two cases react in different ways. **B** undergoes an elimination in the sulfonylation (→ **C,** Figure 11.9) but not in the bromination (Figure 11.10). This difference is less puzzling than it might seem at first. In fact, it is likely that an elimination also occurs intermittently during the course of the bromination (Figure 11.10) but the reaction simply is inconsequential.

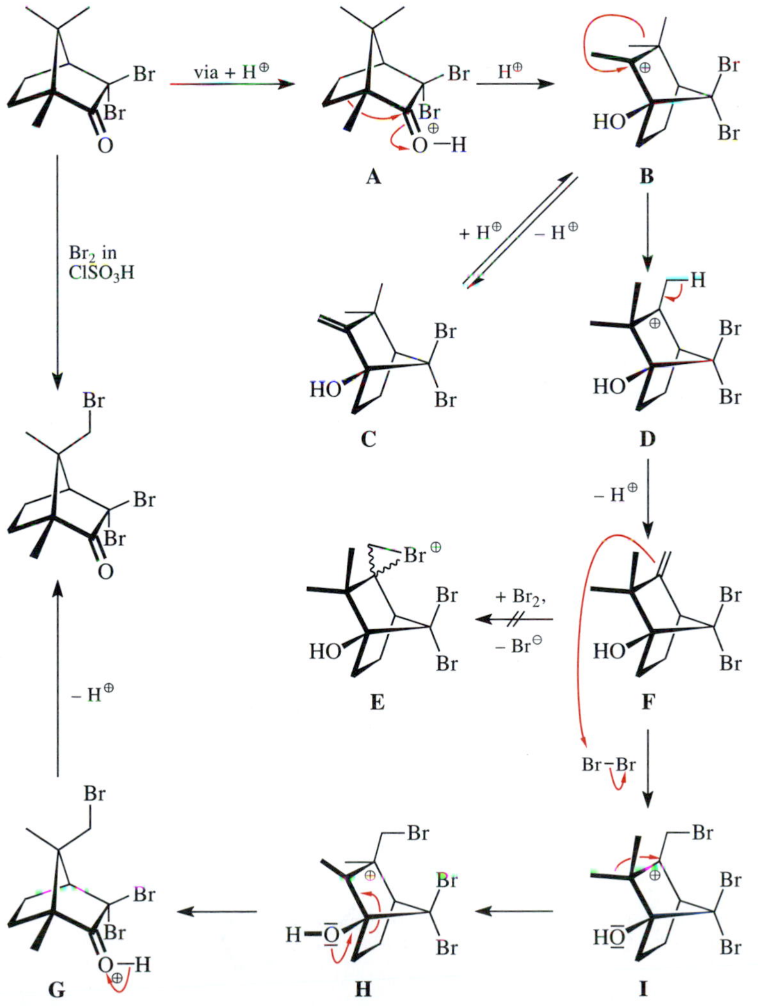

Fig. 11.10. Preparation of optically active tribromocamphor via a path involving three Wagner–Meerwein rearrangements (**A** → **B, B** → **D, I** → **E**) and a semipinacol rearrangement (**H** → **G**).

Molecular bromine, Br_2, is a weak electrophile and does not attack the alkene **C** fast enough. It is for this reason and in contrast to the sulfonylation (Figure 11.9) that the carbenium ion **B** of Figure 11.10, which is in equilibrium with alkene **C**, has sufficient time to undergo another Wagner–Meerwein reaction that converts the β-hydroxycarbenium ion **B** into the carbenium ion **D.** Ion **D** is more stable than **B** because the hydroxy group in **D** is in the γ position relative to the positive charge, while it is in the β position in **B;** that is, the destabilization of the cationic center by the electron-withdrawing group is reduced in **D** compared to **B.** The carbenium ion **D** is now deprotonated to give the alkene **F.** In contrast to **C,** alkene **F** reacts with Br_2. The bromination results in the formation of the bromosubstituted carbenium ion **I.** The bromination mechanism of Section 3.5.1 might rather have suggested the formation of the bromonium ion **E,** but this is not formed. It is known that open-chain intermediates of type **I** also may occur in brominations of C═C double bonds (see commentary in Section 3.5.1 regarding Figures 3.5 and 3.6). In the *present* case, the carbenium ion **I** is presumably formed because it is less strained than the putative polycyclic bromonium ion isomer **E.** In light of the reaction mechanism, one can now understand why Br_2 attacks **F** faster than **C.** The hydroxyl group is one position farther removed from the cationic center in the carbenium ion **I** that is generated from alkene **F** in comparison to the carbenium ion that would be formed by bromination of **C.** Hence, **I** is more stable than the other carbenium ion, so that the formation of **I** is favored by product-development control.

B

The reactions shown in Figures 11.9 and 11.10 exemplify a **tandem rearrangement** and a **cascade rearrangement,** respectively. These terms describe sequences of two or more rearrangements taking place more or less directly one after the other. Cascade rearrangements may involve even more Wagner–Meerwein rearrangements than the one shown in Figure 11.10. The rearrangement shown in Figure 11.11, for example, involves five [1,2]-rearrangements, each one effecting the conversion of a spiro-annulated cyclobutane into a fused cyclopentane.

Every polycyclic hydrocarbon having the molecular formula $C_{10}H_{16}$ can be isomerized to adamantane. The minimum number of [1,2]-rearrangements needed in such rearrangements is so high that it can be determined only with the use of a computer pro-

R OH
$SOCl_2$,
pyridine
R
R
R H N

Fig. 11.11. An E1 elimination involving five Wagner–Meerwein rearrangements.

gram. The rearrangements occur in the presence of catalytic amounts of $AlCl_3$ and *tert*-BuCl:

cat. $AlCl_3$,
cat. *tert*-BuCl

These isomerizations almost certainly involve [1,2]-shifts of H atoms as well as of alkyl groups. One cannot exclude that [1,3]-rearrangements may also play a role. The reaction product, adamantane, is formed under thermodynamic control under these conditions. It is the so-called **stabilomer** (the most stable isomer) of all the hydrocarbons having the formula $C_{10}H_{16}$.

This impressive cascade reaction begins with the formation of a small amount of the *tert*-butyl cation by reaction of $AlCl_3$ with *tert*-BuCl. The *tert*-butyl cation abstracts a hydride ion from the substrate $C_{10}H_{16}$. Thus, a carbenium ion with formula $C_{10}H_{15}^+$ is formed. These carbenium ions $C_{10}H_{15}^+$ are certainly substrates for Wagner–Meerwein rearrangements and also potentially substrates for [1,3]-rearrangements, thereby providing various isomeric cations *iso*-$C_{10}H_{15}^+$. Some of these cations can abstract a hydride ion from the neutral starting material $C_{10}H_{16}$. The saturated hydrocarbons *iso*-$C_{10}H_{16}$ obtained in this way are isomers of the original starting material $C_{10}H_{16}$. Such hydride transfers and [1,2]- and [1,3]-shifts, respectively, are repeated until the reaction eventually arrives at adamantane by way of the adamantyl cation.

Pinacol Rearrangement

B

Di-*tert*-glycols rearrange in the presence of acid to give α-tertiary ketones (Figure 11.12). The trivial name of the simplest glycol of this type is **pinacol,** and this type of reaction therefore is named pinacol rearrangement (in this specific case, the reaction is called a pinacol–pinacolone rearrangement). The rearrangement involves four steps. One of the hydroxyl groups is protonated in the first step. A molecule of water is eliminated in the second step, and a tertiary carbenium ion is formed. The carbenium ion rearranges in the third step into a more stable carboxonium ion via a [1,2]-rearrangement. In the last step, the carboxonium ion is deprotonated and the product ketone is obtained.

For asymmetric di-*tert*-glycols to rearrange under the same conditions to a single ketone, steps 2 and 3 of the overall reaction (Figure 11.12) must proceed chemoselec-

H_2SO_4
$-H_2O$
$-H^{\oplus}$
pinacol
pinacolone

Fig. 11.12. Mechanism of the pinacol rearrangement of a symmetric glycol. The reaction involves the following steps: (1) protonation of one of the hydroxyl groups, (2) elimination of one water molecule, (3) [1,2]-rearrangement, and (4) deprotonation.

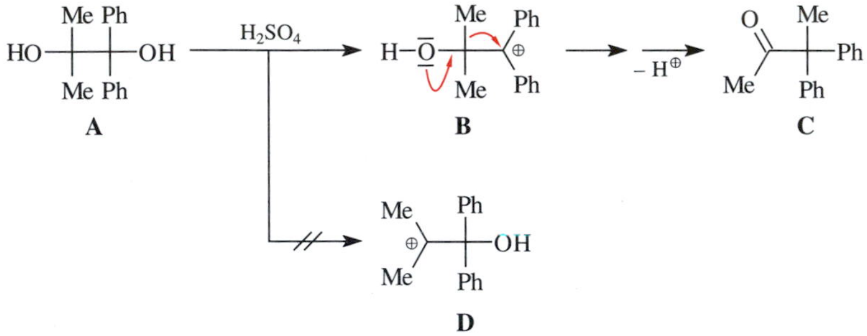

Fig. 11.13. Regioselectivity of the pinacol rearrangement of an asymmetric glycol. The more stable carbenium ion is formed under product-development control. Thus, the benzhydryl cation **B** is formed here, while the tertiary alkyl cation **D** is not formed.

tively. Only one of the two possible carbenium ions can be allowed to form and, once it is formed, only one of the neighboring alkyl groups can be allowed to migrate. Product-development control ensures the formation of the more stable carbenium ion in step 2. In the rearrangement depicted in Figure 11.13, the more stable cation is the benzhydryl cation **B** rather than the tertiary alkyl cation **D.**

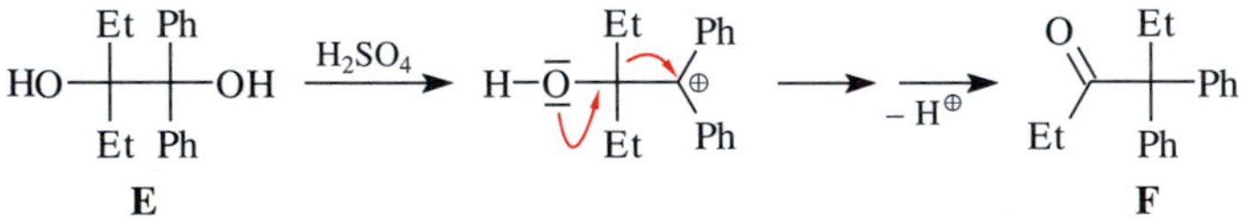

Fig. 11.14. Regioselective pinacol rearrangement of an asymmetric glycol. As in the case depicted in Figure 11.13, the reaction proceeds exclusively via the benzhydryl cation.

The pinacol rearrangement **E** → **F** (Figure 11.14) can be carried out in an analogous fashion for the same reason. As with the reaction shown in Figure 11.13, this reaction proceeds exclusively via a benzhydryl cation. In a crossover experiment (see Section 2.4.3 regarding the "philosophy" of crossover experiments), one rearranged a mixture of di-*tert*-glycols **A** (Figure 11.13) and **E** (Figure 11.14) under acidic conditions. The reaction products were the α-tertiary cations **C** and **F,** already encountered in the respective reactions conducted separately. The absence of the crossover products **G** and **H** proves the intramolecular nature of the pinacol rearrangement.

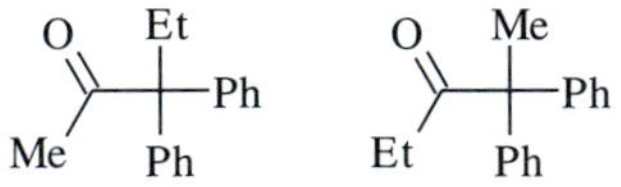

Semipinacol Rearrangements

B

Rearrangements that are not pinacol rearrangements but also involve a [1,2]-shift of an H atom or of an alkyl group from an *oxygenated* C atom to a neighboring C atom, that is, a carbenium ion → carboxonium ion rearrangement, are called semipinacol rearrangements. As shown later, however, there also are some semipinacol rearrangements that proceed without the intermediacy of carboxonium ions (Figures 11.18 and 11.20–11.22).

A

Lewis acids catalyze the ring opening of epoxides. If the carbenium ion that is generated is not trapped by a nucleophile, such an epoxide opening initiates a semipinacol rearrangement (Figures 11.15 and 11.16). Epoxides with different numbers of alkyl

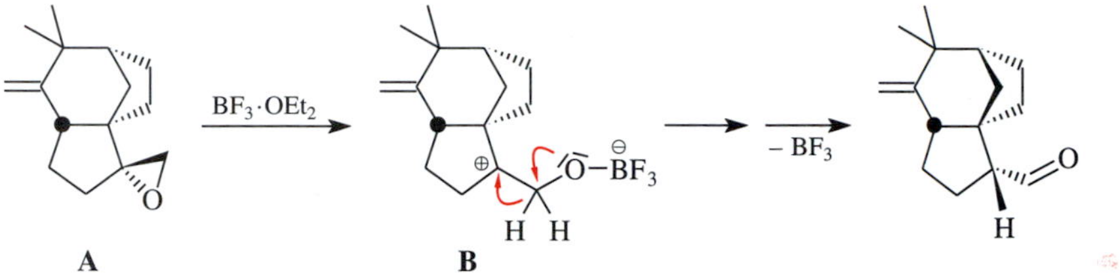

Fig. 11.15. "Accidental" diastereoselectivity in the semipinacol rearrangement of an epoxide. The more substituted carbenium ion is formed exclusively during ring-opening because of product-development control. Only two H atoms are available for possible migrations, and no alkyl groups. In general, diastereoselectivity may or may not occur, depending on which one of the diastereotopic H atoms migrates in which one of the diastereotopic conformers. The present case exhibits diastereoselectivity.

substituents on their ring C atoms are ideally suited for such rearrangements. In those cases, because of product-development control, only one carbenium ion is formed in the ring-opening reaction. It is the more alkylated carbenium ion, hence the more stable one. There is only an H atom and no alkyl group available for a [1,2]-migration in the carbenium ion **B** in Figure 11.15. It is again an H atom that migrates in the carbenium ion **B** formed from the epoxide **A** in Figure 11.16. A carbenium ion with increased ring strain would be formed if an alkyl group were to migrate instead.

The [1,2]-rearrangements shown in Figures 11.15 and 11.16 are stereogenic and proceed stereoselectively (which is not true for many semipinacol rearrangements). In [1,2]-rearrangements, the migrating H atom always is connected to the same face of the carbenium ion from which it begins the migration. It follows that the [1,2]-shift in the carbenium ion **B** of Figure 11.16 *must* proceed stereoselectively. This is because the H atom can begin its migration only on one side of the carbenium ion. On the other hand, an H atom can in principle migrate on either side of the carbenium ion plane of carbenium ion **B** of Figure 11.15. However, only the migration on the top face (in the selected projection) does *in fact* occur.

The semipinacol rearrangements of Figures 11.15 and 11.16 are [1,2]-rearrangements in which the target atoms of the migrations are the higher alkylated C atoms of the 1,2-dioxygenated (here: epoxide) substrates. The opposite direction of migration—toward the less alkylated C atom of a 1,2-dioxygenated substrate—can be realized in *sec,tert*-glycols. These glycols can be toslyated with tosyl chloride at the less hindered

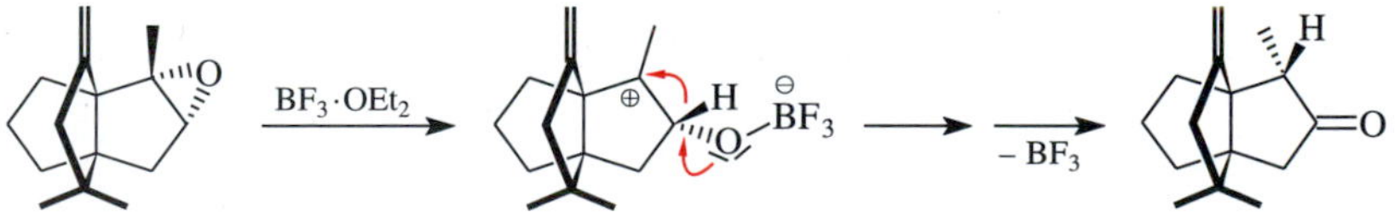

Fig. 11.16. Mechanism-based diastereoselectivity in the semipinacol rearrangement of an epoxide. This rearrangement is stereoselective, since there is only one H atom in the position next to the sextet center and the H atom undergoes the [1,2]-migration on the same face of the five-membered ring.

Fig. 11.17. First semipinacol rearrangement of a glycol monotosylate. The reaction involves three steps in neutral media: formation of a carbenium ion, rearrangement to a carboxonium ion, and deprotonation to the ketone.

secondary OH group. Glycol monosulfonates of the types shown in Figures 11.17 and 11.18 can be obtained in this way. With these glycol monosulfonates, semipinacol rearrangements with a reversed direction of the migration can be carried out.

Glycol monosulfonates of the type discussed may cleave off a tosylate under solvolysis conditions, as illustrated in Figure 11.17. Solvolysis conditions are achieved in this case by carrying out the reaction in a solution of $LiClO_4$ in THF. Such a solution is more polar than water! The solvolysis shown first leads to a carbenium ion **A.** At this point, a rearrangement into a carboxonium ion (and subsequently into a ketone) could occur via the migration of a primary alkyl group or of an alkenyl group. The migration of the alkenyl group is observed exclusively. Alkenyl groups apparently possess a higher intrinsic propensity for migration than alkyl groups.

Figure 11.18 depicts a semipinacol rearrangement that is initiated in a different manner. This reaction proceeds—in contradiction to the title of Section 11.3—*without* the occurrence of a sextet intermediate. We want to discuss this reaction here, nevertheless, because this process provides synthetic access to molecules that are rather similar to molecules that are accessible via "normal" semipinacol rearrangements. The glycol monotosylate **A** is deprotonated by KO*tert*-Bu to give alkoxide **B** in an equilibrium reaction. Under these conditions other glycol monotosylates would undergo a ring closure delivering the epoxide. However, *this* compound cannot form an epoxide because the alkoxide O atom is incapable of a backside attack on the C—OTs bond. Hence,

Fig. 11.18. Second semipinacol rearrangement of a glycol monotosylate. The reaction involves two steps in basic media, since the [1,2]-rearrangement and the dissociation of the tosylate occur at the same time.

the tosylated alkoxide **B** has the opportunity for a [1,2]-alkyl shift to occur with *concomitant* elimination of the tosylate. This [1,2]-shift occurs stereoselectively in such a way that the C—OTs bond is broken by a backside attack of the migrating alkyl group.

B

Other leaving groups in other glycol derivatives facilitate other semipinacol rearrangements. Molecular nitrogen, for example, is the leaving group in

- the Tiffeneau–Demjanov rearrangement of diazotized amino alcohols—a reaction that is of general use for the ring expansion of cycloalkanones to their next higher homologs (for example, see Figure 11.19),
- analogous ring expansions of cyclobutanones to cyclopentanones (for example, see Figure 11.20), or
- the ring expansion of cyclic ketones to carboxylic esters of the homologated cycloalkanone (for example, see Figure 11.21); each of these ring expansions leads to one additional ring carbon, as well as
- a chain-elongating synthesis of β-ketoesters from aldehydes and diazomethyl acetate (Figure 11.22).

A

Figure 11.19 shows how the **Tiffeneau–Demjanov rearrangement** can be employed to insert an additional CH_2 group into the ring of cyclic ketones. Two steps are required to prepare the actual substrate of the rearrangement. A nitrogen-containing C_1 nucleophile is added to the substrate ketone. This nucleophile is either HCN or nitromethane, and the addition yields a cyanohydrin or a β-nitroalcohol, respectively. Both these compounds can be reduced with lithium aluminum hydride to the vicinal amino alcohol **A.** The Tiffeneau–Demjanov rearrangement of the amino alcohol is initiated by a diazotation of the amino group. The diazotation is achieved either with sodium nitrite in aqueous acid or with isoamyl nitrite in the absence of acid and water. The mechanism of these reactions corresponds to the usual preparation of aryldiazonium chlorides from anilines (Figure 14.34, top) or to a variation thereof.

B

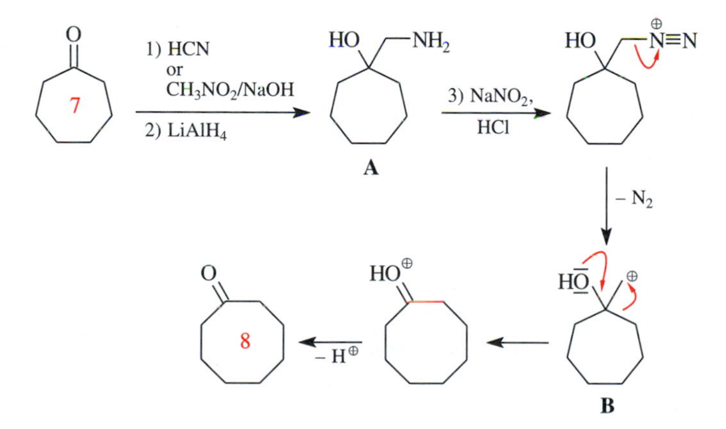

Fig. 11.19. Ring expansion of cyclic ketones via the Tiffeneau–Demjanov rearrangement. The first step consists of the additions of HCN or nitromethane, respectively, to form either the cyanohydrin or the β-nitroalcohol, respectively. The vicinal amino alcohol **A** is formed in the next step by reduction with $LiAlH_4$. The Tiffeneau–Demjanov rearrangement starts after diazotation with the dediazotation.

Aliphatic diazonium salts are much less stable than aromatic diazonium salts (but even the latter tend to decompose when isolated!). The first reason for this difference is that aliphatic diazonium salts, in contrast to their aromatic counterparts, lack stabilization through resonance. Second, aliphatic diazonium salts release N_2 much more readily than their aromatic analogs since they thus react to give relatively stable alkyl cations. Hence, the decomposition of aliphatic diazonium ion is favored by product-development control. Aromatic diazonium salts, on the other hand, form phenyl cations upon releasing N_2 and these are even less stable than alkenyl cations. The electron-deficient carbon atom in an alkenyl cation can be stabilized, at least to a certain degree, because of its linear coordination (which the phenyl cation cannot adopt). This can be rationalized with the MO diagrams of Figure 1.3. The orbital occupancy of a bent carbenium ion $=C^+—R$ resembles that of the bent carbanion $=C^-—R$ except that the n_{sp^2} orbital remains empty in the former. Nevertheless, even linear alkenyl cations are less stable than alkyl cations.

The molecular nitrogen in aliphatic diazonium ions is an excellent leaving group. In fact, nitrogen is eliminated from these salts so fast that an external nucleophile does not stand a chance of actively assisting in the nucleophilic displacement of molecular nitrogen. Only an *internal* nucleophile, that is, a neighboring group, can provide such an assistance in displacing nitrogen (example in Figure 2.24). Therefore, aliphatic diazonium ions *without* neighboring groups always form carbenium ions.

Unfortunately, carbenium ions often undergo a variety of consecutive reactions and yield undesired product mixtures. The situation changes significantly if the carbenium ion carries a hydroxyl group in the β position or if the diazonium salt contains an O^-, an OBF_3^-, or an $OSnCl_2^-$ substituent in the β position with respect to the N_2^+ group. The first of these structural requirements is fulfilled in the carbenium ion intermediate **B** of the Tiffeneau–Demjanov rearrangement of Figure 11.19; it contains a hydroxyl group in the β position. The diazonium ion intermediate **B** of the ring-expansion of Figure 11.10 contains an O^--substituent in the β-position, that of Figure 11.11 an OBF_3^- substituent. The diazonium ion intermediate **A** of Figure 11.22 carries an $OSnCl_2^-$ substituent in the β position. The aforementioned O substituents in the β positions of the diazonium ions and of the resulting carbenium ions allow for [1,2]-rearrangements that generate a carboxonium ion (Figure 11.19) or a ketone (Figures 11.20–11.22), respectively. Accordingly, a favorable all-octet species is formed in both cases, whether charged or neutral.

Let us take a closer look at the ring expansion of Figure 11.20. Cyclobutanones like **A** initially add diazomethane, which is a C nucleophile, and a tetrahedral intermediate **B** is formed. The cyclopentanone **C** is obtained by dediazotation and concomitant regioselective [1,2]-rearrangement. This rearrangement does not belong in the present section in the strictest sense, since the rearrangement **B** → **C** is a one-step process and therefore does not involve a sextet intermediate. Nevertheless, the reaction is described here because of its close similarity to the Tiffeneau–Demjanov rearrangement of Figure 11.19.

The tetrahedral intermediate **B** of Figure 11.20 is not enriched to any significant concentration because the ring expansion reaction, which it undergoes, is at least as fast as its formation. Once the *first* ring-expanded ketone is formed, there obviously still is plenty of unreacted substrate ketone **A** present. The question is now: which ketone re-

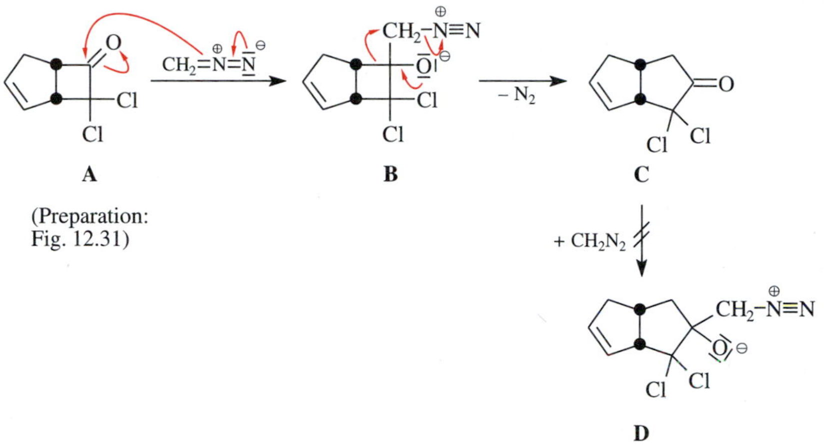

(Preparation: Fig. 12.31)

Fig. 11.20. Ring expansion of cyclobutanones. The cyclobutanones are accessible via [2+2]-cycloadditions of dichloroketene (for the mechanism, see Section 12.4).

acts faster with diazomethane? The answer is that the cyclobutanone reacts faster with diazomethane because of product-development control: the formation of the tetrahedral intermediate **B** results in a substantial reduction of the ring strain in the four-membered ring because of the rehybridization of the carbonyl carbon from sp^2 to sp^3. As is well known, carbon atoms with sp^2 hybridization prefer 120° bond angles, while sp^3-hybridized carbon atoms prefer tetrahedral bond angles (109° 28′). Hardly any ring strain would be relieved in the addition of diazomethane to the newly formed cyclopentanone **C.** It thus follows that the ring expansion of the cyclobutanone occurs fast and proceeds to completion before the product cyclopentanone **C** in turn undergoes its slower ring expansion via the tetrahedral intermediate **D.**

This line of argument also explains why most other cyclic ketones do not undergo chemoselective ring expansions with diazomethane. In the general case, both the substrate ketone and the product ketone would be suitable reaction partners for diazomethane, and multiple, i.e., consecutive ring expansions could not be avoided. Therefore, it is worth remembering that the Tiffeneau–Demjanov rearrangement of Figure 11.19 shows how to accomplish a ring expansion of *any* cycloalkanone by exactly one CH_2 group.

A process is shown in Figure 11.21 that allows for the ring expansion of any cycloalkanone by exactly one C atom, too, using diazoacetic acid ethyl ester. Diazoacetic acid ethyl ester is a relatively weak nucleophile because of its CO_2R substituent, and its addition to unstrained cycloalkanones is possible only in the presence of $BF_3 \cdot OEt_2$. In that case, the electrophile is the ketone–BF_3 complex **A.** The tetrahedral intermediate **B** formed from **A** and diazoacetic acid ethyl ester also is a diazonium salt. Thus it is subject to a semipinacol rearrangement. As in the cases described in Figures 11.18 and 11.20, this semipinacol rearrangement occurs without the intermediacy of a carbenium ion. A carbenium ion would be greatly destabilized because of the close proximity of a positive charge and the CO_2R substituent in the α position.

The diazonium salt **B** of Figure 11.21 contains a tertiary and a primary alkyl group in suitable positions for migration. In contrast to the otherwise observed intrinsic migratory trends, only the primary alkyl group migrates in this case, and the β-ketoester

A

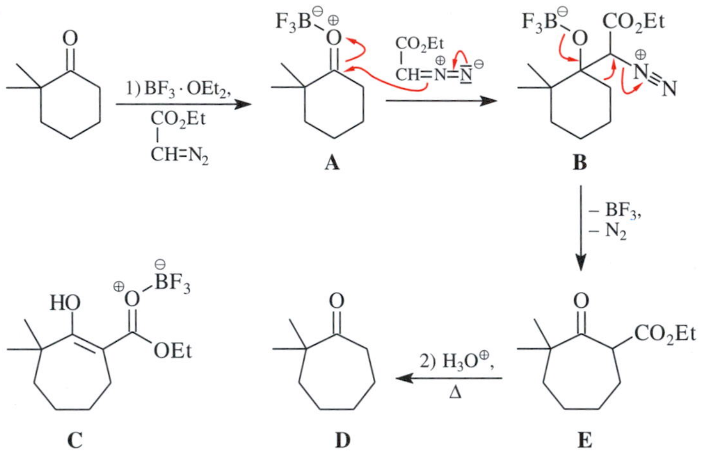

Fig. 11.21. Ring expansion of a cyclohexanone via addition of diazoacetic acid ethyl ester and subsequent [1,2]-rearrangement.

E is formed by the rearrangement. Interestingly, the product does not undergo further ring expansion. This is because the Lewis acid BF_3 catalyzes the enolization of the keto group of **E,** and BF_3 subsequently complexes the resulting enol at the ester oxygen to yield the aggregate **C.** In contrast to the BF_3 complex **A** of the unreacted substrate, this species **C** is not a good electrophile. Hence, only the original ketone continues to react with the diazoacetic acid ethyl ester.

The cyclic β-ketoester **E** subsequently can be saponified in acidic medium. The acid obtained then decarboxylates according to the mechanism of Figure 10.24 to provide the ester-free cycloalkanone **D.** Product **D** represents the product of a CH_2 insertion into the starting ketone. It would be impossible to obtain this product with diazomethane (cf. discussion of Figure 11.20). In Section 11.3.2 (Figure 11.23), a third method will be described for the insertion of a CH_2 group into a cycloalkanone. Again, a [1,2]-rearrangement will be part of that insertion reaction.

Finally, an interesting C_2 elongation of aldehydes to β-ketoesters is presented in Figure 11.22. This elongation reaction involves a semipinacol rearrangement that occurs in complete analogy to the one shown in Figure 11.21. The differences are merely that the Lewis acid $SnCl_2$ is employed instead of BF_3 in the first step, and an H-atom undergoes the [1,2]-migration instead of an alkyl group.

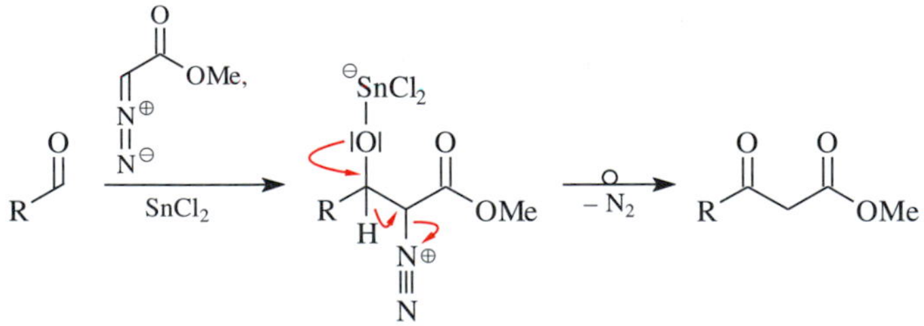

Fig. 11.22. C_2 extension of aldehydes to β-ketoesters via a semipinacol rearrangement.

11.3.2 [1,2]-Rearrangements in Carbenes or Carbenoids

A Ring Expansion of Cycloalkanones

A

Figure 11.23 shows how the ring expansion of cyclic ketones can be accomplished without the liberation of molecular nitrogen (in contrast to the ring expansions of Figures 11.19–11.21). A chemoselective *mono*insertion of CH_2 occurs because the product ketone is never exposed to the reaction condition to which the substrate ketone is subjected. This is a similarity between the present method and the processes described in Figures 11.19 and 11.20, and this feature is in contrast to the method depicted in Figure 11.21.

In the first step of the reaction shown in Figure 11.23, CH_2Br_2 is deprotonated by LDA and the organolithium compound Li-$CHBr_2$ is formed. This reagent adds to the C=O double bond of the ketone substrate and forms an alkoxide. The usual acidic work up yields the corresponding alcohol **A.** The alcohol group of alcohol **A** is deprotonated with one equivalent of *n*-BuLi in the second step of the reaction. A bromine/lithium exchange (mechanism in Figure 13.11, top row) is accomplished in the resulting alkoxide with another equivalent of *n*-BuLi. The resulting organolithium compound **D** is a carbenoid. As discussed in Section 3.3.2, a carbenoid is a species whose reactivity resembles that of a carbene even though there is no free carbene involved. In the VB model, one can consider carbenoid **D** to be a resonance hybrid between an organolithium compound and a carbene associated with LiBr.

The elimination of LiBr from this carbenoid is accompanied or followed (the timing is not completely clear) by a [1,2]-rearrangement. The alkenyl group presumably migrates faster than the alkyl group, as in the case of the semipinacol rearrangement of Figure 11.17. The primary product most probably is the enolate **C,** and it is converted into the ring-expanded cyclooctanone **B** upon aqueous workup. The C=C double bond in **B** is not conjugated with the C=O double bond. This can be attributed to a kinetically controlled termination of the reaction. If thermodynamic control had occurred, some 20% of the unconjugated ketone would have isomerized to the conjugated ketone.

1) LDA, CH_2Br_2 ($\rightarrow$ Li–$CHBr_2$); $H_3O^{\oplus}$ — **A** — 2) 2 *n*-BuLi — **D** — – LiBr — **C** — acidic workup — **B**

Fig. 11.23. Ring expansion of cycloheptenone via a carbenoid intermediate. The elimination of LiBr from the carbenoid occurs with or is followed by a [1,2]-alkenyl shift. The enolate **C** is formed and, upon aqueous workup, it is converted to the ring-expanded cycloalkenone **B.**

(The formation of no more than 20% of the conjugated ketone is due to a medium-sized ring effect. 3-Cyclohexenone contains a normal ring instead of a medium-sized ring, and it would be converted completely to 2-cyclohexenone under equilibrium conditions.)

Wolff Rearrangement

B

Wolff rearrangements are rearrangements of α-diazoketones leading to carboxylic acid derivatives via ketene intermediates. Wolff rearrangements can be achieved with metal catalysis or photochemically. As Figure 11.24 shows, the α-diazoketone **D** initially loses a nitrogen molecule and forms a ketene **G.** Heteroatom-containing nucleophiles add to the latter in uncatalyzed reactions (see Figure 7.1 for the mechanism). These heteroatom-containing nucleophiles must be present during the ketene-forming reaction because only if one traps the ketenes immediately in this way can unselective consecutive reactions be avoided. Thus, after completion of the Wolff rearrangement, only the addition products of the transient ketenes are isolated. These addition products are carboxylic acid derivatives (cf. Section 7.1.2). In the presence of water, alcohols, or amines, the Wolff rearrangement yields carboxylic acids, carboxylic esters, or carboxylic amides, respectively.

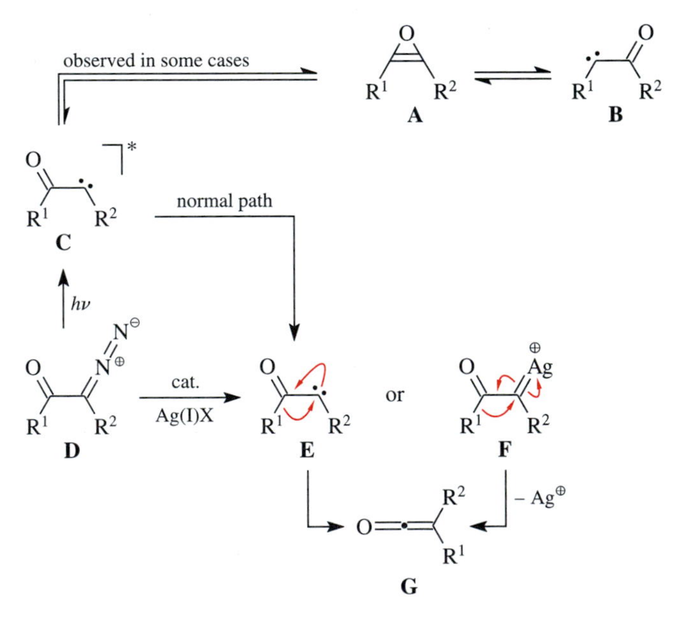

Fig. 11.24. Mechanisms of the photochemically initiated and Ag(I)-catalyzed Wolff rearrangements with formation of the ketocarbene **E** and/or the ketocarbenoid **F** by dediazotation of the diazoketene **D** in the presence of *catalytic amounts of Ag(I)*. **E** and **F** are converted into **G** via a [1,2]-shift of the alkyl group R^1. N_2 and an excited carbene **C** are formed in the *photochemically initiated* reaction. The excited carbene usually relaxes to the normal ketocarbene **E,** and this carbene **E** continues to react to give **G.** The ketocarbene **C** may on occasion isomerize to **B** via an oxacyclopropene **A.** The [1,2-]-shift of **B** also leads to the ketene **G.**

Let us consider the mechanistic details of the Wolff rearrangement (Figure 11.24). If the rearrangement is carried out in the presence of catalytic amounts of silver(I) salts, the dediazotation of the α-diazoketone initially generates the ketocarbene **E** and/or the corresponding ketocarbenoid **F.** A [1,2]-shift of the alkyl group R^1 which stems from the acyl substituent R^1—C═O of the carbene or the carbenoid follows. This rearrangement converts each of the potential intermediates **E** and **F** into the ketene **G.** The same ketene **G** is obtained if the Wolff rearrangement is initiated photochemically. In this case, molecular nitrogen and an excited ketocarbene are formed initially. The excited ketocarbene may undergo relaxation to the normal ketocarbene **E,** which can then undergo the [1,2]-rearrangement discussed earlier. On the other hand, excited ketocarbenes **C** occasionally rearrange into an isomeric ketocarbene **B** via an *anti*-aromatic oxirene intermediate **A.** In that case the [1,2]-rearrangement would occur in **E** and in **B** as well or in **B** alone. One and the same ketene **G** is formed in any case.

The Wolff rearrangement is the third step of the Arndt–Eistert homologation of carboxylic acids. Figure 11.25 picks up an example that was discussed in Section 7.2, that is, the homologation of trifluoroacetic acid to trifluoropropionic acid. The first step of the Arndt–Eistert synthesis consists of the activation of the carboxylic acid via the acid chloride. The C_1 elongation to an α-diazoketone occurs in the second step.

If an alkyl group that migrates in a Wolff rearrangement contains a stereocenter at the C-α-atom, the migration of the alkyl group proceeds with retention of configuration. An example of a reaction that allows one to recognize this stereochemical propensity is provided by the double Wolff rearrangement depicted in Figure 11.26. The bis(diazoketone) **A,** a *cis*-disubstituted cyclohexane, is the substrate of the rearrangement. The bisketene, which cannot be isolated, must have the same stereochemistry, because the dimethyl ester **B** formed from the bisketene by the addition of methanol still is a *cis*-disubstituted cyclohexane.

Cycloalkanones can be converted into the enolate of an α-formyl ketone by reaction with ethyl formate and one equivalent of sodium ethioxide (Figure 10.55). Such a reaction is shown in Figure 11.27 as transformation **A** → **B.** This reaction sets the stage for a **diazo group transfer reaction** (for the mechanism, see Figure 12.39), which results in the cyclic diazoketone **C.** This compound **C** can be converted into ketene **D** *via* a photochemical Wolff rearrangement. If this is done in an aqueous medium, the ketene hydrolyzes *in situ* to give the carboxylic acid **E.** This carboxylic acid contains a nine-membered ring, while its precursor was a ten-membered ring. Hence, these kinds of Wolff rearrangements are ring-contraction reactions.

Fig. 11.25. Wolff rearrangement as the third step in the Arndt–Eistert homologation of carboxylic acids. This example shows the homologation of trifluoroacetic acid to trifluoropropionic acid. The conversion of the acid into the acid chloride is the first step (not shown).

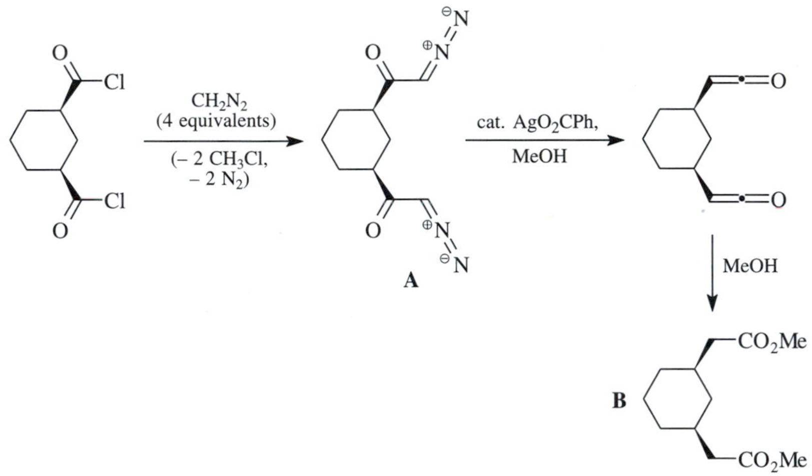

Fig. 11.26. Twofold Wolff rearrangement in the bishomologation of dicarboxylic acids according to Arndt and Eistert. Both Alkyl group migrations occur with retention of configuration.

Aldehyde → Alkyne Elongation via Carbene and Carbenoid Rearrangements

B

Figures 11.28 and 11.29 show a reaction and a sequence of reactions, respectively, that allow for the conversion of aldehydes into alkynes that contain one more C atom. These transformations involve a [1,2]-rearrangement.

The one-step **Seyferth procedure** is shown in Figure 11.28. The reaction begins with the Horner–Wadsworth–Emmons olefination of the aldehyde to the alkene **A.** It is a disadvantage of this reaction that the phosphonate used is not commercially available. The mechanism of this olefination is likely to resemble the mechanism earlier presented in Figure 9.14 for a reaction that involved a different phosphonate anion. As can be seen, alkene **A** also is an unsaturated diazo compound. As soon as the reaction mixture is allowed to warm to room temperature, compound **A** eliminates molecular nitrogen and the vinyl carbene **B** is generated. **A** [1,2]-rearrangement of **B** forms the alkyne. It is presumably the H atom rather than the alkyl group that migrates to the electron-deficient center.

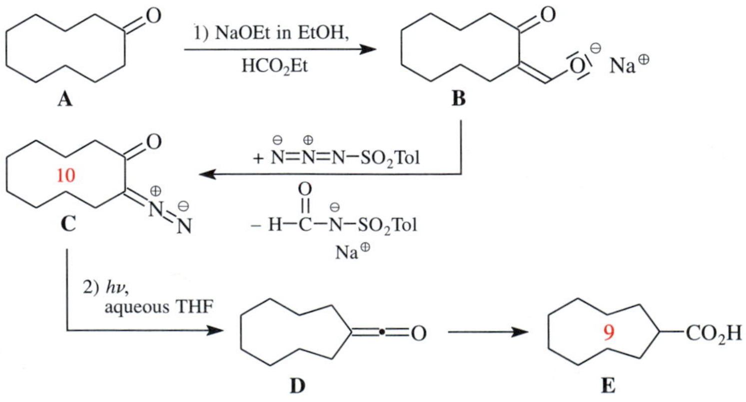

Fig. 11.27. Ring contraction via Wolff rearrangement. The 10-membered cyclic diazoketone **C** rearranges in aqueous media to give the nine-membered ring carboxylic acid **E** via the ketene **D.**

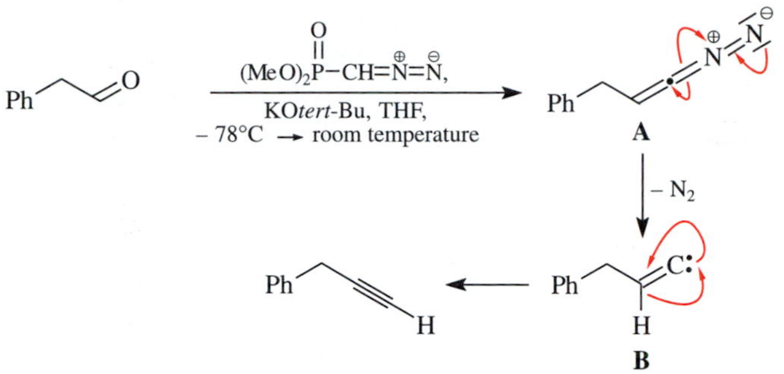

Fig. 11.28. Aldehyde → alkyne chain elongation via [1,2]-rearrangement of a vinyl carbene (Seyferth procedure). First, a Horner–Wadsworth–Emmons olefination of the aldehyde is carried out to prepare alkene **A**. Upon warming to room temperature, alkene **A** decomposes and gives the vinyl carbene **B**. From that, the alkyne is formed by way of a [1,2]-rearrangement.

The two-step **Corey–Fuchs procedure** offers an alternative aldehyde → alkyne elongation (Figure 11.29). In the first step, the dibrominated phosphonium ylide **A** is generated *in situ* by reaction of Ph_3P, CBr_4, and Zn. A Wittig reaction (for the mechanism, see Figure 9.7) between ylide **A** and an added aldehyde elongates the latter to give a 1,1-dibromoalkene. In the second phase of the reaction, the 1,1-dibromoalkene is treated with two equivalents of *n*-BuLi. The *n*-BuLi presumably first initiates a bromine/lithium exchange (for the mechanism, see top row of Figure 13.11)

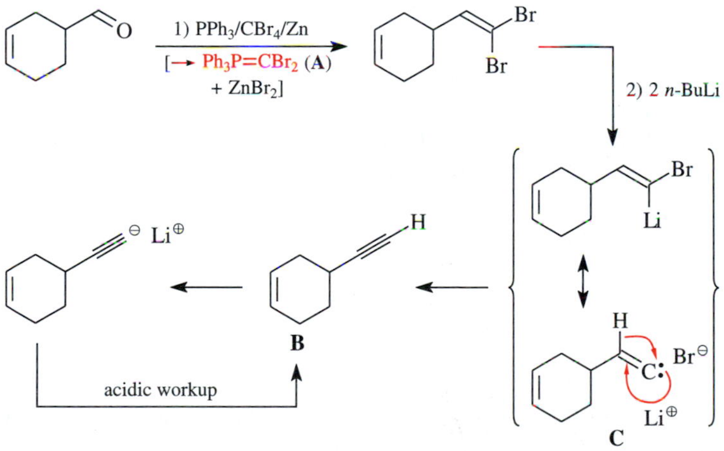

Fig. 11.29. Aldehyde → alkyne chain elongation via [1,2]-rearrangement of a vinyl carbenoid (Corey–Fuchs procedure). The aldehyde and phosphonium ylide **A** generated *in situ* undergo a Wittig reaction and form the 1,1-dibromoalkene. In the second stage, the dibromoalkene is reacted with two equivalents of *n*-BuLi and the vinyl carbenoid **C** is formed. The carbenoid undergoes H migration to form the alkyne **B**. The alkyne **B** reacts immediately with the second equivalent of *n*-BuLi to give the lithium acetylide.

and generates the α-lithiated bromoalkene **C.** Compound **C** is a carbenoid, as indicated by the resonance forms shown in Figure 11.29. It is unknown whether the carbenoid rearranges or whether the free carbene is formed prior to the rearrangement. It is known from analogous experiments in which the carbene carbon was ^{13}C-labeled that only an H atom migrates, not the alkyl group. The alkyne formed **(B)** is so acidic that it immediately reacts with a second equivalent of *n*-BuLi to give the corresponding lithium acetylide. Alkyne **B** is regenerated from the acetylide upon aqueous workup.

When the reaction of Figure 11.29 was carried out with less than two equivalents of *n*-BuLi, the alkyne **B** was deprotonated not only by *n*-BuLi but also by some of the carbenoid **C.** In this way, **C** was converted into a monobromoalkene, which could be isolated. This observation provided evidence that the reaction indeed proceeds via the carbenoid **C** and not by another path.

11.4 [1,2]-Rearrangements *without* the Occurrence of a Sextet Intermediate

B

The reader has already encountered semipinacol rearrangements in which the elimination of the leaving group was not *followed* but instead was *accompanied* by the [1,2]-rearrangement (Figures 11.18 and 11.20–11.22). In this way, the temporary formation of an energetically unfavorable valence electron sextet could be avoided. These rearrangements are summarized in the first two rows of Table 11.1.

It was pointed out in the discussion of the lower part of Figure 11.1 that leaving groups cannot be dissociated from O or N atoms, respectively, if these dissociations resulted in the formation of oxenium or nitrenium ions, respectively. The same is true if a nitrene (R—N:) would have to be formed. These three sextet systems all are highly destabilized in comparison to carbenium ions and carbenes because of the high elec-

Table 11.1. Survey of [1,2]-Rearrangements without Sextet Intermediates

R(H)–a—b–Y $\xrightarrow{-Y^\ominus}$ $\overset{\oplus}{a}$—b(–R(H))

a	b–Y	
$CR(O^\ominus)$	CRH–OTs	Semipinacol rearrangement
$CR(O^\ominus)$	$CH_2-\overset{\oplus}{N}\equiv N$	Semipinacol rearrangement
CR_2	$O-OH_2^\oplus$	Hydroperoxide rearrangement
CR(OH)	O–OC(=O)Ar	Baeyer–Villiger oxidation
$BR_{2-n}(OR)_n^\ominus$	O–OH	Borane oxidation/ boronate oxidation
$C_{sp^2}R$	$N_{sp^2}-OH_2^\oplus$	Beckmann rearrangement
C(=O)	$\overset{\ominus}{N}-\overset{\oplus}{N}\equiv N$	Curtius degradation

tronegativities of O and N. The O- and N-bound leaving groups therefore can be expelled only if *at the same time* either an α-H-atom or an alkyl group undergoes a [1,2]-rearrangement to the O or N atom:

- The entries in rows 3–5 in Table 11.1 refer to one-step eliminations/rearrangements of this type in which oxenium ions are avoided.
- The Beckmann rearrangement (details: Figure 11.38) is a [1,2]-rearrangement in which the occurrence of a nitrenium ion is avoided via the one-step mode of elimination and rearrangement (second to last entry in Table 11.1).
- Curtius rearrangements (details in Figures 11.39 and 11.40) occur as one-step reactions to avoid the intermediacy of nitrenes (last entry in Table 11.1).

11.4.1 Hydroperoxide Rearrangements

B

Tertiary hydroperoxides undergo a hydroperoxide rearrangement in acidic media, as exemplified in Figure 11.30 by the rearrangement of cumene hydroperoxide (for the preparation, see Figure 1.27). The cumene hydroperoxide rearrangement is employed for the synthesis of acetone and phenol on an industrial scale. The OH group of the hydroperoxide is protonated by concentrated sulfuric acid so that the carboxonium ion **A** can be generated by the elimination of water. The ion **A** immediately adds the water molecule again under formation of the protonated hemiacetal **C**. Tautomerization of **C** leads to **B**, and the latter decomposes to phenol and protonated acetone. The last reaction step is merely the deprotonation of the protonated acetone.

Ph O–OH $\xrightarrow{H_2SO_4}$ Ph O–$\overset{\oplus}{O}H_2$ ⟶ **A** + OH_2 ⟶ **C** $\xrightarrow{\sim H^{\oplus}}$ **B** ⟶ =$\overset{\oplus}{O}$H + PhOH $\xrightarrow{-H^{\oplus}}$ =O

Fig. 11.30. Cumene hydroperoxide rearrangement.

11.4.2 Baeyer–Villiger Rearrangements

B

The Baeyer–Villiger rearrangement often is called Baeyer–Villiger oxidation (see the last subsection of Section 14.3.2, Oxidative Cleavage of Ketones). In the Baeyer–Villiger rearrangement a carbonyl compound (ketones are almost always used) and an

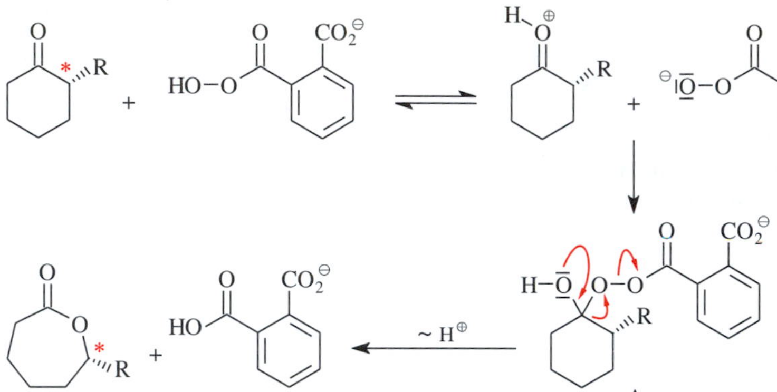

Fig. 11.31. Regioselective and stereoselective Baeyer–Villiger rearrangement of an asymmetric ketone with magnesium monoperoxophthalate hexahydrate (in the drawing, Mg^{2+} is omitted for clarity).

aromatic peracid form esters via insertion of the peroxo-O atom next to the C=O bond of the carbonyl compound. The Baeyer–Villiger rearrangement of *cyclic* ketones results in *lactones,* (as in Figure 11.31).

A Baeyer–Villiger rearrangement starts with the proton-catalyzed addition of the peracid to the C=O double bond of the ketone (Figure 11.31). This affords the α-hydroxyperoxoester **A.** The O—O bond of **A** is labile and breaks even without prior protonation of the leaving group. This is different from the fate of the O—O bond of a hydroperoxide which does not break unless it is protonated. The different behavior is due to the fact that the Baeyer–Villiger rearrangement releases magnesium phthalate, and that this anion is rather stable. The cleavage of a hydroperoxide in the absence of an acid would result in the much more basic and therefore much less stable hydroxide ion.

The O—O bond cleavage of the α-hydroxyperoxoester intermediate of a Baeyer–Villiger rearrangement is accompanied by a [1,2]-rearrangement. One of the two substituents of the former carbonyl group migrates. In the example shown in Figure 11.31, either a primary or a secondary alkyl group in principle could migrate. As in the case of the Wagner–Meerwein rearrangements, the intrinsic propensity toward migration in Baeyer–Villiger rearrangements follows the order $R_{tert} > R_{sec} > R_{prim}$. Hence, the secondary alkyl group migrates exclusively in intermediate **A.** It migrates with complete retention of configuration. This stereochemistry is quite common in [1,2]-rearrangements and was mentioned earlier in connection with the double Wolff rearrangement (Figure 11.26). If the substrate of the Baeyer–Villiger rearrangement of Figure 11.31 is an enantiomerically pure ketone (possible method for the preparation: in analogy to Figures 10.31 or 10.32), an enantiomerically pure lactone is formed.

The Baeyer–Villiger rearrangement of acetophenone, shown in Figure 11.32, also proceeds via the mechanism described in Figure 11.31. The aryl group migrates rather than the methyl group—this is true no matter whether the acetophenone is electron rich or electron poor. The product of this rearrangement is an aryl acetate. The hydrolysis of this aryl acetate occurs quickly (cf. discussion of Table 6.1). The combination of the method for preparing acetophenones (cf. Section 5.2.7) with the Baeyer–Villiger rearrangement allows for the synthesis of phenols from aromatic compounds.

In Baeyer–Villiger rearrangements electron-rich aryl groups migrate faster than H-atoms, and H atoms in turn migrate faster than electron-poor aryl groups. Aldehydes,

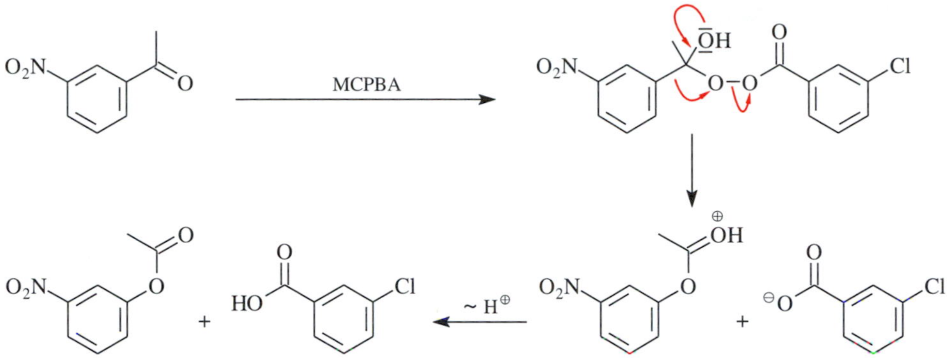

Fig. 11.32. Regioselective Baeyer–Villiger rearrangement of an asymmetric ketone with MCPBA (*meta*-**c**hloro**p**er**b**enzoic **a**cid). The aryl group is [1,2]-shifted in all cases and irrespective of whether the acetophenone is electron rich or electron poor.

benzaldehydes, and electron-poor aromatic aldehydes thus react with peracids under formation of *carboxylic acids* (for example, see Figure 11.33), while electron-rich aromatic aldehydes react with peracids to afford *phenyl formates* (for example, see Figure 11.34). It must be emphasized that, in contrast to Figures 11.31 and 11.32, the transition state of the "Baeyer–Villiger rearrangement" of Figure 11.33 perhaps might not even be that of a rearrangement at all. Instead, it is entirely possible that a β-elimination of benzoic acid from the α-hydroxyperoxoester occurs. This β-elimination might involve a cyclic transition state (cf. *cis*-eliminations in Section 4.2).

Phenyl formates (see Figure 11.34 for synthesis) hydrolyze to phenols more easily than the phenyl acetates shown in Figure 11.32. Overall, these species bring short reaction sequences to an end which may start with any aromatic methyl ketone or with an electron-rich aromatic aldehyde and which end with a phenol with widely variable substituent patterns.

Cyclobutanones are the only ketones that undergo Baeyer–Villiger rearrangements not only with peracids but even with alkaline H_2O_2 or alkaline *tert*-BuOOH (Figure 11.35). In this case, the driving forces of two crucial reaction steps are higher than

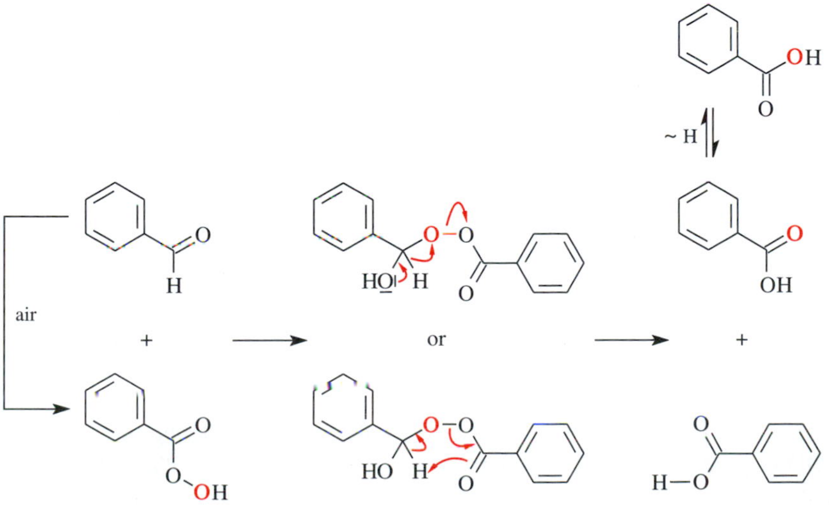

Fig. 11.33. Regioselective Baeyer–Villiger rearrangement of an electron-poor aromatic aldehyde. This reaction is part of the autoxidation of benzaldehyde to benzoic acid. Both alternative reaction mechanisms are shown: the [1,2]-rearrangement (top) and the β-elimination (bottom).

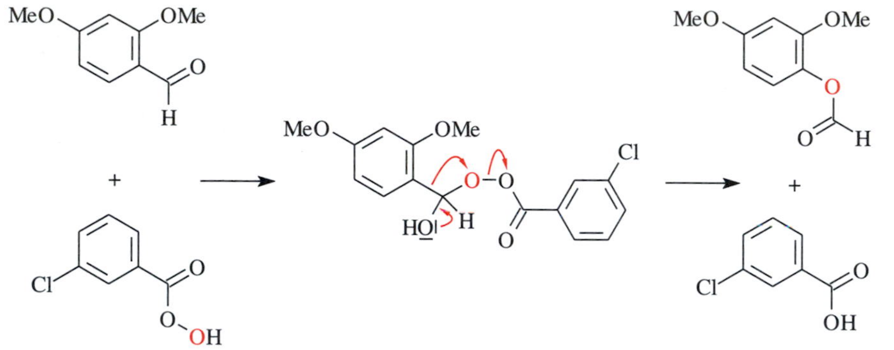

Fig. 11.34. Regioselective Baeyer–Villiger rearrangement of an electron-rich aromatic aldehyde.

normal because of the stepwise release of ring strain. The ring strain in the four-membered ring is reduced somewhat during the formation of the tetrahedral intermediate in step **A** → **B,** since the attacked C atom is rehybridized from sp^2 to sp^3. Accordingly, the preferred bond angle is reduced from 120° to 109° 28′. While still too large for **B** to be strain-free, some relief in comparison to **A** is provided. In the second step **B** → **C** of this particularly fast Baeyer–Villiger rearrangement, the four-membered tetrahedral intermediate is converted into the five-membered rearrangement product **C.** In the process, the ring strain is drastically reduced. The extra exothermicities of these two reaction steps are manifested in lowered activation barriers because of product-development control.

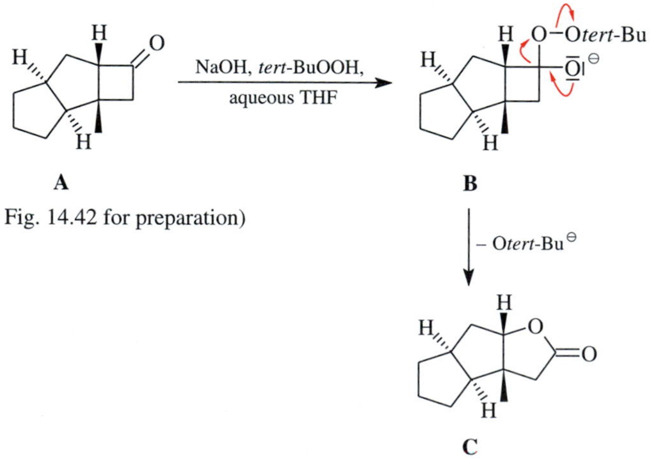

(See Fig. 14.42 for preparation)

Fig. 11.35. Baeyer–Villiger rearrangement of a strained ketone with alkaline *tert*-BuOOH.

11.4.3 Oxidation of Organoborane Compounds

B

Rearrangements also are involved in the oxidations of trialkylboranes with H_2O_2/NaOH (Figure 11.36) and of arylboronic acid esters with H_2O_2/HOAc (Figure 11.37).

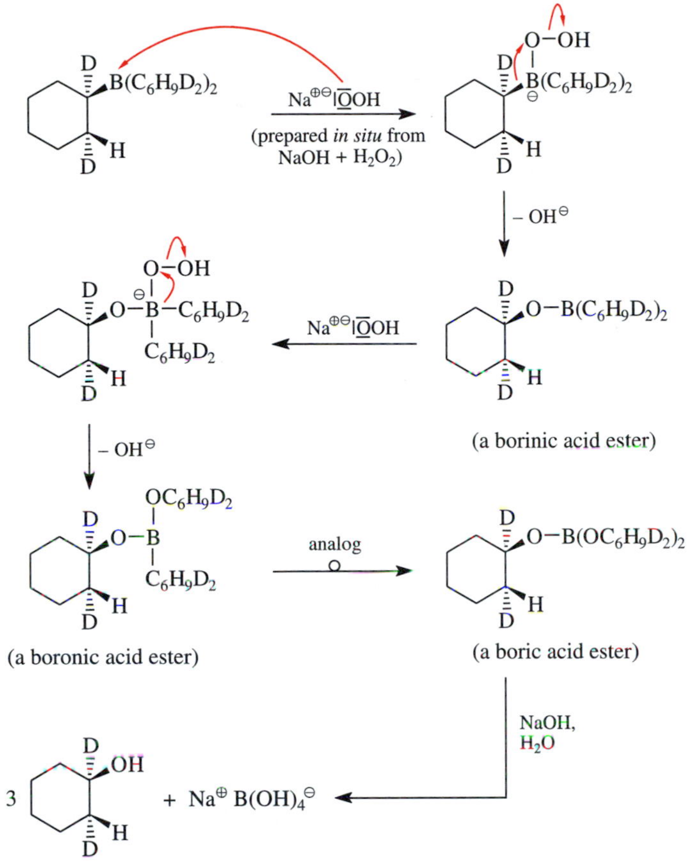

Fig. 11.36. H_2O_2/NaOH oxidation of a trialkylborane (see Figure 3.17 for the preparation of trialkylboranes **D** and **E** and for the mechanism of the hydrolysis of the resulting boric acid ester). Deuterium labeling studies show that the conversion of the C–B into the C–O bonds occurs with retention of configuration.

These rearrangements are driven by three O—O bond cleavages or one O—O bond cleavage, respectively. The formation of energetically unacceptable oxenium ions is strictly avoided. The oxidation of trialkylboranes with H_2O_2/NaOH is the second step of the hydroboration/oxidation/hydrolysis sequence for the hydration of alkenes (cf. Section 3.3.3). The oxidation of arylboronic acid esters with H_2O_2 is the second step of the reaction sequence Ar—Br → Ar—$B(OMe)_2$ → ArOH or of a similar reaction sequence *o*-MDG—Ar → *o*-MDG—Ar—$B(OMe)_2$ → *o*-MDG—ArOH (cf. Section 5.3.3; remember that MDG is the abbreviation for **m**etallation **d**irecting **g**roup).

The mechanisms of these two oxidations are presented in detail in Figures 11.36 and 11.37. They differ so little from the mechanism of the Baeyer–Villiger rearrangement (Section 11.4.2) that they can be understood without further explanations. In the example depicted in Figure 11.36, all the [1,2]-rearrangements occur with complete retention of configuration at the migrating C atom just as we saw in other [1,2]-rearrangements [cf. Figures 11.26 (Wolff rearrangement) and 11.31 (Baeyer–Villiger rearrangement)]. The Curtius rearrangement shown later in Figure 11.40 also occurs with retention of configuration.

Fig. 11.37. H_2O_2/HOAc oxidation of an arylboronic acid ester (for the preparation of this compound, see Figure 5.38).

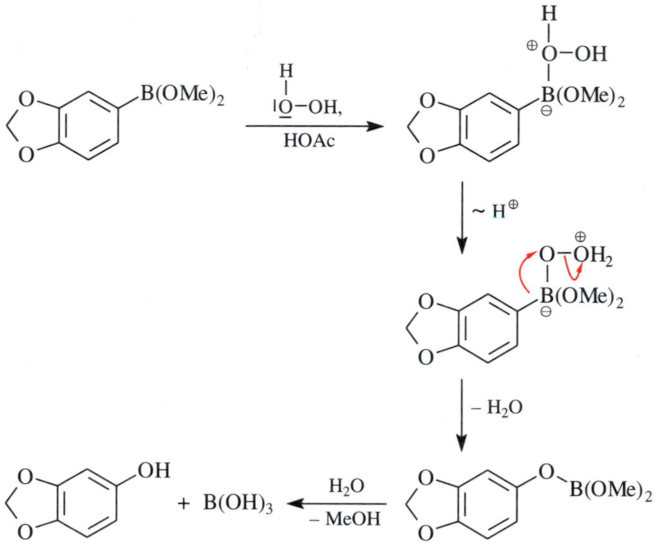

11.4.4 Beckmann Rearrangement

B

The OH group of ketoximes $R^1R^2C(=NOH)$ can become a leaving group. Tosylation is one way to convert this hydroxyl group into a leaving group. The oxime OH group also can become a leaving group if it is either protonated or coordinated by a Lewis acid in an equilibrium reaction. Oximes activated in this fashion may undergo N—O heterolysis. Since the formation of a nitrenium ion needs to be avoided (see discussion of Table 11.1), this heterolysis is accompanied by a simultaneous [1,2]-rearrangement of the group that is attached to the C=N bond in the *trans* position with regard to the O atom of the leaving group. A nitrilium ion is formed initially (see **A** in Figure 11.38). It reacts with water to form an imidocarboxylic acid, which tautomerizes immediately to an amide. The overall reaction sequence is called the Beckmann rearrangement.

The Beckmann rearrangement of cyclic oximes results in *lactams.* This is exemplified in Figure 11.38 with the generation of **ε**-caprolactam, the monomer of nylon-6. The nitrilium ion intermediate cannot adopt the preferred linear structure because it is embedded in a seven-membered ring. Therefore, in this case the intermediate might better be described as the resonance hybride of the resonance forms **A** ($C\equiv N^+$ triple bond) and **B** ($C^+=N$ double bond). The C,N multiple bond in this intermediate resembles the bond between the two C atoms in benzyne that do not carry H atoms.

11.4.5 Curtius Rearrangement

B

The Curtius degradation of acyl azides (Figure 11.39) consists of the thermolysis of the "inner" N=N double bond. This thermolysis expels molecular nitrogen and at the same

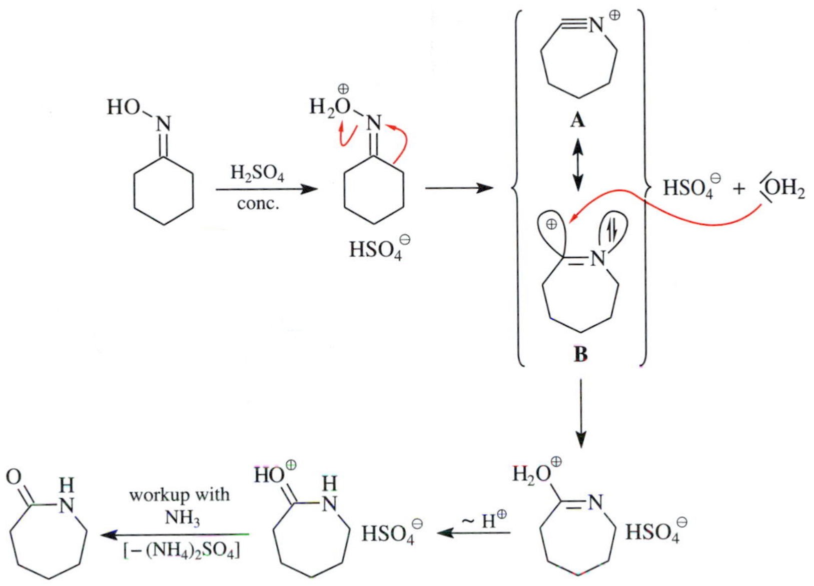

Fig. 11.38. Industrial synthesis of caprolactam via the Beckmann rearrangement of cyclohexanone oxime.

time leads to the [1,2]-rearrangement of the substituent that is attached to the carboxyl carbon. It is the simultaneous occurrence of these two events that prevents the formation of an energetically unacceptably disfavored acylnitrene intermediate. The rearranged product is an isocyanate.

The isocyanate can be isolated if the Curtius degradation is carried out in an inert solvent. The isocyanate also can be reacted with a heteroatom-nucleophile either subsequently or already *in situ.* The heteroatom nucleophile adds to the C=N double bond of the isocyanate according to the uncatalyzed mechanism of Figure 7.1. In this way, the addition of water initially results in a carbamic acid. However, all carbamic acids are unstable and decarboxylate to give amines (cf. Section 7.1.2). Because of *this* consecutive reaction, the Curtius rearrangement presents a valuable amine synthesis. The amines formed contain one C atom less than the acyl azide substrates. It is due to this feature that one almost always refers to this reaction as Curtius *degradation,* not as Curtius *rearrangement.*

A

The reaction sequence of Figure 11.40 shows how a *carboxylic acid* (which can be prepared by saponification of the methyl ester that is accessible according to Figure 9.6) can be subjected to a Curtius degradation *in a one-pot reaction.* This one-pot procedure is convenient because there is no need to isolate the potentially explosive acyl

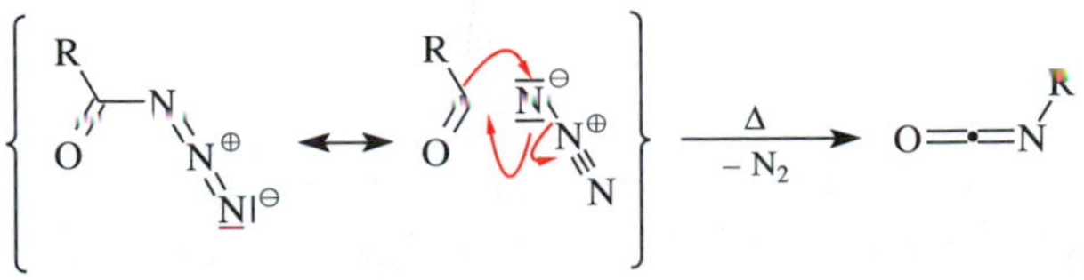

Fig. 11.39. Mechanism of the Curtius degradation.

Fig. 11.40. A one-pot diastereoselective degradation of a carboxylic acid to a Boc-protected amine via a Curtius rearrangement; Boc refers to *tert*-**b**ut**o**xyl**c**arbonyl. The mixed anhydride **B** is formed by a condensation of the phosphorus(V) reagent with the carboxyl group. The anhydride **B** acylates the concomitantly generated azide ion forming the acylazide **A**. A Curtius degradation converts **A** to **C,** and the latter reacts subsequently with *tert*-butanol to the Boc-protected amine.

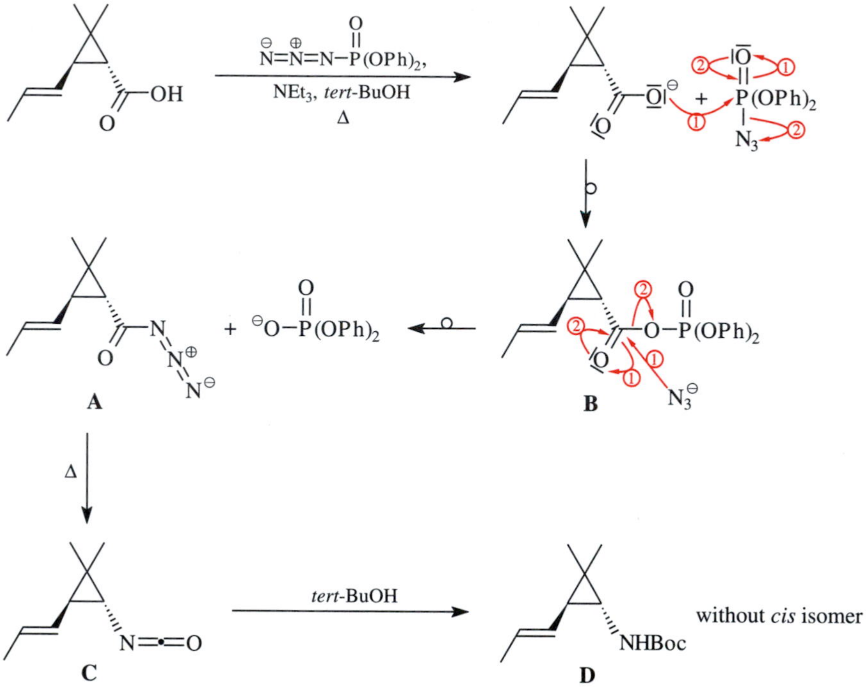

azide. The conversion of the carboxylic acid into the acyl azide occurs in the initial phase of the one-pot reaction of Figure 11.40 by means of a phosphorus(V) reagent. This reagent reacts in a manner analogous to the role of $POCl_3$ in the conversion of carboxylic acids into acid chlorides [and similar to $SOCl_2$ or $(COCl)_2$: cf. Figure 6.10]. Accordingly, a mixed carboxylic acid/phosphoric acid anhydride **B** is formed *in situ.* It acylates the simultaneously formed azide ion, whereupon the acyl azide **A** is obtained.

Compound **A** represents the immediate substrate of the Curtius degradation of Figure 11.40. Compound **A** contains a *trans*-configured cyclopropyl substituent. The *trans*-configuration of this substituent remains unchanged in the course of the [1,2]-rearrangement leading to the isocyanate **C.** Thus it migrates with complete retention of the configuration at the migrating C atom. Since it is possible to synthesize α-chiral carboxylic acids with well-defined absolute configurations (for a possible preparation, see Figure 10.38), the Curtius degradation represents an interesting method for their one-step conversion into enantiomerically pure amines of the structure $R^1R^2CH—NH_2$.

Figure 11.40 also shows how the Curtius degradation of an acyl azide can be combined with the addition of *tert*-butanol to the initially obtained isocyanate. According to Section 7.1.2, this addition gives carbamates. In the present case a *tert*-butoxycarbonyl-protected amine ("Boc-protected amine") is formed.

11.5 Claisen Rearrangement

11.5.1 Classical Claisen Rearrangement

B

The classical Claisen rearrangement is the first and slow step of the isomerization of allyl aryl ethers to *ortho*-allylated phenols (Figure 11.41). A cyclohexadienone **A** is formed in the actual rearrangement step, which is a [3,3]-sigmatropic rearrangement (see Section 11.1 for the nomenclature of sigmatropic rearrangements). Three valence electron pairs are shifted simultaneously in this step. Cyclohexadienone **A,** a nonaromatic compound, cannot be isolated and tautomerizes immediately to the aromatic and consequently more stable phenol **B.**

Fig. 11.41. Preparation of an allyl aryl ether and subsequent Claisen rearrangement. (The rearrangement is named after the German chemist Ludwig Claisen.)

Not only an aryl group—as in Figure 11.41—but also an alkenyl group can participate in the Claisen rearrangement of allyl ethers. (Figure 11.42). Allyl enol ethers are the substrates in this case. Figure 11.42 shows how such an allyl alkenyl ether, **D,** can

Fig. 11.42. Preparation of an allyl enol ether, **D,** from allyl alcohol and a large excess of ethyl vinyl ether. Subsequent Claisen rearrangement **D** → **C** proceeding with chirality transfer.

be prepared from an allyl alcohol in a single operation. The allyl alcohol is simply treated with a large excess of ethyl vinyl ether in the presence of catalytic amounts of $Hg(OAc)_2$.

The preparation involves (kind of) an oxymercuration (cf. Section 3.5.3) of the C=C double bond of the ethyl vinyl ether. The $Hg(OAc)^+$ ion is the attacking electrophile as expected, but it forms an open-chain cation **A** as an intermediate rather than a cyclic mercurinium ion. The open-chain cation **A** is more stable than the mercurinium ion because it can be stabilized by way of carboxonium resonance. Next, cation **A** takes up the allyl alcohol, and a protonated mixed acetal **B** is formed. Compound **B** eliminates EtOH and $Hg(OAc)^+$ in an E1 process, and the desired enol ether **D** results. The enol ether **D** is in equilibrium with the substrate alcohol and ethyl vinyl ether. The equilibrium constant is about 1. However, the use of a large excess of the ethyl vinyl ether shifts the equilibrium to the side of the enol ether **D** so that the latter can be isolated in high yield.

Upon heating, the enol ether **D** is converted into the aldehyde **C** via a Claisen rearrangement as depicted in Figure 11.42. The product **C** and its precursor **D** both are *cis*-substituted cyclohexanes. The σ bond that has migrated connects two C atoms in the product, while it connected a C and an O atom in the substrate. The σ bond remains on the same side of the cyclohexane ring; hence this Claisen rearrangement occurs with *complete* transfer of the stereochemical information from the original, oxygenated stereocenter to the stereocenter that is newly formed. Such a stereocontrolled transformation of an old stereocenter into a new one is called a **chirality transfer.** The Claisen rearrangement **D** $\rightarrow$ **C** of Figure 11.42 represents the special case of a **1,3-chirality transfer,** since the new stereocenter is at position 3 with respect to the old stereocenter, which is considered to be at position 1.

A

11.5.2 Claisen–Ireland Rearrangements

As shown earlier (Figure 10.19), silyl ketene acetals can be prepared at $-78°C$ by the reaction of ester enolates with chlorosilanes. *O*-allyl-*O*-silyl ketene acetals (**A** in Figure 11.43) are formed in this reaction if one employs *allyl* esters. Silyl ketene acetals of type **A** undergo [3,3]-rearrangements already upon thawing and warming to room temperature. This variation of the Claisen rearrangement is referred to as the Claisen–Ireland rearrangement.

Claisen–Ireland rearrangements obviously occur under much milder conditions than the classical Claisen rearrangements of Figures 11.41 and 11.42. Among other things, this is due to product-development control. The rearranged product of a Claisen–Ireland rearrangement is an α-allylated silyl ester, and its C=O bond is stabilized by ester resonance (14 kcal/mol according to Table 6.1). This resonance stabilization provides an additional driving force in comparison to the classical Claisen rearrangement: the primary products of classical Claisen rearrangements are ketones (Figure 11.41) or aldehydes (Figure 11.42), and the C=O double bonds of these species do not benefit from a comparably high resonance stabilization. The additional driving force corresponds, according to the Hammond postulate, to a lowered activation barrier, i.e., to an increased rearrangement rate.

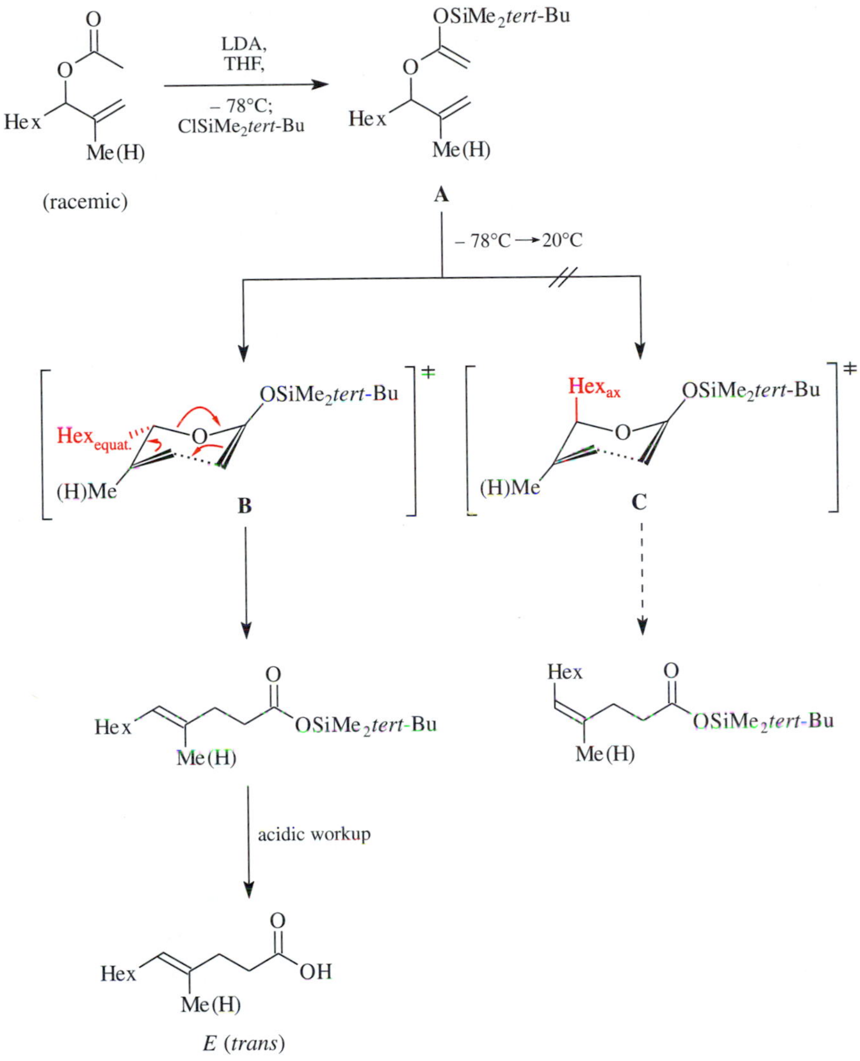

Fig. 11.43. Claisen–Ireland rearrangement of two *O*-allyl-*O*-silyl ketene acetals. *Trans*-selective synthesis of disubstituted and *E*-selective synthesis of trisubstituted alkenes.

The product of a Claisen–Ireland rearrangement essentially is a silyl ester. However, silyl esters generally are so sensitive toward hydrolysis that one usually does not attempt to isolate them. Instead, the silyl esters are hydrolyzed completely during work-up. Thus, Claisen–Ireland rearrangements *de facto* afford carboxylic acids and, more specifically, they afford γ,δ-unsaturated carboxylic acids.

Claisen–Ireland rearrangements are extraordinarily interesting from a synthetic point of view for several reasons. First, the Claisen–Ireland rearrangement is an important C=C bond-forming reaction. Second, Claisen–Ireland rearrangements afford γ,δ-unsaturated carboxylic acids, which are valuable bifunctional compounds. Both of the functional groups of these acids can then be manipulated in a variety of ways.

Claisen–Ireland rearrangements frequently are used for the synthesis of alkenes. This works particularly well if the allyl ester is derived from a secondary allyl alcohol. In this case a stereogenic double bond is formed in the rearrangement. The examples in Figure 11.43 show that the alkene is mostly *trans*-configured if this C=C bond is 1,2-disubstituted and almost completely *E*-configured if it is trisubstituted.

The stereoselectivity of the Claisen–Ireland rearrangement is the result of kinetic control. In other words, the stereoselectivity reflects that the rearrangement proceeds via the lowest-lying transition state. The transition state structure of the Claisen–Ireland rearrangement is a six-membered ring, for which a chair conformation is preferred. In the case of the two Claisen–Ireland rearrangements shown in Figure 11.43, one can imagine two chair-type transition states **B** and **C.** These transition states differ only in the orientation of the substituent at the allyllic center with respect to the chair: the substituent is quasi-equatorial in **B** and quasi-axial in **C.** Hence, the substituent is in a better position in **B** than in **C.** This is true even though in the rearrangement of the *methylated* substrate the allylic substituent and the sp^2-bound methyl group approach each other closer in the transition state **B** than in the transition state **C.**

Figure 11.44 shows how enantiomerically pure allyl alcohols (for a possible preparation, see Figure 3.31) can be converted first into allyl acetates **A** and then at −78°C into *O*-allyl-*O*-silyl ketene acetals **C.** These will undergo Claisen–Ireland rearrangements if they are warmed slowly to room temperature. The ultimately formed unsaturated carboxylic acids then contain a stereogenic C=C bond as well as a stereocenter. Both stereoelements have defined configurations. The double bond is *trans*-configured, and the absolute configuration of the stereocenter depends only on the substitution pattern of the allyl alcohol precursor.

Fig. 11.44. *Trans*-selective Claisen–Ireland rearrangements with 1,3-chirality transfer. DMAP refers to 4-**di**methyl**a**mino**p**yridine; see Figure 6.9 on DMAP-catalyzed ester formation.

Structure **B** corresponds to the most stable transition state of the Claisen–Ireland rearrangement of Figure 11.44. In this transition state, the substituent at the allyllic stereocenter is in a quasi-equatorial orientation with respect to the chair-shaped skeleton. This is the same preferred geometry as in the case of the most stable transition state **B** of the Claisen rearrangement of Figure 11.43. The reason for this preference is as before: that is, an allylic substituent that is oriented in this way experiences the smallest possible interaction with the chair skeleton. The obvious similarity of the preferred transition state structures of the Claisen–Ireland rearrangements of Figures 11.44 and 11.43 causes the same *trans*-selectivity.

The quasi-equatorial orientation of the allylic substituent in the preferred transition state **B** of the Claisen–Ireland rearrangement of Figure 11.44 also is responsible for a nearly perfect 1,3-chirality transfer (the term was explained earlier in connection with Figure 11.42). The absolute configuration of the new chiral center depends on the one hand on the configuration of the chiral center of the allyl alcohol (here *S*-configured) and on the other hand on the allyl alcohol's configuration about the C=C double bond (*cis* or *trans*). The main rearrangement product **D** is an *S*-enantiomer if a *trans*-configured *S*-allyl alcohol is rearranged. However, **D** will be an *R*-enantiomer if a *cis*-configured *S*-allyl alcohol is the starting material.

Claisen–Ireland rearrangements also allow for the realization of 1,4-chirality transfers. Two examples are shown in Figure 11.45, where the usual *trans*-selectivity also is observed (cf. Figures 11.43 and 11.44). In the rearrangements in Figure 11.45, the 1,4-chirality transfer is possible basically because propionic acid esters can be deprotonated in pure THF or in a THF/DMPU mixture, respectively, in a stereoselective fashion to provide "*E*"-configured (see Figure 10.13) or "*Z*"-configured (see Figure 10.14) ester enolates, respectively. This knowledge can be applied to the propionic acid allyl esters of Figure 11.45; that is, the propionic acid ester *syn*-**B** can be converted into an "*E*"-enolate and the ester *anti*-**B** into a "*Z*"-enolate. Each of these enolates can then be silylated with *tert*-$BuMe_2SiCl$ at the enolate oxygen.

The *O*-allyl-*O*-silyl ketene acetals *syn,E*-**C** and *anti,Z*-**C** of Figure 11.45 are thus formed isomerically pure. Each one undergoes a Claisen–Ireland rearrangement upon thawing and warming to room temperature. Again, the allylic substituent is oriented in a quasi-equatorial fashion in the energetically most favorable transition state (**D** and **E**, respectively). It follows that this allylic substituent determines (a) the configuration of the newly formed C=C double bond (similar to the cases in Figures 11.43 and 11.44) and (b) the preferred configuration of the new chiral center. These stereochemical relationships can be recognized if one "translates" the stereo-structures of the transition state structures **D** and **E** into the stereostructures of the respective rearrangement products by way of shifting three valence electron pairs simultaneously. Interestingly, *one and the same rearrangement product* is formed from the *stereoisomeric* silyl ketene acetals of Figure 11.45 via the *stereoisomeric* transition states **D** and **E.**

The rearrangement product of Figure 11.45 is synthetically useful. Heterogeneous catalytic hydrogenation allows for the conversion of this compound into a saturated compound with two methyl-substituted chiral centers with defined relative configurations. The racemic synthesis of a fragment of the vitamin E side chain (for the structure of this vitamin, see Figure 14.66) has been accomplished in this way. Claisen–Ireland rearrangements of this type also play a role in the stereoselective synthesis of other acyclic terpenes.

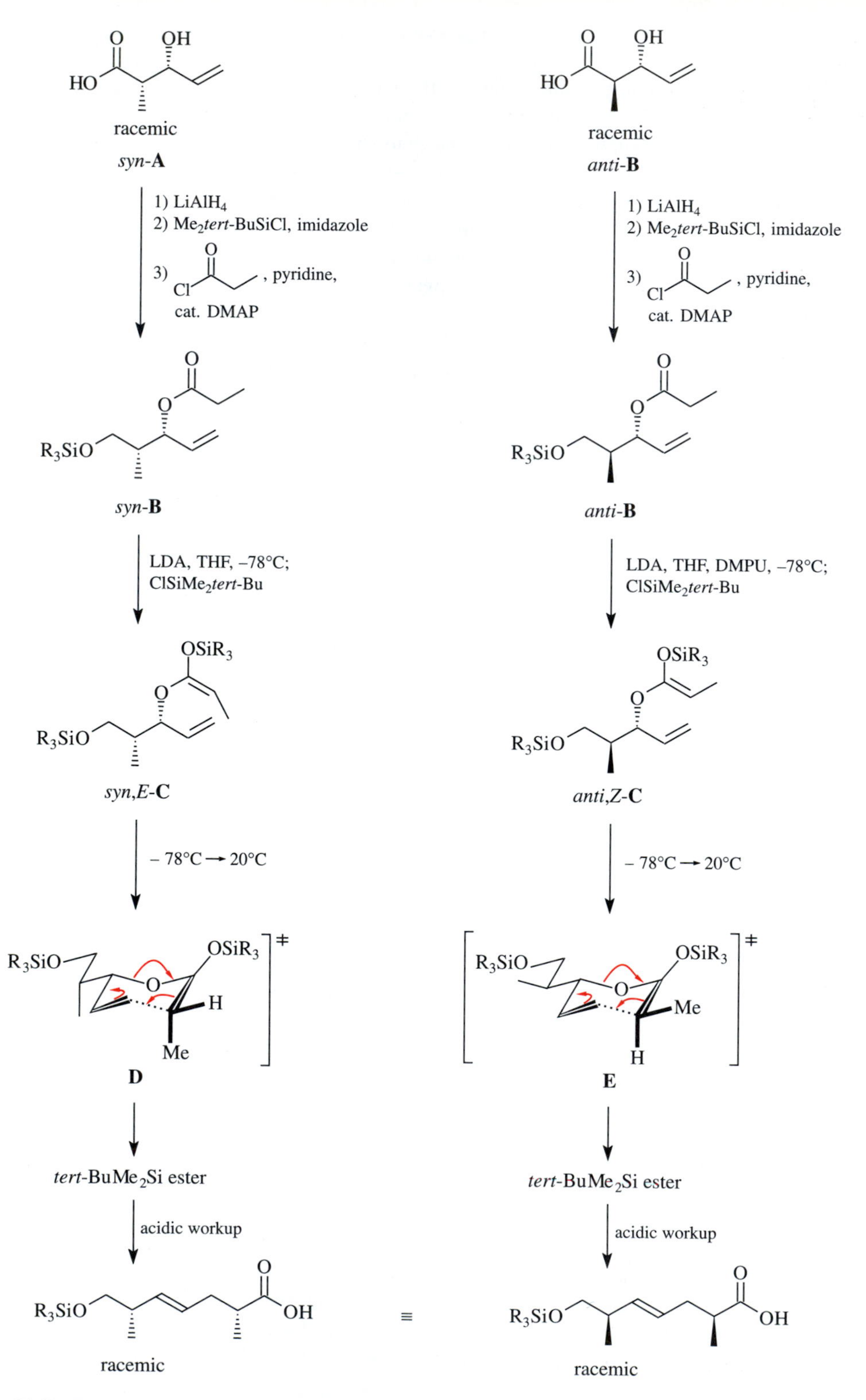

Fig. 11.45. *Trans*-selective Claisen–Ireland rearrangements with 1,4-chirality transfer. (See Figures 10.42 and 10.43, respectively, with R = vinyl in both cases, for preparations of the starting materials *syn*-**A** and *anti*-**A,** respectively.)

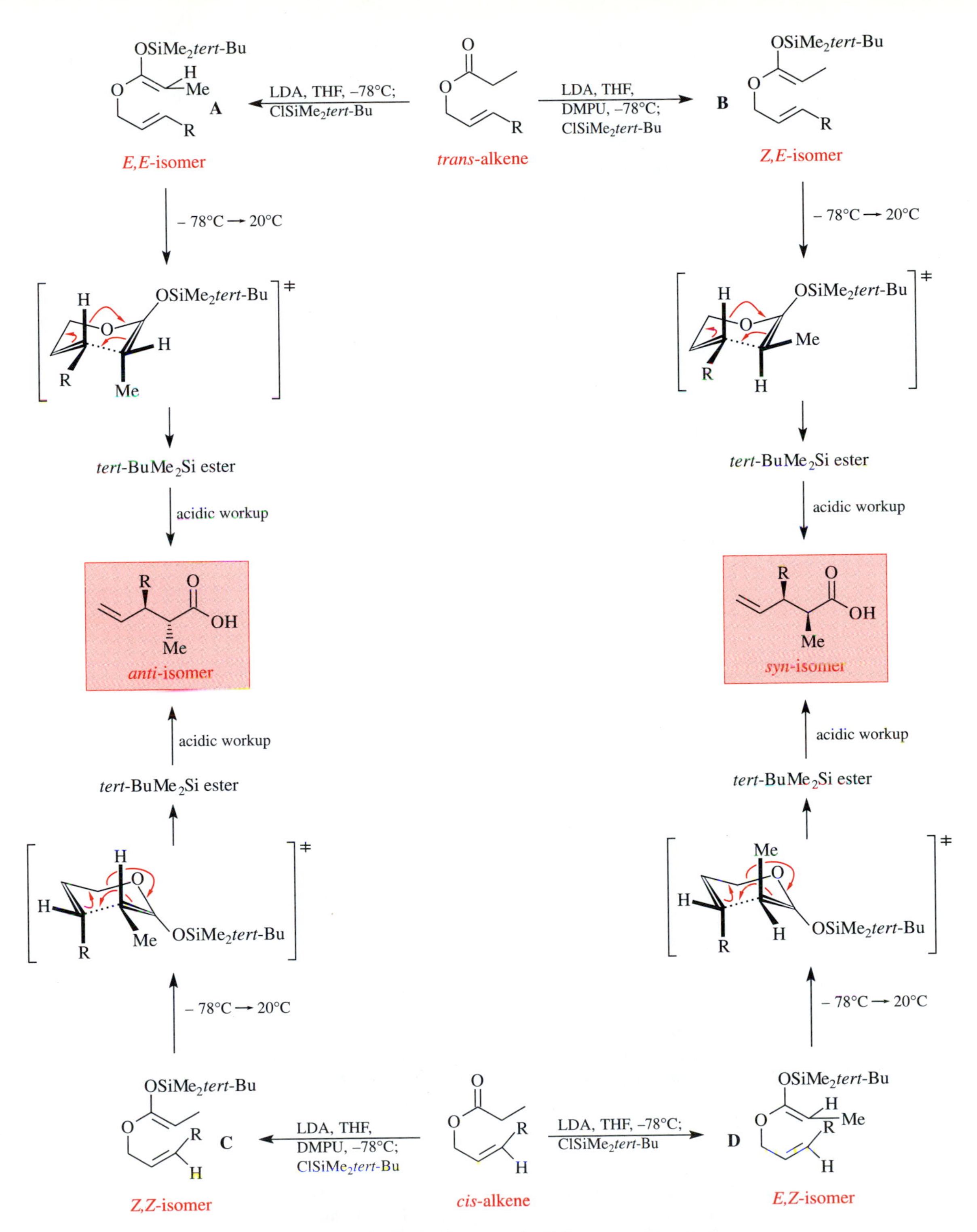

Fig. 11.46. Claisen–Ireland rearrangements with simple diastereoselectivity.

Figure 11.46 shows four Claisen–Ireland rearrangements that exhibit simple diastereoselectivity (see Section 9.3.2 for a definition of the term). The substrates are two *cis, trans*-isomeric propionic acid esters. The propionic acid esters in Figure 11.46 are derived from achiral allyl alcohols. This is different from the situation in Figure 11.45. However, these esters contain a stereogenic C=C double bond. Both the esters in Figure 11.46 can be converted into their "*E*"-enolates with LDA in pure THF (cf. Figure 10.13). Silylation affords the two *E*-configured *O*-allyl-*O*-silyl ketene acetals **A** and **D,** respectively. Alternatively, the two esters of Figure 11.46 can be converted into their "*Z*"-enolates with LDA in a mixture of THF and DMPU (cf. Figure 10.14). Treatment with *tert*-$BuMe_2SiCl$ then leads to the *Z*-isomers **B** and **C** of the *O*-allyl-*O*-silyl ketene acetals **A** and **D,** respectively.

Each of the four *O*-allyl-*O*-silyl ketene acetals **A–D** of Figure 11.46 undergoes a Claisen–Ireland rearrangement between −78°C and room temperature. These reactions all are highly stereoselective. After aqueous workup, only one of the two possible diastereomeric carboxylic acids is formed in each case. These carboxylic acids contain two new stereocenters at the α and β C-atoms. The two stereoisomers are *syn*- or *anti*-disubstituted. The *anti*-configured carboxylic acid is formed stereoselectively if one starts with the silyl ketene acetal **A** or its isomer **C.** In contrast, the Claisen–Ireland rearrangement of the silyl ketene acetal **B** or its isomer **D** gives the *syn*-configured carboxylic acid. These simple diastereoselectivities result from the fact that the transition states of the Claisen–Ireland rearrangements must have a chair conformation (see chair conformations in Figure 11.46).

It does not matter whether it is the *cis*- or the *trans*-isomer of the allyl alcohol that is more easily accessible. According to Figure 11.46, by selecting the appropriate solvent enolate formation can be directed to convert both the *cis*- and the *trans*-allyl alcohols into rearranged products that contain either a *syn*- or an *anti*-arrangement of the vicinal alkyl groups.

References

L. M. Harwood (Ed.), "Polar Rearrangements," Oxford University Press, New York, **1992.**

I. Coldham, "One or More CH and/or CC Bond(s) Formed by Rearrangement," in *Comprehensive Organic Functional Group Transformations* (A. R. Katritzky, O. Meth-Cohn, C. W. Rees, Eds.), Vol. 1, 377, Elsevier Science, Oxford, U.K., **1995.**

H. McNab, "One or More C=C Bond(s) by Pericyclic Processes," in *Comprehensive Organic Functional Group Transformations* (A. R. Katritzky, O. Meth-Cohn, C. W. Rees, Eds.), Vol. 1, 771, Elsevier Science, Oxford, U.K., **1995.**

P. J. Murphy, "One or More =CH, =CC and/or C=C Bond(s) Formed by Rearrangement," in *Comprehensive Organic Functional Group Transformations* (A. R. Katritzky, O. Meth-Cohn, C. W. Rees, Eds.), Vol. 1, 793, Elsevier Science, Oxford, U.K., **1995.**

11.3

M. Saunders, J. Chandrasekhar, P. v. R. Schleyer, "Rearrangements of Carbocations," in *Rearrangements in Ground and Excited States* (P. D. Mayo, Ed.), Vol. 1, 1, Academic Press, New York, **1980.**

J. R. Hanson, "Wagner-Meerwein Rearrangements," in *Comprehensive Organic Synthesis* (B. M. Trost, I. Fleming, Eds.), Vol. 3, 705, Pergamon Press, Oxford, **1991.**

B. Rickborn, "The Pinacol Rearrangement," in Comprehensive Organic Synthesis (B. M. Trost, I. Fleming, Eds.), Vol. 3, 721, Pergamon Press, Oxford, 1991.

D. J. Coveney, "The Semipinacol and Other Rearrangements," in *Comprehensive Organic Synthesis* (B. M. Trost, I. Fleming, Eds.), Vol. 3, 777, Pergamon Press, Oxford, **1991.**

C. D. Gutsche, "The reaction of diazomethane and its derivatives with aldehydes and ketones," *Org. React.* **1954,** *8,* 364–429.

P. A. S. Smith and D. R. Baer, "The Demjanov and Tiffeneau-Demjanov ring expansions," *Org. React.* **1960,** *11,* 157–188.

T. Money, "Remote Functionalization of Camphor: Application to Natural Product Synthesis," in *Organic Synthesis: Theory and Applications* (T. Hudlicky, Ed.), **1996,** 3, JAI, Greenwich, CT.

W. E. Bachmann and W. S. Struve, "The Arndt-Eistert reaction," *Org. React.* **1942,** *1,* 38–62.

G. B. Gill, "The Wolff Rearrangement," in *Comprehensive Organic Synthesis* (B. M. Trost, I. Fleming, Eds.), Vol. 3, 887, Pergamon Press, Oxford, **1991.**

T. Ye and M. A. McKervey, "Organic synthesis with α-diazo carbonyl compounds," *Chem. Rev.* **1994,** *94,* 1091–1160.

M. P. Doyle, M. A. McKervey, T. Ye, "Reactions and Syntheses with α-Diazocarbonyl Compounds," Wiley, New York, **1997.**

C. H. Hassall, "The Baeyer-Villiger oxidation of aldehydes and ketones," *Org. React.* **1957,** *9,* 73–106.

G. R. Krow, "The Baeyer-Villiger Reaction," in *Comprehensive Organic Synthesis* (B. M. Trost, I. Fleming, Eds.), Vol. 7, 671, Pergamon Press, Oxford, **1991.**

G. R. Krow, "The Bayer-Villiger oxidation of ketones and aldehydes," *Org. React.* **1993,** *43,* 251–798.

L. G. Donaruma and W. Z. Heldt, "The Beckmann rearrangement," *Org. React.* **1960,** *11,* 1–156.

D. Craig, "The Beckmann and Related Reactions," in *Comprehensive Organic Synthesis* (B. M. Trost, I. Fleming, Eds.), Vol. 7, 689, Pergamon Press, Oxford, **1991.**

R. E. Gawley, "The Beckmann Reactions: Rearrangements, Elimination- Additions, Fragmentations, and Rearrangement-Cyclisations," *Org. React.* (N.Y.) **1987,** *35,* 1.

P. A. S. Smith, "The Curtis reaction," *Org. React.* **1946,** *3,* 337–449.

11.5

D. S. Tarbell, "The Claisen rearrangement," *Org. React.* **1944,** *2,* 1–48.

S. J. Rhoads and N. R. Raulins, "The Claisen and Cope rearrangements," *Org. React.* **1975,** *22,* 1–252.

F. E. Ziegler, "The thermal aliphatic Claisen rearrangement," *Chem. Rev.* **1988,** *88,* 1423.

P. Wipf, "Claisen Rearrangements," in *Comprehensive Organic Synthesis* (B. M. Trost, I. Fleming, Eds.), Vol. 5, 827, Pergamon Press, Oxford, **1991.**

H.-J. Altenbach, "Diastereoselective Claisen Rearrangements," in *Organic Synthesis Highlights* (J. Mulzer, H.-J. Altenbach, M. Braun, K. Krohn, H.-U. Reißig, Eds.), VCH, Weinheim, New York, **1991,** 111–115.

H.-J. Altenbach, "Ester Enolate Claisen Rearrangements," in *Organic Synthesis Highlights* (J. Mulzer, H.-J. Altenbach, M. Braun, K. Krohn, H.-U. Reißig, Eds.), VCH, Weinheim, New York, **1991,** 116–118.

S. Pereira and M. Srebnik, "The Ireland-Claisen rearrangement," *Aldrichimica Acta* **1993,** *26,* 17–29.

B. Ganem, "The mechanism of the Claisen rearrangement: Déjà vu all over again," *Angew. Chem. Int. Ed. Engl.* **1996,** *35,* 937.

H. Frauenrath, "Formation of C—C Bonds by [3,3] Sigmatropic Rearrangements," in *Stereoselective Synthesis* (Houben-Weyl) 4th ed. 1996, (G. Helmchen, R. W. Hoffmann, J. Mulzer, E. Schaumann, Eds.), **1996,** Vol. E21 (Workbench Edition), *6,* 3301–3756, Georg Thieme Verlag, Stuttgart.

Further Reading

E. L. Wallis and J. F. Lane, "The Hoffmann reaction," *Org. React.* **1946,** *3,* 267–306.

H. Wolff, "The Schmidt reaction," *Org. React.* **1946,** *3,* 307–336.

A. H. Blatt, "The Fries reaction," *Org. React.* **1942,** *1,* 342–369.

G. Magnusson, "Rearrangements of epoxy alcohols and related compounds," *Org. Prep. Proced. Int.* **1990,** *22,* 547.

A. Heins, H. Upadek, U. Zeidler, "Preparation of Aldehydes by Rearrangement with Conservation of Carbon Skeleton," in *Methoden Org. Chem.* (*Houben-Weyl*) *4th ed. 1952-, Aldehydes* (J. Falbe, Ed.), Vol. E3, 491, Georg Thieme Verlag, Stuttgart, **1983.**

W. Kirmse, "Alkenylidenes in organic synthesis," *Angew. Chem.* **1997,** *109,* 1212–1218; *Angew. Chem. Int. Ed. Engl.* **1997,** *36,* 1164–1170.

G. Strukul, "Transition Metal Catalysis in the Baeyer-Villiger Oxidation of Ketones," *Angew. Chem.* **1998,** *110,* 1256–1267; *Angew. Chem. Int. Ed. Engl.* **1998,** *37,* 1198–1209.

M. Braun, "α-Heteroatom substituted 1-alkenyllithium reagents: Carbanions and Carbenoids for C-C bond formation," *Angew. Chem.* **1998,** *110,* 444–465; *Angew. Chem. Int. Ed. Engl.* **1998,** *37,* 430–451.

K. N. Houk, J. Gonzalez, Y. Li, "Pericyclic reaction transition states: Passions and punctilios, 1935-1995," *Acc. Chem. Res.* **1995,** *28,* 81.

R. K. Hill, "Cope, Oxy-Cope and Anionic Oxy-Cope Rearrangements," in *Comprehensive Organic Synthesis* (B. M. Trost, I. Fleming, Eds.), Vol. 5, 785, Pergamon Press, Oxford, **1991.**

F. E. Ziegler, "Consecutive Rearrangements," in *Comprehensive Organic Synthesis* (B. M. Trost, I. Fleming, Eds.), Vol. 5, 875, Pergamon Press, Oxford, **1991.**

D. L. Hughes, "Progress in the Fischer indole reaction," *Org. Prep. Proced. Int.* **1993,** *25,* 607.

G. Boche, "Rearrangements of carbanions," *Top. Curr. Chem.* **1988,** *146,* 1.

R. K. Hill, "Chirality Transfer via Sigmatropic Rearrangements," in *Asymmetric Synthesis. Stereodifferentiating Reactions—Part A* (J. D. Morrison, Ed.), Vol. 3, 502, AP, New York, **1984.**

C. J. Roxburgh, "Syntheses of medium sized rings by ring expansion reactions," *Tetrahedron* **1993,** *49,* 10749–10784.

Thermal Cycloadditions

12

12.1 Driving Force and Feasibility of One-Step [2+4]- and [2+2]-Cycloadditions

A

We dealt with [2+4]-cycloadditions very briefly in Section 3.3.1. As you saw there, a [2+4]-cycloaddition requires two different substrates: one of these is an alkene—or an alkyne—and the other is 1,3-butadiene or a derivative thereof. The reaction product, in this context also called the cycloadduct, is a six-membered ring with one or two double bonds. Some hetero analogs of alkenes, alkynes, and 1,3-butadiene also undergo analogous [2+4]-cycloadditions. In a [2+2]-cycloaddition an alkene or an alkyne reacts with ethene or an ethene derivative to form a four-membered ring. Again, hetero analogs may be substrates in these cycloadditions; allenes and some heterocumulenes also are suitable substrates.

Two new σ bonds are formed in both the [2+4]- and the [2+2]-cycloadditions while two π bonds are broken. It is for this reason that most cycloadditions exhibit a significant driving force, and this remains true even when the cycloadduct is strained. Having realized this, one is a bit surprised that only a few of the cycloadditions just mentioned occur quickly (see Figure 12.1, bottom). Others require quite drastic reaction conditions. The two simplest [2+4]-cycloadditions, the additions of ethene or acetylene to 1,3-butadiene (Figure 12.1, top), belong to the latter group of reactions. Some cycloadditions cannot be carried out in a one-step process at all. The [2+2]-cycloadditions between two alkenes or between an alkene and an alkyne (Figure 12.1, center) belong to this kind of cycloadditions.

[2+2]-Cycloadditions are less exothermic than [2+4]-cycloadditions, since the former result in a strained cycloadduct while the latter give unstrained rings. Thus, according to the Hammond postulate, the [2+2]-cycloadditions should occur more slowly than the [4+2]-cycloadditions. While these cycloadditions would be expected to be *slower,* thermochemistry does *not* explain why one-step cycloadditions between two alkenes or between an alkene and an alkyne—in the absence of light—cannot be achieved at all, *whereas other one-step cycloadditions that lead to four-membered rings do occur remarkably fast at room temperature:* the additions of dichloroketene to alkenes or acetylenes provide striking examples (Figure 12.1, bottom). The latter [2+2]-cycloadditions afford cyclobutanones and cyclobutenones as cycloadducts, which are even more strained than the cyclobutanes and cyclobutenes, which are inaccessible via one-step additions.

Evidently, the ring-size dependent exothermicities of one-step cycloadditions do not explain the differences in their reaction rates. In fact, these differences can be understood only by going beyond the simplistic "electron-pushing" formalism. To really understand these reactions, one needs to compare the transition states of these reactions in the context of molecular orbital (MO) theory. These comparisons—and the presentation of the requisite theoretical tools—are the subjects of Sections 12.2.2–12.2.4.

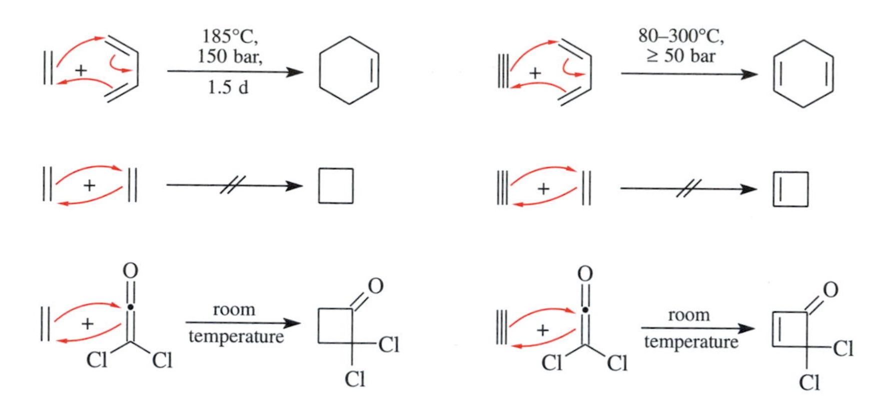

Fig. 12.1. One-step [2+4]- and [2+2]-cycloadditions and their feasibility in the absence of light. The [2+4]-cycloaddition between ethene or acetylene and 1,3-butadiene (top) requires drastic conditions. The one-step [2+2]-cycloaddition between two alkenes or between an alkene and an acetylene (center) cannot be achieved at all. The only [2+2]-cycloadditions that proceed at room temperature are those between ethene or an alkyne and dichloroketene.

12.2 Transition State Structures of Selected One-Step [2+4]- and [2+2]-Cycloadditions

12.2.1 Stereostructure of the Transition States of One-Step [2+4]-Cycloadditions

The combination of powerful computers and modern methods of theoretical chemistry makes it possible to obtain detailed information about transition states and transition state structures. One can compute the Cartesian coordinates of all the atoms involved and all their bond lengths and angles. The energies of the transition states also can be computed. The theoretical estimate for the activation energy of a specific reaction can be determined by subtracting the energies of the starting materials from the energy of the transition state. Yet, we will not be concerned with such numerical data in this section.

The computed transition state of the [2+4]-cycloaddition between ethene and butadiene is shown in Figure 12.2 (top), along with the computed transition state of the [2+4]-cycloaddition between acetylene and butadiene. It is characteristic of the stereochemistry of these transition states that ethene or acetylene, respectively, approach the *cis* conformer of butadiene from a face (and not in-plane). Figure 12.2 also shows that the respective cycloadducts—cyclohexene or 1,4-cyclohexadiene—initially result in the twist–boat conformation.

The transition states of the two [2+4]-cycloadditions in Figure 12.2 are "early" because their geometries are more similar to the starting materials than to the cy-

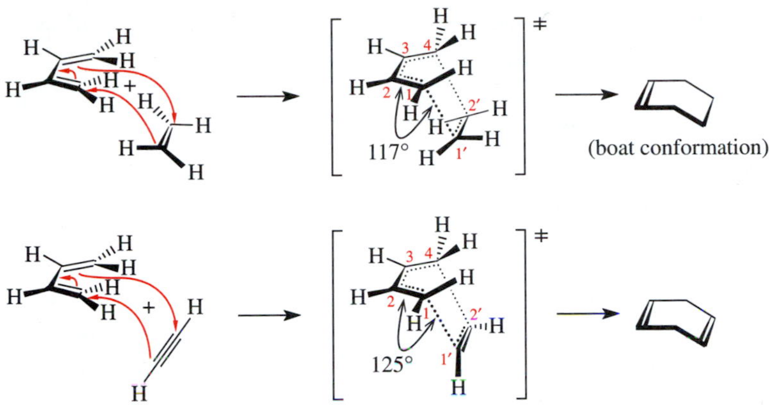

Fig. 12.2. Perspective drawings of the transition state structures of the [2+4]-cycloadditions ethene + butadiene → cyclohexene and acetylene + butadiene → 1,4-cyclohexadiene.

(boat conformation)

[4 + 2]-Addition of 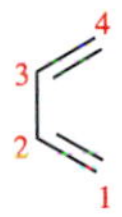 butadiene (C1–C4) and ...	C^1/C^2 or C^3/C^4	C^2/C^3	$C^{1'}/C^{2'}$	$C^1/C^{1'}$ and $C^4/C^{2'}$
	Extent of the bond length changes in the transition state (100% bond length change is realized when the cycloadduct has been reached):			Extent of the bond length elongations in the transition state (in percent of the respective bond length in the cycloadduct):
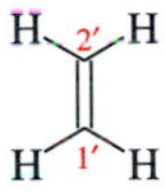 $H_2C^{2'}{=}C^{1'}H_2$	31%	43%	28%	152%
 $HC^{2'}{\equiv}C^{1'}H$	28%	52%	29%	146%

cloadducts (see tabular section of Figure 12.2). To begin with, consider the distances between atoms that are bonded in the starting materials and in the products via bonds that undergo a bond order *change* in the course of the [2+4]-cycloadditions. In going from the starting materials to the transition states, these distances are altered (increased or decreased) by less than half the overall difference between the starting materials and the products. Second, the newly formed σ bonds in both transition states remain about 1.5 times longer than the respective bond lengths in the cycloadducts: the formation of these σ bonds is only just starting in the transition states. Third, the hybridization change of the four C atoms that change their hybridization in the course of the [2+4]-cycloaddition has not progressed much: the bond angles at these C atoms have changed only very little in comparison to the bond angles in the starting materials.

12.2.2 Frontier Orbital Interactions in the Transition States of One-Step [2+4]-Cycloadditions

What Are the Factors Contributing to the Activation Energy of [2+4]-Cycloadditions?

The computation of the activation energy of the cycloaddition of ethene and butadiene requires that one sums up the cumulative effects of all the energy changes that are associated with the formation of the transition state of this reaction (Figure 12.2, top) from the starting materials. To begin with, there are the energy increases due to the changes of bond lengths and bond angles, which already were presented. A second contribution is due to the newly occurring inter- and intramolecular van der Waals repulsions. The energy lowering that is associated with incipient bond formations would be a third contribution to consider.

Especially the last factor often is the one that determines the reaction rates of [2+4]-cycloadditions. This factor allows one to understand, for example, why the cycloadditions of ethene or acetylene with butadiene occur only under rather drastic conditions, while the analogous cycloadditions of tetracyanoethene or acetylenedicarboxylic acid esters "work like a charm." As will be seen, merely an orbital interaction between the reagents at the sites where the new σ bonds are formed is responsible for this advantageous reduction of the activation energies of the latter two reactions.

One can, with surprisingly simple methods, establish whether at all such transition state stabilizations through orbital interactions occur, and one can even estimate their extent. These considerations are based on the knowledge that the transition states are "early" (Section 12.2.1), that is, that the transition states resemble the starting materials both structurally and energetically. It is for this reason that one can *model these transition states using the starting materials* and *discuss the MOs of the transition states by inspection of the MOs of the starting materials.* The stabilization of the transition states as the result of the incipient σ bond formations is thus due to additional orbital overlap, which does not occur in the separated starting materials. Note that these overlaps result from *intermolecular* orbital interactions.

These intermolecular orbital interactions, which are of the σ type, occur between the ends of the π-type MOs that are associated with the respective two reagents. We will see in the next subsection what the ends of these π-type MOs are like. In the subsequent subsection we will deal with the energy effects of the new orbital interactions. The energy effects associated with the special case of [2+4]-cycloadditions will be considered thereafter.Finally, these new orbital interactions will be discussed for [2+2]-cycloadditions.

The LCAO Model of π MOs of Ethene, Acetylene, and Butadiene; Frontier Orbitals

There are a variety of methods for the computation of the MOs that interact in the transition states of [2+4]-cycloadditions. The **LCAO method** (**l**inear **c**ombination of **a**tomic **o**rbitals) is often employed, and the basic idea is as follows. The MOs of the π systems of alkenes, conjugated polyenes, or conjugated polyenyl cations, radicals, or anions all are built by so-called linear combinations of $2p_z$ AOs. In a somewhat casual formulation, one might say that the MOs of these π systems are constructed "with the help of the $2p_z$ AOs." These AOs are centered at the positions of the n C atoms that are part of the π system. LCAO computations describe a conjugated π-electron system that extends over n sp^2-hybridized C atoms by way of n π-type MOs.

MOs that describe a π system and have lower and higher energy, respectively, than a $2p_z$ AO, are called **bonding** and **antibonding MOs,** respectively. Figure 12.3 shows examples. In acyclic π systems with an odd number of centers n, a **nonbonding MO** also occurs. This is illustrated in Figure 12.3 as well. The nonbonding MO has the same energy as the $2p_z$ AO. An **MO diagram** shows the n π-type MOs and their occupation.

The distribution of π electrons over the π MOs is regulated by the *Aufbau* principle and the Pauli rule. The only occupied MO of an MO diagram or the highest occupied MO thereof is called the **HOMO** (**h**ighest **o**ccupied **m**olecular **o**rbital). The only unoccupied MO of an MO diagram or the lowest unoccupied MO thereof is called the **LUMO** (**l**owest **u**noccupied **m**olecular **o**rbital). HOMOs and LUMOs are the so-called **frontier orbitals** since they flank the borderline between occupied and unoccupied orbitals.

The application of the LCAO method to ethene yields two π-type MOs, since two sp^2-hybridized centers build up the π-electron system. Butadiene contains four π-type MOs because four sp^2-hybridized centers build up the conjugated π-electron system. The MOs of ethene and butadiene and their occupations are shown in Figure 12.4. In the π-MO diagrams of ethene and butadiene, all the bonding MOs—one for ethene and two for butadiene—are completely occupied. The antibonding MO of ethene and the antibonding MOs of butadiene are unoccupied.

Frontier Orbital Interactions in Transition States of Organic Chemical Reactions and Associated Energy Effects

Generally, when two originally isolated molecules approach each other as closely as is the case in the transition state of a chemical reaction, interactions will occur between all the MOs of one molecule and all the MOs of the other molecule in those regions

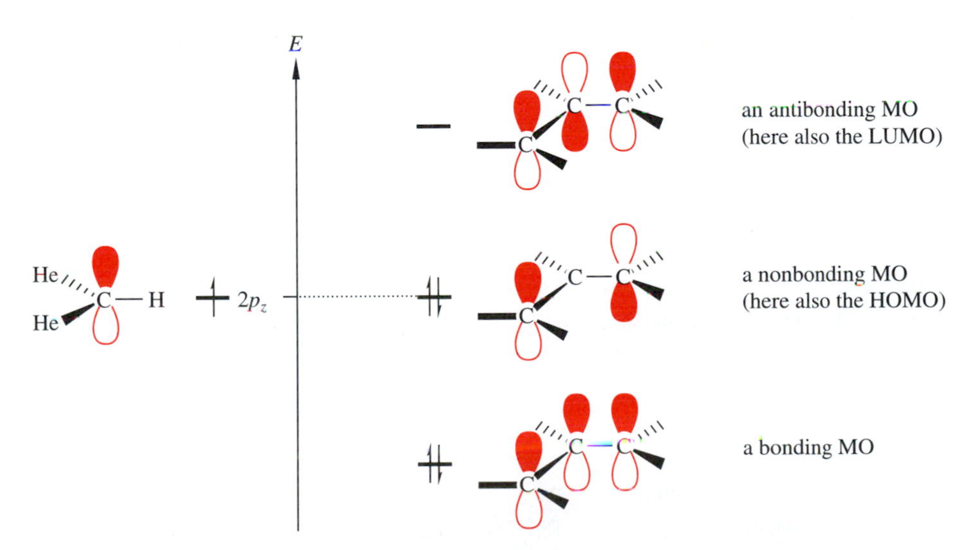

Fig. 12.3. Illustration of the term "MO diagram": left, the π-MO diagram of the methyl radical (the only π MO is identical with the $2p_z$ AO of the trivalent C atom: cf. Section 1.1.2); right, the π-MO diagram of the allyl anion.

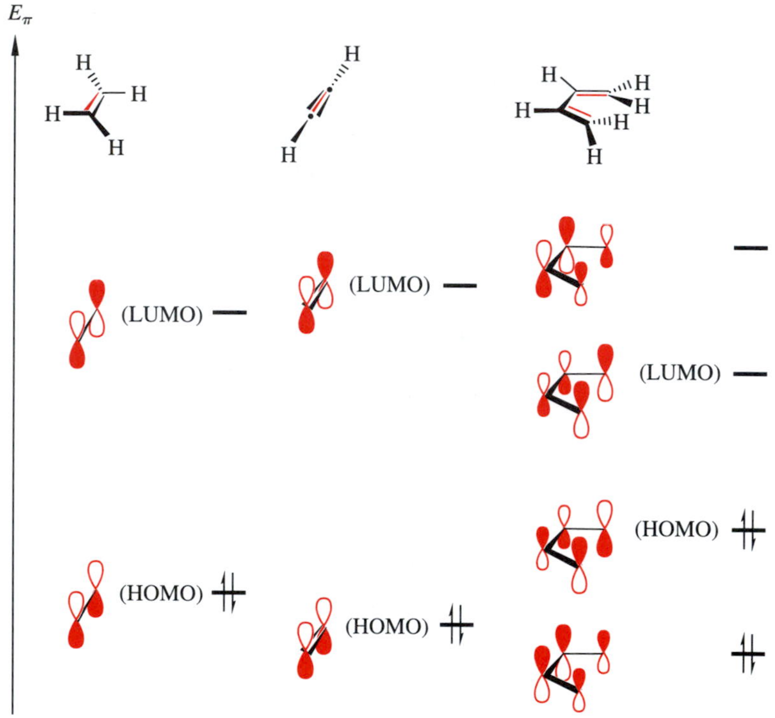

Fig. 12.4. The π MOs of ethene, acetylene, and 1,3-butadiene and the respective energy-level diagrams. The sign of the $2p_z$ AOs is indicated by the open and shaded orbital lobes. The relative importance of each contributing AO is indicated by the size of the atomic orbital.

in which the molecules approach each other most closely. From now on, we shall refer to those moieties of the MOs of the substrates, which are directly involved in the mentioned interactions, as "orbital fragments" of the respective MOs. The MOs that used to be localized on the individual substrates now serve to create new MOs that are delocalized over both substrates. If each substrate has only one MO, which contributes to the orbital interaction, the latter leads to *two* more delocalized MOs. Specifically, the result of this interaction is that one of the delocalized MOs comes to lie below the more stable MO of the isolated substrates while the other delocalized MO comes to lie above the less stable MO of the isolated substrates. The stabilization of the one delocalized MO is proportional to the bonding overlap between the MOs of the substrates. The other delocalized MO becomes destabilized by about the same amount as a result of the antibonding overlap between the MOs of the substrates. Several factors determine whether the interactions between the substrate MOs—or, differently expressed, the concomitant formation of more delocalized MOs in the transition state—will lead to a stabilization or a destabilization of the transition state and what the extent of that (de)stabilization will be. These factors can be summarized as follows.

- The interaction between a fully occupied MO of the substrate and a fully occupied MO of the other starting material at a single center of each component leads neither to stabilization nor to destabilization (Figure 12.5a). The same is true if both interacting MOs are empty (Figure 12.5c).
- The interaction between a fully occupied MO of the substrate and an unoccupied MO of the other starting material at a single center of each component leads to sta-

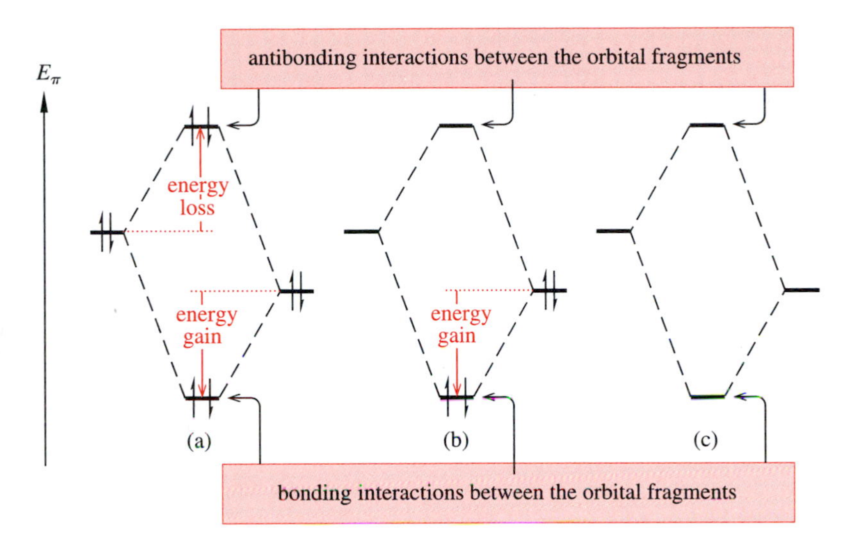

Fig. 12.5. σ-Type interaction between one end of a conjugated π-electron system and one end of another conjugated π-electron system; influence of the orbital occupancy of the π MOs on the electronic energy.

bilization if the interaction is bonding (Figure 12.5b) and to destabilization if the interaction is antibonding.

- The extent of the stabilization or destabilization, respectively, is inversely proportional to the energy difference between the localized MOs involved (Figure 12.6). The stabilization increases with decreasing energy difference.
- For a given energy difference between the interacting MOs, the magnitude of the stabilization or destabilization, respectively, is proportional to the amount of overlap between the orbital fragments (Figure 12.7 provides a plausible example). A large overlap results in a large stabilization or destabilization, respectively, and vice versa.

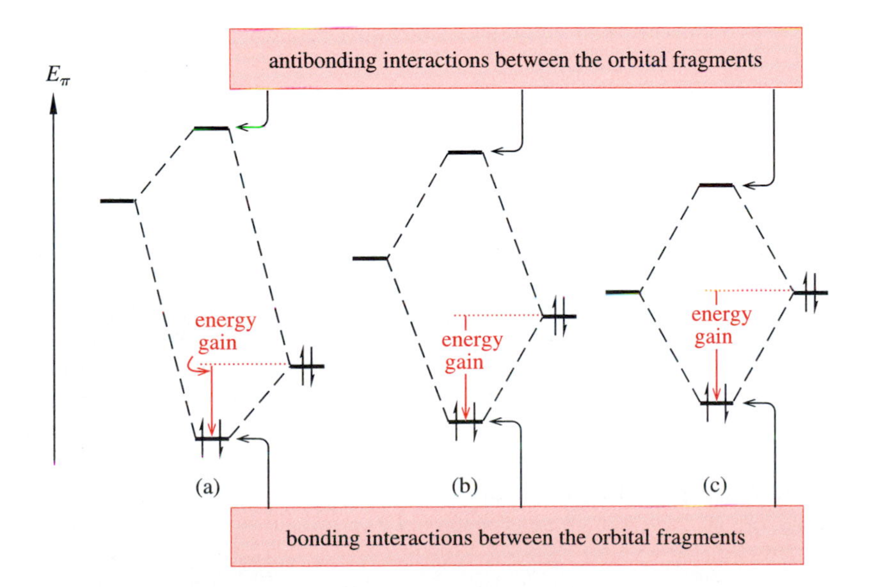

Fig. 12.6. σ-Type interaction between one end of a conjugated π-electron system and one end of another conjugated π-electron system; influence of the energy difference between the π MOs on the energy gain.

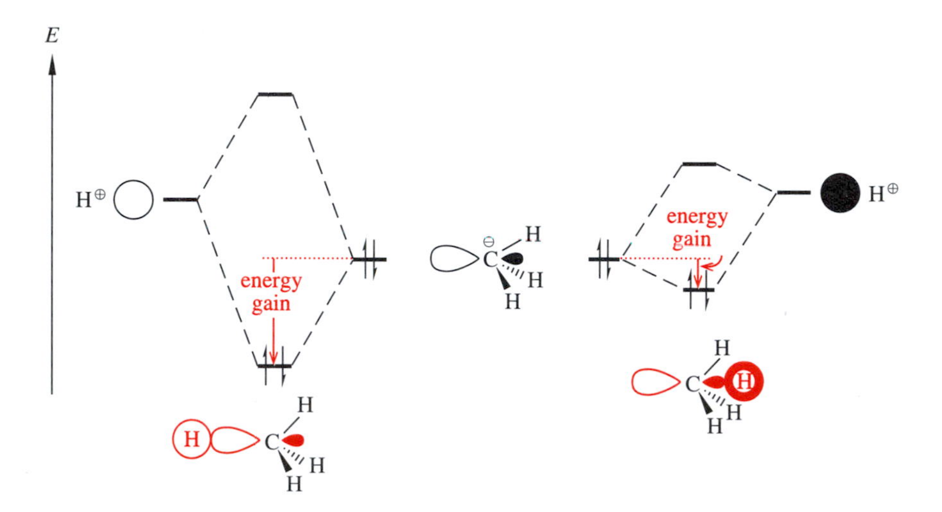

Fig. 12.7. σ-Type interactions between the unoccupied 1*s* AO of a proton and the doubly occupied sp^3 AO of a CH_3 anion; influence of the magnitude of the overlap on the stabilization of the transition states of two bond-forming reactions. Left, formation of tetrahedral methane; right, formation of a fictitious stereoisomer—an asymmetric trigonal bipyramid.

The foregoing statements concerning stabilizing orbital interactions can be extended to transition states. For them, they can be formulated more concisely and more generally using the term "frontier orbitals," as introduced in the discussion of Figure 12.3:

> Essentially the only mechanism for the electronic stabilization of transition states is the bonding interaction between fully occupied and empty frontier orbitals.

Equation 12.1 offers a quantitative formulation of this statement. This equation makes a statement about the stabilization ΔE_{TS} of the transition state of a reaction between substrate I (reacting at its reactive center 1, where its frontier orbitals have the coefficients $c_{1,HOMO_I}$ and $c_{1,LUMO_I}$) and substrate II (reacting at its reactive center 1 as well, where its frontier orbitals have the coefficients $c_{1,HOMO_{II}}$ and $c_{1,LUMO_{II}}$) owing to the two frontier orbital interactions:

$$\Delta E_{TS} \propto \frac{(c_{1,HOMO_I} \cdot c_{1,LUMO_{II}})^2}{E_{HOMO_I} - E_{LUMO_{II}}} + \frac{(c_{1,LUMO_I} \cdot c_{1,HOMO_{II}})^2}{E_{HOMO_{II}} - E_{LUMO_I}} \tag{12.1}$$

Here we are mostly interested in the transition states of one-step cycloadditions between two unsaturated molecules I and II. In this special case, the frontier orbitals will be π-type orbitals, and overlaps at *two* ends of each orbital fragment contribute to each frontier orbital interaction: overlap at the termini $C1_I/C1_{II}$ and another overlap involving the termini $C\omega_I/C\omega_{II}$. (Substrate I reacts at its C atom 1 with C atom 1 of substrate II and at its C atom ω with C atom ω of substrate II. Substrate I possesses the frontier orbital coefficients $c_{1,HOMO_I}$ and $c_{1,LUMO_I}$ at its reactive center C1 and the coefficients $c_{\omega,HOMO_I}$ and $c_{\omega,LUMO_I}$ at $C\omega$. In analogy, the frontier orbital coefficients of substrate II are $c_{1,HOMO_{II}}$ and $c_{1,LUMO_{II}}$ at its reactive center C1 and $c_{\omega,HOMO_{II}}$ and $c_{\omega,LUMO_{II}}$ at reactive center $C\omega$. The stabilization ΔE_{TS} of the transition state of such a one-step cycloaddition can be expressed in terms of the frontier orbitals as Equation 12.2.

$$\Delta E_{\mathrm{TS}} \propto \frac{[(c_{1,\mathrm{HOMO_I}} \cdot c_{1,\mathrm{LUMO_{II}}}) + (c_{\omega,\mathrm{HOMO_I}} \cdot c_{\omega,\mathrm{LUMO_{II}}})]^2}{E_{\mathrm{HOMO_I}} - E_{\mathrm{LUMO_{II}}}} + \frac{[(c_{1,\mathrm{LUMO_I}} \cdot c_{1,\mathrm{HOMO_{II}}}) + (c_{\omega,\mathrm{LUMO_I}} \cdot c_{\omega,\mathrm{HOMO_{II}}})]^2}{E_{\mathrm{HOMO_{II}}} - E_{\mathrm{LUMO_I}}} \tag{12.2}$$

Interestingly, according to Equation 12.2 the HOMO/LUMO interactions in cycloadditions are not necessarily stabilizing; they also can be nonbonding. Whether the interaction is bonding or nonbonding depends on the size and the sign of the fragment orbitals at the reacting centers. In contrast, according to Equation 12.1, the HOMO/LUMO for reactions in which only one bond is formed interaction always is bonding.

Frontier Orbital Interactions in Transition States of One-Step [2+4]-Cycloadditions

Figure 12.2 showed the stereostructures of the transition states of the [2+4]-cycloadditions between ethene or acetylene, respectively, and butadiene. The HOMOs and LUMOs of all substrates involved are shown in Figure 12.4. Figures 12.8 and 12.9 depict the corresponding HOMO/LUMO pairs in the transition states of the respective [2+4]-cycloadditions. Evaluation of Equation 12.2 reveals two new bonding HOMO/LUMO interactions of comparable size in both transition states. Therefore, the transition states of *both* cycloadditions benefit from a significant stabilization. Consequently, these types of cycloadditions can be realized under (fairly) mild conditions.

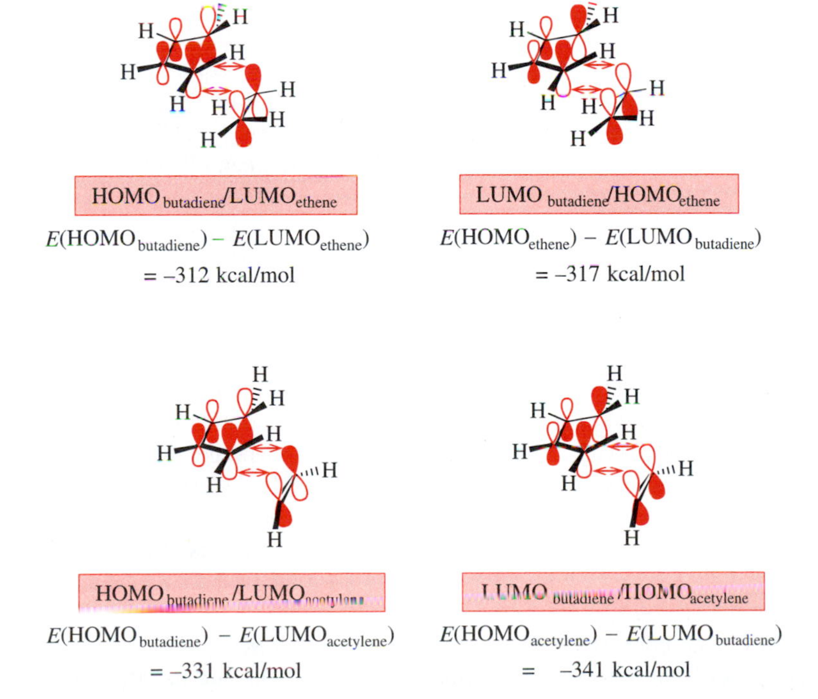

Fig. 12.8. Frontier orbital interactions in the transition state of the one-step [2+4]-cycloaddition of ethene and butadiene.

Fig. 12.9. Frontier orbital interactions in the transition state of the one-step [2+4]-cycloaddition of acetylene and butadiene.

12.2.3 Frontier Orbital Interactions in the Transition States of the Unknown One-Step Cycloadditions of Alkenes or Alkynes to Alkenes

The one-step cycloadditions ethene + ethene → cyclobutane and ethene + acetylene → cyclobutene are unknown (see Figure 12.1). One can understand why this is so by analyzing the frontier orbital interactions in the associated transition states (Figure 12.10). Both HOMO/LUMO interactions are nonbonding. This circumstance contributes to the fact that the respective transition states are energetically out of reach.

Fig. 12.10. Frontier orbital interactions in plausible transition states of the one-step [2+2]-cycloadditions of ethene and ethene (left) and of ethene and acetylene (center and right).

12.2.4 Frontier Orbital Interactions in the Transition State of One-Step [2+2]-Cycloadditions Involving Ketenes

The transition state of the [2+2]-cycloaddition of ketene and ethene is shown in Figure 12.11. In this transition state, the carbonyl C atom (C2) of the ketene approaches the ethene more closely than the methene C atom (C1) does. The two C atoms and the O atom of the ketene fragment no longer are collinear. Yet, all five atoms of the ketene remain in one plane. The structural changes between ethene and the ethene moiety in the transition state are minor. Besides, in the transition state of the [2+2]-cycloaddition, the four atoms that will eventually form the cycloadduct are still far removed from their positions in the cycloadduct. All these structural features characterize this transition state as an early one. Therefore, much as in the case of the transition states of the one-step [2+4]-cycloadditions in Section 12.2.2, the bonding situation can be described by means of the MOs of the separated reagents.

The π MO diagram of the substrate ethene is shown on the left of Figure 12.12; in the center and to the right is shown the MO diagram of the other reactant, the ketene. The HOMO of ethene (HOMO$_A$ in Figure 12.12) is its bonding π MO and the LUMO (LUMO$_A$ in Figure 12.12) is its antibonding π^* MO. The HOMO of the ketene is oriented perpendicular with regard to the plane of the methene group (HOMO$_B$ in Figure 12.12). This MO extends over three centers; it has its largest coefficient at the methene carbon, a small coefficient at the oxygen, and a near-zero coefficient at the central C atom. The LUMO of ketene is the antibonding π^* orbital of the C=O double bond (LUMO$_B$ in Figure 12.12). This MO lies in the plane of the methene group: that is, it is perpendicular to the HOMO$_B$, and its largest coefficient by far is located at the carbonyl carbon.

is identical to

C1 is hidden behind H

Fig. 12.11. Perspective drawing of the structure of the transition state of the [2+2]-cycloaddition ketene + ethene → cyclobutanone.

As with the transition state of the [4+2]-addition of butadiene and ethene (Figure 12.8), *both* HOMO/LUMO interactions are stabilizing in the transition state of the [2+2]-addition of ketene to ethene (Figure 12.13). This explains why [2+2]-cycloadditions of ketenes to alkenes—and similarly to alkynes—can occur in one-step reactions while this is not so for the additions of alkenes to alkenes (Section 12.2.3).

In contrast to the [4+2]-additions of butadiene to ethene or acetylene (Figures 12.8 and 12.9), the two HOMO/LUMO interactions stabilize the transition state of the [2+2]-

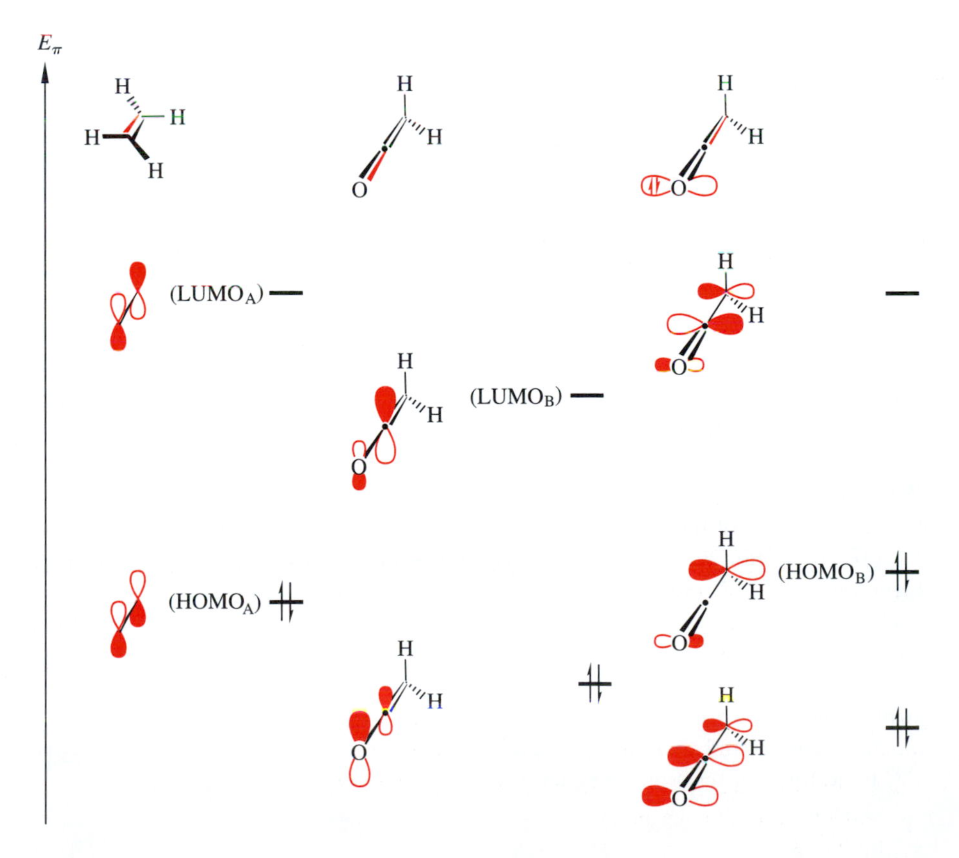

Fig. 12.12. π-Type MOs of ethene (left) and ketene (center and right); the subscripts A and B refer to ethene and ketene, respectively.

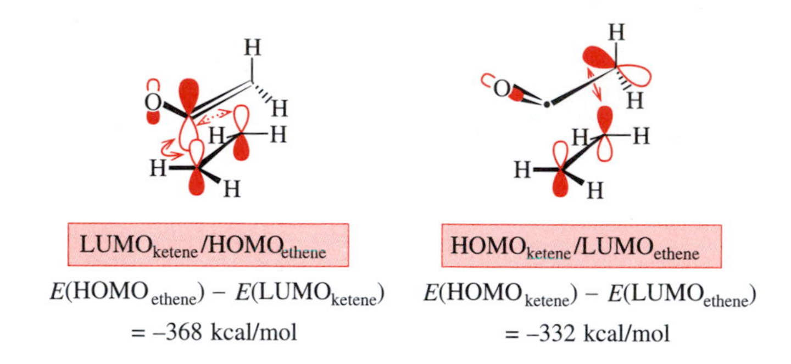

Fig. 12.13. Frontier orbital interactions in the transition state of the one-step [2+2]-cycloaddition of ketene and ethene.

addition of ketenes to alkenes *to a very different extent.* Equation 12.2 reveals that the larger part of the stabilization is due to the $LUMO_{ketene}/HOMO_{ethene}$ interaction. This circumstance greatly affects the geometry of the transition state. If there were *only this one* frontier orbital interaction in the transition state, the carbonyl carbon of the ketene would occupy a position in the transition state that would be perpendicular above the midpoint of the ethene double bond. The Newman projection of the transition state (Figure 12.11) shows that this is almost the case but not quite. The less stabilizing frontier orbital interaction—the one between the HOMO of the ketene and the LUMO of the ethene—is responsible for this small distortion. The big $2p_z$ lobe of the ketene's HOMO—located at the methene carbon—overlaps best with the LUMO of ethene in a way that a banana bond of sorts results between this lobe and one of the $2p_z$ lobes of the ethene LUMO. It is for this reason that the carbonyl carbon of the ketene is slightly moved out of the π-orbital plane of the ethene and at the same time approaches one of the two C atoms of ethene (C2) more closely.

12.3 Diels–Alder Reactions

[2+4]-Cycloadditions are called Diels–Alder reactions in honor of Otto Diels and Kurt Alder, the chemists who carried out the first such reaction. The substrate that reacts with the diene in these cycloadditions is called the **dienophile.** As you saw in Figure 12.1, the simplest Diels–Alder reactions, i.e., the ones between ethene and butadiene and between acetylene and butadiene, respectively, occur only under drastic conditions. Well-designed Diels–Alder reactions, on the other hand, occur much more readily. In the vast majority of those cases acceptor-substituted alkenes serve as dienophiles. In the present section we will be concerned only with such Diels–Alder reactions (see Figures 12.16, 12.17, and 12.22 for exceptions).

Up to four stereocenters may be constructed in one Diels–Alder reaction, and the great number of possible substrates gives the reaction great scope. The enormous value of the Diels–Alder reaction for organic synthesis is thus easy to understand. It is not at all exaggerated to say that this reaction is the most important synthesis for six-membered rings and that, moreover, it is one of the most important stereoselective C,C bond-forming reactions in general.

12.3.1 Stereoselectivity of Diels–Alder Reactions

Essentially all Diels–Alder reactions are one-step processes. If the reactions are stereogenic, they often occur with predictable stereochemistry. For example, *cis,trans*-1,4-disubstituted 1,3-butadienes afford cyclohexenes with a *trans* (!) arrangement of the substituents at C3 and C6 (Figure 12.14, top). In contrast, *trans,trans*-1,4-disubstituted 1,3-butadienes afford cyclohexenes in which the substituents at C3 and C6 are *cis* (!) with respect to each other (Figure 12.14, bottom). Accordingly, Diels–Alder reactions exhibit stereospecificity with regard to the diene.

cis,cis-1,4-Disubstituted 1,3-butadienes undergo Diels–Alder reactions only when they are part of a cyclic diene. Cyclopentadiene is an example of such a diene. In fact it is one of the most reactive dienes in general. In stark contrast, the transition states of Diels–Alder reactions of acyclic *cis,cis*-1,4-disubstituted 1,3-butadienes usually suffer from a prohibitively large repulsion between the substituents in the 1- and 4-positions, which arises when *these* substrates—as any 1,3-diene must—assume the *cis* conformation about the C2—C3 bond.

Stereoselectivity also is observed in Diels–Alder reactions of dienophiles, which contain a stereogenic C═C double bond (Figure 12.15). A *cis,trans* pair of such dienophiles, moreover, react stereospecifically with 1,3- dienes (as long as the latter do not contain any stereogenic double bonds): the *cis*-configured dienophile affords a 4,5-*cis*-disubstituted cyclohexene (Figure 12.15, top), while its *trans* isomer gives the 4,5-*trans*-disubstituted product (Figure 12.15, bottom).

Only very few [4+2]-cycloadditions are known that are not stereoselective with regard to the diene or the dienophile (see, e.g., Figure 12.17), or are stereoselective but not stereospecific (e.g., Figure 12.18). From these stereochemical outcomes, one can then safely conclude that the latter [4+2]-cycloadditions are multistep reactions.

Chloroprene undergoes three different [4+2]-cycloadditions with itself, proceeding as parallel reactions. One of these [4+2]-cycloadditions does not occur in a stereoselective fashion with respect to the dienophile. These cycloadditions are dimerizations

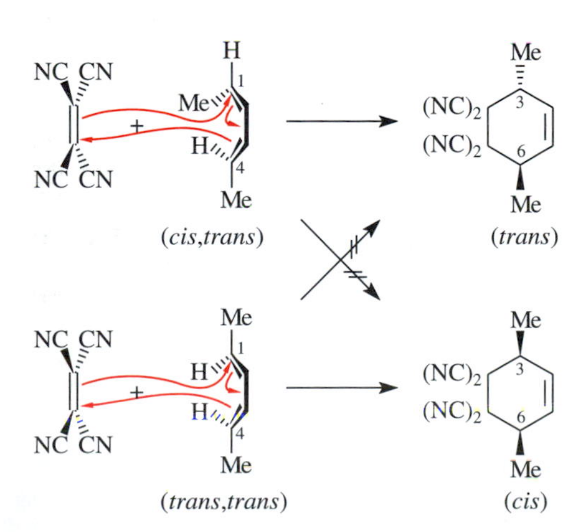

Fig. 12.14. Evidence for stereoselectivity and stereospecificity with regard to the butadiene moiety in a pair of Diels–Alder reactions. The *cis,trans*-1,4-disubstituted 1,3-butadiene forms cyclohexene with a *trans* arrangement of the methyl groups. The *trans,trans*-1,4-disubstituted 1,3-butadiene forms cyclohexene with *cis*-methyl groups.

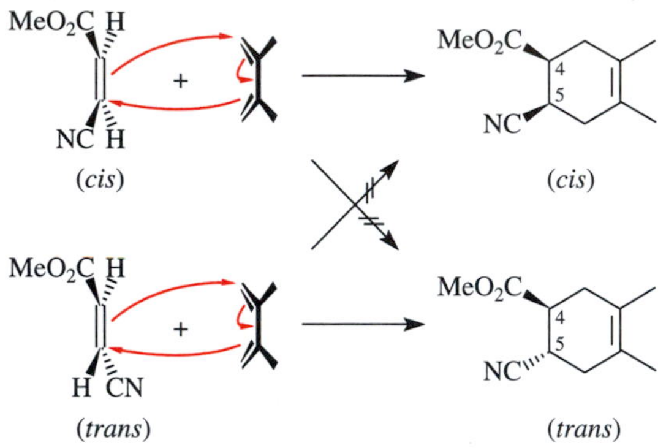

Fig. 12.15. Evidence for stereoselectivity and stereospecificity with regard to the dienophile in a pair of Diels–Alder reactions. The *cis*-configured dienophile affords a 4,5-*cis*-substituted cyclohexene, whereas the *trans* isomer results in a 4,5-*trans*-substituted cyclohexene.

that yield compounds **A**–**C** in Figure 12.16. Chloroprene plays two roles in these [4+2]-cycloadditions: it serves as diene and also as dienophile. In addition, small amounts of chloroprene dimerize (in a multistep process!) to give a [2+2]-cycloadduct **D** and to give a [4+4]-cycloadduct **E** (Figure 12.16).

The dimerization of chloroprene leading to the [4+2]-cycloadduct **C** (Figure 12.16) definitely is a multistep process. This has been demonstrated by analysis of the stereochemistry of a [4+2]-cycloaddition that led to the dideutero analogs of this cycloadduct (Figure 12.17). Instead of chloroprene, a monodeuterated chloroprene (*trans*-[D]-chloroprene) was dimerized. This monodeuterated chloroprene of course also

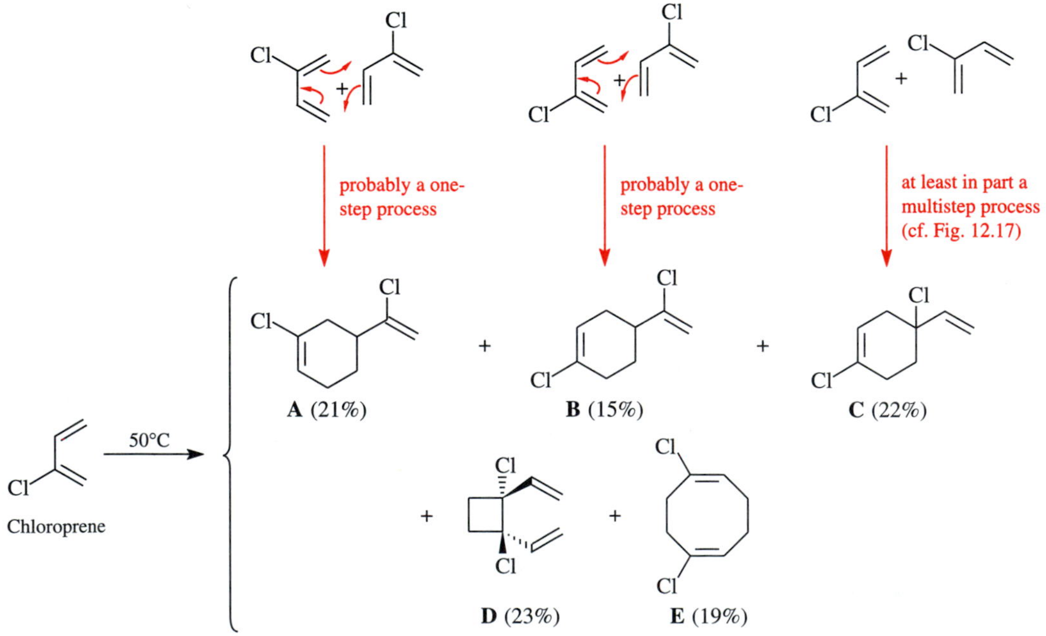

Fig. 12.16. Thermal cycloadditions of chloroprene (2-chlorobutadiene) I: product distribution.

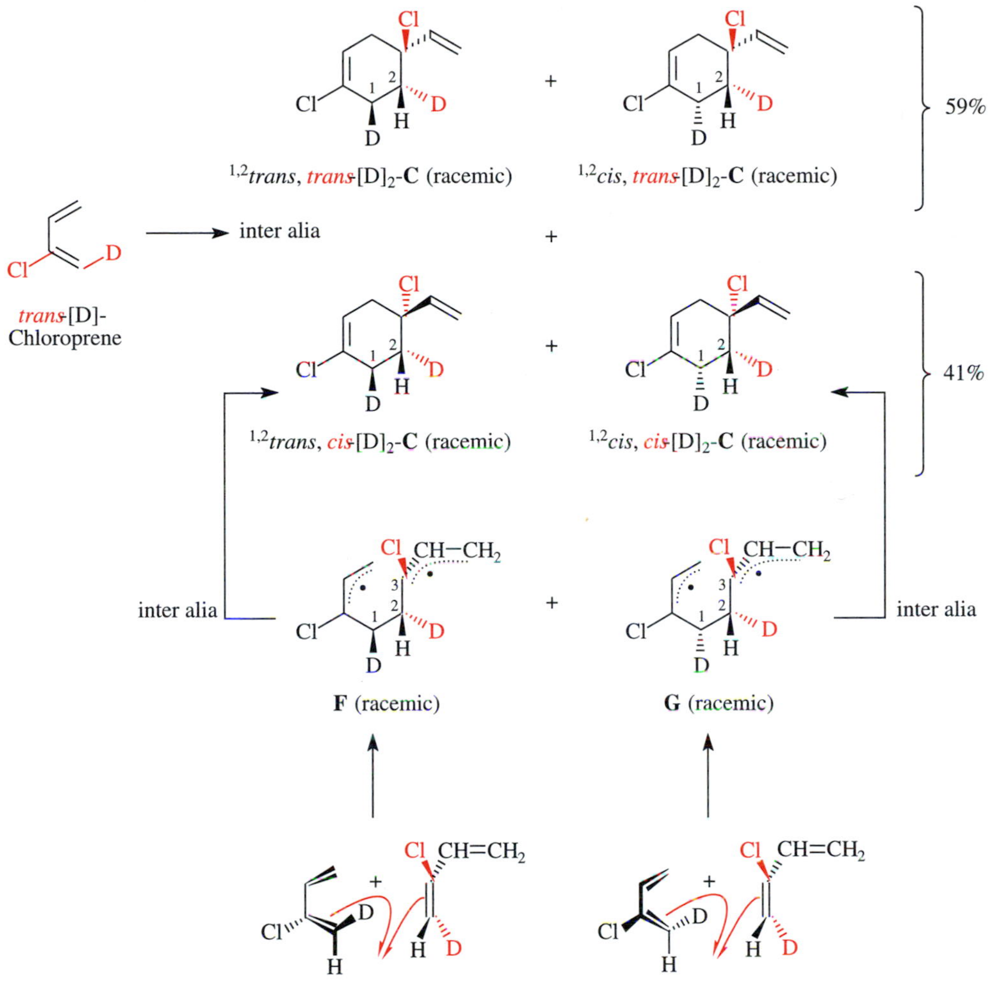

Fig. 12.17. Thermal cycloadditions of chloroprene (2-chlorobutadiene) II: detection and explanation of the two-step nature of the [4+2]-adduct formation.

underwent all five chloroprene dimerization reactions. The elucidation of the stereochemistry of the dideutero analog $[D]_2$-**C** (Figure 12.17) established how the [4+2]-cycloadduct **C** (Figure 12.16) is formed. Compound $[D]_2$-**C** was isolated as a mixture of four racemic diastereomers: ^{1,2}*trans,trans*-$[D]_2$-**C,** ^{1,2}*cis,trans*-$[D]_2$-**C,** ^{1,2}*trans,cis*-$[D]_2$-**C,** and ^{1,2}*cis, cis*-$[D]_2$-**C.**

A one-step dimerization of the *trans*-[D]-chloroprene shown in Figure 12.17 would yield only the first two diastereomers, that is, ^{1,2}*trans, trans*-$[D]_2$-**C** and ^{1,2}*cis, trans*-$[D]_2$-**C.** In these isomers, the *trans* arrangement between the D atom and the proximate Cl-atom in the dienophile is conserved. However, the same two atoms are in a *cis* arrangement in the additionally formed [4+2]-cycloadducts ^{1,2}*trans, cis*-$[D]_2$-**C** and ^{1,2}*cis,cis*-$[D]_2$-**C.** These cycloadducts $[D]_2$-**C** can only have lost the original *trans* arrangement of the atoms in question because they were formed via a multistep mechanism. Specifically, this mechanism must include an intermediate in which the *trans* arrangement of the D and Cl atoms is partly lost. This is only conceivable if in this intermediate a rotation is possible about the C,C bond that connects the deuterated and the chlorinated C atoms that are configurationally stable, i.e., without the possibility of such a rotation, in the dienophile as well as in the cycloadduct.

The most likely multistep mechanism of this type is shown in the lower part of Figure 12.17. It is a two-step mechanism where the diastereomeric diradicals **F** and **G** are the two intermediates that allow for rotation about the configuration-determining C—C bond. Each of the two radical centers is part of a well-stabilized allyl radical (cf. Section 1.2.1). It is unknown whether the formation of biradical **F** is subject to simple diastereoselectivity in comparison to **G** (for the occurrence of simple diastereoselectivity in one-step Diels–Alder reactions, see Section 12.3.4). Biradicals **F** and **G** cyclize *without* diastereocontrol to deliver the [4+2]-cycloadducts: biradical **F** forms a *mixture of* ^{1,2}*trans,cis*-$[D]_2$-**C** and ^{1,2}*trans,trans* $[D]_2$-**C,** since a rotation about the C2—C3 bond is possible but not necessary. For the same reason, biradical **G** forms a mixture of ^{1,2}*cis,cis*-$[D]_2$-**C** and ^{1,2}*cis,trans* $[D]_2$-**C.**

1-(Dimethylamino)-1,3-butadiene and *trans*-dicyanoethenedicarboxylic acid diester react with each other in a stereoselective [4+2]-cycloaddition to give the cyclohexene *trans,*^{2,3}*trans*-**A** (Figure 12.18). 1-(Dimethylamino)-1,3-butadiene also undergoes a stereoselective [4+2]-cycloaddition with *cis*-dicyanoethenedicarboxylic acid diester (Figure 12.18). However, the latter reaction results in the same cyclohexene *trans,* ^{2,3}*trans*-**A** that is formed from the *trans*-configured ester. Thus, Figure 12.18 shows a pair of stereoselective [4+2]-cycloadditions that occur without stereospecificity but with **stereoconvergence** (see Section 3.2.2 for the introduction of this term). 1-(Dimethylamino)-1,3-butadiene and *trans*-dicyanoethenedicarboxylic acid diester thus form the [4+2]-cycloadduct with complete retention of the *trans* relationship between the ester groups. The same would be true if a one-step addition mechanism were operative. 1-(Dimethylamino)-1,3-butadiene and *cis*-dicyanoethenedicarboxylic acid diester, on the other hand, form *trans,*^{2,3}*trans*-**A** with complete inversion of the rela-

Fig. 12.18. [4+2]-Cycloaddition between 1-(dimethylamino)-1,3-butadiene and the two isomeric dicyanoethenedicarboxylic acid diesters I: product distribution.

MeO$_2$C CN / MeO$_2$C CN — *cis* + :NMe$_2$ — concerted (crossed out) → *cis*,^{2,3}*cis*-**A** (racemic) oder *cis*,^{2,3}*trans*-**A** (racemic)

observed exclusively, therefore multistep (cf. Fig. 12.19)

NC CO$_2$Me / MeO$_2$C CN — *trans* + :NMe$_2$ — possibly concerted but thought to be multistep → *trans*,^{2,3}*trans*-**A** (racemic) tive configuration of the two ester groups. This finding can be explained only by a multistep addition mechanism. A one-step mechanism could lead only to the cycloadducts *cis,*^{2,3}*cis*-**A** and *cis,*^{2,3}*trans*-**A.**

Figure 12.19 shows the multistep mechanism of the [4+2]-cycloaddition between 1-(dimethylamino)-1,3-butadiene and *cis*-dicyanoethenedicarboxylic acid diester. The reaction proceeds via an intermediate, which must be conformer **B** of a zwitterion. The anionic moiety of this zwitterion is well stabilized because it represents the conjugate base of a carbon-acidic compound (cf. Section 10.1.2). The cationic moiety of zwitterion **B** also is well stabilized. It represents an iminium ion (i.e., a species with valence electron octet) rather than a carbenium ion (which is a species with valence electron sextet); moreover, the iminium ion is stabilized by conjugation to a C═C double bond.

The zwitterion intermediate of the [4+2]-cycloaddition depicted in Figure 12.19 is formed with stereostructure **B.** Therein the ester groups of the dienophile fragment maintain their original *cis* arrangement. However, this *cis* arrangement is quickly lost owing to rotation about the C2—C1 bond. Zwitterion conformers with stereostructure **C,** i.e., with a *trans* arrangement of the ester groups, are thus formed. Conformer **C** undergoes ring closure to the [4+2]-cycloadduct significantly faster than conformer **B.** In addition, the ring closure occurs with simple diastereoselectivity (cf. Section 12.3.4 for a discussion of simple diastereoselectivity in Diels–Alder reactions). Consequently, the zwitterion **C** leads to the formation of the [4+2]-cycloadduct *trans,*^{2,3}*trans*-**A** only; *trans,*^{2,3}*cis*-**A** does not form.

MeO_2C CN :NMe_2 + MeO_2C CN

MeO_2C NMe_2 NC 3 2 1 MeO_2C CN

trans,^{2,3}*trans*-**A**

Ring closure with simple and with induced stereoselectivity

$\oplus$ NMe_2 NC MeO_2C 2 $\ominus$ 1 MeO_2C CN **B**

Rotation about C2 – C1 bond

$\oplus$ NMe_2 MeO_2C NC 2 $\ominus$ 1 MeO_2C CN **C**

Fig. 12.19. [4+2]-Cycloaddition between 1-(dimethylamino)-1,3-butadiene and *cis*-dicyanoethenedicarboxylic acid diester II: explanation of the inversion of configuration in the dienophile moiety.

12.3.2 Substituent Effects on Reaction Rates of Diels–Alder Reactions

Cyclopentadiene is such a reactive 1,3-diene that it undergoes Diels–Alder reactions with all cyanosubstituted ethenes. The rate constants of these cycloadditions (Table 12.1) show that each cyano substituent increases the reaction rate significantly and that geminal cyano groups accelerate more than vicinal cyano groups.

Table 12.1. Relative Rate Constants $k_{2+4,\mathrm{rel}}$ of Analogous Diels–Alder Reactions of Polycyanoethenes

	NC–CH=CH$_2$	NC/NC (cis)	NC/CN (trans)	NC/CN (1,1)	NC/CN, CN (tri)	NC, CN / NC, CN (tetra)
$k_{2+4,\mathrm{rel}}$	≡ 1	81	91	45 500	480 000	43 000 000

Diels–Alder reactions of the type shown in Table 12.1, that is, Diels–Alder reactions between electron-poor dienophiles and electron-rich dienes, are referred to as **Diels–Alder reactions with normal electron demand.** The overwhelming majority of known Diels–Alder reactions exhibit such a "normal electron demand." Typical dienophiles include acrolein, methyl vinyl ketone, acrylic acid esters, acrylonitrile, fumaric acid esters (*trans*-butenedioic acid esters), maleic anhydride, and tetracyanoethene—all of which are acceptor-substituted alkenes. Typical dienes are cyclopentadiene and acyclic 1,3-butadienes with alkyl-, aryl-, alkoxy-, and/or trimethylsilyloxy substituents—all of which are dienes with a donor substituent.

The reaction rates for the cycloaddition of several of the mentioned dienophiles to electron-rich dienes are significantly increased upon addition of a catalytic amount of a Lewis acid. The $AlCl_3$ complex of methyl acrylate reacts 100,000 times faster with butadiene than pure methyl acrylate (Figure 12.20). Apparently, the C=C double bond in the Lewis acid complex of an acceptor-substituted dienophile is connected to a stronger acceptor substituent than in the Lewis-acid-free analog. A *better* acceptor increases the dienophilicity of a dienophile in a manner similar to the effect *several* acceptors have in the series of Table 12.1.

Fig. 12.20. Diels–Alder reactions with normal electron demand; increase of the reactivity upon addition of a Lewis acid. The $AlCl_3$ complex of the acrylate reacts 100,000 times faster with butadiene than does the uncomplexed acrylate.

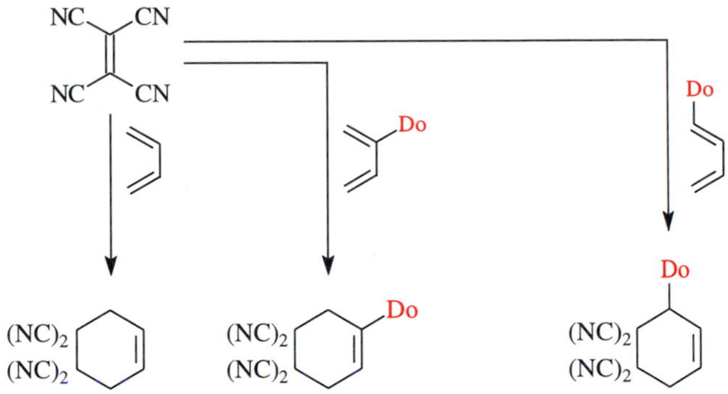

		Do = Me	Ph	OMe	Do = Me	Ph	OMe
$k_{2+4,\ rel}$	≡ 1	44	191	1 720	104	386	50 900

Fig. 12.21. Diels–Alder reactions with normal electron demand; reactivity increase by the use of donor-substituted 1,3-dienes (Do refers to a donor substituent).

An increase in reactivity also can be observed in Diels–Alder reactions with normal electron demand if a given dienophile is reacted with a series of more and more electron-rich dienes. The reaction rates of the Diels–Alder reactions of Figure 12.21 show that the substituents MeO > Ph > alkyl are such reactivity-enhancing donors. The tabulated rate constants also show that a given donor substituent accelerates the Diels–Alder reaction more if located in position 1 of the diene than if located in position 2.

Diels–Alder reactions also may occur when the electronic situation of the substrates is completely reversed, that is, when electron-rich dienophiles react with electron-poor dienes. [4+2]-Cycloadditions of this type are called **Diels–Alder reactions with inverse electron demand.** 1,3-Dienes that contain heteroatoms such as O and N in the diene backbone are the dienes of choice for this kind of cycloaddition. The data in Figure 12.22 show the rate-enhancing effect of the presence of donor substituents in the dienophile.

Why do the Diels–Alder reactions with both normal and inverse electron demand occur under relatively mild conditions? And, in contrast, why can [2+4]-cycloadditions between ethene or acetylene, respectively, and butadiene be realized only under extremely harsh conditions (Figure 12.1)? Equation 12.2 described the amount of transition state stabilization of [4+2]-cycloadditions as the result of HOMO/LUMO interactions between the π MOs of the diene and the dienophile. Equation 12.3 is derived from Equation 12.2 and presents a *simplified* estimate of the magnitude of the stabilization. This equation features a sum of two simple terms, and it highlights the essence better than Equation 12.2.

$$\Delta E_{TS} \propto \frac{1}{E_{HOMO,\ diene} - E_{LUMO,\ dienophile}} + \frac{1}{E_{HOMO,\ dienophile} - E_{LUMO,\ diene}} \quad (12.3)$$

The first term of Equation 12.3 is responsible for most of the transition state stabilization of a Diels–Alder reaction with normal electron demand. In this case, the first term is larger than the second term because the denominator is smaller. The denomi-

Fig. 12.22. Diels–Alder reactions with inverse electron demand; reactivity increase by the use of donor-substituted dienophiles (X refers to a substituent that may be a donor or an acceptor).

(products undergo subsequent reactions)

X	O_2N	H	MeO
$k_{2+4,\ \mathrm{rel}}$	0.13	≡ 1	3.8

nator of the first term is smaller because the HOMO of an electron-rich diene is closer to the LUMO of an electron-poor dienophile than is the LUMO of the same electron-rich diene with respect to the HOMO of the same electron-poor dienophile (Figure 12.23, column 2). Acceptors lower the energy of all π-type MOs irrespective of whether

Fig. 12.23. Frontier orbital interactions in Diels–Alder reactions with varying electron demand.

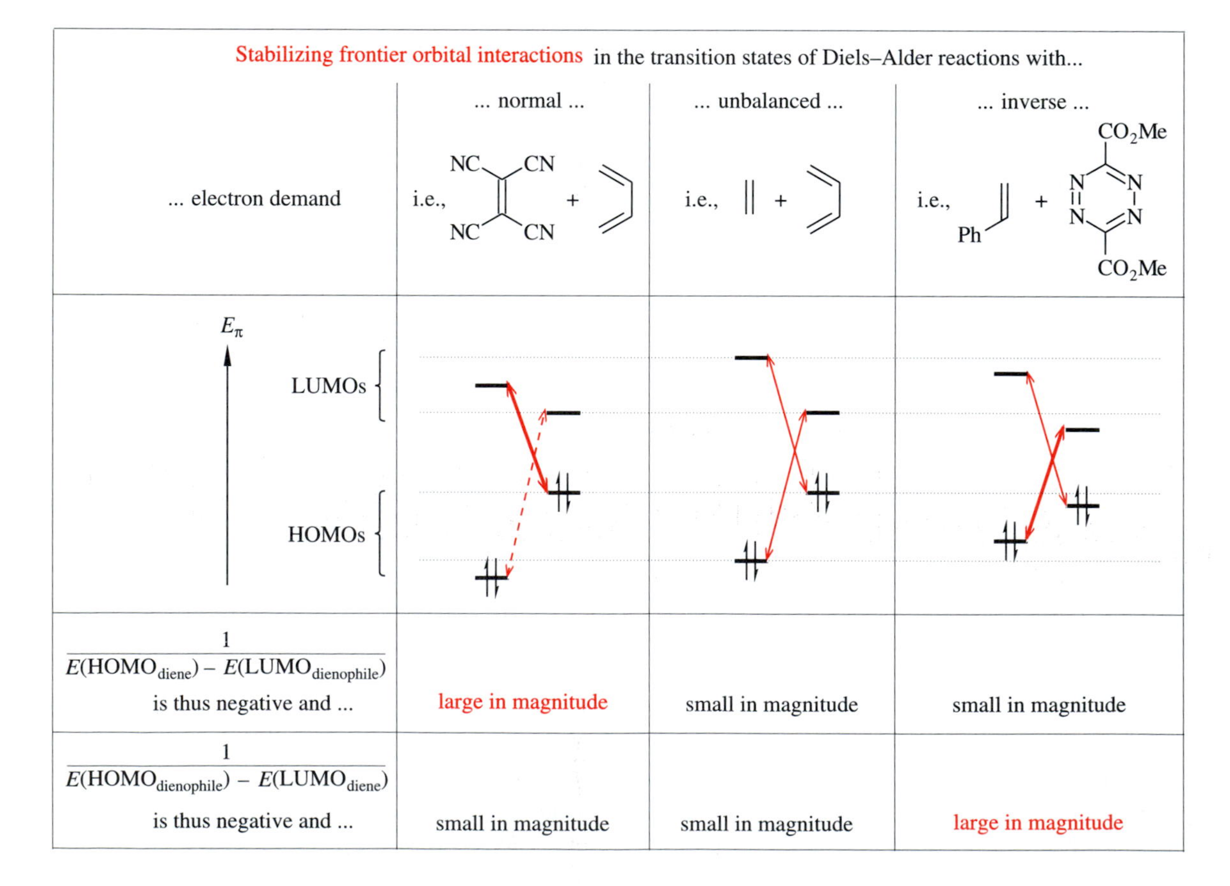

these MOs are bonding or antibonding. This is all the more true the stronger the substituent effects and the more substituents are present.

The most important stabilizing interaction of the transition states of Diels–Alder reactions with inverse electron demand is due to the second term of Equation 12.3. In this case, the denominator of the second term is substantially smaller than that of the first term. This is because the HOMO of an electron-rich dienophile is closer to the LUMO of an electron-poor diene than is the HOMO of the same diene relative to the LUMO of the same dienophile (Figure 12.3, column 4). We saw the reason for this previously: acceptors lower the energies of all π-type MOs; donors increase these energies.

Summary

The transition states of Diels–Alder reactions with either normal or inverse electron demand are substantially stabilized because the HOMO of one reagent lies close to the LUMO of the other reagent. This stabilization of the transition states is responsible for the fast cycloadditions.

In the Diels–Alder reactions between ethene and butadiene and between acetylene and butadiene, respectively, the HOMOs are nearly isoenergetic and they are rather far away from the LUMOs (Figures 12.8 and 12.9). According to Equation 12.3, the transition states of these Diels–Alder reactions experience only a minor stabilization and, for *this* reason, these [2+4]-cycloadditions (Figure 12.23, column 3) are so much slower than the others (columns 2 and 3).

12.3.3 Orientation Selectivity of Diels–Alder Reactions

Diels–Alder reactions with symmetrically substituted dienophiles and/or with symmetrically substituted dienes afford cycloadducts that must be *constitutionally homogeneous*. In contrast, Diels–Alder reactions between an asymmetrically substituted dienophile and an asymmetrically substituted diene may afford two *constitutionally isomeric* cycloadducts. If only one of these isomers is actually formed, the Diels–Alder reaction is said to be **orientation selective.**

1,3-Butadienes with alkyl substituents in the 2-position favor the formation of the so-called *para* products (Figure 12.24, X = H) in their reactions with acceptor-substituted dienophiles. The so-called *meta* product is formed in smaller amounts. This orientation selectivity increases if the dienophile carries two geminal acceptors (Figure 12.24, X = CN). 2-Phenyl-1,3-butadiene exhibits a higher "*para*" selectivity

NC X + 2 —20°C→ NC X "*para* product" + NC X "*meta* product"

	"*para* product"		"*meta* product"
X = H:	70	:	30
X = CN:	91	:	9

Fig. 12.24. Orientation-selective Diels–Alder reactions with a 2-substituted 1,3-diene I: comparison of the effects exerted by one or two dienophile substituents.

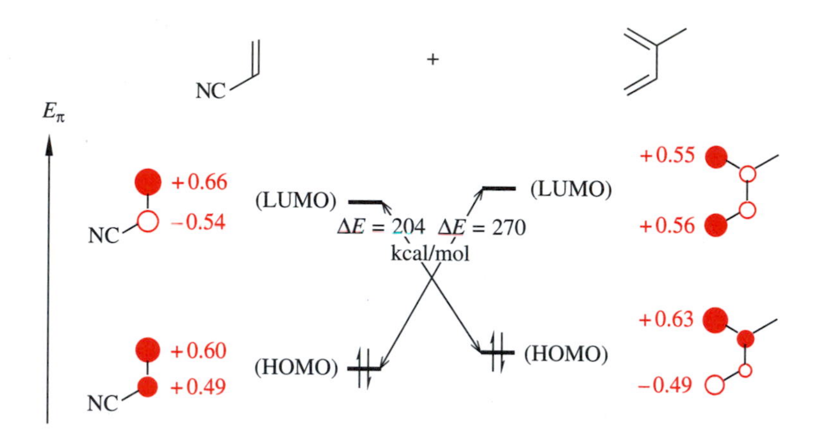

Fig. 12.25. Frontier orbital coefficients and energy difference of the HOMO–LUMO gaps in orientation-selective Diels–Alder reactions (cf. Figure 12.24, X = H).

in its reactions with every asymmetric dienophile than any 2-alkyl-1,3-butadiene does. This is even more true for 2-methoxy-1,3-butadiene and 2-(trimethylsilyloxy)-1,3-butadiene. Equation 12.2, which describes the stabilization of the transition states of Diels–Alder reactions in terms of the frontier orbitals, also explains the "*para*"/"*meta*" orientation. The numerators of both fractions assume different values depending on the orientation, while the denominators are independent of the orientation.

One can compute, for example, the stabilizations ΔE_{TS} for the transition states of the "*para*"- and "*meta*"-selective cycloadditions, respectively, of acrylonitrile and isoprene according to Equation 12.2 with the data provided in Figure 12.25 (HOMO/LUMO gaps, LCAO coefficients at the centers that interact with each other). The result for the Diels–Alder reaction of Figure 12.24 is shown in Equations 12.4 and 12.5:

$$\Delta E_{\text{TS}\rightarrow\text{“}para\text{ product”}} \propto -0.0036 \text{ kcal/mol (63\% of which is due to the HOMO}_{\text{diene}}\text{/LUMO}_{\text{dienophile}} \text{ interaction)} \tag{12.4}$$

$$\Delta E_{\text{TS}\rightarrow\text{“}meta\text{ product”}} \propto -0.0035 \text{ kcal/mol (61\% of which is due to the HOMO}_{\text{diene}}\text{/LUMO}_{\text{dienophile}} \text{ interaction)} \tag{12.5}$$

The transition state leading to the "*para*" product is slightly more stabilized, and accordingly this product is favored. However, the 70 : 30 selectivity shows the preference to be small, as one would anticipate based on the minuscule energy difference computed by means of Equations 12.4 and 12.5.

The same kind of computation can be carried out for the substrate pair consisting of isoprene and 1,1-dicyanoethene. Based on the HOMO/LUMO gaps and with the LCAO coefficients at the centers that interact with each other (see Figure 12.26), Equation 12.2 again gives a higher stabilization ΔE_{TS} for the "*para*" transition state than for the "*meta*" transition state (Equations 12.6 and 12.7).

$$\Delta E_{\text{TS}\rightarrow\text{“}para\text{ product”}} \propto -0.0038 \text{ kcal/mol (67\% of which is due to the HOMO}_{\text{diene}}\text{/LUMO}_{\text{dienophile}} \text{ interaction)} \tag{12.6}$$

$$\Delta E_{\text{TS}\rightarrow\text{“}meta\text{ product”}} \propto -0.0036 \text{ kcal/mol (66\% of which is due to the HOMO}_{\text{diene}}\text{/LUMO}_{\text{dienophile}} \text{ interaction)} \tag{12.7}$$

This difference between the stabilization energies is a bit larger than in the case of the addition of acrylonitrile to isoprene (Equations 12.4 and 12.5). This agrees with the

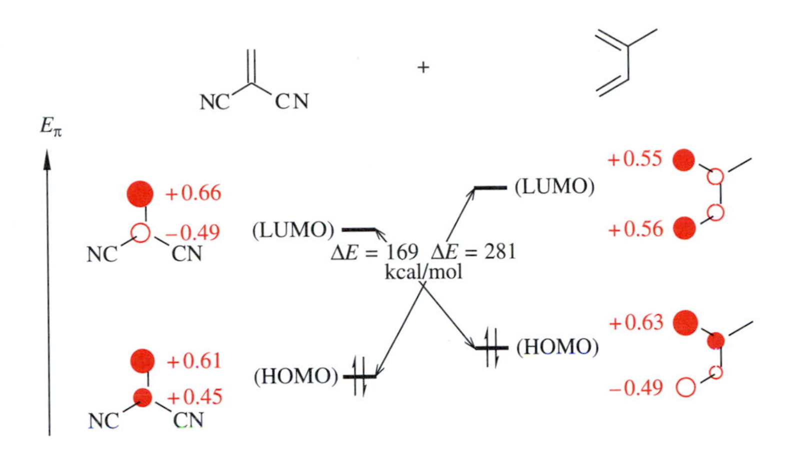

Fig. 12.26. Frontier orbital coefficients and energy difference of the HOMO–LUMO gaps in orientation-selective Diels–Alder reactions (cf. Figure 12.24, X = CN).

data in Figure 12.24, which show a "*para*/*meta*" selectivity of 91 : 9 for the addition of 1,1-dicyanoethene to isoprene—i.e., somewhat higher than the 70 : 30 ratio of the addition involving acrylonitrile.

From Equation 12.2 one may derive the following general rules concerning the orientation selectivity of any one-step cycloaddition.

Rules for the Orientation Selectivity of Any One-Step Cycloaddition

The substrates preferentially bind each other with those atoms that exhibit the largest LCAO coefficients (absolute values) in the closest pair of frontier orbitals. The orientation selectivity generally increases, the larger the relative significance of one HOMO/LUMO interaction compared to the other and the greater in each of the two crucial frontier orbitals the difference in magnitude of the LCAO coefficients (absolute values) at one terminus, compared to the other.

We can customize these general statements specifically for the case of the orientation selectivity of Diels–Alder reactions with normal electron demand and make the following statement right away:

The LCAO coefficient in the HOMO of a 1,3-diene

- increases at C1 compared to C4, the better the donor in position 2,
- increases at C4 compared to C1, the better the donor in position 1, and
- is larger at C4 than at C1 when the same kind of donor is attached both to positions 1 and 2.

The consequences thereof for the orientation selectivity of Diels–Alder reactions can be summarized as follows.

Orientation Selectivity of Diels–Alder Reactions

Asymmetric dienophiles react with 1-donor-, 2-donor-, or 1,2-didonor-substituted 1,3-dienes, preferentially to the "*ortho*," "*para*," or "*ortho,meta*" cycloadduct, respectively. With regard to a given dienophile, the "*ortho*" selectivity is larger than the "*para*" selectivity. Comparison of the upper and the lower pairs of reactions in Figures 12.27 and 12.28 underscores the latter statement.

We go on to state that the LCAO coefficient in the LUMO of a dienophile

- increases at C2 in comparison to C1, the more acceptors are bound to C1 [cf. the LCAO coefficients of acrylonitrile (Figure 12.25) compared to those of 1,1-dicyanoethene (Figure 12.26)] and
- increases at C2 in comparison to C1, the stronger the acceptor that is attached to C1.

What has just been stated regarding the LCAO coefficients of the dienophile LUMO combined with the "rules for the orientation selectivity of any one-step cycloaddition" leads to the following consequences for the Diels–Alder reactions of isoprene:

1) Acrylonitrile shows less "*para*" selectivity in its addition to isoprene than in its addition to 1,1-dicyanoethene (Figure 12.24).
2) Acrylic acid esters show less "*para*" selectivity in their additions to isoprene than do $AlCl_3$-complexed acrylic acid esters (Figure 12.27; the complex formation with $AlCl_3$ converts the CO_2Me group into a better acceptor than the uncomplexed CO_2Me-group).

The increase in orientation selectivity of Diels–Alder reactions upon addition of Lewis acid has a second cause aside from the one which was just mentioned. The reaction conditions described in Figure 12.27 indicate that $AlCl_3$ increases the rate of cycloaddition. The same effect also was seen in the cycloaddition depicted in Figure 12.20. In both instances, the effect is the consequence of the lowering of the LUMO level of the dienophile. According to Equation 12.2, this means that the magnitude of the denominator of the first term decreases and the first term therefore becomes larger than the second term. If, in addition, the numerators of these terms differ by a certain amount for the "*para*" and "*meta*" transition states (as determined by the combinations of the LCAO coefficients), the effect is further enhanced. *This* also increases the "*para*" selectivity.

Fig. 12.27. Orientation-selective Diels–Alder reactions with a 2-substituted diene II: selectivity increase by way of addition of a Lewis acid.

MeO_2C + → MeO_2C + MeO_2C

	"*para* product"	:	"*meta* product"
25°C, 41 d:	70	:	30
10–20°C, 1 mol% $AlCl_3$, 3 h:	95	:	5

Finally, the examples of the two Diels–Alder reactions in Figure 12.27 lead us to a general statement: in Diels–Alder reactions with normal electron demand, the addition of $AlCl_3$ increases the reaction rate and the orientation selectivity. This situation marks one of the most notable exceptions from the reactivity–selectivity principle (Section 1.7.4), which is otherwise so often encountered in organic chemistry.

12.3.4 Simple Diastereoselectivity of Diels–Alder Reactions

We saw in Section 12.3.3 that 1,3-butadienes with a donor in the 1-position react with acceptor-substituted alkenes to form cycloadducts with high "*ortho*" selectivity. The amount

	"*ortho* product"	:	"*meta* product"
25°C, 70 d:	90 (*cis* : *trans* = 57 : 43)	:	10 (*cis* : *trans* = 73 : 27)
10–20°C, 10 mol% $AlCl_3$, 3 h:	98 (*cis* : *trans* = 95 : 5)	:	2 (mainly *cis*)

Fig. 12.28. Orientation selectivity and simple diastereoselectivity of a Diels–Alder reaction with a 1-substituted diene; selectivity increase by way of addition of a Lewis acid.

of "*meta*" products formed is usually less than 10% (see example, Figure 12.28). This is particularly true for Diels–Alder reactions that are carried out in the presence of $AlCl_3$, which has the *same* effect of enhancing the orientation selectivity as seen in Figure 12.27.

Here we are primarily concerned with the fact that this almost exclusively formed "*ortho*" adduct may occur in the form of two diastereomers. The diastereomers are formed as a 57:43 *cis/trans* mixture in the absence of $AlCl_3$, but a 95:5 *cis/trans* mixture is obtained in the presence of $AlCl_3$. In the latter case, thus, one is dealing with a Diels–Alder reaction that exhibits a substantial "simple diastereoselectivity" (see Section 9.3.2 for a definition of the term). Here, the simple diastereoselectivity is due to kinetic rather than thermodynamic control, since the preferentially formed *cis*-disubstituted cyclohexene is less stable than its *trans* isomer.

Simple diastereoselectivity may also occur in Diels–Alder reactions between electron-poor dienophiles and cyclopentadiene (Figure 12.29). Acrylic acid esters or *trans*-crotonic acid esters react with cyclopentadiene in the presence or absence of $AlCl_3$ with substantial selectivity to afford the so-called *endo* adducts. When the bicyclic skeleton of the main product is viewed as a "roof," the prefix "*endo*" indicates that the ester group is below this roof, rather than outside (*exo*). However, methacrylic acid esters add to cyclopentadiene without any *exo,endo* selectivity no matter whether the reaction is carried out with or without added $AlCl_3$.

R^1	R^2	Conditions (30°C)	*endo* product	:	*exo* product
H	H	7.5 h:	78	:	22
		1 equivalent $AlCl_3$, 30 min:	95	:	5
H	Me	7.5 h:	54	:	46
		1 equivalent $AlCl_3$, 30 min:	94	:	6
Me	H	7.5 h:	31	:	69
		1 equivalent $AlCl_3$, 30 min:	60	:	40

Fig. 12.29. Simple diastereoselectivity of the additions of various acrylic acid derivatives to cyclopentadiene.

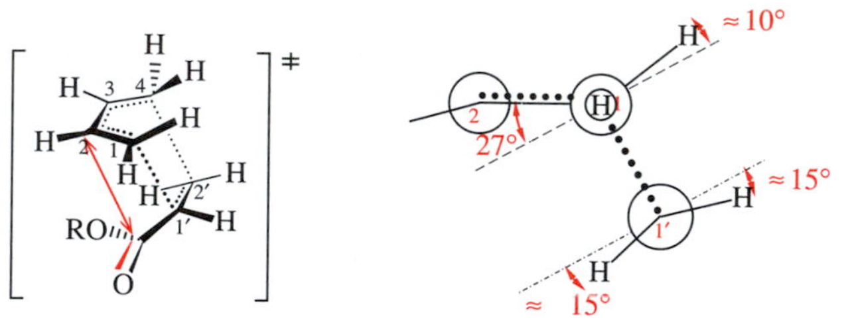

Fig. 12.30. Transition state structures of Diels–Alder additions of butadiene; **A,** side view of the addition of acrylic acid ester and **B,** Newman projection of the addition of ethene.

The high simple diastereoselectivities seen in Figures 12.28 and 12.29 are due to one and the same preferential orientation of the ester group in the transition states. The stereostructure of the cycloadduct shows unequivocally that the ester group points underneath the diene plane in each of the transition states of both cycloadditions and not away from that plane. Figure 12.30 exemplifies this situation for two transition states of simple Diels–Alder reactions of 1,3-butadiene: **A** shows a perspective drawing of the transition state of the acrylic acid ester addition, and **B** provides a side view of the addition of ethene, which will serve as an aid in the following discussion. Both structures were determined by computational chemistry.

One has not identified with certainty the origin(s) for the preference of stereostructure **A** in the acrylic acid ester addition. Possibly a steric effect explains the observation. The bulky acceptor substituent of the dienophile might be less hindered—and this is quite counterintuitive—in the *endo* orientation in the transition state shown in Figure 12.30 than in the alternative *exo* position. One might interpret the better known geometry of the transition state without an acceptor, **B,** to suggest that the substituent of the dienophile in **A** does not try to avoid the C atoms C2 and C3 as much as it tries to stay away from the H atoms *cis*-H1 and *cis*-H4. The increase of *endo* selectivity upon addition of a Lewis acid could then be explained by the premise that the complexing Lewis acid renders the ester group more bulky. This increased steric demand enhances its desire to avoid the steric hinderance in its *exo* position.

12.4 [2+2]-Cycloadditions with Dichloroketene

Only a few ketenes can be isolated, and diphenylketene is one of those. The majority of the other ketenes dimerize quickly, as exemplified by the parent ketene $H_2C{=}C{=}O$. Cycloadditions with *reactive* ketenes therefore can be observed only when they are prepared *in situ* and in the presence of the alkene to which they shall be added. Dichloroketene generated *in situ* is the best reagent for intermolecular [2+2]-cycloadditions. Dichloroketene is poorer in electrons than the parent ketene and therefore more reactive toward the relatively electron-rich standard alkenes. The reason is that the dominating frontier orbital interaction between these reactants involves the LUMO of the ketene, not its HOMO (see Section 12.2.4).

Fig. 12.31. Orientation-selective [2+2]-cycloaddition with *in situ* generated dichloroketene I: the dichloroketene is generated by way of an NEt_3-mediated β-elimination of HCl from dichloroacetyl chloride.

The first of the two common preparations of dichloroketene, shown in Figure 12.31, consists of an NEt_3-mediated β-elimination of HCl from dichloroacetyl chloride. Interestingly, dichloroketene does not participate as a dienophile in a Diels–Alder reaction with cyclopentadiene (Figure 12.31), preferring instead a [2+2]-cycloaddition. The [2+2]-cycloaddition occurs with perfect orientation selectivity. The preferred transition state is in line with the rule of thumb formulated in Section 12.3.3 for the orientation selectivity of one-step cycloadditions in general. The preferred transition state is the one in which the atoms that have the largest LCAO coefficients (absolute values) in the closest frontier orbitals are connected together. Figure 12.13 shows that the relevant frontier orbital pair includes the LUMO of dichloroketene and the HOMO of cyclopentadiene. The carbonyl carbon possesses the largest LCAO coefficient in the LUMO of dichloroketene (Figure 12.12). The largest LCAO coefficient in the HOMO of cyclopentadiene is located at C1, not at C2, since cyclopentadiene is a 1,3-butadiene that bears an alkyl substituent both at C1 and C4 (cf. discussion of the effects of alkyl substituents on the HOMO coefficients of 1,3-dienes in Section 12.3.3). Hence, the carbonyl carbon of the dichloroketene binds to C1 of cyclopentadiene in the most stable transition state of the [2+2]-cycloaddition.

Figure 12.32 shows the second commonly employed method for the generation of dichloroketene, which involves the reductive β-elimination of chlorine from trichloroacetyl chloride by zinc (cf. Sections 4.7.1 and 14.4.1 for mechanistic considerations). The addition of the dichloroketene to the trisubstituted alkene **A** (Figure 12.32) exhibits ori-

Fig. 12.32. Orientation-selective and diastereoselective [2+2]-cycloaddition with *in situ* generated dichloroketene II: the dichloroketene is generated by way of a reductive β-elimination of chlorine from trichloroacetyl chloride.

entation selectivity. The carbonyl carbon of the dichloroketene is connected to that C atom of the olefinic C=C double bond, which has the larger LCAO coefficient in the alkene HOMO. This C atom is marked with a β. This [2+2]-cycloaddition also shows diastereoselectivity. The alkene can add to the bicyclic alkene **A** only from the more accessible convex side (cf. Section 8.3.1 regarding the kinetic advantage of reactions on the convex sides of convex/concave molecules).

12.5 1,3-Dipolar Cycloadditions

12.5.1 1,3-Dipoles

A 1,3-dipole is a compound of the type a—Het—b that may undergo 1,3-dipolar cycloadditions with multiply bonded systems and can best be described with a zwitterionic all-octet Lewis structure (**"Huisgen ylid"**). An unsaturated system that undergoes 1,3-dipolar cycloadditions with 1,3-dipoles is called **dipolarophile.** Alkenes, alkynes, and their diverse hetero derivatives may react as dipolarophiles. Since there is a considerable variety of 1,3-dipoles—Table 12.2 shows a small selection—1,3-dipolar cycloadditions represent not only a general but also the most universal synthetic approach to five-membered heterocycles.

Table 12.2. Important 1,3-Dipoles

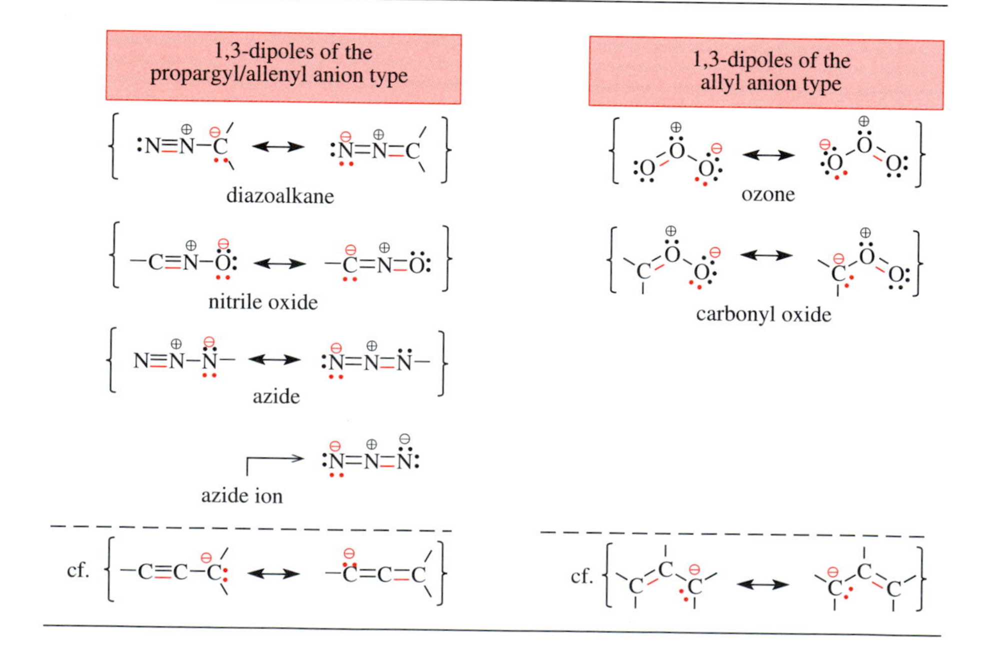

1,3-Dipoles are isoelectronic either to both propargyl and allenyl anions or to allyl anions. One may thus group these 1,3-dipoles into 1,3-dipoles of the propargyl/allenyl anion types (Table 12.2, left) and 1,3-dipoles of the allyl anion type (Table 12.2, right). The 1,3-dipoles of the propargyl/allenyl anion type contain a linearly coordinated central atom, as is the case for propargyl and allenyl groups in all sorts of compounds. The linearly coordinated central atom of these 1,3-dipoles is an N atom. This N atom carries a formal positive charge in both all-octet resonance forms. This central N atom is connected to C and/or to other heteroatoms, and these carry—one in each of the resonance forms—the formal negative charge. The azide ion is the only 1,3-dipole of the propargyl/allenyl anion type with a *net* negative charge.

1,3-Dipoles of the allyl anion type are bent just like allyl anions. The central atom can be an N, O, or S atom, and this atom carries a formal positive charge in both all-octet resonance forms. Centers 1 and 3 of the 1,3-dipole of the allyl anion type carry the formal negative charge, again one in each of the resonance forms. As with the dipoles of the propargyl/allenyl anion type, this negative charge may be located on C atoms and/or on heteroatoms.

12.5.2 Frontier Orbital Interactions in the Transition States of One-Step 1,3-Dipolar Cycloadditions; Sustmann Classification

Diazomethane adds to ethene to form Δ^1-pyrazoline (**A** in Figure 12.33). Its addition to acetylene first leads to the formation of the nonaromatic 3-*H*-pyrazole (**B** in Figure 12.33), which subsequently is converted into the aromatic 1-*H*-pyrazole (**C**) by way of a fast 1,5-hydrogen migration.

Do the transition states of the 1,3-dipolar cycloadditions with diazomethane benefit from a stabilizing frontier orbital interaction? Yes! Computations show that the $HOMO_{diazomethane}/LUMO_{ethene}$ interaction (orbital energy difference, −229 kcal/mol) stabilizes the transition state of the 1,3-dipolar cycloaddition to ethene (Figure 12.34) by about 11 kcal/mol. Moreover, computations also show that the $HOMO_{ethene}/LUMO_{diazomethane}$ interaction (orbital energy difference, −273 kcal/mol) contributes a further stabilization of 7 kcal/mol.

H₂C ⊕N ⊖N + || room temperature → A; H₂C ⊕N ⊖N + ||| room temperature → B; ~H → C (HN)

Fig. 12.33. The simplest 1,3-dipolar cycloadditions with diazomethane.

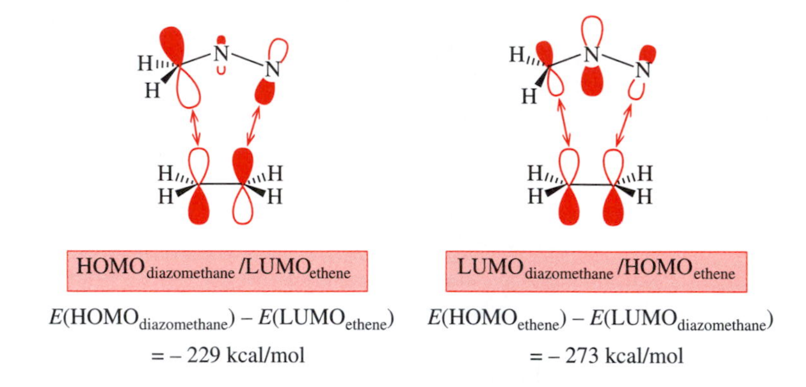

Fig. 12.34. Frontier orbital interactions in the transition state of the 1,3-dipolar cycloaddition of diazomethane to ethene.

Frontier orbital interactions are stabilizing the transition states of all the 1,3-dipolar cycloadditions. It is for this reason that one-step 1,3-dipolar cycloadditions are generally possible and, aside from some exotic exceptions, one does indeed observe one-step reactions.

As in the case of Diels–Alder reactions or of [2+2]-cycloadditions of ketenes, the rate of 1,3-dipolar cycloadditions is affected by donor and acceptor substituents in the substrates. Again, Equation 12.2 can be used to obtain a good approximation of their effects, since this equation applies to *any* one-step cycloaddition. We restated this equation once before as the approximation expressed in Equation 12.3. In that case, it was our aim to understand the rate of Diels–Alder reactions, and we are now faced with the task of making a statement concerning the rate of 1,3-dipolar cycloadditions. To this end it is advantageous to employ a different approximation to Equation 12.2, and that approximation is expressed in Equation 12.9.

$$\Delta E_{\mathrm{TS}} \propto \frac{1}{E_{\mathrm{HOMO,dipole}} - E_{\mathrm{LUMO,dipolarophile}}} + \frac{1}{E_{\mathrm{HOMO,dipolarophile}} - E_{\mathrm{LUMO,dipole}}} \tag{12.9}$$

This approximation again is a crude one, but it allows one to recognize the essentials more clearly.

- According to Equation 12.9, the transition state of a 1,3-dipolar cycloaddition is more stabilized and the reaction proceeds faster, the closer the occupied and the unoccupied frontier orbitals are to each other.
- Especially fast 1,3-dipolar cycloadditions can be expected whenever the $\mathrm{HOMO_{dipole}}$/$\mathrm{LUMO_{dipolarophile}}$ interaction is particularly strong. In this case, the denominator of the first term in Equation 12.9 will be rather small, which makes the first term large. This scenario characterizes the so-called Sustmann type I additions (Figure 12.35, column 2).
- Especially fast 1,3-dipolar cycloadditions also can be expected whenever the denominator of the second term is very small, so that the magnitude of the second term is high. This scenario characterizes the so-called Sustmann type III additions. Therein, it is essentially the $\mathrm{HOMO_{dipolarophile}}$/$\mathrm{LUMO_{dipole}}$ interaction which stabilizes the transition state (Figure 12.35, column 4).
- Sustmann type II additions occur whenever both terms of Equation 12.9 contribute to a similar (and small) extent to the stabilization of the transition state of 1,3-cycloadditions. These reactions correspondingly represent a reactivity minimum.

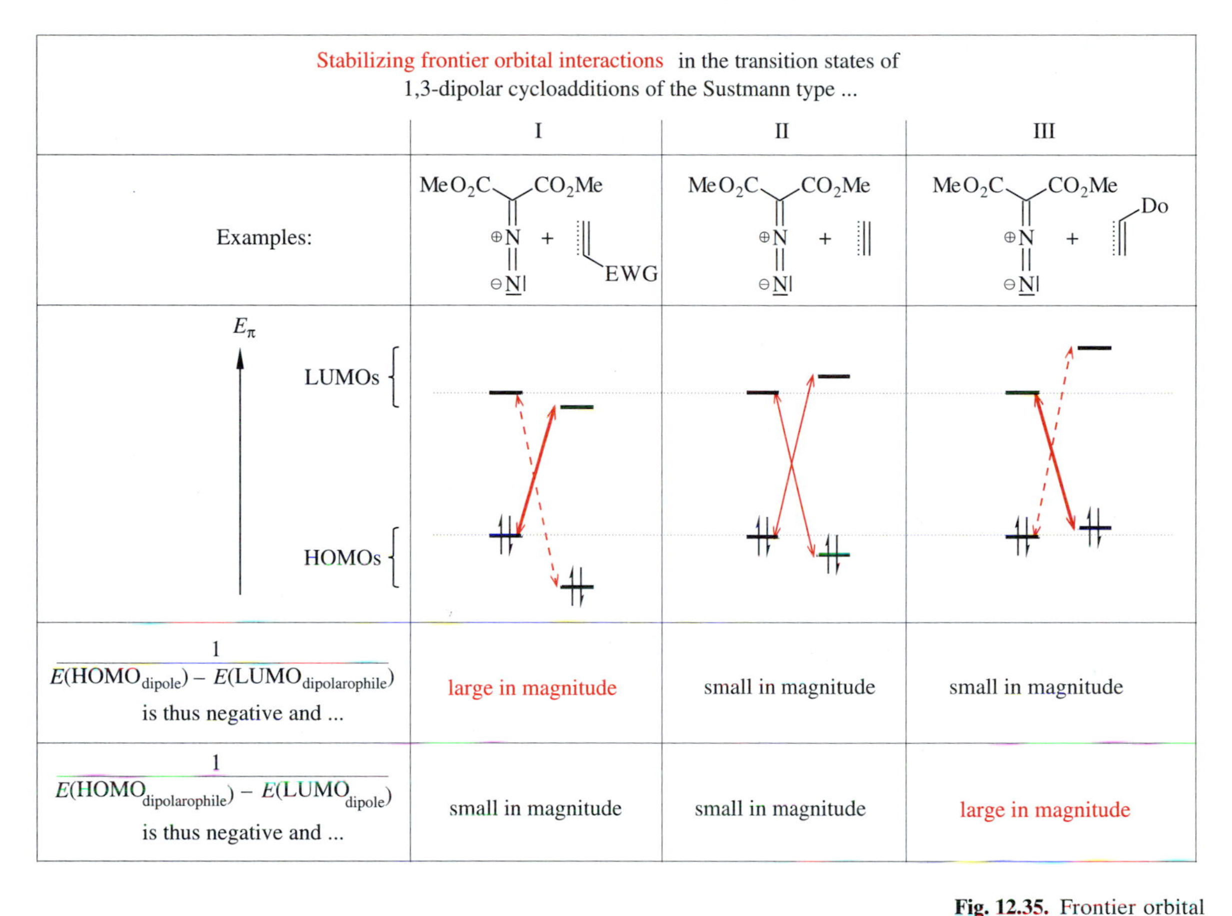

Fig. 12.35. Frontier orbital interactions of 1,3-dipolar cycloadditions with varying electron demands.

Some 1,3-dipoles possess HOMO and LUMO energies that allow for fast Sustmann type I additions with electron-poor dipolarophiles and for fast Sustmann type III additions with electron-rich dipolarophiles. In reactions with dipolarophiles with intermediate electron density, such dipoles merely are substrates of the much slower Sustmann type II additions. A plot of the rate constants of the 1,3-dipolar cycloadditions of such dipoles as a function of the HOMO energy of the dipolarophiles—or as a function of their LUMO energy that varies with the same trend—then has a U-shape. One such plot is shown in Figure 12.36 for 1,3-dipolar cycloadditions of diazomalonates.

12.5.3 1,3-Dipolar Cycloadditions of Diazoalkanes

The simplest 1,3-dipolar cycloadditions of **diazomethane** were presented in Figure 12.33. Diazomethane is generated from sulfonamides or alkyl carbamates of *N*-nitrosomethylamine. The preparation shown in Figure 12.37 is based on the commercially available *para*-toluenesulfonylmethylnitrosamide (Diazald). In a basic medium, this amide forms a sulfonylated diazotate **A** by way of a [1,3]-shift, which then undergoes a base-mediated 1,3-elimination (cf. Section 4.1.1).

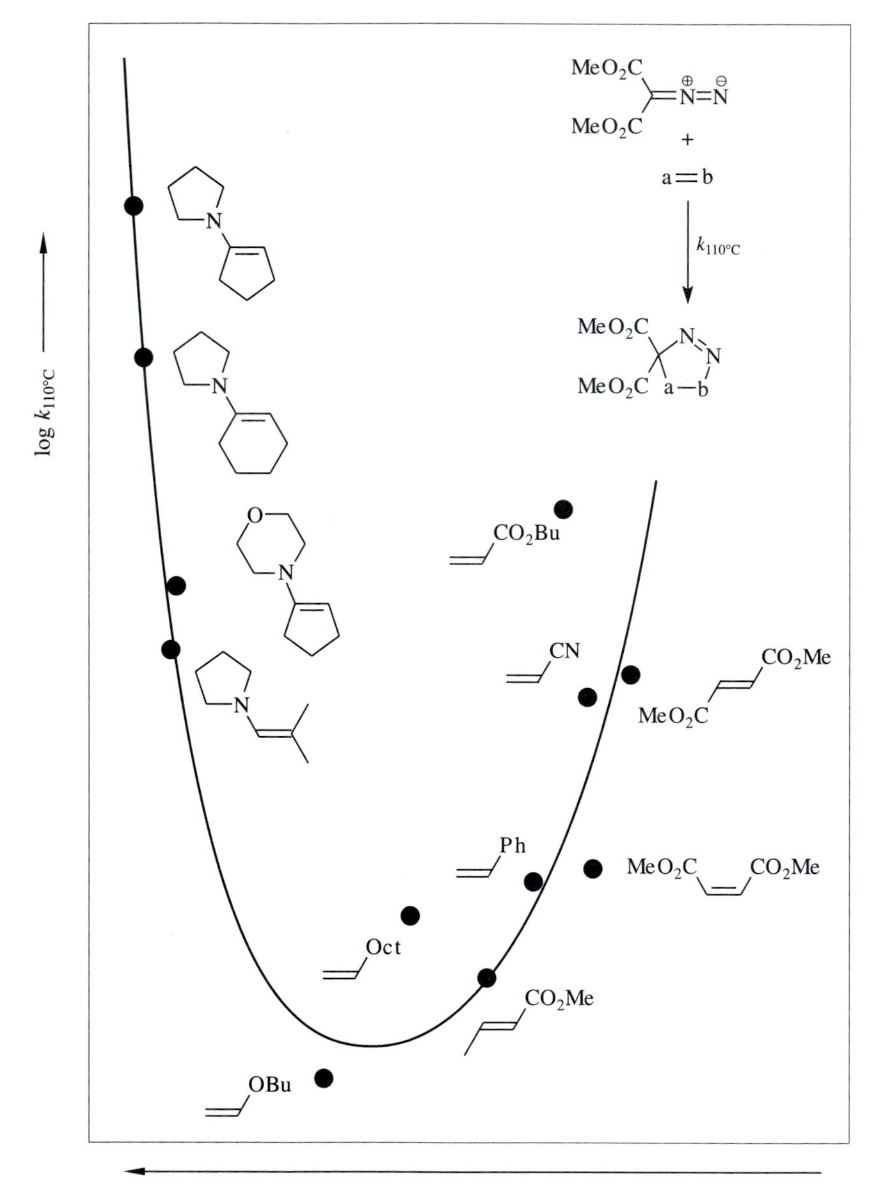

Fig. 12.36. Rate constants of 1,3-dipolar cycloadditions of diazomalonic ester as a function of the HOMO or LUMO energies, respectively, of the dipolarophile.

Diazomethane is an electron-rich 1,3-dipole, and it therefore engages in Sustmann type I 1,3-dipolar cycloadditions. In other words, diazomethane reacts with acceptor-substituted alkenes or alkynes (e.g., acrylic acid esters and their derivatives) much faster than with ethene or acetylene (Figure 12.33). Diazomethane often reacts with asymmetric electron-deficient dipolarophiles with orientation selectivity, as exempli-

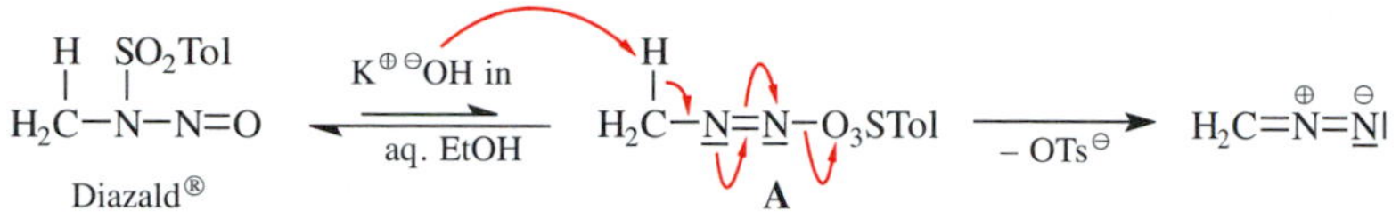

Fig. 12.37. Preparation of diazomethane by way of a 1,3-elimination.

fied in Figure 12.38 for the case of the 1,3-dipolar cycloaddition between diazomethane and the methyl ester of 2-methyl-2-butenoic acid.

This 1,3-dipolar cycloaddition also shows stereoselectivity. From the *trans*-configured ester the *trans*-configured cycloadduct is formed with a diastereoselectivity exceeding 99.997 : 0.003. This finding provides compelling evidence that this cycloaddition occurs in a single step. If the reaction were a two-step process, either dipole **A** or biradical **C** would occur as an intermediate. None of the intermediates would have to possess exactly the conformations shown in Figure 12.38, *but the conformation depicted certainly is correct in one detail:* the *trans* configuration of the 2-methyl-2-butenoic acid moiety is initially conserved.

The reactive intermediates **A** and **C** could be so short-lived that they cyclize extremely fast. They certainly could cyclize so fast that the *trans* configuration of the 2-methyl-2-butenoic acid moiety *by and large* is carried over into the cycloadduct. Intermediates **A** and **C** would therefore hardly have enough time to isomerize to conformers **B** and **D,** respectively, by way of a rotation about the C—C single bond between the two C atoms that used to be connected by a configurationally stable C═C

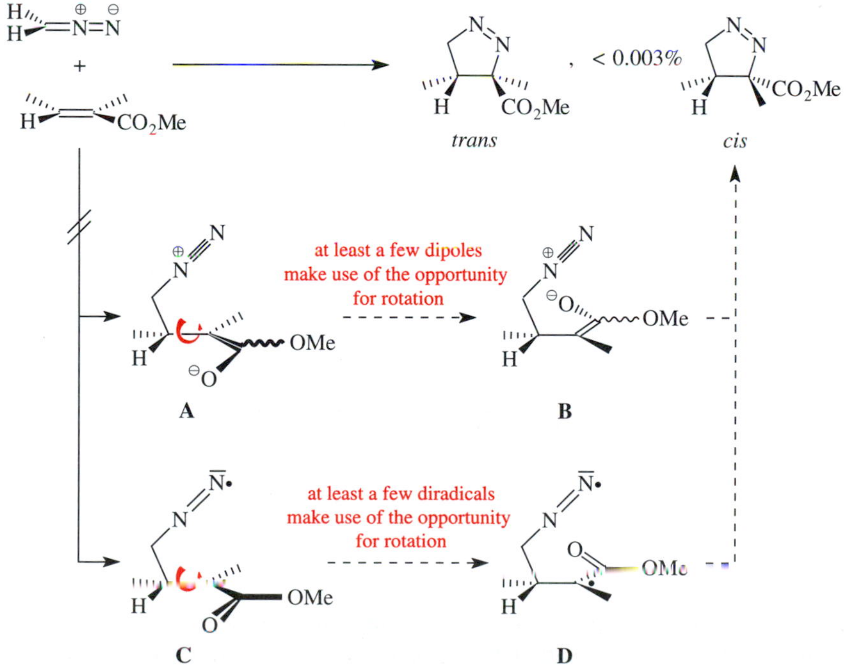

Fig. 12.38. Orientation-selective and stereoselective 1,3-dipolar cycloaddition of diazomethane. The *trans*-configured 2-methyl-2-butenoic acid is converted to the *trans*-configured cycloadduct with a diastereoselectivity of better than 99.997 : 0.003.

double bond in the starting ester. However, it seems unbelievable that not even as little as 0.003% of either of the potential intermediates would have an opportunity to rotate. The resulting isomerized intermediates **B** and **D** would then have cyclized to give the *cis*-cycloadduct about as fast as the original intermediates **A** and **C** would be assumed to give the *trans*-cycloadduct. Hence, for a multistep course of the 1,3-dipolar cycloaddition of Figure 12.38, one would expect more *cis*-cycloadduct to form than the amount (< 0.003%) that actually occurs.

Diazomalonic ester is another important 1,3-dipole for synthesis. We saw the kinetics of 1,3-dipolar cycloadditions of diazomalonic ester earlier, in the discussion of Figure 12.36. The preparation of this 1,3-dipole is accomplished most conveniently with the procedure shown in Figure 12.39: **diazo group transfer according to Regitz.**

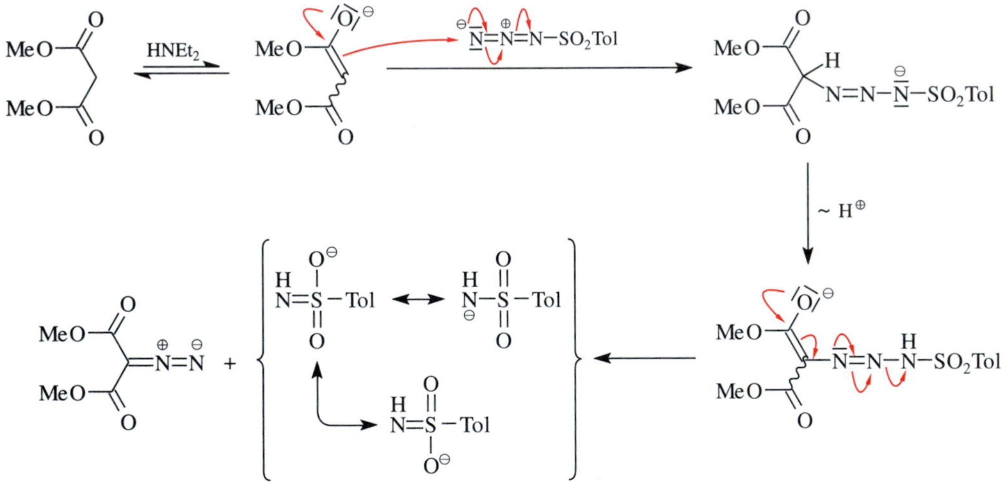

Fig. 12.39. Preparation of diazomalonic ester via diazo group transfer according to Regitz.

12.5.4 1,3-Dipolar Cycloadditions of Nitrile Oxides

Nitrile oxides have the structure R—C≡N$^+$—O$^-$ or Ar—C≡N$^+$—O$^-$. Aliphatic nitrile oxides usually can be prepared only *in situ,* while the analogous aromatic compounds, which are resonance-stabilized, generally can be isolated. The most common preparation of nitrile oxides is the dehydration of aliphatic nitro compounds. Figure 12.40 shows in detail how this dehydration can be effected by the reaction of nitroethane with a mixture of NEt_3 and Ph—N=C=O.

NEt_3 deprotonates a fraction of the nitro compound to give the nitronate **A.** One of its negatively charged O atoms adds to the C=O double bond of the isocyanate and the negatively charged adduct **D** is formed. This mode of reaction is reminiscent of the addition of carboxylic acids, alcohols, or water to isocyanates (Section 7.1). Adduct **D** undergoes a β-elimination of phenyl carbamate. This elimination proceeds via a cyclic transition state that resembles the transition state of the Chugaev reaction (Figure 4.13). The nitrile oxide is formed and immediately adds to the *trans*-butene, which must be

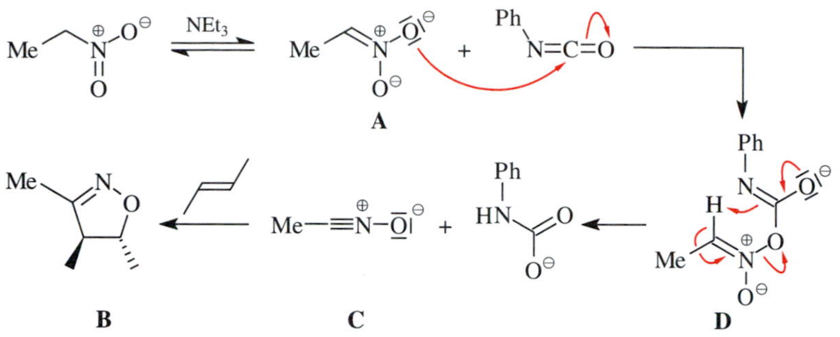

Fig. 12.40. Isoxazoline formation by way of a 1,3-dipolar addition of an *in situ* generated nitrile oxide to *trans*-butene.

added to the reaction mixture before the nitrile oxide preparation is begun. The isoxazoline **B** is formed with complete *trans*-selectivity in accordance with a one-step 1,3-cycloaddition.

Benzonitrile oxide (**C** in Figure 12.41) is an isolable 1,3-dipole. It can be generated from benzaldoxime and an NaOH/Cl_2 solution. Under these reaction conditions initially the oxime/nitroso anion (**A** ↔ **B**) is formed and chlorine disproportionates into Cl—O^- and chloride. An S_N reaction of the negatively charged C atom of the anion **A**↔**B** at the Cl atom of Cl—O^- or of Cl—O—H affords the α-chlorinated nitroso compound **E,** which tautomerizes to the hydroxamic acid chloride **D.** From that species, the nitrile oxide **C** is generated via a base mediated 1,3-elimination. Isoxazoles are formed in the reactions of **C** with alkynes (Figure 12.41), while isoxazolines would be formed in its reactions with alkenes.

The sequence on the left in Figure 12.42 shows how the method described in Figure 12.40 can be used to convert unsaturated nitro compounds into nitrile oxides containing C=C double bonds. The sequence on the right in Figure 12.42 shows how unsaturated oximes can be converted into unsaturated nitrile oxides with the method described in Figure 12.41. The dipole and the dipolarophile contained in C=C containing nitrile oxides can undergo an intramolecular 1,3-dipolar cycloaddition if these functional groups are located at a suitable distance—that is, if the possibility exists

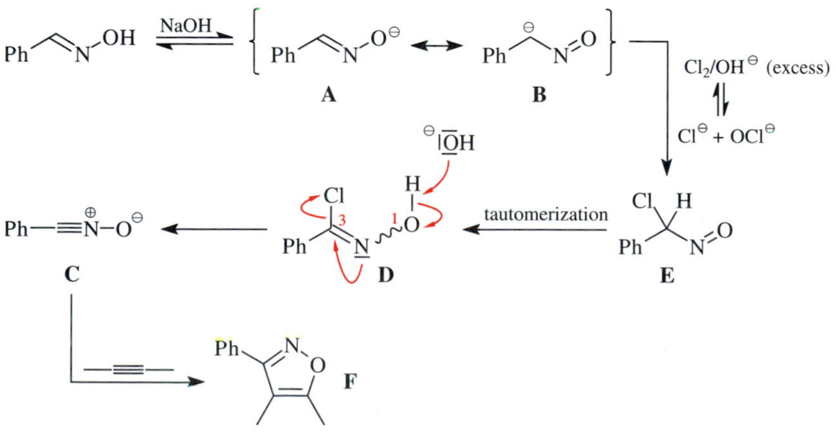

Fig. 12.41. Isoxazole formation by way of a 1,3-dipolar addition of an isolable nitrile oxide.

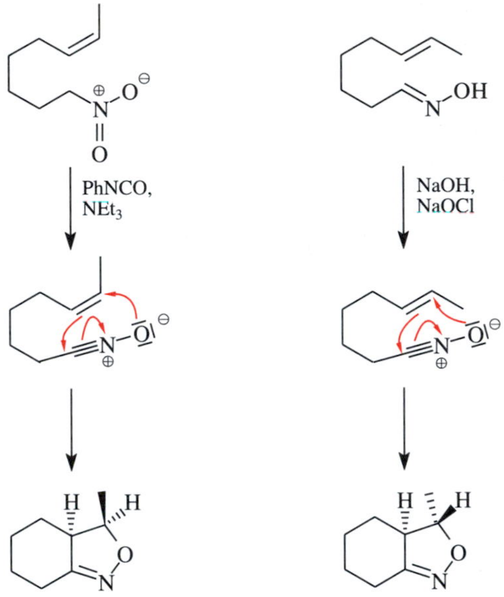

Fig. 12.42. Intramolecular 1,3-dipolar additions of stereoisomeric nitrile oxides to form stereoisomeric isoxazolines.

for the formation of a five- or six-membered ring. Two such cycloadditions are shown in Figure 12.42. Each is stereoselective and, when viewed in combination, the two reactions are stereospecific.

The N—O bond of isoxazolines can be cleaved easily via reduction. It is for this reason that isoxazolines are interesting synthetic intermediates. γ-Amino alcohols are formed by reduction with $LiAlH_4$ (for an example, see Figure 12.43, left). Hydrogenation of isoxazolines catalyzed by Raney nickel yields β-hydroxyimines, which undergo hydrolysis to β-hydroxycarbonyl compounds in the presence of boric acid (Figure 12.43, right). Figures 12.42 and 12.43 illustrate impressively that the significance of 1,3-dipolar cycloadditions extends beyond the synthesis of five-membered heterocycles. In fact, these reactions can provide a valuable tool in the approach to entirely different synthetic targets. In the cases at hand, one can view the 1,3-dipolar cycloaddition of nitrile oxides to alkenes as a ring-closure reaction and more specifically, as a

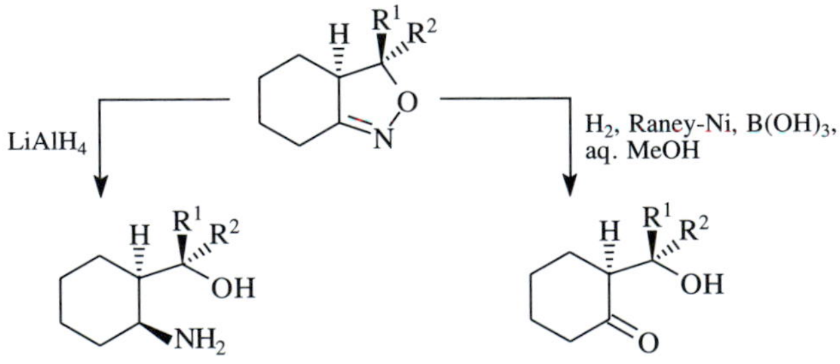

Fig. 12.43. Isoxazolines (for their preparation, see Figure 12.42) are synthetic equivalents for γ-amino alcohols and β-hydroxyketones.

means of generating interestingly functionalized five- and six-membered rings in a stereochemically defined fashion.

12.5.5 1,3-Dipolar Cycloadditions and 1,3-Dipolar Cycloreversions as Steps in the Ozonolysis of Alkenes

The reaction of ozone with a C=C double bond begins with a 1,3-dipolar cycloaddition. It results in a 1,2,3-trioxolane, the so-called **primary ozonide:**

in CH_2Cl_2 or in MeOH

primary ozonide

The presence of two O—O bonds renders primary ozonides so unstable that they decompose immediately (Figures 12.44 and 12.45). The decomposition of the permethylated symmetric primary ozonide shown in Figure 12.44 yields acetone and a carbonyl oxide in a one-step reaction. The carbonyl oxide represents a 1,3-dipole of the allyl an-

a hydroperoxide

$\sim H^{\oplus}$

if the ozonolysis is carried out in Me—OH

a primary ozonide

a carbonyl oxide

if the ozonolysis is carried out in CH_2Cl_2: × 2

a tetroxane

Fig. 12.44. Solvent dependence of the decomposition of the symmetric primary ozonide formed in the ozonolysis of tetramethylethene (2,3-dimethyl-2-butene). The initially formed carbonyl oxide forms a hydroperoxide in methanol, while it dimerizes to give a 1,2,4,5-tetroxane in CH_2Cl_2.

ion type (Table 12.2). However, it is known that a correct valence bond description of *this* dipole also requires the consideration of a biradical resonance form. When acetone is viewed as a dipolarophile, then the decomposition of the primary ozonide into acetone and a carbonyl oxide is recognized as the reversion of a 1,3-cycloaddition. Such a reaction is referred to as a **1,3-dipolar cycloreversion.**

The carbonyl oxide, a valence-unsaturated species, is not the final product of an ozonolysis. Rather, it will react further in one of two ways. Carrying out the ozonolysis in methanol leads to the capture of the carbonyl oxide by methanol under formation of a hydroperoxide, which is structurally identical to the "ether peroxide" of isopropyl methyl ether. However, if the same carbonyl oxide is formed in the absence of methanol (e.g., if the ozonolysis is carried out in dichloromethane) the carbonyl oxide undergoes a cycloaddition. If the carbonyl oxide is formed along with a

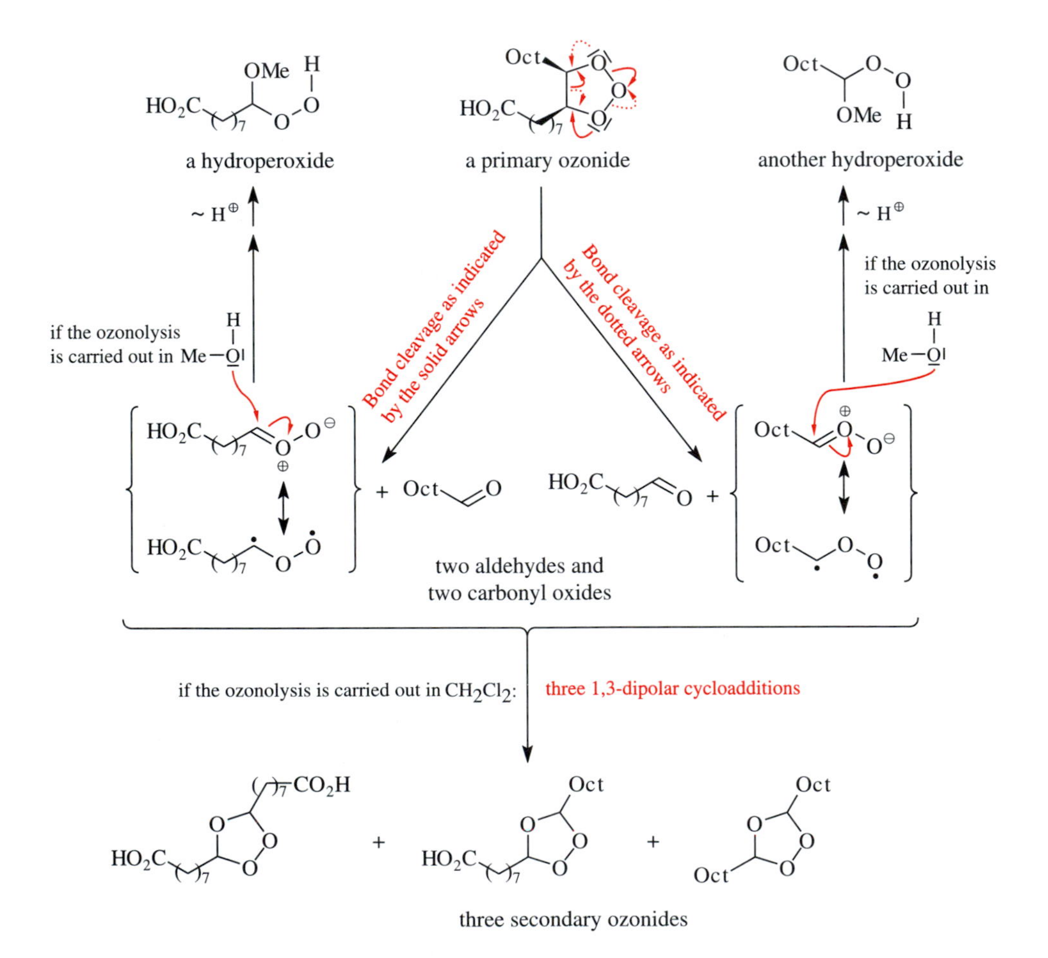

Fig. 12.45. Decomposition of the asymmetric primary ozonide formed in the ozonolysis of oleic acid. Two different aldehydes and two different carbonyl oxides are formed. In CH_2Cl_2 these molecules react with each other to form three secondary ozonides. In methanol, on the other hand, the carbonyl oxides react with the solvent to form hydroperoxides.

ketone (Figure 12.44), as opposed to an aldehyde, it preferentially participates in a cycloaddition with another carbonyl oxide, i.e., a dimerization to a 1,2,4,5-tetroxane occurs. Such tetroxanes are extremely explosive compounds that one may not even attempt to isolate. Instead these compounds must be destroyed via reduction while they remain in solution. We will see later (Figures 14.21 and 14.22) how this is accomplished.

The chemistry of primary ozonides is more varied if they are less highly alkylated than the primary ozonide of Figure 12.44. This is particularly true if the primary ozonide is asymmetric, like the one shown in Figure 12.45. This is because its decay may involve two different 1,3-dipolar cycloreversions. Both of them result in one carbonyl oxide and one carbonyl compound. If the reaction is carried out in methanol, the two carbonyl oxides can react with the solvent (as in Figure 12.44) whereby each of them delivers a hydroperoxide (an "ether peroxide" analog).

Obviously, this reaction channel is not an option for carbonyl oxides generated in dichloro-methane (Figure 12.45). Instead, they can continue to react by way of a cycloaddition. In contrast to the carbonyl oxide of Figure 12.44, they do not undergo a cycloaddition with each other. Instead, they undergo a 1,3-dipolar cycloaddition to the C=O double bond of the concomitantly formed aldehyde(s). The orientation selectivity is such that the trioxolane formed differs from the primary ozonide; the 1,2,4-trioxolane products are the so-called **secondary ozonides.**

Secondary ozonides are significantly more stable than primary ozonides, since the former contain only *one* O—O bond while the latter contain *two.* Still, these compounds are reasonably stable only in solution, being explosive when neat. Therefore, secondary ozonides must be reduced to compounds without O—O bonds prior to reaction workup, too. Figures 12.44 and 12.45 show how that is accomplished.

12.5.6 A Tricky Reaction of Inorganic Azide

Aryldiazonium ions and sodium azide react to form aryl azides. We did not mention *this* reaction in the discussion of the S_N reactions of Ar—N_2^+ (Section 5.4) because it belongs, at least in part, to the 1,3-dipolar cycloadditions.

Phenyl azide is formed from phenyldiazonium chloride and sodium azide by way of two competing reactions (Figure 12.46). The reaction path to the right begins with a 1,3-dipolar cycloaddition. At low temperature, this cycloaddition affords phenylpentazole, which decays above 0°C via a 1,3-dipolar cycloreversion. This cycloreversion produces the 1,3-dipole phenyl azide as the *desired* product, and molecular nitrogen as a side product.

The rest of the phenyl azide produced in this reaction comes from an intermediate called 1-phenylpentazene, which is formed in competition with phenylpentazole (Figure 12.46, left). 1-Phenylpentazene is a compound that features a phenyl group at the end of a chain of five N atoms. The unavoidably ensuing reaction of this pentazene is an α-elimination of molecular nitrogen from the central N atom. It leads to the same organic product—phenyl azide—as the phenylpentazole route. That these two mechanisms are realized concomitantly was established by the isotopic labelling experiments reviewed in Figure 12.46.

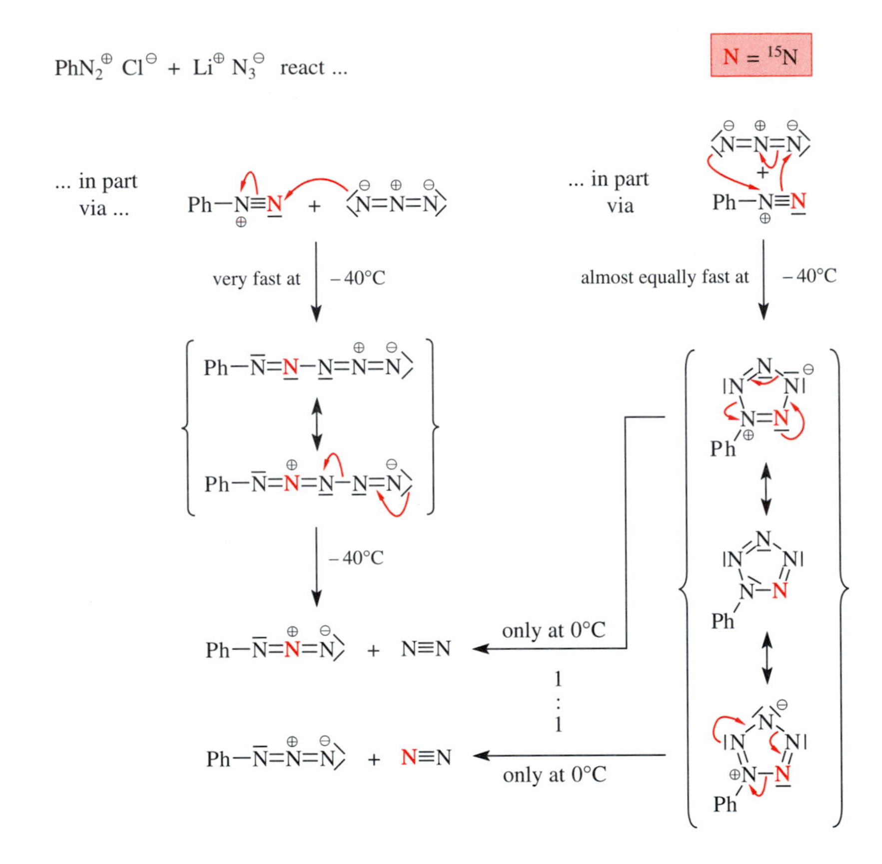

Fig. 12.46. The competing mechanisms of the "S_N reaction" $Ar—N_2^+ + N_3^- \rightarrow Ar—N_3 + N_2$. The reaction proceeds following two reaction channels: 1,3-dipolar cycloaddition via the intermediate phenylpentazole (right) and via the alternative phenylpentazene (left).

References

W. Carruthers, "Cycloaddition Reactions in Organic Synthesis," Pergamon Press, Elmsford, NY, **1990.**

J. Jurczak, T. Bauer, C. Chapuis, D. Craig, M. Cinquini, F. Cozzi, W. Sander, P. Binger, D. Fox, B. B. Snider, J. Mattay, R. Conrads, H.-U. Reißig, "Formation of C-C Bonds by Pericyclic Reactions—Cycloadditions," in *Methoden Org. Chem.* (*Houben-Weyl*) *4th ed. 1952-, Stereoselective Synthesis* (G. Helmchen, R. W. Hoffmann, J. Mulzer, E. Schaumann, Eds.), Vol. E21c, 2735, Georg Thieme Verlag, Stuttgart, **1995.**

H. McNab, "One or More C=C Bond(s) by Pericyclic Processes," in *Comprehensive Organic Functional Group Transformations* (A. R. Katritzky, O. Meth-Cohn, C. W. Rees, Eds.), Vol. 1, 771, Elsevier Science, Oxford, U.K., **1995.**

12.1

F. Bernardi, M. Olivucci, M. A. Robb, "Predicting forbidden and allowed cycloaddition reactions: Potential surface topology and its rationalization," *Acc. Chem. Res.* **1990,** *23,* 405.

12.2

K. N. Houk, J. Gonzalez, Y. Li, "Pericyclic reaction transition states: Passions and punctilios, 1935–1995," *Acc. Chem. Res.* **1995,** *28,* 81–90.

O. Wiest and K. N. Houk, "Density-functional theory calculations of pericyclic reaction transition structures," *Top. Curr. Chem.* **1996,** *183,* 1–24.

K. N. Houk, Y. Li, J. D. Evanseck, "Transition structures of hydrocarbon pericyclic reactions," *Angew. Chem.* **1992,** *104,* 711–739; *Angew. Chem. Int. Ed. Engl.* **1992,** *31,* 682–708.

12.3

H. L. Holmes, "The Diels-Alder reaction: Ethylenic and acetylenic dienophiles," *Org. React.* **1948,** *4,* 60–173.

M. C. Kloetzel, "The Diels-Alder reaction with maleic anhydride." *Org. React.* **1948,** *4,* 1–59.

L. L. Butz and A. W. Rytina, "The Diels-Alder reaction: Quinones and other cyclenones," *Org. React.* **1949,** *5,* 136–192.

F. Fringuelli and A. Taticchi, "Dienes in the Diels-Alder Reaction," Wiley, New York, **1990.**

V. D. Kiselev and A. I. Konovalov, "Factors that determine the reactivity of reactants in the normal and catalyzed Diels-Alder reaction," *Russ. Chem. Rev.* (*Engl. Transl.*) **1989,** *58,* 230.

W. Oppolzer, "Intermolecular Diels-Alder Reactions," in *Comprehensive Organic Synthesis* (B. M. Trost, I. Fleming, Eds.), Vol. 5, 315, Pergamon Press, Oxford, **1991.**

F. Fringuelli, A. Taticchi, E. Wenkert, "Diels-Alder reactions of cycloalkenones in organic synthesis. A review," *Org. Prep. Proced. Int.* **1990,** *22,* 131–165.

F. Fringuelli, L. Minuti, F. Pizzo, A. Taticchi, "Reactivity and selectivity in Lewis acid-catalyzed Diels-Alder reactions of 2-cyclohexenones," *Acta Chem. Scand.* **1993,** *47,* 255–263.

E. Ciganek, "The intramolecular Diels-Alder reaction," *Org. React.* (*N.Y.*) **1984,** *32,* 1.

W. R. Roush, "Stereochemical and Synthetic Studies of the Intramolecular Diels-Alder Reaction," in *Advances in Cycloaddition* (D. P. Curran, Ed.) **1990,** *2,* Jai Press, Greenwich, CT.

12.4

S. Patai (Ed.), "The Chemistry of Ketenes, Allenes, and Related Compounds," Wiley, New York, **1980.**

W. T. Brady, "Synthetic applications involving halogenated ketenes," *Tetrahedron* **1981,** *37,* 2949.

P. W. Raynolds, "Ketene," in *Acetic Acid and Its Derivatives* (V. H. Agreda, J. R. Zoeller, Eds.), 161, Marcel Dekker, New York, **1993.**

E. Lee-Ruff, "New Synthetic Pathways from Cyclobutanones," in *Advances in Strain in Organic Chemistry* (B. Halton, Ed.) **1991,** *1,* Jai Press, Greenwich, CT.

12.5

A. Padwa (Ed.), "1,3-Dipolar Cycloaddition Chemistry," Wiley, **1984.**

R. Huisgen, "Steric Course and Mechanism of 1,3-Dipolar Cycloadditions," in *Advances in Cycloaddition* (D. P. Curran, Ed.) **1988,** *1,* 11–31, Jai Press, Greenwich, CT.

R. Sustmann, "Rolf Huisgen's contribution to organic chemistry, emphasizing 1,3-dipolar cycloadditions," *Heterocycles* **1995,** *40,* 1–18.

A. Padwa, "Intermolecular 1,3-Dipolar Cycloadditions," in *Comprehensive Organic Synthesis* (B. M. Trost, I. Fleming, Eds.), Vol. 4, 1069, Pergamon Press, Oxford, **1991.**

A. Padwa and A. M. Schoffstall, "Intramolecular 1,3-Dipolar Cycloaddition Chemistry," in *Advances in Cycloaddition* (D. P. Curran, Ed.), Vol. 2, 1, Jai, Press Inc., Greenwich, CT, **1990.**

P. A. Wade, "Intramolecular 1,3-Dipolar Cycloadditions," in *Comprehensive Organic Synthesis* (B. M. Trost, I. Fleming, Eds.), Vol. 4, 1111, Pergamon Press, Oxford, **1991.**

T. H. Black, "The preparation and reactions of diazomethane," *Aldrichimica Acta* **1983,** *16,* 3.

K. B. G. Torsell, "Nitrile Oxides, Nitrones and Nitronates in Organic Synthesis. Novel Strategies in Synthesis," VCH Verlagsgesellschaft, Weinheim, **1988.**

S. Kanemasa and O. Tsuge, "Recent advances in synthetic applications of nitrile oxide cycloaddition (1981–1989)," *Heterocycles* **1990,** *30,* 719–736.

D. P. Curran, "The Cycloaddition Approach to α-Hydroxy Carbonyls: An Emerging Alternative to the Aldol Strategy," in *Advances in Cycloaddition* (D. P. Curran, Ed.) **1988,** *1,* 129–189, Jai Press, Greenwich, CT.

R. L. Kuczkowski, "Formation and structure of ozonides," *Acc. Chem. Res.* **1983,** *16,* 42.
R. L. Kuczkowski, "The structure and mechanism of formation of ozonides," *Chem. Soc. Rev.* **1992,** *21,* 79–83.
O. Horie and G. K. Moortgat, "Gas-phase ozonolysis of alkenes. Recent advances in mechanistic investigations," *Acc. Chem. Res.* **1998,** *31,* 387–396.
W. H. Bunnelle, "Preparation, properties, and reactions of carbonyl oxides," *Chem. Rev.* **1991,** *91,* 335–362.
K. Ishiguro, T. Nodima, Y. Sawaki, "Novel aspects of carbonyl oxide chemistry," *J. Phys. Org. Chem.* **1997,** *10,* 787–796.

Further Reading

S. Danishefsky, "Cycloaddition and cyclocondensation reactions of highly functionalized dienes: Applications to organic synthesis," *Chemtracts: Org. Chem.* **1989,** *2,* 273–297.
T. Kametani and S. Hibino, "The synthesis of natural heterocyclic products by hetero Diels-Alder cycloaddition reactions," *Adv. Heterocycl. Chem.* **1987,** *42,* 246.
J. A. Coxon, D. Q. McDonald, P. J. Steel, "Diastereofacial Selectivity in the Diels-Alder Reaction," in *Advances in Detailed Reaction Mechanisms* (J. M. Coxon, Ed.), Vol. 3, Jai Press, Greenwich, CT, **1994.**
J. D. Winkler, "Tandem Diels-Alder cycloadditions in organic synthesis," *Chem. Rev.* **1996,** *96,* 167–176.
C. O. Kappe, S. S. Murphree, A. Padwa, "Synthetic applications of furan Diels-Alder chemistry," *Tetrahedron* **1997,** *53* 14179–14231.
D. Craig, "Stereochemical aspects of the intramolecular Diels-Alder reaction," *Chem. Soc. Rev.* **1987,** *16,* 187.
G. Helmchen, R. Karge, J. Weetman, "Asymmetric Diels-Alder Reactions with Chiral Enoates as Dienophiles," in *Modern Synthetic Methods* (R. Scheffold, Ed.), Vol. 4, 262, Springer, Berlin, **1986.**
L. F. Tietze and G. Kettschau, "Hetero Diels-Alder reactions in organic chemistry," *Top. Curr. Chem.* **1997,** *189,* 1–120.
K. Neuschuetz, J. Velker, R. Neier, "Tandem reactions combining Diels-Alder reactions with sigmatropic rearrangement processes and their use in synthesis," *Synthesis* **1998,** 227–255.
D. L. Boger and M. Patel, "Recent Applications of the Inverse Electron Demand Diels-Alder Reaction," in *Progress in Heterocyclic Chemistry* (H. Suschitzky, E. F. V. Scriven, Eds.) **1989,** *1,* 36–67, Pergamon Press, Oxford, U.K.
M. J. Tashner, "Asymmetric Diels-Alder Reactions," in *Organic Synthesis: Theory and Applications* (T. Hudlicky, Ed.) **1989,** *1,* Jai Press, Greenwich, CT.
H. B. Kagan and O. Riant, "Catalytic asymmetric Diels-Alder reactions," *Chem. Rev.* **1992,** *92,* 1007.
T. Oh and M. Reilly, "Reagent-controlled asymmetric Diels-Alder reactions," *Org. Prep. Proced. Int.* **1994,** *26,* 129.
I. E. Marko, G. R. Evans, P. Seres, I. Chelle, Z. Janousek, "Catalytic, enantioselective, inverse electron-demand Diels-Alder reactions of 2-pyrone derivatives," *Pure Appl. Chem.* **1996,** *68,* 113.
H. Waldmann, "Asymmetric hetero-Diels-Alder reactions," *Synthesis* **1994,** 635–651.
H. Waldmann, "Asymmetric Aza-Diels-Alder Reactions," in *Organic Synthesis Highlights II* (H. Waldmann, Ed.), VCH, Weinheim, New York, **1995,** 37–47.
L. C. Dias, "Chiral Lewis acid catalysts in Diels Alder cycloadditions: Mechanistic aspects and synthetic applications of recent systems," *J. Braz. Chem. Soc.* **1997,** *8,* 289–332.
J. Mulzer, "Natural Product Synthesis via 1,3-Dipolar Cycloadditions," in *Organic Synthesis Highlights* (J. Mulzer, H.-J. Altenbach, M. Braun, K. Krohn, H.-U. Reißig, Eds.), VCH, Weinheim, New York, **1991,** 77–95.
K. V. Gothelf and K. A. Jürgensen, "Metal-catalyzed asymmetric 1,3-dipolar cycloaddition reactions," *Acta Chem. Scand.* **1996,** *50,* 652.
P. N. Confalone and E. M. Huie, "The [3+2] nitrone-olefin cycloaddition reaction," *Org. React. (N.Y.)* **1988,** *36,* 1.

Transition Metal–Mediated Alkenylations, Arylations, and Alkynylations

13

A

As a rule, halogens, OH groups, and all other O substituents that are attached to an sp^2-hybridized C atom of an alkene or an aromatic compound cannot be substituted by nucleophiles alone. One exception has already been discussed and that was the addition/elimination mechanism of the nucleophilic substitution. This mechanism occurs, among other substrates, with alkenes that carry a strong electron acceptor on the neighboring C atom (see Figures 8.36–8.38 for examples). The same mechanism is known to also operate for aromatic compounds—under the condition, again, that they carry an acceptor substituent at an appropriate position (Section 5.5). The benzyne mechanism of nucleophilic substitution, that is, an elimination/addition mechanism (Section 5.6), presents a second mode of nucleophilic substitution of a halogen at an sp^2-hybridized C atom.

Halogens or O-bound leaving groups, however, also can be detached from the sp^2-hybridized C atom of an alkene or an aromatic compound

- if organometallic compounds act as nucleophiles, and
- if a transition metal is present in at least catalytic amounts. These S_N reactions are designated as **C,C-coupling reactions**.

The most important substrates for substitutions of this type are alkenyl and aryl triflates, bromides, or iodides (Sections 13.1–13.3). The most important organometallic compounds to be introduced into the substrates contain Cu, Mg, B, or Zn. The metal-bound C atom can be sp^3-, sp^2-, or sp-hybridized in these compounds, and each of these species, in principle, is capable of attacking unsaturated substrates. Organocopper compounds most usually (Section 13.1), but not always (Section 13.3.4), substitute *without* the need for a catalyst. Grignard compounds substitute in the presence of catalytic amounts of Ni complexes (Section 13.2), while organoboron (Section 13.3.2) and organozinc (Section 13.3.3) compounds typically are reacted in the presence of $Pd(PPh_3)_4$.

All these C,C-coupling reactions can be carried out in an analogous fashion at sp-hybridized carbons, too, as long as this carbon binds to a Br or I as a leaving group. However, we will present this type of reaction only briefly, in Section 13.4.

In Section 13.5, a few *other* C,C-coupling reactions of alkenes and of aromatic compounds, which contain an sp^2—OTf, an sp^2—Br, or an sp^2—Cl bond, will be discussed because these C,C couplings and the preceding ones are closely related mechanistically. These substrates, however, react with *metal-free* alkenes. Palladium-complexes again serve as the catalysts.

13.1 Alkenylation and Arylation of Copper-Bound Organyl Groups

Me_2CuLi couples with a variety of alkenyl triflates (bromides, iodides) giving methyl derivatives (Table 13.1), and, in complete analogy, with aryl triflates (bromides, iodides) giving toluenes.

Table 13.1. Product Spectrum of C,C-Coupling Reactions with Me_2CuLi

Substrate	Me, OSO_2CF_3 (cyclohexenyl triflate)	R–CH=CH–Br(I)	R, Br(I) (cis-alkenyl)	$ArOSO_2CF_3$ or Ar–I(Br)
Preparation according to ...	Fig. 10.20	Fig. 13.10	Fig. 13.10 Fig. 13.12	i.e., as in Section 5.2.1
Reaction with Me_2CuLi yields	Me, Me (cyclohexene)	R–CH=CH–Me	R, Me (cis-alkenyl)	Ar–Me

In reactions of alkenyl triflates with stereogenic C=C double bonds, coupling reactions of these kinds convert the C_{sp^2}—X bond of the alkenyl triflate into the C_{sp^2}—C bond of the substitution product with complete retention of configuration. The stereoselective synthesis of a 1,3-diene from an alkenyl triflate and $(vinyl)_2CuLi$ provides an example (Figure 13.1).

Gilman cuprates also convert the C_{sp^2}—Br and C_{sp^2}—I bonds of stereogenic haloalkenes into a C_{sp^2}—C bond with complete retention of configuration (Table 13.1,

Fig. 13.1. Stereoselective synthesis of a trisubstituted ethylene by way of C,C coupling of a Gilman reagent. The C_{sp^2}–triflate bond is converted into the C_{sp^2}—C bond of the product with complete retention of configuration.

columns 3 and 4). Pairs of coupling reactions of stereoisomeric haloalkenes thus show stereospecificity (cf. columns 3 and 4).

Figure 13.2 depicts the presumed reaction mechanism for such C,C-coupling reactions between Gilman cuprates and the C atom of suitable C_{sp^2}—X bonds. Some of the reaction steps (1–4) might well be comprised of more than one elementary reaction. It is very important to understand these four steps as a general reaction concept. In step 1, the heterosubstituted alkene or the aromatic compound enters the coordination sphere of copper as a two-electron π donor. The reactive π complex will have at least one vacancy at the metal. Step 2 consists of the oxidative addition of the triflate or the halide, respectively, to the metal. In this step the metal inserts into the C_{sp^2}—X bond and the metal's oxidation state is increased from +1 to +3. The reaction types of steps 1 and 2 reoccur in reverse order in steps 3 and 4. Step 3 is complementary to step 2 and represents a reductive elimination. The reaction product is a coordinatively unsaturated π complex of—again—Cu(I).The C,C-coupling product functions as a π donor in this complex until the metal assumes a coordination number suitable for further reaction. The final step, 4, is complementary to step 1 and entails the dissociation of the π-bonded ligand. This step yields the final coupling product and an organocopper compound (in place of the cuprate employed originally).

We saw earlier (Figure 8.37) that MeMgBr/MeCu allows for stereospecific substitution reactions in a pair of stereoisomeric enolphosphates. In the discussion of the mechanism of these substitution reactions, we mentioned that they presumably do not involve an addition/elimination mechanism. Instead, it is more likely that the mechanism of these reactions is analogous to the C,C couplings discussed in the present chapter.

Arylcopper compounds are accessible from aryllithium compounds by way of transmetallation with a *full* equivalent of CuI (Figure 13.3). [This transmetallation also is a step in the formation of Gilman cuprates from organolithium compounds and *half* an equivalent of CuI (cf. Figure 8.34). In the latter case, the transmetallation is followed by the addition of the excess half-equivalent of the organolithium compound to the ini-

X + Me–Cu–Me⊖ Li⊕ (+1) —(1) (see text)→ [π complex: X, Cu–Me (+1, ⊖), Me, Li⊕]
↓ (2)
X–Cu–Me (+3, ⊖), Me, Li⊕ —(3)→ Me, [π complex: X–Cu–Me (+1, ⊖), Li⊕] —(4)→ Me + MeCu (+1) + LiX

Fig. 13.2. Presumed elementary steps of a C,C coupling between a Gilman cuprate and an alkenyl or aryl triflate (X = O_3S—CF_3), bromide (X = Br), or iodide (X = I). The four elementary steps of the reaction, discussed in the text, are (1) complexation, (2) oxidative addition of the substrate to the metal, (3) reductive elimination, and (4) dissociation of the π-bound ligand.

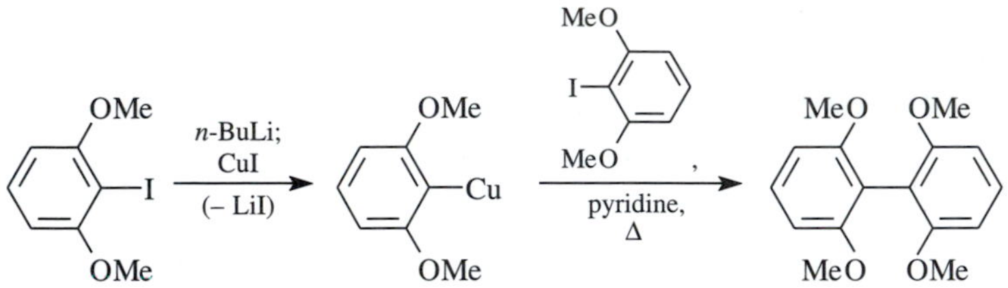

Fig. 13.3. Biaryl syntheses via arylcopper compounds I: preparation of the nucleophile in a separate reaction step.

tially formed half equivalent of RCu ($\rightarrow R_2CuLi$).] The arylcopper compound shown in Figure 13.3 can subsequently couple with an aryl iodide as the nucleophilic reagent. Substituted biphenyls are accessible in this way. The mechanism of this C,C coupling probably resembles the mechanism of the coupling shown in Figure 13.4. It is thus also similar to the mechanism of the Cadiot–Chodkiewicz coupling shown later (Figure 13.25).

The **Ullmann reaction** (Figure 13.4) represents another synthesis of substituted biphenyls. In this process an aryl iodide or—as in the present case—an aryl iodide/aryl chloride mixture is heated with Cu powder. It is presumed that under standard conditions the aryl iodide reacts *in situ* with Cu to form the aryl copper compound. Usually, the latter couples with the remaining aryl iodide and a symmetric biphenyl is formed. In a few instances it is also possible to generate asymmetric biaryls via a **crossed Ullmann reaction.** In these cases one employs a mixture of an aryl iodide and another aryl halide (not an iodide!); the other aryl halide *must* exhibit a higher propensity than the aryl iodide to couple to the arylcopper intermediate. It is presumed that the mechanism of the Ullmann reaction parallels the mechanism of the Cadiot–Chodkiewicz coupling, which we will discuss in Section 13.4.

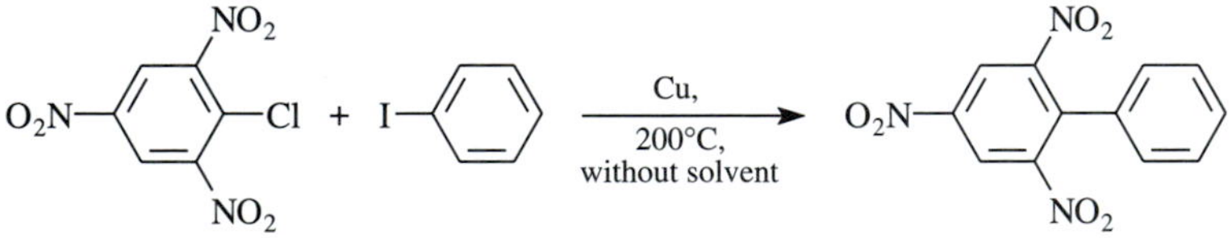

Fig. 13.4. Biaryl syntheses via arylcopper compounds II: *in situ* preparation of the nucleophile in the Ullmann procedure.

13.2 Alkenylation and Arylation of Grignard Compounds

Alkenyl bromides and iodides as well as aryl triflates, bromides, and iodides can undergo substitution reactions with Grignard compounds in the presence of catalytic amounts of a nickel complex. Even though the catalytically active species is a Ni(0) complex, these reactions can be initiated by Ni(II) complexes, which are easier to handle. Starting with a different oxidation state is possible since the organometallic compound reduces the Ni(II) complex to Ni(0) *in situ*. The addition of about 1 mol% $NiCl_2$(dppe) to the reaction mixture suffices to catalyze the alkenylation (Figure 13.5) or arylation (Figure 13.6) of Grignard compounds.

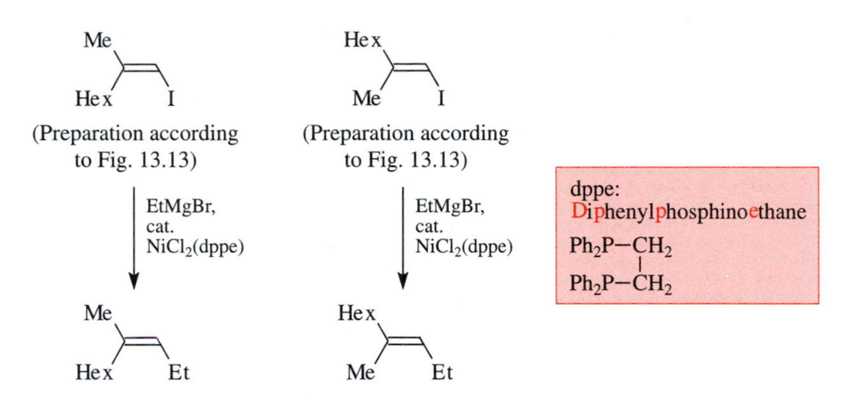

Fig. 13.5. Nickel-catalyzed alkenylation of Grignard compounds; occurrence of stereoselectivity and stereospecificity.

These reactions provide representative examples and the polarity of the substrates is interchangeable:

- Primary alkyl Grignard compounds can be alkenylated (Figure 13.5) and arylated (in analogy to Figure 13.6).
- Aromatic Grignard compounds can be arylated (Figure 13.6) and alkenylated (in analogy to Figure 13.5).
- Alkenyl Grignard compounds, too, can be alkenylated as well as arylated.

In all these reactions, any existing configuration of any stereogenic double bond—whether it be in the alkenyl bromide or iodide or in the alkenyl Grignard compound—is completely retained. The reactions in Figure 13.5 provide good examples.

A plausible reaction mechanism for such couplings is presented in Figure 13.7 for the specific example of the transformation of Figure 13.6. We do not specify the number n and the nature of the ligands L of the intermediate Ni complexes in Figure 13.7. Little is known about either one. Also, it is quite possible that more than one elementary reaction is involved in some of the steps 1–5. In any case, these five steps are *certainly* involved in the overall reaction. In step 1, the aryl bromide enters the coordination sphere of the Ni(0) compound as a π ligand. At least one metal coordination site must be vacated before an oxidative addition of the aryl bromide to the Ni atom may occur in step 2. The nickel inserts into the C_{sp^2}—Br bond, and its oxidation

MeO H — *tert*-BuLi (– *tert*-BuH); MgBr$_2$ (– LiBr) → MeO MgBr — Br (1-bromonaphthalene), cat. NiCl$_2$(dppe) → MeO

Fig. 13.6. Nickel-catalyzed arylation of Grignard compounds. The Grignard compound can be prepared *via* a substituent-directed *peri*-lithiation of a substituted naphthalene (see Section 5.3.1 for the analogous *ortho*-lithiation) and subsequent transmetallation with $MgBr_2$ (cf. Table 8.1 regarding the method).

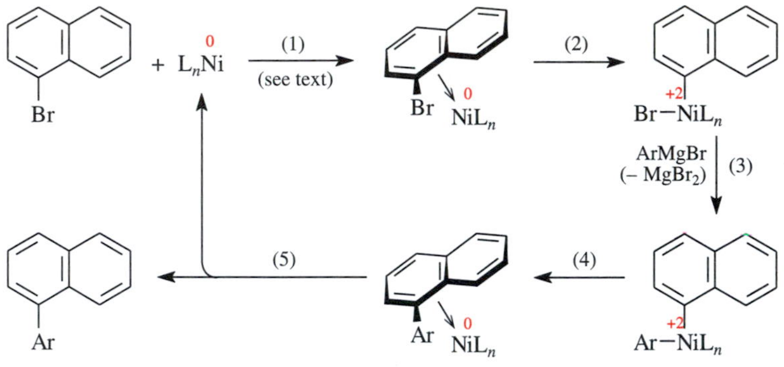

Fig. 13.7. Presumed elementary steps of a Ni-catalyzed C,C coupling between an aryl Grignard reagent and an alkyl halide—using the example of the reaction in Figure 13.6. The elementary steps, discussed in the text, are (1) complexation, (2) oxidative addition, (3) transmetallation, (4) reductive elimination, and (5) dissociation.

number is increased from +0 to +2. Step 3 is a transmetallation. The aryl Grignard compound is converted into an aryl–nickel compound (a diaryl–nickel compound in the specific example) by way of replacement of a bromide ion by a Mg-bound aryl group. Step 4 is a reductive elimination of the coupling product. It yields a coordinatively unsaturated Ni(0) complex to which the product is attached as a π ligand. Step 5 completes the reaction: the coupled aromatic compound leaves the complex, and the remaining Ni(0) complex is ready to enter the next passage through the catalytic cycle.

C,C couplings of secondary alkyl Grignard reagents may occur in analogy to the reactions of the primary Grignard reagents (Figure 13.5), but they may also lead to unexpected reactions (Figure 13.8 and 13.9). Depending on the catalyst, the Ni-catalyzed C,C-coupling reactions of *sec*-BuMgCl and chlorobenzene result in either the expected *sec*-butylbenzene or the unexpected *n*-butylbenzene. How does such an isomerizing coupling occur?

The next to last step of the respective catalytic cycle—the one that would correspond to step 4 in Figure 13.7—apparently is sterically hindered. In Figure 13.8, this step would be the reductive elimination of *i*-BuPh from the Ni complex. For this elimination to occur as desired, a *secondary* C atom would have to be connected to the aromatic compound. As the result of steric hindrance and the associated slowing down of this reaction step a competing reaction becomes feasible. This competing reaction consists of a β-elimination and leads to the hydrido–Ni complex **B** and 1-butene. This Ni complex can add again to 1-butene in a reversal of its formation reaction. This addition corresponds to a hydronickelation of the olefinic C=C double bond (cf. the hydroboration of the olefinic C=C double bonds, Section 3.3.3). The hydronickelation of 1-butene (Figure 13.8) has two regioselectivity options. The original Ni complex **A** can be reformed as the "Markovnikov adduct," in which case the overall reaction is not affected because the *expected* coupling product with the branched side chain is still obtained. However, the hydrido–Ni complex **B** and 1-butene also may add to each other with the opposite regioselectivity to form an "*anti*-Markovnikov adduct," in which case the isomeric Ni complex **C** with its unbranched butyl group is formed. The reductive elimination of an alkylbenzene from **C** then leads to the unexpected substitution product with the isomerized, i.e., unbranched side chain.

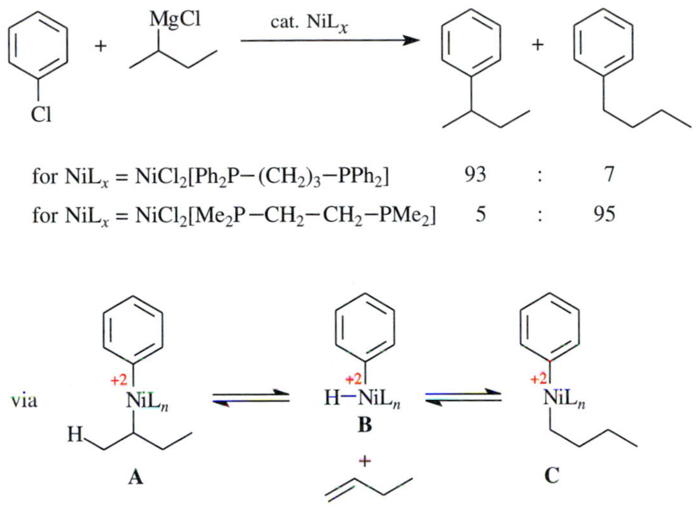

for $NiL_x = NiCl_2[Ph_2P{-}(CH_2)_3{-}PPh_2]$ 93 : 7

for $NiL_x = NiCl_2[Me_2P{-}CH_2{-}CH_2{-}PMe_2]$ 5 : 95

Fig. 13.8. Normal and isomerizing C,C couplings of a Grignard compound.

Figure 13.9 shows another case, namely, the reaction of isopropylmagnesium chloride with aryl triflates, in which a Ni complex does not effect the expected C,C coupling. Instead, it initiates a reduction and the deoxygenated aromatic compound is isolated (this is a valuable method for the deoxygenation of phenols!). This reaction mode again is a consequence of steric hindrance. The latter causes the interruption of the catalytic cycle of the Ni-catalyzed C,C coupling (Figure 13.7) prior to the reductive elimination of the coupling product (step 4 in Figure 13.7). This elimination would have to occur in intermediate **A** in Figure 13.9, but it is slowed down because a *secondary* C atom would have to bind to the aromatic compound. Thus, Ni complex **A**—as an alternative—undergoes a β-elimination. This leads to a hydrido–Ni complex, **B,** and propene. In contrast to the hydrido–Ni complex **B** of Figure 13.8, the

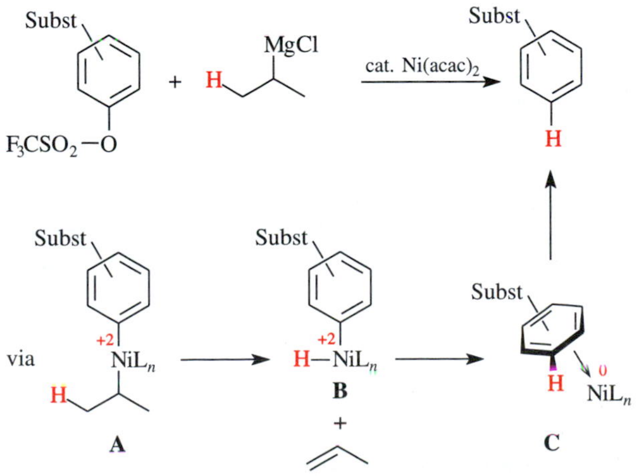

Fig. 13.9. Nickel-catalyzed reduction of an aryl triflate by a Grignard compound; acac refers to acetylacetonate, which is the enolate of pentane-2,4-dione.

analogous complex **B** of Figure 13.9 will *not* react back and forth with free propene to (re)form an alkyl–Ni complex. Instead, a reductive elimination occurs in which the aryl group picks up the H atom from the Ni. The elimination product is a π complex of the *reduced* aromatic compound. Finally, this aromatic compound dissociates off the metal.

13.3 Palladium-Catalyzed Alkenylation and Arylation of Organometallic Compounds

The stereoselective synthesis of alkenes is basically a solved problem. Nowadays, all kinds of alkenes can be synthesized irrespective of whether their double bond is isolated or conjugated with another C═C double bond, a C≡C triple bond, or an aromatic ring. This state of affairs is largely due to the discovery and the development of a number of palladium-catalyzed alkenylation and arylation reactions of organometallic compounds.

13.3.1 A Prelude: Preparation of Haloalkenes and Alkenylboronic Acid Derivatives, Important Building Blocks for Palladium-Mediated C,C Couplings

According to Section 3.3.3, borane and its mono- and dialkyl derivatives add to C═C double bonds in a *cis*-selective fashion. These reagents also add to C≡C triple bonds *cis*-selectively, but the primary products formed, alkenylboranes, may react with the boranes once more. The second reaction almost always is faster than the first one. Consequently, alkenylboranes are not accessible in this way.

Fortunately, though, there is *one* borane, catecholborane (**A** in Figure 13.10), that adds to C≡C triple bonds but not to C═C double bonds (in the absence of transition metal catalysts). This reagent adds to alkynes with *cis*-selectivity, so that the reaction stops at the stage of *trans*-alkenylboronic acid esters (**B** in Figure 13.10).

These boronic esters are easily hydrolyzed to give *trans*-alkenylboronic acids with complete retention of their stereochemistry (**C** in Figure 13.10). Alkenylboronic esters and alkenylboronic acids are organometallic compounds that can be alkenylated and arylated in Pd-catalyzed reactions (Section 13.3.2). Aside from this, the *trans*-alkenylboronic acid esters as well as the *trans*-alkenylboronic acids are valuable precursors of haloalkenes (Figure 3.10).

Alkenylboronic acid esters **B** react with bromine to give the *trans* addition product initially. This primary product is not isolated but is immediately reacted with a solution of NaOMe in MeOH. Addition of MeO^- ion to the B atom of the bromine adduct forms the borate complex **D.** This borate complex is converted into a *cis*-configured bromoalkene and a mixed boronic acid ester by way of a β-elimination.

trans-Alkenylboronic acids (**C** in Figure 13.10) exhibit the complementary diastereoselectivity in their reactions with elemental iodine: they form iodoalkenes with *trans*-selectivity. The conversion of the C—B bond into the C—I bond occurs with complete retention of configuration. This stereoselectivity is typically encountered in reactions of no matter which organometallic compounds with electrophiles, i.e., the C—M bond almost always is converted into a C—E bond with retention of configuration. The electrophile (here: I_2) apparently attacks at the location of highest electron density, that is, at the center of the C—M bond [here: C—$B(OH)_3^-$ bond].

Bromoalkenes—accessible, for instance, according to the procedure outlined in Figure 13.10—can be employed for the preparation of alkenylboronic acid esters and alkenylboronic acids. To do so, one first carries out a bromine/lithium exchange by way of treating the bromide with *n*-BuLi (Figure 13.11). This exchange is analogous to the bromine/lithium exchange reaction of aryl bromides (Figure 5.34). In agreement with a foregoing statement, a key step consists of the attack of Li^+ at the center of the

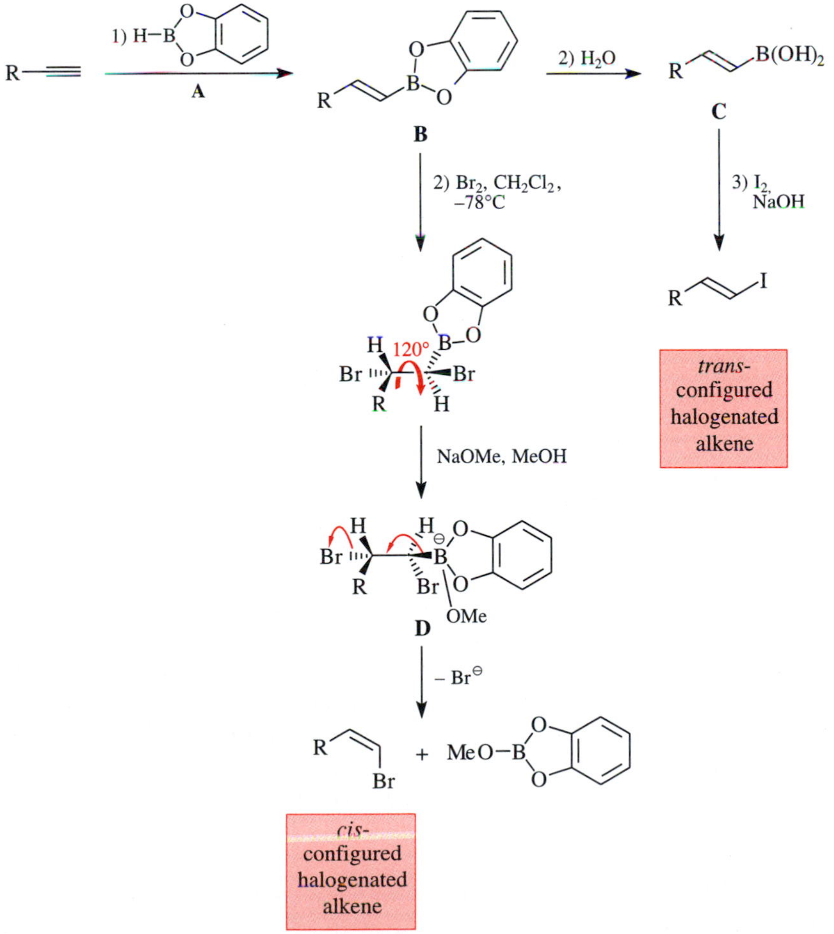

Fig. 13.10. Stereoselective preparations of *trans*-alkenylboronic acid esters **(B)** and *trans*-alkenylboronic acids **(C)** and their stereoselective conversion into *cis*-bromoalkenes and *trans*-iodoalkenes, respectively.

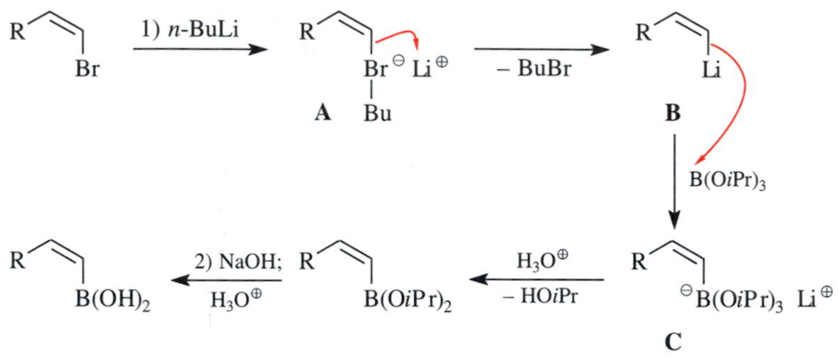

Fig. 13.11. Stereoselective preparations of *cis*-alkenylboronic acids and the corresponding diisopropyl ester starting with *cis*-bromoalkenes. The first step involves a Br/Li exchange to form the alkenyllithium compound **B.** This organolithium compound is subsequently transmetallated to give complex **C** by using $B(OiPr)_3$.

C—Br$^-$—*n*-Bu bond of the transient complex **A.** Thus, complete retention of configuration is observed in the conversion of the C_{sp^2}—Br bond into the C_{sp^2}—Li bond of the lithioalkene **B.** The transmetallation of this lithioalkene with boric acid triisopropyl ester again proceeds with complete retention of configuration. Again, the reason for this retention is that the electrophile, $B(OiPr)_3$, attacks the C—M bond—that is, the C_{sp^2}—Li bond—at the center. Hence, a *cis*-configured complex **C** with tetravalent and negatively charged boron is obtained. Complex **C** can be hydrolyzed under mildly acidic conditions to give a *cis*-configured alkenylboronic acid ester. This ester in turn can be hydrolyzed—now preferentially with base catalysis—to give the parent alkenylboronic acid with retention of the *cis* geometry.

The basicity of Gilman cuprates is so low that they do not undergo acid/base reactions with acetylene or higher terminal alkynes. Instead, Gilman cuprates effect the carbocupration of the C≡C triple bond (Figures 13.12 and 13.13). This reaction formally resembles the hydroboration of a C≡C triple bond (see example in Figure 13.10). The regioselectivity also is the same; hence, the metal is connected to the C1 center of a terminal alkyne. Finally, the reaction shows the same stereoselectivity as in the case of the hydroboration of a C≡C triple bond: carbocupration occurs as a *cis* addition.

Gilman cuprates successively transfer their two methyl or primary alkyl groups, respectively, in the carbocupration of C≡C triple bonds (Figures 13.12 and 13.13). The resulting products are Gilman cuprates, too. Depending on the structure of the alkyne used, the newly obtained Gilman cuprates contain two *cis*-configured *mono*substituted alkenyl groups (Figure 13.12) or two stereochemically homogeneous *di*substituted alkenyl groups (Figure 13.13). These dialkenyl cuprates can be treated with elemental iodine. Without affecting the C=C double bonds, the respective C_{sp^2}—Cu bonds thereby are converted into the C_{sp^2}—I bonds of iodoalkenes. This occurs stereoselectively,

Fig. 13.12. Stereoselective preparation of *cis*-iodoalkenes after a carbocupration.

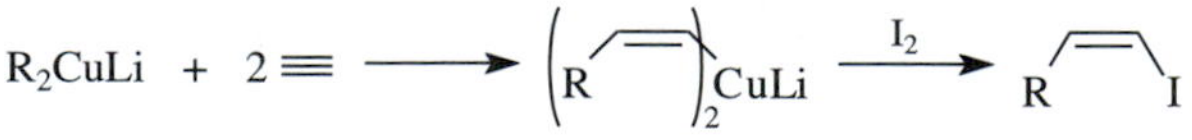

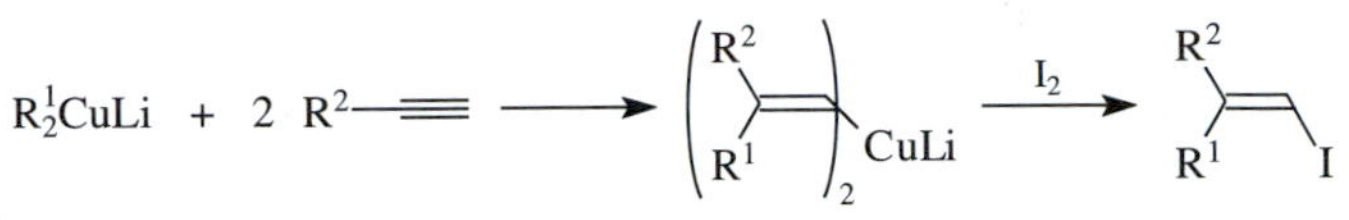

Fig. 13.13. Stereoselective synthesis of disubstituted iodoalkenes after carbocupration.

namely, with complete retention of configuration. The mechanism of this transformation is exactly the same as that of the conversion of *trans*-alkenylboronic acid **C** into the *trans*-iodoalkene in Figure 13.10.

13.3.2 Alkenylation and Arylation of Boron-Bound Groups

Alkenylations of organoboron compounds can be realized with alkenyl triflates, bromides, and iodides. Arylations of organoboron compounds are possible in complete analogy with aryl triflates, bromides, and iodides. Suitable organoboron compounds are alkenylboronic acid esters and alkenylboronic acids, arylboronic acid esters and arylboronic acids, and 9-BBN derivatives with a primary alkyl group. However, all these reactions succeed only in the presence of catalytic amounts of Pd complexes. $Pd(PPh_3)_4$ is most commonly used, that is, a Pd(0) complex. In solution, one or two PPh_3 ligands dissociate off to form the electron-deficient complexes $Pd(PPh_3)_3$ or $Pd(PPh_3)_2$, and these complexes initiate the respective reactions.

All the alkenylations and arylations discussed in this section follow the common mechanism exemplified in Figure 13.14 by the arylation of a *trans*-configured alkenyl-

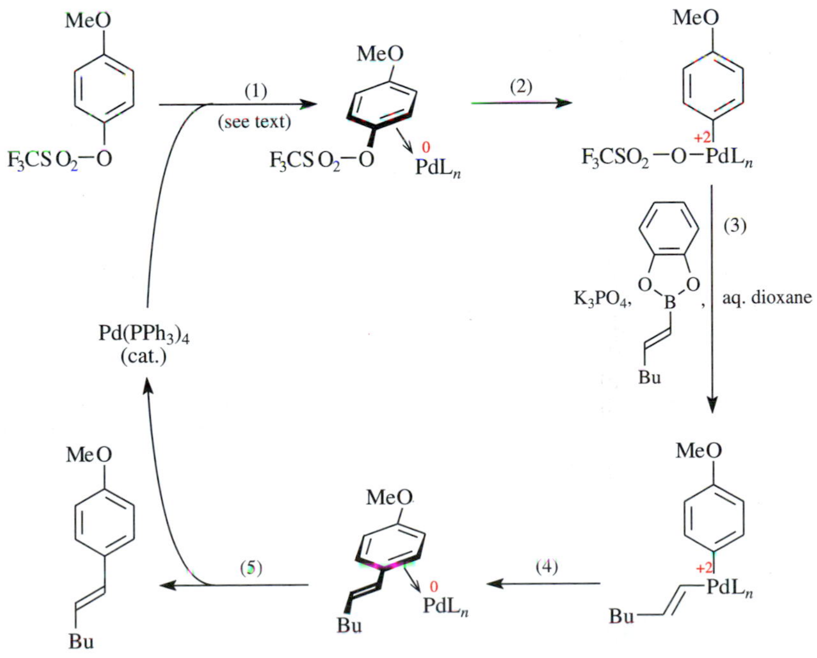

Fig. 13.14. Representative mechanism of the Pd-catalyzed C,C coupling of an organoboron compound. The elementary steps, discussed in the text, are (1) complexation, (2) oxidative addition, (3) transmetallation of the alkenylboron compound to afford an alkenylpalladium compound, (4) reductive elimination, and (5) dissociation of the coupled product from the metal.

boronic acid ester. The steps shown in Figure 13.14 may involve more than one elementary reaction, as in the case of the presentation of the mechanistic course of the Ni-catalyzed C,C coupling with Grignard compounds (Figure 13.7). It should be noted that the *basic* sequence of steps is very much the same in Figures 13.14 and 13.7.

In step 1 of Figure 13.14 one of the above-mentioned electron-deficient Pd(0) complexes, $Pd(PPh_3)_3$ or $Pd(PPh_3)_2$, forms a π complex with the aryl triflate. Step 2 is an oxidative addition. The Pd inserts into the C_{sp^2}—O bond of the aryl triflate, and the oxidation number of Pd increases from 0 to +2. A transmetallation occurs in step 3: the alkenyl–B compound is converted into an alkenyl–Pd compound. This step corresponds to a ligand exchange reaction at Pd; concomitantly the former leaving group of the electrophile—here the triflate—is replaced by the alkenyl group. The two σ-Pd-bonded organic moieties combine in step 4, which, since the oxidation number of Pd is reduced from +2 to 0, is a reductive elimination. At that stage, the coupling product remains bound to Pd as a π complex. It dissociates off the metal in step 5. This step reconstitutes the valence-unsaturated Pd(0) complex which can start another passage through the catalytic cycle.

Arylalkenes are accessible not only by way of an arylation of alkenylboron compounds (example in Figure 13.14) but also via an alkenylation of arylboron compounds. Figure 13.15 exemplifies this for a Pd-catalyzed reaction of an arylboronic acid with iodoalkenes with widely variable substitution patterns. The addition of KOH increases the reactivity of the arylboronic acid in this coupling and in similar ones. The base converts the boronic acid into a negatively charged boronate ion **A.** This ion **A** is transmetallated faster than the neutral boronic acid by the Pd(II) intermediate; the boronate ion is a superior nucleophile and replaces the iodide ion in the Pd(II) complex particularly fast.

Alkenylboronic acid ester can be prepared with *cis* configuration (e.g., according to Figure 13.11) or with *trans* configuration (e.g., according to Figure 13.10). Iodoalkenes also are easily synthetically accessible in both configurations (preparation with *cis* and *trans* configuration possible according to Figures 13.12 and 13.10). Palladium-catalyzed C,C coupling reactions between two of these building blocks lead stereoselectively to

Fig. 13.15. Palladium-catalyzed, stereoselective alkenylation of an arylboronic acid (preparation according to Figure 5.32) with a variety of iodoalkenes. The boronic acid is converted into the boronate anion **A.** The ion **A** reacts with the Pd(II) intermediate **B** via transmetallation: subsequent reductive elimination leads to the coupling products.

NEt2
O O
B(OH)2
+
I R2 R1
Pd(PPh3)4 (cat.),
aq. KOH,
dioxane
NEt2
O O
R2
R1

NEt2
O O
$B(OH)_3^{\ominus}$
I Pd L L R2 R1
A
B

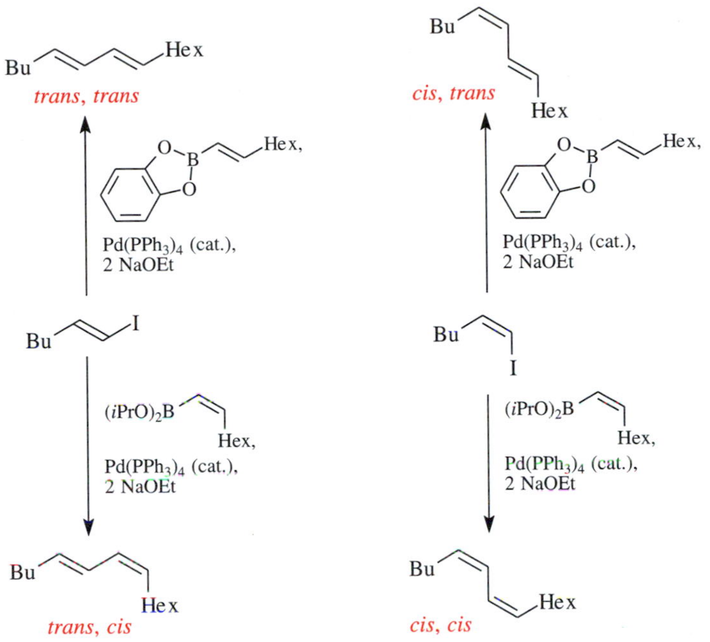

Bu Hex
trans, trans
Bu Hex
cis, trans
Hex,
Hex,
$Pd(PPh_3)_4$ (cat.),
2 NaOEt
$Pd(PPh_3)_4$ (cat.),
2 NaOEt
Bu I
Bu I
$(iPrO)_2B$ Hex,
$Pd(PPh_3)_4$ (cat.),
2 NaOEt
$(iPrO)_2B$ Hex,
$Pd(PPh_3)_4$ (cat.),
2 NaOEt
Bu Hex
trans, cis
Bu Hex
cis, cis

Fig. 13.16. Alkenylation of isomeric alkenylboronic acid esters with isomeric iodoalkenes; stereoselective synthesis of isomeric 1,3-dienes.

each of the four geometrical isomers of the 1,3-diene depicted in Figure 13.16. The couplings shown in this figure are best carried out in the presence of NaOEt. Thereby, the nucleophilicity of the organometallic compound can be increased just as in the case of the coupling of boronic acids by way of KOH addition (Figure 13.15). In both cases, one generates *in situ* negatively charged and therefore more nucleophilic tetravalent boron complexes. These complexes have the structure alkenyl–$B(OR)_3^-$ in the reactions shown in Figure 13.16, while such a complex **A** of composition aryl–$B(OH)_3^-$ occurs in Figure 13.15.

Side Note 13.1
Stereoselective Synthesis of Vitamin A

As discussed, in the absence of catalysts catecholborane adds to C≡C triple bonds but not to C=C double bonds (cf. Figure 13.10). Accordingly, if enynes are reacted with catecholborane, the latter will add chemoselectively to their C≡C triple bond (Figure 13.17). As expected, the resulting boronic ester features the same regio- and stereochemistry seen in Figure 13.10. The acid-catalyzed hydrolysis of this boronic ester affords a *trans*-configured dienylboronic acid (**A** in Figure 13.17). This acid undergoes a base-catalyzed reaction (cf. reaction conditions in Figure 13.15) with the iodotriene **B** (see Figure 13.13 for a method of preparation) to provide a conjugated pentaene. The latter contains nothing but stereodefined C=C double bonds and it is none other than vitamin A.

Fig. 13.17. Alkenylation of a dienylboronic acid with an iodinated triene; stereoselective synthesis of vitamin **A.** The enyne (top left) is added to catecholborane to prepare the *trans*-configured boronic ester in a chemoselective fashion. The latter affords *trans*-dienylboronic acid **A** upon acid-catalyzed hydrolysis.

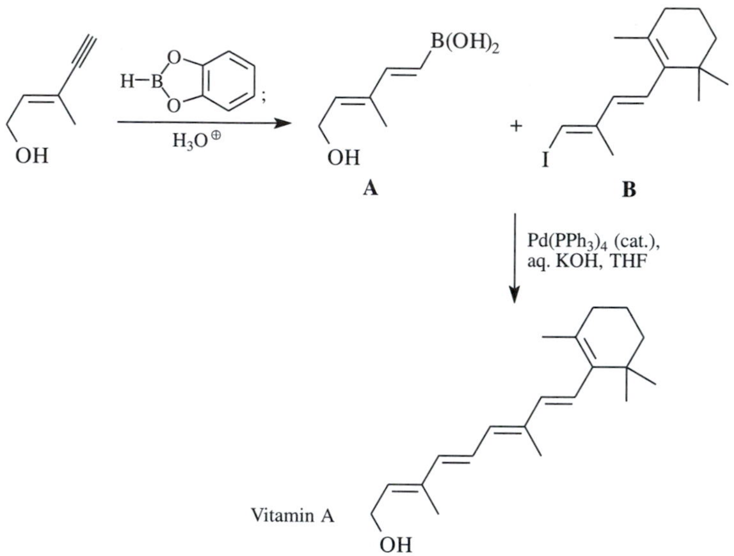

In the case of aromatic compounds with two triflate groups and/or bromine or iodine substituents, it is sometimes possible to achieve chemoselective and sequential Pd-catalyzed substitutions of these leaving groups by organoboron compounds. In the ideal scenario, this reaction sequence is even possible in a one-pot reaction. One first adds *one* equivalent of *one* unsaturated boronic acid to the substrate/$Pd(PPh_3)_4$/base mixture (cf. discussion of Figure 13.15 with regard to the reaction conditions). The leaving groups are cleaved from the aromatic compound in the order I > Br > triflate. Hence, the first boronic acid replaces the iodide in bromoiodobenzene (Figure 13.18). Subsequently, one adds a *second* unsaturated boronic acid *in excess* to replace the remaining bromine. Alkenylboronic acids are just as suitable for such tandem coupling reactions as the arylboronic acids shown in Figure 13.18.

In Figures 13.14–13.18 we introduced the reaction principle of the Pd(0)-catalyzed

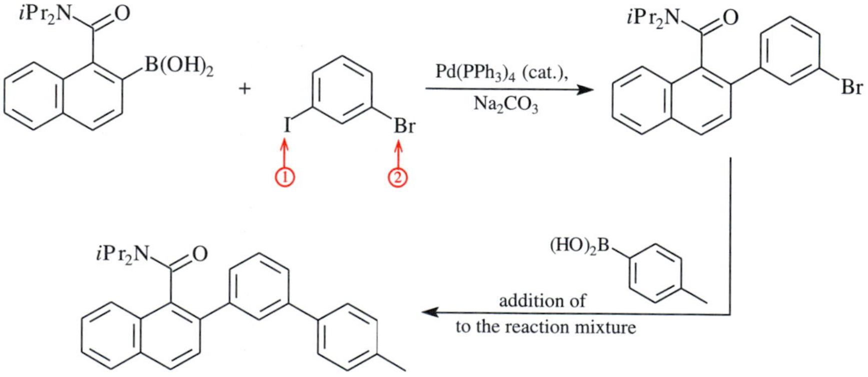

Fig. 13.18. Palladium-catalyzed C,C coupling with bromoiodobenzene: regioselective tandem coupling I.

A

1) MCPBA (1.0 equivalent)
2) 9-BBN

B

Br ; Pd(PPh_3)$_4$ (cat.), K_3PO_4

Fig. 13.19. Suzuki coupling I for the synthesis of functionalized aromatic compounds.

alkenylation or arylation of *unsaturated* boronic esters and boronic acids, respectively. This reaction principle can be extended to another class of organoboron compounds, namely, to certain *trialkylboranes*. Specifically, the trialkylboranes resulting from the addition of 9-BBN to terminal alkenes can be combined with aryl–X or alkenyl–X compounds (X = OTf, Br, I) via Pd-catalyzed **Suzuki couplings**. 9–BBN adds to terminal alkenes in a regioselective fashion (cf. Section 3.3.3) and, interestingly, this reaction proceeds uneventfully even if the alkene contains a series of other functional groups (cf. also first two rows of Figures 8.30 and 8.38). In fact, a number of functional groups interfere neither with the formation of a 9-BBN adduct nor with the subsequent Pd(0)-catalyzed C,C coupling. Such arylating (Figures 13.19 and 13.20) or alkenylating coupling reactions

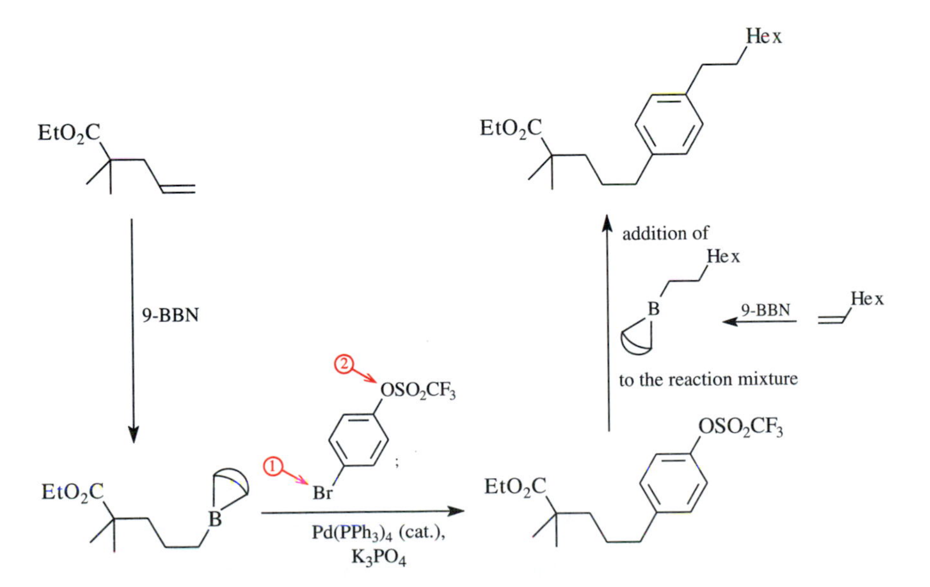

Fig. 13.20. Suzuki coupling II for the synthesis of functionalized aromatic compounds: regioselective tandem coupling II.

of hydroborated olefins are starting points for the synthesis of aromatic compounds and alkenes, respectively, that may contain a variety of functional groups in different parts of the molecules. The figures cited show compatibility with epoxides, ketones, and esters.

Finally, Figure 13.20 shows a particularly interesting one-pot reaction consisting of two consecutive Suzuki coupling reactions of an aromatic compound bearing two sp^2-bonded leaving groups. The Ar—Br moiety reacts faster than the Ar—OTf moiety. This is quite expected in the context of what was said regarding the regioselectivity observed in Figure 13.18.

13.3.3 Alkenylation and Arylation of Zinc-Bound Functionalized Groups

Arylzinc and alkylzinc iodides can be arylated by aryl triflates (bromides, iodides) and alkenylated by alkenyl triflates (bromides, iodides) in the presence of catalytic amounts of $Pd(PPh_3)_4$. These arylzinc and alkylzinc iodides are particularly interesting because they are accessible even if they contain other functional groups (cf. discussions in connection with Figures 8.34 and 14.38). The organozinc compound shown in Figure 13.21 contains, for example, an ester group. The 9-BBN analog of this organozinc compound—potentially a substrate for a Suzuki coupling (cf. Figures 3.19 and 13.20)—would *not*

Cl
F_3CSO_2-O
(1) (see text)
Cl
F_3CSO_2-O $\overset{0}{Pd}L_n$
(2)
Cl
$F_3CSO_2-O-\overset{+2}{Pd}L_n$
LiCl (3)
Cl
$Cl-\overset{+2}{Pd}L_n$
$\overset{0}{Pd}(PPh_3)_4$ (cat.)
I
EtO_2C
Zn
ZnI
EtO_2C
(4)
Cl
$\overset{+2}{Pd}L_n$
EtO_2C
(5)
Cl
$\overset{0}{Pd}L_n$
EtO_2C
(6)
Cl
EtO_2C

Fig. 13.21. Representative mechanism of the Pd(0)-catalyzed arylation and alkenylation of organozinc iodides. Steps 1, 2, and 4–6 correspond to steps that can also be found—in some cases with different numbers—in Figures 13.7 and 13.14. Step 3, which is new, represents a ligand exchange reaction of the aryl–Pd(II) complex.

be accessible in a straightforward manner, that is, by the hydroboration of the C=C bond of ethyl acrylate. This is because 9-BBN and ethyl acrylate would react with each other in a 1,4-addition to yield a boron enolate of ethyl propionate.

The mechanism of the Pd(0)-catalyzed coupling of functionalized organozinc compounds is exemplified in Figure 13.21 by the arylation of the above-mentioned ester-functionalized alkylzinc iodide. The catalytic cycle consists of six reaction steps. Steps 1, 2, and 4–6 correspond exactly to the five steps of the C,C-coupling reactions already discussed, that is, the Ni-catalyzed coupling of Grignard reagents (Figure 13.7) and the Pd-catalyzed coupling of organoboron compounds (Figure 13.14). Everything that was said in those discussions applies fully to the sequence of events Figure 13.21.

The only difference between the present mechanism and the mechanisms to which it was compared is that it requires the addition of LiCl. Some Pd-catalyzed C,C-coupling reactions between organotin compounds and arylating or alkenylatng reagents also require such an addition of LiCl. The role of LiCl is accounted for by step 3 of the reaction mechanism. This step consists of a ligand exchange reaction of an aryl–Pd(II) complex, namely, the exchange of a triflate group for a chloride. The Cl-containing complex subsequently (step 4) reacts *much* faster with the organozinc compound than the triflate precursor ever could have.

13.3.4 Alkenylation and Arylation of Copper Acetylides

Terminal alkynes can be alkenylated by alkenyl triflates (bromides, iodides) and arylated by aryl triflates (bromides, iodides). These reactions are called **Cacchi coupling** reactions if the reaction is catalyzed by Cu(I) and Pd(0) and if *triflate* reagents are employed, **Sonogashira–Hagihara coupling** reactions if the reaction is catalyzed by Cu(I) and Pd(0) and *halides* are employed as substrates, and **Stephens–Castro coupling** reactions for the more specialized case of the noncatalyzed coupling of copper acetylides with aryl halides.

Cacchi and Sonogashira–Hagihara coupling reactions occur only if a primary, secondary, or tertiary amine is present, and it is best to have the amine present in large excess. Under these conditions the acetylene will form at least a small equilibrium amount of an ammonium acetylide. Aside from this, a substoichiometric amount of CuI is almost always added to capture the small equilibrium concentration of the ammonium acetylide as a copper acetylide. The copper acetylide represents a substantially improved nucleophile in comparison to the free acetylene. Without the CuI addition, the acetylide content of the reaction mixture is so small that a reaction occurs only at higher temperatures (for an example, see Figure 13.23).

Furthermore, Cacchi and Sonogashira–Hagihara coupling reactions require catalysis by a complex of zero-valent Pd. $PdCl_2(PPh_3)_2$, a Pd(II) complex, can be handled more conveniently and is usually employed, since under the reaction conditions this complex is reduced immediately to provide the Pd(0) complex. The amine as well as the terminal alkyne or its respective acetylide ion can act as the reducing reagent. The amine as a reductant would be oxidized to give an iminium ion while the acetylede as a reductant would be converted into a 1,3-diyne, that is, the product of an oxidative dimerization (**Glaser coupling**).

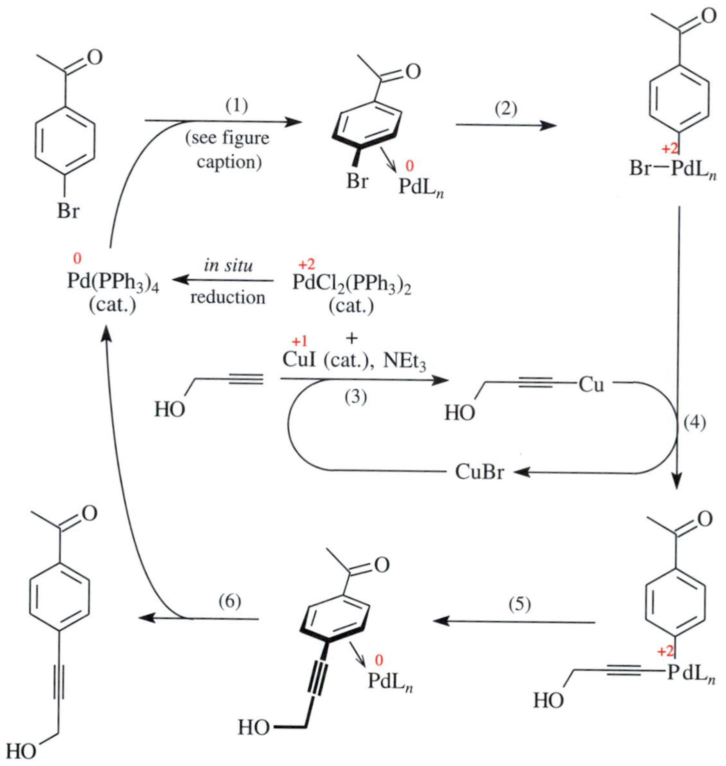

Fig. 13.22. Mechanism of the Pd(0)-catalyzed arylation of a copper acetylide. Step 1: formation of a π complex between the catalytically active Pd(0) complex and the arylating agent. Step 2: oxidative addition of the arylating agent and formation of a Pd(II) complex with a σ-bonded aryl moiety. Step 3: formation of a Cu-acetylide. Step 4: transmetallation; the alkynyl–Pd compound is formed from the alkynyl–Cu compound via ligand exchange. Step 5: reductive elimination to form the π complex of the arylated alkyne. Step 6: decomposition of the complex into the coupling product and the unsaturated Pd(0) species, which reenters the catalytic cycle anew with step 1.

The mechanism of the arylation of a terminal acetylene is illustrated in Figure 13.22 as a representative example. The catalytic cycle starts with the already mentioned formation of the Cu-acetylide (step 3 in Figure 13.22) and comprises five other steps. These steps are basically the same as the steps in the Ni-catalyzed coupling of Grignard reagents (Figure 13.7) and the Pd-catalyzed coupling of organoboron compounds (Figure 13.14). An equally pronounced similarity exists between this mechanism and the Pd-catalyzed coupling of organozinc compounds (Figure 13.21). Only step 3 of the latter mechanism is missing. The series of steps listed in the caption of Figure 13.22 thus suffices completely to describe the new reaction mechanism.

Alkenylations of alkynes (Figures 13.23 and 13.24) are the methodological supplement to the arylation of alkynes (Figure 13.22). The substrates of the coupling in Figure 13.23 are an alkenyl triflate (preparation analogous to Figure 10.20) and an alkyne

C_8H_{17} $F_3CSO_2{-}O$ H + $PdCl_2(PPh_3)_2$ (cat.), NEt_3, DMF, Δ C_8H_{17} H O O

Fig. 13.23. Alkenylation of an iminium acetylide catalyzed by Pd(0).

and the coupling product is a 1,3-enyne. The C≡C triple bond of such an enyne can be hydrogenated in a *cis*-selective fashion by using Lindlar Pd as a catalyst (cf. Figure 14.70). In this way 1,3-dienes are formed that contain at least one *cis*-configured C=C double bond.

The reactions of Cu-acetylides with configurationally homogenous *cis*-iodoalkenes (accessible via the procedure in Figure 13.12) or with stereopure *trans*-iodoalkenes (e.g., preparation according to Figure 3.11), respectively, result in 1,3-enynes with retention of the respective double bond geometry (Figure 13.24).

The coupling of an iodoalkene with a trimethylsilylacetylene and a subsequent desilylation result in the formation of 1,3-enynes with a terminal C≡C triple bond. The parent acetylene usually does not react to give such an enyne under these reaction conditions, since Cu_2C_2 is formed. The solubility of this carbide is very low so that it precipitates. The very small amount of this copper species that remains in solution—if it couples at all—couples at both C atoms. Hence, the major coupling product then is the bis-coupling product of acetylene, a 3-ene-1,5-diyne, but in any case this product is formed only in small amounts because of the low solubility of Cu_2C_2.

R I + $SiMe_3$ 1) $Pd(PPh_3)_4$ (cat.), CuI (cat.), $PrNH_2$ $SiMe_3$ R 2) $NH_4^{\oplus}$ $F^{\ominus}$, MeOH (– $FSiMe_3$, – NH_3) R

R I + $SiMe_3$ 1) $Pd(PPh_3)_4$ (cat.), CuI (cat.), $PrNH_2$ R $SiMe_3$ 2) $NH_4^{\oplus}$ $F^{\ominus}$, MeOH (– $FSiMe_3$, – NH_3) R

Fig. 13.24. Stereoselective and stereospecific Pd(0)-catalyzed alkenylations of copper acetylides.

As briefly mentioned, aryl bromides or aryl iodides and stoichiometric amounts of copper acetylide also can be coupled in the absence of a palladium catalyst by refluxing the components in pyridine solution. These so-called **Stephens–Castro couplings** follow a mechanism that is analogous to the coupling of aryl copper compounds with aryl iodides (Figure 13.3) or the coupling of alkynyl copper compounds with bromoalkynes (Figure 13.25).

13.4 Alkynylation of Copper Acetylides

For the preparation of conjugated alkynes, one can alkenylate or arylate alkynes according to Section 13.3.4. Alternatively, metallated alkenes or metallated aromatic compounds also may be alkynylated, but this option will not be pursued further. We merely mention in passing that bromoalkynes and iodoalkynes are suitable alkynylating agents and that these can be obtained in a one-step reaction from terminal alkynes:

$$\text{H–C}\equiv\text{C–Ph} \xrightarrow[\text{HN}\diagup\!\!\diagdown\text{O (morpholine)}]{Br_2 \text{ or } I_2,} \text{Br–C}\equiv\text{C–Ph} \quad \text{or} \quad \text{I–C}\equiv\text{C–Ph}$$

More often such bromo- and iodoalkynes are employed with another synthetic goal in mind, namely, in the **Cadiot–Chodkiewicz** reaction for the formation of symmetric or asymmetric 1,3-diynes by reaction of the haloalkyne with a terminal alkyne (Figure 13.25). Additional reagents essential for the success of this reaction are one equivalent or more of an amine and a substoichiometric amount of CuI. As with the Cacchi and Stephens–Castro coupling reactions of Section 13.3.4, a Cu-acetylide is the reactive species in the Cadiot–Chodkiewicz coupling. It is formed in step 1 of the mechanism illustrated in Figure 13.25.

The remaining steps 2 and 3 of the Cadiot–Chodkiewicz coupling and the mechanisms discussed in earlier sections of Chapter 13 show obvious analogies. Step 2 consists of the oxidative addition of the haloalkyne to a Cu(I) species. This addition may involve a transient π complex or it may be a one-step process. The C,C coupling occurs in step 3 as a reductive elimination and leads to the coupling product and CuBr. This step may proceed indirectly via an intermediate π complex or directly.

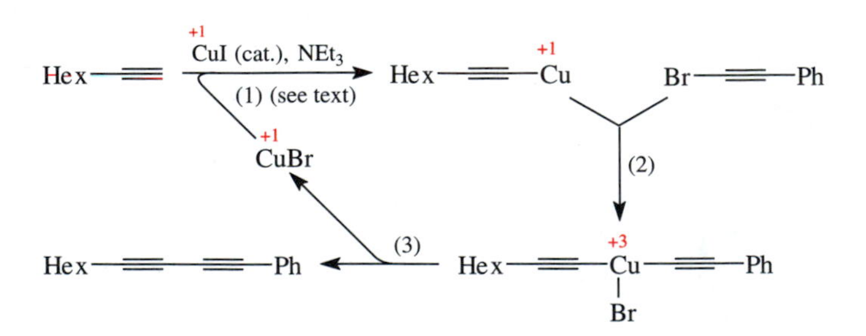

Fig. 13.25. Mechanism of a Cadiot–Chodkiewicz coupling.

13.5 Heck Reactions

Certain alkenes can be alkenylated by alkenyl triflates (bromides, iodides) and arylated by aryl triflates (bromides, iodides) even though they do not contain a metal. For these so called **Heck reactions** to occur, catalytic amounts of palladium(II) acetate and triphenylphosphine, as well as stoichiometric amounts of triethylamine, need to be added to the mixture of the starting materials. The amine serves to reduce Pd(II) to the catalytically active Pd(0) complex (cf. Section 13.3.4). The amine also has an important second role in that it neutralizes the strong acid formed in the reaction (TfOH, HBr, and HI). Apart from that, there also exists a variation of the Heck reaction that works without triphenylphosphine (examples in Figures 13.27 and 13.29).

Alkenes without allylic H atoms—such as ethylene, acceptor-substituted ethylenes, and styrene—can be alkenylated and arylated by Heck reactions in a clearly predictable fashion. These alkenes can be *alkenylated* to provide 1,3-dienes, $\alpha,\beta,\gamma,\delta$-unsaturated carbonyl compounds (Figures 13.27 and 13.28), $\alpha,\beta,\gamma,\delta$-unsaturated carboxyl compounds (Figures 13.27 and 13.28), as well as aryl-substituted 1,3-dienes. Moreover, the same alkenes can be *arylated* to give styrenes, α,β-unsaturated β-arylated carbonyl compounds, or α,β-unsaturated β-arylated carboxyl compounds (Figure 13.26) and stilbenes (Figure 13.29).

Figure 13.26 elucidates in detail a representative reaction mechanism of one of these Heck reactions. The arylating reagent is a benzene derivative that contains both a good leaving group (triflate group) and a poor leaving group (chlorine). The good leaving group is cleaved off, the poor one is not. The alkene that is being functionalized is acrylic acid methyl ester, and the coupling product is a *trans*-configured cinnamic ester.

Steps 1–3 of the catalytic cycle correspond to various steps of other catalytic cycles already discussed in Chapter 13. Step 1: π-complex formation by combination of the aryl triflate and a sufficiently valence-unsaturated and thus sufficiently reactive Pd(0) species. Step 2: oxi-dative addition of the aryl triflate to Pd with formation of a C_{sp^2}—Pd(II) bond. Steps 3a and 3b: exchange of a PPh_3 ligand by an acrylic acid methyl ester via a dissociation/addition mechanism. The newly entered acrylic acid ester is bound as a π complex.

Step 4 of the mechanism shown in Figure 13.26 is new. This step consists of the *cis*-selective addition of the aryl–Pd complex to the C=C double bond of the acrylic acid methyl ester. This is a **carbopalladation** of the double bond. A related reaction, the *cis*-selective **carbocupration** of C≡C triple bonds was mentioned in connection with Figures 13.12 and 13.13. The regioselectivity of the **carbopalladation** of Figure 13.26 is such that the organic moiety is bonded to the methylene carbon and Pd to the methyne carbon of the C=C double bond. The addition product is an alkyl–Pd(II) complex.

Prior to the next bond formation/bond cleavage event, a rotation about the newly formed C—C bond of this complex is required; this occurs in step 5. The rotation brings an H-β atom and the PdL_x group into a *syn* relation. This conformation is essential for the formation of a hydrido–Pd(II) complex to occur by a β-elimination of this β-H atom and the PdL_x group in step 6. This complex contains the arylated acrylic ester as

Fig. 13.26. Mechanism of a Heck coupling with acrylic acid methyl ester. The elementary steps 1–3 correspond to those in Figures 13.7 and 13.14, and the further course of the reaction is described in the text.

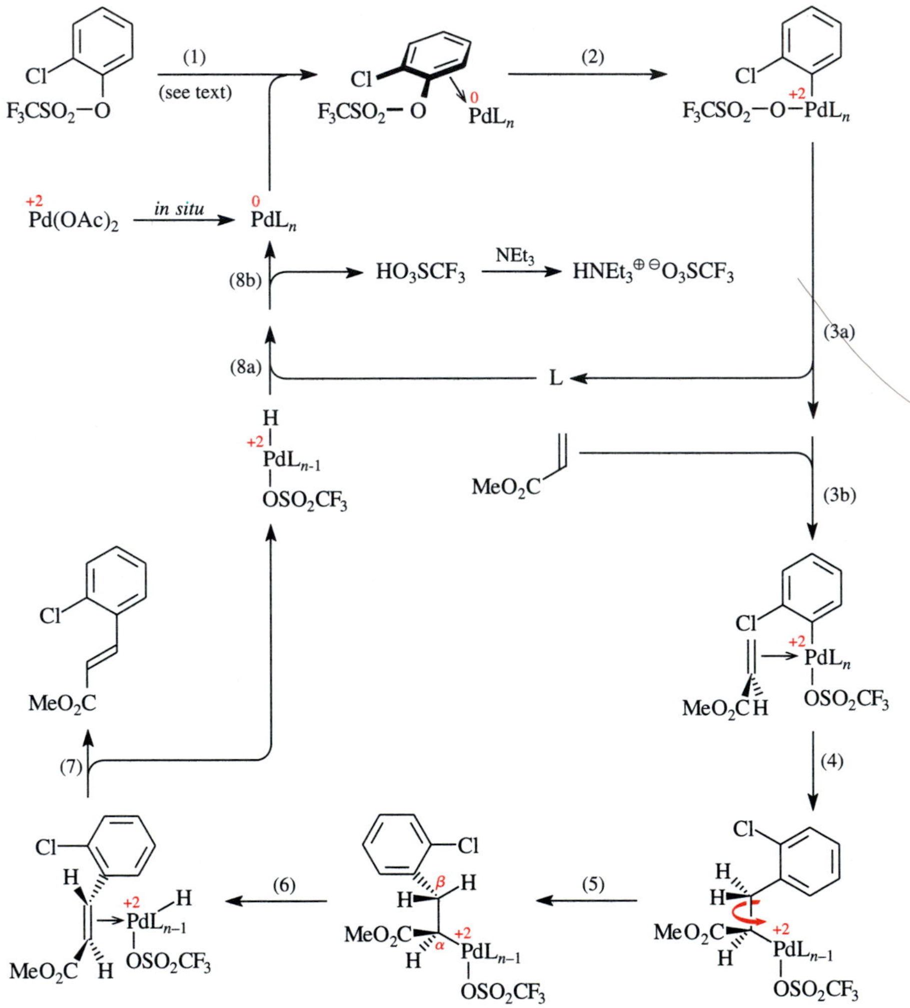

a π-bonded ligand. Mechanistically related β-eliminations were encountered earlier in Figures 13.8 and 13.9 where hydrido–Ni(II) complexes were formed from alkyl–Ni(II) complexes.

The π-bonded coupling product dissociates from the hydrido–Pd(II) complex in step 7, and another hydrido–Pd(II) complex is formed (Figure 13.26). It loses trifluoromethane-sulfonic acid in step 8 of the catalytic cycle. Thereby, the same valence-unsaturated Pd(0) complex that initiated the reaction in step 1 is formed, so this complex is now available to begin another cycle.

In the Heck reaction, stereogenic C=C double bonds always are formed in a *trans*-selective fashion; a few examples are shown in Figures 13.26–13.29. This *trans*-selectivity is due to product-development control in the β-elimination of the hydrido–Pd(II) complex (cf. step 6 of Figure 13.26): the more stable *trans*-alkene is formed faster than the

Bu–CH=CH–I + CH₂=CH–EWG → Pd(OAc)$_2$ (cat.), K_2CO_3, $Bu_4N^{\oplus}Cl^{\ominus}$ → Bu–CH=CH–CH=CH–EWG

(EWG = CH=O, C(=O)Me, CO_2Me)

Fig. 13.27. Stereoselectivity and stereospecificity of Heck coupling reactions with isomeric iodoalkenes.

cis isomer. But there also is another reason. Step 6 can be reversible, and the hydrido–Pd(II) complex may react back to the alkyl–Pd(II) complex (an analogous option occurred in the case of the hydrido–Ni(II) complex **B** in Figure 13.8). The coupling product and its C=C bond in particular thus disappear again. Subsequently, step 6 will occur again in the forward direction, that is, as a β-elimination. Thus, the coupling product and its C=C double bonds are formed again—*but perhaps not with the same configuration as when formed for the first time.* A sufficiently high number of such readditions and renewed β-eliminations assures that the double bond geometry will be determined by thermodynamic control, and the *trans* configuration is of course the thermodynamically preferred geometry. Therefore, the synthetic importance of the Heck reaction is not just to make possible the alkenylation or arylation of alkenes, but also to guarantee that these alkenylations and arylations occur in a *trans*-selective fashion.

A haloalkene that already contains a stereogenic C=C double bond usually can be coupled with alkenes via the Heck reaction without isomerization. The reaction pair in Figure 13.27 provides two sets of examples. As can be seen, both the *cis*- and the *trans*-configured iodoalkenes react with acceptor-substituted alkenes with complete retention of the C=C double bond configuration. These coupling reactions thus are *stereoselective* and—when considered as a pair—*stereospecific.*

Alkenyl *triflates* are capable of Heck reactions, too, as can be seen in Figure 13.28.

–O–SO_2CF_3 + EWG → $PdCl_2(PPh_3)_2$ (cat.), NEt_3 → EWG

(EWG = CH=O, C(=O)Me, CO_2Me)

Fig. 13.28. Heck coupling of an enol triflate without isomerization, that is, without change of the location of the C=C double bond. (For a preparation of the triflate, see Figure 10.20.)

Substrates with several sp^2-bound triflate groups, bromine, or iodine atoms, respectively, may undergo several Heck reactions in a row. A threefold Heck reaction of 1,3,5-tribromobenzene with styrene is shown in Figure 13.29.

One of the things the Heck reaction *cannot* do, at least not in an intermolecular fashion, is couple alkenyl triflates (bromides, iodides) with metal-free *aromatic* compounds,

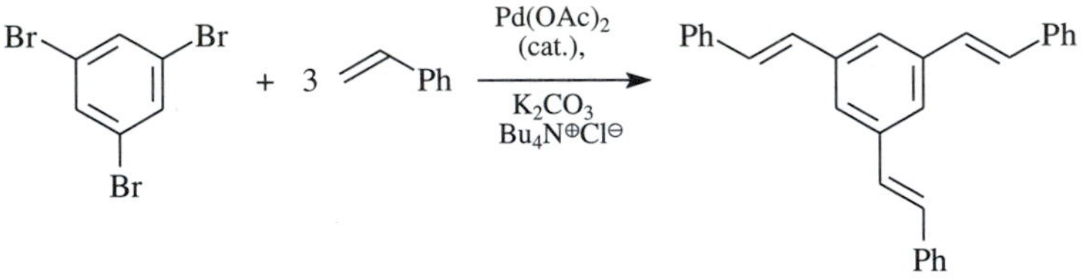

Fig. 13.29. Three "one-pot," *trans*-selective Heck coupling reactions.

while similar couplings with metal-free *alkenes* would occur. Step 4 of the mechanism in Figure 13.26 reveals why. If an aromatic compound were the substrate instead of an alkene, the aromaticity would have to be destroyed temporarily in the carbopalladation of a C=C double bond. Heck reactions typically already require temperatures of 100°C, and an additional energy-consuming disruption of aromaticity is therefore out of the question.

References

K. Tamao, "Coupling Reactions between sp^3 and sp^2 Carbon Centers," in *Comprehensive Organic Synthesis*" (B. M. Trost, I. Fleming, Eds.), Vol. 3, 435, Pergamon Press, Oxford, **1991.**

D. W. Knight, "Coupling Reactions between sp^2 Carbon Centers," in *Comprehensive Organic Synthesis* (B. M. Trost, I. Fleming, Eds.), Vol. 3, 481, Pergamon Press, Oxford, **1991.**

R. F. Heck, "Palladium Reagents in Organic Synthesis," Academic Press, **1985.**

A. de Meijere and F. E. Meyer, "Kleider machen Leute: Die Heck-Reaktion in neuem Gewand," Angew. Chem. 1994, 106, 2473-2506; "Fine feathers make fine birds: The Heck reaction in modern garb," *Angew. Chem. Int. Ed. Engl.* **1994,** *33,* 2379–2411.

H.-J. Altenbach, "Regio- and Stereoselective Aryl Coupling," in *Organic Synthesis Highlights* (J. Mulzer, H.-J. Altenbach, M. Braun, K. Krohn, H.-U. Reißig, Eds.), VCH, Weinheim, New York, **1991,** 181–185.

K. Ritter, "Synthetic transformations of vinyl and aryl triflates," *Synthesis* **1993,** 735–762.

P. J. Stang and F. Diederich (Eds.), "Modern Acetylene Chemistry," VCH, Weinheim, Germany, **1995.**

J. Tsuji, "Palladium Reagents and Catalysis: Innovations in Organic Synthesis," Wiley, New York, **1995.**

S. P. Stanforth, "Catalytic cross-coupling reactions in biaryl synthesis," *Tetrahedron* **1998,** *54,* 263–303.

G. Bringmann, R. Walter, R. Weirich, "Biaryls," in *Stereoselective Synthesis* (Houben-Weyl) 4th ed. 1996, (G. Helmchen, R. W. Hoffmann, J. Mulzer, E. Schaumann, Eds.), **1996,** Vol. E21 (Workbench Edition), *1,* 567–588, Georg Thieme Verlag, Stuttgart.

F. Diederich and P. J. Stang (Eds.), "Metal-Catalyzed Cross-Coupling Reactions," Wiley-VCH, Weinheim, Germany, **1998.**

13.1

P. E. Fanta, "The Ullmann synthesis of biaryls," *Synthesis* **1974,** 9.

13.3

R. G. Jones and H. Gilman, "The halogen-metal interconversion reaction with organolithium compounds," *Org. React.* **1951,** *6,* 339–366.

V. Snieckus, "Combined directed ortho metalation-cross coupling strategies: Design for natural product synthesis," *Pure Appl. Chem.* **1994,** *66,* 2155–2158.

D. S. Matteson, "Boronic esters in stereodirected synthesis," *Tetrahedron* **1989,** *45,* 1859–1885.

A. R. Martin and Y. Yang, "Palladium catalyzed cross-coupling reactions of organoboronic acids with organic electrophiles," *Acta Chem. Scand.* **1993,** *47,* 221–230.

A. Suzuki, "New synthetic transformations via organoboron compounds," *Pure Appl. Chem.* **1994,** *66,* 213–222.

N. Miyaura and A Suzuki, "Palladium-catalyzed cross-coupling reactions of organoboron compounds," *Chem. Rev.* **1995,** *95,* 2457–2483.

S. Suzuki, "Cross-coupling reactions of organoboron compounds with organic halides," in *Metal-Catalyzed Cross-Coupling Reactions,* (F. Diederich and P. J. Stang, Eds.), Wiley-VCH, Weinheim, **1998,** 49–89.

N. Miyaura, "Synthesis of Biaryls via the Cross-Coupling Reaction of Arylboronic Acids" in *Advances in Metal-Organic Chemistry* (L. S. Liebeskind, Ed.), JAI, Greenwich, CT, **1998,** *6,* 187–243.

E. Erdik, "Use of activation methods for organozinc reagents," *Tetrahedron* **1987,** *43,* 2203.

E. Erdik, "Transition metal catalyzed reactions of organozinc reagents," *Tetrahedron* **1992,** *48,* 9577-9648.

E. Erdik (Ed.), "Organozinc Reagents in Organic Synthesis," CRC Press, Boca Raton, FL, **1996.**

Y. Tamaru, "Unique Reactivity of Functionalized Organozincs," in *Advances in Detailed Reaction Mechanism. Synthetically Useful Reactions* (J. M. Coxon, Ed.), **1995,** *4,* JAI Press, Greenwich, CT.

P. Knochel, "Organozinc, Organocadmium and Organomercury Reagents," in *Comprehensive Organic Synthesis* (B. M. Trost, I. Fleming, Eds.), Vol. 1, 211, Pergamon Press, Oxford, **1991.**

P. Knochel, M. J. Rozema, C. E. Tucker, C. Retherford, M. Furlong, S. A. Rao, "The chemistry of polyfunctional organozinc and copper reagents," *Pure Appl. Chem.* **1992,** *64,* 361–369.

P. Knochel, "Zinc and Cadmium: A Review of the Literature 1982–1994," in *Comprehensive Organometallic Chemistry II* (E. W. Abel, F. G. A. Stone, G. Wilkinson, Eds.), Vol. 11, 159, Pergamon, Oxford, UK, **1995.**

P. Knochel, "Preparation and Application of Functionalized Organozinc Reagents," in *Active Metals* (A. Fürstner, Ed.), 191, VCH, Weinheim, Germany, **1996.**

P. Knochel, J. J. A. Perea, P. Jones, "Organozinc mediated reactions," *Tetrahedron,* **1998,** *54,* 8275–8319.

P. Knochel, "Carbon-Carbon Bond Formation Reactions Mediated by Organozinc Reagents," in *Metal-Catalyzed Cross-Coupling Reactions,* (F. Diederich, P. J. Stang, Eds.), Wiley-VCH, Weinheim, **1998,** 387–416.

P. Knochel and P. Joned, (Eds.) "Organozinc Reagents: A Practical Approach," Oxford University Press, Oxford, U.K., **1999.**

I. B. Campbell, "The Sonogashira Cu-Pd-Catalyzed Alkyne Coupling Reactions," in *Organocopper Reagents: A Practical Approach* (R. J. K. Taylor, Ed.), 217, Oxford University Press, Oxford, U.K., **1994.**

13.4

P. Siemsen, R. C. Livingston, F. Diederich, "Acetylenic coupling: A powerful tool in molecular construction," *Angew. Chem.* **2000,** *112,* 2740–2767; *Angew. Chem. Int. Ed. Engl.* **2000,** *39,* 2632–2657.

13.5

R. F. Heck, "Vinyl Substitutions with Organopalladium Intermediates," in *Comprehensive Organic Synthesis* (B. M. Trost, I. Fleming, Eds.), Vol. 4, 833, Pergamon Press, Oxford, **1991.**

H.-U. Reißig, "Palladium-Catalyzed Arylation and Vinylation of Olefins," in *Organic Synthesis Highlights* (J. Mulzer, H.-J. Altenbach, M. Braun, K. Krohn, H.-U. Reißig, Eds.), VCH, Weinheim, New York, etc., **1991,** 174–180.

L. E. Overman, "Application of intramolecular Heck reactions for forming congested quaternary carbon centers in complex molecule total synthesis," *Pure Appl. Chem.* **1994,** *66,* 1423–1430.

A. de Meijere and F. E. Meyer, "Fine feathers make fine birds: The Heck reaction in modern garb," *Angew. Chem.* **1994,** *106,* 2473–2506; *Angew. Chem. Int. Ed. Engl.* **1994,** *33,* 2379–2411.

W. Cabri and I. Candiani, "Recent developments and new perspectives in the Heck reaction," *Acc. Chem. Res.* **1995,** *28,* 2–7.

S. E. Gibson and R. J. Middleton, "The intramolecular Heck reaction," *Contemp. Org. Synth.* **1996,** *3,* 447–472.

T. Jeffery, "Recent Improvements and Developments in Heck-Type Reactions and Their Potential in Organic Synthesis," in *Advances in Metal-Organic Chemistry* (L. S. Liebeskind, Ed.), **1996,** *6,* JAI, Greenwich, CT.

M. Shibasaki, C. D. J. Boden, A. Kojima, "The asymmetric Heck reaction," *Tetrahedron* **1997,** *53,* 7371–7393.

J. T. Link and L. E. Overman, "Forming cyclic compounds with the intramolecular Heck reaction," *Chem. Br.* **1998,** *28,* 19–26.

S. Bräse and A. de Meijere, "Palladium-Catalyzed Coupling of Organyl Halides to Alkenes—The Heck Reaction," in *Metal-Catalyzed Cross-Coupling Reactions,* (F. Diederich, P. J. Stang, Eds.), Wiley-VCH, Weinheim, **1998,** 99–154.

J. T. Link and L. E. Overman, "Intramolecular Heck Reactions in Natural Product Chemistry," in *Metal-Catalyzed Cross-Coupling Reactions,* (F. Diederich, P. J. Stang, Eds.), Wiley-VCH, Weinheim, **1998,** 231–266.

J. T. Link and L. E. Overman, "Forming cyclic compounds with the intramolecular Heck reaction," *Chemtech* **1998,** *28,* 19–26.

I. P. Beletskaya and A. V. Cheprakov, "The Heck reaction as a sharpening stone of palladium catalysis," *Chem. Rev.* **2000,** *100,* 3009–3066.

G. T. Crisp, "Variations on a theme: Recent developments on the mechanism of the Heck reaction and their implications for synthesis," *Chem. Soc. Rev.* **1998,** *27,* 427–436.

C. Amatore and A. Jutand, "Anionic Pd(0) and Pd(II) intermediates in palladium-catalyzed Heck and cross-coupling reactions," *Acc. Chem. Res.* **2000,** *33,* 314–321.

Further Reading

V. Fiandanese, "Sequential cross-coupling reactions as a versatile synthetic tool," *Pure Appl. Chem.* **1990,** *62,* 1987–1992.

R. Rossi, A. Carpita, F. Bellina, "Palladium- and/ or copper-mediated cross-coupling reactions between 1-alkynes and vinyl, aryl, 1-alkynyl, 1,2-propadienyl, propargyl and allylic halides or related compounds: A review," *Org. Prep. Proced. Int.* **1995,** *27,* 127–160.

R. Rossi and F. Bellina, "Selective transition metal-promoted carbon-carbon and carbon-heteroatom bond formation: A review," *Org. Prep. Proced. Int.* **1997,** *29,* 139–176.

V. Farina and G. P. Roth, "Recent Advances in the Stille Reaction," in *Advances in Metal-Organic Chemistry* (L. S. Liebeskind, Ed.), **1996,** *6,* JAI, Greenwich, CT.

V. Farina, V. Krishnamurthy, W. J. Scott, "The Stille reaction," *Org. React.* **1997,** *50,* 1–652.

A. G. Davies, "Organotin Chemistry," Wiley-VCH, Weinheim, Germany, **1997.**

C. Amatore and A. Jutand, "Role of DBA in the reactivity of palladium(0) complexes generated in situ from mixtures of $Pd(DBA)_2$ and phosphines," *Coord. Chem. Rev.* **1998,** *178–180,* 511–528.

K. Osakada and T. Yamamoto, "Mechanism and relevance of transmetalation to metal promoted coupling reactions," *Rev. Heteroatom Chem.* **1999,** *21,* 163–178.

Oxidations and Reductions

14

14.1 Oxidation States of Organic Chemical Compounds, Oxidation Numbers in Organic Chemical Compounds, and Organic Chemical Redox Reactions

B

Everybody learns early on how to determine the oxidation states of inorganic compounds by assigning oxidation numbers to the atoms of which they are composed. Let us review the examples of H_2O (Figure 14.1), H_2O_2 (Figure 14.3), and NH_4^+ (Figure 14.5). After dealing with the inorganic molecules, we will analyze the oxidation numbers of comparable organic molecules, i.e., the oxidation numbers in CH_4 (Figure 14.2), C_2H_6 (Figure 14.4), and $CH_3—NH_3^+$ (Figure 14.6). It will be seen that the organic chemist's approach is entirely the same as the familiar approach taken in inorganic chemistry.

The principles for the assignment of oxidation numbers to atoms in covalently bound molecules are as follows:

Determination of the Oxidation Numbers in Molecules That Contain Covalent Bonds Only

1) In the case of a covalent bond between *different* atoms A and B, one assumes that all the bonding electrons are localized on the more electronegative atom B. Therefore, a diatomic molecule AB or a diatomic substructure AB of a molecule is considered
   - as A^+B^-, if it contains an A—B single bond,
   - as $A^{2+}B^{2-}$, if it contains an A═B double bond, and
   - as $A^{3+}B^{3-}$, if it contains an A≡B triple bond.
2) In the case of a covalent bond between *identical* atoms A, one assumes that 50% of the bonding electrons are localized on each atom A. In a diatomic molecule AA or in a diatomic substructure AA of a molecule, one therefore assigns
   - the two bonding electrons of an A—A single bond to the bonded atoms in the form $A^{\bullet}$ $A^{\bullet}$,
   - the four bonding electrons of an A═A double bond to the bonded atoms in the form $A^{\bullet\bullet}$ $A^{\bullet\bullet}$, and
   - the six bonding electrons of an A≡A triple bond to the bonded atoms in the form $A^{\bullet\bullet\bullet}$ $A^{\bullet\bullet\bullet}$.
3) For each atom of the molecule under consideration, one determines the formal charge of the atom based on the foregoing electron assignments. This formal charge is the **oxidation number.** The sum of the oxidation numbers of all atoms must be zero for an uncharged molecule, and the sum of the oxidation numbers of all atoms equals the total charge of an ion.

Figure 14.1 shows how the oxidation numbers of the atoms in H_2O are obtained in this way. According to the foregoing rules, the bonding electrons of both O—H bonds are assigned to the more electronegative oxygen, not to the hydrogens. The O atom thus possesses oxidation number −2, and both H atoms have oxidation number +1.

Fig. 14.1. Determination of the oxidation numbers in H_2O.

1. Step: Set oxidation number = + 1

H−O−H

2. Step: Set oxidation number = – 2

The same procedure is used in Figure 14.2 to determine the oxidation numbers in methane, the simplest organic molecule. Since carbon has a higher electronegativity than hydrogen, the two bonding electrons of each C—H bond are assigned to carbon. Hence, the oxidation number of carbon is −4 and that of all H atoms is +1.

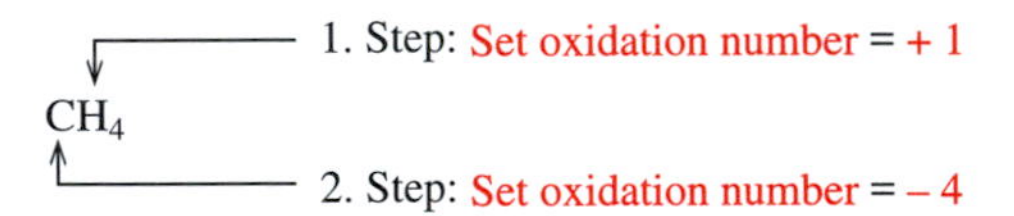

Fig. 14.2. Determination of the oxidation numbers in CH_4.

For the assignment of the oxidation numbers in hydrogen peroxide (Figure 14.3), the two electrons of each O—H bond count only for the O atoms, and the two bonding electrons of the O—O bond count 50% to each O atom. In this way, one finds the oxidation numbers in +1 for both H atoms and −1 for both O atoms.

Fig. 14.3. Determination of the oxidation numbers in H_2O_2.

1. Step: Set oxidation number = + 1

H−O−O−H

2. Step: Set oxidation number = – 1

Applying the analogous approach to ethane (Figure 14.4) results in the oxidation number of +1 for each H atom and −3 for each C atom.

Fig. 14.4. Determination of the oxidation numbers in C_2H_6.

1. Step: Set oxidation number = + 1

$H_3C{-}CH_3$

2. Step: Set oxidation number = – 3

Figure 14.5 reminds us that the oxidation numbers of the atoms in ions and molecules are determined in the same way as in inorganic chemistry. In the ammonium ion, the four H atoms again possess the oxidation number +1, and the N atom has the oxidation number −3.

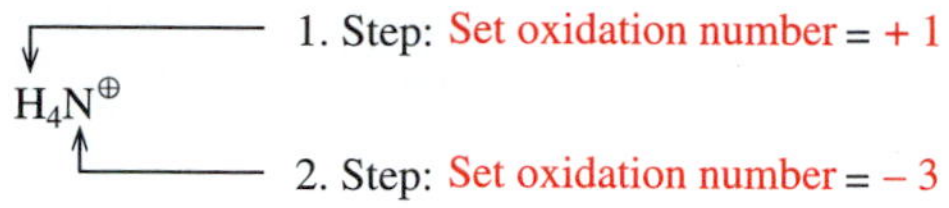

$H_4N^{\oplus}$

1. Step: Set oxidation number = + 1

2. Step: Set oxidation number = – 3

Fig. 14.5. Determination of the oxidation numbers in NH_4^+.

The simplest organic analog of the ammonium ion is the methylammonium ion (Figure 14.6). If one assigns the bonding electrons of the C—H bonds to carbon and those of the N—H and C—N bonds to nitrogen, one obtains oxidation numbers of +1 for each of the H atoms, −3 for the N atom, and −2 for the C atom.

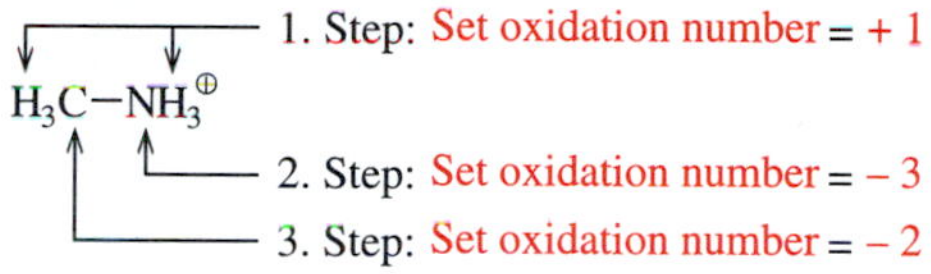

$H_3C{-}NH_3^{\oplus}$

1. Step: Set oxidation number = + 1

2. Step: Set oxidation number = – 3

3. Step: Set oxidation number = – 2

Fig. 14.6. Determination of the oxidation numbers in $CH_3{-}NH_3^+$.

The procedure exemplified for ethane (Figure 14.4) can be employed to assign oxidation numbers to the C and H atoms of all hydrocarbons. The oxidation number of every H atom is +1. However, the oxidation numbers in the C atoms depend on the structure, and they are summarized in Figure 14.7. The C atom of a methyl group always possesses the oxidation number −3 in any hydrocarbon. The C atom of a methylene group always possesses the oxidation number −2 in any hydrocarbon, the C atom of a methyne group always possesses −1, and every quaternary C atom possesses the oxidation number 0.

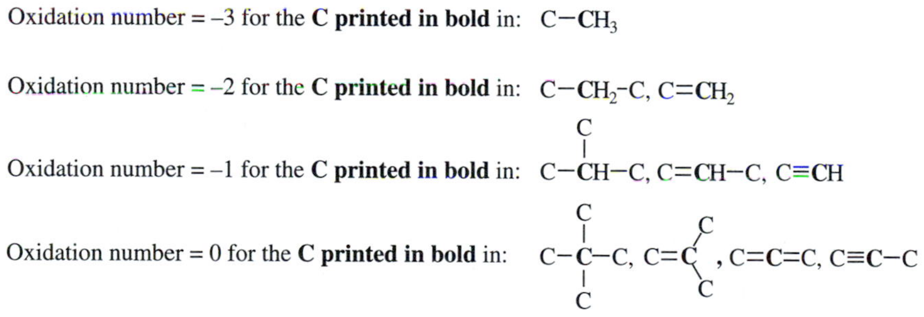

Oxidation number = –3 for the **C printed in bold** in: C−**C**H_3

Oxidation number = –2 for the **C printed in bold** in: C−**C**H_2-C, C=**C**H_2

Oxidation number = –1 for the **C printed in bold** in: C−**C**H(−C)−C, C=**C**H−C, C≡**C**H

Oxidation number = 0 for the **C printed in bold** in: C−**C**(C)(C)−C, C=**C**(C)(C), C=**C**=C, C≡**C**−C

Fig. 14.7. Oxidation numbers of C atoms in selected hydrocarbon substructures.

With the data in Figure 14.7, oxidation numbers can be assigned to the C atoms of the two isomeric butenes of Figure 14.8. Hence, 1-butene possesses the oxidation number −3 at one C atom, the oxidation number −2 at two C atoms, and the oxidation number −1 at one C atom. On the other hand, 2-butene consists of two sets of two C atoms with oxidation numbers −3 and −1, respectively.

$$\overset{-3}{H_3C}-\overset{\boxed{-2}}{CH_2}-\overset{-1}{CH}=\overset{\boxed{-2}}{CH_2} \longrightarrow \overset{-3}{H_3C}-\overset{\boxed{-1}}{CH}=\overset{-1}{CH}-\overset{\boxed{-3}}{CH_3}$$

Fig. 14.8. Oxidation numbers of the C atoms of 1- and 2-butene.

This difference has one irritating consequence: the isomerization of 1-butene to 2-butene would change the oxidation numbers of two atoms. This isomerization thus would constitute a redox reaction or, more specifically, a redox disproportionation. That result, however, is not compatible with "good common sense."

What causes this problem? The assignment of oxidation numbers in organic chemistry should not be overly burdened by questions of whether the procedure really makes sense. The important feature of the butenes of Figure 14.8 lies with the fact that *the C atoms in the butenes on average possess the same oxidation number.* The average oxidation numbers are $(-3 - 2 - 1 - 2)/4 = -2$ for 1-butene and $(-3 - 1 - 1 - 3)/4 = -2$ for 2-butene. The isomerization 1-butene → 2-butene leaves the average oxidation numbers of the atoms invariant, and the isomerization of butene rightly no longer needs to be viewed as a redox reaction. It is best to remember the following:

Two organic chemical compounds possess the *same* oxidation state if the average oxidation numbers of their C atoms are the same *and* if any heteroatoms that might be present possess their usual oxidation numbers (Li, +1; Mg, +2; B, +3; N and P, −3; O and S, −2; and −1 for halogen atoms).

The six C_3 skeletons shown in Figure 14.9 all have an average oxidation number of −1 1/3 of their C atoms. Accordingly, all these compounds are representatives of the same oxidation state.

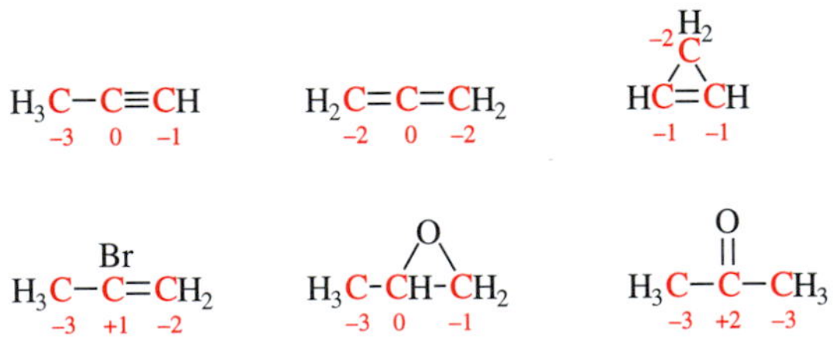

Fig. 14.9. A selection of compounds with the same average oxidation number of −1 1/3 at every atom: oxidation number of the O atoms, −2; oxidation number of the Br atoms, −1.

$H_3C{-}C{\equiv}CH$ (−3, 0, −1)

$H_2C{=}C{=}CH_2$ (−2, 0, −2)

cyclopropene $HC{=}CH$ / CH_2 (−1, −1, −2)

$H_3C{-}C(Br){=}CH_2$ (−3, +1, −2)

$H_3C{-}CH{-}CH_2$ epoxide (−3, 0, −1)

$H_3C{-}C({=}O){-}CH_3$ (−3, +2, −3)

Based on the preceding rule, one can state the following definition:

The Terms "Oxidation" and "Reduction" in Organic Chemistry

Reactions that increase the average oxidation number of the C atoms of a substrate are **oxidations.** The same is true if the oxidation number of one of the heteroatoms increases. Conversely, reactions that decrease the average oxidation number of the C atoms in the substrate or decrease the oxidation number of one of the heteroatoms are **reductions.**

The columns of Table 14.1 contain characteristic substructures ordered by common average carbon oxidation numbers. Also, the average carbon oxidation number increases in going from left to right in this table. Accordingly, reactions are oxidations if they

Table 14.1. Organic Chemical Redox Reactions I: Change of the Average Oxidation Numbers of C Atoms*

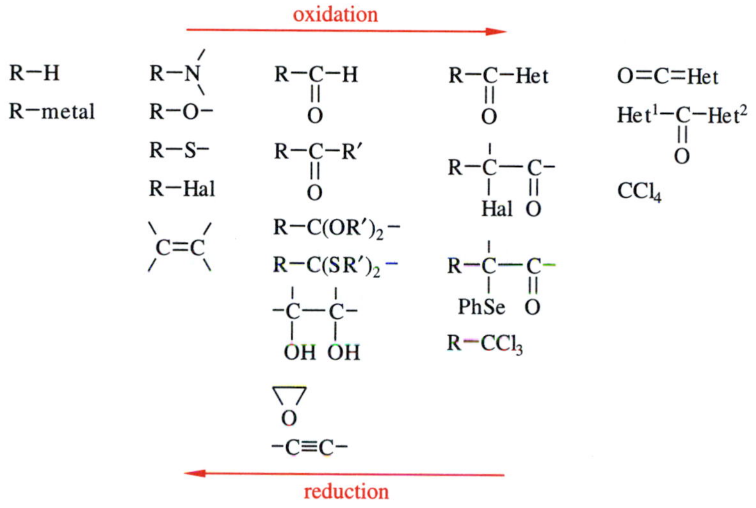

oxidation →

R−H	R−N<	R−C(=O)−H	R−C(=O)−Het	O=C=Het
R−metal	R−O−	R−C(=O)−R′	R−C(Hal)−C(=O)−	Het^1−C(=O)−Het^2
	R−S−	R−C(OR′)$_2$−	R−C(PhSe)−C(=O)−	CCl_4
	R−Hal	R−C(SR′)$_2$−	R−CCl_3	
	>C=C<	−C(OH)−C(OH)−		
		epoxide (three-membered ring with O)		
		−C≡C−		

← reduction

*Selenium is considered to be more electronegative in organic compounds than carbon, even though the Pauling electronegativities are about the same.

Table 14.2. Organic Chemical Redox Reactions II: Change of the Average Oxidation Numbers of N Atoms

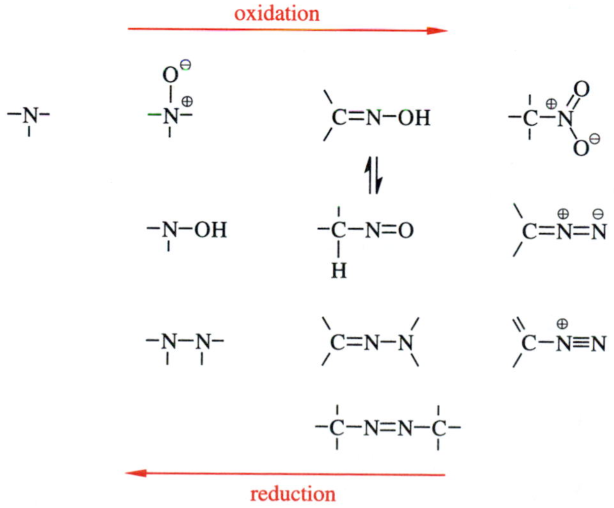

oxidation →

−N<	$−N^{\oplus}(−O^{\ominus})<$	>C=N−OH ⇅ H−C−N=O	$−C−N^{\oplus}(=O)−O^{\ominus}$
	−N(−)−OH	>C=N−N<	$>C=N^{\oplus}=N^{\ominus}$
	−N(−)−N(−)−	−C−N=N−C−	$^{\ominus}>C−N^{\oplus}≡N$

← reduction

convert a substructure of one column into a substructure of a column that is further to the right. The opposite is true for reductions.

An analogous listing of N-containing substructures of organic compounds is given in Table 14.2. Again, these substructures are organized in such a way that the oxidation number in the N atom or the average oxidation number of the N atoms, respectively, increases from left to right. From this it follows that here, too, oxidations are reactions that convert a substrate into a compound that occurs further to the right in the table, and vice versa for reductions.

14.2 Cross-References to Redox Reactions Already Discussed in Chapters 1–13

A B

Many reactions that fit the definition of an organic chemical redox reaction have already been presented in Chapters 1–13. The presentation of these reactions in various other places—without alluding at all to their redox character—was reasonable because they follow mechanisms that were discussed in detail in the respective chapters or because these reactions showed chemical analogies to reactions discussed there. Tables 14.3 and 14.4 provide cross-references to all oxidations and reductions discussed thus far.

The plethora of entries in Tables 14.3 and 14.4 emphasizes that organic chemical redox reactions are not limited to one mechanism and are not even based on a small number of mechanistic principles. Hence, one should not expect any mechanistic homogeneity among the reactions to be discussed in Chapter 14. Sections 14.3 (oxida-

Table 14.3. Compilation of Oxidation Reactions Presented Elsewhere in This Book

Reaction	Reference
$R{-}Hg{-}O_2CCF_3 \xrightarrow[NaBH_4]{O_2,} R{-}OH$	Fig. 1.14
$R{-}H \xrightarrow{Hal_2} R{-}Hal$	Section 1.7
$R{-}H \xrightarrow{O_2} R{-}O{-}O{-}H$	Fig. 1.27, 1.28
R_x $\xrightarrow{ArCO_3H}$ O, R_x	Section 3.14
OH, R_x $\xrightarrow[Ti(OiPr)_4]{tert\text{-}BuOOH,}$ OH, O, R_x	Section 3.4.6

Table 14.3. Continued

Substrate	Reagents	Product	Reference
alkene (R_x)	Br_2	$-C(Br)-C(Br)-$	Section 3.5.1
	NBS, H_2O	$-C(Br)-C(OH)-$	Fig. 3.33, 3.34
	Chloramine T, H_2O	$-C(Cl)-C(OH)-$	Fig. 3.35
HO–C(=O)– (unsaturated acid)	I_2	O= (lactone), I	Fig. 3.36
OH, H	NBS	O, Br	Fig. 3.36
R–Se–Ph	H_2O_2	R–Se(=O)–Ph	Fig. 4.10
Ar–H	Hal_2, $MHal_x$	Ar–Hal	Section 5.2.1
	H_2SO_4	Ar–SO_3H	Section 5.2.2
	HNO_3, H_2SO_4	Ar–NO_2	Section 5.2.3
	$N{\equiv}N^{\oplus}$–Ar′	Ar–N=N–Ar′	Fig. 5.18, 5.19
Ar–Metal	$E^{\oplus}$	Ar–E	Section 5.3.1, 5.3.2
Ar–$B(OMe)_2$	H_2O_2, HOAc	Ar–OH	Fig. 5.39, 11.37
O	H_2SO_4, Ac_2O	O, SO_3H	Fig. 11.9
Br, Br, O	Br_2, $ClSO_3H$	Br, Br, Br, O	Fig. 11.10

Table 14.3. Continued

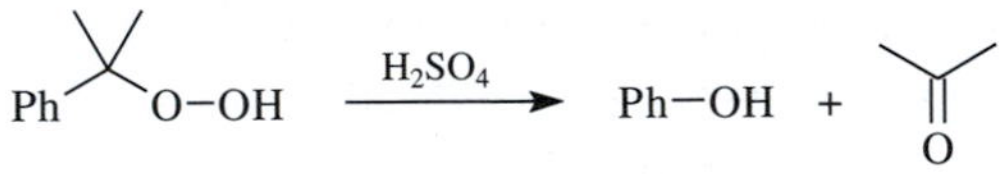

Starting material	Reagent	Product	Reference
Ph–C(CH_3)$_2$–O–OH	H_2SO_4	Ph–OH + acetone	Fig. 11.30
R^1–C(=O)–R^2	$ArCO_3H$	R^1–C(=O)–OR^2	Section 11.4.2
	H_2O_2 or *tert*-BuOOH		Section 11.4.2
BR_3	NaOOH	$B(OR)_3$	Fig. 11.36
	H_2SO_4		Fig. 11.38
	TsN_3, $HNEt_2$		Fig. 12.39
	O_3		Section 12.5.5
	Br_2; NaOMe		Fig. 13.10
	I_2		Fig. 13.10
	I_2		Fig. 13.12, 13.13
R–≡–H	Br_2 or I_2	R–≡–Hal	Section 13.4

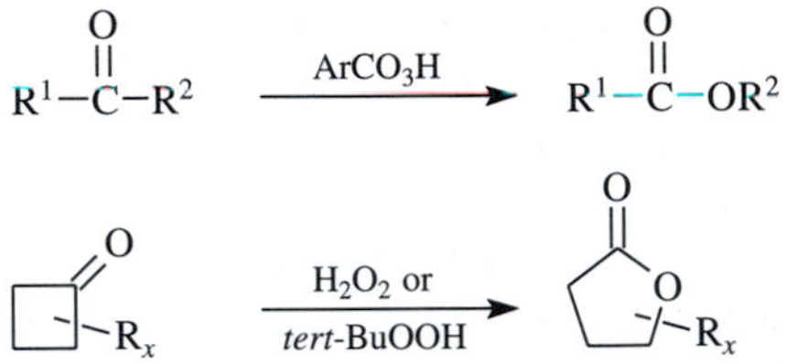

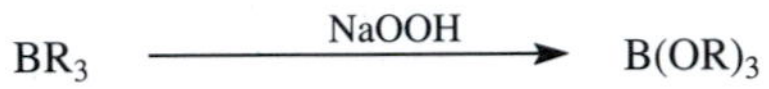

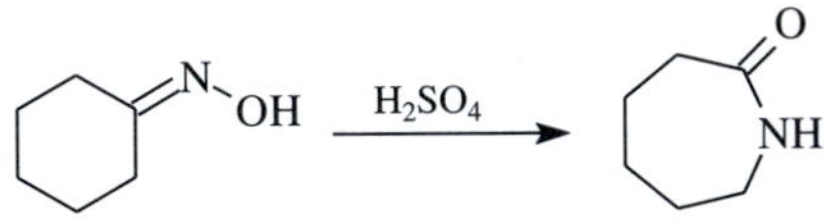

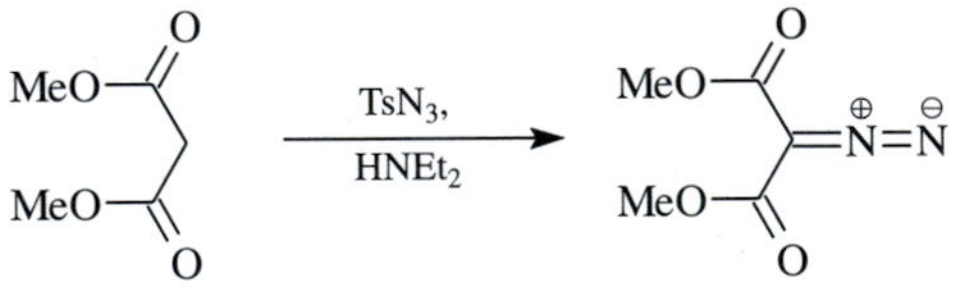

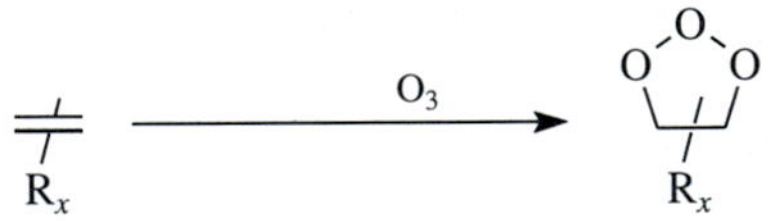

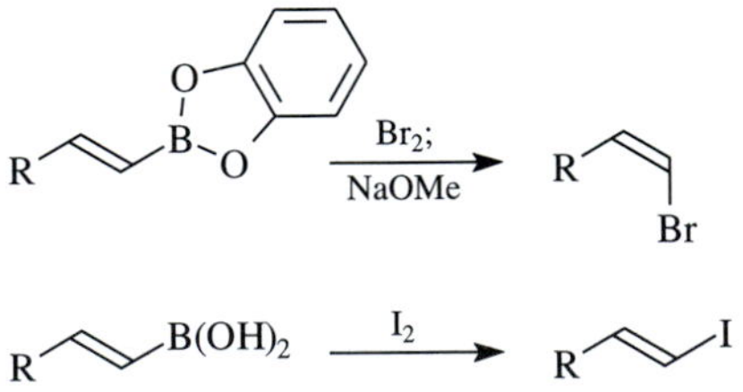

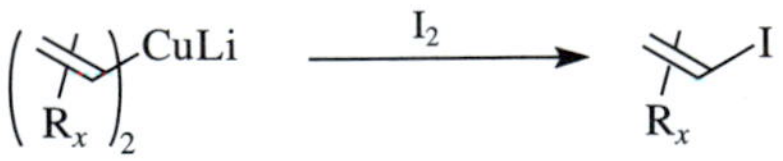

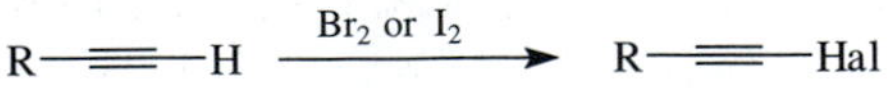

Table 14.4. Compilation of Reduction Reactions Presented Elsewhere in This Book

Substrate	Reagent	Product	Reference
R−Hal; R−O−C(=S)−X	Bu_3SnH or $(Me_3Si)_3SiH$	R−H	Section 1.9.1
alkene (R_x)	H−BL_n	−C(H)−C(BL_n)−	Fig. 3.15
alkene (R_x)	H_2, Pd/C	−C(H)−C(H)−	Fig. 3.22
N=N (1-pyrazoline)	Δ	cyclopropane	Fig. 4.1
Br, Br (1,3-dibromopropane)	Mg	cyclopropane	Fig. 4.1
Br−CH₂−C=C−CH₂−Br (R_x)	Zn	1,3-diene (R_x)	Fig. 4.1
Br−CH₂CH₂−Br	Mg	ethylene	Section 4.7.1
Cl_3C−C(=O)−Cl	Zn	Cl_2C=C=O (see Fig. 12.31)	Section 4.7.1
Br−C−C−OR (R_x)	Mg or Li	alkene (R_x)	Section 4.7.1
R^1−CH($PhSO_2$)−CH(OAc)−R^2	$NaHg_x$	R^1−CH=CH−R^2	Fig. 4.37
cyclic thiocarbonate (O−C(=S)−O, R_x)	$P(OMe)_3$	alkene (R_x)	Fig. 4.42
$Ar-\overset{\oplus}{N}\equiv N\ X^{\ominus}$	H_3PO_2 or EtOH	Ar−H	Fig. 5.42
R−C(=O)−Het	"$H^{\ominus}$"	R−C(=O)−H	Section 6.5.2

Table 14.4. Continued

Starting material	Reagent	Product	Reference
$R^1{-}C({=}O){-}R^2$	"$H^{\ominus}$"	$R^1{-}CH(OH){-}R^2$	Section 8.2–8.4
O	*i*PrMgBr	OH	Fig. 8.22
$R_{func}{-}I$	Zn	$R_{func}{-}ZnI$	Fig. 8.29
R	$HBEt_2$; $ZnEt_2$	$(R{-}CH_2CH_2)_2Zn$	Fig. 8.30, 8.38
n, OH, Br, Br	2 *n*-BuLi	*n*+1, O or *n*+1, O	Fig. 11.23
R, Br, Br	2 *n*-BuLi	$R{-}C{\equiv}CH$	Fig. 11.29
R_x, O, N	$LiAlH_4$	R_x, OH, NH_2	Fig. 12.43
R_x, O, N	H_2, Raney-Ni, $B(OH)_3$	R_x, OH, O	Fig. 12.43
$Ph{-}\overset{\oplus}{N}{\equiv}N\ X^{\ominus}$	NaN_3	$Ph{-}N{=}\overset{\oplus}{N}{=}\overset{\ominus}{N}$	Fig. 12.46
OTf, R_x	*i*PrMgCl, $Ni(acac)_2$	H, R_x	Fig. 13.9
$R{-}C{\equiv}CH$	H–B, O, O (catecholborane)	R, B, O, O	Fig. 13.10
R, Br	n-BuLi	R, Li	Fig. 13.11

tions) and 14.4 (reductions) are thus not organized on the basis of mechanistic considerations. Instead, the ordering principle reflects preparative aspects: Which classes of compounds can be oxidized or reduced into which other classes of compounds, and how can these transformations be accomplished?

14.3 Oxidations

14.3.1 Oxidations in the Series Alcohol → Aldehyde → Carboxylic Acid

Survey

B

For the oxidation of primary or secondary alcohols on a laboratory scale, one usually employs one of the reagents listed in Table 14.5. Rows 1–5 of this table list Cr(VI) oxidizing agents, namely, aqueous $K_2Cr_2O_7$, the **Jones reagent,** the **Collins reagent,** PCC, and PDC. Row 6 of Table 14.5 lists a mixture of reagents that leads to the so-called **activated dimethyl sulfoxide** as an oxidant. The oxidation potential of these six reagents toward alcohols and aldehydes can be summarized as follows:

1) *All* reagents listed in Table 14.5 can be used for the oxidation of secondary alcohols to ketones.
2) Primary alcohols can be oxidized to the respective carboxylic acids; aqueous $K_2Cr_2O_7$ and the Jones reagent allow for this possibility.
3) Alternatively it is possible to oxidize a primary alcohol no further than to give the aldehyde. This is the domain of the Collins reagent, PCC, PDC, or activated dimethyl sulfoxide. The oxidation of primary alcohols *with $K_2Cr_2O_7$ in aqueous solution* to nothing but the aldehyde, (i.e., without further oxidation to the carboxylic acid) is possible only if a volatile aldehyde results and is distilled off as it is formed. This is the only way to prevent the further oxidation of the aldehyde in the (aqueous) reaction mixture. Selective oxidations of primary alcohols to aldehydes with the Jones reagent succeed only for allylic and benzylic alcohols. Otherwise, the Jones reagent directly converts alcohols into carboxylic acids (see above).
4) Aside from aqueous $K_2Cr_2O_7$ and the Jones reagent (Table 14.5), there also exists a chromium-free method for the oxidation of aldehydes to carboxylic acids. This method is of interest because Cr(VI) compounds are recognized to be carcinogens. Conjugated aldehydes in particular can be oxidized "chromium free" to carboxylic acids with a mixture of sodium chlorite and hydrogen peroxide. Sodium chlorite is a weaker (!) oxidizing agent than the sodium hypochlorite, which it delivers through reduction by the substrate. To prevent destruction of this substrate by sodium hypochlorite at the expense of the desired oxidation by sodium chlorite, hydrogen peroxide is added to oxidize sodium hypochlorite back to sodium chlorite as soon as it is formed. The mechanism of this alcohol → carboxylic acid oxidation is not known.

A

Table 14.5. Standard Reagents for Oxidations in the Series Alcohol → Aldehyde → Carboxylic Acid and Alcohol → Ketone

Oxidizing agent	Can be used in selective oxidations: $R^1R^2CHOH \rightarrow R^1C(=O)R^2$	$RCH_2OH \rightarrow RCHO$	$RCH_2OH \rightarrow RCOOH$
$K_2\overset{+6}{Cr}_2O_7$, dilute H_2SO_4	✓	no	✓
$\overset{+6}{Cr}O_3$, dilute H_2SO_4, acetone [1]	✓	sometimes [6]	✓
Pyridine $\cdot \frac{1}{2}\,\overset{+6}{Cr}O_3$ [2]	✓	✓	no
Pyridinium($\oplus$)N–H $Cl-\overset{+6}{Cr}O_3^{\ominus}$ (≡ PCC [3])	✓	✓	no
(Pyridinium($\oplus$)N–H)$_2$ $\overset{+6}{Cr}_2O_7^{2\ominus}$ (≡ PDC [4])	✓	✓	no
$H_3C-\overset{\pm 0}{S}(=O)-CH_3$, $Cl-C(=O)-C(=O)-Cl$; NEt_3 [5]	✓	✓	no

[1] Jones reagent.
[2] Collins reagent.
[3] PCC, **p**yridinium **c**hloro**c**hromate.
[4] PDC, **p**yridinium **d**i**c**hromate.
[5] Swern oxidation.
[6] For R = aryl or alkenyl (in the latter case, *cis* → *trans* and *E* ⇌ *Z* isomerizations are possible).

Cr(VI) Oxidation of Alcohol and Aldehydes

B

The oxidation of alcohols to carbonyl compounds with Cr(VI) occurs via chromium(VI) acid monoesters (**A** in Figure 14.10). These esters yield chromium(IV) acid by way of a β-elimination via a cyclic transition state (Figure 14.10). Alternatively, one could also imagine that an acyclic transition state might be involved. The chromium(IV) acid could then disproportionate giving Cr(III) and Cr(VI) (see Figure 14.10, center); that is, the inorganic Cr(III) and Cr(VI) products would be obtained without the participation of an organic molecule. On the other hand, the chromium(IV) acid itself also is capable of oxidizing the alcohol while being reduced to Cr(III), presumably via a

Cr(VI) Chemistry:

A

$$R^1{-}CH(OH){-}R^2 + CrO_3 \longrightarrow \mathbf{A} \xleftarrow{-\,X^{\ominus}} R^1{-}CH(OH){-}R^2 + [CrO_3X]^{\ominus}$$

$$\mathbf{A} \longrightarrow R^1{-}C(=O){-}R^2 + \overset{+4}{Cr}O(OH)_2$$

+ Follow-up chemistry of Cr(IV):

Either: $3\ \overset{+4}{Cr}O(OH)_2 \xrightarrow[\text{chemistry"}]{\text{"Inorganic}} \overset{+3}{Cr_2O_3} + \overset{+6}{Cr}O_3 + 3\ H_2O$

or:

$$R^1{-}CH(OH){-}R^2 \xrightarrow{\overset{+4}{Cr}O(OH)_2 \ \rightarrow\ \frac{1}{2}\overset{+3}{Cr_2O_3} + 1\frac{1}{2}H_2O} R^1{-}\dot{C}(OH){-}R^2\ (\mathbf{B}) \xrightarrow{\overset{+6}{Cr}O_3 \ \rightarrow\ \overset{+5}{Cr}O_3H} R^1{-}C(=O){-}R^2$$

$$\overset{+5}{Cr}O_3H + R^1{-}CH(OH){-}R^2 \longrightarrow \mathbf{C} \longrightarrow R^1{-}C(=O){-}R^2 + \overset{+3}{Cr}(OH)_3$$

C

*) Or abstraction by water

Fig. 14.10. Mechanism of the Cr(VI) oxidation of alcohols to carbonyl compounds. The oxidation proceeds via the chromium(VI) acid ester **A** ("chromic acid ester") and yields chromium(IV) acid. The chromium(IV) acid may either disproportionate in an "inorganic" reaction or oxidize the alcohol to the hydroxy-substituted radical **B.** This radical is subsequently oxidized to the carbonyl compound by Cr(VI), which is reduced to Cr(V) acid in the process. This Cr(V) acid also is able to oxidize the alcohol to the carbonyl compound while it is undergoing reduction to a Cr(III) compound.

radical mechanism (Figure 14.10, bottom). The hydroxy-substituted radical **B** is formed from the alcohol by a one- or multistep transfer of an H atom from the α-position onto the Cr(IV) species. Subsequently, this radical is oxidized further to the carbonyl compound. Cr(VI) might act as the oxidizing reagent, and it would be reduced to a chromium(V) acid. This species is the third oxidizing agent capable of attacking the alcohol in Cr(VI) oxidations; presumably, the alcohol and the Cr(V) acid form a chromium(V) acid monoester **C.** This ester may undergo a β-elimination, much like the Cr(VI) analog **A,** and thereby form the carbonyl compound and a Cr(III) compound.

Oxidations of alcohols with *water-free* Cr(VI) reagents, such as the ones in rows 3–5 in Table 14.5, always result in the formation of carbonyl compounds. In particular, a carbonyl compound is obtained even if it is an aldehyde and therefore in principle could be oxidized to give the carboxylic acid. On the other hand, aldehydes *are* oxidized further (cf. rows 1–2 in Table 14.5) if one uses water-containing Cr(VI) reagents (e.g., oxidations with H_2SO_4–$K_2Cr_2O_7$) or the Jones reagent (unless the latter is used under really mild conditions). Figure 14.11 provides an explanation of this effect of water. Cr(VI) compounds cannot attack aldehydes at all. Only the hydrates of these aldehydes can be attacked, and these hydrates are formed in an equilibrium reaction from the aldehydes in the presence of water (Figure 7.7). The hydrate of an aldehyde behaves toward the Cr(VI) compound just like any ordinary alcohol. Hence, the hydrate of the aldehyde is oxidized by the mechanism discussed for alcohols (Figure 14.10) after the hydrate has been converted into a chromium(VI) acid monoester **A** (Figure 14.11). This ester undergoes the same reactions described in Figure 14.10; that is, a maximum of two steps gives the carboxylic acid, and a maximum of three steps leads to the final inorganic Cr(III) product.

Oxidations of Alcohols with Activated Dimethyl Sulfoxide

B

In this subsection we want to consider oxidations that employ dimethyl sulfoxide (DMSO) as the oxidizing reagent. These oxidations, which almost always are carried out in the presence of first oxalyl chloride and then NEt_3, are referred to as **Swern oxidations.** The mechanism of this reaction is known in detail (Figure 14.12). In the prelude, the O atom of DMSO acts as a nucleophile and, following the mechanism of Figure 6.2, undergoes an S_N reaction at one of the carboxyl carbons of the oxalyl chloride to form the sulfonium ion **B.** This is *one* form of the so-called activated DMSO. Addition of a chloride ion yields a sulfurane intermediate. Loss of the —O—C(=O)—C(=O)—Cl group, which fragments, results in the formation of the sulfonium ion **D.** This ion **D** represents *another* "activated DMSO."

Irrespective of whether the initially formed sulfonium ion **B** or the subsequently formed sulfonium ion **D** reacts with the alcohol, the alcohol is taken up by such a sulfonium ion with formation of sulfuranes **A** (first case) or **C** (second case). Any of these sulfuranes would yield the sulfonium salt **E** after dissociation. Once this sulfonium salt has formed,

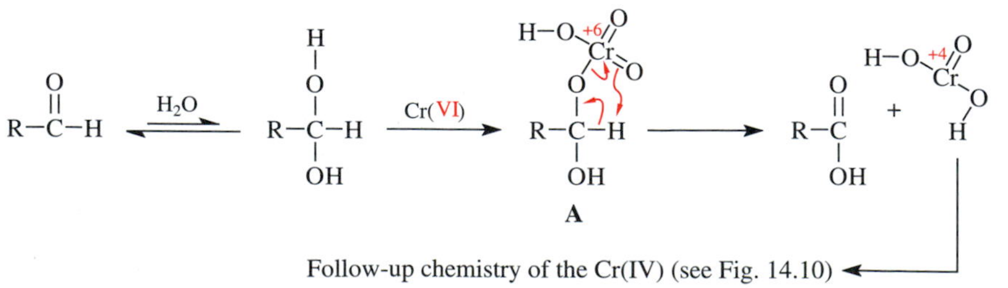

Fig. 14.11. Mechanism of the Cr(VI) oxidation of an aldehyde. As in the case described in Figure 14.10, the oxidation can proceed via three different paths.

Follow-up chemistry of the Cr(IV) (see Fig. 14.10)

Fig. 14.12. Mechanism of the Swern oxidation of alcohols. The actual reagent is an "activated DMSO" (compound **B** or **D**), which reacts with an alcohol with formation of **A** or **C**, respectively. Dissociation leads to the sulfonium salt **E**, which is then converted into the sulfonium ylide **F** after NEt_3 addition and raising the temperature from −60 to −45°C. β-Elimination via a cyclic transition state generates the carbonyl compound and dimethyl sulfide from **F**.

five equivalents of NEt_3 are added to the reaction mixture, which then is allowed to warm up from −60 to −45°C. Under these conditions, the sulfonium salt **E** is deprotonated to give the sulfonium ylide **F**. This ylide undergoes a β-elimination via a cyclic transition state to form the desired carbonyl compound and dimethyl sulfide as a side-product.

Special Oxidation Methods for R—CH_2OH → R—CH(=O)

It might seem that the methods presented thus far for the oxidation of alcohols to aldehydes offer more than enough options. In practice, however, situations may well arise in which *none* of these methods works. It is because of such failures that there has been and still is a constant need for the development of alternatives. Three of these alternative procedures are used nowadays as "advanced routine procedures" for such R—CH_2OH → R—CH(=O) transformations (Figures 14.13–14.15).

The oxidation of alcohol **A** of Figure 14.13 to give aldehyde **E** fails under Swern conditions because NEt_3 catalyzes an ensuring E2 elimination leading to the conjugated aldehyde **D.** In contrast, the oxidation of the same aldehyde **A** with the **Dess–Martin reagent** (**B** in Figure 14.13; preparation, Figure 14.33) occurs without base. **B** is a mixed anhydride of an aryliodo(III) acid and two different carboxylic acids. One

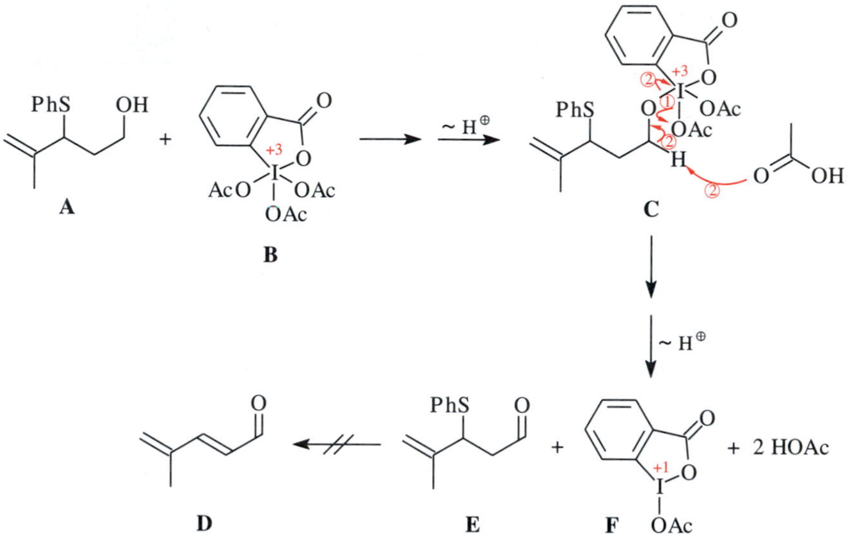

Fig. 14.13. Mechanism of the Dess–Martin oxidation of alcohols to aldehydes. The aryliodo(III) ester **C** is formed from the Dess–Martin reagent **B** and the alcohol. This ester undergoes a β-elimination and forms the aldehyde **E** along with the iodo(I) compound. **F,** presumably in two steps.

of the three acetoxy groups of reagent **B** is replaced by the substrate alcohol via a dissociation/association mechanism. The aryliodo(III) acid ester **C** is formed in this way. **C** undergoes a β-elimination, presumably in two steps. The first step consists of the cleavage of an acetate ion from the iodine. The second step is an E2-elimination. The products of elimination are the desired aldehyde **E** and the iodo(I) derivative **F.**

The enantiomerically pure epoxy alcohols **A** shown in Figure 14.14 (see Figure 3.29 for preparation) can be oxidized with a combination of two oxidizing reagents. This combination consists of a stoichiometric amount of *N*-methylmorpholine-*N*-oxide and a catalytic amount of tetrapropylammonium perruthenate (TPAP). This oxidation works much better than either the standard oxidizing agent PDC or the Swern reagent. The water formed is removed by added molecular sieves to prevent the formation of an aldehyde hydrate and, via the latter, a progression of the oxidation toward the carboxylic acid (cf. the mechanism in Figure 14.11).

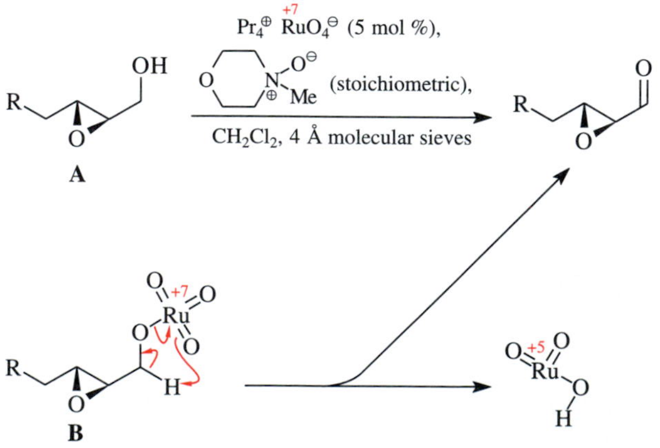

Fig. 14.14. TPAP oxidation of an alcohol to an aldehyde; TPAP stands for **t**etra**p**ropyl**a**mmonium **p**erruthenate.

The effective oxidant in the TPAP oxidation of alcohols is the perruthenate ion, a Ru(VII) compound. This compound is employed in catalytic amounts only but is continuously replenished (see below). The mechanism of the alcohol → aldehyde oxidation with TPAP presumably corresponds to the nonradical pathway of the same oxidation with Cr(VI) (Figure 14.10, top). Accordingly, the key step of the TPAP oxidation is a β-elimination of the ruthenium(VII) acid ester **B.** The metal is reduced in the process to ruthenium(V) acid.

If no *N*-methylmorpholine-*N*-oxide were added the ruthenium(V) acid would be converted into RuO_2. In that case, Ru(VII) would be a three-electron oxidizing agent just like Cr(VI) (Figure 14.10). Such a conversion of Ru(V) into Ru(IV) could in principle occur, since Ru(V) also oxidizes alcohols. This oxidation presumably would proceed via an α-hydroxylated radical as discussed for the Cr(IV) oxidation of alcohols (Fig 14.10, center). Yet, there is no indication for such a radical pathway to occur when the reaction is carried out in the presence of *N*-methylmorpholine-*N*-oxide. Hence, it appears that *N*-methylmorpholine-*N*-oxide reoxidizes the ruthenium(V) acid to perruthenate faster than the ruthenium(V) acid could attack an alcohol molecule.

Figure 14.15 illustrates a third "advanced" procedure for the alcohol → aldehyde oxidation with the example of a racemization-free oxidation of an enantiomerically pure alcohol. Two oxidizing agents are employed, a stoichiometric amount of NaOCl

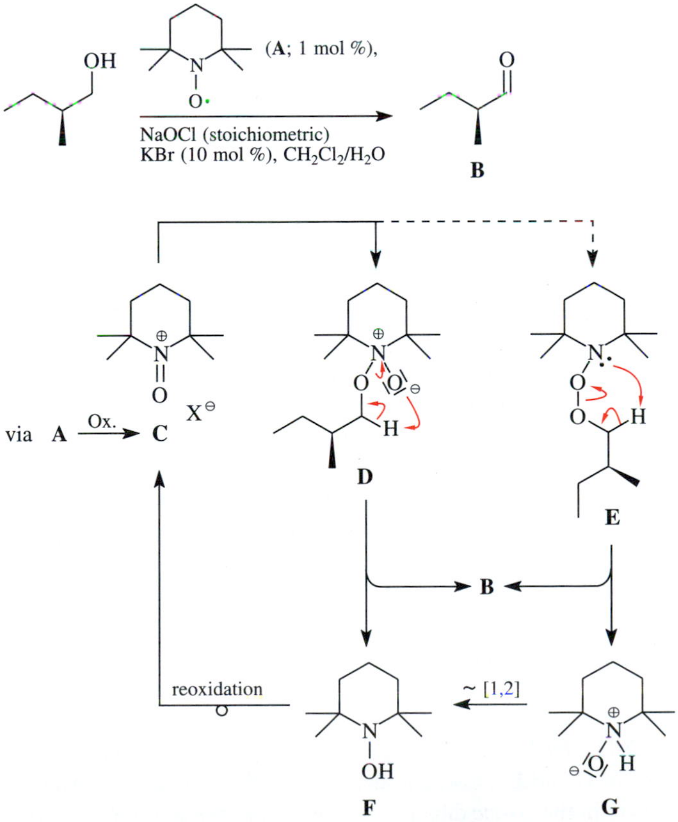

Fig. 14.15. Mechanism of the TEMPO oxidation of alcohols to aldehydes; TEMPO stands for **te**tra**m**ethyl**p**iperidine nitr**o**xyl.

and a catalytic amount of the nitroxyl **A.** These two components form the actual oxidizing reagent, the nitrosonium ion **C** (X = Cl). A small amount of KBr is added as a third component to the mixture. KBr serves to increase the solubility of the nitrosonium ion in the organic phase, where the bromide salt **C** (X = Br) is more soluble than the chloride salt **C** (X = Cl). In the organic phase, the nitrosonium ion combines with the alcohol, which acts as a nucleophile and adds to the $N^+{=}O$ double bond. This attack of the alcohol is more likely to occur at the N atom (→ **D**) than at the O atom (→ **E**) of the $N^+ = 0$ bond. However, each of these intermediates would produce the same desired aldehyde **B** via a *β*-elimination. The accompanying product could be the hydroxylamine **F** or its tautomer, the *N*-oxide **G.** Certainly, **G** would immediately isomerize to give hydroxylamine **F** via a [1,2]-rearrangement (see Section 11.1 for a definition of this term). Subsequently, **F** is reoxidized by NaOCl to restore, via the TEMPO radical **A,** the nitrosonium salt **C.** This completes the catalytic cycle.

14.3.2 Oxidative Cleavages

The *cis-vic* Dihydroxylation of Alkenes: No Oxidative Cleavage, but an Important Prelude

B

Alkenes can be dihydroxylated *cis*-selectively by reaction with a stoichiometric amount of *N*-methylmorpholine-*N*-oxide (NMO; for a preparation, see Figure 14.30), a catalytic amount of a suitable Os(VIII) reagent, and in the presence of water. This reaction, the *cis-vic* dihydroxylation, was not discussed in the section on *cis*-selective additions to C=C double bonds in alkenes (Section 3.3), and it also was not discussed in the context of cycloadditions (Chapter 12). From a preparative point of view, this reaction is closely related to oxidative cleavages, and for this reason it is introduced now (Figure 14.16).

OsO_4 is the reactive Os(VIII) species in the *cis-vic* dihydroxylation of alkenes. This compound is a solid but is not easy to handle because it has a rather high vapor pressure. Therefore, it is often preferable to prepare the compound *in situ.* Such a preparation involves the oxidation of the potassium salt $K_2OsO_4 \cdot 2H_2O$ of osmium(VI) acid with NMO as the oxidizing agent. This salt also is a solid but has a much lower vapor pressure than OsO_4. The co-oxidant NMO also—and mainly—effects the *regeneration* of Os(VIII) from Os(VI), which otherwise would be the final inorganic product of the dihydroxylation of alkenes by OsO_4.

It is not completely clear whether the addition of OsO_4 to the alkene is a [2+2]- or a 1,3-dipolar cycloaddition (Figure 14.16). Depending on the mode of attack, the first intermediate would be either the four-membered ring **A**—an Os(VIII) derivative—or the five-membered ring **B**—an Os(VI) derivative. The four-membered ring **A** would react further via the five-membered ring **B,** which would require a fast ring expansion. The intermediate **B,** a cyclic diester of osmium(VI) acid, could be hydrolyzed by the water present (Figure 14.16, center). This hydrolysis would liberate the *cis*-diol and osmium(VI) acid. The latter would be reoxidized to OsO_4 by the co-oxidant NMO.

The lower part of Figure 14.16 shows that the osmium(VI) derivatives **C** and **E** are formed when OsO_4 reacts with alkenes in the absence both of a co-oxidant and water. The occurrence of **C** and **E** allowed to deduce that the osmium(VI) acid ester **B** (also) is an intermediate in the *cis-vic* dihydroxylation of alkenes when water is present. In the

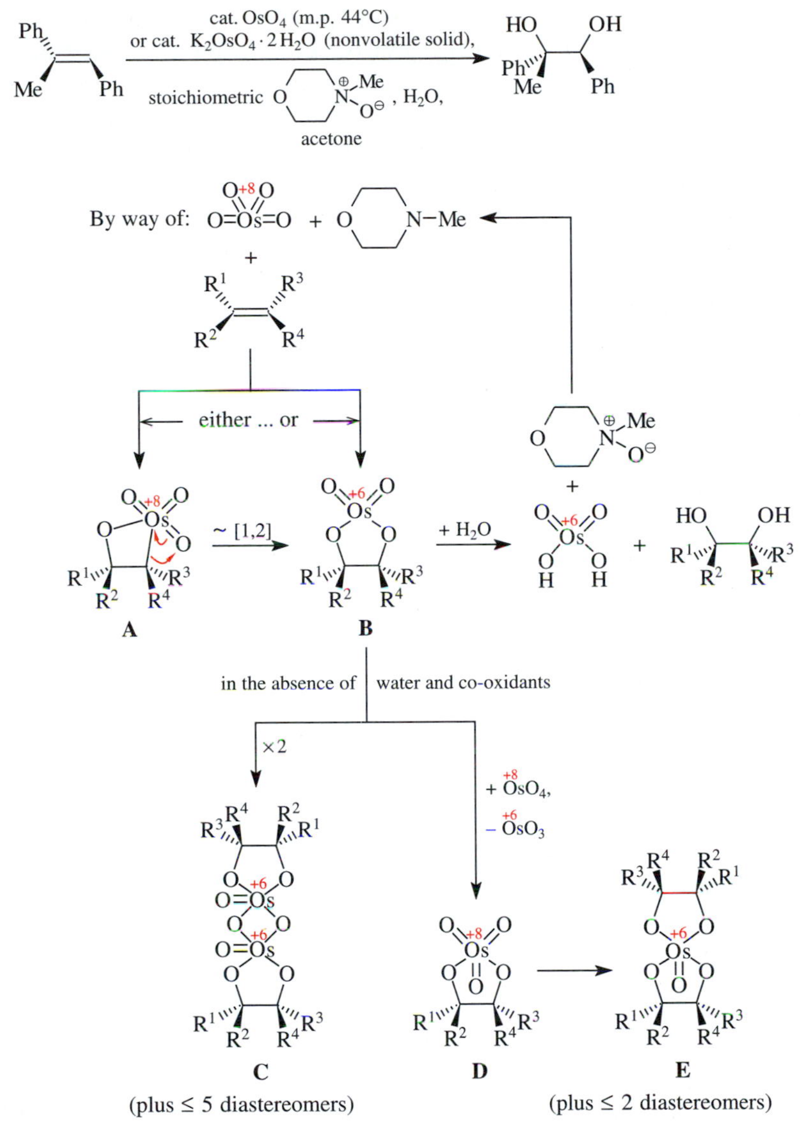

Fig. 14.16. *cis-vic* Dihydroxylation of alkenes with Os(VIII) and pertinent mechanistic insights obtained thus far.

absence of water, of course, the Os(VI) ester **B** cannot hydrolyze to the *cis-vic*-diol, and dimerization to **C** presents a possible alternative reaction path. On the other hand, the Os(VI) ester **B** can be oxidized to the Os(VIII) analog **D** by some of the remaining OsO_4. Then **D** adds to the double bond of another alkene and the tetraester **E** is formed. This addition presumably is similar to the addition of OsO_4 to a C=C double bond.

The possibility cannot be excluded that the *cis-vic* dihydroxylation of alkenes with OsO_4 in the presence of water proceeds via intermediate **D** and perhaps even via **E.** Hydrolysis of either of these species would namely yield the same diol as the hydrolysis of **B.**

Often *cis-vic* dihydroxylations of alkenes similar to the ones effected by OsO_4/NMO/H_2O or K_2OsO_4/NMO/H_2O can be achieved by using **potassium permanganate** as a stoichiometric oxidizing reagent. In this case, careful control of the reaction condition is called for. In contrast to OsO_4, potassium permanganate is capable of effecting a subsequent oxidation of the diols to give two carbonyl and/or carboxyl compounds. Hence, potassium permanganate by itself on occasion may have the same effect as the $KMnO_4$/$NaIO_4$ mixture in the Lemieux–von Rudloff oxidation of alkenes (see Figure 14.20, right).

Oxidative Cleavage of Glycols

B

$NaIO_4$ or H_5IO_6 can be used to cleave vicinal glycols into two carbonyl compounds or into a single compound with two carbonyl groups (Figure 14.17). The formation of a diester of iodo(VII) acid as an intermediate is decisive for the success of this reaction. Such a diester formation is even possible starting from *trans*-configured 1,2-cyclohexanediols. The diester intermediate decomposes in a one-step reaction in which three valence electron pairs are shifted simultaneously. One of these shifted electron pairs ends up as a lone pair on iodine, and the iodine(VII) initially present is thereby reduced to iodine(V).

Fig. 14.17. Mechanism of the glycol cleavage with $NaIO_4$ or H_5IO_6, respectively. A diester of iodo(VII) acid (periodic acid) is formed initially. The ester decomposes in a one-step reaction in which three valence electron pairs are shifted simultaneously.

Glycols are essential structural elements of sugars. Figure 14.18 shows the oxidative cleavages of two sugar derivatives **A** and **B,** both monoglycols, with $Pb(OAc)_4$ (mechanism below). Glycol **A** is formed in acidic environment from D-mannitol and acetone; it is formed chemoselectively and regioselectively as the least sterically hindered, five-membered (cf. discussion of Figure 7.19) D-mannitol bisacetonide. Glycol **B** is obtained in an analogous fashion in the presence of acid as the most stable monoacetonide of the dithioacetal of L-arabinose. This dithioacetal can be prepared starting from L-arabinose with the procedure described in Section 7.3.3 for the preparation of the dithioacetal of D-glucose. The oxidative cleavages of glycols **A** and **B** (Figure 14.18) result in acetonides of the *R*- and *S*-configured glycerol aldehydes. Taking into account their origin, these compounds belong to the "chiral pool," that is, the pool of naturally occurring chiral compounds.

The glycol cleavages shown in Figure 14.18 are carried out with $Pb(OAc)_4$ in an anhydrous solvent. If one wanted to use $NaIO_4$ as the oxidizing agent, the same cleav-

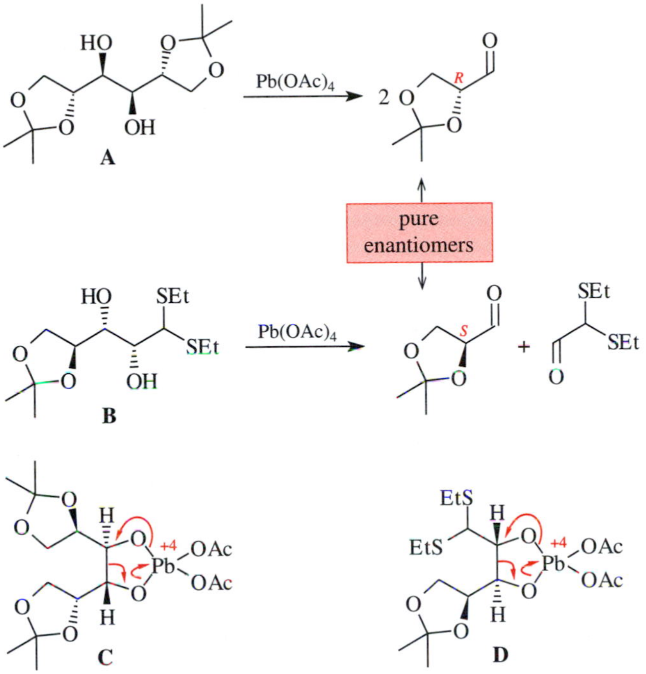

Fig. 14.18. Standard mechanism of the glycol cleavage with $Pb(OAc)_4$. The reaction proceeds preferentially via a cyclic diester of Pb(IV) acid, which decomposes in a one-step reaction to $Pb(OAc)_2$ and two equivalents of the carbonyl compound.

ages would have to be carried out in water-containing solvents for reasons of solubility. In such water-containing media, however, the desired aldehydes—α-oxygenated aldehydes—tend to form aldehyde hydrates (Figure 7.7), from which the aldehydes can be recovered only by way of an azeotropic (i.e., water-removing) distillation. With H_5IO_6, though, glycol cleavages can be achieved in the absence of water. Hence, α-oxygenated aldehydes such as the ones in Figure 14.18 could be obtained *without hydrate formation* in principle by way of the H_5IO_6-mediated glycol cleavage. Yet, this procedure would be prone to fail because of the acid sensitivity of the acetal groups in the aldehydes shown as well as in their precursors.

Glycol cleavages with $Pb(OAc)_4$ preferentially proceed via five-membered lead(IV) acid diesters. Such esters are shown in Figure 14.18 as **C** and **D.** Each one of these esters decomposes in a one-step reaction to furnish $Pb(OAc)_2$ and two equivalents of the carbonyl compound. The cleavage is caused by the concerted shift of three valence electron pairs. One of these becomes a lone pair at Pb. The oxidation number of lead is thereby reduced from +4 to +2.

A

Neither $NaIO_4$ nor H_5IO_6 can convert a conformationally fixed *trans*-glycol (such as compound **A** in Figure 14.19) into a cyclic iodo(VII) acid ester (of the type shown in Figure 14.17). Such glycols therefore cannot be cleaved oxidatively by these reagents (Figure 14.19, top left). $Pb(OAc)_4$, on the other hand, does cleave the same diol oxidatively even though a cyclic lead(IV) acid diester of the type shown in Figure 14.18 also cannot be formed. The reason is that $Pb(OAc)_4$ can react with glycols by way of a second cleavage mechanism that involves the formation of a lead(IV) acid monoester

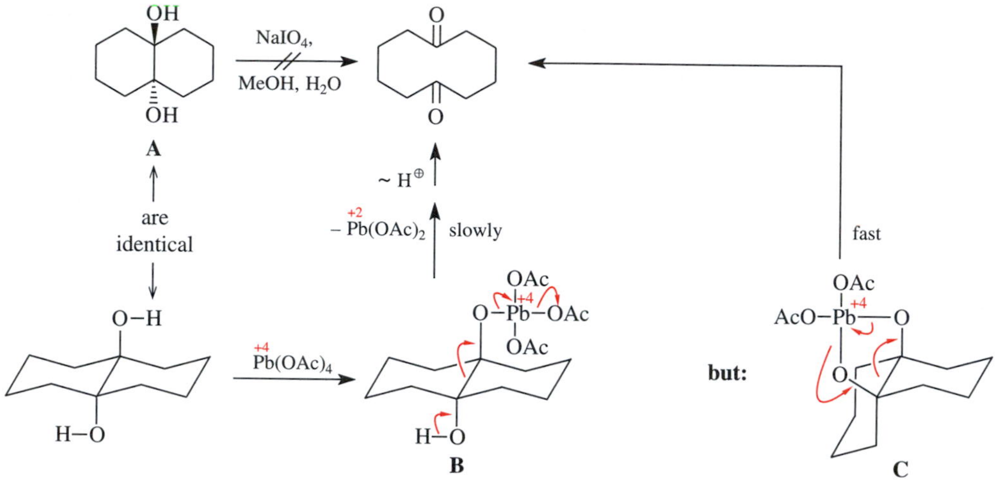

Fig. 14.19. The standard mechanism (transition state **C**) and the alternative mechanism (transition state **B**) of glycol cleavages with $Pb(OAc)_4$. The *trans*-glycol **A** reacts slowly via the monoester **B** of Pb(IV) acid, while the isomeric *cis*-glycol reacts fast via the cyclic diester **C** of lead(IV) acid.

(**B** in Figure 14.19). The cleavage products are formed by a fragmentation reaction in which four electron pairs are shifted simultaneously.

The second mechanism described for the $Pb(OAc)_4$ cleavage of glycols is only "second rate." Accordingly, $Pb(OAc)_4$ cleaves the *cis* isomer of *trans*-glycol **A** (Figure 14.19) much faster because the *cis*-glycol can react with $Pb(OAc)_4$ via the mechanism shown in Figure 14.18 involving a cyclic intermediate (**C** in Figure 14.19).

Oxidative Cleavage of Alkenes

A

The glycol formation from alkenes (Figure 14.16) and the glycol cleavage (Figure 14.17) can be combined in a one-pot reaction. As mentioned in the discussion of Figure 14.16, such a one-pot oxidation/cleavage of alkenes can be accomplished with $KMnO_4$. Under these conditions, however, controlling the oxidation state of the cleavage products might pose problems. Aldehydes *or* carboxylic acids may be formed, and ketones might perhaps even be subject to further oxidation (cf. Figure 14.17).

Well-defined cleavage products will *definitely* be obtained if one employs the pairs of oxidizing reagents shown in Figure 14.20 instead of $KMnO_4$ alone. The stoichiometric oxidizing reagent is still $NaIO_4$, and catalytic amounts of an appropriate co-oxidant are added. The basic approach to achieving the various cleavages in Figure 14.20 always is the same: the alkenes first are converted into *cis*-configured diols, which are then subjected to glycol cleavage. The further oxidation of an aldehyde might be a consecutive reaction.

In the **Lemieux–Johnson oxidation** of alkenes (Figure 14.20, left) the dihydroxylation of the C=C double bond is achieved by using catalytic OsO_4. $NaIO_4$, employed in stoichiometric amounts, plays two roles: it effects the reoxidation of Os(VI) to Os(VIII), and it cleaves the glycols to the aldehydes and/or ketones (mechanism in Figure 14.17).

In the **Lemieux–von Rudloff oxidation** of alkenes (Figure 14.20, right) stoichiometric amounts of the oxidizing reagent $NaIO_4$ are used. In the classical procedure, a catalytic amount of $KMnO_4$ is employed as the co-oxidizing reagent; more modern vari-

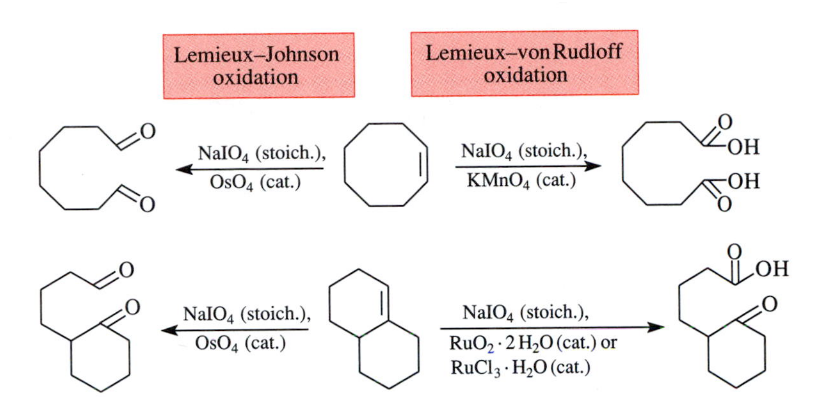

Fig. 14.20. Oxidative cleavages of alkenes with symmetrically (top) and asymmetrically (bottom) substituted C=C double bonds.

ants employ RuO_2 or $RuCl_3$. The *cis* hydroxylation is affected by the MnO_4^- anion in the first case and by RuO_4, formed *in situ,* in the other two cases. Both reactions formally resemble the corresponding OsO_4 reaction (Figure 14.16) and are mechanistically rather similar. The periodate subsequently cleaves the formed glycol to aldehydes and/or ketones (for the mechanism, see Figure 14.17). The MnO_4^- anion or RuO_4 further oxidizes any initially formed aldehydes to carboxylic acids via the aldehyde hydrate (cf. the mechanism shown in Figure 14.11).

Alkene C=C double bonds also can be cleaved by ozone. The mechanism of this reaction was discussed in connection with Figures 12.44 and 12.45, and it has been explained that

- mixtures of hydroperoxides and carbonyl compounds are formed in MeOH, and
- mixtures of tetroxane(s) and ketone(s) or mixtures of secondary ozonides are formed in CH_2Cl_2.

These findings are summarized in the upper part of Figure 14.21.

The ozonolyses need to be *terminated* with a redox reaction (or by the β-elimination of an acylated hydroperoxide as shown, e.g., in Figure 14.22, right). Each of these reactions breaks the weak O—O single bonds of the mentioned primary oxidation products, which are the hydroperoxides, the tetroxanes, or the secondary ozonides. The usual methods for the workup of an ozonolysis with a redox reaction are shown in the lower part of Figure 14.21:

- The use of $NaBH_4$ or $LiAlH_4$ results in the reduction of the O—O and C=O bonds in the primary products to alcohol groups.
- The use of Me_2S or Ph_3P or Zn/HOAc leaves the molecules containing a C=O bond unaffected, while the molecules containing an O—O bond are reduced to carbonyl compounds.
- The use of H_2O_2 leaves ketones intact and converts all initially formed aldehydes and their derivatives into carboxylic acids.

The ozonolysis of cyclohexene to 1,6-dioxygenated compounds is shown in Figure 14.22. Other cycloalkenes similarly deliver other 1,ω-dioxygenated cleavage products. With the three methods for workup (Figure 14.21), this ozonolysis provides access to

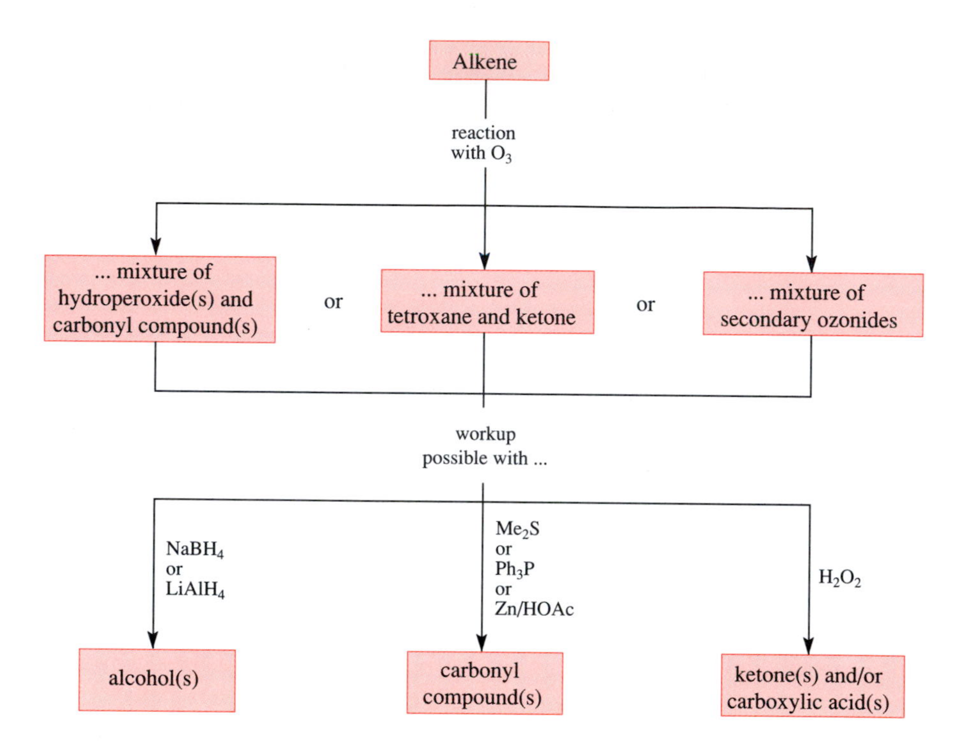

Fig. 14.21. Intermediates (center row) and final products (bottom row) of the ozonolysis of alkenes.

1,6-hexanediol, 1,6-hexanedial, or to 1,6-hexanedicarboxylic acid. Each of these compounds contains two functional groups of the same kind.

A

Cyclohexene also can be cleaved into a 1,6-dioxygenated C_6 chain with *different* termini, and this option is shown at the far right in Figure 14.22. Cyclohexene is subjected to ozonolysis in methanol to afford the hydroperoxide **A.** The peroxoacetate **B** is obtained by treatment of **A** with acetic anhydride/NEt_3 (mechanism analogous to Figure 6.2). NEt_3 might cause the elimination of HOAc from **B** in an E2-type fashion. One could also imagine that the peroxoacetate **B** undergoes a *cis*-elimination via the cyclic transition state shown (this transition state resembles the one of the Baeyer–Villiger rearrangement shown in Figure 11.33). In any case, the aldehyde ester **C** is formed as the desired elimination product aside from HOAc.

Oxidative Cleavage of Aromatic Compounds

A

The reagents that effect the oxidative cleavage of *alkene* C═C double bonds (Figures 14.20–14.22) in principle also are suitable for the cleavage of *aromatic* C═C double bonds (Figures 14.23–14.25). The mechanism is unchanged.

Ru(VIII) cleaves monoalkylbenzene to alkanecarboxylic acid (Figure 14.23). The Ru(VIII) is generated *in situ* by using a stoichiometric amount of NaOCl or $NaIO_4$ and a catalytic amount of RuO_2. The Ru(VIII) is continuously regenerated. The α-ketoaldehyde **A** is formed as the first key intermediate. Its two C═O bonds stem from the oxidative cleavage of the C═C bonds which the substrate contained in their

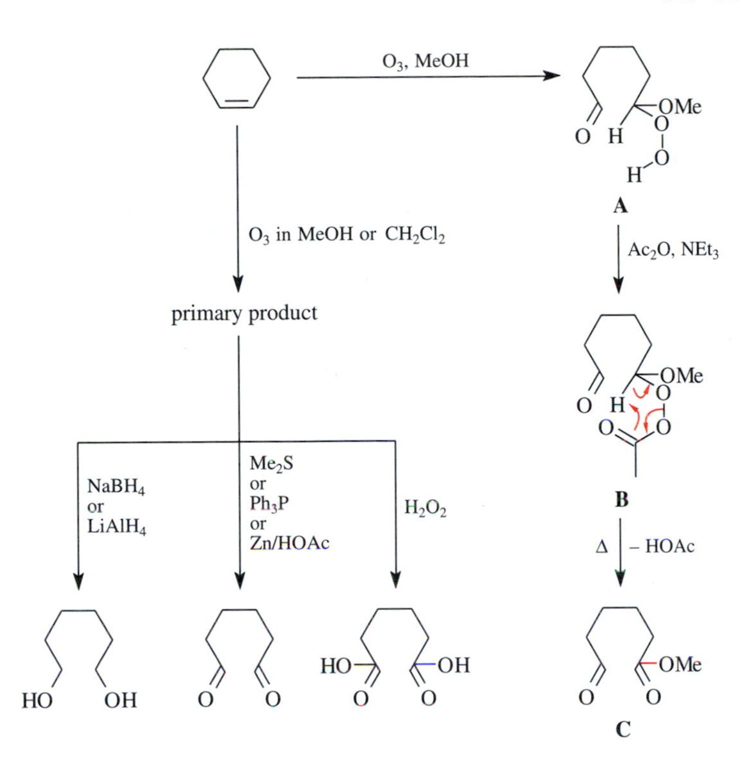

Fig. 14.22. Conversion of cyclohexene into symmetric cleavage products (left side) and into one asymmetric product (right side).

place. The further reaction of the α-ketoaldehyde **A** is not clear. It is possible that the dianion of the ruthenium(VIII) acid engages in a nucleophile addition to the C═O double bonds of **A.** In this way, the dianion of a ruthenium(VIII) acid diester would be formed. It next would be converted into the neutral ruthenium(VIII) acid diester **D** by twofold protonation. A plausible mechanistic alternative (Figure 14.23, right) to the conversion ketoaldehyde **A** $\rightarrow$ diester **D** starts with the energetically favorable formation of the aldehyde hydrate **B** by addition of water. The aldehyde hydrate **B** might then add to a Ru═O "double bond" of RuO_4 and form the acyclic ruthenium(VIII) acid *mono*ester **C.** The cyclic ruthenium(VIII) acid *di*ester **D** would be formed subsequently by ring-closure of the monoester.

The Ru(VIII) acid diester **D,** a key intermediate of the oxidative cleavage shown in Figure 14.23, fragments by way of a concerted shift of three valence electron pairs in analogy to the reactions of cyclic iodo(VII) acid diesters (Figure 14.17) and of cyclic lead(IV) acid diesters (Figure 14.18). One of the participating valence electron pairs becomes a nonbonding electron pair at Ru, and the oxidation number of Ru therefore is lowered from +8 to +6. A substituted acetic acid and formic acid are the organic products of this cleavage. The formic acid is readily oxidized further to carbon dioxide. Hence, an aliphatic carboxylic acid is the only oxidation product formed from the alkyl-substituted aromatic compound that remains in the reaction mixture.

Only *one* ring of condensed aromatic systems is cleaved with either of the two mixtures of oxidizing agents shown in Figure 14.23—stoichiometric $NaIO_4$/catalytic NaOCl

Fig. 14.23. The RuO_4 cleavage of a phenyl ring is exemplified by the case of an alkylated aromatic compound using the modified Lemieux–von Rudloff conditions (Figure 14.20). The reaction involves two key intermediates: the α-ketoaldehyde **A** and the ruthenium(VIII) acid diester **D**. The mechanism for the reaction of **A** to **D** is not known. Diester **D** fragments and forms a substituted acetic acid, RuO_3, and formic acid in the last step via a cyclic transition state. Under the reaction conditions, formic acid is oxidized further to CO_2.

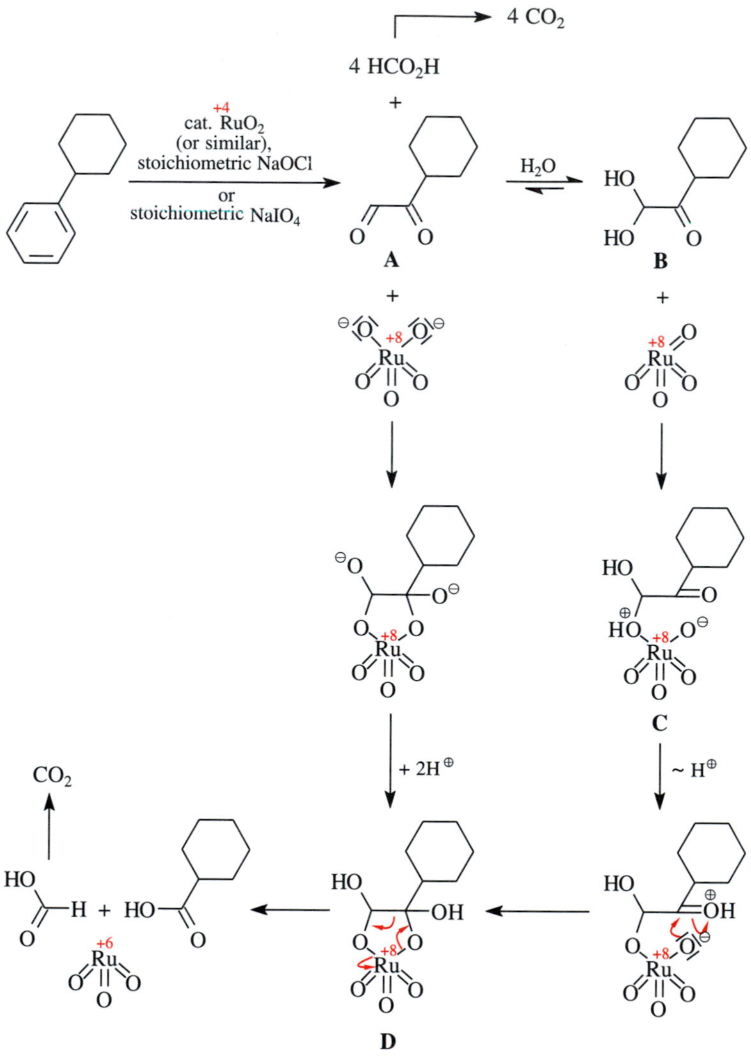

or stoichiometric $NaIO_4$/catalytic RuO_2. In the laboratory, naphthalene, for example, can be oxidized to obtain phthalic acid anhydride by using such a catalytic oxidation that features RuO_4 as the actual oxidizing reagent (Figure 14.24). The industrial variant of this oxidation uses air as the stoichiometric oxidant and a V_2O_5 catalyst. It is unknown whether the mechanisms of these two oxidations are analogous.

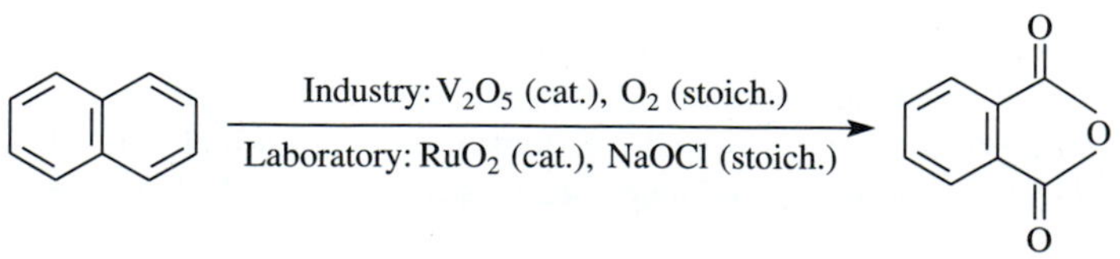

Fig. 14.24. Transition metal–mediated cleavage of naphthalene.

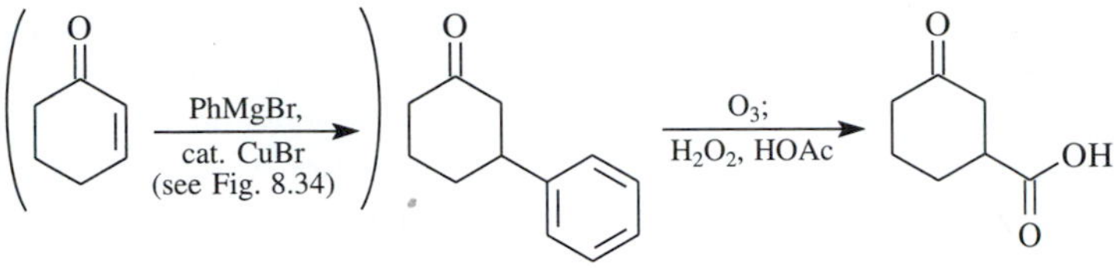

Fig. 14.25. Ozonolysis of a phenyl ring.

Ozone, too, can be used to break down a benzene ring into a carboxylic acid (Figure 14.25). In this case, the primary product of the ozonolysis is worked up under oxidizing conditions (cf. Figure 14.21). The fact that phenyl rings undergo ozonolysis is synthetically useful because it allows one to introduce a comparatively inert phenyl ring into a synthetic intermediate instead of a more reactive carboxylic acid. The carboxyl group can be generated subsequently, as illustrated by the example shown in Figure 14.25. In this context, the phenyl ring plays the role of a **masked** or **latent** carboxyl group.

Oxidative Cleavage of Ketones

B

The C_α—C(=O) bond of ketones can be cleaved with peracids, and esters are formed via a Baeyer–Villiger rearrangement and the insertion of an O atom (for mechanism, see Figures 11.31–11.34). Cyclic ketones produce lactones in this way (Figure 11.31). Cyclobutanones react in a rather similar fashion with H_2O_2 or *tert*-BuOOH and form γ-butyrolactones (example in Figure 11.35). The Baeyer–Villiger rearrangement of asymmetric ketones breaks the C_α—C(=O) bond that leads to the higher substituted C atom in the α-position (cf. Section 11.4.2). If this atom is a stereocenter, one observes complete retention of configuration (cf. Figure 11.31). The top reaction in Figure 14.26 illustrates the corresponding synthetic potential of the Baeyer–Villiger oxidation with the example of the regioselective and stereoselective cleavage of menthone.

The bottom reaction of Figure 14.26 shows an oxidative cleavage of the same menthone with the opposite regiochemistry. In a preparatory step, menthone is converted into the silyl enol ether **B** via its kinetic enolate (cf. Figure 10.9). The C=C double

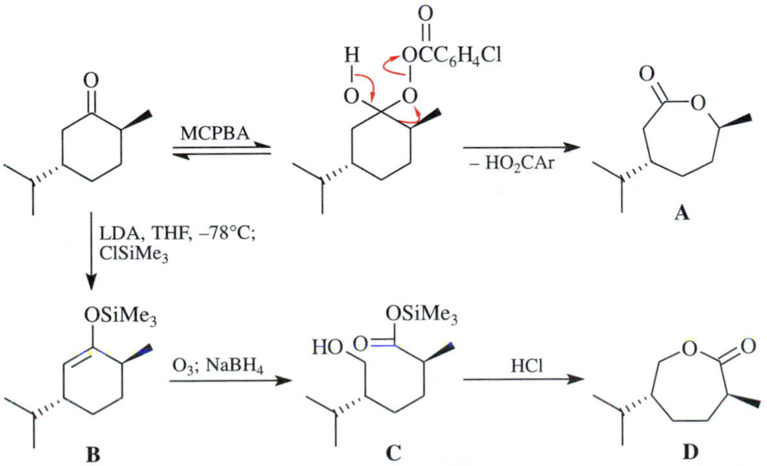

Fig. 14.26. Oxidative cleavage of an asymmetric ketone with complementary regioselectivities. Lactone **A** is obtained by Baeyer–Villiger oxidation of menthone [2-methyl-5-(1-methylethyl)cyclohexanone]. Alternatively, one may first convert menthone into the silyl enol ether **B** and cleave its C=C double bond with ozone to obtain a bifunctional compound, which exhibits a trimethylsilyl ester and an α-methoxyhydroperoxide. The latter is reduced with $NaBH_4$ to the hydroxylated silyl ester **C.** The hydroxycarboxylic acid is obtained therefrom by acid-catalyzed hydrolysis. It cyclizes spontaneously to give lactone **D.**

bond of this compound is subsequently cleaved with ozone. This cleavage (cf. Figures. 12.44 and 12.45) proceeds in MeOH as if the Me_3SiO substituent was not even present, and the ozonolysis results in an α-methoxyhydroperoxide that is a part of a trimethylsilyl ester. $NaBH_4$ is added to reduce the hydroperoxide to an alcohol. Of course, $NaBH_4$ does not attack the trimethylsilyl ester. The *reduced* cleavage product is thus the hydroxy-substituted silyl ester **C.** Upon acidic workup, ester **C** hydrolyzes, and the hydroxycarboxylic acid that is formed cyclizes rapidly to provide lactone **D** (cf. Figure 6.22). **D** is a constitutional isomer of lactone **A,** which was obtained earlier via the Baeyer–Villiger oxidation of menthone (cf. Figure 14.26, top).

A

Cyclic ketones also can be cleaved by oxidation with $KMnO_4$ (Figure 14.27). Under these conditions, 1,ω-ketocarboxylic acids or 1,ω-dicarboxylic acids are formed. The mechanism of this reaction is not known in detail. It is known only that $KMnO_4$ serves as a three-electron oxidizing reagent and that it is reduced to MnO_2.

The surmised mechanism is shown in Figure 14.27. It is assumed that the enolate, not the ketone itself, reacts with the permanganate. The MnO_4^- anion presumably attacks the enolate carbon electrophilically with one of its doubly bonded O atoms. The manganese(V) acid *mono*ester **C** is formed and is oxidized by a permanganate ion to

Fig. 14.27. Mechanism of the $KMnO_4$ cleavage of cyclohexanone to adipic acid.

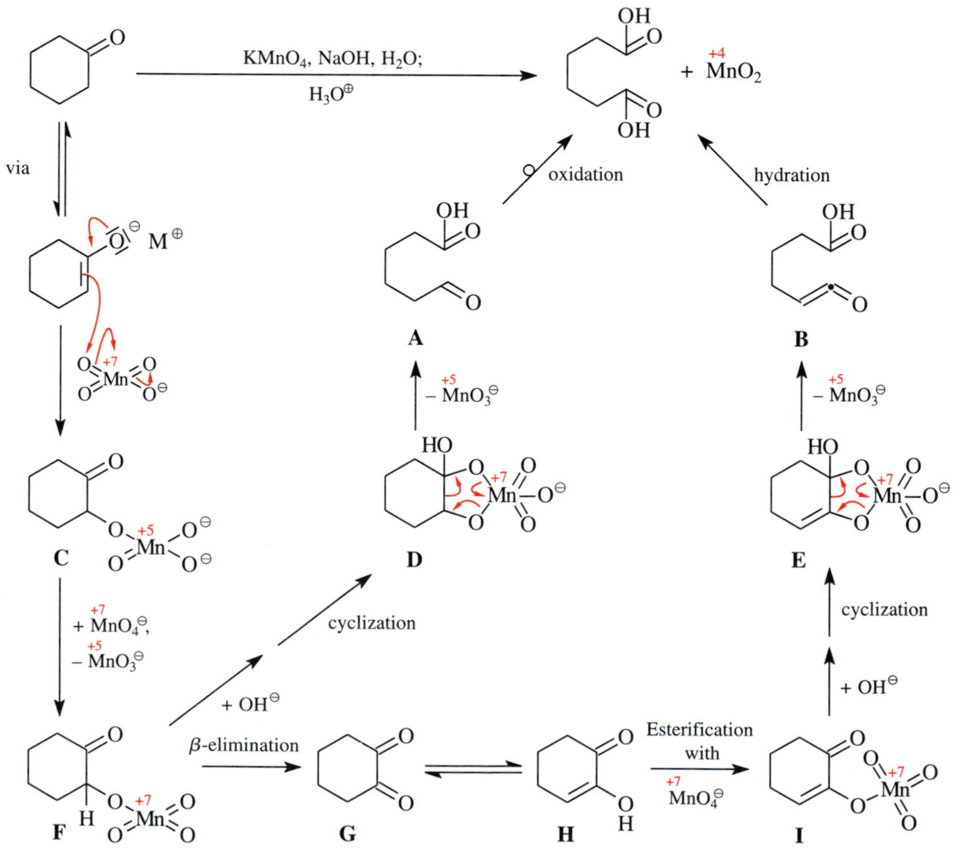

$KMnO_4$, NaOH, H_2O; $H_3O^{\oplus}$

+ $\overset{+4}{MnO_2}$

via

$M^{\oplus}$

oxidation

hydration

A

B

$-\overset{+5}{MnO_3^{\ominus}}$

$-\overset{+5}{MnO_3^{\ominus}}$

C

D

E

$+\overset{+7}{MnO_4^{\ominus}}$, $-\overset{+5}{MnO_3^{\ominus}}$

cyclization

cyclization

+ $OH^{\ominus}$

+ $OH^{\ominus}$

β-elimination

Esterification with $\overset{+7}{MnO_4^{\ominus}}$

F

G

H

I

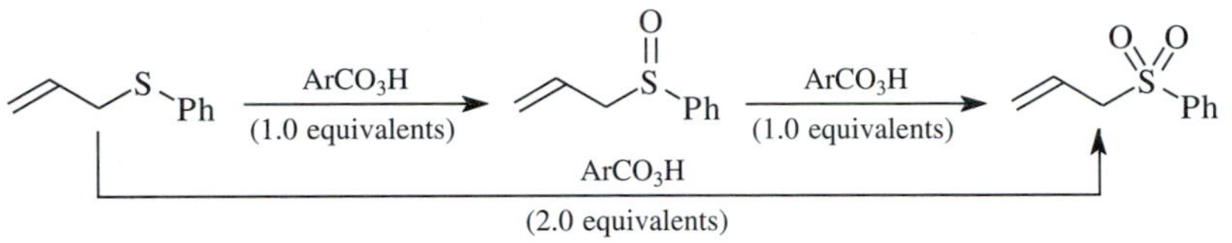

Fig. 14.28. Sulfide → sulfoxide → sulfone oxidation(s).

give the analogous manganese(VII) acid *mono*ester **F**. The subsequent chemistry of **F** is difficult to predict.

- **Variant 1.** The manganese(VII) acid monoester **F** is hydrated at the carbonyl group or at the metal. Subsequently the ring closes and leads to the manganese(VII) acid diester **D.** Its heterocyclic ring fragments by a concerted shift of three valence electron pairs, as in the cases of the esters of iodo(VII) acid, lead(VI) acid, and ruthenium(VIII) acid (Figures 14.17, 14.18, and 14.23). The aldehyde acid **A** is the organic product of this fragmentation reaction, and it is oxidized subsequently to the dicarboxylic acid.
- **Variant 2.** The manganese(VII) acid monoester **F** undergoes a β-elimination of a manganese(V) acid, and diketone **G** is formed. Compound **G** is in equilibrium with enol **H.** Enol **H** and a permanganate ion form the manganese(VII) enol ester **I** by way of a condensation reaction. Ester **I** reacts further to form the manganese(VII) acid diester **E.** Then **E** might undergo fragmentation to yield the ketene carboxylic acid **B.** The sequence **I** → **E** → **B** is step for step analogous to the sequence **F** → **D** → **A** of variant 1. Finally, the ketene carboxylic acid **B** would have to be hydrated to the dicarboxylic acid (for the mechanism, see Figure 7.2).

14.3.3 Oxidations at Heteroatoms

The oxidations shown in Figures 14.28–14.31 and the primary reaction in Figure 14.32 all involve the same pattern. In all cases, the heteroatom reacts as the nucleophile in an S_N2-type substitution reaction with an O atom of the O—O bond of H_2O_2 or a peracid.

Sulfoxides most commonly are prepared via the oxidation of sulfides with peracids, *tert*-BuOOH, or H_2O_2 (Figure 14.28). Sulfoxides are obtained in a chemoselective fashion if one equivalent of the respective oxidizing reagent is employed. Sulfones also can be obtained chemoselectively from the same sulfides if two equivalents of the oxidizing reagent are employed (Figure 14.28). Obviously, sulfoxides can be oxidized to give sulfones by any of the same oxidizing agents. All these reactions can be carried out in the presence of C=C double bonds.

The oxidation of selenides to selenium oxides (Figure 14.29) is faster than the oxidation of sulfides to sulfoxides. The former reaction also succeeds in the presence of

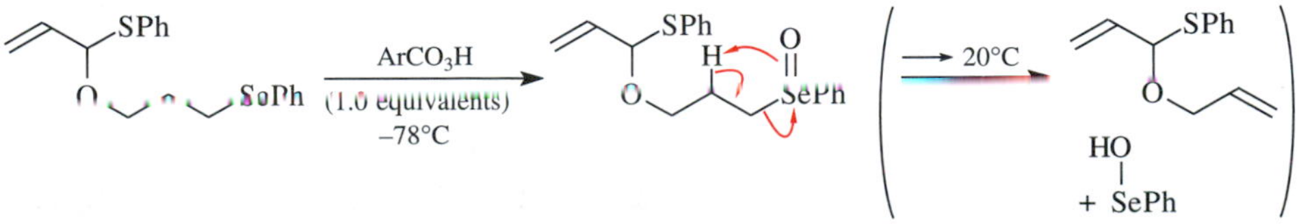

Fig. 14.29. A one-pot combination of a selenide → selenoxide oxidation and a selenoxide "pyrolysis" (see Figures 4.10–4.12 for the mechanism).

Fig. 14.30. A *tert*-amine → amine *N*-oxide oxidation as preparation of the oxidizing reagent NMO (see Figures. 14.14 and 14.16 for synthetic applications).

C=C double bonds, and even a sulfide group that might be present in the substrate will not be attacked.

Tertiary amines yield amine *N*-oxides in a similar fashion (Figure 14.30).

A

Hydrazone **A** in Figure 14.31 (the procedure in Figure 10.31 shows one possibility for the preparation) undergoes oxidation just like an amine, and amine *N*-oxide **B** is formed. Immediately, **B** undergoes a β-elimination via a cyclic transition state, and nitrile **C** and a hydroxylamine are formed. Since the hydrazone precursor is accessible as a pure enantiomer, the nitrile also can be generated as a pure enantiomer.

Fig. 14.31. A hydrazone → hydrazone *N*-oxide oxidation as part of a one-pot reaction for the conversion of SAMP hydrazones into enantiomerically pure nitriles.

Side Note 14.1
Preparation of TEMPO

Peroxo compounds initially oxidize secondary amines to the corresponding hydroxylamines (e.g., **A** in Figure 14.32) but the oxidation generally does not stop at this stage. The hydroxylamine loses the H atom of the hydroxyl group and a nitroxyl radical, a rather resonance-stabilized radical, is formed. The nitroxyl radical shown in Figure 14.32 is so stable that it can be stored. It is commercially available as the oxidizing agent TEMPO (**te**tra**m**ethyl**p**iperidine nitr**o**xyl free radical).

Fig. 14.32. A *sec*-amine → nitroxyl radical oxidation exemplified by the preparation of the oxidizing agent TEMPO (see Figure 14.15 for a synthetics application).

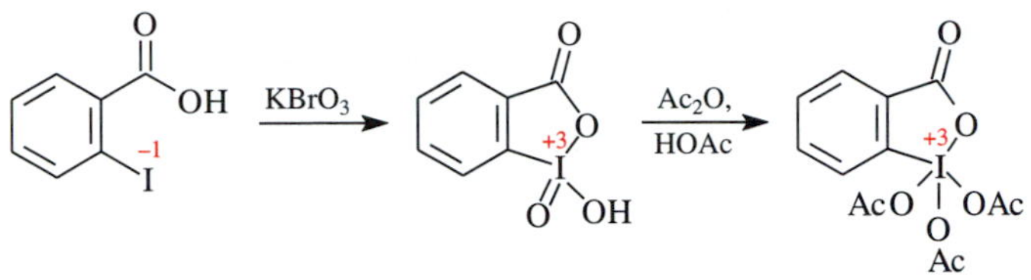

Fig. 14.33. A iodide → periodinane oxidation exemplified by the preparation of the Dess–Martin reagent (see Figure 14.13 for a synthetics application).

The oxidation of *ortho*-iodobenzoic acid to give the **Dess–Martin reagent** is shown in Figure 14.33. This oxidation is a synthetically important oxidation of an iodine atom. The mechanism of the oxidation is not yet understood.

As our last examples of oxidation reactions of heteroatoms, we consider the reactions depicted in Figure 14.34. These reactions are oxidations and, in contrast to the reactions in Figure 14.28–14.32, these oxidations also are condensation reactions, since the oxidizing reagent remains in the product. The mechanistic details outlined in Figure 14.34 are so familiar by now that no further explanation is needed.

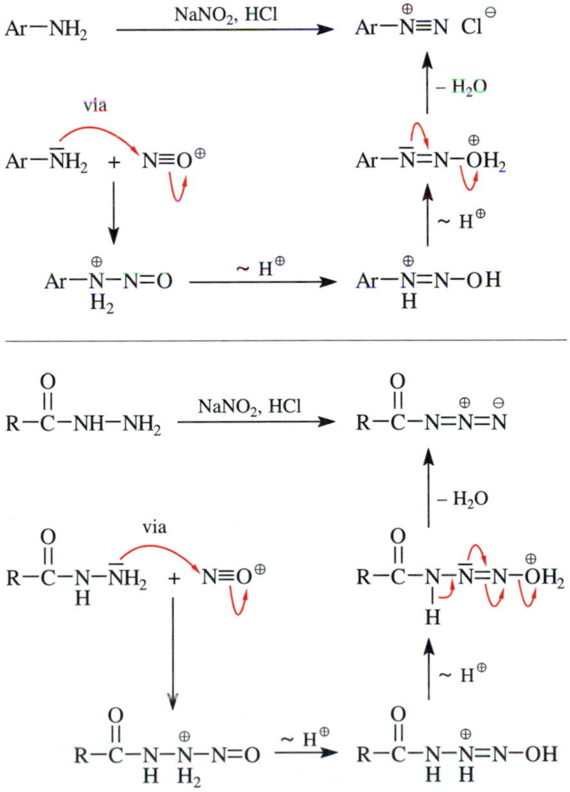

B

Fig. 14.34. Comproportionation of two nitrogen compounds to give a diazonium ion (top) or an acyl azide (bottom), respectively.

14.4 Reductions

B

In inorganic chemistry, the term **reduction** indicates a process in which a substrate absorbs electrons. Of course, the same is true in organic chemistry as well. Additional orientation is provided by the classes of compounds compiled in Tables 14.1 and 14.2. Because of their particular order, reductions can be described as "transformations that convert any given compound into a compound in a column further to the left." As the tables reveal, reductions of organic chemical compounds often are associated with a net-uptake of hydrogen. In accordance with these statements, there are the following reducing reagents in organic chemistry:

- electron donors (metals, which dissolve in suitable solvents in the presence or absence of a proton donor);
- elemental hydrogen (in catalytic hydrogenations or hydrogenolyses, respectively);
- H-atom transfer reagents [Bu_3SnH, $(Me_3Si)_3SiH$; see Section 1.9]; and
- reagents that transfer nucleophilic hydrogen.

The kinds of reagents that belong to the last group of reducing reagents were mostly already discussed in Section 8.1:

- covalent neutral metal hydrides such as BH_3, DIBAL, or Et_3SiH (in the special case of the reduction of carboxonium ions or benzyl cations; not yet discussed reductant);
- soluble ionic complex metal hydrides derived from tetravalent boron or from tetravalent aluminum; and
- organometallic compounds that contain a β H-atom that can be transferred onto an organic substrate.

14.4.1 Reductions R_{sp^3}—X → R_{sp^3}—H or R_{sp^3}—X → R_{sp^3}—M

B

Primary and secondary alkyl bromides, iodides, and sulfonates can be reduced to the corresponding alkanes with $LiBHEt_3$ **(superhydride)** or with $LiAlH_4$. If such a reaction occurs at a stereocenter, the reaction proceeds with substantial or often even complete stereoselectivity via backside attack by the hydride transfer reagent. The reduction of alkyl chlorides to alkanes is much easier with superhydride than with $LiAlH_4$. The same is true for sterically hindered halides and sulfonates:

$$\text{Bu}-\text{C}(\text{Me})_2-\text{CH}_2-\text{OTs} \xrightarrow{\text{LiBHEt}_3} \text{Bu}-\text{C}(\text{Me})_2-\text{CH}_3$$

Several options can be considered when it comes to the mechanisms of these reductions and of other reductions with complex metal hydrides. It is convenient to imagine that a hydrogen atom with hydride character is detached from the reducing agent in the transition state. However, $LiAlH_4$ seemingly is also capable of effecting a single electron transfer onto organic substrates.

Fig. 14.35. Reduction of epoxides. The sterically less hindered C—O bond is attacked regioselectively.

The same reducing agents $LiBHEt_3$ and $LiAlH_4$ also react with epoxides in S_N2-type reactions converting them into alcohols (Figure 14.35). The sterically less hindered C—O bond is attacked in asymmetric epoxides regioselectively.

A

Epoxy alcohols (**A** in Figure 14.36) are accessible in enantiomerically pure form via the Sharpless oxidation (Figure 3.29). The epoxide substructure of these compounds can be reduced to obtain an alcohol in a regiocontrolled fashion being part of either an enantiomerically pure 1,3-diol, **B**, or an enantiomerically pure 1,2-diol, **C** (Figure 14.36). The 1,3-diols **B** and the 1,2-diols **C** are best accessible via reduction of **A** with Red-Al [$NaAlH_2(OCH_2CH_2\,OCH_3)_2$] or DIBAL, respectively.

Red-Al first generates one equivalent of hydrogen gas from such epoxy alcohols **A** (Figure 14.36). An O—Al bond forms in the resulting trialkoxyaluminate **D.** The epoxy fragment in **D** then is reduced via an intramolecular reaction. The transfer of a hydride ion from aluminum leads selectively to the formation of a 1,3-diol, since the approach path that would lead to the 1,2-diol cannot be collinear to the C—O bond which would have to be broken (cf. Section 2.4.3).

The treatment of epoxy alcohols **A** with DIBAL also first liberates one equivalent of H_2 (Figure 14.36, right). An O—Al bond is formed, which in the resulting interme-

Red-Al ⟶	99.3	:	0.7
DIBAL ⟶	7	:	93

Fig. 14.36. Regioselective reduction of enantiomerically pure epoxy alcohols to enantiomerically pure diols. 1,3-Diols are formed with Red-Al and 1,2-diols are formed with DIBAL.

diate **E** is part of a chelating ring. This intermediate can be reduced only in an *intermolecular* fashion, since it does not contain any hydridic hydrogen. The binding of the epoxy oxygen to the Al atom in the chelate should remain intact as much as possible during this reaction, and it is for this reason that the 1,2-diol is formed preferentially.

B

Tertiary iodides and bromides usually are reduced (defunctionalized) with Bu_3SnH or $(Me_3Si)_3SiH$ via a radical chain reaction, as discussed in Section 1.9. Primary and secondary alkyl iodides, bromides, and xanthogenates can be reduced with the same reagents under the same conditions via the same radical mechanism.

Primary, secondary, and tertiary alkyl halides also can be reduced with dissolving metals. The primary reduction product is an organometallic compound. Whether the latter is formed quantitatively or whether prior to that, it is converted into the corresponding hydrocarbon by protonation depends on the solvent. The organometallic compound is stable in aprotic solvents (hexane, ether, THF), while it is protonated in protic solvents (HOAc, alcohols).

The reduction of alkyl halides to organometallic compounds in aprotic solvents involves a heterogeneous reaction on the metal surface. This metal surface must be pure metal if it is to react efficiently in the desired way. If the surface of the metal has reacted with oxygen (Li, Mg, Zn) or even with nitrogen (Li), one must first remove the metal oxide or lithium nitride layer. This can be accomplished in the following ways:

- mechanically (stirring of Mg shavings overnight; pressing of Li through a fine nozzle into an inert solvent),
- chemically (etching of Mg shavings with I_2 or 1,2-dibromoethylene; etching of Zn powder with Me_3SiCl),
- or by an *in situ* preparation of the respective metal via reduction of a solution of one of its salts under an inert atmosphere [$MgCl_2$ + lithium naphthalenide → "Rieke-Mg"].

The mechanism of the dissolving metal reduction of alkyl halides presumably is the same for reductions with Li, Mg, or Zn. Alkenyl and aryl halides generally also can be reduced in this fashion via the corresponding alkenyl or aryl organometallic compounds. The mechanism shown in Figure 14.37 for the reduction of MeI by Mg to give a Grignard compound is representative; the individual steps of this reaction are described in the figure caption.

A

Alkyllithium compounds can be prepared in a heterogeneous reaction between the organic halide and lithium metal (Figure 14.38, center). This is similar to the preparations of Grignard compounds (Figure 14.38, top) and alkylzinc iodide (Figure 14.38, bottom). However, a heterogeneous reaction is not the only way to make alkyllithium compounds. They also can be prepared by homogeneous reactions of alkyl phenyl sulfides or alkyl chlorides (Figure 14.39). This method for the preparation of alkyllithium compounds (and other organolithium compounds as well, cf. Figure 14.40) is known as **reductive lithiation.** Independent of the substrate, the reducing agent is the soluble lithium salt of a radical anion derived from naphthalene, 1-(dimethylamino)naphthalene, or 4,4′-di-*tert*-butylbiphenyl.

The first step in the reductive lithiation of alkyl phenyl sulfides **(Cohen process)** consists of preparing the reducing reagent in stoichiometric amounts. Lithium naphthalenide, for example, is made from lithium and naphthalene. The sulfide is added

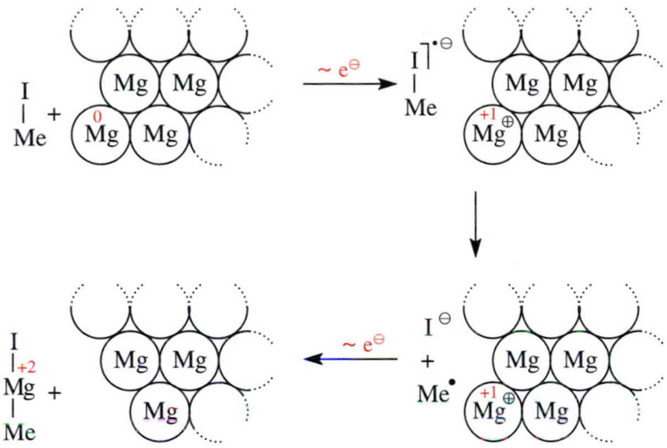

Fig. 14.37. Mechanism of the formation of a Grignard compound; $\sim e^-$ indicates electron migration. The reaction is initiated by an electron transfer from the metal to the substrate. The extra electron occupies the σ^*(C—I) orbital, whereby the C—I bond is weakened and breaks. This cleavage leads to the formation of a methyl radical and an iodide ion on the metal surface. In the third step of the reaction, the valence electron septet of the methyl radical is converted into an octet by formation of a covalent bond between the methyl radical and a metal radical. The Grignard reagent is thus formed.

dropwise to this reducing reagent. The mechanism of reduction corresponds step by step to the one outlined in Figure 14.37 except that the dissolved radical anion is the source of the electrons, which are transferred as opposed to a metal surface. Furthermore, a C_{sp^3}—S bond breaks instead of a C_{sp^3}—I bond. In the reductive lithiation of alkyl chlorides **(Screttas–Yus process),** the radical anion is generated *in situ* and only in catalytic amounts, starting with a stoichiometric amount of Li powder and a few molepercent of di-*tert*-butylbiphenyl (possibility for preparation: Section 5.2.5). The continuously generated lithium di-*tert*-butylbiphenylide reduces the alkyl chloride. This reaction and the reductive lithiation of alkyl phenyl sulfides follow the same mechanism.

Lithium di-*tert*-butylbiphenylide in homogeneous solution is a *very* strong reducing reagent. This reagent allows for easy metallations in cases that are more difficult with metallic lithium and would be impossible to accomplish with magnesium. The reduc-

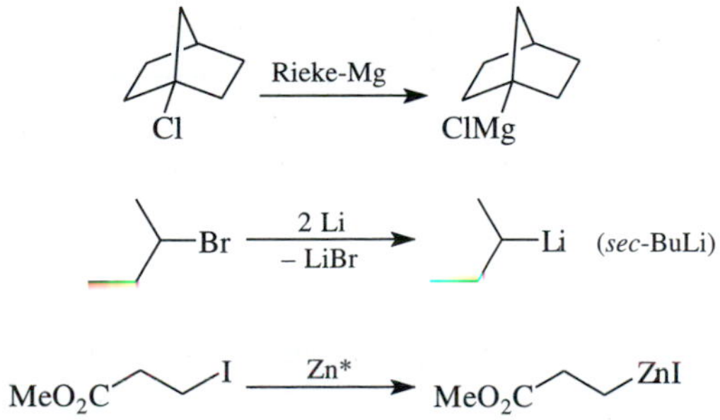

Fig. 14.38. Heterogeneous reduction of alkyl halides to Grignard compounds, organolithium compounds, and—possibly functionalized—organozinc compounds; Zn* refers to surface-activated metallic zinc.

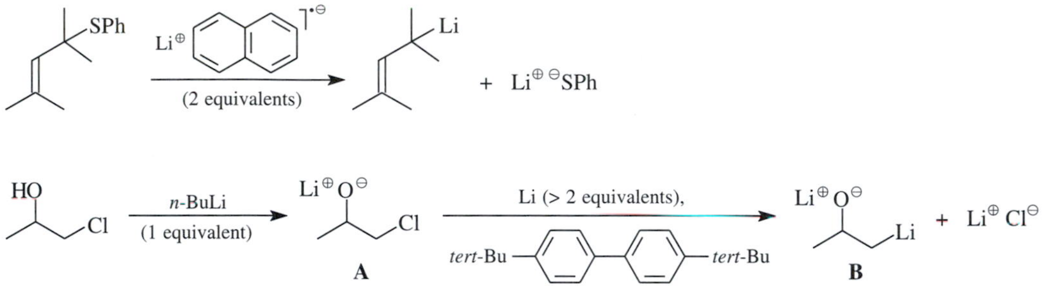

Fig. 14.39. Reductive lithiation of an alkyl aryl sulfide (Cohen process) and an alkyl chloride (Screttas–Yus process).

tion of chloride **A** in Figure 14.39 provides a good example. **A** is a rather weak electron acceptor because it is an alkoxide, that is, an anion.

The reducing power of lithium di-*tert*-butylbiphenylide is so high that it is even capable of the reductive lithiation of a C_{sp^2}—Cl bond (Figure 14.40; C_{sp^2}—Cl bonds are stronger than C_{sp^3}—Cl bonds according to Section 1.2). In the example given, a lithium (dialkylamino)carbonyl compound **A** is formed that *is not accessible in any other way*. If one generates the organolithium compound **A** in this way and in *the presence of a carbonyl compound,* the former adds immediately to the C=O double bond of the latter. For example, **A** reacts *in situ* with benzaldehyde to give alcohol **B.** The generation of organometallic compounds from halides in the presence of a carbonyl compound followed by an *in situ* reaction of these species with each other is referred to as the **Barbier reaction.**

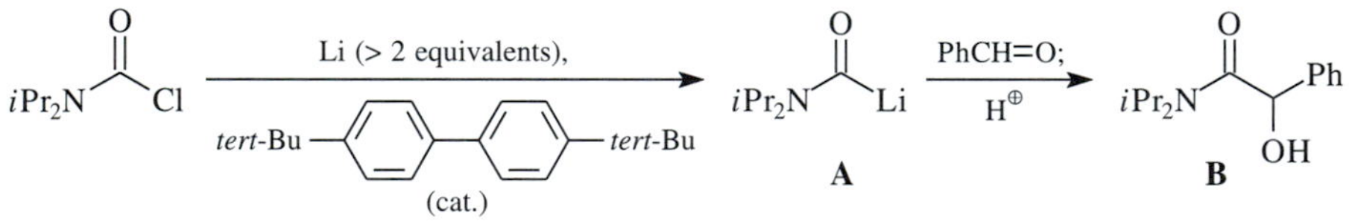

Fig. 14.40. Reductive lithiation of a carbamoyl chloride to the (dialkylamino)carbonyl lithium compound **A** and its immediately following reaction with a carbonyl compound (Barbier reaction) leading to alcohol **B.**

In some cases, a metal reduces a C_{sp^3}—heteroatom bond faster to a C_{sp^3}-M bond than it reduces the hydrogen of an OH group to H_2 and, in these cases, protic solvents can be used. Instead of the organometallic compound, one then obtains the product resulting from its protonation. This kind of reduction is used, for example,

- in the defunctionalization of alkyl phenyl sulfones with sodium amalgam in MeOH. (Figure 14.41; another reaction of the intermediate formed, an organo-sodium compound, occurs in the Julia–Lythgoe olefination because there is a leaving group in the β-position relative to the metallated center (cf. Figure 4.37)];
- in the dehalogenation of 2,2-dichlorocyclobutanones (Figure 14.42, top) with Zn/HOAc;
- in the dehalogenation of dichlorocyclopropane (Section 3.3.1) with Na/*tert*-BuOH;

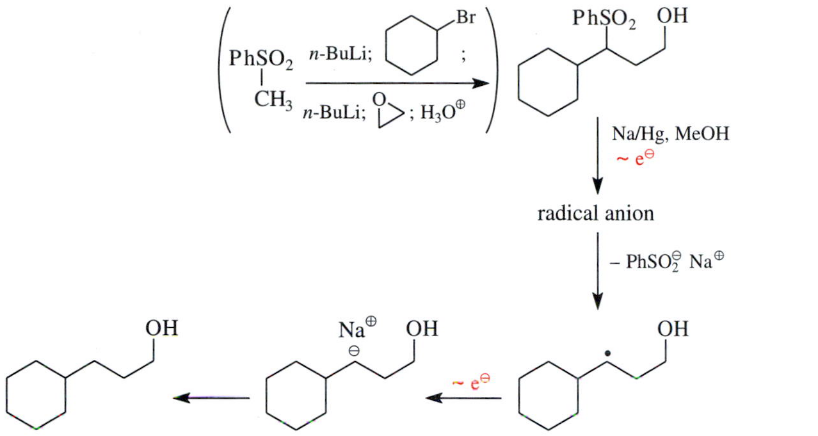

PhSO2 / CH3 — n-BuLi; (bromocyclohexane); n-BuLi; (ethylene oxide); $H_3O^{\oplus}$ → PhSO2, OH

Na/Hg, MeOH ~ $e^{\ominus}$

radical anion

– $PhSO_2^{\ominus}$ $Na^{\oplus}$

OH ← ~ $e^{\ominus}$ ← $Na^{\oplus}$ ⊖ OH ← OH

Fig. 14.41. The reduction of an alkyl aryl sulfone to an alkane. The preparation of the alkyl aryl sulfone (top) is followed by a reduction with Na amalgam (right and bottom).

- and in the deoxygenation of acyloin to ketones (Figure 14.42, bottom) with Zn/HCl.

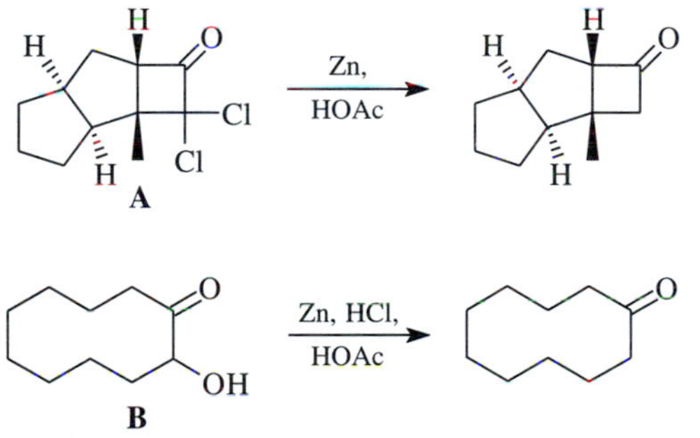

H, H, O, Cl, Cl, H — **A** — Zn, HOAc → H, H, O, H

O, OH — **B** — Zn, HCl, HOAc → O

Fig. 14.42. Reductions of α-heterosubstituted ketones to α-unsubstituted ketones. See Figures 12.32 and 14.51 for the preparation of compounds **A** and **B**, respectively.

The C_{sp^3}—O bonds of benzyl alcohols, benzyl ethers, benzyl esters, and benzyl carbamates also can be reduced to a C—H bond (Figures 14.43 and 14.44). Lithium or

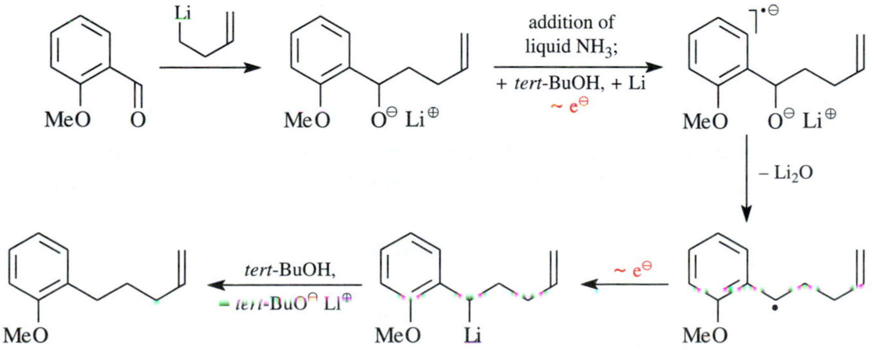

MeO, O — Li (butenyl) → MeO, $O^{\ominus}$ $Li^{\oplus}$ — addition of liquid NH_3; + *tert*-BuOH, + Li ~ $e^{\ominus}$ → MeO, $O^{\ominus}$ $Li^{\oplus}$ (radical anion $]^{\bullet\ominus}$)

– Li_2O

MeO ← *tert*-BuOH, – *tert*-$BuO^{\ominus}$ $Li^{\oplus}$ ← MeO, Li ← ~ $e^{\ominus}$ ← MeO (radical)

Fig. 14.43. One-pot synthesis of an alkyl-substituted aromatic compound that involves a dissolving lithium reduction of a benzyl alkoxide.

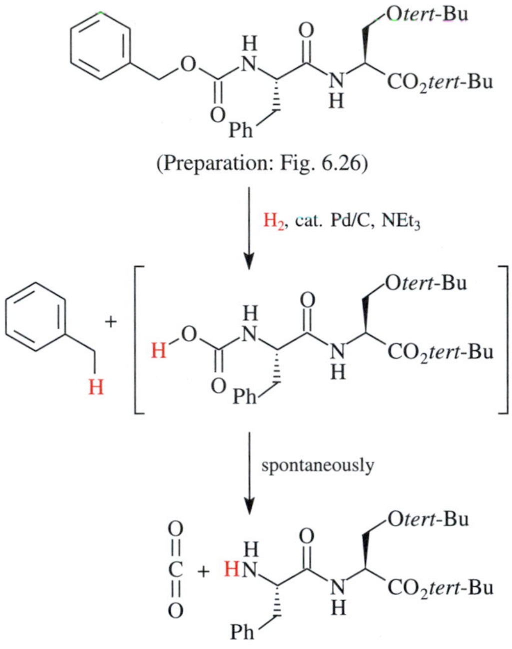

Fig. 14.44. The *O*-benzylcarbamate → toluene reduction for the removal of a protecting group.

(Preparation: Fig. 6.26)

H_2, cat. Pd/C, NEt_3

spontaneously

sodium in liquid ammonia are good reducing agents for this purpose. One usually adds an alcohol, such as *tert*-BuOH, as a weak proton source. Lithium and sodium dissolve in liquid ammonia. The resulting solutions contain metal cations and solvated electrons. Hence, it is assumed that the electrons are not transferred from the solid metal onto the substrate but that instead the electron transfer occurs in homogeneous solution.

Figure 14.43 shows how a benzyl alkoxide can be reduced in this way. Here, the reduction is part of the synthesis of an alkylated aromatic system. The target molecule contains a primary alkyl group. It is not possible to introduce this alkyl group by way of a Friedel–Crafts alkylation into the *ortho*-position of anisol in a regioselective fashion and without rearrangements (cf. Section 5.2.5).

B

The reduction of a benzylic C—O bond to a C—H bond often is employed to remove benzyl-containing protecting groups in benzyl ethers, benzyl carbonates, or *O*-benzyl carbamates. One can use the procedure shown in Figure 14.43 and react the respective compound with Li or Na in an NH_3/*tert*-BuOH mixture. On the other hand, benzylic C—O bonds of the substrates mentioned can be cleaved via Pd-catalyzed hydrogenolyses. Figure 14.44 presents an example of the removal of a benzyloxycarbonyl group from an amino group in this way. The cleavage product is a carbamic acid. Such an acid decarboxylates spontaneously (cf. Section 7.1.2) and produces the amine as the final cleavage product.

14.4.2 One-Electron Reductions of Carbonyl Compounds and Esters; Reductive Coupling

Dissolving metals initially convert aldehydes, ketones, and esters into radical anions. Subsequently, proton donors may react with the latter, which leads to neutral radicals. This mode of reaction is used, for example, in the drying of THF or ether with potassium in the presence of the indicator benzophenone. Potassium and benzophenone react to give the deep-blue potassium ketyl radical anion **A** (Figure 14.45). Water then protonates ketyl **A** to the hydroxylated radical **B** as long as traces of water remain. Further potassium reduces **B** via another electron transfer to the hydroxysubstituted organopotassium compound **C. C** immediately tautomerizes to the potassium alkoxide **D.** Once all the water has been consumed, no newly formed ketyl **A** can be protonated so that its blue color indicates that drying is complete.

In the drying of THF or ether (Figure 14.45), the sequence ketone → ketyl → hydroxylated radical → hydroxylated organometallic compound → alkoxide is of course not intended to convert *all* the ketone into "product." The reaction depicted in Figure 14.46 features the same sequence of steps as Figure 14.45 (and there is thus no need to discuss their mechanism again). In the reaction of Figure 14.46, however, the reaction is intended to run to *completion* until all of the ketone has been consumed. The reason for this is that it is the purpose of this reaction to reduce the ketone to the alcohol. The

Fig. 14.45. Chemistry of the drying of THF or ether with potassium and benzophenone featuring the ketone → ketyl reduction and the trapping reaction of the ketyl with residual water.

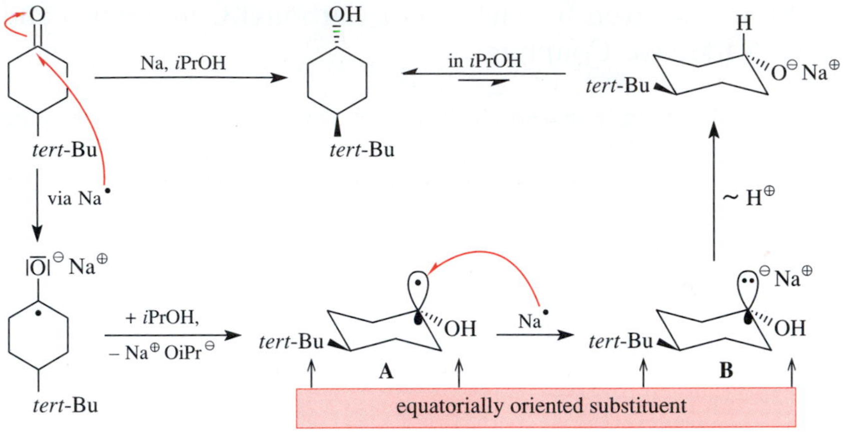

Fig. 14.46. Diastereoselective reduction of a cyclohexanone with dissolving sodium.

substrate in Figure 14.46 is a conformationally fixed cyclohexanone and the reducing agent is sodium, which dissolves in isopropanol. Playing a dual role, this solvent also acts as a proton source. Since the supply of this proton source is unlimited in this case, *all* of the ketone is converted into alkoxide.

Interestingly, the reduction shown in Figure 14.46 is highly diastereoselective. Only the *trans*-configured cyclohexanol is formed, that is, the equatorial alcohol. Such a level of diastereoselectivity cannot be achieved with hydride transfer reagents (cf. Figure 8.8).

The diastereoselectivity of the reduction depicted in Figure 14.46 is determined when the hydroxylated radical **A** is reduced to the hydroxylated organosodium compound **B.** For steric reasons, the OH group assumes a pseudo-equatorial position in the trivalent and moderately pyramidalized C atom of the radical center of the cyclohexyl ring of intermediate **A.** Consequently, the unpaired electron at that C atom occupies a pseudo–axially oriented AO. This preferred geometry is fostered and settled once and for all with the second electron transfer. It gives rise to the organosodium compound **B. B** isomerizes immediately to afford the equatorial sodium alkoxide.

In reactions of carbonyl compounds other than the ones shown in Figures 14.45 and 14.46 with nonprecious metals, initially, ketyls are formed as well. However, the Mg ketyl of acetone (Figure 14.47) is sterically much less hindered than the K ketyl of benzophenone (Figure 14.45). The former therefore dimerizes while the latter does not.

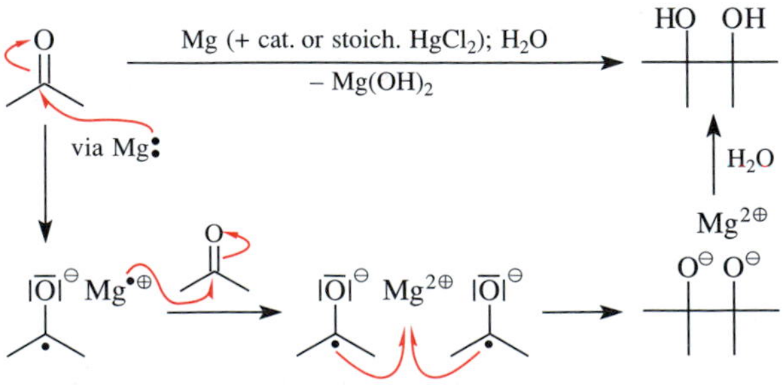

Fig. 14.47. Reductive dimerization of acetone (pinacol coupling).

The magnesium salt of glycol is thus formed. It yields a glycol with the trival name pinacol upon workup with weak acid. Hence, this kind of reductive coupling of carbonyl compounds is called **pinacol coupling**.

A

Carbonyl compounds also can undergo a reductive coupling with so-called low-valent titanium, that is, titanium compounds in which Ti possesses an oxidation number between 0 and +2 (Figure 14.48). Depending on the temperature, different products are formed. A pinacol coupling occurs at *low* temperature and leads to a heterocyclic monometal salt (as shown in Figure 14.49 [top]) or to an acyclic dimetal salt (not shown) of a glycol. The glycol can be isolated after aqueous workup.

The special significance of the Ti-mediated pinacol coupling of carbonyl compounds is that intramolecular ring-closure reactions succeed with outstanding yields (e.g., Figure 14.48, top). Rings with sizes from 3 to more than 20 C atoms can be formed equally well. These coupling reactions have only one disadvantage, namely, that the glycols are formed without simple diastereoselectivity. That is, they are formed as mixtures of diastereoisomers. The coupling shown in Figure 14.48, for example, yields the *cis*- and *trans*-dihydroxylated 14-membered rings in a 30:70 ratio.

At *higher* temperatures, the same carbonyl compounds and the same "low-valent titanium" yield alkenes instead of glycols. Dicarbonyl compounds cyclize under these conditions giving cycloalkenes (e.g., Figure 14.48, bottom). This type of reductive coupling is called a **McMurry reaction.** As with the Ti-mediated glycol formation, this reaction generally is not diastereoselective. The 14-membered cycloalkenes of Figure 14.48, for example, are formed in a 10:90 *cis*:*trans* ratio.

The diastereoselectivities for the Ti-mediated glycol vs alkene formations from the dialdehyde of Figure 14.48 are rather different. *Yet, the Ti glycolate intermediates of the first coupling mode are converted into the alkenes of the second coupling mode upon heating.* That nonetheless this discrepancy of diastereoselectivities is observed is in

OH OH 14 + 14 OH OH
trans 70 : 30 *cis*
lower temperatures
O 14 O + $TiCl_3$ (prereduced by Zn/Cu couple)
higher temperatures
14 + 14
cis *trans*
10 : 90

Fig. 14.48. Reductive coupling of a dicarbonyl compound to afford diastereomeric glycols or diastereomeric alkenes (McMurry reaction).

perfect agreement with the mechanism of the part of the McMurry reaction shown in Figure 14.49. Remember that upon protonation, the titanium glycolates **A** and **B** of Figure 14.49 furnish the *trans*- and *cis*-diols, respectively. Therefore, according to the relative yields given in Figure 14.48, these glycolates **A** and **B** must be formed in a 30 : 70 ratio. At sufficiently high temperatures, the same glycolates decompose via the homolytic cleavage of *one* of their C—O bonds. Both glycolates react in this fashion to deliver one and the same radical intermediate **C.** The only C—O bond left in this radical also breaks homolytically, and spin pairing leads to the formation of the C═C double bond of the resulting alkene. The rotation about the C—C(O) bond in the intermediate **C** is essentially free. This explains why the alkene, the McMurry product, is not formed as a pure stereoisomer and why its double bond configuration is independent of the stereostructure of the initially present titanium glycolates.

Even C═O groups of esters (Figure 14.50, top) and amides (Figure 14.50, bottom) can undergo intramolecular McMurry reactions with ketone carbonyl groups if they are exposed to "low-valent titanium" prepared in a slightly different way. The mechanism of these reactions corresponds by and large to the mechanism of the aldehyde/aldehyde coupling (Figures 14.48 and 14.49).

The McMurry cyclizations shown in Figure 14.50 are quite interesting from a synthetic point of view. The cyclization product **B** of ketoester **A** is a 10-membered enol

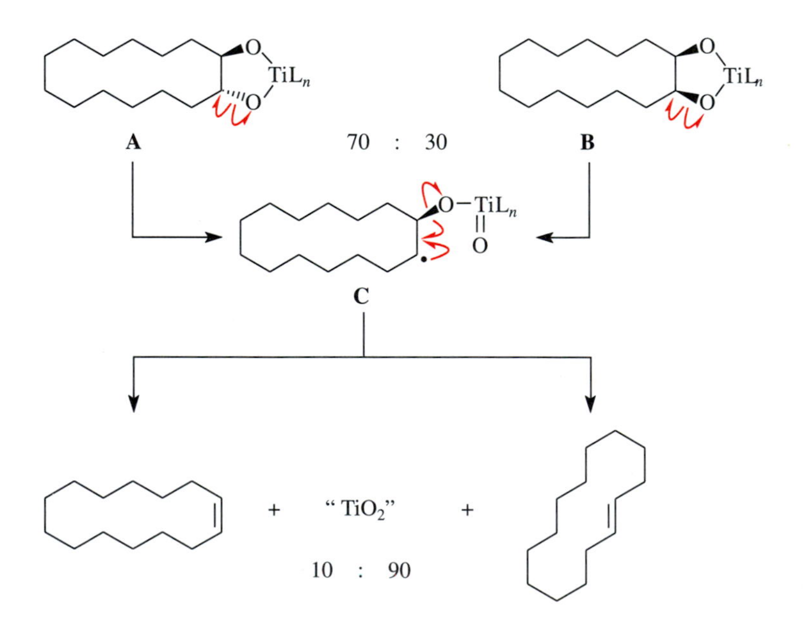

Fig. 14.49. Mechanism of the McMurry reaction. The heterocyclic monotitanium glycolates **A** and **B** or analogous dititanium glycolates decompose at higher temperatures via heterolytic cleavage of one of their C—O bonds and form the radical intermediate **C.** The alkene is formed by cleavage of the second C—O bond. This alkene is not obtained as a single stereoisomer because of the free rotation about the C—C(O) bond in the radical intermediate **C.** Note that the alkene is formed with a *cis,trans*-selectivity that is independent of the configuration of the titanium glycolate precursor(s).

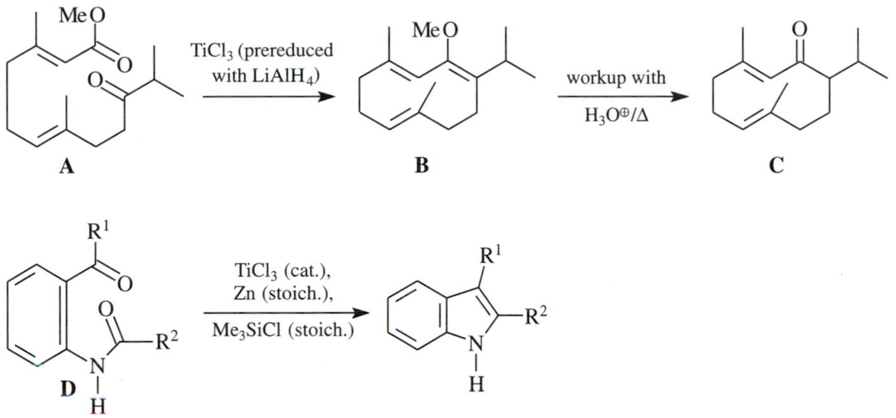

Fig. 14.50. Crossed McMurry reactions: ketone/ester coupling (top) and ketone/amide coupling (bottom).

ether. It can be hydrolyzed by acid to give ketone **C,** which itself *cannot* be the product of a McMurry reaction. The cyclization of ketoamide **D** illustrates one of the most versatile syntheses of 2,3-disubstituted indoles (**Fürstner indole synthesis**).

B

Carboxylic esters also can be reduced with dissolving sodium (Figure 14.51). Very different products are obtained depending on whether the reduction is carried out in ethanol or xylene. The reaction of esters with sodium in ethanol is referred to as the **Bouveault–Blanc reaction.** Prior to the discovery of complex metal hydrides, this reaction was the only method for the reduction of esters to alcohols. The *diester* shown in Figure 14.51 produces a *diol* in this way.

The mechanism of the Bouveault–Blanc reduction is shown in rows 1 and 2 of Figure 14.51. It starts with the sequence ester → radical anion **C** → hydroxylated radical **D** → hydroxylated organosodium compound **B** → hemiacetal anion **A.** This sequence is completely analogous to the sequence ketone → ketyl → hydrox-ylated radical **A** → hydroxylated organosodium compound **B** → sodium alkoxide that occurs in the reduction of a ketone with Na in *i*PrOH (Figure 14.46).

Intermediate **A** of the Bouveault–Blanc reduction of Figure 14.51 is not a simple alkoxide but rather the anion of a hemiacetal. Accordingly, it decomposes into an alkoxide anion and an aldehyde. In the further course of the Bouveault–Blanc reduction, this aldehyde is reduced by Na/EtOH just as the ketone of Figure 14.46 is reduced by Na/*i*PrOH.

The so-called **acyloin condensation** consists of the reduction of esters—and the reduction of diesters in particular—with sodium in xylene. The reaction mechanism of this condensation is shown in rows 2–4 of Figure 14.51. Only the first of these intermediates, radical anion **C,** occurs as an intermediate in the Bouveault–Blanc reduction as well. In xylene, of course, the radical anion **C** cannot be protonated. As a consequence, it persists until the second ester also has taken up an electron while forming the bis(radical anion) **F.** The two radical centers of **F** combine in the next step to give the sodium "glycolate" **G.** Compound **G,** the dianion of a bis(hemiacetal), is converted into the 1,2-diketone **J** by elimination of two equivalents of sodium alkoxide. This diketone is converted by two successive electron transfer reactions into the enediolate **I,** which is stable in xylene until it is converted into the enediol **H** during acidic aqueous workup. This enediol tautomerizes subsequently to furnish the α-hydroxyketone—or

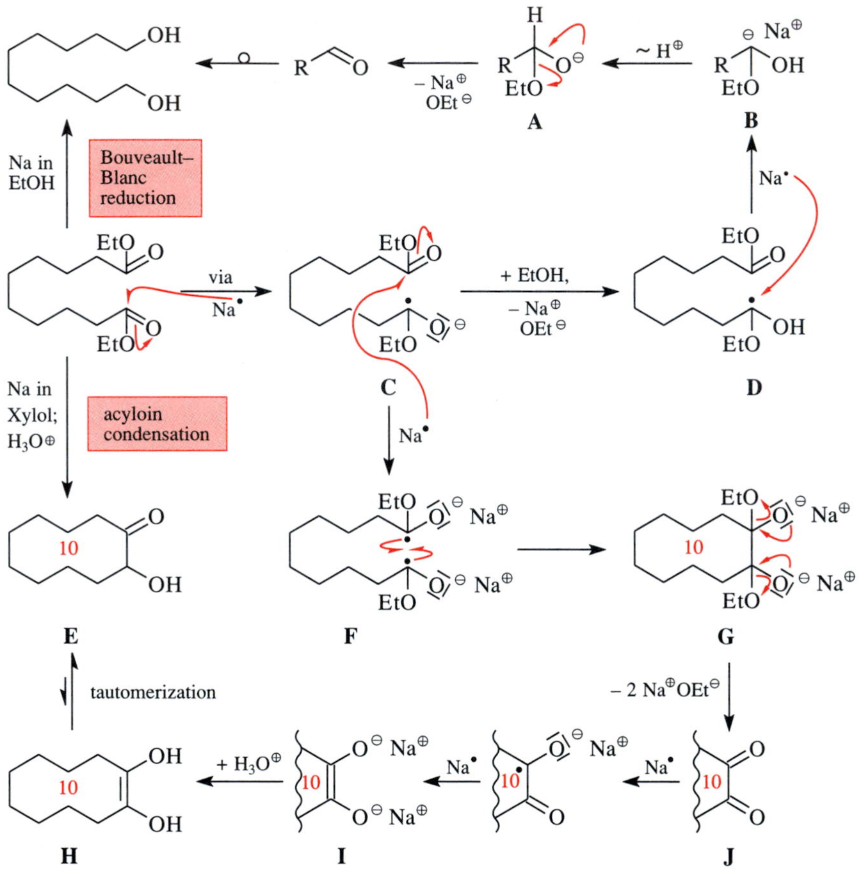

Fig. 14.51. Reduction of a carboxylic ester with dissolving sodium. Branching of the reduction paths in the presence (Bouveault–Blanc reduction) and absence (acyloin condensation) of protons.

acyloin—**E.** The acyloin condensation is of special value because this method allows for the synthesis of medium and large rings in good yields without the need for high dilution techniques.

14.4.3 Reductions of Carboxylic Acid Derivatives to Alcohols or Amines

B

Complex or soluble neutral metal hydrides are usually employed for the reduction of carboxylic acid derivatives to alcohols or amines. The standard reagents for the most important transformations are shown in Table 14.6. For completeness, various reagents also are listed for the reduction of carboxylic acid derivatives to aldehydes. The latter mode of reduction was discussed in Section 6.5.2.

The mechanism of the $LiAlH_4$ reduction of carboxylic esters to alcohols (Figure 14.52) can be readily understood based on the discussions of Section 6.5.1. A tetrahedral intermediate **A** is formed by addition of a hydride ion to the ester. Intermediate

Table 14.6. Survey of Reductions of Carboxylic Acid Derivatives to Alcohols, Amines, or Aldehydes

Reduction	Typically used reagent(s)
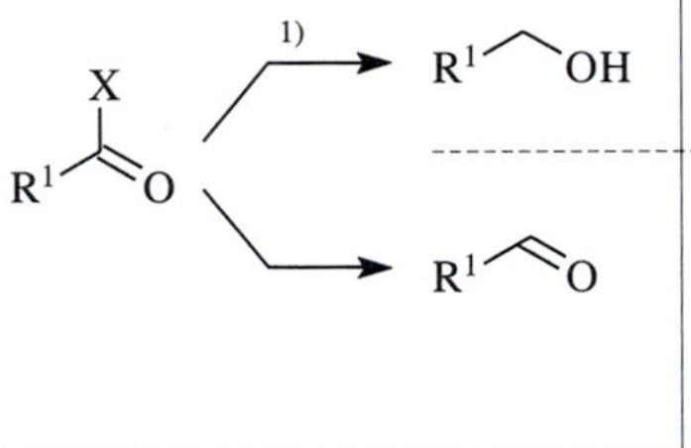 $R^1C(=O)X$ —[1)]→ R^1CH_2OH	for X = OR: $LiAlH_4$; DIBAL[2] in THF or Et_2O for X = OH: BH_3[3]; $LiAlH_4$
$R^1C(=O)X$ → R^1CHO	for X = Cl: $NaBH_4$; LiAlH(O*tert*-Bu)$_3$ for X = NMe–OMe: $LiAlH_4$; DIBAL for X = NMePh: add $^1/_4$ $LiAlH_4$ for X = OR: 1 DIBAL in toluene or hexane or CH_2Cl_2
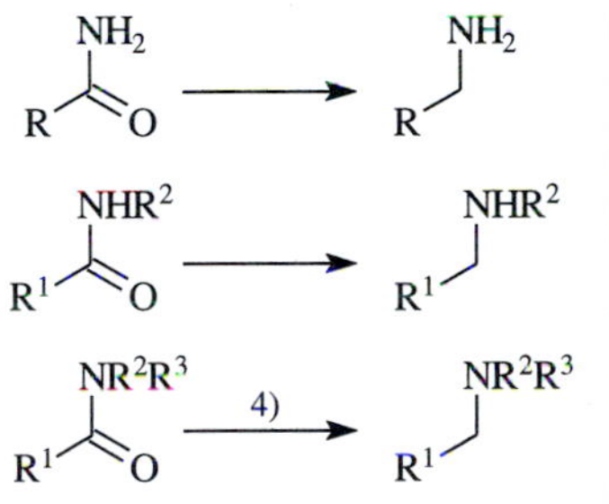 $RC(=O)NH_2$ → RCH_2NH_2 $R^1C(=O)NHR^2$ → $R^1CH_2NHR^2$ $R^1C(=O)NR^2R^3$ —[4)]→ $R^1CH_2NR^2R^3$	$LiAlH_4$; DIBAL
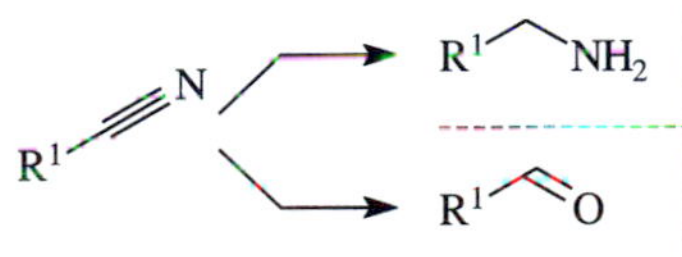 $R^1C≡N$ → $R^1CH_2NH_2$	$LiAlH_4$; H_2, NH_3, Pd/C
$R^1C≡N$ → R^1CHO	1 DIBAL in toluene or hexane or CH_2Cl_2

[1] Bouveault–Blanc reduction, not commonly used nowadays except for the chemoselective reduction of X = OR in the presence of X = OH.
[2] Preferred reagent for the reduction of α,β-unsaturated esters to allyl alcohols.
[3] Can be employed only if the substrate does not contain C=C or C≡C bonds; reduces carboxylic acids (X = OH) chemoselectively in the presence of carboxylic acid esters (X = OR).
[4] Only a few special tertiary amides can be reduced to alcohols, e.g., with superhydride.

A decomposes, faster than it is formed, into an aldehyde and a lithium alkoxide. The aldehyde is a better electrophile than the remaining ester, and therefore the aldehyde is attacked faster by the reductant than the ester. Consequently, $LiAlH_4$ effectively reduces esters to alcohols without there being a way of scavenging the aldehyde in this process.

A similar mechanism is operative in the reduction of carboxylic esters with DIBAL (Figure 14.53). The tetrahedral intermediate **A** is formed by addition of an Al—H bond of the reducing agent to the ester C=O bond. This tetrahedral intermediate **A** does *not necessarily* decompose immediately to an aldehyde and ROAl(iBu)$_2$. In *nonpolar* media **A** definitely decomposes quite slowly. In fact, at very low temperatures **A** remains unchanged until it is protonated to a hemiacetal during aqueous workup. The latter eliminates water to give the aldehyde.

Fig. 14.52. Mechanism of the $LiAlH_4$ reduction of carboxylic esters to alcohols via aldehydes.

In polar solvents, however, the tetrahedral intermediate **A** of Figure 14.53 decomposes faster than it is generated, with formation of an aldehyde and $ROAl(iBu)_2$. This solvent effect is explained in Figure 14.53, using THF as an example. The tetrahedral intermediate **A** contains a trivalent Al and forms a Lewis acid–Lewis base complex with the solvent. The intermediate **A** is thus converted into the aluminate complex **B,** which contains tetravalent Al. The Al atom in **A** is bonded to only one O atom, which

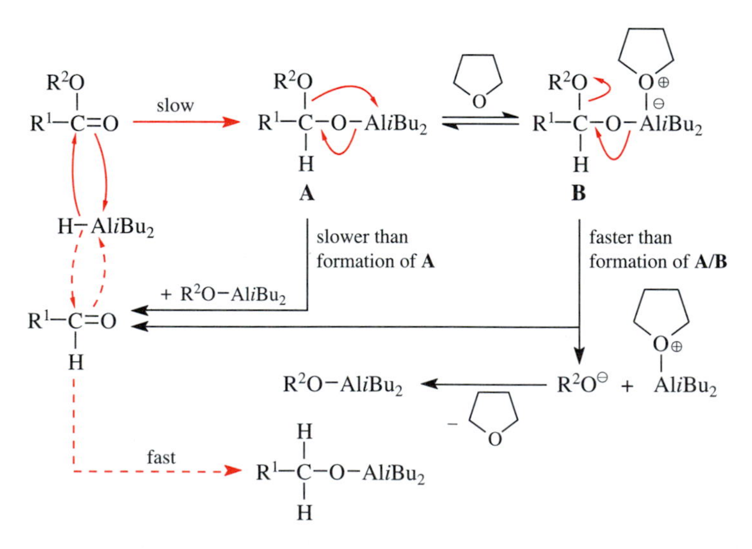

Fig. 14.53. Mechanism of the DIBAL reduction of carboxylic esters to aldehydes and further to alcohols. In nonpolar solvents the reaction stops with the formation of the tetrahedral intermediate **A.** During aqueous workup, **A** is converted into the aldehyde via the hemiacetal. In polar solvents, however, the tetrahedral intermediate **A** quickly decomposes forming the aldehyde via complex **B.** In the latter situation the aldehyde successfully competes with unreacted ester for the remaining DIBAL. The aldehyde is reduced preferentially, since the aldehyde is the stronger electrophile, and it is converted into the alcohol.

used to be the carbonyl oxygen of the ester. In complex **B,** the Al atom still binds to the same O atom, but it also binds weakly to a THF molecule. The bond between Al and the O atom that originated from the ester is stronger in **A** than in **B** because of the additional contact with THF in the latter. The Al—O bond thus breaks relatively easily in **B** and gives rise to a situation in which the aldehyde is formed faster from **B** than the tetrahedral intermediate **A** is formed from DIBAL and the ester. The already formed aldehyde and the still unconsumed ester therefore compete for the remaining DIBAL, and the aldehyde wins because of its higher electrophilicity. Therefore, DIBAL reductions of carboxylic esters in polar solvents go "all the way" to the alcohol.

There is one type of ester → alcohol reduction for which one always employs DIBAL (in a polar solvent) rather than $LiAlH_4$ (in ether of THF). This reduction is the reduction of α,β-unsaturated esters to allyl alcohols (example in Figure 14.54). The reaction of this kind of substrate with $LiAlH_4$ sometimes results in a partial reduction of the C=C double bond to a C—C single bond in addition to the desired transformation —C(=O)OR → —CH_2OH.

DIBAL (2 equivalents), Et_2O

OMe OH

Fig. 14.54. DIBAL reduction of an α,β-unsaturated ester to an allylic alcohol. (See Figure 9.11 for a preparation of the substrate.)

The two reducing agents considered so far in this section—$LiAlH_4$ and DIBAL—also are the reagents of choice for the reduction of nitriles (Figure 14.55). The mechanistic details of these reactions can be gathered from the figure, and the result can be summarized as follows.

Products of the Reduction of Nitriles with $LiAlH_4$ and DIBAL

- The reduction of nitriles with $LiAlH_4$ leads to iminoalanate **B** via iminoalanate **A.** The hydrolytic workup affords an amine.
- The reduction of nitriles with DIBAL can be stopped at this stage of the iminoalane **C.** The hydrolysis of this iminoalane gives the aldehyde. The iminoalane **C** also can be reacted—more slowly—with another equivalent of DIBAL to the aminoalane **D.** The latter yields the amine upon addition of water.

$LiAlH_4$ or DIBAL reduces carboxylic amides at *low* temperatures only to such an extent that aldehydes are obtained after aqueous workup (Figure 14.56). This works most reliably, according to Figure 6.33, if the amides are Weinreb amides. At *higher* temperatures, the treatment of carboxylic amides with either $LiAlH_4$ or DIBAL results in amines. Accordingly, *N,N*-disubstituted amides give tertiary amines, monosubstituted amides give secondary amines, and unsubstituted amides give primary amines (Figure 14.56).

It is noteworthy that all of the latter reductions yield amines rather than alcohols. This is due to the favored decomposition path of the initially formed tetrahedral intermediates. Specifically, the tetrahedral intermediates **B** and **D** are formed in amide

Fig. 14.55. Mechanism of the $LiAlH_4$ reduction (top) and the DIBAL reduction (bottom) of nitriles.

$R{-}CH{=}N^{\ominus}\ Li^{\oplus}$ + AlH_3 ⇌ A → B; B → ($H_3O^{\oplus}$; $OH^{\ominus}$) → $R{-}CH_2{-}NH_2$

$H{-}AlH_3^{\ominus}\ Li^{\oplus}$ + $R{-}C{\equiv}N$ + $H{-}AliBu_2$ — reduction; hydrolysis → either $R{-}CH{=}O$ or $R{-}CH_2{-}NH_2$

$R{-}CH{=}N{-}AliBu_2$ (C) + $H{-}AliBu_2$ → $R{-}CH_2{-}N(AliBu_2)_2$ (D); C → ($H_3O^{\oplus}$) → $R{-}CH{=}O$; D → ($H_3O^{\oplus}$; $OH^{\ominus}$) → $R{-}CH_2{-}NH_2$

reductions with $LiAlH_4$ and DIBAL, respectively. Their decomposition in principle could affect the C—O bond (→ → → amine) or the C—N bond (→ → → alcohol). There are two factors that provide an advantage for the C—O bond cleavage:

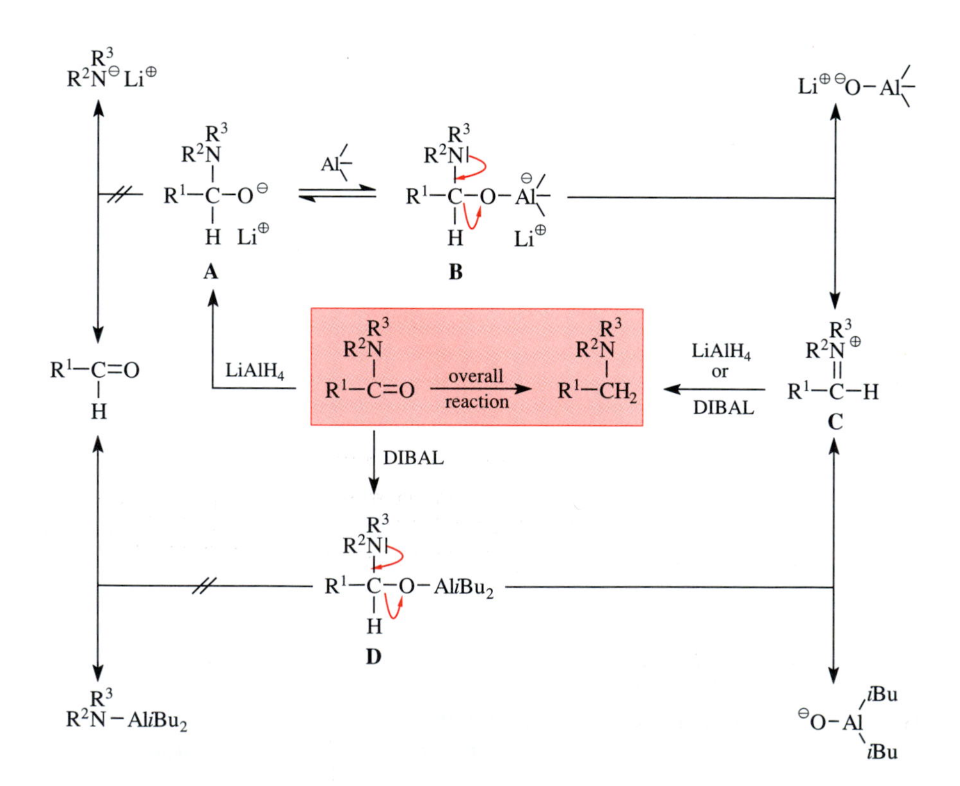

Fig. 14.56. Chemoselectivity of the reduction of amines.

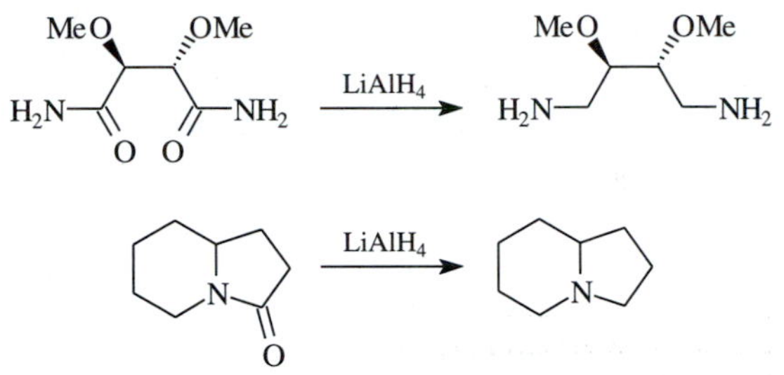

Fig. 14.57. Amide → amine reduction (top) and lactam → cycloamine reduction (bottom) with $LiAlH_4$.

1) The strong O—A bond of the respective tetrahedral intermediate is replaced by a still stronger O—Al bond in the leaving group ($R_2Al—O^-$ or $R_3Al^-—O^-$).
2) The amino group, a π-donor substituent, facilitates loss of the leaving group ($R_2Al—O^-$ or $R_3Al^-—O^-$), because an iminium ion is formed.

The competing C—N bond cleavage of the tetrahedral intermediates **B** and **D** does not occur for the reasons just given and because C—N bond cleavage is disadvantaged *in addition* because of the following factors:

1) The amide anion is a much poorer leaving group than the aluminum alkoxide ($R_2Al—O^-$ or $R_3Al^-—O^-$), and this remains true even if the amide group is complexed by $i Bu_2Al$.
2) The putative cation that would be generated by C—N bond cleavage in **D** would contain an O atom with a very weak +M effect. This is because the $i Bu_2Al$ group, which binds to this O atom, acts as a Lewis acid. Consequently, the cation in question would be an acceptor-substituted carbenium ion (with a C—O single bond) rather than an oxocarbenium ion (with a C═O double bond).

All four factors push in the same direction. Hence, *only* the iminium ion **C** is formed from the tetrahedral intermediates **B** and **D** of Figure 14.45. This iminium ion finally takes up a second hydride ion from the reducing agent and yields the Al derivative of the product amine.

The two specific examples shown in Figure 14.57 illustrate the general concept of the amide → amine reduction depicted in Figure 14.56. The reduction of a diamide to a diamine is shown on top, and the reduction of a lactam to a bicyclic amine is shown on bottom.

14.4.4 Reductions of Carboxylic Acid Derivatives to Aldehydes

These reductions were discussed in Section 6.5.2.

14.4.5 Reductions of Carbonyl Compounds to Alcohols

The reader already is familiar with the reductions of carbonyl compounds to alcohols, since they were described in Sections 8.2–8.4.

14.4.6 Reductions of Carbonyl Compounds to Hydrocarbons

B

The Wolff–Kishner reduction is an old and still often used method for the reduction of a ketone to the corresponding alkane. The **Huang–Minlon modification** of this reduction is commonly employed. It entails the treatment of the ketone with hydrazine hydrate and KOH in diethylene glycol, first at low temperature and then at reflux (200°C). The assumed reaction mechanism is exemplified in Figure 14.58 with the second step **A** → **E** of the Haworth naphthalene synthesis (steps 1 and 3 of this synthesis: Figure 5.28). This reduction includes the following steps: (1) formation of the hydrazone **B** (for the mechanism, see Table 7.2), (2) tautomerization of hydrazone **B** to the azo compound **C,** (3) E2 elimination with the liberation of nitrogen and formation of the benzylpotassium **D,** and (4) protonation of **D** in the benzylic position, leading to the formation of the carboxylate **F.** The carboxylic acid **E** is formed by protonation of this carboxylate during acidic workup of the reaction mixture.

A

Starting from previously isolated hydrazones, it turns out that they can be reduced to the corresponding hydrocarbons by treatment with base in an *aprotic* solvent at temperatures significantly below the 200°C of the Huang–Minlon modification of the Wolff–Kishner reduction. However, hydrazones cannot be prepared in a one-step reaction between a ketone and hydrazine, since usually azines

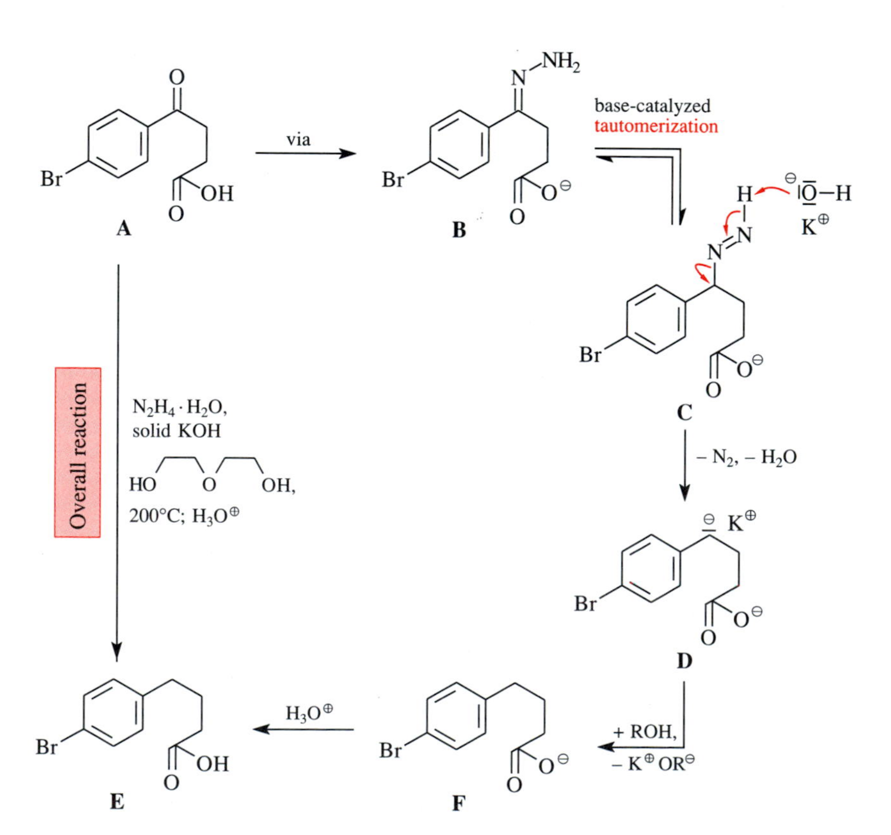

Fig. 14.58. Wolff–Kishner reduction of a ketone. The example shows the second step of the five-step Haworth synthesis of naphthalene.

($R^1R^2C{=}N{-}N{=}CR^1R^2$) are formed instead. However, semicarbazones are hydrazone derivatives that are easily accessible by the reaction of a ketone with semicarbazide (for the mechanism, see Table 7.2). **Semicarbazones** can be converted into alkanes with KO*tert*Bu in toluene at temperatures as low as 100°C. This method provides an alternative to the Wolff–Kishner reduction when much lower than usual reduction temperatures are desirable.

The mechanism of such a semicarbazone reduction is exemplified in Figure 14.59 by the deoxygenation of the α,β-unsaturated ketone **A** (possible preparation: in analogy to compound **A** in Figure 10.62). Since the substrate contains a C=C double bond—which is, however, not a prerequisite for the feasibility of such a reaction—one obtains a product with a C=C double bond. Hence, an *alkene* is formed, not an alkane. The first two steps **A** → **B** → **C** of the semicarbazone reduction of Figure 14.59 correspond to the introductory steps of the Wolff–Kishner reduction (Figure 14.58). They comprise a hydrazone formation—here the semicarbazone formation as a special case of a hydrazone formation—and a tautomerization to an azo compound **C** (the position of the C=C double bond in **C** is not known). The *tert*-butoxide adds to the C=O double bond of the carbonic acid moiety of the azo compound **C** and forms the tetrahedral intermediate **F** (the position of the C=C double bond remains unknown). The fragmentation of **F** yields a carbamate, N_2, and the allylpotassium compound **E.** The protonation of **F**—presumably under kinetic control—occurs in a regioselective fashion and results in the alkene **D.**

Tosylhydrazones can be reduced to the corresponding alkanes under milder conditions compared to the reduction of carbonyl compounds by the Wolff–Kishner method. This is illustrated in Figure 14.60 by the reduction of the aldhydrazone **A** (for a possible preparation, see Table 7.2) to the alkane **C.** The reduction is carried out with $NaBH_4$ in MeOH. The effective reducing agent, formed *in situ,* is $NaBH(OMe)_3$. This reductant delivers a hydride ion for addition to the C=N double bond of the tosylhydrazone **A.** Thereby the hydrazide anion **B** is formed. Much as in the second step

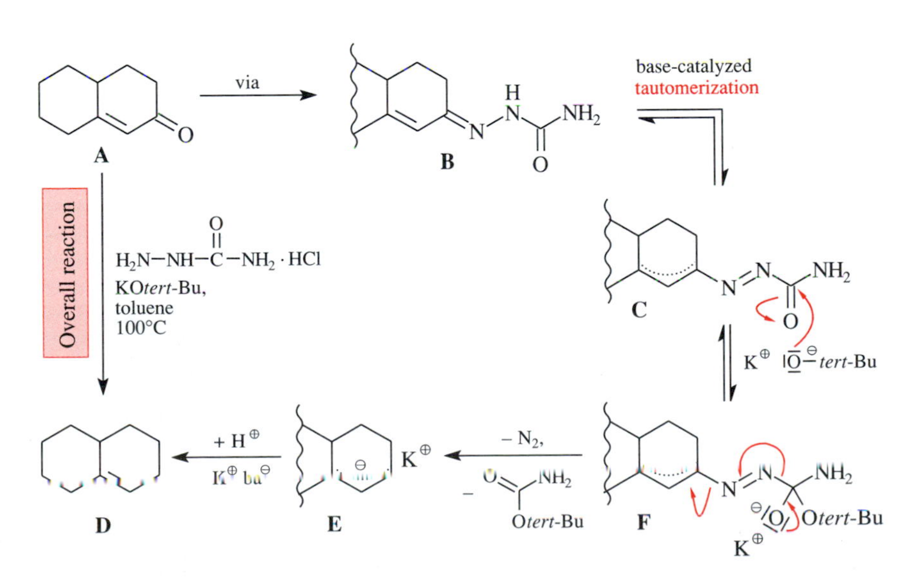

Fig. 14.59. Alternative I to the Wolff–Kishner reduction: reductive decomposition of a semicarbazone.

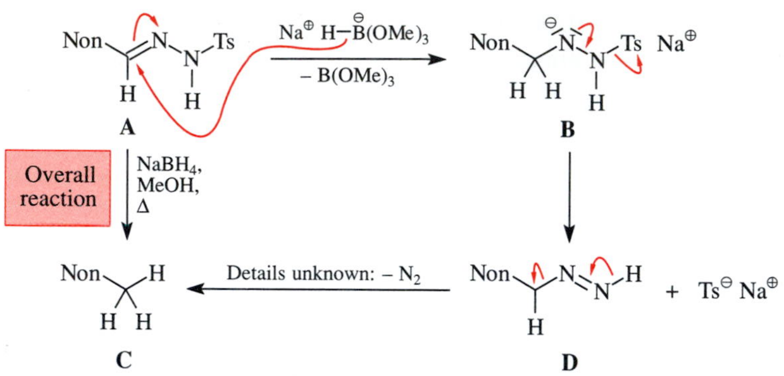

Fig. 14.60. Alternative II to the Wolff–Kishner reduction: reduction of a tosylhydrazone; "Non" refers to a nonyl group.

of an $E1_{cb}$ elimination, the anion **B** eliminates a *para*-tolylsulfinate and a diazo compound **D** is generated. The conversion of the diazo compound **D** into the hydrocarbon product **C** requires the same structural changes as the conversion of the diazo compound **C** of the Wolff–Kishner reduction (Figure 14.58) into the reduction product. These two transformations could therefore be mechanistically analogous.

Conjugated tosylhydrazones also can be reduced to hydrocarbons with the method depicted in Figure 14.60, that is, with $NaBH(OMe)_3$. The C=C double bond is retained but it is shifted. Figure 14.61 exemplifies this situation for tosylhydrazone **A.** The sequence of initial steps **A** → **B** → **D** resembles the one shown in Figure 14.60. However, the diazo compound **D** undergoes a different reaction, namely, **a retro-ene reaction.** A retro-ene reaction is a one-step fragmentation reaction of

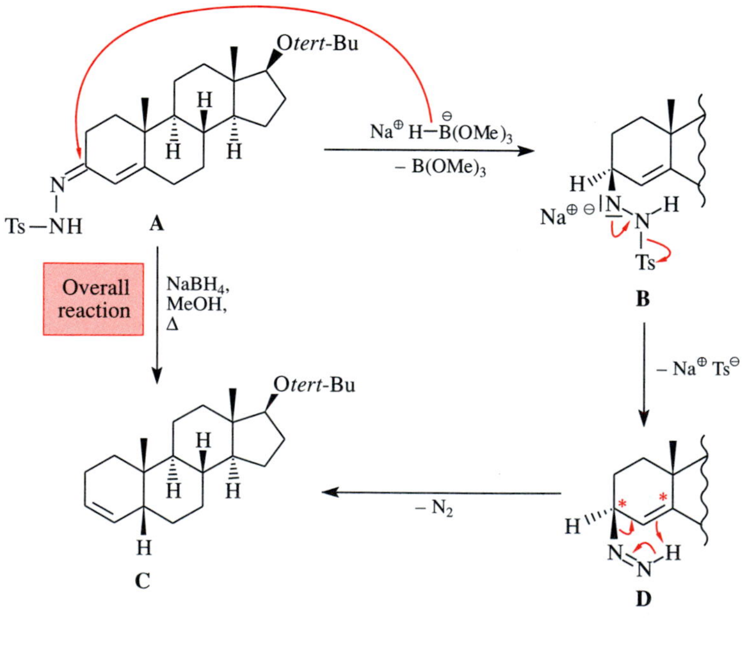

Fig. 14.61. Reduction of a conjugated tosylhydrazone.

an unsaturated compound of type **A** or **B,** which affords two unsaturated compounds according to the following pattern:

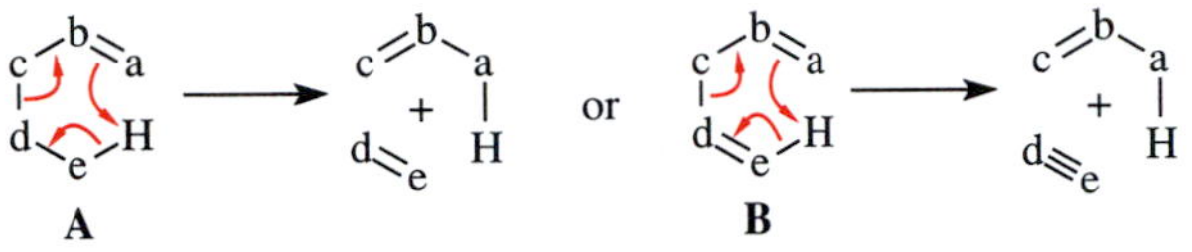

The retro-ene reaction of Figure 14.61 directly leads to the reaction product **C.**

Aromatic aldehydes and aromatic ketones also can be reduced to hydrocarbons in a completely different manner, namely via the so-called **ionic hydrogenation** followed by an **ionic hydrogenolysis.** This kind of reduction is possible only if it can proceed via resonance-stabilized cationic intermediates. This resonance stabilization is readily achieved in a benzylic position, and it is therefore advantageous to employ aromatic carbonyl compounds in this kind of reduction. The carboxonium ion **A,** formed

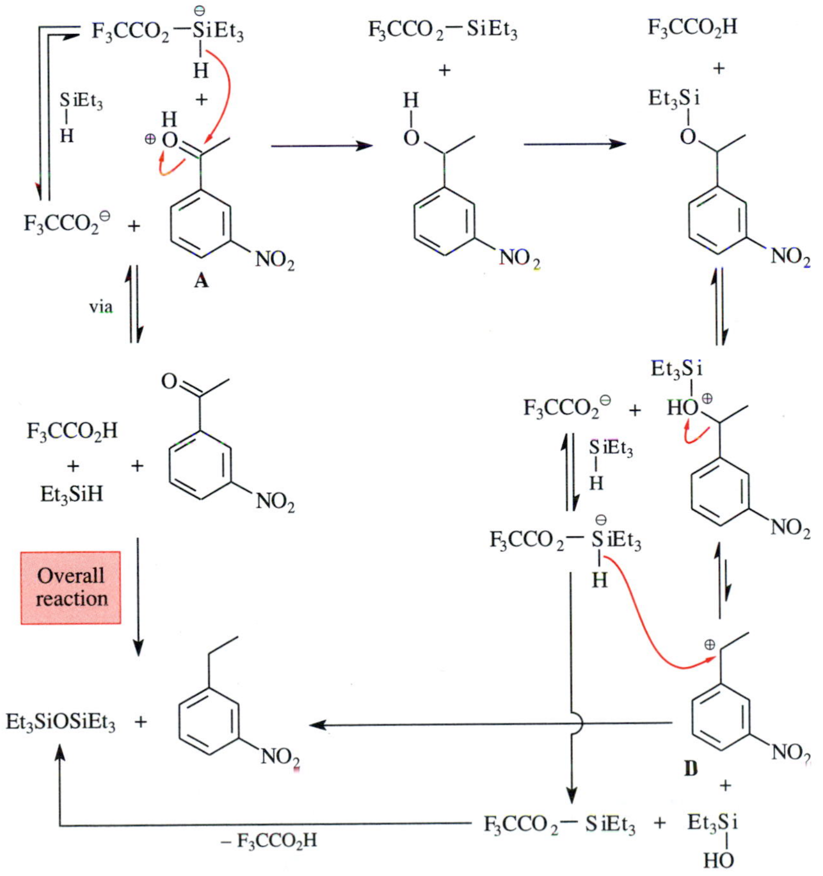

Fig. 14.62. Polar hydrogenation/hydrogenolysis of an aromatic ketone (*meta*-nitroacetophenone). CF_3COOH causes a reversible protonation of the ketone to the carboxonium ion **A.** The reducing agent triethylsilane then transfers a hydride ion onto **A** to form a benzylic alcohol. This alcohol presumably is silylated, protonated, and converted into the benzyl cation **B.** A second hydride transfer yields the final product.

in a protonation equilibrium with CF_3CO_2H, is the first cationic intermediate in the reduction of *meta*-nitroacetophenone (Figure 14.62). Cation **A** abstracts a hydride ion from the first equivalent of the reducing agent, triethylsilane. A benzyl alcohol is thus obtained as a transient intermediate. It is presumably silylated, protonated, and eventually converted into the benzyl cation **B.** The latter abstracts another hydride ion from the second equivalent of the triethylsilane, which leads to the final product.

The reduction of a ketone to an alkene is feasible not only for unsaturated ketones (Figures 14.59 and 14.61) but for saturated ketones as well. To this end the latter must be converted into enol phosphonoamidates or enol dialkylphosphonates via suitable lithium enolates. One substrate of each type is shown in Figure 14.63 (**A** and **B**). Lithium dissolved in $EtNH_2$/*tert*-BuOH mixtures is a suitable reducing agent for both these compounds. Their C_{sp^2}—O bond is cleaved by a sequence of three elementary steps with which you are familiar from the formation of methylmagnesium iodide (Figure 14.37): (1) electron transfer, (2) dissociation of the radical anion obtained to a vinyl radical and a negatively charged phosphoric acid derivative, and (3) electron transfer onto the vinyl radical and formation of an alkenyllithium compound. In the final and unavoidable fourth reaction step, the alkenyllithium compound is protonated by *tert*-BuOH to furnish the alkene product. The C=C double bond remains in the same position as in the precursors **A** and **B,** respectively (Figure 14.63). This means that two different alkenes are formed, since the double bonds were in different positions in compounds **A** and **B.**

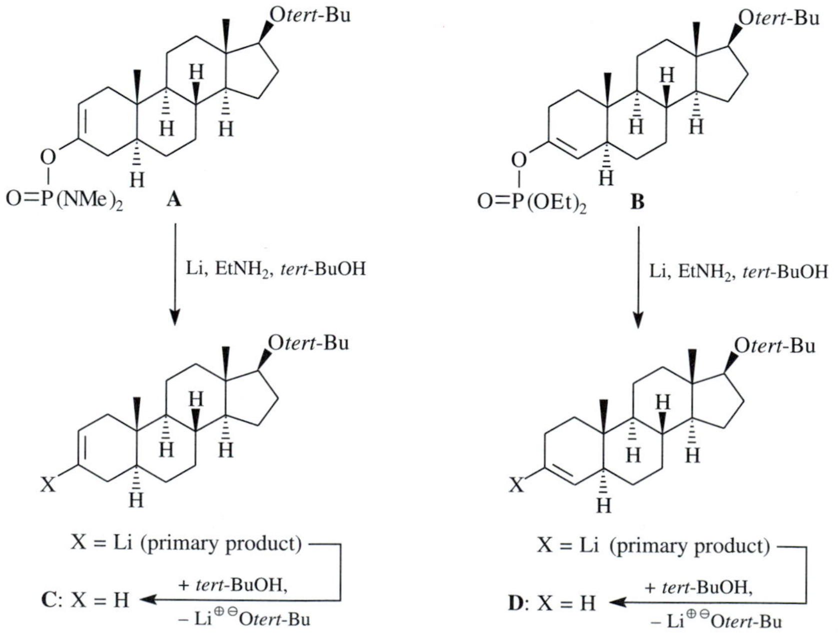

Fig. 14.63. Ketone → alkane reduction via enol phosphonoamidates (for one way to prepare **A,** see Figure 10.21) and enol dialkylphosphates (one way to prepare **B** is to use a combination of the methods depicted in Figures 10.17 and 10.22). The cleavage of the C_{sp^2}—O bond of the substrates occurs in analogy to the electron transfers in the formation of methylmagnesium iodide (Figure 14.37). The alkenyllithium intermediates are protonated in the terminating step to afford the target alkenes.

14.4.7 Hydrogenation of Alkenes

B

It was mentioned in Section 3.3.4 that alkenes react with H_2 on the surface of elemental Pd or elemental Pt to form alkanes. Similar hydrogenations occasionally also can be accomplished using Raney nickel as a catalyst. Raney nickel is prepared from a 1:1 NiAl alloy and aqueous KOH and has a high specific surface.

The examples of Figure 3.22 illustrate standard scenarios of heterogeneous catalytic hydrogenations of alkenes. Such reductions usually are highly stereo-selective *cis* additions. Exceptions can be observed occasionally; for example, the hydrogenations shown in Figure 14.64 produce up to 30% of the *trans*-hydrogenated product. The mechanism outlined in Figure 14.65 shows that hydrogenations involve *several* steps. This fact explains why heterogeneous catalytic hydrogenations usually are *cis*-selective (part I) and why *trans* product may occur occasionally (part II).

Part I of Figure 14.65 shows the beginning and the *cis*-selective path of the hydrogenation. First, an H_2 molecule dissolved in the liquid phase is bonded covalently to the surface of the metal catalyst. The alkene **B** also is bonded to that surface. This bonding is accomplished by reversible π-complex formation. Occasionally, an alkene will thus be bonded to the catalyst surface in proximity to a Pd—H bond as shown in **C.** What follows is a kind of *cis*-selective hydropalladation of the alkene. A Pd atom binds to one end of a C=C bond and an H atom that was bonded to a proximate Pd adds to the other end of the C=C bond. The hydropalladation product is described by stereostructure **E.** Compound **E** can react further only if an H atom migrates to a Pd atom that is right next to the Pd atom that was involved in the hydropalladation. This migration (of an H atom that is already bonded elsewhere on the surface) occurs by way of surface diffusion. The intermediate **D** is then formed. It releases alkane **A,** the product of a stereoselective *cis*-hydrogenation.

A

Part II of Figure 14.65 shows the side reactions that occur when the Pd-catalyzed hydrogenation is not completely *cis*-selective. The start is the formation of the π-complex **F** from the hydropalladation product **E** by a β-hydride elimination of sorts (see Figures 13.8, 13.9, 13.26). In a way, this reaction is the reverse of the reaction type that formed **E** from the other π-complex **C** (part I of Figure 14.65). In an equilibrium reaction, the isomerized π-complex **F** subsequently releases the alkene *iso*-**B,** which is a double bond isomer of the substrate alkene **B.**

Subsequently, the new alkene *iso*-**B** is hydrogenated. In principle it may add hydrogen from either of its diastereotopic faces. However even if the addition were 100%

Me, Me — H_2, Pt/C, EtOH → H, Me, Me, H + H, Me, Me, H

at 25°C → 70 : 30

at 0°C → 84 : 16

Fig. 14.64. Examples of heterogeneous catalytic hydrogenations of C=C double bonds that occur with incomplete *cis*-selectivity.

Fig. 14.65, Part I
Mechanism of the *cis*-selective heterogeneous Pd-catalyzed hydrogenation of C=C double bonds.

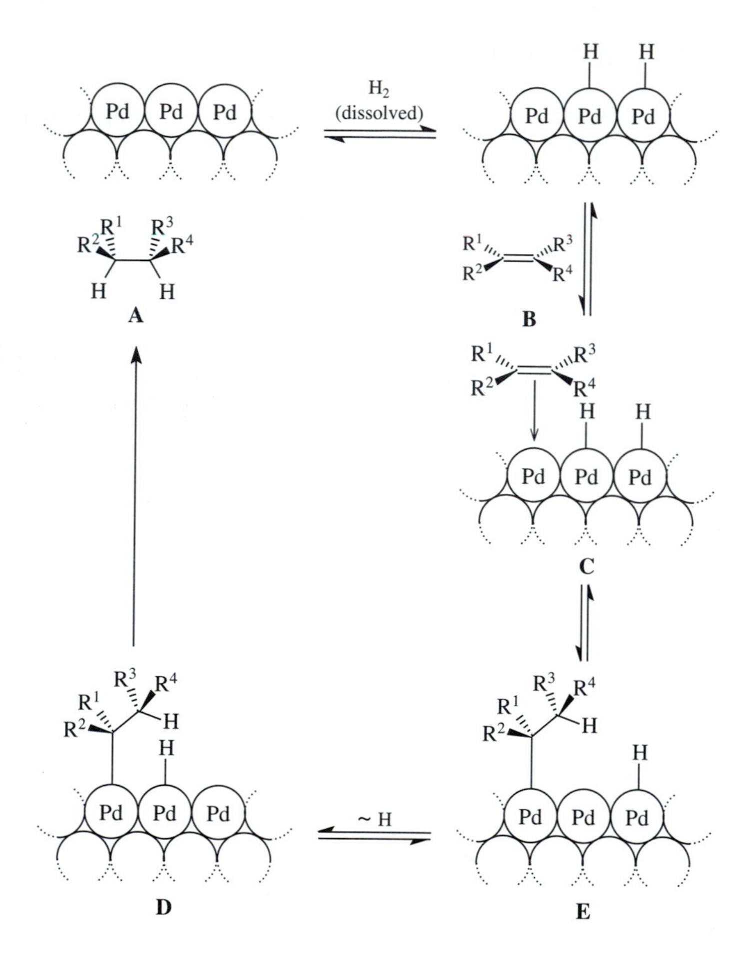

cis-selective, a mixture of alkane **A** and *iso*-**A** could be formed, as shown by the structures depicted in the bottom half of Figure 14.65 (part II). With respect to the alkene plane of the original substrate **B** the newly added hydrogen atoms in *iso*-**A** are oriented in the way that would have resulted from a *trans*-hydrogenation of **B.**

The alkene isomerization **B** → *iso*-**B** also may have another stereochemical consequence (part II of Figure 14.65): the destruction of the configurational homogeneity of a stereocenter $C^{*}HR^5R^6$ in the allyl position of the substrate. In that case, the hydrogenation results in a mixture containing the stereoisomers **A** and *iso*-**A.**

Catalytic hydrogenations of C=C double bonds can be carried out not only in a heterogeneous fashion—on metal surfaces—but also in **homogeneous phase** using soluble metal complexes as catalysts. This possibility is of special significance if the hydrogenation is carried out with an enantiomerically pure chiral catalyst. It allows for the enantioselective addition of hydrogen to certain functionalized alkenes. Highly efficient catalysts for such **chiral hydrogenations** are derived from BINAP, a phosphine derivative of 2,2-binaphthyl. The synthesis of this ligand from BINOL is described in

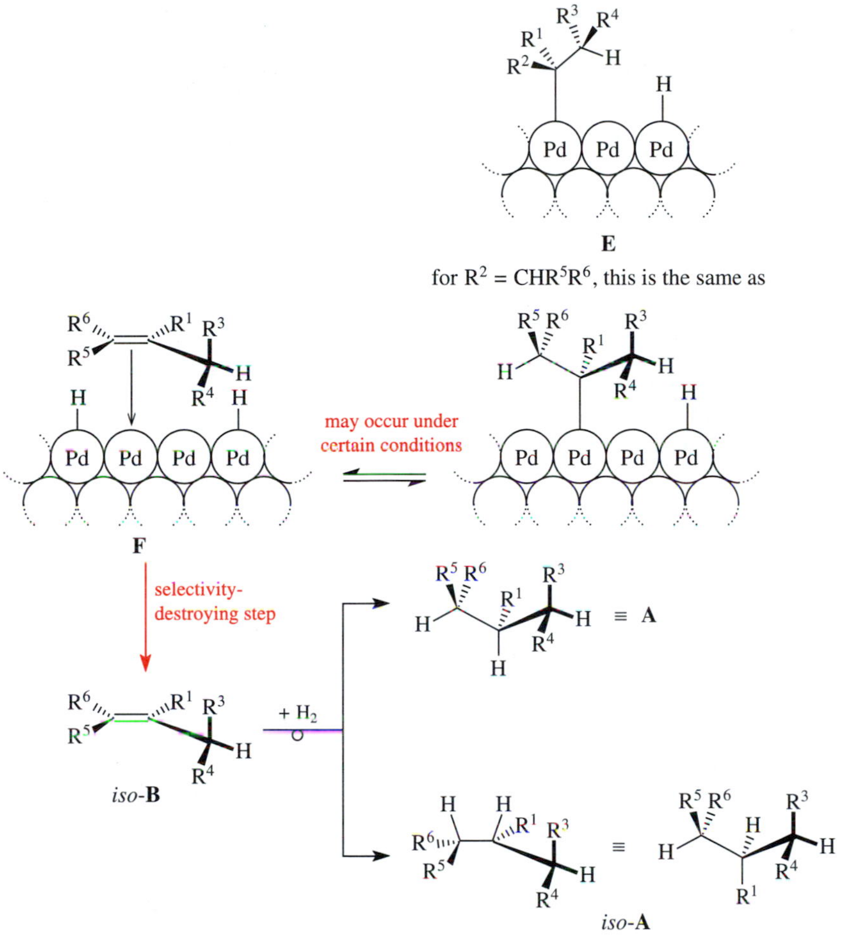

Fig. 14.65, Part II Mechanism of the stereo-unselective heterogeneous Pd-catalyzed hydrogenation of C═C double bonds.

Figures 5.50 and 5.37. BINAP-containing precious metal complexes catalyze—among others—the following hydrogenations:

- enantioselective hydrogenations of the C═C double bonds of certain allyl alcohols (example in Figure 14.66),
- enantioselective hydrogenations of the C═C double bonds of α-(acylamino)acrylic acids (example in Figure 14.67), and
- enantioselective hydrogenations of the C═O double bond of β-ketoesters.

The asymmetric hydrogenation of geraniol or nerol is the most elegant method for the stereoselective construction of the side chains of vitamins E and K_1 (Figure 14.66). In this manner, not only the stereocenter of these vitamins highlighted in red can be established with the correct configuration by the incorporation of the analogously configured precusor molecule **A.** One can achieve comparably high stereocontrol also at the neighboring stereocenter by means of hydrogenations that are analogous to the ones

Fig. 14.66. Enantioselective homogeneous catalytic hydrogenations of nerol (left) and geraniol (right); OPiv refers to the pivaloyloxy group *tert*-Bu—C(=O)—O—.

shown, namely, which start with trisubstituted allylic alcohols. Such further-leading allylic alcohols can be constructed from **A** by way of chain-elongating syntheses.

The enantioselective hydrogenation of α-(acylamino)acrylic acids (Figure 14.67) is mainly used for the preparation of *R*- (unnatural) and *S*-configured (natural) amino acids. Such enantiomerically pure amino acids are needed, among others, for the synthesis of peptide-based antibiotics and peptide mimetics.

Homogeneous catalytic hydrogenations proceed via catalytic cycles that involve many steps. Not all the species discussed in these catalytic cycles have been proven completely, but it is very likely that these cycles are basically correct.

For enantioselectivity to occur, the unsaturated substrate must bind to the catalytic center in such a way that a complex with well-defined stereostructure is formed. Accordingly, a highly enantioselective hydrogenation is assured—at least in most cases—

(S-BINAP)Rh(MeOH)$_2^{\oplus}$ $BF_4^{\ominus}$ (1 mol %)
H_2 (3 atm), room temperature
(100% *ee*)

Fig. 14.67. Enantioselective homogeneous catalytic hydrogenations of an α-(acylamino)acrylic acid to an *R*-amino acid.

if the substrate forms *two* bonds to the metal. The substrate is π-bonded to the metal via the C=C double bond that is to be hydrogenated. It is also σ-bonded to the metal via a heteroatom that is close enough to this C=C double bond.

Achieving successful enantioselective hydrogenations of C=C double bonds requires that one more prerequisite be met. It must be possible to convert the substrate/metal complex into a complex that contains a hydrido ligand as well. Hydrometallation of the π-bonded C=C double bond is possible only in such a hydrido complex. The hydrometallation reaction in turn is the prerequisite for a subsequent second C—H bond formation. The latter completes what is an overall "hydrogenation" of the substrate. Figure 14.68 substantiates this general description of the species involved in asymmetric hydrogenations by depicting the *specific* intermediates of the hydrogenations of Figures 14.66 and 14.67.

The rhodium/substrate complex **A** in Figure 14. 68 is the key intermediate in the hydrogenation of α-(acylamino)acrylic acids (Figure 14.67). The reaction starts with the equilibrium dissociation of two MeOH ligands from the Rh(I) atom. The two now-vacant coordination sites allow for the oxidative addition of a hydrogen molecule. The dihydrido Rh(III) complex **C** is formed. The hydrometallation of the C=C double bond, that is, the *cis*-addition of L_nRh—H to that double bond, occurs *in this complex*. This reaction delivers the monohydrido Rh(III) complex **E**. The next step consists of the reductive elimination of the Rh(III)-bonded H atom jointly with the Rh(III)-bonded alkyl group. The C—Rh bond in **E** thereby is converted into a C—H bond with retention of configuration. Two MeOH molecules add to the new Rh(I) complex, and the hydrogenated product—the *N*-benzoylphenylalanine—dissociates.

Structure **B** in Figure 14.68 describes the initially formed metal/substrate complexes in the hydrogenation of allylic alcohols according to Figure 14.66. A molecule of hydrogen can add only if both the carboxyl oxygen of the pivaloate group and the C=C double bond of the allylic alcohol free their coordination site. The dihydrido complex of Ru(IV) formed by this addition (not shown in the figure) undergoes a reductive elimination: the monohydrido Ru(II) complex **D** is formed via elimination of pivaloic acid. It is *this complex* in which the hydrometallation of the substrate occurs via a *cis*-addition of L_nRu—H to the C=C double bond. Complex **F** is produced, which still contains Ru(II). The neutral ligand L of this complex **F** dissociates in the next step, leaving vacant a total of two metal coordination sites. An oxidative addition of pivaloic acid fills these coordination sites by a pivaloate and a hydride. At this point, the oxidation number of the Ru once again is +4 (in a complex, which is not shown in the

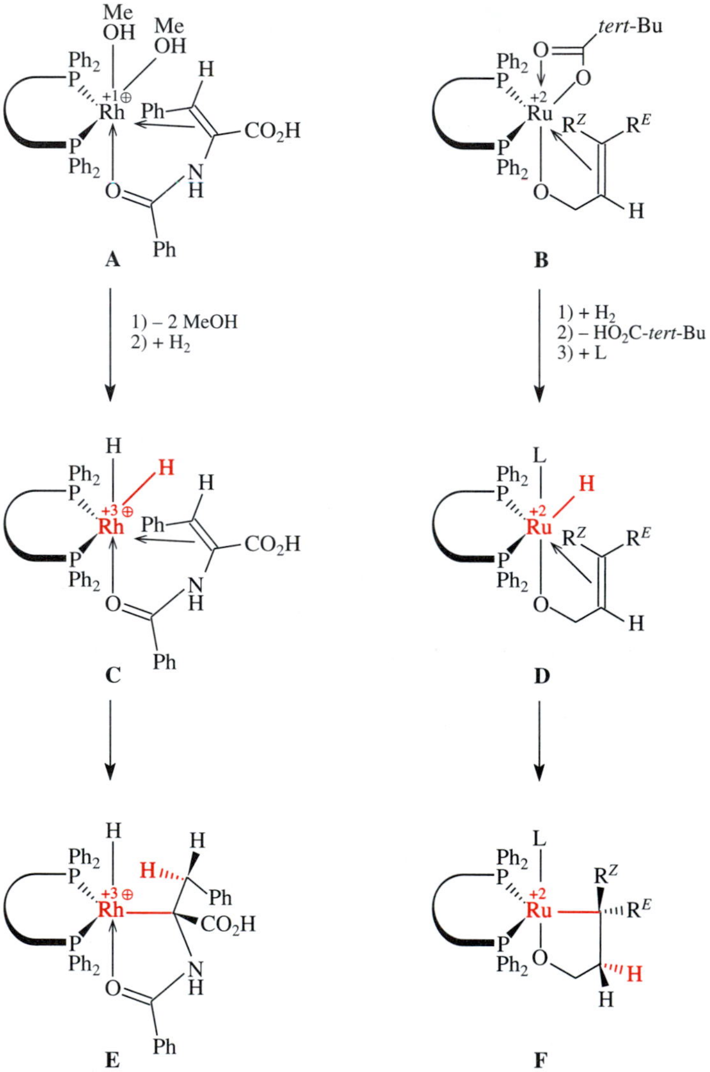

Fig. 14.68. Key intermediates in the enantioselective hydrogenations of Figure 14.66 (right) and Figure 14.67 (left). The BINAP ligand is shown schematically as U-shaped, with two PPh_2 substituents.

figure). Its substitution pattern initiates the last step of the catalytic cycle: the reductive elimination of the hydrogenation product. As with the conversion of the Rh—C bond into the C—H bond in complex **E,** the C—Ru bond is converted into the C—H bond with complete retention of configuration. Therefore, this H_2 addition to the C═C double bond again occurs with *cis*-selectivity.

14.4.8 Reductions of Aromatic Compounds and Alkynes

B

Benzene and its derivatives can be hydrogenated on the surface of precious metals only under pressure, and even then generally only at elevated temperatures (Figure

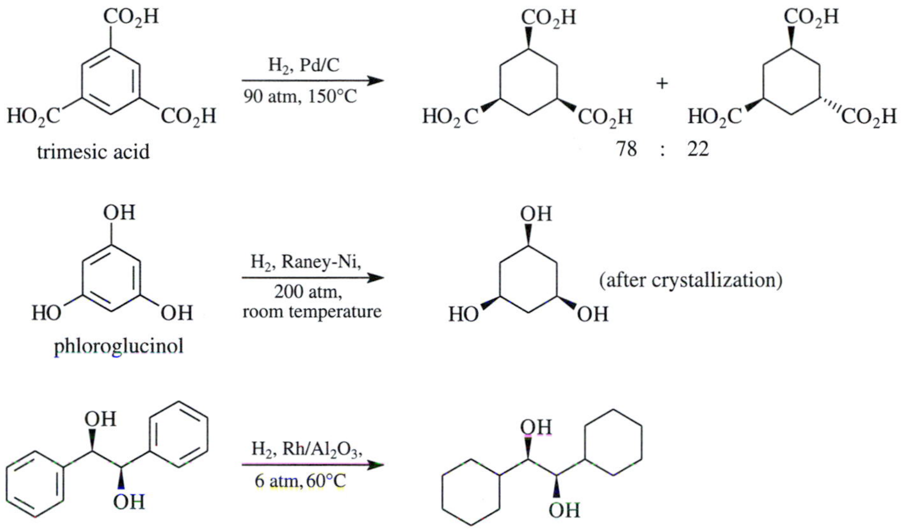

Fig. 14.69. Heterogeneous catalytic hydrogenations of aromatic compounds.

14.69). Obviously, these hydrogenations require more drastic conditions than the hydrogenations of alkenes—as one would expect, based on the Hammond postulate. The first step in the hydrogenation of an aromatic compound is the conversion into a cyclohexadiene. This step is associated with the loss of the benzene resonance energy of 36 kcal/mol. It is for this reason, and in contrast to the hydrogenation of alkenes, that the hydrogenation of the first aromatic C=C double bond is endothermic. Once the benzene has been converted into the cyclohexadiene, the consecutive hydrogenations via cyclohexene on to cyclohexane are exothermic reactions. In agreement with Hammond's postulate, the latter two hydrogenations are faster than the initial hydrogenations of the aromatic system. Consequently, benzene derivatives are hydrogenated "all the way" to cyclohexanes.

The two examples on top of Figure 14.69 show that these kinds of hydrogenations have a certain significance for the stereoselective synthesis of *cis*-1,3,5-trisubstituted cyclohexanes. The third example in Figure 14.69 shows that aromatic compounds can be hydrogenated with metallic Rh as the catalyst in a *chemoselective* fashion even though C—O bonds that are rather sensitive to hydrogenolysis are present in the molecule (cf. Figure 14.44).

Alkynes are hydrogenated "all the way" to alkanes if the usual heterogeneous catalysts (Pd, Pt, Raney Ni) are used. If a suitable deactivated catalyst is used, however, it is possible to stop these reactions after monohydrogenation. The so-called **Lindlar palladium** is a commonly used deactivated catalyst of this type (Figure 14.70). To prevent an overhydrogenation, it is still necessary to monitor the rate of hydrogen consumption and to interrupt the reaction after one equivalent of hydrogen gas has been absorbed even when the deactivated catalyst is used. The Lindlar hydrogenation depicted in Figure 14.70 shows that C=C double bonds already present in the substrate also are not hydrogenated under these conditions.

The mechanism of the Lindlar hydrogenation is analogous to the mechanism of the heterogeneously catalyzed alkene hydrogenation (Figure 14.65). It is for this similarity

Fig. 14.70. Lindlar hydrogenation of a C≡C triple bond. (The starting material can be prepared in analogy to Figure 13.24.)

Bu Bu H_2, Lindlar-Pd (i.e., either Pd/$BaSO_4$ or Pd/PbO/$CaCO_3$/ quinoline) Bu H H Bu

that hydrogen additions to C≡C triple bonds usually also exhibit high *cis*-selectivity. Lindlar hydrogenations present a good synthetic route to *cis*-alkenes. Only occasionally, this *cis*-selectivity is not perfect, for the reasons discussed in the case of the heterogeneously catalyzed hydrogenations of C═C double bonds (Figure 14.65, part II).

A

The **Birch reduction** is another method for the conversion of benzene and its derivatives into nonaromatic six-membered rings (Figure 14.71). Solvated electrons in liquid NH_3 or in liquid $EtNH_2$ are the effective reducing agent. The solvated electrons are generated by dissolving Li or Na in these media. Birch reductions of aromatic compounds can be carried out in the presence of several equivalents of alcohol without the formation of hydrogen gas. This is important because some of the later steps of the Birch reduction require an alcohol as a proton donor.

One of the solvated electrons is transferred into an antibonding π orbital of the aromatic compound, and a radical anion of type **C** is formed (Figure 14.71). The alcohol protonates this radical anion in the rate-determining step with high regioselectivity. In the case under scrutiny, and starting from other *donor-substituted* benzenes as well, the protonation occurs in the *ortho* position relative to the donor substituent. On the other hand, the protonation of the radical anion intermediate of the Birch reduction of *acceptor-substituted* benzenes occurs in the *para* position relative to the acceptor substituent.

Protonation of the radical anion **C** of Figure 14.71 results in the radical **D.** This radical regains its valence electron octet by capturing another solvated electron to form the carbanion **E.** Carbanions of this type are protonated regioselectively by the second equivalent of alcohol. The regioselectivity is independent of the substituents: protonation forms a 1,4-dihydroaromatic compound rather than a 1,2-dihydroaromatic compound. Hence, the 1,4-dihydrobenzene **A** is the reduction product in the example of Figure 14.71.

1,4-dihydroaromatic compounds obtained in this way can be used for the preparation of six-membered rings that would be difficult to synthesize otherwise. For example, the dihydrobenzene **A** of Figure 14.71 can be converted into the 4-substituted 2-cyclohexenone **B** via an acid-catalyed hydrolysis and an acid-catalyzed migration of the remaining C═C double bond.

The Birch reduction of aromatic compounds in NH_3/ROH or $EtNH_2$/ROH mixtures includes the following steps (Figure 14.71): electron transfer to the substrate, protonation of the resulting radical anion by the alcohol, electron transfer to the resulting radical, and protonation of the formed organometallic compound. The same four elementary reactions also can be used for the reduction of disubstituted alkynes to give alkenes (Figure 14.72). In comparison to aromatic compounds, alkynes are poorer electron acceptors, and the solvated electrons would relatively quickly reduce any added alcohol, with formation of hydrogen gas. Hence, alcohol is omitted in the reduction of alkynes. Protonation of the radical anion intermediate **A** or of the alkenylsodium compound **B** is accomplished in this case by the extremely weakly acidic solvent NH_3.

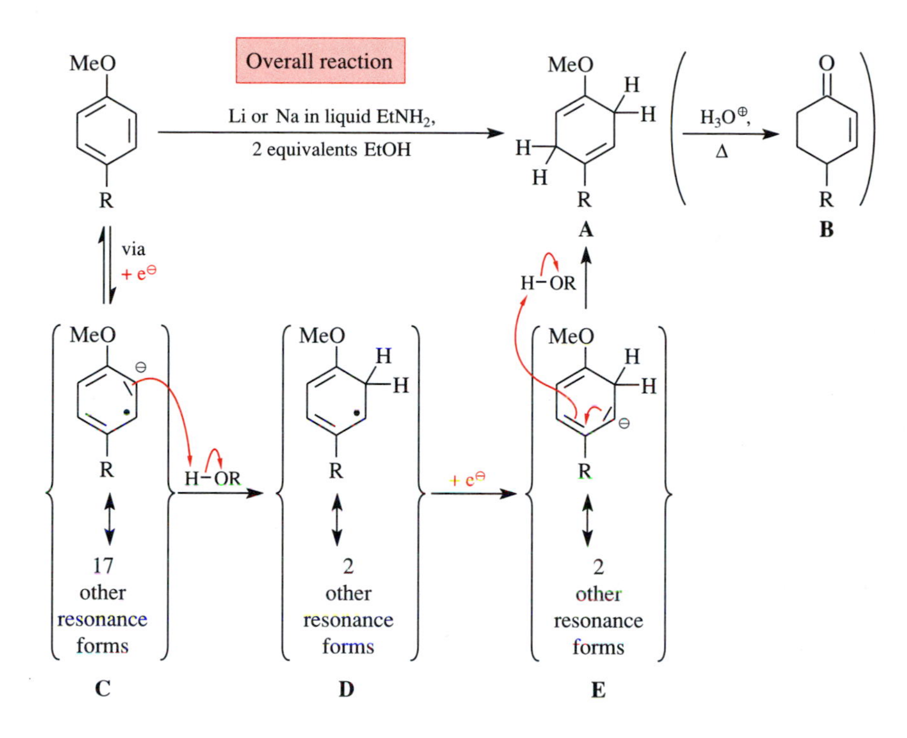

Fig. 14.71. Birch reduction of benzenes give 1,4-cyclohexadienes. The radical anion **C** is formed by capture of a solvated electron in an antibonding π^* orbital of an aromatic compound. The alcohol protonates this radical anion to the radical **D,** which captures another electron from the solution to form the carbanion **E.** The carbanion is protonated by a second equivalent of the alcohol, and the 1,4-dihydroaromatic compound results.

The rate-determining step in the Na/NH_3 reduction of alkynes is the protonation of the radical anion **A.** The next step, the reaction of the alkenyl radical **C** to the alkenylsodium intermediate **B,** determines the stereochemistry. The formation of **B** occurs such that the substituents of the C═C double bond are in *trans* positions. This *trans*-selectivity can be explained by product-development control in the formation of **B** or perhaps also by the preferred geometry of radical **C** provided it is nonlinear at the radical carbon. The alkenylsodium compound **B** is protonated with retention of configuration, since alkenylsodium compounds are configurationally stable (cf. Section 1.1.1). The Na/NH_3 reduction of alkynes therefore represents a synthesis of *trans*-alkenes.

It is difficult to determine directly whether the electron transfer to an alkyne is a reversible or an irreversible reaction (cf. reaction in Figure 14.72, top). Compounds containing two C≡C triple bonds that are not too far apart can be reduced chemo-

Fig. 14.72. *trans*-Selective reduction of C≡C triple bonds.

Fig. 14.73. Chemoselective and *trans*-selective reduction of exactly one C≡C triple bond of a diyne.

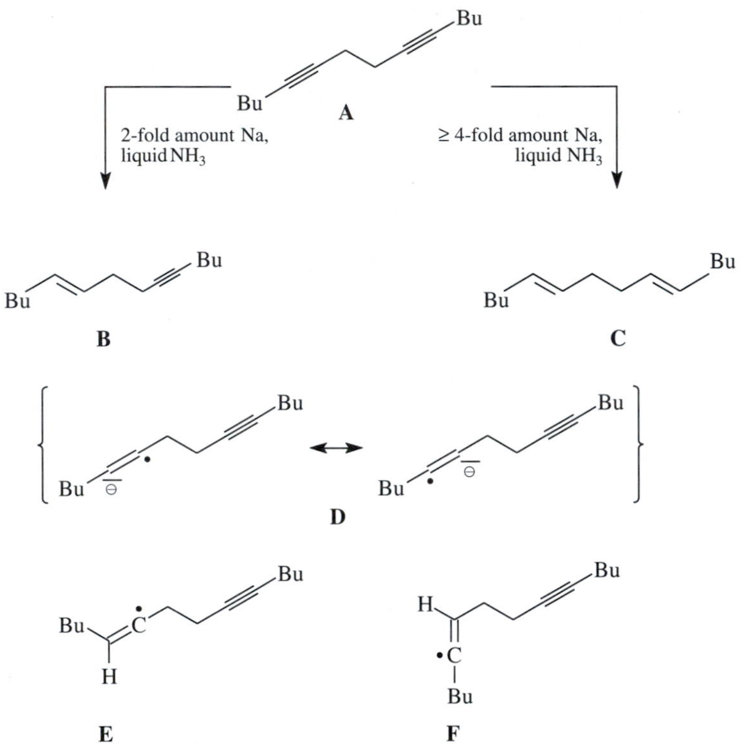

selectively with Na in liquid ammonia such that only one of the C≡C triple bonds is reduced. Figure 14.73 depicts the example of the dialkyne **A.** Such a reduction to the monoalkyne can be accomplished if one employs only the amount of sodium required for monoreduction. If one employs twice the amount of sodium, one obtains the dialkene **C,** as expected. The first reduction **A** → **B** consequently must be faster than the subsequent reduction **B** → **C.** This can be understood immediately if one assumes that the electron transfer to the C≡C triple bond is irreversible. If that is true, then the radical anion **D** would be formed quantitatively upon consumption of exactly 50% of the reducing reagent. After protonation in the rate-determining step, the radical anion would yield a mixture of the isomeric alkenyl radicals **E** and **F.** These radicals would quickly consume the remaining 50% of the reducing reagent. Thus, their reaction to form the monoalkyne **B** would be faster than the reduction of the monoalkyne to the dialkene **C.**

References

14.3

A. H. Haines, "Methods for the Oxidation of Organic Compounds: Alkanes, Alkenes, Alkynes, Arenes," Academic Press, New York, **1985.**

A. H. Haines, "Methods for the Oxidation of Organic Compounds: Alcohols, Alcohol Derivatives, Alkyl Halides, Nitroalkanes, Alkyl Azides, Carbonyl Compounds, Hydroxyarenes, and Aminoarenes," Academic Press, New York, **1988.**

M. Hudlicky, "Oxidations in Organic Chemistry," American Chemical Society, Washington, DC, **1990.**

H. Bornowski, D. Döpp, R. Jira, U. Langer, H. Offermans, K. Praefcke, G. Prescher, G. Simchen, D. Schumann, "Preparation of Aldehydes by Oxidation," in *Methoden Org. Chem. (Houben-Weyl) 4th ed. 1952-, Aldehydes* (J. Falbe, Ed.), Vol. E3, 231, Georg Thieme Verlag, Stuttgart, **1983.**

A. J. Mancuso and D. Swern, "Activated dimethyl sulfoxide: Useful reagents for synthesis," *Synthesis* **1981,** 165.

T. T. Tidwell, "Oxidation of alcohols by activated dimethyl sulfoxide and related reactions: an update," *Synthesis* **1990,** 857–870.

T. T. Tidwell, "Oxidation of alcohols to carbonyl compounds via alkoxysulfonium ylides: The Moffatt, Swern and related reactions," *Org. React.* **1990,** *39,* 297–572.

E. J. de Nooy, A. C. Besemer, H. van Bekkum, "On the use of stable organic nitroxyl radicals for the oxidation of primary and secondary alcohols," *Synthesis,* **1996,** 1153–1174.

R. L. Kuczkowski, "The structure and mechanism of formation of ozonides," *Chem. Soc. Rev.* **1992,** *21,* 79–83.

G. R. Krow, "The Baeyer-Villiger oxidation of ketones and aldehydes," *Org. React. (N.Y.)* **1993,** *43,* 251–798.

14.4

M. Hudlicky, "Reductions in Organic Chemistry," The Royal Society of Chemistry, Cambridge, U.K., **1996.**

A. F. Abdel-Magid (Ed.), "Reductions in Organic Synthesis: Recent Advances and Practical Applications," ACS Symposium Series, The Royal Society of Chemistry, Cambridge, U.K., **1996.**

A. Hajos, "Reduction with Inorganic Reducing Agents—Metal Hydrides and Complex Hydrides," in *Methoden Org. Chem. (Houben-Weyl) 4th ed. 1952-, Reduction Part II* (H. Kropf, Ed.), Vol. 4/1d, 1, Georg Thieme Verlag, Stuttgart, **1981.**

J. Malek, "Reduction by metal alkoxyaluminum hydrides," *Org. React. (N.Y.)* **1985,** *34,* 1.

J. Malek, "Reductions by metal alkoxyaluminum hydrides. Part II. Carboxylic acids and derivatives, nitrogen compounds, and sulfur compounds," *Org. React.* **1988,** *36,* 249–590.

J. Seyden-Penne, "Reductions by the Alumino- and Borohydrides in Organic Synthesis," VCH, New York, 1991.

A. J. Downs and C. R. Pulham, "The hydrides of aluminum, gallium, indium, and thallium—A reevaluation," *Chem. Soc. Rev.* **1994,** *23,* 175.

N. M. Yoon, "Selective reduction of organic compounds with aluminum and boron hydrides," *Pure Appl. Chem.* **1996,** *68,* 843.

J. Seyden-Penne, "Reductions by the Alumino- and Borohydrides in Organic Synthesis," 2nd ed., Wiley, New York, **1997.**

L. K. Keefer and G. Lunn, "Nickel-aluminum alloy as a reducing agent," *Chem. Rev.* **1989,** *89,* 459–502.

T. Imamoto, "Reduction of Saturated Alkyl Halides to Alkanes," in *Comprehensive Organic Synthesis* (B. M. Trost, I. Fleming, Eds.), Vol. 8, 793, Pergamon Press, Oxford, **1991.**

S. W. McCombie, "Reduction of Saturated Alcohols and Amines to Alkanes," in *Comprehensive Organic Synthesis* (B. M. Trost, I. Fleming, Eds.), Vol. 8, 811, Pergamon Press, Oxford, **1991.**

A. G. Sutherland, "One or More CH Bond(s) Formed by Substitution: Reduction of C-Halogen and C-Chalcogen Bonds," in *Comprehensive Organic Functional Group Transformations* (A. R. Katritzky, O. Meth-Cohn, C. W. Rees, Eds.), Vol. 1, 1, Elsevier Science, Oxford, U.K., **1995.**

C. Blomberg, "The Barbier Reaction and Related One-step Processes," Springer-Verlag, Heidelberg, **1994.**

C. G. Screttas and B. R. Steele, "Organometallic carboxamidation: A review," *Org. Prep. Proced. Int.* **1990,** *22,* 269–314.

T. Cohen and M. Bhupathy, "Organoalkali compounds by radical anion induced reductive metalation of phenyl thioethers," *Acc. Chem. Res.* **1989,** *22,* 152–161.

M. Yus, "Arene-catalyzed lithiation reactions," *Chem. Soc. Rev.* **1996,** *25,* 155–162.

M. Yus and F. Foubelo, "Reductive opening of saturated oxa-, aza- and thia-cycles by means of an arene-promoted lithiation: Synthetic applications," *Rev. Heteroatom Chem.* **1997,** *17,* 73–108.

Y. H. Lai, "Grignard reagents from chemically activated magnesium," *Synthesis* **1981,** 585.

C. Walling, "The nature of radicals involved in Grignard reagent formation," *Acc. Chem. Res.* **1991,** *24,* 255.

H. M. Walborsky, "Mechanism of Grignard reagent formation. The surface nature of the reaction," *Acc. Chem. Res.* **1990,** *23,* 286–293.

J. F. Garst, "Grignard reagent formation and freely diffusing radical intermediates," *Acc. Chem. Res.* **1991,** *24,* 95–97.

R. D. Rieke, "Preparation of organometallic compounds from highly reactive metal powders," *Science* **1989,** *246,* 1260–1264.

J. S. Thayer, "Not for synthesis only: The reactions of organic halides with metal surfaces," *Adv. Org. Chem.* **1995,** *38,* 59–78.

J. W. Huffman, "Reduction of C=X to CHXH by Dissolving Metals and Related Methods," in *Comprehensive Organic Synthesis* (B. M. Trost, I. Fleming, Eds.), Vol. 8, 107, Pergamon Press, Oxford, **1991.**

H. M. R. Hoffmann and A. M. El-Khawaga, "Formation of C—H Bonds by Reduction of Carbonyl Groups (C=O)—Reduction of Carbonyl Groups with Metals," in *Methoden Org. Chem. (Houben-Weyl) 4th ed. 1952-, Stereoselective Synthesis* (G. Helmchen, R. W. Hoffmann, J. Mulzer, E. Schaumann, Eds.), Vol. E21d, 3967, Georg Thieme Verlag, Stuttgart, **1995.**

J. J. Bloomfield, D. C. Owsley, J. M. Nelke, "The Acyloin Condensation," *Org. React. (N.Y.)* **1976,** *23,* 259.

R. Brettle, "Acyloin Coupling Reactions," in *Comprehensive Organic Synthesis* (B. M. Trost, I. Fleming, Eds.), Vol. 3, 613, Pergamon Press, Oxford, **1991.**

G. M. Robertson, "Pinacol Coupling Reactions," in *Comprehensive Organic Synthesis* (B. M. Trost, I. Fleming, Eds.), Vol. 3, 563, Pergamon Press, Oxford, **1991.**

D. Lenoir, "The application of low-valent titanium reagents in organic synthesis," *Synthesis* **1989,** *12,* 883–897.

J. E. McMurry, "Carbonyl-coupling reactions using low-valent titanium," *Chem. Rev.* **1989,** *89,* 1513–1524.

T. Lectka, "The McMurry Recation," in *Active Metals* (A. Fürstner, Ed.), 85, VCH, Weinheim, Germany, **1996.**

A. Fürstner and B. Bogdanovic, "New developments in the chemistry of low-valent titanium," *Angew. Chem.* **1996,** *108,* 2582–2609; *Angew. Chem. Int. Ed. Engl.* **1996,** *35,* 2442–2469.

A. G. M. Barrett, "Reduction of Carboxylic Acid Derivatives to Alcohols, Ethers and Amines," in *Comprehensive Organic Synthesis* (B. M. Trost, I. Fleming, Eds.), Vol. 8, 235, Pergamon Press, Oxford, **1991.**

J. S. Cha, "Recent developments in the synthesis of aldehydes by reduction of carboxylic acids and their derivatives with metal hydrides," *Org. Prep. Proced. Int.* **1989,** *21,* 451–477.

A. P. Davis, "Reduction of Carboxylic Acids to Aldehydes by Other Methods," in *Comprehensive Organic Synthesis* (B. M. Trost, I. Fleming, Eds.), Vol. 8, 283, Pergamon Press, Oxford, **1991.**

N. Greeves, " Reduction of C=O to CHOH by Metal Hydrides ," in *Comprehensive Organic Synthesis* (B. M. Trost, I. Fleming, Eds.), Vol. 8, 1, Pergamon Press, Oxford, **1991.**

H. Brunner, "Formation of C—H Bonds by Reduction of Carbonyl Groups (C=O)—Hydrogenation," in *Methoden Org. Chem. (Houben-Weyl) 4th ed. 1952-, Stereoselective Synthesis* (G. Helmchen, R. W. Hoffmann, J. Mulzer, E. Schaumann, Eds.), Vol. E21d, 3945, Georg Thieme Verlag, Stuttgart, **1995.**

A. P. Davis, M. M. Midland, L. A. Morell, "Formation of C—H Bonds by Reduction of Carbonyl Groups (C=O), Reduction of Carbonyl Groups with Metal Hydrides," in *Methoden Org. Chem. (Houben-Weyl) 4th ed. 1952-, Stereoselective Synthesis* (G. Helmchen, R. W. Hoffmann, J. Mulzer, E. Schaumann, Eds.), Vol. E21d, 3988, Georg Thieme Verlag, Stuttgart, **1995.**

M. M. Midland, L. A. Morell, K. Krohn, "Formation of C—H Bonds by Reduction of Carbonyl Groups (C=O)—Reduction with C—H Hydride Donors," in *Methoden Org. Chem. (Houben-Weyl) 4th ed. 1952-, Stereoselective Synthesis* (G. Helmchen, R. W. Hoffmann, J. Mulzer, E. Schaumann, Eds.), Vol. E21d, 4082, Georg Thieme Verlag, Stuttgart, **1995.**

R. O. Hutchins, "Reduction of C=X to CH_2 by Wolff-Kishner and Other Hydrazone methods," in Comprehensive Organic Synthesis (B. M. Trost, I. Fleming, Eds.), Vol. 8, 327, Pergamon Press, Oxford, 1991.

P. N. Rylander, "Hydrogenation Methods," 216, Academic Press, **1990.**

U. Kazmaier, J. M. Brown, A. Pfaltz, P. K. Matzinger, H. G. W. Leuenberger, "Formation of C—H Bonds by Reduction of Olefinic Double Bonds—Hydrogenation," in *Methoden Org. Chem. (Houben-Weyl) 4th ed. 1952-, Stereoselective Synthesis* (G. Helmchen, R. W. Hoffmann, J. Mulzer, E. Schaumann, Eds.), Vol. E21d, 4239, Georg Thieme Verlag, Stuttgart, **1995.**

V. A. Semikolenov, "Modern approaches to the preparation of 'palladium on charcoal' catalysts," *Russ. Chem. Rev.* **1992,** *61,* 168–174.

N. M. Dmitrievna and K. O. Valentinovich, "Heterogeneous catalysts of hydrogenation," *Russ. Chem. Rev.* **1998,** *67,* 656–687.

H. Takaya, "Homogeneous Catalytic Hydrogenation of C=C and Alkynes," in *Comprehensive Organic Synthesis* (B. M. Trost, I. Fleming, Eds.), Vol. 8, 443, Pergamon Press, Oxford, **1991.**

R. Noyori, H. Takaya, "BINAP: An efficient chiral element for asymmetric catalysis," *Acc. Chem. Res.* **1990,** *23,* 345–350.

J. M. Hook, and L. N. Mander, "Recent developments in the Birch reduction of aromatic compounds: Applications to the synthesis of natural products," *Nat. Prod. Rep.* **1986,** *3,* 35.

P. W. Rabideau, "The metal-ammonia reduction of aromatic compounds," *Tetrahedron* **1989,** *45,* 1579–1603.

L. N. Mander, "Partial Reduction of Aromatic Rings by Dissolving Metals and by Other Methods," in *Comprehensive Organic Synthesis* (B. M. Trost, I. Fleming, Eds.), Vol. 8, 489, Pergamon Press, Oxford, **1991.**

P. W. Rabideau and Z. Marcinow, "The Birch reduction of aromatic compounds," *Org. React.* **1992,** *42,* 1–334.

A. J. Birch, "The Birch reduction in organic synthesis," *Pure Appl. Chem.* **1996,** *68,* 553–556.

Further Reading

A. B. Jones, "Oxidation Adjacent to X=X Bonds by Hydroxylation Methods," in *Comprehensive Organic Synthesis* (B. M. Trost, I. Fleming, Eds.), Vol. 7, 151, Pergamon Press, Oxford, **1991.**

P. T. Gallagher, "The synthesis of quinones," *Contemp. Org. Synth.* **1996,** *3,* 433–446.

V. D. Filimonov, M. S. Yusubov, Ki-WhanChi, "Oxidative methods in the synthesis of vicinal di- and poly-carbonyl compounds," *Russ. Chem. Rev.* **1998,** *67,* 803–826.

S. Akai and Y. Kita, "Recent progress in the synthesis of *p*-quinones and *p*-dihydroquinones through oxidation of phenol derivatives," *Org. Prep. Proced. Int.* **1998,** *30,* 603–629.

J. Tsuji, "Synthetic applications of the palladium-catalysed oxidation of olefins to ketones," *Synthesis* **1984,** 369.

J. Tsuji, "Addition Reactions with Formation of Carbon-Oxygen Bonds—The Wacker Oxidation and Related Reactions," in *Comprehensive Organic Synthesis* (B. M. Trost, I. Fleming, Eds.), Vol. 7, 469, Pergamon Press, Oxford, **1991.**

A. Fatiadi, "The classical permanganate ion: Still a novel oxidant in organic chemistry," *Synthesis* **1987,** 85.

H. Kropf, E. Müller, A. Weickmann, "Ozone as an Oxidation Agent," in *Methoden Org. Chem. (Houben-Weyl) 4th ed. 1952-, Oxidation Part I* (H. Kropf, Ed.), Vol. 4/1a, 3, Georg Thieme Verlag, Stuttgart, **1981.**

H. Heaney, "Oxidation reactions using magnesium monoperphthalate and urea hydrogen peroxide," *Aldrichimica Acta* **1993,** *26,* 35–45.

W. P. Griffith and S. V. Ley, "TPAP: Tetra-n-propylammonium perruthenate, a mild and convenient oxidant for alcohols," *Aldrichimica Acta* **1990,** *23,* 13–19.

S. V. Ley, J. Norman, W. P. Griffith, S. P. Maraden, "Tetrapropylammonium Perruthenate, $Pr_4N^+RuO_4^-$, TPAP: A catalytic oxidant for organic synthesis," *Synthesis* **1994,** 639–666.

H. Waldmann, "Hypervalent Iodine Reagents," in *Organic Synthesis Highlights II* (H. Waldmann, Ed.), VCH, Weinheim, New York, etc., **1995,** 223–230.

A. Varvoglis (Ed.), "Hypervalent Iodine in Organic Synthesis," Academic Press, San Diego, CA, **1996.**

M. Nishizawa and R. Noyori, "Reduction of C=X to CHXH by Chirally Modified Hydride Reagents," in Comprehensive Organic Synthesis (B. M. Trost, I. Fleming, Eds.), Vol. 8, 159, Pergamon Press, Oxford, **1991.**

R. M. Kellogg, "Reduction of C=X to CHXH by Hydride Delivery from Carbon," in *Comprehensive Organic Synthesis* (B. M. Trost, I. Fleming, Eds.), Vol. 8, 79, Pergamon Press, Oxford, **1991.**

R. O. Hutchins, "Reduction of C=N to CHNH by Metal Hydrides," in *Comprehensive Organic Synthesis* (B. M. Trost, I. Fleming, Eds.), Vol. 8, 25, Pergamon Press, Oxford, **1991.**

H. Kumobayashi, "Industrial application of asymmetric reactions catalyzed by BINAP–metal complexes," *Rec. Trav. Chim. Pays-Bas* **1996,** *115,* 201–210.

K. Inoguchi, S. Sakuraba, K. Achiwa, "Design concepts for developing highly efficient chiral bisphosphine ligands in rhodium-catalyzed asymmetric hydrogenations," *Synlett* **1992,** 169–178.

C. Rosini, L. Franzini, A. Raffaelli, P. Salvadori, "Synthesis and applications of binaphthylic C_2-symmetry derivatives as chiral auxiliaries in enantioselective reactions," *Synthesis* **1992,** 503–517.

D. J. Ager and S. A. Laneman, "Reductions of 1,3-dicarbonyl systems with ruthenium-biarylbisphosphine catalysts," *Tetrahedron Asymmetry* **1997,** *8,* 3327–3355.

S. Otsuka, K. Tani, "Catalytic asymmetric hydrogen migration of allylamines" *Synthesis* **1991,** 665–680.

E. Block, "Olefin synthesis via deoxygenation of vicinal diols," *Org. React. (N.Y.)* **1984,** *30,* 457.

M. M. Midland, "Asymmetric reduction with organoborane reagents," *Chem. Rev.* **1989,** *89,* 1553.

H. C. Brown and P. V. Ramachandran, "Asymmetric reduction with chiral organoboranes based on α-pinene," *Acc. Chem. Res.* **1992,** *25,* 16–24.

V. Ponec, "Selective de-oxygenation of organic compounds," *Rec. Trav. Chim. Pays-Bas* **1996,** *115,* 451–455.

V. K. Singh, "Practical and useful methods for the enantioselective reduction of unsymmetrical ketones," *Synthesis* **1992,** 607–617.

M. Wills and J. R. Studley, "The asymmetric reduction of ketones," *Chem. Ind.* **1994,** 552–555.

S. Itsuno, "Enantioselective reduction of ketones, *Org. Prep. Proced Int.,* **1998,** 52, 395–576.

Subject Index

A

C

E

F

G

H

I

J

K

L

M

N

O

S

T

U